Clouds in the Perturbed Climate System

Strüngmann Forum Reports

Julia Lupp, series editor

The Ernst Strüngmann Forum is made possible through
the generous support of the Ernst Strüngmann Foundation,
inaugurated by Dr. Andreas and Dr. Thomas Strüngmann.

Additional support for this Forum was received from the
Deutsche Forschungsgemeinschaft (German Science Foundation).

Clouds in the Perturbed Climate System

Their Relationship to Energy Balance, Atmospheric Dynamics, and Precipitation

Edited by

Jost Heintzenberg and Robert J. Charlson

Program Advisory Committee:

Jean-Louis Brenguier, Robert J. Charlson, Jim Haywood, Jost Heintzenberg, Teruyuki Nakajima, and Bjorn Stevens

The MIT Press

Cambridge, Massachusetts
London, England

Series Editor: J. Lupp
Assistant Editor: M. Turner
Photographs: U. Dettmar
Typeset by BerlinScienceWorks

MIT Press books may be purchased at special quantity discounts for business or sales promotional use. For information, please email special_sales@mitpress.mit.edu or write to Special Sales Department, The MIT Press, 55 Hayward Street, Cambridge, MA 02142.

The book was set in TimesNewRoman and Arial.
Printed and bound in the United States of America.

Library Congress Cataloging-in-Publication Data

Ernst Strüngmann Forum (2008 : Frankfurt, Germany)
Clouds in the perturbed climate system : their relationship to energy balance, atmospheric dynamics, and precipitation / edited by Jost Heintzenberg and Robert J. Charlson.
p. cm. — (Strüngmann Forum reports)
Forum held March 2–7, 2008 in Frankfurt, Germany.
Includes bibliographical references and indexes.
ISBN 978-0-262-01287-4 (hardcover : alk. paper)
1. Cloud physics—Congresses. 2. Clouds—Dynamics—Congresses. 3. Climatic changes—Congresses. I. Heintzenberg, J. (Jost) II. Charlson, Robert. J. III. Title.
QC921.48.E76 2009
551.57'6—dc22

2008049592

Contents

Ernst Strüngmann Forum ix

List of Contributors xiii

1 **Introduction** 1
Jost Heintzenberg and Robert J. Charlson

2 **Trends in Observed Cloudiness and Earth's Radiation Budget: What Do We Not Know and What Do We Need to Know?** 17
Joel R. Norris and Anthony Slingo

3 **Climatologies of Cloud-related Aerosols Part 1: Particle Number and Size** 37
Stefan Kinne

Part 2: Particle Hygroscopicity and Cloud Condensation Nucleus Activity 58
Ulrich Pöschl, Diana Rose, and Meinrat O. Andreae

4 **Cloud Properties from *In-situ* and Remote-sensing Measurements: Capability and Limitations** 73
George A. Isaac and K. Sebastian Schmidt

5 **Clouds and Precipitation: Extreme Rainfall and Rain from Shallow Clouds** 107
Yukari N. Takayabu and Hirohiko Masunaga

6 **Temporal and Spatial Variability of Clouds and Related Aerosols** 127
Theodore L. Anderson, Rapporteur
Andrew Ackerman, Dennis L. Hartmann, George A. Isaac, Stefan Kinne, Hirohiko Masunaga, Joel R. Norris, Ulrich Pöschl, K. Sebastian Schmidt, Anthony Slingo, and Yukari N. Takayabu

7 **Laboratory Cloud Simulation: Capabilities and Future Directions** 149
Frank Stratmann, Ottmar Möhler, Raymond Shaw, and Heike Wex

8 **Cloud-controlling Factors: Low Clouds** 173
Bjorn Stevens and Jean-Louis Brenguier

9 **Deep Convective Clouds** 197
Wojciech W. Grabowski and Jon C. Petch

10 **Large-scale Controls on Cloudiness** 217
Christopher S. Bretherton and Dennis L. Hartmann

11 **Cloud-controlling Factors of Cirrus** 235
Bernd Kärcher and Peter Spichtinger

12 **Cloud-controlling Factors** 269
A. Pier Siebesma, Rapporteur
Jean-Louis Brenguier, Christopher S. Bretherton, Wojciech W. Grabowski, Jost Heintzenberg, Bernd Kärcher, Katrin Lehmann, Jon C. Petch, Peter Spichtinger, Bjorn Stevens, and Frank Stratmann

13 **Cloud Particle Precursors** 291
Sonia M. Kreidenweis, Markus D. Petters, and Patrick Y. Chuang

14 **Cloud–Aerosol Interactions from the Micro to the Cloud Scale** 319
Graham Feingold and Holger Siebert

15 **Weather and Climate Engineering** 339
William R. Cotton

16 **Air Pollution and Precipitation** 369
Greg Ayers and Zev Levin

17 **What Do We Know about Large-scale Changes of Aerosols, Clouds, and the Radiation Budget?** 401
Teruyuki Nakajima and Michael Schulz

18 **The Extent and Nature of Anthropogenic Perturbations of Clouds** 433
Patrick Y. Chuang and Graham Feingold, Rapporteurs
Greg Ayers, Robert J. Charlson, William R. Cotton, Sonia M. Kreidenweis, Zev Levin, Teruyuki Nakajima, Daniel Rosenfeld, Michael Schulz, and Holger Siebert

19 **Global Indirect Radiative Forcing Caused by Aerosols: IPCC (2007) and Beyond** 451
Jim Haywood, Leo Donner, Andy Jones, and Jean-Christophe Golaz

20 **Simulating Global Clouds: Past, Present, and Future** 469
William D. Collins and Masaki Satoh

21 **Observational Strategies from the Micro- to Mesoscale** 487
Jean-Louis Brenguier and Robert Wood

22 **Observational Strategies at Meso- and Large Scales to Reduce Critical Uncertainties in Future Cloud Changes** 511
Anthony Illingworth and Sandrine Bony

23 **Aerosols and Clouds in Chemical Transport Models and Climate Models** 531
Ulrike Lohmann and Stephen E. Schwartz

24 Current Understanding and Quantification of Clouds in the Changing Climate System and Strategies for Reducing Critical Uncertainties **557**
Johannes Quaas, Rapporteur
Sandrine Bony, William D. Collins, Leo Donner, Anthony Illingworth, Andy Jones, Ulrike Lohmann, Masaki Satoh, Stephen E. Schwartz, Wei-Kuo Tao, and Robert Wood

Abbreviations **575**

Name Index **581**

Subject Index **591**

The Ernst Strüngmann Forum

The Ernst Strüngmann Forum is founded on the tenets of scientific independence and the inquisitive nature of the human mind. Through its innovative communication process, the Ernst Strüngmann Forum facilitates the continual expansion of knowledge by providing a creative environment within which experts can scrutinize high-priority issues from multiple vantage points.

The themes that are selected transcend classic disciplinary boundaries; they address topics that are problem-oriented and of high-priority interest to science and society. Proposals are submitted by leading scientists active in their field and selected by the Scientific Advisory Board.

"Clouds in the Perturbed Climate System" began to take form in 2004, when Jost Heintzenberg and Bob Charlson approached me with the idea of convening a follow-up meeting to one held almost a decade earlier.[1] That meeting focused on aerosol forcing: the nature and effects of aerosols, the evidence for aerosol effects, the uncertainties in climate forecasts attributable to aerosol forcing, and policy implications. As Heintzenberg and Charlson reflected, major international field campaigns and basic research were stimulated by that meeting, but many issues remained unresolved. Subsequent progress had been made in understanding various parts (and their interconnections) of the Earth system. However, cloud feedbacks still pose the largest source of uncertainty in the climate system, and the resulting implications are immense.

It was decided, therefore, to convene this Ernst Strüngmann Forum to assess the temporal and spatial variability of clouds in an effort to further understanding of the physical processes that control cloud evolution. In addition, the extent and nature of anthropogenic perturbations of clouds and cloud-related processes would be examined and strategies formulated to reduce critical uncertainties.

Jean-Louis Brenguier, Jim Haywood, Terry Nakajima, and Bjorn Stevens joined Heintzenberg and Charlson on the steering committee. Together they refined the key topics for discussion and selected the experts to be invited to the central meeting, which was held in Frankfurt from March 2–7, 2008.

An Ernst Strüngmann Forum is a communication process—a carefully orchestrated dialog between diverse participants, each of whom assumes an active role. It is a dynamic process that unfolds over an extended period of time. The results have now been set down in this volume for further consideration.

This volume is not meant to be a consensus document. Its intent is to communicate the scope of the topics as well as the essence of the discourse that evolved. Chapters 6, 12, 18, and 24, for example, are the reports of the individual discussion groups. To draft such a report and bring it into final

[1] Charlson, R. J., and J. Heintzenberg. 1995. Aerosol Forcing of Climate. Chichester: John Wiley & Sons.

form is no simple matter, and for this, I wish to extend a particular word of thanks to Tad Anderson, Pier Siebesma, Patrick Chuang, Graham Feingold, and Johannes Quaas. As rapporteurs of the discussion groups, their task was certainly not easy. Neither was the job of the moderators, so ably assumed by Dennis Hartmann, Jost Heintzenberg, Zev Levin, and Bill Collins. A dialog of this kind creates its own unique dynamics; invitees contributed not only their congenial personalities but also a willingness to question and probe beyond that which is already evident.

Long-held views or opinions are never easy to put aside. However, achieving this allows boundaries of the unknown to become clearer, gaps in knowledge to be recognized, and response strategies to be formulated. This is what we set out to accomplish.

A communication process of this nature requires institutional stability and an environment that guarantees free thought. Through the generous support of the Ernst Strüngmann Foundation, which was founded by Dr. Andreas and Dr. Thomas Strüngmann in honor of their father, the Ernst Strüngmann Forum is able to conduct its work in the service of science. The involvement of the Scientific Advisory Board ensures the scientific independence of the Forum, and their work is gratefully acknowledged. Additional support was received from the German Science Foundation.

Together, these partnerships have enabled this unique gathering. It is our hope that the resulting insights will stimulate further debate and direct future enquiry, as science strives to reduce the uncertainties that cloud feedbacks pose to the perturbed climate system.

Julia Lupp, Program Director
Ernst Strüngmann Forum
Frankfurt Institute for Advanced Studies (FIAS)
Ruth-Moufang-Str. 1, 60438 Frankfurt am Main, Germany
http://fias.uni-frankfurt.de/esforum/

In recognition of his contributions to science and to this Ernst Strüngmann Forum, and with deep respect and admiration, we dedicate this volume to our colleague and friend, Anthony Slingo.

During the final preparation of this book, Professor Anthony Slingo died suddenly, and it was a shock to all who knew him. On October 14, 2008, the University of Reading lost a leading academic; his wife, Julia, lost her soul mate and greatest fan; his daughters, Mary and Anna, lost a proud and loving father; I lost a personal friend; and the scientific community lost a great scientist who made lasting contributions to the environmental and climate sciences over many years.

Tony's work on observing and modeling the Earth's radiation budget added fundamental knowledge that advanced our understanding of how the radiative processes influence the Earth system, and how the greenhouse effect of our planet shapes our climate. More recently, Tony worked on pioneering new observations that have begun to quantify the effects of dust on the radiation balance of desert regions.

There were a number of character traits that endeared Tony to many people. His enormous sincerity spilled over into his attitudes on science. He was incredibly careful and thorough in everything he did and did not tolerate anything less than excellence. Tony pursued what he considered important with sincerity and unwavering resolve. He was direct, honest, and adept at shunning the ever-growing non-essential activities that are sadly part of an expanding overhead in science today. I admired his dogma and the way he remained true to himself over the years.

The dedication of this particular volume to the memory of Tony seems most appropriate. He clearly valued the topics addressed in this book. As was his style, he would not have attended this Ernst Strüngmann Forum otherwise. Its focal topics, in many ways, mirror the important contributions that he made to science.

During the 12 years that he headed the Clouds and Climate group and then Model Parameterization group at the Hadley Center, UK Meteorological Office, Tony led many research activities that are reflected in the chapters contained within this book. He led the development of the Edward–Slingo radiative parameterization that is still in use today and which is obviously central to the cloud radiation feedbacks of the current UK Met Office model. He also developed new diagnostic approaches that eventually forged into his more recent work involving systematic comparison of GERB observations with models. Tony was involved in early aerosol indirect studies using global models, also a topic central to the theme of this book. However, he also sought more from his science and longed to become more engaged with observations and the real world than was possible, given his duties, at the Met Office. The move to the University of Reading gave him greater freedom, which made the highly successful Radagast project of today possible.

To Tony, you are missed by many.

—Graeme L. Stephens
University Distinguished Professor, Colorado State University

List of Contributors

Ackerman, Andrew NASA Goddard Institute for Space Studies, 2880 Broadway, New York, NY 10025, U.S.A.

Anderson, Theodore L. Department of Atmospheric Sciences, University of Washington, Box 35164 / 408 ATG, Seattle, WA 98195–1640, U.S.A.

Andreae, Meinrat O. Biogeochemistry Department, Max Planck Institute for Chemistry, Becherweg 27/29, 55128 Mainz, Germany

Ayers, Greg CSIRO, Marine and Atmospheric Research, 107–121 Station Street, Private Bag 1, Aspendale VIC 3195, Australia

Bony, Sandrine Laboratoire de Météorologie Dynamique, CNRS/UPMC, 4 place Jussieu, Box 99, 75252 Paris cedex 05, France

Brenguier, Jean-Louis Météo-France CNRM, 42 avenue Coriolis, 31057 Toulouse cedex 1, France

Bretherton, Christopher S. Department of Atmospheric Sciences, University of Washington, Box 351640, Seattle, WA 98195–1640, U.S.A.

Charlson, Robert J. Department of Atmospheric Sciences, University of Washington, Box 351640, Seattle, WA 98195–1640, U.S.A.

Chuang, Patrick Y. Department of Earth and Planetary Sciences, University of California Santa Cruz, EMS A254, Santa Cruz, CA 95064, U.S.A.

Collins, William D. Lawrence Berkeley Laboratory, Earth Sciences Division, 1 Cyclotron Road Mail Stop 90R1116, Berkeley, CA 9470–8126, U.S.A.

Cotton, William R. Department of Atmospheric Science, Colorado State University, 200 West Lake Street, Fort Collins, CO 80523–1371, U.S.A.

Crutzen, Paul Department of Atmospheric Chemistry, Max Planck Institute for Chemistry, 55020 Mainz, Germany

Donner, Leo Geophysical Fluid Dynamics Laboratory/NOAA, Princeton University, Forrestal Campus, 201 Forrestal Rd., Princeton, NJ 08540, U.S.A.

Feingold, Graham NOAA Earth System Research Laboratory, Chemical Sciences Division R/CSD3, 325 Broadway, Boulder, CO 80305, U.S.A.

Golaz, Jean-Christophe Geophysical Fluid Dynamics Laboratory/NOAA, Princeton University Forrestal Campus, 201 Forrestal Road, Princeton, NJ 08540, U.S.A.

Grabowski, Wojciech W. Mesoscale and Microscale Meteorology Division, National Center for Atmospheric Research, P.O. Box 3000, Boulder, CO 80307–3000, U.S.A.

Hartmann, Dennis L. Department of Atmospheric Sciences, University of Washington, Seattle, WA 98195–1640, U.S.A.

Haywood, Jim Met Office Hadley Centre, FitzRoy Road, Exeter EX1 3PB, U.K.

Heintzenberg, Jost Leibniz Institute for Tropospheric Research, Permoserstr. 15, 04318 Leipzig, Germany

Illingworth, Anthony Department of Meteorology, University of Reading, Earle Gate, Reading RG6 6BB, U.K.

Isaac, George A. Cloud Physics and Severe Weather Research Section, Environment Canada, 4905 Dufferin Street, Toronto, Ontario M3H 5T4, Canada

Jones, Andy Met Office Hadley Centre, FitzRoy Road, Exeter EX1 3PB, U.K.

Kärcher, Bernd Deutsches Zentrum für Luft- und Raumfahrt, Institut für Physik der Atmosphäre, Oberpfaffenhofen, 82234 Wessling, Germany

Kinne, Stefan Max Planck Institute for Meteorology, Bundesstraße 53, 20146 Hamburg, Germany

Kreidenweis, Sonia M. Department of Atmospheric Science, Colorado State University, 200 West Lake Street, Fort Collins, CO 80523–1371, U.S.A.

Lehmann, Katrin Scripps Institution for Oceanography, Center for Atmospheric Sciences, 9500 Gilman Dr., La Jolla, CA 92093, U.S.A.

Levin, Zev Department of Geophysics and Planetary Sciences, Tel Aviv University, Ramat Aviv 69978, Israel

Lohmann, Ulrike Institute for Atmospheric and Climate Science, ETH Zürich, Universitätsstr. 16, 8092 Zurich, Switzerland

Masunaga, Hirohiko Hydrospheric Atmospheric Research Center, Nagoya University, Furocho, Chikusa-ku, Nagoya 464–8601, Japan

Möhler, Ottmar Forschungszentrum Karlsruhe in der Helmholtz-Gemeinschaft, Technik und Umwelt IMK Bau 326, Bereich Atmosphärische Aerosolforschung, Weberstraße 5, 76133 Karlsruhe, Germany

Nakajima, Teruyuki Center for Climate System Research, University of Tokyo, 5-1-5, Kashiwanoha, Kashiwa, Chiba 277–8568, Japan

Norris, Joel R. Nierenberg Hall 440, Scripps Institution of Oceanography, 9500 Gilman Drive, Department 0224, La Jolla, CA 92093–0224, U.S.A.

Petch, Jon C. UK Met Office, FitzRoy Road, Exeter EX1 3PB, U.K.

Petters, Markus D. Department of Atmospheric Science, Colorado State University, 200 West Lake Street, Fort Collins, CO 80523–1371, U.S.A.

Pöschl, Ulrich Biogeochemistry Department, Max Planck Institute for Chemistry, Becherweg 27/29, 55128 Mainz, Germany

Quaas, Johannes Max Planck Institute for Meteorology, Bundesstr. 53, 20146 Hamburg, Germany

Rose, Diana Biogeochemistry Department, Max Planck Institute for Chemistry, Becherweg 27/29, 55128 Mainz, Germany

Rosenfeld, Daniel Institute of Earth Sciences, The Hebrew University of Jerusalem, Jerusalem 91904, Israel

Satoh, Masaki Center for Climate System Research, University of Tokyo, 5-1-5 Kashiwanoha, Kashiwa, Chiba 277–8568, Japan
Schmidt, K. Sebastian Institute for Atmospheric Physics, University of Mainz, Becherweg 21, 55099 Mainz, Germany
Schulz, Michael Laboratoire des Sciences du Climat et de l'Environnement, CEA/CNRS, LSCE L'Orme des Merisiers, Bat. 709, 91191 Gif-sur-Yvette cedex, France
Schwartz, Stephen E. Atmospheric Sciences Division, Brookhaven National Laboratory, Upton NY, 11973, U.S.A.
Shaw, Raymond Michigan Technological University, Cloud Physics Laboratory, 1400 Townsend Drive, Houghton, MI 49931–1295, U.S.A.
Siebert, Holger Leibniz Institute for Tropospheric Research, Permoserstrasse 15, 04318 Leipzig, Germany
Siebesma, A. Pier Atmospheric Research Division, Royal Netherlands Meteorological Institute, 3730 AE De Bilt, The Netherlands
Slingo, Anthony Environmental Systems Science Centre, University of Reading, Harry Pitt Building, Whiteknights, Reading, RG6 6AL, U.K.
Spichtinger, Peter Institute for Atmospheric and Climate Science, ETH Zürich, Universitätsstr. 16, 8092 Zurich, Switzerland
Stevens, Bjorn Max Planck Institute for Meteorology, 20146 Hamburg, Germany
Stratmann, Frank Leibniz Institute for Tropospheric Research, Permoserstrasse 15, 04318 Leipzig, Germany
Takayabu, Yukari N. Center for Climate System Research, University of Tokyo, 5-1-5 Kashiwanoha, Kahiwa, Chiba 277–8568, Japan
Tao, Wei-Kuo Mesoscale Atmospheric Processes Branch Code 613.1, NASA's Goddard Space Flight Center, Greenbelt, Maryland 20771, U.S.A.
Wex, Heike Leibniz Institute for Tropospheric Research, Permoserstrasse 15, 04318 Leipzig, Germany
Wood, Robert Atmospheric Sciences, University of Washington, 711 Atmospheric Science–Geophysics Building, Seattle, WA 98195–1640, U.S.A.

1

Introduction

Jost Heintzenberg[1] and Robert J. Charlson[2]

[1]Leibniz Institute for Tropospheric Research, Leipzig, Germany
[2]Department of Atmospheric Sciences, University of Washington, Seattle, WA, U.S.A.

Clouds populate the Earth's atmosphere from the surface, as fog, to the mesosphere, as noctilucent clouds (cf. Table 1.1). They form whenever air is cooled sufficiently for its relative humidity to exceed 100%. Such cooling occurs, for example, when air is lofted upward or when a volume of air loses energy by radiating longwave radiation. Clouds can form at temperatures greater than 0°C (so-called "warm clouds") or below 0°C, and they can exist for lengthy periods of time as supercooled water droplets or as ice, either frozen droplets or crystals grown from the vapor phase. Since cooling can occur at almost any altitude and in myriad meteorological circumstances, clouds take on a nearly indescribable range of physical appearances, ranging from massive cumulus that dominate the sky to wispy veils that may even be too thin to be seen with the naked eye.

Clouds, however, constitute the largest source of uncertainty in the climate system, and there are solid reasons why our knowledge of clouds and their related processes is very limited. To approach these issues, this *Ernst Strüngmann Forum* was convened to assess the limits of current knowledge and to offer new approaches to the understanding of cloud-related issues in the Earth system.

Perturbations of Clouds and Related Aerosols

Humankind is perturbing the Earth's cloud system through its actions (e.g., emissions and surface changes). Contrails, which result from aircraft emissions, represent the most obvious (but not necessarily most relevant) and easily perceived evidence of regional perturbations. Other anthropogenic cloud perturbations in the form of ship tracks, found in persistent low marine clouds, are clearly visible from space. Table 1.2 lists the primary mechanisms of

Table 1.1 Range of typical cloud properties. LWC/IWC = liquid water content/ice water content; $N_{hydrometeors}$ = number of cloud particles per volume of air.

Location	Height (km)	Type	Temperature (°C)	LWC/IWC ($mg\ m^{-3}$)	$N_{hydrometeors}$ (cm^{-3})
Surface	0	Fog	≈ 0	10–100	1–100
Lower troposphere	1–5	Cumulus	10 to ≈ –35	100–1000	10–1000
Lower troposphere	1–3	Stratus	10 to ≈ –35	100–500	10–1000
Troposphere	1–15	Cumulo-nimbus	0 to –60	1000–10,000	100–1000
Upper troposphere	7–15	Cirrus	–40 to –90	1–10	0.01–10
Stratosphere	15–25	Polar strato-spheric clouds	< –80	0.001–0.01	1–10
Mesosphere	80–85	Noctilucent clouds	≈ –120	0.00001–0.0001	25–500

anthropogenic perturbations of clouds recognized today (for a detailed discussion, see Chapters 6, 15–17).

Anderson et al. (Chapter 6) devoted considerable time to the discussion of confounding meteorological influences, which makes it difficult to test hypotheses of anthropogenic effects on clouds. They offer strategies for separating aerosol and meteorological effects in view of the classical "null" hypothesis combined with specific atmospheric settings in which potential anthropogenic cloud changes should be sought.

The observed and hypothesized perturbations of clouds listed in Table 1.2 require a comparison to long-term trends in observed clouds over the past several decades—a period marked by rapidly rising temperatures and changes in the Earth's radiation budget. The radiation budget controls the formation of clouds and is also strongly influenced by their existence, as discussed in Chapter 2. Here, Norris and Slingo review multidecadal variations in various cloud and radiation parameters, documented in previous studies; they argue that no conclusive results are yet available. Problems include the lack of global and quantitative surface measurements, the shortness of the available satellite record, the inability to determine correctly cloud and aerosol properties from satellites, many different kinds of inhomogeneities in data, and insufficient precision to measure the small changes in cloudiness and radiation, which nevertheless can significantly impact the Earth's climate. Their recommendations to improve this situation include (a) processing the available historical measurements as a means of mitigating inhomogeneities, (b) providing better

Table 1.2 Observed and hypothesized anthropogenic perturbations of clouds. CCN = cloud condensation nuclei; IN = ice nuclei; LWC = liquid water content; PBL = planetary boundary layer.

Cloud type	Perturbation	Potential mechanism
Contrails	+ Albedo	Water vapor and anthropogenic CCN/IN[1]
Contrails	– Daily temperature range	Change in air traffic in connection with 9/11[2]
Ship trails	+ Albedo	Anthropogenic water vapor, and CCN[3]
Continental stratocumulus	+ Albedo	Anthropogenic CCN[4]
Continental stratocumulus	+ Cloud-top temperature	Anthropogenic CCN[5]
Continental stratocumulus	– Precipitation	Anthropogenic CCN[6]
Global PBL stratocumulus	+ Albedo	Anthropogenic CCN[7]
Continental rain clouds	– Precipitation	Anthropogenic CCN[8]
Continental deep convection	+ Freezing level	Anthropogenic CCN[9]
Continental low clouds	+ Precipitation	Cloud seeding with CCN or IN[10]
Continental low clouds	± Cloudiness	Surface flux change attributable to vegetation change[11]
Marine PBL clouds	– "Effective radius"	Anthropogenic CCN[12]
PBL stratocumulus	– LWC	Anthropogenic soot[13]
Cloud formation	+ Atmospheric heating	Anthropogenic greenhouse gases[14]
Global cloud cover	+ Cloudiness	Cosmic radiation, ions, anthropogenic CCN[15]
Regional weather	± Synoptic weather systems	Anthropogenic energy release or redirection[16]

[1]Scorer 1955, Meerkötter et al. 1999; [2]Travis et al. 2002; [3]Twomey 1974, Coakley et al. 1987; [4]Twomey 1974, Krüger and Graßl 2002; [5]Devasthale et al. 2005; [6]Albrecht 1989, Rosenfeld 1999, Rosenfeld 2000; [7]Twomey 1974, Nakajima et al. 2003, Sekiguchi et al. 2003; [8]Bell et al. 2008; [9]Andreae et al. 2004; [10]Garstang et al. 2004; [11]Pitman et al. 1999, Ray et al. 2003; [12]Twomey 1974, Albrecht 1989, Han et al. 1994; [13]Ackerman et al. 2000; [14]Douville et al. 2002, Wetherald and Manabe 2002; [15]Marsh and Svensmark 2000; [16]Hoffman 2002

retrievals of cloud and aerosol properties, and (c) extending the record farther back in time. In addition, they advocate an observation system with sufficient stability and longevity to measure long-term variations in cloudiness and the radiation budget with improved precision and accuracy. Unfortunately, as they note, there is currently little prospect in enhancing the present system, which is, moreover, in danger of deterioration since there are no definite commitments to replace several critical instruments when the current satellite missions end.

Clouds consist of particles of condensed water that have grown from either a cloud condensation nucleus (CCN) or an ice nucleus (IN), which caused either a supercooled water droplet to freeze by means of several possible mechanisms or water vapor to deposit directly to form solid water ice. Because CCN and IN are found in the form of aerosol particles, and because almost all aerosol particles can become CCN and some of them are inherently IN, understanding how and why clouds form and what properties they have requires us first to understand the nature and amounts of aerosol particles. The atmospheric aerosol spans a range of four orders of magnitude in particle size and seven orders of magnitude in number concentration (cf. Figure 1.1); CCNs are a subpopulation of this aerosol. Figure 1.1 illustrates the size and concentration ranges that typically act as CCN. Again, complexity arises because of the myriad sorts of aerosol particles, deriving from a host of natural and anthropogenic aerosol sources, that produce the starting material for the formation of cloud particles.

In Chapter 3, Kinne (Part 1) and Pöschl et al. (Part 2) discuss climatologies of cloud-related aerosols in terms of particle number, size, and hygroscopic properties. To date, the high temporal and spatial variability of concentration, size, and composition of atmospheric aerosols has been mapped, based largely on insufficiently evaluated datasets of model simulations or satellite retrievals. Their approach merges data from ground-based remote-sensing networks into multi-model, median background fields that yield global monthly maps of columnar particle properties. The vertical distribution of aerosol characteristics is derived from global modeling. Applying the argument that hygroscopic growth of atmospheric aerosol particles is relatively well-constrained, global

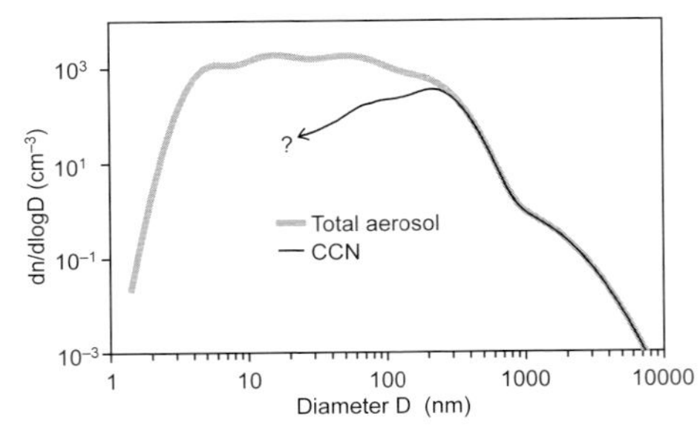

Figure 1.1 Typical near-surface nonurban continental number-size distribution of atmospheric particles (Birmili et al. 1999; Heintzenberg et al. 1998). The typical size distribution of cloud condensation nuclei (CCN) is illustrated by the drop-scavenged fraction according to counterflow virtual impactor (CVI) data from Mertes et al. (2005). The CVI-scavenging data are extrapolated from their upper limit at D = 900 nm to the value of one at 10,000 nm. The question mark indicates the lack of data for smaller particles.

monthly maps for concentrations of CCN are presented. The uncertainty of these results is not known.

Cloud characteristics have vertical and geographical variations, which are important but poorly constrained by present experimental methods. Isaac and Schmidt (Chapter 4) describe the *in-situ* and remote-sensing instrumentation currently available, as well as potential problems in discerning cloud properties. They discuss the necessity to measure parameters on the scales of interest and to present those measurements in proper units. Recommendations for future action include improvements in the accuracy of cloud measurements, global cloud data sets, and better collaborations between those who make and those who use *in-situ* and remote-sensing measurements.

Variability and potential trends of cloud properties affect not only the global radiation budget but also the global hydrological cycle through precipitation (e.g., rain and snow). Precipitation is difficult to assess on large scales. Based on recent developments in passive and active remote sensing, Takayabu and Masunaga (Chapter 5) review current understanding of extreme rainfall, as well as the statistics of light rain and rain from shallow clouds. They find a "butterfly" geographical pattern of shallow rainfall across the equator over both the tropical Pacific and Atlantic oceans. It is not fully understood why this quasi-symmetric pattern appears, as the tropical convergence zones, which geographically constrain deep convective rainfall, are highly asymmetric around the equator. The nature of extreme precipitation varies, depending on the timescale of interest, and is discussed in terms of hourly and daily extremes. Satellite observations imply that the global distribution of extreme precipitation shows a systematic difference from the total rainfall map in terms of, for example, the contrast between land and ocean. Results suggest that the realistic reproduction in models of synoptic systems as well as proper representations of shallow convection and its interaction with the synoptic-scale systems are indispensable for adequate reproduction of extreme daily precipitation.

Anderson et al. (Chapter 6) confirm the findings of the authors of Chapter 5 and emphasize that the most uncertain aspects in current knowledge concern ice microphysics and ice nucleation. Particular difficulties exist because of the confounding effects of built-in correlations of aerosols, clouds, and the meteorological fields in which they are found. These discussions strongly confirm the necessity of understanding and quantifying aerosol and cloud effects as a prerequisite to a full explanation of the climatic records of the twentieth century. Of considerable interest to the entire Forum was the conclusion that observational evidence for large-scale impacts of aerosols on cloud albedo, cloud amount, and precipitation remain ambiguous. Despite this ambiguity, participants were convinced that the emerging trend of warming over the past few decades makes it imperative to look for and quantify coincident changes in clouds. They emphasize the serious need for long-term planning of satellites to monitor the Earth's radiation budget and propose suggestions for new technologies and new orbits (e.g., at the Lagrange point L1 in space).

Cloud-controlling Factors

Since the atmosphere allows only the observation of clouds and does not permit us to control the initial and/or boundary conditions, laboratory studies are important tools for the examination and understanding of microphysical cloud processes under well-defined and repeatable conditions. Stratmann et al. (Chapter 7) provide an overview of the capabilities and limitations of laboratory facilities (ranging in scales from bench-top instruments to vertical mine shafts), wherein clouds are generated artificially and studied under controlled conditions. In this context, hygroscopic growth and activation of aerosol particles, droplet dynamic growth, ice nucleation, and droplet–turbulence interactions can be investigated. Stratmann et al. offer suggestions for future research topics, including investigations into particle hygroscopic growth and activation, the accommodation coefficients of water vapor on liquid water and ice, aerosol effects on primary ice formation in clouds, aerosol-based parameterizations of cloud ice formation, secondary ice formation/multiplication, the production and characterization of particles suitable for cloud simulation experiments, and experiments which combine turbulence and microphysics. The latter is emphasized because of the potential importance of interactions between the microphysical (activation, growth, freezing) and turbulent transport processes within clouds, and the difficulty of studying these through any other approach.

Stevens and Brenguier (Chapter 8) review how meteorological and aerosol factors determine the statistics and climatology of layers of shallow (boundary layer) clouds, with an emphasis on factors that may be expected to change in a fluctuating climate. They identify the paramount role of theory, both to advance our understanding and to improve our modeling and attribution of specific cause and effects. In particular, they argue that limits to current understanding of meteorological controls on cloudiness make it difficult, and in many situations perhaps impossible, to attribute changes in cloudiness to perturbations in the aerosol. Suggestions for advancing our understanding of low cloud-controlling processes include renewing our focus on theory, model craftsmanship, and increasing the scope and breadth of observational efforts.

In Chapter 9, Grabowski and Petch address deep convection, which plays a key role in the Earth's atmospheric general circulation and is often associated with severe weather. They argue that an understanding of the role of deep convection in the climate system, as well as in predictions of climate change, necessitates modeling efforts across all scales, from the micro- to global scale, using a variety of models. Traditional atmospheric general circulation models, in which representation of deep convection and how it may change in the perturbed climate is highly uncertain, are not sufficient. Grabowski and Petch review the relevant aspects of the problem, highlight limitations of current modeling and observational approaches, and suggest areas for future research.

Bretherton and Hartmann (Chapter 10) emphasize current limits in modeling accurately the interaction of clouds and dynamics in the present-day climate.

To guide thinking about the real atmosphere, they demonstrate that horizontal gradients in top-of-the-atmosphere cloud radiative forcing act as atmospheric circulation feedbacks, and that cloud shading helps regulate sea surface temperatures. This relates closely to our lack of fundamental understanding of the empirical controls on tropical deep and low cloud forcing. Bretherton and Hartmann advocate the use of new high-resolution modeling tools, discussed elsewhere in the volume (Chapters 8, 9, and 18) and suggest that new observations may lead to progress if cleverly applied. They caution that scientists should tread carefully and test comprehensively when adding components to general circulation models, such as aerosols and soluble trace gases which interact closely with clouds.

Clouds in the upper troposphere and tropopause region that lack a liquid water phase are called cirrus. They represent a special cloud type, not only because of their formation mechanisms and characteristics but also because of their evident anthropogenic perturbations in terms of contrails. Factors controlling cirrus clouds comprise small- and large-scale atmospheric dynamics, ice nucleation behavior of natural and anthropogenic particles, and interaction with terrestrial and solar radiation. Current understanding of these factors is summarized by Kärcher and Spichtinger in Chapter 11. Key uncertainties in this active area of research are outlined, along with viable approaches to minimize them. These areas of concern include relative humidities in the cirrus regions, vertical velocities, the understanding of ice initiation and growth processes, and accurate data on small ice particles.

Siebesma et al. (Chapter 12) address the shortcomings that arise in atmospheric models attributable to the interactions between resolved and unresolved (i.e., parameterized) cloud-related processes. These problems occur because it is necessary to consider simultaneously a wide range of scales of cloud-related processes, from molecular to global (cf. Figures 12.1 and 12.2). One way to do this is to use smaller-scale process models to improve the representation of clouds in climate models. However, problems and questions arise in deciding just what level of complexity is needed in global models. Siebesma et al. suggest three different pathways to improve the representation of cloud-related processes in future climate models. They recognize that there are many open issues concerning the description of cloud particle formation, right down to questions about the behavior of the water molecule during phase transitions. They echo statements made by Anderson et al. (Chapter 6), in terms of the difficulties in describing the ice phase; however, they add the case of mixed-phase clouds as another cloud process with a serious knowledge deficit. In terms of observations, Siebesma et al. note that polar-orbiting satellites, which are necessary for global, high-resolution coverage with some classes of instruments, do not provide adequate information about the diurnal variations of clouds (e.g., the mid-day convection maximum).

Extent and Nature of Anthropogenic Perturbations of Clouds

Starting with the characteristic parameters of cloud particle precursors and the sensitivities of warm and cold cloud formation to these characteristics, Kreidenweis et al. (Chapter 13) discuss changes to these parameters and propose recommendations for future closure exercises between modeled and experimentally characterized cloud formation processes.

Feingold and Siebert (Chapter 14) extend the microphysical interaction of aerosols and cloud processes to the cloud scale, which involves vertical motions and turbulent mixing processes inside clouds as well as at their borders. Many uncertainties remain on this scale. In particular, there is scant observational evidence of aerosol effects (positive or negative) on surface precipitation. In addition, clouds and precipitation modify the amount of aerosol through both physical and chemical processes so that a three-way interactive feedback between aerosol, cloud microphysics, and cloud dynamics must be considered. Feingold and Siebert demonstrate the dubious utility of simple constructs to separate aerosol effects from the rest of the cloud system. Both observations and modeling suggest that the magnitude (as well as perhaps the sign) of these effects depend on the larger-scale meteorological context in which aerosol–cloud interactions are embedded. They also consider alternate approaches and the possibility of self-regulation processes, which may act to limit the range over which aerosol significantly affects clouds.

The most obvious, yet controversial, perturbation of clouds through willful human intervention—cloud seeding—is addressed in Chapter 15 by Cotton. Here, he reviews research that confirms or refutes the existing concepts for increasing rainfall, decreasing hail damage, and reducing hurricane intensity, and provides a critical overview of the existing atmospheric concepts for climate engineering to counter greenhouse warming.

The physical hypothesis that air pollution in the form of small particles should lead to less efficient formation of precipitation has been established for several decades and is considered by some to be scientifically sound. Ayers and Levin (Chapter 16) provide strong arguments that there is as yet no convincing proof that such a microphysical control of precipitation efficiency has been the prime cause of rainfall reduction in any area of the globe. They emphasize the need for new experimental designs to test this hypothesis in a holistic way, taking into account all possible confounding influences on rainfall trends in a climate that is clearly nonstationary in the face of global warming and natural decadal variability.

Nakajima and Schulz (Chapter 17) broaden the scope of anthropogenic perturbations of clouds to the global scale, using satellite data and global models. Recent observations have detected what appear to be signatures of large-scale changes in the atmospheric aerosol amount and associated changes in cloud fraction and microphysical structures on a global scale. Models can simulate these signatures fairly well, but problems still exist, thus necessitating further improvements. Fields of anthropogenic aerosol optical depth from several atmospheric models have been found to be consistent with the spatial pattern obtained from satellite-derived products. Further studies are needed (a) to improve our ability to differentiate between natural and anthropogenic aerosols, (b) to interpret observed temporal and regional trends in aerosol parameters, and (c) to interpret the extent to which the covariation of satellite-derived aerosol and cloud characteristics can be utilized to advance understanding of aerosol–cloud interactions.

Chuang et al. (Chapter 18) emphasize the daunting task of identifying the myriad effects which must be considered, and the consequences of these for relevant cloud-related processes. These effects include those on microphysics, radiation (both reflected short wave and emitted longwave), precipitation (both rain and snow), dynamics (attributable to the redistribution of energy by clouds), and on chemical processes in clouds and on the composition of precipitation. Three sorts of perturbations are noted: those attributable to aerosols, perturbations of greenhouse gases (which involve changes in dynamics), and changes in the land surface. Three categories of gaps in understanding are identified: conceptual gaps, knowledge or data gaps (which are deficits that could be filled using present-day instruments and data, but for some reason, e.g., lack of resources, have not), and tool gaps or deficits in our ability to make relevant measurements. However, some points seem clear. For example, Chuang et al. emphasize the need to consider multiple scales. The constraints imposed by limitations of available observations were exemplified by the present impossibility to measure small supersaturations in the field (cf. Grabowski and Petch, Chapter 9, who state that it is possible to generate accurately known supersaturations in the laboratory). In addition, Chuang et al. discuss the apparent constancy of global albedo over the past ca. 10 millennia in the context that this stability implies constancy of cloud properties. The possibility exists that as yet unidentified feedbacks might be responsible for such stasis. The difficulty of understanding and quantifying cloud fraction (i.e., the fractional area of a region or the globe covered by clouds) was highlighted as a key problem. Once again, Chuang et al. identified the need for longevity of satellite observations of 30+ years, as well as the need for new sorts of instruments in new orbits (e.g., L1 satellite).

Current Understanding and Quantification of the Effects of Clouds in the Changing Climate System and Strategies to Reduce the Critical Uncertainties

Anthropogenic aerosols are thought to exert a significant indirect radiative forcing because they act as CCN in warm cloud formation and as ice nuclei in cold cloud-forming processes. Haywood et al. (Chapter 19) address this issue by comparing the radiative forcing from the indirect effect of aerosols with those from other radiative forcing components, such as that from changes in well-mixed greenhouse gases. They highlight problems in assessing the effect of anthropogenic aerosols upon clouds under the strict definitions of radiative forcing provided by the IPCC (2007). Straightforward scaling between forcing and the temperature change it induces is significantly compromised in the case of aerosols, where feedbacks from indirect aerosol effects are responses to both radiative and cloud microphysical perturbations. Haywood et al. argue that additional characterization, such as climate efficacy, is required when comparing indirect aerosol effects with other radiative forcings. They suggest using the *radiative flux perturbation* associated with a change from preindustrial to present-day composition, calculated in a global climate model with fixed sea-surface temperature and sea ice, as a supplement to IPCC's definition of forcing.

Collins and Satoh (Chapter 20) discuss the differences of cloud responses to increasing greenhouse gas concentrations using global cloud-resolving models (GCRMs) in comparison to conventional global climate models with cloud parameterization. They demonstrate that high clouds behave differently within these models, suggesting the questions: How is high cloud amount sensitive to cloud processes such as cloud generation, precipitation efficiency, or sedimentation of cloud ice? How are model results of high clouds comparable to current satellite observations such as CloudSat and CALIPSO? How can we understand the change in dynamic fields such as narrowing the precipitation regions, increase in transport of water, and relative humidity?

Strategies to reduce critical uncertainties in our understanding of inadvertent anthropogenic perturbations of clouds are discussed on micro- to mesoscales by Brenguier and Wood (Chapter 21). They emphasize that the challenge is to establish the links between two contrasting forcings, i.e., to understand how clouds respond to changes in the general circulation in order to quantify how this response might be modulated by changes in their microphysical properties. The two generic classes of micro- to mesoscale observational strategies, the Eulerian column closure and the Lagrangian cloud system evolution approaches, are described using examples of low-level cloud studies, and recommendations are made on how they should be combined with large-scale information to address this issue.

Illingworth and Bony (Chapter 22) extend this strategic discussion from the mesoscale to larger scales, where the response of clouds to climate change remains very uncertain because of an incomplete knowledge of the cloud physics and the difficulties in simulating the different properties of clouds. They propose an observational strategy to improve the representation of clouds in large-scale models and reduce uncertainties in the future change of cloud properties. This consists first in determining what key aspects of the simulation of clouds are the most critical with respect to future climate changes, and then in using specific methodologies and new datasets to improve the simulation of these aspects in large–scale models.

A critical review of the representation of clouds in large-scale models by Lohmann and Schwartz (Chapter 23) reveals a major unresolved problem. This is attributable to the high sensitivity of radiative transfer and water cycle to cloud properties and processes, an incomplete understanding of these processes, and the wide range of scales over which these processes occur. Small changes in the amount, altitude, physical thickness, and/or microphysical properties of clouds which result from human influences can exert changes in the Earth's radiation budget that are comparable to the radiative forcing by anthropogenic greenhouse gases, thus either partly offsetting or enhancing greenhouse warming. Because clouds form on aerosol particles, changes in the amount and/or composition of aerosols affect clouds in a variety of ways. Because of the forcing of the radiation balance that results from aerosol–cloud interactions, major uncertainties exist and must be addressed before accurate results can be obtained.

Quaas et al. (Chapter 23) focus on the necessity of models at all scales, especially global, and note the apparent lack of progress in quantifying the cloud–albedo–climate feedback, even though this problem has been identified for more than two decades. Substantial discussion centers on our need for present-day observational proxies to extrapolate future cloud perturbations. In addition, Quaas et al. emphasize the role of small-scale models to describe processes in large-scale models. Substantial effort seems to be required if we are to be able to identify and isolate key cloud-related processes. Quass et al. discuss the problem of applying the concept of climate forcing (Wm^{-2}) to systems in which the fast response of the system via feedbacks changes the initial forcing itself. They recommend the use of a different terminology (i.e., the term *radiative flux perturbation*) to avoid misapplication of the concept of forcing. This new forcing concept, however, could be defined in an even more rigorous way, with explicit statements about the maximum response time of system adjustments in the models that are allowed. Quaas et al. note the necessity of developing process-based evaluation of large-scale models: What aspects of clouds do we need to represent to achieve an accurate assessment of aerosol–cloud interactions? Is it possible to design an observational program to detect and quantify aerosol indirect effects?

Describing the Response of Clouds to Changing Climate: The Need for Multiple Indices of Climate Change

Although there is no question that clouds must have changed as a result of forced climate change and because of the impositions of anthropogenic aerosol on the atmosphere, major questions and uncertainties remain in terms of the details: How have clouds changed? How much have they changed? How will they change in the future? The simplest climate models of the 1960s, as well as the zero-dimensional model of Arrhenius (1896), projected increased

Table 1.3 Indices of climate change.

Index of change	Symbol, unit
Global mean surface temperature	Δ T
Ocean heat content	Joules
Change in regional-scale surface temperature	Δ T
Change in global- or regional-scale atmospheric water content	B_{H2O}, g m^{-2} B_{H2O} region
Total greenhouse absorption	LW_{abs} W m^{-2}
Global or regional mean radiative forcing	Δ F, W m^{-2}
Global or regional mean precipitation	mm
Atmospheric GHG concentration or concentration change	e.g., Δ CO_2
Ocean pH	–
Global or regional mean albedo	Δ A
Sea level change	meters
Global or regional change in solar irradiance at the surface	W m^{-2}
Change in cloud cover, type of cloud, height of cloud, etc.	–

anthropogenic water vapor as a result of global warming. Yet the simplicity of this phenomenon (attributable to consideration of the accurately known and strong dependence of water vapor pressure on temperature change) belies the complexity of responses via a multitude of cloud processes and feedbacks. Indeed, *temperature change* is a misleadingly simple index of the known and suspected changes in clouds, cloud processes, and cloud functions.

Temperature change is the only "gold standard" index of forced climate change and natural variability. However, because this parameter does not capture the essence of changes in clouds and cloud functions (e.g., their role in planetary albedo or the amount, location, and timing of precipitation), we suggest that other indices can and should be used to describe more fully and quantify the consequences of change in the atmosphere caused by human activity (see Table 1.3). Of these, several indices pertain to the known or suspected changes of clouds in the perturbed climate system. In Table 1.3, we include regional-scale variables because regional changes are generally more important to society than global mean changes and because regional-scale changes in cloud-related parameters may, in some cases, be easier to detect and attribute than global-scale changes.

Context of this Forum: The Urgency of Current Demands by the Policy Community on the Scientific Community and the Need for High Scientific Standards

Given the various forecasts of impending climatic catastrophes, there can be little doubt that the issue of "global warming" has captured world attention. Such forecasts range from modest increases of global mean temperature to severe climatic shifts, flooding coastlines, crop failures, and beyond. Yet the term "global warming" is also a source of some ambiguity insofar as the verb "to warm" has both transitive and intransitive meanings.

There is no scientific doubt that the increase in manmade greenhouse gases (e.g., CO_2) has created a warming of the lower layers of the atmosphere in the sense that these gases have caused heat to be added to the air (the transitive meaning). However, substantial uncertainty exists as to how much warming (in the intransitive sense) can be expected as a result of this additional heat energy, not least because the sensitivity of the climate to such perturbations is itself uncertain. Indeed, the sensitivity of global mean surface temperature to a given change in the content of greenhouse gases is uncertain to at least a factor of two and perhaps a factor of three (Schwartz 2008). A large portion of this uncertainty in climate sensitivity, and hence uncertainty in the climate forecast, stems from the uncertainty in the numerous effects of clouds and associated aerosols.

This high degree of uncertainty regarding clouds, combined with the urgency for societies to make firm decisions on the emissions of greenhouse gases (most especially on the continued combustion of fossil carbon fuels), places a great burden on our scientific community. Because we are the only group trained to study the details of clouds and climate, we must do our utmost to reduce the uncertainties and clarify the details of the climate forecast. In doing so, we assume the awesome obligation to communicate our research to the policy community in ways that are impeccably honest and forthright, so that the uncertainties that will always remain and which will, by nature, constrain the confidence that can be taken regarding policy decisions are understood. Just a few decades ago, our fields of science contributed far less to policy making, and we enjoyed the freedom to speculate openly about the physics of clouds and aerosols. Today, however, what we say does count, and a very attentive audience is listening. We must therefore hold ourselves and our findings to an ever-higher standard of scientific proof and be candid about what we have and have not found.

Reducing the uncertainty of climate sensitivity requires vast improvements in the ways that clouds and aerosols are understood and described in the models used by decision makers. What is literally at stake is the ability of the global society to plan rationally ways to conduct its business. As stated by Schwartz (2008):

> This uncertainty in climate sensitivity, which gives rise to a comparable uncertainty in the shared global resource of the amount of fossil fuel that can be burned consonant with a given increase in global mean surface temperature, greatly limits the ability to effectively formulate strategies to limit climate change while meeting the world's energy requirements.

It is of crucial importance for us to find answers to the many puzzles posed by clouds in the perturbed climate system.

References

Ackerman, A. S., Toon, O. B., Stevens, D. E., Heymsfield, A. J., Ramanathan, V. and Welton, E. J. 2000. Reduction of tropical cloudiness by soot. *Science* **288**:1042–1047.

Albrecht, B. A. 1989. Aerosols, cloud microphysics, and fractional cloudiness. *Science* **245**:1227–1230.

Andreae, M. O., D. Rosenfeld, P. Artaxo et al. 2004. Smoking rain clouds over the Amazon. *Science* **303(5662)**:1337–1342.

Arrhenius, S. A. 1896. Nature's heat usage. *Nordisk Tidskrift* **14**:121–130.

Bell, T. L., D. Rosenfeld, K.-M. Kim et al. 2008. Midweek increase in U.S. summer rain and storm heights suggests air pollution invigorates rainstorms. *J. Geophys. Res.* **113**:D02209.

Birmili, W., J. Heintzenberg, and A. Wiedensohler. 1999. Representative measurement and parameterization of the submicron continental particle size distribution. *J. Aerosol Sci.* **30**:S229–S230.

Coakley, Jr., J. A., R. L. Bernstein, and P. A. Durkee. 1987. Effect of ship-stack effluents on cloud reflectivity. *Science* **237**:1020–1022.

Devasthale, A., O. Krüger, and H. Graßl. 2005. Change in cloud-top temperatures over Europe. *IEEE Trans. Geosci. Rem. Sens.* **2(3)**:333–336.

Douville, H., F. Chauvin, J.-F. Royer, D. Salas-Melia, and S. Tyteca. 2002. Sensitivity of the hydrological cycles to increasing amounts of greenhouse gases and aerosols. *Climate Dyn.* **20(1)**:45–68.

Garstang, M., R. Bruintjes, R. Serafin et al. 2004. Weather modification: Finding common ground. *Bull. Amer. Meteor. Soc.* **86**:647–655.

Han, Q., W. B. Rossow, and A. A. Lacis. 1994. Near-global survey of effective droplet radii in liquid water clouds using ISCCP data. *J. Climate* **7**:465–497.

Heintzenberg, J., K. Müller, W. Birmili, G. Spindler, and A. Wiedensohler. 1998. Mass-related aerosol properties over the Leipzig Basin. *J. Geophys. Res.* **103(D11)**:13,125–13,135.

Hoffman, R. N. 2002. Controlling the global weather. *Bull. Amer. Meteor. Soc.* **83(2)**:241–248.

IPCC. 2007. Climate Change 2007: The Physical Science Basis. Contribution of Working Group I to the Fourth Assessment Report of the Intergovernmental Panel on Climate Change, ed. S. Solomon, D. Qin, M. Manning et al. New York: Cambridge Univ. Press.

Krüger, O., and H. Graßl. 2002. The indirect aerosol effect over Europe. *Geophys. Res. Lett.* **29(19)**:1925.

Marsh, N. D., and H. Svensmark. 2000. Low cloud properties influenced by cosmic rays. *Phys. Rev. Lett.* **85(23)**:5004–5007.

Meerkötter, R., U. Schumann, D. R. Doelling et al.1999. Radiative forcing by contrails. *Ann. Geophysicae* **17**:1080–1094.

Mertes, S., K. Lehmann, A. Nowak, A. Maßling, and A. Wiedensohler. 2005. Link between aerosol hygroscopic growth and droplet activation observed for hill-capped clouds at connected flow conditions during FEBUKO. *Atmos. Environ.* **39(23–24)**:4247–4256.

Nakajima, T., M. Sekiguchi, T. Takemura et al. 2003. Significance of direct and indirect radiative forcings of aerosols in the East China Sea region. *J. Geophys. Res.* **108(D23)**:8658.

Pitman, A., R. Pielke, Sr., R. Avissar et al. 1999. The role of the land surface in weather and climate: Does the land surface matter? *IGBP Newsletter* **39**:4–11.

Ray, D. K., U. S. Nair, R. M. Welch et al. 2003. Effects of land use in Southwest Australia: 1. Observations of cumulus cloudiness and energy fluxes. *J. Geophys. Res.* **108(D14)**:4414.

Rosenfeld, D. 1999. TRMM observed first direct evidence of smoke from forest fires inhibiting rainfall. *Geophys. Res. Lett.* **26**:3105–3108.

Rosenfeld, D. 2000. Suppression of rain and snow by urban and industrial air pollution. *Science* **287**:1793–1796.

Schwartz, S. E. 2008. Uncertainty in Climate Sensitivity: Causes, Consequences, Challenges. *Energy Environ. Sci.* **1**:430

Scorer, R. S. 1955. Condensation trails. *Weather* **10(9)**:281–287.

Sekiguchi, M., T. Nakajima, K. Suzuki et al. 2003. A study of the direct and indirect effects of aerosols using global satellite data sets of aerosol and cloud parameters. *J. Geophys. Res.* **108(D22)**:4699.

Travis, D. J., A. M. Carleton, and R. G. Lauritsen. 2002. Contrails reduce daily temperature range. *Nature* **418**:601.

Twomey, S. A. 1974. Pollution and the planetary albedo. *Atmos. Environ.* **8**:1251–1256.

Wetherald, R. T. and S. Manabe. 2002. Simulation of hydrologic changes associated with global warming. *J. Geophys. Res.* **107(D19)**:4379.

2

Trends in Observed Cloudiness and Earth's Radiation Budget

What Do We Not Know and What Do We Need to Know?

Joel R. Norris[1] and Anthony Slingo[2]

[1]Scripps Institution of Oceanography, San Diego, La Jolla, CA, U.S.A.
[2]Environmental Systems Science Centre, University of Reading, Reading, U.K.

Abstract

The response of clouds and their radiative effects to global warming represents a long-standing and considerable area of uncertainty in our understanding of climate change. At present, it is not known whether changes in cloudiness will exacerbate, mitigate, or have little effect on the increasing global surface temperature caused by anthropogenic greenhouse radiative forcing. Another substantial uncertainty is the magnitude of radiative forcing resulting from the modification of cloud properties by anthropogenic aerosols. Global climate models provide scant reliable insight regarding these issues because of their inability to parameterize correctly or otherwise represent the small-scale convective, turbulent, and microphysical processes that control cloud properties. It is therefore crucial to document and assess global and regional low-frequency variations in clouds and radiation flux that have occurred over the past several decades, a period marked by rapidly rising temperature and changes in anthropogenic aerosol emissions. This will enable us to estimate from observations how clouds and their impacts on the radiation budget are responding to global warming and aerosol changes. Moreover, a trustworthy observational record will provide a good constraint on global climate model simulations.

Previous investigators have documented multidecadal variations in various cloud and radiation parameters, but no conclusive results are yet available. Problems include the lack of global and quantitative surface measurements, the shortness of the available satellite record, the inability to determine correctly cloud and aerosol properties from satellite data, many different kinds of inhomogeneities in the data, and insufficient precision to measure the small changes in cloudiness and radiation that nevertheless can have large impacts on the Earth's climate. Many of these deficiencies emerge from the absence of an observing system designed to monitor variations in clouds and radiation on timescales relevant to climate; to compensate, observations must be assembled from a system originally designed for purposes of weather forecasting. Although we cannot go back in time to make more and better observations, we need to improve our processing of the available historical measurements to mitigate inhomogeneities, provide better retrievals of cloud and aerosol properties, and extend the record farther back in time. Furthermore, it is essential to construct an observation system with sufficient stability and longevity to measure long-term variations in cloudiness and the radiation budget with improved precision and accuracy. Our present observing system has unfortunately little prospect of enhancement at this time and, moreover, is in danger of future deterioration since there are no definite commitments to replace several critical instruments when current satellite missions end.

Introduction

Clouds and Earth's Radiation Budget

Clouds greatly impact the Earth's radiation budget (ERB) because they reflect solar or shortwave (SW) radiation back to space and restrict the emission of thermal or longwave (LW) radiation to space. One conventional measure of this radiative effect is "cloud radiative forcing" (CRF); that is, the difference between actual radiation flux and what it would be if clouds were absent (Ramanathan et al. 1989). A more proper term might be "cloud radiative effect," since "forcing" is usually reserved for external perturbations to the climate system rather than internal components. Regardless of etymology, we will follow common usage and employ CRF in this chapter. Shortwave cloud radiative forcing (SWCRF) is almost always negative, since reflection by clouds usually causes a loss of energy from the climate system, and LWCRF is almost always positive, since clouds usually reduce the amount of radiation emitted to space and thus cause a gain of energy to the climate system.

Several factors influence SWCRF, including the incident solar flux, the horizontal extent of clouds, the vertically integrated amount of cloud condensate (cloud water path), and cloud particle size, phase, and shape. Cloud water path and cloud particle characteristics contribute both to cloud optical thickness and cloud albedo, with the water path being the dominant factor. Variations in cloud cover generally affect SWCRF much more than do variations in cloud albedo. The magnitude of SWCRF varies linearly with insolation such that cloud changes occurring during daytime, the summer season, and at low latitude have the greatest impact on ERB. The dominance of clouds in the SW portion of the spectrum is illustrated by the fact that if clouds were instantaneously removed, the reflectivity of the planet would be approximately halved.

In addition, cloud condensate and cloud particle characteristics affect cloud emissivity in the LW part of the spectrum, and little upwelling radiation passes through clouds with high emissivity (low transmissivity). Usually, clouds emit less radiation than does the surface due to their colder temperatures, and extensive coverage by clouds with tops high in the atmosphere reduces substantially the amount of LW flux escaping to space. Variability in the horizontal extent of high-level clouds generally has a greater influence on LWCRF than does variability in the specific height or emissivity of high-level clouds. Outgoing LW radiation would increase by about 15% if clouds were instantaneously removed. The loss of radiation by SWCRF is greater than the retention of radiation by LWCRF in the global average, and clouds consequently produce a net cooling effect in the current climate.

The basic geographical distributions and seasonal variations of reflected SW and outgoing LW radiation are well observed and well understood. Information is also available on interannual variability, particularly the large perturbations to ERB resulting from the El Niño/Southern Oscillation (ENSO) phenomenon (Figure 2.1). These perturbations are closely related to shifts in the spatial distribution of optically thick clouds with high tops, which are primarily generated by deep convection. One interesting feature, which may have far-reaching consequences for climate stability, is the tendency for a cancellation between the SWCRF and LWCRF over tropical deep convection regions. It

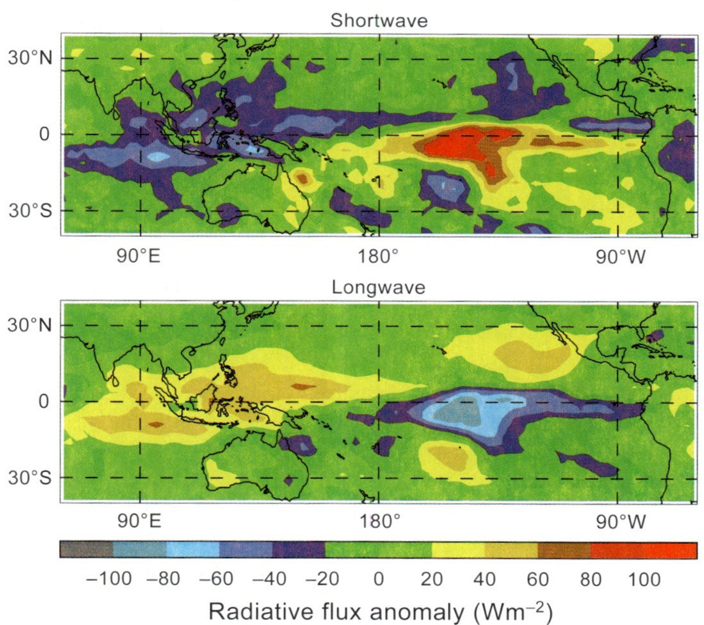

Figure 2.1 Differences in outgoing shortwave (top) and longwave (bottom) radiation fluxes between CERES data for January 1998 (the peak of a very strong ENSO event) and the January average of ERBE data for 1985–1989.

has been argued that this is a fundamental property of the tropical atmosphere, which ensures that the net cloud radiative feedback from deep tropical convection during climate change must be close to zero (Hartmann et al. 2001; cf. also Bretherton and Hartmann, this volume).

Importance of Small Cloud and ERB Changes and the Difficulties in Measuring Them

Global climate is very sensitive to even slight changes in clouds and radiation, hence requiring very precise measurements of variations in these parameters. For example, an instantaneous doubling of CO_2 (100% increase) would produce about a 4 W m^{-2} reduction in outgoing radiation flux, which is less than 2% of the 235 W m^{-2} in global mean outgoing radiation. Since the trend in CO_2 concentration over the span of a few decades is much smaller than a doubling, the trend in ERB is also commensurately much smaller and thus even more difficult to detect. Such seemingly minor departures from radiative balance, when sustained over decades, could nevertheless produce changes in global temperature and climate sufficient to have substantial effects on human society and natural ecosystems. A second measurement obstacle is that radiation flux, unlike long-lived greenhouse gases, has little spatial coherence and must be measured from space at many locations to obtain a reasonable global average for ERB. This must be done by satellites, which suffer the additional disadvantage of being able to make observations of upwelling radiation intensity from only a single direction and spot on the Earth. Radiative modeling is required to convert this into a hemispheric flux.

Small cloud changes are important because they can exert more leverage over ERB than equivalent changes in greenhouse gases. For example, a 15–20% relative increase in low-level cloud amount is presumed to counteract the radiative forcing caused by a doubling of CO_2 (Slingo 1990). Low-level marine stratocumulus and summertime midlatitude frontal clouds have an especially large negative net CRF and impact on ERB because, unlike the case for tropical deep convective clouds, their SWCRF is much greater than their LWCRF (Ramanathan et al. 1989). In contrast, optically thin high-level clouds have a positive net CRF because they strongly reduce LW emission while transmitting SW radiation. Since various cloud types have strikingly different radiative effects, it is not a simple matter to determine the overall global impact of cloud changes; each cloud type and climate regime must be examined in particular. Moreover, alterations of cloud albedo, cloud emissivity, and cloud height can affect ERB even when cloud amount remains the same. Since insolation strongly effects the magnitude of SWCRF, shifts in the latitude or seasonal cycle of clouds can change ERB even if mean cloud amount and cloud properties do not change. To detect long-term variations in their properties and radiative effects, clouds must be monitored with precision everywhere around the globe.

Cloud Simulations and Climate Sensitivity in Global Climate Models

Over the past two decades, generations of global climate models (GCMs) have been tuned to ERB, CRF, and cloud data to ensure that they represent at least the current effects of clouds as well as possible. Despite this empirical contribution, many studies comparing simulated clouds with observed clouds have found that GCMs poorly represent clouds when evaluated on terms for which they were not explicitly tuned. In fact, GCMs frequently reproduce the observed time-averaged ERB only by compensating for errors (e.g., cloud fraction that is too small and cloud optical thickness that is too large). Observed relationships between clouds and dynamical parameters are especially poorly simulated by GCMs (e.g., Norris and Weaver 2001; Tselioudis and Jakob 2002) even though these are more likely to be relevant to climate change questions than is the production of realistic cloud and radiation climatologies. One of the main reasons for such incorrect and inconsistent cloud and radiation simulations is that GCMs lack sufficient spatial resolution to represent properly the nonlinear small-scale convective, turbulent, and microphysical processes that control cloud properties, which instead must be crudely parameterized.

Climate sensitivity has been conventionally defined as the equilibrium change in global mean surface temperature in response to the radiative forcing caused by a doubling of CO_2 in the atmosphere. Recent GCM intercomparisons definitively indicate that low-level marine clouds provide the greatest contribution to the spread in climate sensitivity between models (Bony and Dufresne 2005). This was confirmed in the findings of the recent IPCC Fourth Assessment Report (Randall et al. 2007). Some GCMs suggest that the horizontal extent of low-level clouds over low-latitude oceans will increase with higher global temperature, whereas other GCMs suggest that low-level cloud amount will decrease. Changes in trade cumulus (small cloud fraction and weak negative net CRF) may be more important than changes in marine stratocumulus (large cloud fraction and strong negative net CRF) because the former occur over a much larger area of the ocean. The amount of coverage by trade cumulus or stratocumulus is notoriously difficult to predict in large-scale models because the dynamical forcing is weak and small changes in this forcing and in the boundary layer properties can have a large impact on the mean relative humidity and hence on the cloud cover. Note that the importance of marine boundary layer clouds to the climate sensitivity of GCMs does not necessarily imply that such clouds are equally important to the climate sensitivity of the Earth, nor is it necessarily the case that changes in CRF will be greatest in regions where climatologic CRF is currently largest. Nevertheless, marine boundary layer clouds can potentially exert greater radiative leverage on the climate system than any other cloud type.

Two Key Questions

How Are Clouds Responding to Global Warming?

One long-standing and critical issue for our understanding of climate change is to determine how clouds have changed in response to global warming and how they will change in the future, and whether the cloud radiative response will exacerbate, mitigate, or have little effect on the increase in Earth's temperature caused by anthropogenic greenhouse gases (Randall et al. 2007). The enduring problems with GCMs motivate an investigation of the observational record made over the past several decades. Despite the presence of natural variability in the historical cloud and ERB time series, a climate change signal may be apparent, especially when we consider that the last forty years have been a time period of substantially increasing anthropogenic radiative forcing and global temperature. Relating multidecadal changes in cloud properties and ERB to changes in temperature and atmospheric circulation will provide insight into cloud feedbacks on the climate system and a useful constraint on GCM climate simulations. Our ability to document and interpret the observational record is unfortunately hampered by limitations and weaknesses in the global observational system, which was developed to monitor weather rather than climate.

How Are Clouds Responding to Anthropogenic Aerosols?

A second key question is how clouds and CRF have been altered in response to changes in the concentration and composition of anthropogenic aerosol particles. Without a dynamic feedback on cloud macrophysics, an increase in the number of hygroscopic particles will produce a larger number of smaller cloud droplets and hence increase cloud albedo. Changes in cloud particle size, phase, and number concentration have furthermore been hypothesized to have a substantial influence on cloud microphysical and precipitation processes, thus potentially affecting cloud condensate, cloud lifetime, and cloud fraction. Although local changes in cloud albedo and regional changes in cloud droplet radius have been observed, it has been difficult to demonstrate unambiguously a significant large-scale influence of anthropogenic aerosol on cloud albedo and cloud fraction due to the very large confounding impact of meteorology. Since aerosol transport is highly correlated with atmospheric dynamics, it is no simple matter to distinguish the effects of anthropogenic aerosol from natural meteorological influences on cloud properties. Another complication is that strong atmospheric radiative heating and surface radiative cooling by aerosols may possibly change regional circulation patterns and thus affect cloudiness through a non-microphysical mechanism (e.g., Menon et al. 2002). Investigations of long-term trends in cloud properties in regions that have experienced trends in aerosols may reveal the true aerosol influence on cloudiness and provide constraints for GCM simulations. It is necessary, however, to control carefully for changes in meteorological conditions and to determine whether aerosol radiative heating may have modified regional circulation[4]. The specific radiative impacts of aerosols, cloudiness, or aerosol–cloud interactions must also be correctly identified and distinguished in long-term satellite measurements, which is difficult because thin cloudiness appears optically similar to aerosol haze (e.g., Charlson et al. 2007; Koren et al. 2007).

The Observational Record

Visual Cloud Observations

The longest record of cloudiness comes from synoptic reports made by human observers at weather stations over land and on ships over the ocean. These include the fraction of sky dome covered by all clouds and clouds in the lowest layer as well as morphological cloud types at low, mid, and high levels. The largest archive of synoptic cloud observations is the Extended Edited Cloud Report Archive (EECRA) (Hahn and Warren 1999). Several investigations have found that interannual variability and trends in surface-observed cloudiness are physically consistent with trends and variability in related meteorological parameters in some regions of the world (e.g., Norris 2000a, 2005b; Park and Leovy 2000; Sun et al. 2001). Although visual cloud observations lack quantitative radiative information, the application of locally derived linear empirical coefficients can nonetheless provide useful (but incomplete) information on changes in ERB. The reason for this is that monthly anomalies in cloud cover, the most quantitative parameter in synoptic cloud reports, are the largest contributors to monthly anomalies in ERB. The smaller radiative impacts of variations in cloud albedo, cloud emissivity, and cloud height (aside from differing changes in low- and upper-level cloud cover) cannot be derived from visual cloud observations.

Short- and longwave anomalies estimated from gridded surface cloud cover observations correspond substantially to satellite measurements of SWCRF and LWCRF on interannual and decadal timescales for many locations of the world (Norris 2005a). In regions with inadequate sampling, such as some tropical land areas and most tropical and southern hemisphere ocean areas, there is much less agreement. The limited information obtained from surface cloud observations is nonetheless valuable because it provides the only near-global record of cloud cover and related ERB changes during the pre-satellite era. Satellite measurements of upper-level cloud cover and LW radiation are consistent with the surface-observed record during the period of overlap, suggesting that it may also be reliable in the pre-satellite era (Norris 2005a). For example, the time series of upper-level cloud cover anomalies obtained from the EECRA and from the International Satellite Cloud Climatology Project (ISCCP) (Rossow and Schiffer 1999) coincide well (top of Figure 2.2), apart from the 1991–1993

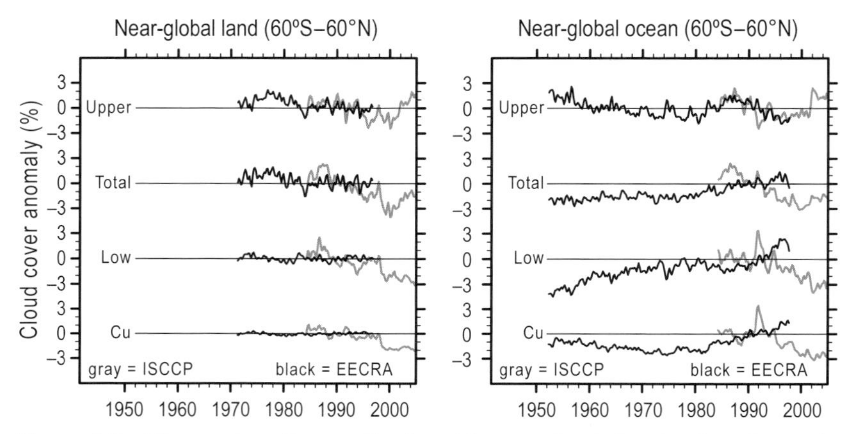

Figure 2.2 Time series of anomalies in total, upper-level, low-level, and cumulus cloud cover from ISCCP satellite (gray) and EECRA surface (black) observations averaged between 60°S and 60°N over land and ocean.

time period when aerosols resulting from the Mt. Pinatubo volcanic eruption caused upper-level cloudiness to be misidentified as low-level cloudiness in ISCCP (Luo et al. 2002).

Surface-observed, upper-level cloud cover has declined globally over the past several decades (Figure 2.2), and this has weakened LWCRF over time. Although it appears that this long-term decrease in upper-level cloud cover was abruptly reversed in October 2001, the increase reported by ISCCP is actually an artifact. Use of an incorrect calibration coefficient for the NOAA-16 polar orbiter produced an underestimate of cloud-top temperature (Knapp 2008), which correspondingly caused an overestimate of upper-level cloud cover. Surface and satellite records disagree on trends in total cloud cover, low-level cloud cover (Figure 2.2), and reflected SW radiation. The surface data indicate that in recent decades there has been a large increase in low-level cloud amount, especially from cumulus clouds over tropical oceans. Such an enhancement of cloud cover would lead to a physically implausible reduction in absorbed SW radiation unless there was a large commensurate decrease in cloud albedo (a parameter that cannot be quantified in visual cloud reports) (Norris 2005a). The origin of this apparent artifact has not yet been identified.

Surface Radiation Measurements

The Global Energy Balance Archive (GEBA) contains measurements of downwelling solar radiation at the surface that were made at various locations around the world during the past several decades, primarily in North America and Eurasia and only over land (Gilgen and Ohmura 1999). Not all stations are trustworthy; many report unrealistically large interannual and decadal variations in solar radiation flux, particularly in the United States and in developing countries. Furthermore, sensor degradation may lead to a spurious reduction in observed solar radiation (M. Wild, pers. comm.). Nonetheless, several studies have made use of high-quality stations and documented a geographically widespread decrease in surface solar radiation from the 1960s until the mid-1980s, followed by an increase from the 1990s onward (Wild et al. 2005). These trends are popularly known as "global dimming" and "global brightening," even though they do not necessarily occur around the entire world. The reported trends probably represent regional rather than global mean changes in downwelling solar radiation, and may suffer from an urban bias (Alpert et al. 2005).

Several explanations for the observed "global dimming" and "global brightening" have been offered, including trends in cloud cover and cloud optical thickness. The most plausible general cause, however, is a rise followed by a decline in anthropogenic aerosol burden. According to this scenario, haze from fossil fuel and biomass burning has increased since the middle of the 20th century and has, as a result, enhanced atmospheric reflection and absorption of solar radiation ("global dimming"). Air pollution control laws enacted during the 1980s and later in certain countries reduced this haze, thus allowing more solar radiation to reach the surface ("global brightening"). Another likely factor contributing to "brightening" in Europe was the shutdown of fossil fuel combustion sources following the collapse of communism in Eastern Europe. In certain developing countries such as India, "global dimming" continues and there has been no reversal to "brightening" (Wild et al. 2005).

It has been difficult to distinguish impacts of cloud changes from aerosol changes using GEBA data because the values typically exist only as monthly averages of the direct plus diffuse flux. Thus, cloudy intervals cannot be separated from cloudless intervals in the time series so as to distinguish cloud radiative effects and potential aerosol impacts on clear-sky solar radiation. This shortcoming is alleviated in the Baseline Surface Radiation Network (BSRN) dataset, which provides data on downwelling SW and LW radiation every minute at a small number of stations as of 1990 (Ohmura et al. 1998). For the larger and lengthier GEBA record, the radiative effects of cloud cover variations on surface solar radiation can be reliably estimated from surface synoptic cloud reports. The recent study of Norris and Wild (2007) demonstrated that cloud cover variability, associated with weather and atmospheric circulation regimes, was the dominant cause of interannual anomalies in surface solar radiation over Europe but not of long-term trends. Removal of the cloud cover contribution from the solar radiation time series revealed more distinct "dimming" and "brightening" trends than were apparent in previous studies.

Retrievals of the components of the surface radiation budget are available from several satellite programs (e.g., Gupta et al. 1999, Zhang et al. 2004); however, care must be taken in using them because a significant amount of modeling is used to create such products. A score or so of well-instrumented surface sites (e.g., BSRN) provide a reference both for satellite retrievals and for the much larger number of surface sites with limited instrumentation. It must also be remembered that the signals of an enhanced greenhouse effect are much larger at the surface than at the top of the atmosphere, so direct surface measurements should play an important part in the detection and attribution of global warming.

One intriguing, new potential method of monitoring changes in the Earth's global albedo is that of "earthshine" observations (Pallé et al. 2004). This approach would use a network of surface telescopes to measure (a) variations in the brightness of the non-sunlit part of the Moon and thus (b) how much solar radiation is reflected by the Earth to the Moon and back again.

Satellite Measurements of the Earth's Radiation Budget

The arrival of the satellite era provided the first global direct measurements of clouds and ERB, starting with the basic instruments mounted on Explorer 7 in 1959 and TIROS 1 in 1960. The Nimbus series of experimental satellites, culminating in the important scanning radiometer observations from Nimbus 7, laid the foundation for the comprehensive Earth Radiation Budget Experiment (ERBE) from 1985–1990. ERBE placed broadband scanning instruments on two operational weather satellites and on the dedicated satellite, ERBS, which flew in a precessing low-inclination orbit to sample all times of day through its 72-day orbital repeat cycle. After ERBE, further scanning radiometers flew on the ScaRaB missions from 1994–1995 and 1998–1999. The CERES instruments on the TRMM, Terra, and Aqua satellites of the NASA Earth Observing System set the highest standards for accuracy (Wielicki et al. 1996).

Many of the satellites mentioned above were, or are, in sun-synchronous orbits with limited diurnal sampling. CERES overcomes this problem by incorporating additional data from the geostationary satellites used also by ISCCP. The ERBS and TRMM satellites, as well as the Russian satellites that carried the ScaRaB instruments, are in precessing orbits, providing coverage of all local times over the orbital repeat period; however, at any one location, the temporal sampling is still limited. To date, the only broadband radiometers in geostationary orbit are the GERB instruments on Meteosat-8 and Meteosat-9, which provide 15-minute data for the whole visible disk, albeit over a limited, although climatically important, geographical region (Harries et al. 2005). The broadband data from all of these instruments provide a wealth of information on the geographical distributions and the diurnal, seasonal, and interannual variability of the ERB, including the effect of clouds. This is possible because the signals are large (typically tens of W m^{-2} or more, as is clear from Figure 2.1) and thus are well observed relative to the instrument noise and absolute calibration.

Whereas multiyear changes in ERB due to El Niño/La Niña and the 1991 Mt. Pinatubo volcanic eruption are clearly apparent in the satellite tropical mean time series of outgoing LW radiation (Figure 2.3), it is much more challenging to study decadal and longer period variability and to search for trends associated with climate change. The latter topics place much more stringent demands on the measurements, which were seldom designed for such work. Some evidence has been put forward for the existence of decadal variations in the tropical radiation budget, particularly from the ERBS wide-field-of-view (WFOV) instrument; however, detection of such changes is hampered by various observational artifacts. Several of these were discovered after the original publication by Wielicki et al. (2002), including the existence of a spurious semiannual cycle due to aliasing of the diurnal cycle by the precessing satellite orbit. The first correction to the ERBS data adjusted for a spurious change in radiation flux that resulted from a downward trend in satellite altitude; the second correction adjusted for a previously unidentified degradation of solar transmission through the instrument dome. These two small corrections to the ERBS data reduced the change in LW flux by a factor of four and reversed the sign of the change in the net radiation, compared with the original study (Wong et al. 2006). The most recent results indicate a slight trend towards more LW emission (consistent with a weakening of LWCRF) but a larger trend towards less SW reflection so as to produce a net gain of energy to the Earth. Measurements of ocean heat content are consistent with a net gain of energy during the past two decades (Wong et al. 2006). It is interesting to note

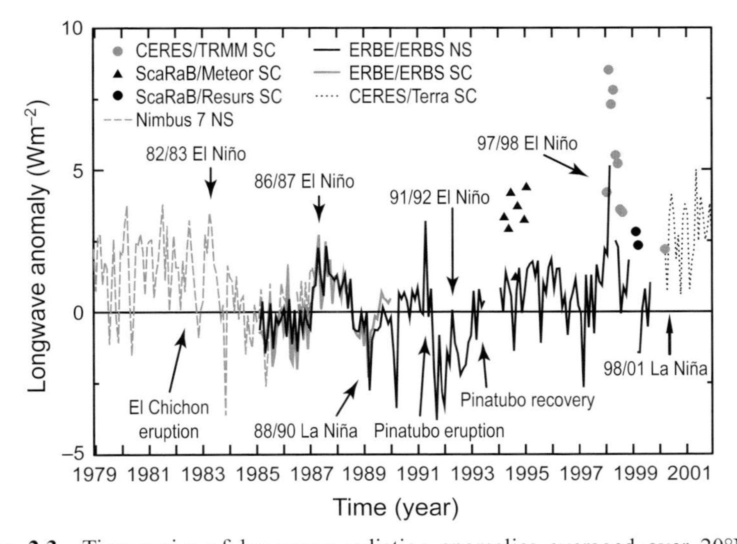

Figure 2.3 Time series of longwave radiation anomalies averaged over 20°N to 20°S based on the new ERBS Nonscanner WFOV Edition3_Rev1 (solid black line), Nimbus 7 Nonscanner (gray dashed line), ERBS Scanner (solid gray line), CERES/Terra FM1 Scanner ES4 Edition2_Rev1 (dotted line), CERES/TRMM Scanner Edition2 (gray circle), ScaRaB/Meteor Scanner (triangle), and ScaRaB/Resurs Scanner (black circle) dataset. Anomalies are defined with respect to the 1985–1989 period (from Wong et al. 2006).

that the GCMs, rightly or wrongly, do not reproduce the magnitude of decadal change in SW and net radiation flux between the 1985–1990 and 1994–1999 periods that are reported by ERBS (Wong et al. 2006). Investigation of changes in ERB over longer periods of time than that covered by ERBS WFOV must concatenate multiple satellite records (Figure 2.3).

Further evidence of the acute observational challenges presented by climate change is illustrated by the accuracy requirements for climate datasets listed by Ohring et al. (2005) and the error analysis of the current CERES sensors performed by Loeb et al. (2007). The stability requirement for measurements of the net SW flux at the top of the atmosphere necessary to identify the feedbacks from low clouds during climate change is listed by Ohring et al. (2005) as 0.3 Wm^{-2} per decade. Loeb et al. (2007) show that from the current CERES instrumentation, 10–15 years of data would be needed before a trend of this magnitude could be detected against the background of natural variability. Clearly, this is only achievable if such well-characterized sensors continue to be flown to provide an unbroken time series from which the cloud feedback could be inferred. The main problem is identifying such a small trend in the face of natural interannual and interdecadal variability, which Figure 2.3 shows can be much larger in magnitude than 0.3 Wm^{-2}, at least regionally. Greater natural variability extends the time period needed to detect a trend of a given magnitude.

The strength of ERB measurements (i.e., that they provide information on the net effect of all the radiatively active constituents on the heating and cooling of the planet) can also be viewed as a weakness. Without additional information, it is difficult to unravel these effects. Retrievals of cloud properties have been combined with ERBE and later data to study the radiative effects of different cloud types and how they contribute to the changes in ERB that take place (e.g., during ENSO events). It is clearly important that complementary information should be available for aerosols.

Satellite Cloud Measurements

International Satellite Cloud Climatology Project

The most commonly used cloud dataset in climate studies is the ISCCP (Rossow and Schiffer 1999). Originally intended to provide a cloud climatology and data for studying cloud processes at synoptic scales, ISCCP uses narrowband radiance data obtained from weather satellites to retrieve cloud fraction and other cloud properties in 280 km grid boxes around the globe every three hours. A radiative transfer model has been applied to these cloud data along with observed stratospheric aerosol and GCM-derived tropospheric aerosol to produce broadband SW and LW fluxes at the top of the atmosphere, the surface, and several levels in between (Zhang et al. 2004). It is interesting to note that interannual and decadal variations in the ISCCP flux dataset have some resemblance to those reported by ERBS (Figure 2.4), considering that they are independently derived datasets.

ISCCP faces much greater difficulty in maintaining homogeneous and stable data over time than do the ERB investigations since it makes use of a long series of weather satellites that were not designed to maintain onboard calibration or interface with other satellites. The ISCCP processing intercalibrates geostationary satellites (the main source of data) using a polar orbiter that flies below each of them, which itself must be calibrated over time. One problem involves calibrating successive polar orbiters so that artificial changes in global cloud properties do not appear at the transition between satellites, as was the case in the original ISCCP C-series dataset (Klein and Hartmann 1993). Since the surface of the Earth happens to be more radiatively stable than any current satellite-observing system, the processing of the ISCCP D-series

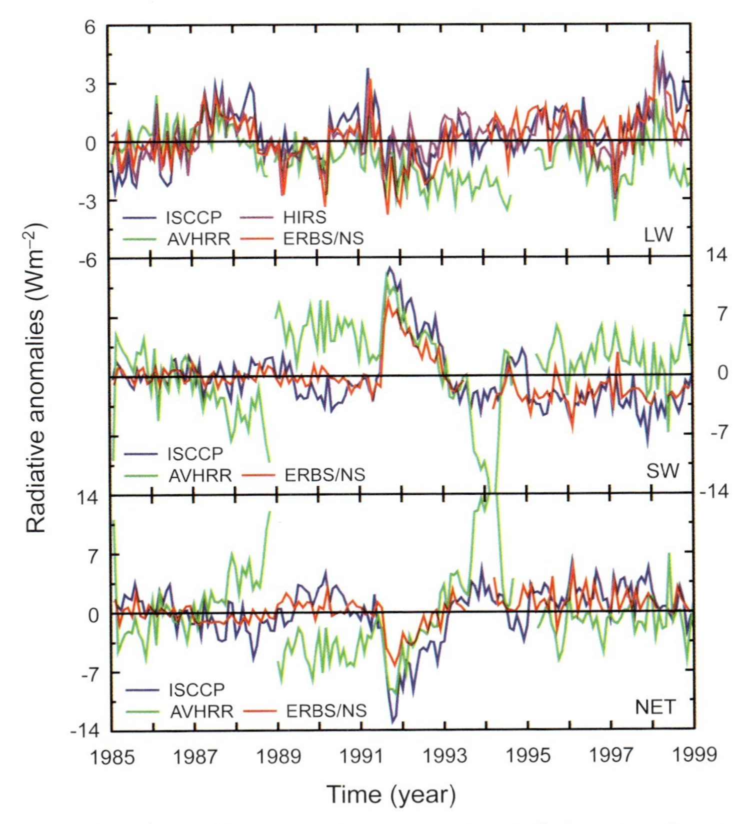

Figure 2.4 Time series of longwave, shortwave, and net radiation anomalies averaged over 20°N to 20°S from the ERBS Nonscanner WFOV Edition3_Rev1 (red), ISCCP FD (blue), HIRS Pathfinder OLR (pink), and AVHRR Pathfinder ERB (green) data records. Anomalies are defined with respect to the 1985–1989 period (from Wong et al. 2006).

dataset included the adjustment of satellite measurements such that global mean surface temperature and reflectance (including the effects of tropospheric aerosol) did not change over time (Brest et al. 1997). Although this method substantially reduced inhomogeneities in the ISCCP time series, it obviously precludes the use of the data to investigate changes in the global surface properties. Moreover, the method will introduce an artifact into the cloud data to the extent that the Earth's surface has changed in the global mean. Nonetheless, it is nominally possible to examine changes in global cloud properties, and ISCCP reports that a general decrease in global mean total cloud amount has occurred over the past couple of decades (Figure 2.5), due primarily to a decrease in low-level cloud amount at low latitudes.

Despite the multiple layers of calibration and adjustment in ISCCP, this trend and other variations in global mean cloud amount appear to be entirely spurious. The pattern of correlation between the global mean time series and the time series in each grid box resembles the circular geostationary satellite fields of view rather than any geophysical pattern, and the spatial pattern of local trends becomes strongly negative near the edges of the geostationary satellite fields of view. Evan et al. (2007) attribute this to a systematic change in the average satellite view angle over time. A geostationary satellite sees pixels near the outside boundary of its view area with a greater slant path than it sees pixels in the center of its view area (i.e., near-nadir). At visible wavelengths, thin low-level clouds appear optically thicker for greater slant path and thus are easier to detect by the threshold methods employed by ISCCP. Since only three geostationary satellites were available at the beginning of the ISCCP record and five geostationary satellites were available at the end, locations that were once seen at high satellite view angle now were seen at low satellite view angle. This means that very optically thin clouds that were once detected were no longer detected, and reported cloud amount consequently decreased.

The view angle artifact has less impact on the calculated ISCCP short- and longwave radiative fluxes than it does on cloud amount because the artificial reduction in cloud amount is compensated by a corresponding artificial enhancement of cloud optical thickness. Moreover, the clouds that are no longer identified are those that are closest to the threshold of detection and thus least radiatively important.

Even when view angle artifacts associated with changes in the number and position of geostationary satellites are statistically removed, other apparently spurious variability remains. For example, coincident variations in cloud amount are seen across the entire view area of a geostationary satellite (Norris 2000b). There is no reason why cloud anomalies in opposite hemispheres and seasons (and often land vs. ocean) should be so closely correlated unless there were some unidentified artifacts in the satellite measurements or application of incorrect calibration (e.g., Norris and Wild 2007).

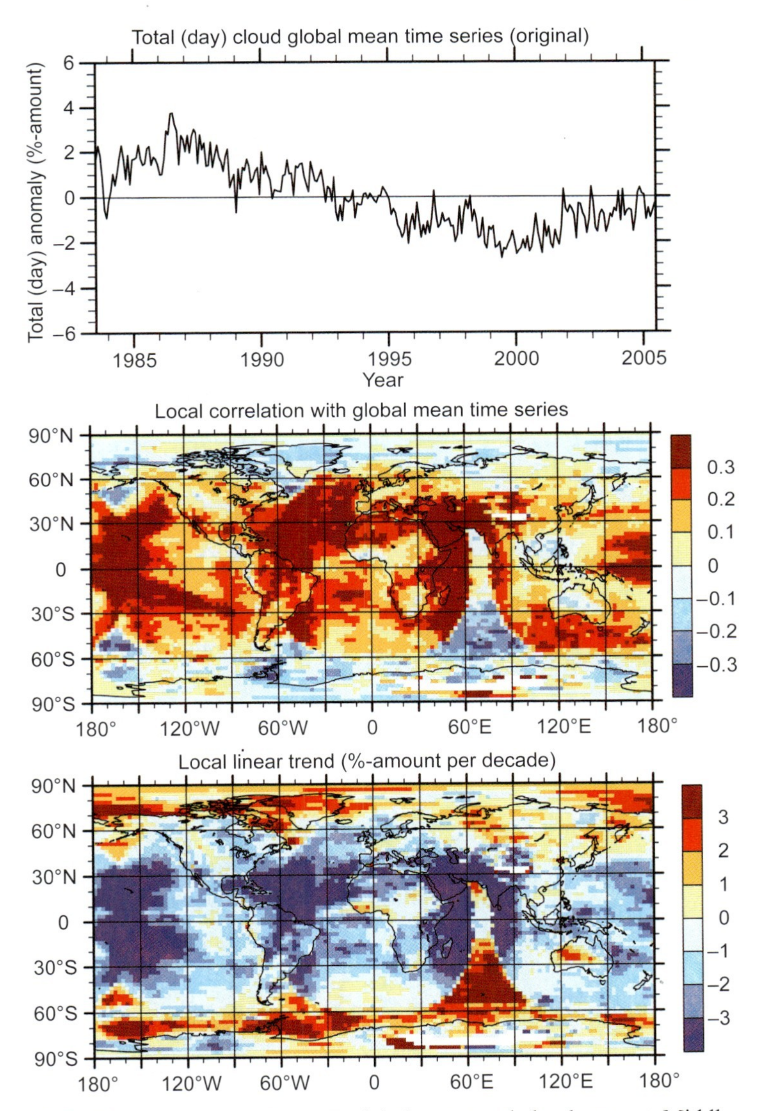

Figure 2.5 Top: time series of ISCCP global mean total cloud amount. Middle: correlation between the global mean time series and the time series at each grid box. Bottom: local linear trend in total cloud amount for each grid box.

Additional Cloud Datasets

Another cloud dataset with a multidecadal record is the AVHRR pathfinder atmosphere (PATMOS) dataset (Jacobowitz et al. 2003). PATMOS cloud and radiation properties are derived from the AVHRR instrument on polar-orbiter weather satellites. Although they do not suffer from shifts in geostationary satellite view angle, as does ISCCP, the nominally sun-synchronous polar orbiters contributing to PATMOS experience substantial orbital drift during their tenure, causing the local equatorial crossing time to occur ever later in the day. This means that the diurnal cycle of cloud properties gets aliased into a long-term trend (i.e., if more clouds occur in the morning than in the afternoon, it will look like cloud amount has declined over time). PATMOS corrects for diurnal drift by statistically regressing out the long-term trend, which must be repeated for each successive satellite in the series. Removing trends in this manner limits the use of these data for long-term climate studies since any real cloud trend may get removed as well. A better method would be to observe the diurnal cycle from a geostationary satellite and then adjust the diurnal drift based on that. PATMOS time series also exhibit extremely large variations that appear to be related to transitions between satellites (Figure 2.4).

A third multidecadal cloud dataset is based on the high resolution infrared radiometer sounder (HIRS) on polar-orbiting weather satellites (Wylie et al. 2005). The HIRS dataset operates by comparing radiances from several infrared channels that view different layers of the atmosphere, and it is especially good at detecting high-level optically thin clouds. Like PATMOS, HIRS suffers from orbital drift and inhomogeneities at satellite transitions. The time series of tropical outgoing LW radiation derived from HIRS (Mehta and Susskind 1999), however, does appear to agree with the ISCCP flux dataset and ERBS time series (Figure 2.4).

Challenges for the Future

It is crucial to avoid treating cloud, radiation, and aerosol observations in isolation. Instead, we must study them in their meteorological context (i.e., temperature, humidity, and winds) because clouds are formed by dynamic and thermodynamic processes, and aerosols provide a perturbation to the formation and dissipation processes as well as to the physical and radiative properties of clouds once they have formed. Numerical weather prediction reanalyses provide the most consistent source of such data, and now they include output for the ERB and cloud fields as well. Reanalyses are thus an important resource for climate studies; however, although the representation of the clear-sky ERB can be good, there are still significant errors in the simulation of clouds and of the all-sky ERB (Allan et al. 2004). A climate-quality reanalysis using a coupled ocean–atmosphere model, paying attention to the radiative and hydrological cycles, would be a major step forward and should be a high priority (Bengtsson et al. 2007).

Several independent cloud and ERB datasets suggest that there might have been a decrease in high-level cloud cover during the past several decades that has reduced LWCRF over time. ERBS, ISCCP flux dataset, and ocean heat content measurements indicate that this slight cooling effect has recently been opposed by a stronger warming effect caused by a weakening of SWCRF. Considering all the corrections that have been applied to the data, can we confidently conclude that these trends are real? If they truly exist, are they related to anthropogenic climate change or are they merely produced by some natural decadal cycle? What implications do the cloud and ERB trends have on our understanding of climate sensitivity?

Answering these questions with confidence will require a thorough reexamination and reprocessing of the different types of available data to reconstruct our best understanding of past variations in cloudiness and ERB and how they are associated with other parameters of the climate system. We should especially use knowledge gained from more recent detailed and comprehensive satellite measurements to improve older records. Better known quantities should be employed to constrain less known quantities when empirical calibration is unavoidable, and improved physical models should be developed to address issues like the perceived variation of cloud optical thickness with satellite view angle. Since the magnitude of systematic artifacts relative to real cloud and radiation variability is typically greatest at the largest scales, patterns of regional changes may be more robust than global changes. In particular, investigations should search for regional trends in cloudiness and ERB that are likely to be associated with anthropogenic global warming or aerosol.

Considering the many inadequacies in the historical data record, it may seem preferable to forget about it and instead to devote our efforts solely to making more precise, comprehensive, and stable measurements in the future. This would be a mistake, we believe, since many years of observation will be required to distinguish a climate change signal from natural variability. We do not have the luxury of waiting for those years to come, but rather must make better use of the record that we do have.

Since there have been damaging gaps in the past record of ERB measurements, which have made it difficult to quantify decadal variability or to identify climate trends, it is essential that we maintain continuity in the future. Any trends in ERB are expected to be small and thus require careful cross-calibration of the various instruments flown at different periods, and without overlaps. Such cross-calibration is very difficult, if not impossible, to achieve. Independent parameters, such as ocean heat content, should also be monitored to provide additional constraints on ERB. As global warming signals begin to emerge clearly above the level of natural variability, one of the greatest challenges facing attempts to ensure continuity of ERB and other measurements is the difficulty of obtaining adequate funding for climate monitoring missions, despite their obvious importance. It is relatively straightforward to justify the continuity of missions to support numerical weather prediction, because of the requirements of operational agencies. However, no satellite agency has yet been able to commit to continuous funding of climate monitoring. Thus, there is a real danger of losing the continuous record of ERB measurements because of future funding constraints. The numbers presented above suggest that the minimum requirement for monitoring changes in the ERB associated with global warming would be continuation of the coverage from the CERES or similar instruments at the current level. Other complementary measurements that might be useful include radiation budget measurements of the sunlit portion of the globe from a satellite at the Lagrange point L1. A particularly valuable mission that has been proposed is a climate calibration observatory, which would be used to cross-calibrate instruments on other satellites to make better use of their data by bringing them together to a common reference scale that is compatible with the CERES data.

Acknowledgments

We are grateful to Richard Allan for providing Figure 2.1, the data for which were retrieved from the NASA Langley DAAC. The data for Figures 2.2 and 2.5 were obtained from the ONRL Carbon Dioxide Information Analysis Center and the NASA Goddard Institute for Space Studies. We thank Takmeng Wong for providing Figures 2.3 and 2.4. An NSF CAREER award, ATM02-38527, supported the work by Joel Norris.

References

Allan, R. P., M. A. Ringer, J. A. Pamment, and A. Slingo. 2004. Simulation of the Earth's radiation budget by the European Centre for Medium-Range Weather Forecasts 40-year reanalysis (ERA40). *J. Geophys. Res.* **109**:D18107.

Alpert, P., P. Kishcha, Y. J. Kaufman, and R. Schwarzbard. 2005. Global dimming or local dimming? Effect of urbanization on sunlight availability. *Geophys. Res. Lett.* **32**:L17802.

Bengtsson, L., P. Arkin, P. Berrisford et al. 2007. The need for a dynamical climate reanalysis. *Bull. Amer. Meteor. Soc.* **88**:495–501.

Bony, S., and J.-L. Dufresne. 2005. Marine boundary layer clouds at the heart of tropical cloud feedback uncertainties in climate models. *Geophys. Res. Lett.* **32**:L20806.

Brest, C. L., W. B. Rossow, and M. D. Roiter. 1997. Update of radiance calibrations for ISCCP. *J. Atmos. Ocean. Technol.* **14**:1091–1109.

Charlson, R. J., A. S. Ackerman, F. A. M. Bender, T. L. Anderson, and Z. Liu. 2007. On the climate forcing consequences of the albedo continuum between cloud and clear air. *Tellus* **59**:715–727.

Evan, A. T., A. K. Heidinger, and D. J. Vimont. 2007. Arguments against a physical long-term trend in global ISCCP cloud amounts. *Geophys. Res. Lett.* **34**:L04701.

Gilgen, H., and A. Ohmura. 1999. The global energy balance archive. *Bull. Amer. Meteor. Soc.* **80**:831–850.

Gupta, S. K., N. A. Ritchey, A. C. Wilber et al. 1999. A climatology of surface radiation budget derived from satellite data. *J. Climate* **12**:2691–2710.

Hahn, C. J., and S. G. Warren. 1999. Extended edited synoptic cloud reports from ships and land stations over the globe, 1952–1996. Report NDP026C. Oak Ridge, TN: Carbon Dioxide Information Analysis Center, Oak Ridge National Lab.

Harries, J. E., J. E. Russell, J. A. Hanafin et al. 2005. The Geostationary Earth Radiation Budget (GERB) project. *Bull. Amer. Meteor. Soc.* **86**:945–960.

Hartmann, D. L., L. A. Moy, and Q. Fu. 2001. Tropical convection and the energy balance at the top of the atmosphere. *J. Climate* **14**:4495–4511.

Jacobowitz, H., L. L. Stowe, G. Ohring et al. 2003. The Advanced Very High Resolution Radiometer Pathfinder Atmosphere (PATMOS) climate dataset: A resource for climate research. *Bull. Amer. Meteor. Soc.* **84**:785–793.

Klein, S. A., and D. L. Hartmann. 1993. Spurious changes in the ISCCP dataset. *Geophys. Res. Lett.* **20**:455–458.

Knapp, K. R. 2008. Calibration assessment of ISCCP geostationary infrared observations using HIRS. *J. Atmos. Ocean. Technol.* **25**:183–195.

Koren I., L. A. Remer, Y. J. Kaufman, Y. Rudich, and J. V. Martins. 2007. On the twilight zone between clouds and aerosols. *Geophys. Res. Lett.* **34**:L08805.

Loeb, N. G., B. A. Wielicki, W. Su et al. 2007. Multi-instrument comparison of top-of-atmosphere reflected solar radiation. *J. Climate* **20**:575–591.

Luo, Z., W. B. Rossow, T. Inoue, and C. J. Stubenrauch. 2002. Did the eruption of Mt. Pinatubo volcano affect cirrus properties? *J. Climate* **15**:2806–2820.

Mehta, A., and J. Susskind. 1999. Outgoing longwave radiation from the TOVS Pathfinder Path A dataset. *J. Geophys. Res. Atmos.* **104**:12,193–12,212.

Menon, S., J. Hansen, L. Nazarenko, and Y. Luo. 2002. Climate effects of black carbon aerosols in China and India. *Science* **297**:2250–2253.

Norris, J. R. 2000a. Interannual and interdecadal variability in the storm track, cloudiness, and sea surface temperature over the summertime North Pacific. *J. Climate* **13**:422–430.

Norris, J. R. 2000b. What can cloud observations tell us about climate variability? *Space Science Rev.* **94**:375–380.

Norris, J. R. 2005a. Multidecadal changes in near-global cloud cover and estimated cloud cover radiative forcing. *J. Geophys. Res. Atmos.* **110**:D08206.

Norris, J. R. 2005b. Trends in upper-level cloud cover and surface divergence over the tropical Indo-Pacific Ocean between 1952 and 1997. *J. Geophys. Res. Atmos.* **110**:D21110.

Norris, J. R., and C. P. Weaver. 2001. Improved techniques for evaluating GCM cloudiness applied to the NCAR CCM3. *J. Climate* **14**:2540–2550.

Norris, J. R., and M. Wild. 2007. Trends in aerosol radiative effects over Europe inferred from observed cloud cover, solar "dimming," and solar "brightening." *J. Geophys. Res. Atmos.* **112**:D08214.

Ohmura, A., H. Gilgen, H. Hegner et al. 1998. Baseline Surface Radiation Network (BSRN/WCRP): New precision radiometry for climate research. *Bull. Amer. Meteor. Soc.* **79**:2115–2136.

Ohring, G., B. Wielicki, R. Spencer, B. Emery, and R. Datla. 2005. Satellite instrument calibration for measuring global climate change. *Bull. Amer. Meteor. Soc.* **86**:1303–1313.

Pallé, E., P. R. Goode, P. Montañés Rodriguez, and S. E. Koonin. 2004. Changes in Earth's reflectance over the past two decades. *Science* **304**:1299–1301.

Park, S., and C. B. Leovy. 2000. Winter North Atlantic low cloud anomalies associated with the Northern Hemisphere annular mode. *Geophys. Res. Lett.* **27**:3357–3360.

Ramanathan, V., R. D. Cess, E. F. Harrison et al. 1989. Cloud-radiative forcing and climate: Results from the Earth Radiation Budget Experiment. *Science* **243**:57–63.

Randall, D. A., R. A. Wood, S. Bony et al. 2007. Climate models and their evaluation. In: Climate Change 2007: The Physical Science Basis. Contribution of Working Group I to the Fourth Assessment Report of the Intergovernmental Panel on Climate Change, ed. S. Solomon, D. Qin, M. Manning et al., pp. 591–662. New York: Cambridge Univ. Press.

Rossow, W. B., and R. A. Schiffer. 1999. Advances in understanding clouds from ISCCP. *Bull. Amer. Meteor. Soc.* **80**:2261– 2287.

Slingo, A. 1990. Sensitivity of the Earth's radiation budget to changes in low clouds. *Nature* **343**:49–51.

Sun, B., P. Y. Groisman, and I. I. Mokhov. 2001. Recent changes in cloud-type frequency and inferred increases in convection over the United States and the former USSR. *J. Climate* **14**:1864–1880.

Tselioudis, G., and C. Jakob. 2002. Evaluation of midlatitude cloud properties in a weather and a climate model: Dependence on dynamic regime and spatial resolution. *J. Geophys. Res.* **107**:4781.

Wielicki, B. A., B. R. Barkstrom, E. F. Harrison et al. 1996. Clouds and the earth's radiant energy system (CERES): An earth observing system experiment. *Bull. Amer. Meteor. Soc.* **77**:853–868.

Wielicki, B. A., T. Wong, R. P. Allan et al. 2002. Evidence for large decadal variability in the tropical mean radiative energy budget. *Science* **295**:841–844.

Wild, M., H. Gilgen, A. Roesch et al. 2005. From dimming to brightening: Decadal changes in surface solar radiation, *Science* **308**:847–850.

Wong, T., B. A. Wielicki, R. B. Lee et al. 2006. Re-examination of the observed decadal variability of the earth radiation budget using altitude-corrected ERBE/ERBS nonscanner WFOV data. *J. Climate* **19**:4028–4040.

Wylie, D., D. L. Jackson, W. P. Menzel, and J. J. Bates. 2005. Trends in global cloud cover in two decades of HIRS observations. *J. Climate* **18**:3021–3031.

Zhang, Y., W. B. Rossow, A. A. Lacis, V. Oinas, and M. I. Mishchenko. 2004. Calculation of radiative fluxes from the surface to top of atmosphere based on ISCCP and other global data sets: Refinements of the radiative transfer model and the input data. *J. Geophys. Res.* **109**:D19105.

3

Climatologies of Cloud-related Aerosols

Part 1: Particle Number and Size

Stefan Kinne

Max Planck Institute for Meteorology, Hamburg, Germany

Abstract

A proper representation of aerosol properties is an essential first step in addressing potential impacts of aerosols on climate and clouds. In the context of temporal and spatial variability of concentration, size, and composition, maps of aerosol properties are usually based on insufficiently evaluated datasets of model simulations or satellite retrievals. Here, a new approach is offered. Quality data from ground-based remote-sensing networks are merged into multi-model median background fields. Global monthly maps are created (at a $1° \times 1°$ horizontal resolution) for aerosol column properties of aerosol optical depth (AOD), single-scattering albedo (ω_0), and Ångström parameter (AnP, an easier to measure substitute for the asymmetry factor, *g*). Adopting the commonly observed bimodal size distribution shape for aerosol, AOD is partitioned into contributions from smaller (accumulation mode, radii: 0.05–0.50 μm) and larger (coarse mode, radii >0.50 μm) particle sizes. This simplifies the spectral extension of mid-visible optical properties and allows anthropogenic AOD estimates to be independent of interannual variations of (larger size) natural aerosol. The AOD is partitioned into its natural and anthropogenic components using global model estimates for the anthropogenic AOD fraction of all small particles, assuming that only small aerosol particles can be anthropogenic. Global modeling also provides the vertical distribution of AOD. Applying the arguments of Pöschl et al. (see Part 2 of this chapter), namely, that hygroscopic growth of atmospheric aerosol particles is relatively well constrained, all necessary ingredients are assembled to derive global monthly maps for concentrations of cloud condensation nuclei (CCN) and to assess regional CCN enhancements as a result of anthropogenic activity.

Introduction

Climate change simulations are constrained by large uncertainties associated with aerosols. Progress has been slow, primarily as a result of temporal and spatial variability of tropospheric aerosols and our limited understanding of the multiple feedback processes involving aerosols. Potential impacts of aerosols on clouds and on the hydrological cycle constitute particular areas of concern. To quantify the global impact requires temporal information on concentration, size, composition, and altitude of aerosols on a global scale. For this purpose, dedicated regional field campaigns have been conducted (e.g., TARFOX, SAFARI, SCAR-B, ACE-Asia, AMMA) to study specific aerosol types, advanced sensors for extra detail and accuracy on aerosol properties have been placed on satellites, and ground-based aerosol monitoring stations have grown in number and equipment.

In the first part to this chapter, global maps of aerosol properties are provided that are also relevant to clouds. Beginning with the status quo of the last decade, recent advances in aerosol global modeling and aerosol remote sensing are presented. An overview of the major ground-based remote-sensing networks for aerosols is included, since this ground monitoring adds essential complementary detail and establishes a reference for remote sensing from space. The preferential use of statistics from these ground-based networks in combination with spatially complete data by global modeling yields new global monthly maps for aerosol properties. These maps include distributions of properties of anthropogenic aerosols and for CCN changes attributable to anthropogenic activity.

Data of Interest

To characterize aerosols properly requires information on particle amount, size, composition, as well as shape in space and time. The relevant optical properties are AOD, ω_0, and g. These properties are associated with overall extinction, the probability for scattering (as opposed to absorption), and the scattering behavior. AOD is the vertically integrated attenuation of radiation caused by scattering and absorption by particles; ω_0 is the single-scattering albedo and captures the scattering probability of attenuation event. The asymmetry factor, g, captures the angular probability of scattering by the cosine (of the scattering angle) weighted integral value. All three properties are a function of location, time, altitude, and (at least solar) wavelength. Because of limitations in aerosol measurements, the requirements for aerosols in global applications are usually relaxed to maps of mid-visible properties of vertically integrated aerosols, and often only to that of the AOD. The justification here is that (a) aerosol amount (captured by AOD) is more variable than size or composition, (b) aerosols are located predominantly in the lower troposphere, and (c) mid-visible aerosol properties are most important, because in that solar spectral region, the product of particle cross section and available (solar) energy is at a maximum. Still, there is a need to account for non-negligible variations in particle size and composition. This requires additional maps of the mid-visible column properties for ω_0 and g. The necessary data for g can be substituted by using a parameter that is easier to measure: mid-visible AnP, which is defined by the AOD spectral dependence as the negative slope in log/log space. This substitution is possible, as both properties are sensitive to particle size. Thus, monthly global maps for AOD, ω_0, and AnP are the minimum that are needed.

Status: Ten Years Ago

Ten years ago, aerosols were treated quite simplistically in global models. Tropospheric aerosols were characterized primarily by coarse size ($r > 0.5$ μm) dust, as a natural component, and by accumulation mode size ($0.05 > r > 0.5$ μm) (non-absorbing) sulfate, as the anthropogenic component (where r is the particle radius). Even by ignoring organic aerosol sources, it was not easy to convert industrial sulfur emissions via chemical and transport processes to somewhat realistic sulfate aerosol mass distributions to exact impact assessments of anthropogenic pollution (Langner and Rodhe 1991). Early evaluations of model simulations using satellite remote-sensing data were difficult. Estimates for mid-visible AOD by AVHRR sensor data (Stowe et al. 2002) have been available since 1981. However, AVHRR retrievals were limited to (deep) ocean regions, and they suffered from calibration problems, overpass time drifts, and sensor discontinuity on consecutive platforms. In contrast, since 1979, TOMS sensor data have provided estimates for AOD and ω_0 even over (snow-free) land regions (Torres et al. 2002). However, a coarse pixel size increased the potential for cloud contamination, and data were provided for an energetically less interesting ultraviolet spectral region. In addition, accuracy of TOMS aerosol data depends strongly on an assumed aerosol altitude. Comparisons of wavelength-normalized AOD annual averages (Figure 3.1) between TOMS and AVHRR reveal, at best, similar overall patterns with evidence of significant differences in magnitude. Detailed comparisons are not possible because of data inconsistencies (e.g., overpass time, pixel size, sample) and differences in retrieval assumptions for aerosols (e.g., particle size or absorption) and environment (e.g., solar surface albedo, definition of cloud-free conditions). Some aspects of these quantitative limitations can be overcome using reference values from complementary ground-based remote sensing. Solar transmission measurements (from the ground) have several advantages over solar reflection measurements (from space): they have a well-defined background and are insensitive to aerosol absorption. Ten years ago, however, the number of monitoring sites that provided AOD and AnP by sun photometry

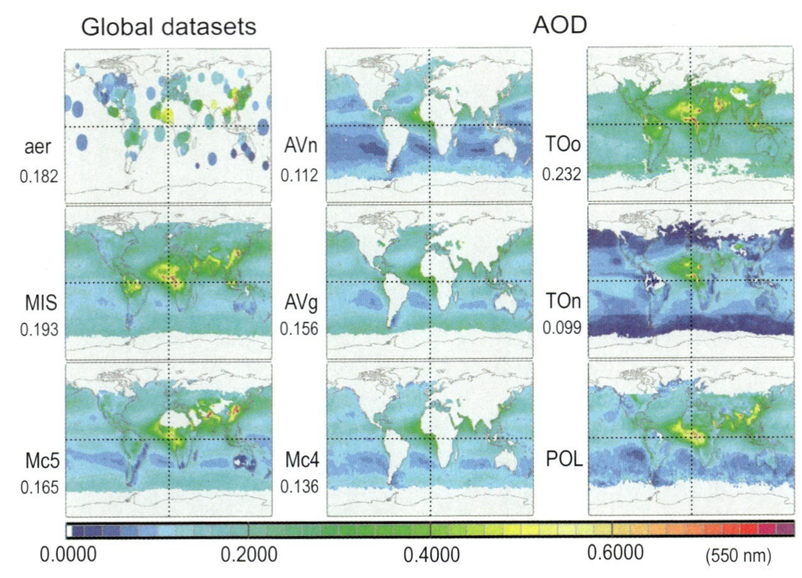

Figure 3.1 Annual mid-visible aerosol optical depth (AOD) fields at 550 nm wavelength from remote sensing. Ground-based remote sensing AOD data by AERONET (aer) have been combined into a gridded product and expanded for better viewing; they establish an AOD reference. In comparison, annual AOD fields of different multiannual satellite retrievals are given (MIS: MISR; Mc5: MODIS-collection5; Mc4: MODIS-collection4; AVn: AVHRR-noaa; AVg: AVHRR-gacp; TOo: TOMS-old; TOn: TOMS-new; POL: POLDER). Times with enhanced stratospheric aerosol after major volcanic eruptions have been excluded, and listed values are global averages of non-zero data.

was small, and measurements at different ground sites lacked coordination and comparison of calibrations.

The Global Aerosol Data Set (GADS) was the first serious attempt to establish a global climatology for aerosol (microphysical and) optical properties. Using the available data, GADS combined the sampling information on aerosols from different (mainly surface *in-situ*) sites with modeling (Koepke et al. 1997). Aerosols in GADS are classified by size and composition into ten main components: water soluble, water insoluble, soot, sulfate, two types of sea salt, and four types of minerals. Each component is given a particular composition, size, and humidification factor (WMO 1983; d'Almeida et al. 1991). Based on assumptions of the ambient relative humidity, Mie simulations (assuming spherical shapes) are applied to determine the spectral aerosol properties for AOD, ω_0, and g, via the OPAC software package (Hess et al. 1998). GADS aerosols are represented locally by a mixture of up to four components. GADS distinguishes between seven different altitude profiles, and horizontal resolution is 5°×5° latitude longitude grid. Data are provided for two months (January and July) to represent a winter and a summer season. The major shortcomings of GADS include the limited seasonality (e.g., it misses the biomass burning maxima of the southern hemisphere), the restriction to a few classes with predefined values for particle size and spectral absorption (e.g., there is now a consensus on a much weaker mid-visible absorption for dust), and the necessity to provide vertically resolved data for the ambient relative humidity to account for the water uptake of the particles.

Present-day Status

Over the last decade, atmospheric research began to concentrate on aerosols, because they were identified as a key regulator of atmospheric change, including clouds. From a modeling perspective, increased computing capacities permitted better characterization of aerosols in global models. Elaborate aerosol submodules now process emissions from different aerosol components (e.g., sulfate, organic matter, black carbon, dust, and sea salt). Particle size is resolved into several size classes; at the very least, there is an accumulation mode and one or several coarse modes for dust and sea salt. Some aerosol modules even allow added complexities of component mixtures. Notwithstanding the limitations to evaluation data (often component-integrated data with non-negligible uncertainty, such as satellite AOD), there is now a great deal of freedom in modeling within the aerosol module. For example, the AeroCom aerosol module exercise demonstrated significant model diversity in compositional attribution (Kinne et al. 2006). This diversity was partly related to prewired assumptions in aerosol processing (e.g., wet deposition), as most model biases remained even though the emission input to all models had been harmonized (Textor et al. 2006).

From the perspective of measurements, ground-monitoring networks were established in parallel with the launching of new aerosol-dedicated satellite sensors. This included multispectral sensors such as MODIS (Tanre et al. 1997; Kaufman et al. 1997), multidirectional sensors such as MISR (Kahn et al. 1998; Martonchik et al. 1998), and even sensors with detection of polarized signals such as POLDER (Deuzé et al. 1999; Deuzé et al. 2001). Onboard calibrations, more detailed retrieval models, and simultaneous sampling with supporting sensors on the same or consecutive platforms (e.g., A-Train; Anderson et al. 2005) provide better detail of aerosol properties than had been previously available. Currently, satellite sensors can supply AOD maps over land (even over bright desert regions). In addition, active remote sensing by the CALIPSO lidar (Winker et al. 2004) and CloudSat radar (Stephens and Kummerow 2007) supplies samples of aerosol vertical distribution and altitude placement for clouds (i.e., essential information for more accurate aerosol direct forcing estimates). Accuracy issues remain, however, with respect to retrieval assumptions (e.g., surface properties, predefined aerosol types) and subpixel contamination (e.g., clouds, snow). Data from ground-based monitoring networks are expected to serve as quality reference. The most important networks for columnar aerosol optical properties are AERONET and SKY-NET; their sun/sky photometers derive all particle properties simultaneously: amount, size, absorption, and shape. Additional sun photometer networks (e.g., GAW) increase the data volume on particle amount and size. In parallel, lidar sampling (for vertical profiling) has organized itself into regional networks, as EARLINET for Europe or NIES for Eastern Asia. Especially attractive are sites with complementary instrumentation: lidar deployments by MPL-Net at AERONET sun/sky photometer sites or sun photometer deployments at lidar sites. Since ground remote sensing establishes a needed reference in the characterization of aerosols, an overview of some of the major networks follows.

Major Ground Networks

AERONET (AErosol RObotic NETwork; http://aeronet.gsfc.nasa.gov/) is a federation of ground-based remote-sensing aerosol networks that was established in the mid-1990s by NASA and CNRS; it was expanded thereafter through the efforts of national agencies, universities, and individuals (Holben et al. 1998). Using standards for instruments, calibration techniques, processing, and data distribution, individual sites are connected into a network. The basic instrument is a CIMEL sun/sky photometer. Data are transmitted via satellite or internet to the central processing facility at NASA, where the retrieved products from almost all sites are available within hours. Later, quality-assured products are released after the instruments have been recalibrated. Retrieved products are aerosol column properties for the mid-visible AOD and the AnP; an inversion algorithm employing sky radiances (Dubovik and King 2000) provides the particle size distribution and data on particle absorption (with the caveat that estimates for ω_0 are only reliable, if the mid-visible AOD value exceeds 0.3). Now more than 150 instruments are simultaneously in operation worldwide. Since many instruments have been moved during their lifetime, column data on annual cycles of aerosol optical properties are available for more than about 400 sites.

SKYNET (http://atmos.cr.chiba-u.ac.jp/) is an observation network that is based in Eastern Asia. It was established to further an understanding of the atmospheric interactions between particles, clouds, and radiation (Aoki 2006). Basic instrumental configurations include a PREDE sun/sky photometer at ten sites. The retrieved aerosol column properties are mid-visible AOD, AnP, and, by processing sky radiances in an inversion routine, ω_0 (whose estimate can only be trusted at larger AOD values, as for AERONET). Data obtained are locally collected and then transferred to a central site at the Chiba University in Japan, where data are archived and made accessible to the public. Side-by-side measurements next to CIMEL instruments of AERONET demonstrated good agreement and comparability of different network data. Thus, even the relatively few SKYNET sites (7 sites in 2003 and 2004) complement AERONET statistics in a CIMEL-sparse region.

GAW (Global Atmospheric Watch) is a WHO/WMO-sponsored program combining 22 global and more than 300 regional sites into a measurement network. GAW is geared primarily toward the background monitoring of surface concentrations of trace gases and compositional properties of the near-surface aerosol. It extends a similar regional network monitoring by EMEP (Klein et al. 2007) in Europe and IMPROVE (Malm and Hand 2003) in the U.S. Some GAW stations operate a PRF sun photometer for estimates of aerosol column properties. Retrieved properties are mid-visible values for AOD and AnP. Observations from each site are sent to Davos, Switzerland, and released after quality assurance to the World Data Centre for Aerosol at Ispra (http://wdca.jrc.it/). The PFR sun photometer differs in design and spectral capabilities from CIMEL (of AERONET) or PREDE (of SKYNET) photometers, but retrieved properties compare well in side-by-side sampling. The relatively few GAW sites (15 sites in 2003 and 2004) complement AERONET by providing aerosol statistics in remote regions.

EARLINET (European Aerosol Research LIdar NETwork, http://www.earlinet.org/) was established in 2000 to provide comprehensive, quantitative, and statistically significant information on the aerosol vertical distribution over Europe (Bösenberg et al. 2003). It involves a coordinated probing of the atmosphere by more than twenty lidar systems at sites that are primarily located in central and southern Europe. About half of the sites have Raman capabilities for more accurate estimates on aerosol extinction profiles. Time-coordinated sampling is conducted on a regular and event basis; a quality assurance program addresses both instrumentation and evaluation algorithms; and efforts are underway to harmonize data exchange with format and processing standards.

NIES (National Institute for Environmental Studies) has established a network of automated two wavelength dual-polarization lidars (http://www.nies.go.jp/gaiyo/index-e.html) in cooperation with various universities and research organizations. This network provides aerosol vertical distribution monitoring (Nishizawa et al. 2007) at about 15 sites in Eastern Asia.

MPL-Net (http://mplnet.gsfc.nasa.gov/) is a network of backscattering lidars, which are co-located with CIMEL sun/sky photometers at 11 AERONET sites (Welton et al. 2001). Such a co-location of aerosol instrumentations elevates the site importance. Thus, many sites of other (aerosol) lidar networks are (and will be) instrumented with co-located sun photometers.

EARLINET, NIES, and MPL-Net are major but certainly not the only existing regional (aerosol) lidar networks. Efforts are underway to merge sampling in individual regions into one global effort, under the GALION initiative, which is supported by the WMO.

Data Sources

To develop a climatology for monthly global fields of mid-visible aerosol properties of AOD, ω_0, and AnP, input from three different sources must be considered: (a) remote sensing from space, (b) remote sensing from the ground, and (c) global modeling. Strengths and weaknesses of these data sources are briefly reviewed to justify the selected choice: a synergetic approach combining the accuracy of ground-based remote-sensing statistics with the completeness from global modeling.

Remote Sensing from Space

Remote sensing from space is only able to retrieve a subset of aerosol properties, mainly AOD and sometimes its spectral dependence, for AnP estimates. These retrievals require a priori assumptions of other aerosol properties (at least aerosol absorption) and environmental properties. Without sufficient accuracy of these assumed properties (or choices in lookup tables), aerosol retrievals are too uncertain to be useful. An example of this is the common failure to retrieve AOD from visible reflectance data over brighter surfaces (e.g., desert, sun glint). In addition, sampling is spatially or temporally limited, depending on the satellite orbit. Over the equator, (geo)stationary platforms have the potential for high temporal sampling; however, because of the geometric aspects of their position, remote sensing is restricted to lower latitudes, and at least five different satellites with five different sensors are needed to achieve global coverage. Polar-orbiting platforms, in contrast, achieve global coverage with a single sensor and thus have been a preferred choice for aerosol sensor platforms. With limitations to the swath, polar orbiters provide at best a single daytime overpass. Temporal sampling frequency is further reduced by the aerosol retrieval requirement for a cloud-free scene. This is particularly a problem if the region associated with the smallest pixel size is large. For polar orbiters, the number of successful samples becomes often so sparse that statistical significance is not guaranteed, even for monthly averages at a spatial resolution of $1° \times 1°$ longitude/latitude. Thus, for typical AOD maps from polar-orbiting remote sensing, only the available multiyear sensor datasets are considered and compared to AERONET statistics (Figure 3.1). For the displayed multiannual averages, periods of enhanced stratospheric aerosols are excluded.

Figure 3.1 demonstrates the significant differences that exist among regional annual values for AOD. The retrieved dataset is sensitive to retrieval assumptions, as demonstrated by comparisons of different processing versions for AVHRR, MODIS, and TOMS. For example, a change to the ocean reflectance in the TOMS retrieval causes a switch from formerly the largest to now the smallest background AOD over oceans. Obviously, the retrieval performance in reference to AERONET varies regionally. To compensate, a satellite retrieval composite for AOD has been developed based on the best overall scores obtained regionally for bias, spatial variability, and seasonality. The combined product appears superior over any individual satellite dataset. The resulting seasonal AOD distribution is presented in Figure 3.2.

Even the best regional AOD satellite retrievals of this composite are far from perfect. Moreover, next to AOD, satellite retrievals provide only limited information on particle size and almost never on particulate absorption. In retrievals developed for different satellite sensors, these assumed aerosol properties as well as their environmental properties are often inconsistent. This complicates comparisons of retrieved AOD values and can be responsible for qualitative differences (e.g., different spatial patterns). Improvements are expected with sensor standards, sensor side-by-side deployments on the same platform, and consistency in assumptions for ancillary and/or a priori data. Useful insights on retrieval capabilities and limitations are expected from careful comparisons to quality reference data (e.g., sun/sky photometry). However, this requires a commitment to compare, evaluate, and reprocess the data plus the necessary time.

Remote Sensing from the Ground

Compared to remote sensing from space, remote sensing from the ground has well-defined backgrounds (sun or dark space), and for AOD data, no assumptions on particulate absorption are required. Ground remote sensing can, in fact, provide data for all three particle characteristics: AOD, ω_0, and

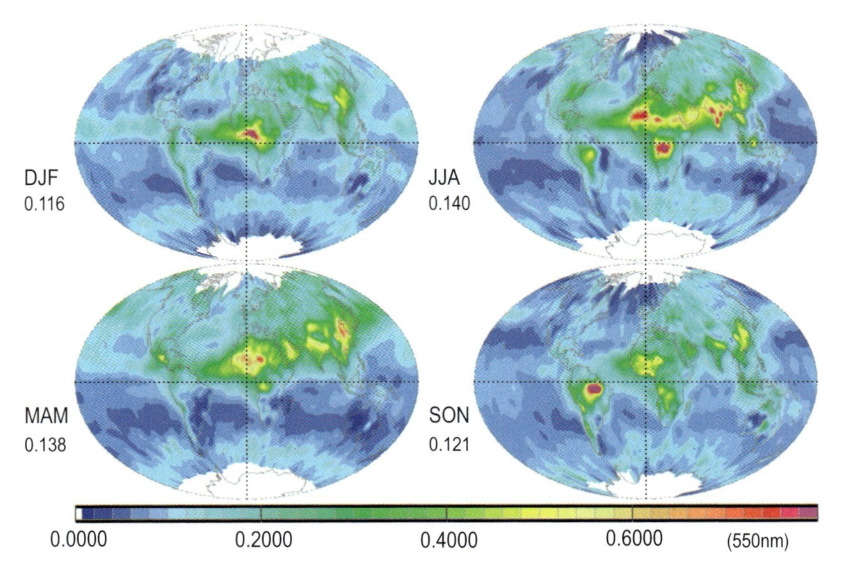

Figure 3.2 Seasonal AOD patterns at 550 nm wavelength of a satellite remote-sensing composite tied to AERONET data.

AnP. Direct solar attenuation measurements with sun photometers provide highly accurate data for AOD and AnP. Additional sky radiance data can provide (via inversions) consistent columnar data on all aerosol properties, including ω_0. However, accurate ω_0 estimates can only be expected if the associated (mid-visible) AOD exceeds 0.3. A higher (compared to satellite retrievals) temporal resolution allows for much better statistics, but the requirement for a cloud-free scene remains. There are major drawbacks in trying to derive global AOD fields from sun/sky photometer sites: (a) site distribution is sparse and uneven; (b) there is potential for local bias, which makes statistics at a site unfit to represent the site's surrounding region. However, if statistics from sky/sun photometer sites are available and if a good regional representation applies (e.g., demonstrated by invariant satellite retrievals for different spatial domains around that site), then sky/sun photometer data are preferable for that region over spaceborne remote-sensing data.

Global Modeling

Global modeling can provide (spatially, temporally, and spectrally) complete and consistent datasets for all aerosol properties. Concerns exist, however, as to the accuracy of the underlying assumptions (e.g., emissions, transport, water uptake) and parameterizations (e.g., aerosol processing, interactions with clouds). With rather general constraints (e.g., column and component-integrated data from remote sensing), there is significant diversity in aerosol global modeling, especially at modeling substeps (Textor et al. 2006). To counteract this diversity and to establish characteristic particle properties from global modeling, aerosol simulations of more than twenty different models were considered. All of these models employed advanced aerosol modules, which distinguished between aerosol components of dust, sulfate, sea salt, organic carbon, and black carbon (Kinne et al. 2006). Simulated monthly averages were regridded to a common $1° \times 1°$ latitude/longitude horizontal resolution. The local median value of monthly averages suggested by all models at any grid point was picked; thereafter, these median values were combined to define monthly fields from global modeling for AOD, ω_0, and AnP. These median fields have the advantage that extreme behavior (e.g., outliers) of individual models is suppressed. In addition, these median model fields tend to score better when evaluated than individual models (Schulz et al. 2006).

New Climatology

The new climatology is defined by monthly maps for (mid-visible) the aerosol characteristics of AOD, ω_0, and AnP. The climatology prioritizes quality data of local statistics, which are merged onto globally and temporally complete background fields, from modeling. To be more specific, monthly maps of

AeroCom model median values (Kinne et al. 2006) define complete and consistent background fields, and monthly statistics from sun/sky photometers are the quality data. Thus, this climatology does not directly involve data from satellite remote sensing. The quality data involve sky/sun photometer monthly statistics of more than 150 AERONET and 7 SKYNET sites, with a complete annual cycle and more than 200 AERONET sites with limited annual coverage. The merging of the sun-/sky photometer data onto the model median background fields occurs in five consecutive steps (Kinne 2008):

1. Rate all sun photometer sites individually in terms of quality and regional representation.
2. Combine sun photometer local monthly statistics onto the regular grid (1°×1°) of modeling.
3. Define a global ratio map (backward field) based on globally spreading available ratios between sun photometer data and model data.
4. Establish local domain fields around each sun photometer site.
5. Establish "effective weights" (apply the domain weight to the ratio map).

To define the new aerosol climatology, the effective weights fields are applied to the background fields from modeling to yield modified monthly maps for AOD, ω_0, and AnP.

Figures 3.3–3.5 compare maps of annual averages between the modeling background, AERONET samples, and the merged product. The data of the sky or sun photometer samples have been enlarged in the figures for legibility, and as a site's domain varies according to its assigned regional representation, changing colors are possible in the case of nearby sites. With respect to the merged product of ω_0 in Figure 3.4, some of the lower values from sky photometry are not shown, because their associated AOD fell below the confidence threshold (mid-visible AOD < 0.3) for reliable ω_0 inversions.

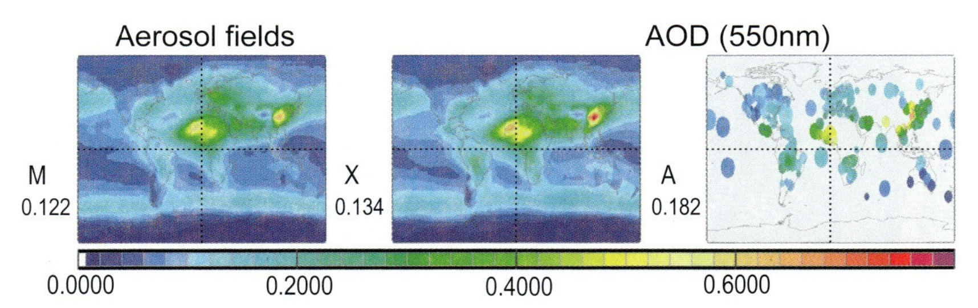

Figure 3.3 Mid-visible aerosol optical depth (AOD) at 550 nm wavelength. Comparison between annual maps from global modeling (M), the merged product (X), and sun photometer gridded samples (A). Deviations to model simulations suggest AOD underestimates in global modeling.

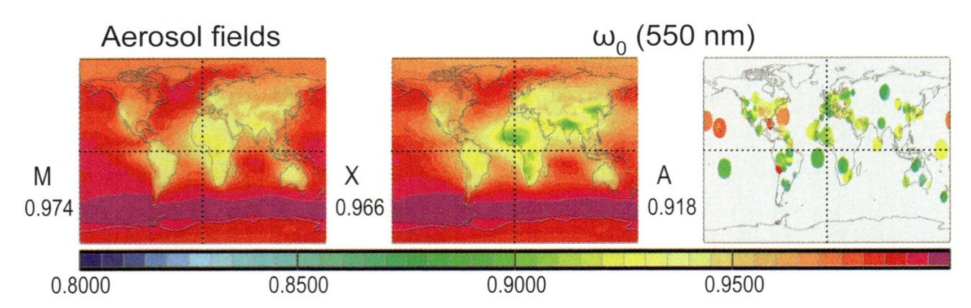

Figure 3.4 Mid-visible single-scattering albedo (ω_0) at 550 nm wavelength. Comparison between annual maps from global modeling (M), the merged product (X), and sky photometer gridded samples (A). Deviations to model simulations suggest absorption underestimates in models.

Spectral Extension

The three particle properties of interest in radiative transfer simulations are AOD, ω_0, and g, which is estimated from AnP. All three properties are spectrally dependent. Thus, for solar and infrared broadband simulations, which are required in climate applications, the merged monthly global fields of the new climatology for the mid-visible spectral region must be extended to other wavelengths of the solar and infrared spectrum. This spectral extension involves the following strategy:

First, *assume bimodality to stratify AOD in accumulation and coarse mode fractions*. The spectral extension takes advantage of observational evidence that almost all particle size distributions are bimodal in shape, with a concentration minimum at radii of 0.5 μm, separating smaller accumulation mode sizes from larger coarse mode sizes. Coarse mode (predominantly natural aerosol) particles are assumed to be large enough so that they do not display any significant spectral AOD dependence in the solar spectrum. Thus, for coarse size mode particles, the Ångström parameter, AnP_C (based on extinctions at 0.44 μm and 0.87 μm), is assumed to be zero ($AnP_C = 0$). For particles of the accumulation mode, the associated Ångström parameter, AnP_A, is prescribed but allowed to

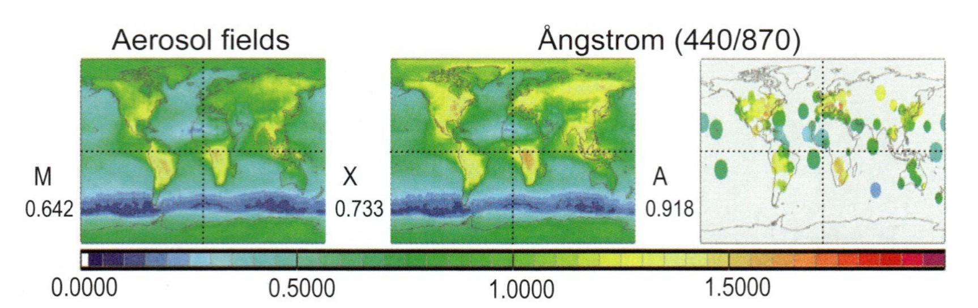

Figure 3.5 Mid-visible Ångström parameter combining 440 and 870 nm wavelength data. Comparison between annual maps from global modeling (M), the merged product (X), and sun-photometer gridded samples (A). Deviations to model simulations suggest size overestimates in modeling.

vary with ambient humidity. At larger ambient relative humidity, particles are expected to increase in size, thus lowering AnP_A. Based on statistics with truncated (accumulation mode only) size distributions of AERONET, AnP_A varies between 2.2 and 1.6. It is assumed that the larger value, 2.2, refers to completely dry conditions and that as the ambient relative humidity increases, this value decreases until, at completely wet conditions, a value of 1.6 is reached. Data on ambient relative humidity are approximated by the scaled low-level cloud cover,

$$C_{LOW}, s = \frac{C_{LOW}}{1 - C_{MID} - C_{HIGH}}, \tag{3.1}$$

by applying cloud cover data (C_{LOW}, C_{MID}, C_{HIGH}) of the ISCCP cloud climatology (Rossow et al. 1993). This defines the locally applicable value for AnP_A ($AnP_A = 2.2 - 0.6 \times C_{LOW}, s$). If the local AnP of the new climatology that needs to be matched falls below the local threshold AnP, this indicates not only the presence of coarse mode particles but also quantifies the AOD fractions attributed to each mode. For a local AnP outside the (two) coarse and accumulation threshold values, the AOD would thus be assigned entirely to the coarse size mode (if $AnP < 0$) or to the accumulation size mode (if $AnP > AnP_A$, whereby AnP_A then adopts the value of AnP).

Second, *apply* ω_0 *to determine the composition and size of the coarse mode.* Coarse mode particles are assumed to be sea salt, dust, or a combination of both. The startup configuration assumes dust over land and sea salt over water. Both aerosol types are defined by log-normal size distributions with effective radii (r_{eff} is the ratio of the third and second moment of the particle size distribution) of 1.25 μm for dust and of 2.5 μm for sea salt. Both components are assumed to be associated with rather wide distributions as a standard deviation (σ) of 2.0 was assumed. The adopted refractive indices for sea salt (Nilsson et al. 1979) and dust (Sokolik, pers. comm.) translated for the mid-visible (spectral) region in no absorption (= 1) for sea salt and in weak absorption for dust. The initially assumed composition is then modified to match the local value for (mid-visible) ω_0. If (over land) the dust-associated single-scattering albedo, $\omega_{0,DU}$, is larger (less absorption) than the local ω_0, then part of the absorbing dust is replaced by non-absorbing sea salt. A possible compensation by a less-absorbing fine mode is not permitted, because it is assumed that the absorption potential of the accumulation size mode ($1 - \omega_{0,A}$) cannot be smaller than the absorption potential of the coarse size mode ($1 - \omega_{0,C}$). If the local ω_0 is smaller than the suggested coarse mode background value (which is quite common), then excess absorption is assumed to have been caused by the smaller (accumulation mode) aerosol. However, to avoid unrealistic low values for $\omega_{0,A}$, a minimum $\omega_{0,Amin}$ is defined as a function of both coarse and accumulation mode AOD ($\omega_{0,Amin} = 0.80 \times AOD_A + 0.95 \times AOD_C$). In case the required $\omega_{0,A}$ falls below that threshold ($\omega_{0,Amin}$), the dust size is doubled (now $r_{eff} = 2.5$ μm), because increasing the size of an absorbing aerosol (here coarse dust) is an alternate way to lower the overall ω_0. Working now with a smaller $\omega_{0,C}$ of larger dust sizes, $\omega_{0,A}$ is recalculated. If the required $\omega_{0,A}$ continues to falls below the revised $\omega_{0,Amin}$ threshold, the doubling of the dust size is repeated to $r_{eff} = 5$ µm and if necessary even to $r_{eff} = 10$ µm. Thus, the ω_0 constraint not only defines the coarse mode type, it also defines the single-scattering albedo of the accumulation mode $\omega_{0,A}$ and provides information on the dust size.

Third, *solar extension of accumulation mode and AnP conversion into g.* Overall spectral aerosol properties are defined by combining spectral properties of coarse size mode and accumulation size mode. Precalculated spectral properties for the relevant dust size and/or sea salt are applied for the coarse mode. For the accumulation mode, the spectral dependence of the AOD is defined by AnP_A, which, as explained above, varies between 2.2 and 1.6 depending on local low cloud cover. Even for the smallest possible AnP_A value of 1.6 (at wet conditions), the associated effective radius of 0.27 µm remains small enough, so that far-infrared impacts (at wavelengths > 4 µm) can be neglected. Thus, accumulation mode properties of $\omega_{0,A}$ and g_A need only to be defined for the solar spectrum. The mid-visible single-scattering albedo of the accumulation mode $\omega_{0,F}$ is assumed to apply for all UV and visible wavelength. However, with size parameters significantly smaller than one in the near-infrared, the single-scattering albedo is reduced. The asymmetry factor g_A is based on the accumulation mode AnP_A using a relationship derived from AERONET statistics. Using truncated AERONET size distributions (no coarse size mode concentrations), scatter plots at wavelengths of 0.44, 0.55, and 1.0 µm suggest an anticorrelation between g_A and AnP_A which increases at longer wavelengths $g_A = 0.7 - 0.4 \times (AnP_A - 1) \times \ln(w^{0.5} + 0.5)$, *when wavelengths (w) are between 0.25 and 3.0* µm. This quite general relationship is associated with significant scatter, but it is certainly better than assuming the mid-visible value g_A (0.58 for dry and 0.64 for wet conditions) for the entire solar spectrum.

Finally, *apply mixing rules to combine coarse mode and accumulation mode properties*. The single-scattering properties of the coarse and accumulation size modes are combined via the common mixing rules, whereby AOD is additive, ω_0 needs to apply an AOD weight, and g is weighted by the product of AOD and ω_0. Figure 3.6 presents samples of annual maps at four wavelengths for AOD, ω_0, and g.

Anthropogenic Fraction

Quantifying anthropogenic AOD is difficult. Today's measurements may provide estimates for total AOD distributions, but they cannot distinguish natural from anthropogenic contributions because measurements of a natural reference (e.g., preindustrial conditions) are extremely rare. Thus, the distinction between natural and anthropogenic aerosols must be based on modeling. Model output from two specific AeroCom experiments (Schulz et al. 2006)

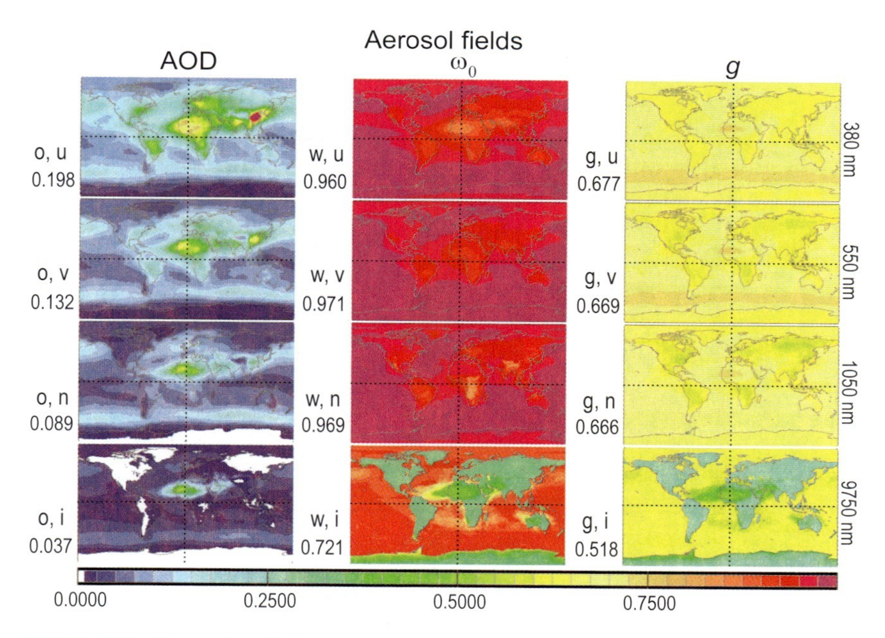

Figure 3.6 Spectral variations for single-scattering properties of total aerosol. Annual global fields for the column-integrated properties of extinction (AOD, left), single-scattering albedo (ω_0, center) and asymmetry factor (g, right) are presented at UV (380 nm), visible (550 nm), near-IR (1050 nm), and far-infrared (9750 nm) wavelengths.

serves as the basis: one uses current (year 2000) emissions while the other applies preindustrial (year 1750) emissions for primary aerosol and precursor gases (Dentener et al. 2006). As these emission inventories were processed using different global models, it was assumed that anthropogenic aerosols only populated the smaller accumulation sizes as a result of added sulfate and carbonaceous particles. Anthropogenic dust, which may have enhanced the coarse size concentrations because of changes in land use, was neglected. Thus, an anthropogenic fraction was defined that relates to AOD_A, not to the total AOD. This means that the anthropogenic fraction is independent of interannual variations in coarse mode sizes. From simulated monthly maps for the AOD_A with current and preindustrial emissions, today's anthropogenic fraction $f_{A,ANT}$ was determined from the normalized difference:

$$f_{A,ANT} = \frac{AOD_{A,2000} - AOD_{A,1750}}{AOD_{A,2000}}. \tag{3.2}$$

Monthly fields for $f_{A,ANT}$ are presented in Figure 3.7.

The anthropogenic AOD is defined by the product of the anthropogenic fraction and the accumulation mode AOD: $AOD_{ANT} = f_{A,ANT} \times AOD_A$. Single-scattering albedo and asymmetry factor are those of the accumulation mode

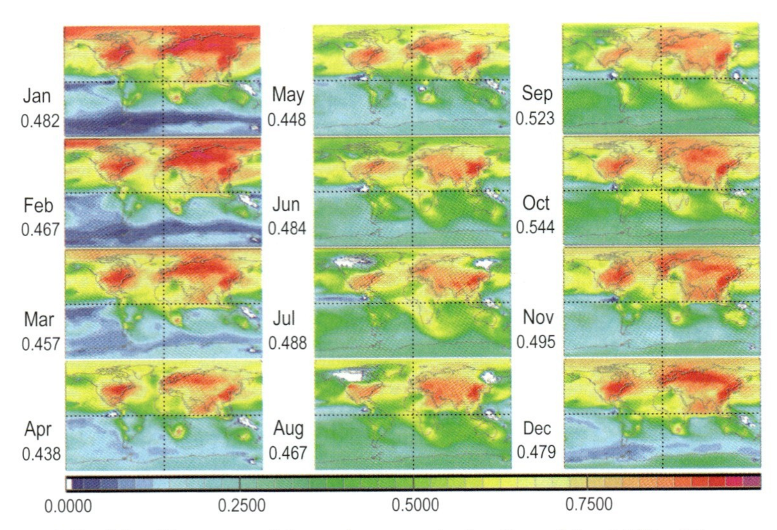

Figure 3.7 Monthly maps of the anthropogenic fraction of the AOD of the accumulation size mode (for aerosol particles smaller than 0.5 μm in radius) based on global model simulations.

($\omega_{0,A}$, g_A); the solar spectral dependencies follow those of the accumulation mode as well. Annual global maps at selected solar wavelengths are shown in Figure 3.8.

The natural aerosol properties are defined by the difference between total and anthropogenic aerosol. Thus, the AOD for natural aerosols is defined by the sum of the accumulation mode AOD, which is not anthropogenic, and the entire coarse mode AOD: $AOD_{NAT} = (1 - f_{ANT}) \times AOD_A + AOD_C$. Values for ω_0 and g of natural aerosols apply a similar relationship with the appropriate mixing weights.

CCN and Their Enhancements

Aerosols can act as CCN and modify as a result cloud microphysics with potential impacts on cloud macrophysical properties (e.g., cloud structure, frequency, or lifetime) and the hydrological cycle (e.g., precipitation intensity, frequency, or distribution; cf. Stratmann et al. and Ayers and Levin, both this volume). Thus, there is a need to quantify the number of available aerosol particles that can serve as nuclei. In the context of a changing climate, there is particular interest in anthropogenic CCNs.

Aerosol particles, acting as CCN, allow atmospheric water vapor to condense and form cloud droplets. This condensation occurs preferably on larger particle sizes, because lower supersaturations are required to overcome their surface curvature effect. Thus, only the larger particle sizes are of interest

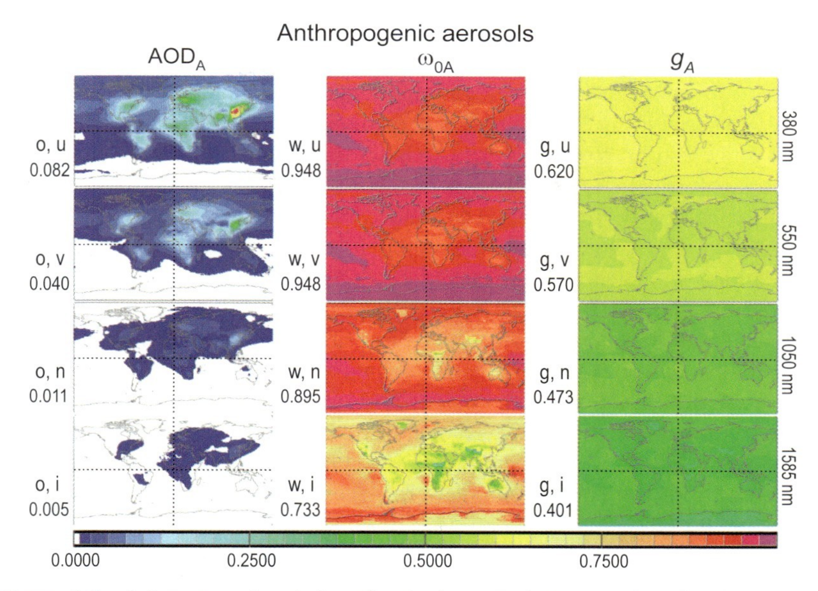

Figure 3.8 Solar spectral variations for single-scattering properties of anthropogenic aerosols. Annual global fields for the column-integrated properties of extinction (AOD_A), single-scattering albedo (ω_{0A}), and asymmetry factor (g_A) are presented at an UV (380 nm), a VISIBLE (550 nm) and two near-IR (1050 nm, 1585 nm) wavelengths.

(here, as well as for radiative transfer impacts). Potential CCN are all aerosol particles of the coarse size mode and the larger sizes of the accumulation size mode. Differences in the hygroscopic particle properties are a complicating factor. However, recent studies (see discussion in Part 2) have concluded that hygroscopic particle properties cluster around characteristic values over land and ocean. Assuming this simplification, the critical size, above which particles can serve as CCN, is primarily a function of the water vapor supersaturation. Typical values for supersaturation are near 0.1% and up to 0.5% for more convective cloud systems. An analytical formula (Rose et al. 2008; see also below, Equation 3.3) places the critical radii at 80 nm (over land) and 30 nm (over oceans) for a supersaturation of 0.1% and at 60 nm (over land) and 20 nm (over oceans) for a supersaturation of 0.5%. Thus, over land as much as 50% of the pollution or biomass (accumulation size) aerosol may be too small to serve as CCN, whereas over the ocean almost all accumulation (and of course coarse) particle sizes can be potential CCNs.

Aside from data on hygroscopicity and supersaturation, estimates for CCN require information on size and (local) aerosol amount. This information is supplied by concepts and data from the new aerosol climatology. Aerosol amount is based on global AOD maps, where sun photometer data are merged onto a modeling background. For the necessary vertical distribution of the AOD, monthly average data from global model simulations were applied. As part of

AeroCom model evaluations (Giubert, pers. comm.), general agreement has been demonstrated between simulated aerosol vertical distributions and lidar profiles. However, given the diversity in modeling and a growing database by active remote sensing from space, it is recognized that the use of model data at this stage is a pragmatic choice to satisfy data needs. At a later stage, data on aerosol vertical distribution will certainly be better constrained by active remote-sensing data from ground and space. Particle size is based on a bimodal, log-normal distribution, and a distinction is made between coarse and accumulation size modes. Apportionment of AOD to each mode (AOD_A, AOD_C) is tied to the mid-visible AOD spectral dependence, combining the spectral insensitivity of the coarse mode with a prescribed spectral dependence for the accumulation mode. The coarse mode composition (either dust or sea salt) as well as the dust size are defined by the mid-visible ω_0. More specifically, the coarse mode assumes a log-normal distribution with a fixed distribution width (standard deviation 2.0). The assumed mode radii are 0.75 μm for sea salt and 0.375 μm for dust. It should be noted, however, that larger dust sizes are successively chosen (0.75, 1.5, or 3.0 μm), if (small) particle absorption alone is unlikely to explain locally the low ω_0 of the climatology. The smaller accumulation mode also assumes a log-normal size distribution with a fixed (though narrower) distribution width (standard deviation 1.7). In conjunction with the prescribed AnP (completely dry: 2.2; completely wet: 1.6) the mode radius is defined to lie between 0.085 μm under completely dry and 0.135 μm under completely wet conditions (where low cloud cover of a cloud climatology is applied to define wetness).

Three types of CCN concentrations are considered. In the first scenario, all aerosols of the coarse and accumulation modes are considered. In the second scenario, supersaturations of up to 0.5% are permitted, which requires that the log-normal distribution of the accumulation mode needed to be truncated from sizes smaller than 60 nm over land and smaller than 20 nm over oceans, since these particles were too small to be activated. In the third scenario, the maximum supersaturation was set to 0.1%, with cutoff sizes at 80 nm over land and 30 nm over oceans. Simulated CCN concentrations at about 1 km above the ocean or land surface are displayed in Figure 3.9 (in log10 space) separately for total, natural, and anthropogenic aerosol.

The CCN concentrations of Figure 3.9 are given for a logarithmic scale and annual averages. Monthly CCN fields for total (natural and anthropogenic) aerosol at 0.1% and 0.5% supersaturations indicate seasonal variations (e.g., increases during the tropical biomass burning season) and demonstrate that CCN concentrations increase (on a global average basis) by about 30%, when relaxing the supersaturation from 0.1% to 0.5%. It should be noted that CCN concentrations of the accumulation mode are about one order of magnitude larger than CCN concentrations of the coarse mode. Thus, it is not surprising that monthly maps for the CCN anthropogenic fraction resemble the monthly maps for the anthropogenic fraction for AOD_A of Figure 3.7.

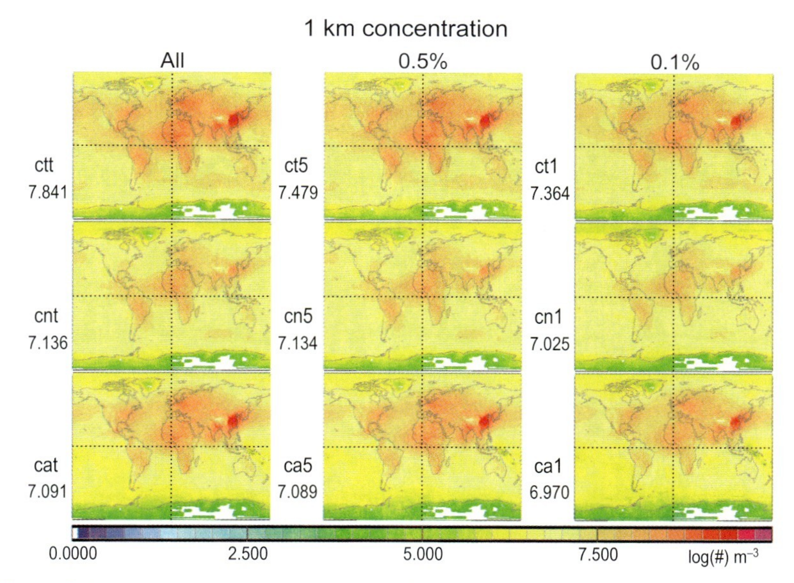

Figure 3.9 Annual average global maps for CCN concentrations without cutoffs to sizes of the accumulation mode (left), at 0.5% supersaturation (center), and at 0.1% supersaturation (right). Concentrations are displayed (in log10 m^{-3}) separately for total aerosols (top), natural aerosols (center), and anthropogenic aerosols (bottom).

CCN concentration changes with altitude. Thus, to obtain information on cloud development at higher altitudes, CCN concentrations were determined at two additional altitudes: 3 km and 8 km. Simulated CCN concentrations for a 0.1% supersaturation at these three altitudes are compared in Figure 3.10 (in log10 space), again separately for natural, anthropogenic, and total (anthropogenic and natural) aerosol.

The simulated total (natural and anthropogenic) CCN concentrations are in general agreement with Glomap simulations (Spracklen et al. 2008). Their global annual average at the surface and for a 0.2% supersaturation is about twice as large as the estimate of this study for 1 km altitude and a 0.1% supersaturation. There is also agreement on the seasonal cycle, as July CCN concentrations (northern hemispheric summer) are higher than for December. Most maxima match (e.g., industrial regions). The largest differences are Glomap-simulated CCN sinks in the ITCZ, which are likely caused by cloud-processing in the Glomap model.

Of particular interest are the CCN enhancement factors, defined by the ratio between anthropogenic and natural concentration. Since the annual ratios resemble each other at different supersaturations and altitudes, enhancement factors are presented on a monthly basis in Figure 3.11 for the most interesting case of low-level water clouds: lower altitude and 0.1% supersaturation.

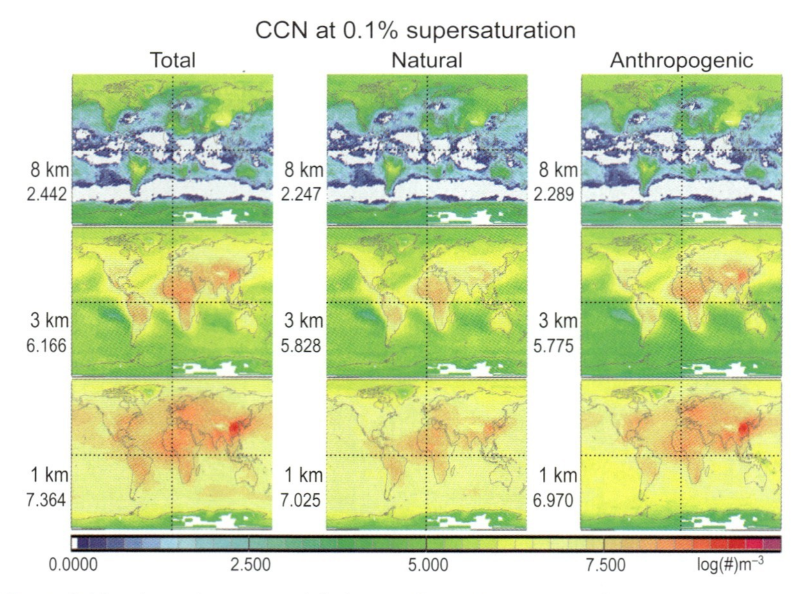

Figure 3.10 Annual average global maps for CCN concentrations at 0.1% supersaturation at different altitudes: 1 km above the ground, at 3 km, and at 8 km. Concentrations (in log10 m^{-3}) are shown for total, natural, and for anthropogenic aerosols.

The monthly maps in Figure 3.11 illustrate that anthropogenic enhancements occur primarily in the Northern Hemisphere, predominantly near industrial areas. Anthropogenic CCN enhancements are greater during the winter season, at which time they significantly impact the Arctic.

Conclusion

Monthly global maps have been developed (a) to define aerosol optical properties (also as a function of wavelength), (b) to distinguish between natural and anthropogenic contributions, and (c) to derive cloud-relevant estimates for CCN and CCN enhancement as a result of anthropogenic activities. In developing these maps, the preferential use of quality measurements from ground-based monitoring networks over pure model simulations was promoted. This study demonstrated the value of coordinated ground-based aerosol monitoring, especially when data were extended with modeling. Modeling and satellite retrievals will also benefit from complementary ground network data on aerosol, more accurate satellite datasets, or better model parameterization and simulations. Many sources have had to rely completely on modeling, because data were either unavailable or lacking in accuracy. Future efforts should explore new measurement-based dataset sources to constrain the freedom in

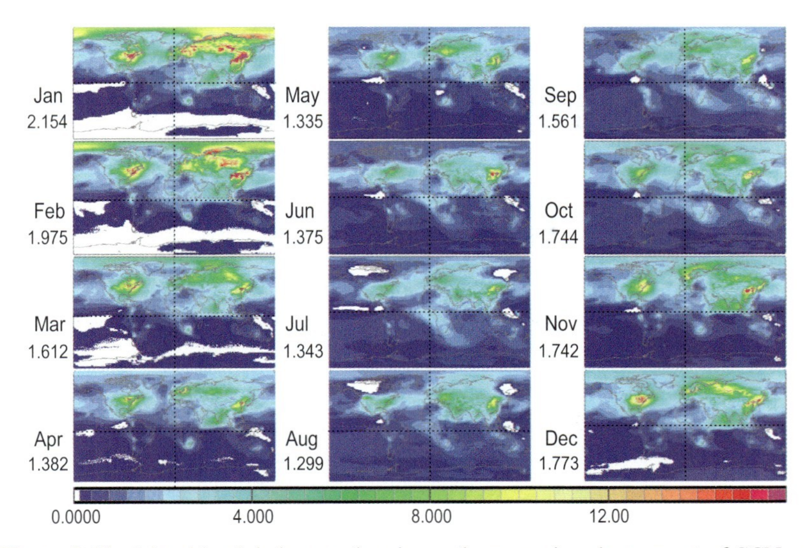

Figure 3.11 Monthly global maps showing anthropogenic enhancement of CCN concentrations over natural CCN at 0.1% supersaturation at 1 km above the ground.

modeling or to replace simulations. Examples are vertical profiling from active remote sensing.

The various global maps created for this chapter relied on many assumptions and applied often rather simple methods. There is obviously room for improvement and fine tuning; the values indicated for (tropospheric) aerosols were intended to provide general numbers on amount, spatial distribution, and seasonality. Despite providing estimates for CCN concentrations, the cloud-related aspect of aerosols remains unclear. Little is known about what happens in cloudy environments to aerosol or cloud particles as a result of interactions, as almost all aerosol data are associated with cloud-free conditions. New measurements (e.g., active remote sensing, high temporal resolution data) and modeling concepts (e.g., bridging modeling scales) are needed to move beyond questionable statistical associations.

Part 2: Particle Hygroscopicity and Cloud Condensation Nucleus Activity

Ulrich Pöschl, Diana Rose, and Meinrat O. Andreae

Max Planck Institute for Chemistry, Biogeochemistry
Department, 55128 Mainz, Germany

Abstract

Aerosol particles that act as cloud condensation nuclei (CCN) play a central role in the formation of clouds. Here, the basic concepts and key aspects that link the CCN activity of aerosol particles to their size, chemical composition, and hygroscopicity (i.e., their ability to absorb water vapor) are discussed. Literature data and recent field measurements suggest that the influence of chemical composition can be efficiently described by a single effective hygroscopicity parameter that relates the dry particle diameter to the so-called critical water-vapor supersaturation (i.e., the minimum supersaturation required to form a cloud droplet). This hygroscopicity parameter, κ, can be easily calculated from chemical composition data and is typically in the range of ~0.1 for pyrogenic and secondary organic aerosols to ~1 for sea spray aerosols. Continental and marine boundary layer aerosols tend to cluster into relatively narrow ranges of effective hygroscopicity (continental $\kappa = 0.3 \pm 0.1$; marine $\kappa = 0.7 \pm 0.2$).

Thus the influence of aerosol chemical composition and hygroscopicity appears to be less variable and less uncertain than other factors that determine the effects of aerosols on warm cloud formation in the atmosphere (e.g., particle number concentration, size distribution, sources, sinks, and meteorological conditions). Nevertheless, more detailed investigations and representations of the hygroscopic properties of aerosol particles, as a function of chemical composition, are needed to elucidate fully aerosol–cloud interactions, especially for low water-vapor supersaturations, low aerosol concentrations, and organic components. Even for simple and well-defined inorganic reference substances, such as ammonium sulfate and sodium chloride, the critical supersaturations or critical dry particle diameters of CCN activation calculated with different Köhler models can deviate as much as 20% from the most accurate available models. To ensure that measurement and model results can be compared properly, CCN studies should always report exactly which Köhler model equations and parameters have been applied. In addition, potential kinetic limitations of water uptake appear to be one of the most crucial open questions of CCN properties and activation.

Introduction

Atmospheric aerosol particles that enable the condensation of water vapor and formation of cloud droplets are called cloud condensation nuclei. Elevated concentrations of CCN tend to increase the concentration and decrease the size of droplets in a cloud. In addition to changing the optical properties and radiative effects of clouds on climate, this could lead to the suppression of precipitation in shallow and short-lived clouds, but also to greater convective overturning and more precipitation in deep convective clouds. The response of cloud characteristics and precipitation processes to increasing anthropogenic aerosol concentrations represents one of the largest areas of uncertainty in the current understanding of climate change. One of the crucial challenges is to determine the ability of aerosol particles to act as CCN under relevant atmospheric conditions, an issue that has received increasing attention over the past years (McFiggans et al. 2006; IAPSAG 2007; IPCC 2007; Andreae and Rosenfeld 2008; and references therein).

Cloud Droplet Formation and Köhler Theory

In the Earth's atmosphere, cloud droplets do not generally form by homogeneous nucleation of supersaturated water vapor (i.e., condensation of water molecules from the gas phase in the absence of a preexisting condensation nucleus). This would require the initial formation of droplet embryos (clusters of water molecules) with a very small radius of curvature. Because of surface tension, however, the equilibrium vapor pressure over such a strongly curved surface is much greater than over a flat surface ("Kelvin effect" or "curvature effect"). Thus water-vapor supersaturations (the relative difference between the actual vapor pressure and the equilibrium vapor pressure over a flat surface) of several hundred percent are needed for homogeneous nucleation of water droplets (Pruppacher and Klett 2000; Andreae and Rosenfeld 2008).

In the atmosphere, such large supersaturations are not reached, because aerosol particles facilitate the condensation of water vapor. The equilibrium water-vapor pressure over an aqueous solution is generally lower than over pure water ("Raoult effect" or "solute effect"; reduction of water activity), and thus water vapor can condense and form solution droplets on particles composed of soluble material (deliquescence and hygroscopic growth). Insoluble, but wettable particles can also facilitate droplet formation by decreasing the curvature effect for water adsorbed on the surface (depending on hydrophilicity and contact angle), and the uptake of water vapor on insoluble particles can be enhanced by soluble materials that are ubiquitous in the atmosphere (e.g., sulfuric acid).

By accounting for curvature and solute effects, Köhler theory describes the hygroscopic growth and CCN activation of soluble aerosol particles as a function of relative humidity or water-vapor supersaturation, respectively (Seinfeld and Pandis 1998; Pruppacher and Klett 2000; McFiggans et al. 2006): For a given dry particle diameter, it enables the critical water-vapor supersaturation to be calculated; that is, the minimum supersaturation required to form an aqueous droplet that can freely grow by further condensation (cloud droplet).

For a given water-vapor supersaturation, it permits the critical dry particle diameter to be determined; that is, the minimum dry particle diameter required to form a cloud droplet.

A wide range of different Köhler models have been applied in experimental and theoretical studies of aerosol–cloud interactions. Depending on the equations, parameterizations, and approximations used to quantify the Raoult effect and describe water activity in the aqueous solution droplet, they can be broadly classified as activity parameterization models, osmotic coefficient models, van't Hoff factor models, effective hygroscopicity parameter models, and analytical approximation models. Some of these, however, yield substantially different results even for simple and well-defined standard aerosols and reference substances (Rose et al. 2008a).

Figure 3.12 illustrates the range of critical supersaturations calculated as a function of dry particle diameter (20–200 nm) with different Köhler models for ammonium sulfate and sodium chloride. As discussed by Rose et al. (2008a), activity parameterization models based on the Aerosol Inorganics Model (AIM; Clegg et al. 1998a, b) can be regarded as the most accurate Köhler models available for these substances. The relative deviations of alternative models range up to 20% for ammonium sulfate and 10% for sodium chloride.

As outlined below, such deviations may be negligibly small compared to other uncertainties in current investigations of the interactions between aerosol, cloud, and climate using regional and global atmospheric models. For detailed mechanistic studies of cloud processes, however, and compared to the high precision of state-of-the-art measurement techniques (e.g., CCN counters), the

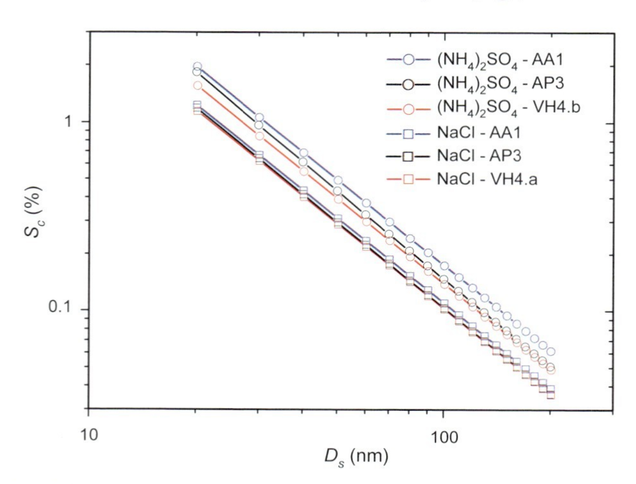

Figure 3.12 Critical supersaturations (S_c) calculated for ammonium sulfate and sodium chloride particles with dry particle mass or volume equivalent diameters (D_s) in the range of 20–200 nm at 298 K using selected Köhler models: an activity parameterization model based on the AIM (AP3, black); a van't Hoff factor model (VH4, red); and an analytical approximation model (AA1, blue). For details see Rose et al. (2008a).

deviations between different Köhler models can be very substantial and exceed other sources of uncertainty. To ensure that measurement and model results can be properly compared, CCN studies should always report exactly which Köhler model equations and parameters have been applied.

CCN Activity of Aerosol Particles

The ability of an aerosol particle to act as a CCN ("CCN activity")—that is, its ability to induce water-vapor condensation and cloud droplet formation under a given set of conditions—depends primarily on (a) water-vapor supersaturation, (b) dry particle size, and (c) hygroscopicity (soluble particles) or wettability (insoluble particles) (Seinfeld and Pandis 1998; Pruppacher and Klett 2000; McFiggans et al. 2006; Andreae and Rosenfeld 2008).

In science and engineering, hygroscopicity is generally defined as the ability of a substance to absorb or adsorb water vapor. In atmospheric and aerosol research, however, the term hygroscopicity is usually reserved for the absorption of water by soluble substances; the term wettability is used for the adsorption of water on the surface of insoluble particles. In the atmosphere, most aerosol particles are at least partly soluble, and thus the following discussion will focus on the hygroscopicity of fully or partially soluble particles and particle components rather than on the wettability of insoluble particles.

In general, the hygroscopicity of an aerosol particle depends on its chemical composition and the following factors:

- the mass or volume fraction of soluble components (water-soluble fraction) and their actual solubility (saturation equilibrium concentration in aqueous solution);
- the efficiency of soluble components in reducing water activity ("Raoult efficiency"), which can be expressed by an equivalent number of ions or molecules that go into solution per unit mass or unit volume of the particle material (Raoult's law);
- the surface activity of soluble components (i.e., their ability to reduce the surface tension and Kelvin effect of an aqueous solution).

To describe and model the CCN activity of atmospheric particles, the influence of chemical composition and the factors listed above can be efficiently summarized and approximated using a single hygroscopicity parameter, as will be detailed below.

The actual solubility of water-soluble components (i.e., their saturation equilibrium concentration in aqueous solution) is usually of minor importance for the CCN activity of atmospheric aerosol particles. Most chemical components can effectively be regarded as either fully soluble or completely insoluble. In Köhler model calculations, the effects of limited solubility need to be considered only for a relatively narrow range of sparingly soluble organic compounds and appear to be negligible for real atmospheric multicomponent particles (Petters and Kreidenweis 2008; Kreidenweis et al., this volume).

Nevertheless, sparingly soluble organic compounds can influence CCN activation as surfactants that reduce droplet surface tension and/or form a surface film that inhibits water uptake (mass transfer kinetics). The practical importance of sparingly soluble as well as of fully soluble surface-active substances and their effects on the CCN activity of atmospheric aerosol particles has not yet been well characterized and needs to be further elucidated (McFiggans et al. 2006). Relative to particle size and the fraction and Raoult efficiency of water-soluble components, however, changes of surface tension appear to be a second-order effect. It is unlikely to exceed the ~10% relative uncertainty that is more or less inherent to most state-of-the-art field measurements and model calculations of atmospheric aerosol properties (McMurry 2000; Raes et al. 2000; Kanakidou et al 2005; Pöschl 2005; Fuzzi et al. 2006; Textor et al. 2006; Rose et al 2008a,b; Kreidenweis et al., this volume).

In the scientific literature and discussion, aerosol particles that are more hygroscopic and CCN-active than others are frequently called "more soluble." This is, however, misleading because the solubility and the hygroscopicity of water-soluble compounds are not directly proportional. In fact, compounds with higher solubility can be less hygroscopic and less CCN-active than compounds with lower solubility. For example, NH_4NO_3 is very highly soluble (~27 mol kg^{-1}) but only moderately hygroscopic ($\kappa \approx 0.7$), whereas NaCl is moderately soluble (~6 mol kg^{-1}) but very highly hygroscopic ($\kappa \approx 1.3$; see discussion in next section). Solubility determines the relative humidity of deliquescence (i.e., the threshold for the transformation of a dry particle into a saturated aqueous solution droplet). The extent of water uptake (hygroscopic growth factor) and the critical supersaturation of CCN activation, however, are primarily governed by the hygroscopicity and Raoult efficiency of the soluble substance (effective molar density of soluble ions or molecules; see below). Accordingly, the deliquescence of NH_4NO_3 particles occurs at lower relative humidity than that of NaCl particles (~60% vs. ~75% RH; Mikhailov et al. 2004), but the critical supersaturation for the CCN activation of NH_4NO_3 particles is higher than that of NaCl particles (e.g., 0.20% vs. 0.15% for particles with a diameter of ~80 nm; cf. Equation 3.3).

In addition to the composition and properties of particulate matter, the gas phase composition can also influence the CCN activity of aerosol particles: Water-soluble gases like ammonia and nitric acid can facilitate droplet formation by dissolving in nascent droplets, adding to the solute amount in the droplet, and thus enhancing the Raoult effect. On the other hand, the timescales of gas-particle interaction can influence the activation of CCN in the atmosphere (clouds) as well as in measurement instruments (HTDMA and CCN counters) and laboratory experiments (aerosol chambers and flow tubes) as a result of the kinetic limitations of mass transport and phase transitions (Mikhailov et al. 2004; McFiggans et al. 2006; Andreae and Rosenfeld 2008;

Ruehl et al. 2008). Potential kinetic limitations of water uptake go beyond equilibrium model calculations and appear to be one of the most crucial open questions of CCN activation and cloud droplet formation.

Note, however, that the understanding of surface and multiphase processes in aerosols and clouds is limited not only by the available kinetic and thermodynamic data but also by a "Babylonian confusion" of terms and parameters (e.g., inconsistency of mass accommodation coefficients for the uptake water vapor on liquid water: bulk vs. surface accommodation?). This problem will hopefully be resolved through the use of a comprehensive kinetic model framework with consistent and unambiguous terminology and universally applicable rate equations and parameters (Pöschl et al. 2007).

Thus it follows that CCN are not a fixed, special type of particle, but rather a highly variable subset of the ambient aerosol population. Size, solute content, the presence of surface-active or slightly soluble substances, the wettability and shape of insoluble particles, the presence of soluble gases, and the timescales of gas-particle interaction all influence whether a particle can act as a CCN at a given supersaturation. Consequently, the abundance and properties of CCN are influenced by all of the mechanisms and processes that lead to the emission, formation, and transformation of atmospheric aerosols. Note, however, that the actual influence of aerosol chemical composition on CCN activation and cloud droplet growth appears to be limited by the dynamics of cloud formation and evolution under real atmospheric conditions (McFiggans et al. 2006).

The atmospheric abundance, sources, properties, and effects of CCN as well as the relations to anthropogenic pollution and global change have recently been reviewed by Andreae and Rosenfeld (2008). For a review of the characteristics and variability of atmospheric aerosol size distributions, the reader is referred to Heintzenberg et al. (2000) and Raes et al. (2000); recent results from remote sensing studies are presented by Kinne (see Part 1).

Characterization and Parameterization of CCN Activity

The CCN activity of an aerosol particle depends on its size and chemical composition. According to Raoult's law, the ability of the water-soluble fraction to absorb water and reduce its activity is determined primarily by the number of ions or molecules that can go into a solution per unit volume of the particulate material (Raoult efficiency of the solute).

For particles with a fully known composition, the effective number of solute molecules or ions per unit volume can be derived from the basic physicochemical properties of the components. The relevant properties, such as dissociation constants and activity coefficients, are, however, not well known for many atmospherically relevant substances (e.g., as secondary organic aerosol components). Ambient particles contain a vast number of compounds many of which are unknown, not individually measurable, and without available thermodynamic information.

Consequently, empirical techniques have been developed and used to characterize the hygroscopicity of ambient aerosol particles. They can be broadly classified in two categories:

- Techniques that measure the water uptake and hygroscopic growth of aerosol particles in the subsaturated regime (RH < 100%), which can be extrapolated to CCN activation under supersaturated conditions.
- Techniques that directly measure the activation of CCN and formation of cloud droplets in the supersaturated regime (RH > 100 %).

Most information can be obtained with online and size-resolved techniques, and the most widely used instruments are hygroscopicty tandem differential mobility analyzers (HTDMA) for subsaturated conditions and various types of CCN counters (mostly thermal gradient diffusion chambers and flow tubes) for supersaturated conditions. HTDMAs are usually more robust and simple to construct and operate than CCN counters, but obviously CCN counters provide more direct insight into CCN activation. The reliability of CCN measurement results has been a subject of continuing debate. Recently, CCN counters with high and well-documented measurement precision and accuracy have become commercially available and are now widely used in laboratory and field studies (McFiggans et al. 2006; Rose et al. 2008a,b).

Measurement results on the hygroscopic growth and CCN activation of atmospheric aerosol particles have often been reported in terms of an (equivalent) soluble fraction, which can be defined as the volume fraction of a model salt (e.g., $(NH_4)_2SO_4$ or NaCl) in a dry particle consisting of the model salt and an insoluble core, such that the model particle would exhibit the same hygroscopicity and CCN activity as the actual particle (Andreae and Rosenfeld 2008a,b).

Petters and Kreidenweis (2007) have proposed an effective hygroscopicity parameter, κ, which relates the volume of water taken up by a particle to the activity of water in the formed aqueous droplet. Measured or calculated values of κ and similar parameters (e.g., "ion densities" as defined and used by Rissler et al. 2006 and Wex et al. 2007) enable efficient calculation of critical supersaturations or critical dry particle diameters according to the Köhler theory (effective hygroscopicity parameter models; Rose et al. 2008a).

A simple analytical approximation model enables the calculation of the critical supersaturation S_c as a function of κ and of the dry particle mass or volume equivalent diameter D_s, respectively (Rose et al. 2008a)

$$S_c \approx \left(\exp\left(\left(\tfrac{4}{27} \right)^{\frac{1}{2}} \cdot A^{\frac{3}{2}} \cdot \kappa^{-\frac{1}{2}} \cdot D_s^{-\frac{3}{2}} \right) - 1 \right) \cdot 100\%. \tag{3.3}$$

Here the Kelvin effect is described by the parameter A as a function of the surface tension of the aqueous solution droplet (σ_{sol}), the absolute temperature (T), the ideal gas constant (R), and the density and molar mass of water (ρ_w, M_w):

$$A = \frac{(4\sigma_{sol} M_W)}{(RT\rho_w)}. \tag{3.4}$$

Under the assumption that the surface tension of the aqueous solution droplets formed upon CCN activation equals the surface tension of water, the Kelvin parameter can be approximated by $A \approx (0.66 \times 10^{-6}\ \mathrm{K\ m})/T$.

Under the assumption of linear additivity of the influence of individual chemical components on the activity of water in a multicomponent aqueous solution (Zdanovski-Stokes-Robinson or ZSR assumption), the effective hygroscopicity parameter of an aerosol particle (κ) can be calculated by linear combination of the effective hygroscopicity parameters of the individual chemical components (κ_j) weighted by their volume fraction (ε_j) in the dry particle:

$$\kappa = \sum_j \varepsilon_j \kappa_j. \tag{3.5}$$

Insoluble components can be described by $\kappa_j = 0$, and limited solubility can be taken into account as described by Petters and Kreidenweis (2008) and Kreidenweis et al. (this volume). For fully soluble components, κ_j is directly related to the molar mass (M_j), density (ρ_j), and van't Hoff factor (i_j) or stoichiometric dissociation number and osmotic coefficient ($v_j\ \varphi_j$) of the dry component and of water (subscript w), respectively (Rose et al. 2008):

$$\kappa_j = \frac{\rho_j / M_j}{\rho_w / M_w} i_j \approx \frac{\rho_j / M_j}{\rho_w / M_w} v_j \varphi_j. \tag{3.6}$$

Accordingly, κ and κ_j can be regarded as an "effective Raoult parameter"; that is, as an effective or equivalent molar density of soluble ions or molecules in the dry particle or dry particle component respectively, normalized by the molar density of water molecules in liquid water (~55 mol $\mathrm{l^{-1}}$). For compounds with low molecular mass (including most inorganic salts like sulfates, nitrates, chlorides as well as mono- or dicarboxylic acids, monosaccharides, etc.), the effective molar density of ions or molecules is usually close to the actual molar density of ions/molecules (i.e., the osmotic coefficients are close to unity). For macromolecular organic compounds like proteins, however, the osmotic coefficient can increase up to >100, and the effective Raoult parameter can only be regarded as an equivalent parameter (Mikhailov et al. 2004). The assumption of a constant value of κ_j for a soluble particle component is essentially equivalent to the assumption of a concentration-independent van't Hoff factor or osmotic coefficient, respectively.

Note, however, that Petters and Kreidenweis (2007) have suggested folding surface tension effects into the hygroscopicity parameter κ as well; that is, to assume that the surface tension of aqueous solution droplets formed upon CCN activation equals the surface tension of water ($\sigma_{sol} = \sigma_w = 0.072\ \mathrm{J\ m^{-2}}$ at 298K) both upon deriving κ values from experimental data as well as upon applying them for model predictions. To avoid confusion, we suggest that a distinction be made between "effective Raoult parameters," as defined in the above equations and in the basic equations of Petters and Kreidenweis (2007), and "effective hygroscopicity parameters" that have been derived and should be applied under the assumption of $\sigma_{sol} = \sigma_w$. For most practical applications, this discrimination will be of minor importance, and in many cases it may also not be possible to determine independently effective Raoult parameters (water activity reduction) and Kelvin parameters (surface tension reduction) for atmospheric aerosol particles. Neverthless, establishing unambiguous terms and definitions should be helpful for consistent and efficient communication between different scientific communities (aerosol and cloud; field, laboratory, and model).

Characteristic values of κ_j are in the range of ~1.3 for NaCl, ~0.6 for $(NH_4)_2SO_4$, ~0.2 for levoglucosan and various organic acids, and ~0.1 for secondary organic aerosols. For biomass-burning particles, κ values range from about 0.01 for very fresh smoke containing mostly soot particles to 0.55 for aerosols from grass burning. The available data suggest that after short aging on a timescale of hours, most pyrogenic aerosols will have κ values in the range of 0.1 to 0.3 (Petters and Kreidenweis 2007; Andreae and Rosenfeld 2008; Rose et al. 2008b; Kreidenweis et al., this volume). Externally mixed aerosol populations can be described by separate κ values, and the effects of internal mixing (coagulation/condensation) can be described according to Equation 3.5. Based on aerosol and CCN field measurements and laboratory experiments, it will be possible to establish an inventory of effective hygroscopicity and CCN activity of particles from various sources and to describe efficiently the influence of atmospheric aging.

A summary of κ values derived from various sources is presented in Figure 3.13. The data plotted in this figure show that continental aerosols fall in a narrow range of κ values around 0.3, consistent with the suggestion by Dusek et al. (2006) that, for such aerosols, the composition can be treated to a good approximation as invariant, and that the CCN activity of particles is controlled mostly by particle size. In that study, the observed range of κ was 0.15–0.30. Although Hudson (2007) correctly points out that field data cover a larger range of κ values, his polluted continental data also give an average of $\kappa = 0.33 \pm 0.15$. As expected, his clean marine data indicate higher values, with $\kappa = 0.87 \pm 0.24$ (colored bands in Figure 3.13). Measurements at the coast of Puerto Rico also showed CCN activation diameters corresponding to κ values in the range 0.6 ± 0.2 (Allan et al. 2007). A large number of field data has been compiled by Kandler and Schütz (2007) and expressed in the form of soluble fractions. When converted to κ values, these data are also consistent with urban and continental values of κ around 0.2–0.3 and marine values around 0.6. Recent measurements of CCN, as a function of particle size and water-vapor supersaturation at background and polluted continental locations in Europe, Asia, and America, confirm fairly uniform average hygroscopic properties and CCN activities of continental aerosols ($\kappa = 0.3 \pm 0.1$). Higher variability was

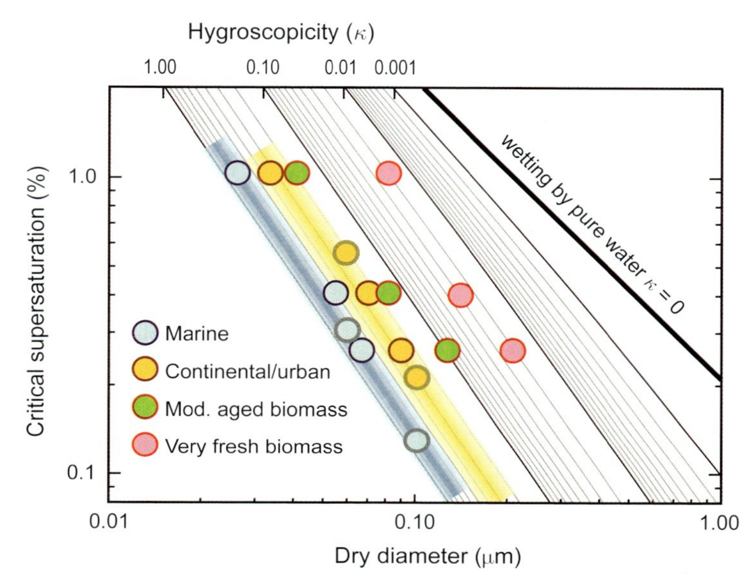

Figure 3.13 Average relations between critical supersaturation and aerosol dry diameter (Andreae and Rosenfeld 2008). The colored bands reflect polluted continental and clean marine data from Hudson (2007); the colored dots with colored borders are from Dusek et al. (2006) and Andreae and Rosenfeld (2008). The colored dots with gray borders have been recalculated from Kandler und Schütz (2007). The lines representing constant effective hygroscopicity parameters κ are from Petters and Kreidenweis (2007).

observed only at low supersaturation ($\leq 0.1\%$) and for freshly emitted or newly formed aerosols dominated by organic components (Rose et al. 2007, 2008b).

Overall, κ values of 0.3±0.1 and 0.7±0.2 appear to be representative for average continental and marine aerosols, respectively. The assumption of fairly uniform effective hygroscopicity parameters is also supported by studies indicating that the influence of aerosol chemical composition is limited by the dynamics of cloud droplet growth. Moreover, the influence of the composition-dependent hygroscopicity parameter is, according to Köhler theory, weaker by the power of 3 than that of particle diameter (see Equation 3.3: $\kappa^{-1/2}$ vs. $D_s^{-3/2}$). Accordingly, Dusek et al. (2006) and Andreae and Rosenfeld (2008) have argued and presented evidence that particle size is the dominant property in controlling the variability in CCN activity of atmospheric aerosols. Moreover, Andreae (2008) found a surprisingly close correlation between the average CCN concentrations observed in field measurements around the globe and the corresponding aerosol optical thickness (AOT) data from collocated or nearby remote sensing instruments (mostly AERONET sun photometers). Aerosol particle number concentrations and size distributions obtained by remote sensing and their implications for the abundance of CCN on global and regional scales have also been addressed by Kinne (see Part 1).

Nevertheless, more detailed investigations and representations of aerosol particle hygroscopicity and composition are desirable for a full elucidation of aerosol–cloud interactions, especially for low water-vapor supersaturations, low aerosol concentrations, and organic components (McFiggans et al. 2006; Rose et al. 2007, 2008b; Kreidenweis et al., this volume).

Summary and Conclusions

The subset of particles from the overall aerosol population that can act as CCN is determined mostly by the number of soluble ions or molecules a particle contains, which in turn is a function of particle size and chemical composition. The complexity and variability of the composition of ambient aerosols make an explicit treatment of the chemical properties of CCN in atmospheric models difficult. Recent studies have shown, however, that for the purpose of assessing CCN activity, the influence of aerosol composition can be efficiently represented by a single parameter: the hygroscopicity parameter κ. Knowledge of this parameter and of the ambient aerosol size distribution enables efficient calculation of CCN spectra (CCN concentration as a function of water-vapor supersaturation).

Measurements of CCN as a function of water-vapor supersaturation and particle size suggest that continental and marine aerosols tend to cluster on average into relatively narrow ranges of effective hygroscopicity (continental $\kappa=0.3\pm0.1$; marine $\kappa=0.7\pm0.2$), which should facilitate their treatment in atmospheric models. Thus the influence of aerosol chemical composition and hygroscopicity appears to be less variable and less uncertain than other factors that determine the effects of aerosols on warm cloud formation in the atmosphere (e.g., particle number concentration, size distribution, sources, sinks, and meteorological conditions) (see Anderson et al., Kinne, and Kreidenweis et al., all this volume).

Nevertheless, more detailed investigations and representations of the hygroscopic properties of aerosol particles as a function of chemical composition are needed before aerosol–cloud interactions—especially for low water-vapor supersaturations, low aerosol concentrations, and organic components—are fully explained. Even for simple and well-defined inorganic reference substances, such as ammonium sulfate and sodium chloride, the critical supersaturations or critical dry particle diameters of CCN activation calculated with different Köhler models can deviate by up to 20%. To ensure that measurement and model results can be properly compared, CCN studies should always report exactly which Köhler model equations and parameters have been applied.

Looking beyond equilibrium model calculations, potential kinetic limitations of water uptake appear to be one of the most crucial open questions of CCN properties and activation. Moreover, knowledge of the chemical composition of aerosol particles is essential to determine aerosol sources and production mechanisms and to assess the role of anthropogenic versus natural contributions to the atmospheric aerosol burden.

Acknowledgments

The authors would like to thank M. Petters, S. Kreidenweis, A. Prenni, G. Roberts, P. Chuang, G. Feingold, S. Kinne, E. Mikhailov, S. Martin, J.-L. Brenguier, R. Charlson, J. Heintzenberg, J. Quaas, and the participants and organizers of this Forum for stimulating comments, discussions, and support.

References

Allan, J. D., D. Baumgardner, G. B. Raga et al. 2007. Clouds and aerosols in Puerto Rico: A new evaluation. *Atmos. Chem. Phys. Discuss.* **7**:12,573–12,616.

Anderson, T. L., R. J. Charlson, N. Bellouin et al. 2005. An “A-train” strategy for quantifying direct climate forcing by anthropogenic aerosols. *Bull. Amer. Meteor. Soc.* **86**:1795–1809.

Andreae, M. O. 2008. Correlation between cloud condensation nuclei concentration and aerosol optical thickness in remote and polluted regions. *Atmos. Chem. Phys. Discuss.* **8**:11,293–11,320.

Andreae, M .O., and D. Rosenfeld. 2008. Aerosol–cloud–precipitation interactions. Part 1: The nature and sources of cloud active aerosols. *Earth Sci. Rev.* **89(1-2)**:13–41.

Aoki, K. 2006. Aerosol optical characteristics in Asia from measurements of SKYNET sky radiometers. In: IRS 2004 – Current Problems in Atmospheric Radiation, ed. H. Fischer and B-J. Sohn, pp. 311–313. Hampton, VA: Deepak Publ.

Bösenberg, J., V. Matthias et al. 2003. EARLINET – A European Aerosol Research Lidar Network to Establish an Aerosol Climatology. Hamburg: MPI Report No. 348.

Clegg, S. L., P. Brimblecombe, and A. S. Wexler. 1998a. A thermodynamic model of the system H+ -NH+4 –SO42−–NO3−H2O at tropospheric temperatures. *J. Phys. Chem. A* **102**:2137–2154.

Clegg, S. L., P. Brimblecombe, and A. S. Wexler. 1998b. A thermodynamic model of the system H+-NH4+-Na+-SO42–NO3–Cl–H2O at 298.15 K. *J. Phys. Chem. A* **102**:2155–2171.

d’Almeida, G. A., P. Koepke, and E. P. Shettle. 1991. Atmospheric Aerosols: Global Climatology and Radiative Characteristics. Hampton, VA: Deepak Publ.

Dentener, F., S. Kinne, T. Bond et al. 2006. Emissions of primary aerosol and precursor gases in the years 2000 and 1750, prescribed datasets for AeroCom. *Atmos. Chem. Phys* **6**:4321–4344.

Deuzé, J. L, F. M. Breon, C. Devaux et al. 2001. Remote sensing of aerosol over land surfaces from POLDER/ADEOS-1 polarized measurements. *J. Geophys. Res.* **106**:4912–4926.

Deuzé, J. L., M. Herman, and P. Goloub. 1999. Characterization of aerosols over ocean from POLDER/ADEOS-1. *Geophys. Res. Lett.* **26**:1421–1424.

Dubovik, O., and M. D. King. 2000. A flexible inversion algorithm for retrieval of aerosol optical properties from Sun and sky radiance measurements. *J. Geophys. Res.* **105**:20,673–20,696.

Dusek, U., G. P. Frank, L. Hildebrandt et al. 2006. Size matters more than chemistry for cloud-nucleating ability of aerosol particles. *Science* **312**:1375–1378.

Fuzzi, S., M. O. Andreae, B. J. Huebert et al. 2006. Critical assessment of the current state of scientific knowledge, terminology, and research needs concerning the role of organic aerosols in the atmosphere, climate, and global change. *Atmos. Chem. Phys*. **6**:2017–2038.

Heintzenberg, J. D., C. Covert, and R. Van Dingenen. 2000. Size distribution and chemical composition of marine aerosols: A compilation and review. *Tellus* **52**:1104–1122.

Hess, M., P. Koepke, and I. Schult. 1998. Optical Properties of Aerosols and Clouds: The software package OPAC. *Bull. Amer. Meteor. Soc.* **79**:831–844.

Holben, B. N., T. F. Eck, I. Slutsker et al. 1998. AERONET: A federated instrument network and data archive for aerosol characterization. *Rem. Sens. Environ.* **66**:1–16.

Hudson, J. G. 2007. Variability of the relationship between particle size and cloud-nucleating ability. *Geophys. Res. Lett.* **34**:L08801.

IAPSAG. 2007. International aerosol precipitation science assessment group (IAPSAG): Aerosol pollution impact on precipitation: A scientific review. Geneva:WMO.

IPCC. 2007. Climate Change 2007: The Physical Science Basis. Contribution of Working Group I to the Fourth Assessment Report of the Intergovernmental Panel on Climate Change, ed. S. Solomon, D. Qin, M. Manning et al. New York: Cambridge Univ. Press.

Kahn, R., P. Banerjee, D. McDonald, D. J. Diner. 1998. Sensitivity of multi-angle imaging to aerosol optical depth and to pure particle size distribution and composition over ocean. *J. Geophys. Res.* **103**:32,195–32,213.

Kanakidou, M., J. H. Seinfeld, and S. N. Pandis et al. 2005. Organic aerosol and global climate modelling: A review. *Atmos. Chem. Phys.* **5**:1053–1123.

Kandler, K., and L. Schütz. 2007. Climatology of the average water-soluble volume fraction of atmospheric aerosol. *Atmos. Res.* **83**:77–92.

Kaufman, Y., D. Tanre, L. Remer et al. 1997. Operational remote sensing of tropospheric aerosol over the land from EOS-MODIS. *J. Geophys. Res.* **102**:17,051–17,061.

Kinne, S. 2008. Aerosol direct radiative forcing with an AERONET touch. *Atmos.Environ.,* in press.

Kinne, S., M. Schulz, C.Textor et al. 2006. An AeroCom initial assessment: Optical properties in aerosol component modules of global models. *Atmos. Chem. Phys.* **6**:1–22.

Klein, H., A. Benedictow, and H. Fagerli. 2007. Transboundary data by main pollutants (S, N, O_3) and PM. *MSC-W Data Note* 1/07.

Koepke, P., M. Hess, I. Schult, and E. P. Shettle. 1997. Global aerosol dataset. Hamburg: MPI Report No. 243.

Langner, J., and H. Rodhe. 1991. A global three-dimensional model of the tropospheric sulfur cycle *J. Atmos. Chem.* **13**:225–263.

Malm, W., and J. Hand. 2007. An examination of the physical and optical properties of aerosols collected in the IMPROVE program. *Atmos. Environ.* **11**:3407–3427.

Martonchik, J. V., D. J. Diner, R. A. Kahn et al. 1998. Techniques for the retrieval of aerosol properties over land and ocean using multi-angle imaging. *IEEE Trans. Geosci. Remt. Sensing* **36**:1212–1227.

McFiggans, G., P. Artaxo, U. Baltensperger et al. 2006. The effect of physical and chemical aerosol properties on warm cloud droplet activation. *Atmos. Chem. Phys.* **6**:2593–2649.

McMurry, P. H. 2000. A review of atmospheric aerosol measurements. *Atmos. Environ.* **34**:1959–1999.

Mikhailov, E., S. Vlasenko, R. Niessner, and U. Pöschl. 2004. Interaction of aerosol particles composed of protein and salts with water vapor: Hygroscopic growth and microstructural rearrangement. *Atmos. Chem. Phys.* **4**:323–350.

Nilsson, B. 1979. Meteorological influence on aerosol extinction in the 0.2–40 μm wavelength range. *Appl. Opt.* **18**:3457–3473.

Nishizawa, T., H. Okamoto, N. Sugimoto et al. 2007. An algorithm that retrieves aerosol properties from dual-wavelength polarized lidar measurements. *J. Geophys. Res.* **112**:D06212.

Petters, M. D., and S. M. Kreidenweis. 2007. A single parameter representation of hygroscopic growth and cloud condensation nucleus activity. *Atmos. Chem. Phys.* **7**:1961–1971.

Petters, M. D., and S. M. Kreidenweis. 2008. A single parameter representation of hygroscopic growth and cloud condensation nucleus activity. Part 2: Including solubility. *Atmos. Chem. Phys. Discuss.* **8**:5939–5955.

Pöschl, U. 2005. Atmospheric aerosols: Composition, transformation, climate and health effects. *Angew. Chemie Intl. Ed.* **44**:7520–7540.

Pöschl, U., Y. Rudich, and M. Ammann. 2007. Kinetic model framework for aerosol and cloud surface chemistry and gas-particle interactions. Part 1: General equations, parameters, and terminology. *Atmos. Chem. Phys.* **7**:5989–6023.

Pruppacher, H. R., and J. D. Klett. 2000. Microphysics of Clouds and Precipitation. Dordrecht: Kluwer Acad. Publ.

Raes, F., R. Van Dingenen, E. Vignati et al. 2000. Formation and cycling of aerosols in the global troposphere. *Atmos. Environ.* **34**:4215–4240.

Rissler, J., A. Vestin, E. Swietlicki et al. 2006. Size distribution and hygroscopic properties of aerosol particles from dry-season biomass burning in Amazonia. *Atmos. Chem. Phys.* **6**:471–491.

Rose, D., G. P. Frank, U. Dusek, M. O. Andreae, and U. Pöschl. 2007. Are the cloud condensation nuclei (CCN) properties in polluted air different from those in a remote region? *Geophys. Res. Abstr.* **9**:09452.

Rose, D., S. S. Gunthe, E. Mikhailov et al. 2008a. Calibration and measurement uncertainties of a continuous-flow cloud condensation nuclei counter (DMT–CCNC): CCN activation of ammonium sulfate and sodium chloride aerosol particles in theory and experiment. *Atmos. Chem. Phys.* **8**:1153–1179.

Rose, D., A. Nowak, P. Achert et al. 2008b. Cloud condensation nuclei in polluted air and biomass burning smoke near the mega-city Guangzhou, China. Part 1: Size-resolved measurements and implications for the modeling of aerosol particle hygroscopicity and CCN activity. *Atmos. Chem. Phys. Discuss.* **8**:17,343–17,392.

Rossow, W., A. Walker, and C. Garder. 1993. Comparison of ISCCP and other cloud amounts. *J. Climate* **6**:2394–2418.

Ruehl, C. R., P. Y. Chuang, and A. Nenes. 2008. How quickly do cloud droplets form on atmospheric particles? *Atmos. Chem. Phys.* **8**:1043–1055.

Schulz, M., C.Textor, S.Kinne et al. 2006. Radiative forcing by aerosols as derived from the AeroCom present-day and preindustrial simulations. *Atmos. Chem. Phys.* **6**:5225–5346.

Seinfeld, J. H., and Pandis, S. N. 1998. Atmospheric Chemistry and Physics: From Air Pollution to Climate Change. New York: Wiley.

Spracklen, D. V., K. J. Pringle, K. S. Carslaw, M. P. Chipperfield, and G. W. Mann. 2005. A global off-line model of size-resolved aerosol microphysics: I. Model development and prediction of aerosol properties. *Atmos. Chem. Phys.* **5**:2227–2252.

Stephens, G. L, and C. Kummerow. The remote sensing of clouds and precipitation from space: A review. *J. Atmos. Sci.* **64(11)**:3742.

Stowe, L., H. Jacobowitz, G. Ohring, K. Knapp, and N. Nalli. 2002. The advanced very high resolution radiometer pathfinder atmosphere (PATMOS) dataset: Initial analyses and evaluations. *J. Climate* **15**:1243–1260.

Tanre, D., Y. Kaufman, M. Herman et al. 1997. Remote sensing of aerosol properties over ocean using the MODIS/EOS spectral radiances *J. Geophys. Res.* **102**:16,971–16,988.

Textor, C. 2006. Analysis and quantification of the diversities of aerosol life cycles within AeroCom. *Atmos. Chem. Phys.* **6**:1777–1811.

Textor, C., Schulz, M., Guibert et al. 2006. Analysis and quantification of the diversities of aerosol life cycles within AeroCom. *Atmos. Chem. Phys.* **6**:1777–1813.

Torres, O., P. K. Barthia, J. R. Herman et al. 2002. A long-term record of aerosol optical depth from TOMS observationsand comparisons to AERONET measurements. *J. Atmos. Sci.* 59:398–413.

Welton, E. J., J. R. Campbell, J. D. Spinhirne and V. S. Scott. 2001. Global monitoring of clouds and aerosols using a network of micro-pulse lidar systems. *Proc. SPIE* 4153:151–158.

Wex, H., T. Hennig, I. Salma et al. 2007. Hygroscopic growth and measured and modeled critical super-saturations of an atmospheric HULIS sample. *Geophys. Res. Lett.* **34**:L02818.

Winker, D. M., W. H. Hunt, and C. A. Hostetler. 2004. Status and performance of the CALIOP lidar. *Proc. SPIE* **5575**:8–15.

WMO. 1986. World climate research program: A preliminary cloudless standard atmosphere for radiation computation. Geneva: WCP-112, WMO/TD-NO. 24.

4

Cloud Properties from *In-situ* and Remote-sensing Measurements

Capability and Limitations

George A. Isaac[1] and K. Sebastian Schmidt[2]

[1]Cloud Physics and Severe Weather Research Section, Environment Canada, Toronto, Canada
[2]Institute for Atmospheric Physics, University of Mainz, Mainz, Germany

Abstract

The ability to measure cloud properties has advanced considerably over recent years. This chapter reviews the *in-situ* and remote-sensing instrumentation currently available and discusses some of the associated potential problems. The requirements for calibrations and use of well-documented software are outlined. Aerosol and cloud characteristics have vertical and geographical variations that are important, but poorly defined. The necessity to measure parameters on the scales of interest and to present those measurements in proper units is discussed. Recommendations are made regarding future actions that would be beneficial to the community. These include improvements required in the accuracy of the measurements, global datasets, and collaborations between those who make and use *in-situ* and remote-sensing measurements.

Introduction

The ability to measure cloud parameters has advanced considerably over the past fifty years, in terms of instrumentation and analysis techniques. Earlier (e.g., Squires 1958), drops were captured on a media such as glass slides exposed to the airstream from aircraft platforms; after suitable corrections, this gave a measure of cloud droplet number concentrations. Those measurements were used to develop the concepts that maritime clouds had droplet concentrations smaller than continental clouds and typical values were assigned to each. That type of characterization still exists today, very often using textbook-type values of microphysical parameters. In this chapter, we explain why it is necessary to get away from such simple approaches and measurements and start to consider clouds as more complex, stressing the use of the latest *in-situ* and remote-sensing probes.

First, it is necessary to consider the accuracy to which cloud properties must be measured to simulate them properly in numerical model simulations of our climate. There have been a few sensitivity studies, such as those by Slingo (1990) and Rotstayn (1999).

> The top of the atmosphere radiative forcing by doubled carbon dioxide concentrations can be balanced by modest relative increases of ~15–20% in the amount of low clouds and 20–35% in liquid–water path, and by decreases of 15–20% in mean drop radius. This indicates that a minimum relative accuracy of ~5% is needed....to simulate these quantities in climate models (Slingo 1990, pp. 49).

> The total indirect forcing is –2.1 W m^{-2}...*resulting*...from a 1% increase in cloudiness, a 6% increase in liquid water path, and a 7% decrease in droplet effective radius (Rotstayn 1999, pp. 9369).

These studies suggest that cloud properties need to be measured with significant precision; Slingo suggested to within 5%. However, this is a very demanding, if not unrealistic goal with today's instruments.

Because of the increasing complexity in obtaining measurements and analyzing them, often the user obtains data from institutional data banks and is only vaguely aware of all the associated problems. Problems such as instrument calibrations, expected accuracy, software analysis problems, and scale effects, for example, are rarely understood. This chapter will hopefully provide a short primer to allow appropriate questions to be asked before the provided data is used.

Current *In-situ* Instrumentation and Its Limitations

In-situ Measurements

The airborne platform used for *in-situ* measurements must be suitable. For example, in some applications a tethered balloon might be adequate and economical, whereas in others a long-range aircraft is essential. For campaigns requiring only a few instruments, a light aircraft or an unmanned airborne vehicle (UAV) might be the best option; in other projects, a large capacity aircraft is necessary to carry all of the technical gear as well as the many operators and scientists involved. Some applications require an aircraft that can fly at high, or low, altitudes.

There is a host of *in-situ* instruments available to measure the properties of aerosols, in particular cloud condensation nuclei (CCN), ice-forming nuclei

(IN), and cloud properties such as cloud droplets, ice particles, cloud water content, precipitation-sized particles, and special properties related to extinction and scattering of the particles themselves. Often it is necessary to measure gas phase chemical composition and the chemical particle properties to simulate the interaction of clouds with their environment properly. It is not possible to address all of the available instruments in this chapter, nor can we successfully forecast new advances in instrumentation. Thus, the reader will be referred to some general papers, and an outline of some of the available instruments will be provided.

In the 1970s, Particle Measuring System (PMS) and, specifically, Bob Knollenberg effected a new era of cloud *in-situ* measurements. Knollenberg (1981) describes some of these new probes; for a more recent summary of *in-situ* measurement techniques, see Baumgardner et al. (2002). Figure 4.1 (see next section) indicates the size ranges of some standard PMS probes for size distribution measurements. Lawson (1998, 2001, 2006) describes new instrumentation that can help image and discriminate small ice particles in the atmosphere, which is a very difficult problem at the moment (see below). Korolev et al. (1998b) introduce a new probe that can simultaneously measure cloud liquid water content (LWC) and cloud total water content (TWC). Rogers et al. (2001) and Rose et al. (2007) discuss some of the problems of measuring IN and CCN.

Table 4.1 shows some cloud microphysical parameters that are measured in all-liquid, all-ice, and mixed-phase clouds, along with their nominal sample volumes and possible accuracy. Many problems exist in creating such a simplified overview; however, the intent is to provide useful information to the nonexpert. Hallett (2003) discusses some of the errors brought about by sampling statistics, which are not adequately addressed in Table 4.1. Some *in-situ* measurement parameters, which are important but not listed, include in-cloud temperature and relative humidity (or supersaturation), as well as updraft and turbulent velocities. Out-of-cloud dewpoint is also an important variable. Temperature in-cloud is an interesting case, because it is often assumed that it can be measured within 1°C, providing the probe does not become wet (Jensen and Raga 1993). Using a fast response probe, however, Haman et al. (2001) demonstrated rapid fluctuations of 1–2°C over distances of one or two centimeters, which could significantly influence in-cloud superaturation. Unfortunately, methods to measure in-cloud supersaturation are limited, an exception being the technique described by Gerber (1991). Korolev and Isaac (2006) show measurements of supersaturation within all-liquid, all-ice, and mixed-phase clouds, where the all-liquid and mixed-phase clouds are close to water saturation. There are several techniques to measure in-cloud turbulence; for a summary of such measurements made in cumulus clouds, see MacPherson and Isaac (1977) and Siebert et al. (2006).

In-situ data must be collected with instruments that perform the tasks required. Isaac et al. (2005) describe instrumentation necessary for obtaining

Table 4.1 *In-situ* measurements normally made in all-liquid, all-ice, and mixed-phase clouds. Nominal sample volume rate is shown at 100 m s^{-1}; source of the nominal accuracies are indicated in the footnotes. Individual datasets should provide their own numbers. The accuracies assume that probes are properly mounted on the aircraft and calibrated. CCN and IN measurements are not necessarily done in-cloud. SS: supersaturation; N: number concentration; D: particle diameter; MVD: median volume diameter.

Parameter	Size Range	Nominal sample (vol. at 100 ms^{-1})	Possible accuracy	Comments
All-liquid Clouds				
CCN [1, 2, 3]	0.1–1.3% SS	1 l min^{-1}	10–40%	No known calibration standards
LWC [4, 5, 6]	0.01–3 g m^{-3}	4 l s^{-1}	15%	Errors higher for low LWC; only accurate for MVD < 40 μm.
Cloud droplets[7, 8]	2–50 μm	30 cm^3 s^{-1}	N: 20%, D: 1–2 μm	Accuracies depend on airspeed, size, and concentration
Large droplets[9, 10,11]	50–500 μm	5–15 l s^{-1}	N: 25%, D: 10%	Accuracies depend on airspeed and size; 50–100 μm is poorly measured for all phases
Precipitation drops[7]	>500 μm	200 l s^{-1}	N: 10%, D: 10%	
All-ice Clouds				
IN[12]	0 to –40°C, ice saturation to 20% SS water	1 l min^{-1}	Unknown	Very difficult measurement
Ice water content[13]	0.01–2 g m^{-3}	4 l s^{-1}	25%	
Small ice particles[14]	2–100 μm			Controversy about measurements by imaging and scattering probes related to shattering off probe tips.

Parameter	Size Range	Nominal sample (vol. at 100 ms^{-1})	Possible accuracy	Comments
Large ice[9, 10, 11]	>100 μm	50–200 l s^{-1}	N: 25%, D: –15%	
Particle shape[11, 15]	Requires 5–10 pixels			Automated recognition software is available but not standardized
Mixed-phase Clouds				
TWC[13]	0.01–3 g m^{-3}	4 l s^{-1}	50%	Accuracy depends on ice/liquid fraction.
LWC[4, 5, 6]	0.01–3 g m^{-3}	4 l s^{-1}	30%	
Ice water content[13]	0.01–2 g m^{-3}	4 l s^{-1}	50%	
Small size distribution[16]	2–100 μm		unknown	No validated technique for phase segregation
Large sizes[9, 10, 11, 16]	>100 μm	5–200 l s^{-1}	N: 25%, D: 15%	Need good liquid/ice discrimination software
Particle shape[11, 15]	Requires 5–10 pixels			Can be accomplished by assuming all circular 2-D images represent liquid drops

Sources:
[1]Lance et al. (2006)
[2]Rose et al. (2008)
[3]Roberts et al. (2006)
[4]Biter et al. (1987)
[5]Korolev et al. (1998a)
[6]Strapp et al. (2003)
[7]Knollenberg (1981)
[8]Baumgardner & Korolev (1997)
[9]Strapp et al. (2001)
[10]Gayet et al. (1993)
[11]Korolev et al. (1998b)
[12]Rogers et al. (2001)
[13]Korolev et al. (2008)
[14]Heymsfield (2007)
[15]Korolev et al. (2000)
[16]Cober et al. (2001b)

data for aircraft icing certification tests, and much of the information in that paper is relevant here. When selecting and evaluating datasets, it is necessary to know (a) the parameters that are required, (b) the accuracy required for those measurements, (c) the range of conditions over which the measurements must be made (e.g., temperature, altitude, LWC, drop size, length scale), and (d) the types of clouds to be examined (convective, stratiform). In general, the dataset must be sufficiently large to obtain a representative sample. Generally, a few case studies are not good enough to characterize clouds for climate models or to evaluate remote-sensing techniques, although they may be exceptionally useful for examining physical processes.

It should be stressed that it is very difficult to measure IN because they can activate through many different mechanisms (e.g., deposition, contact, immersion, condensation freezing; cf. Kreidenweis et al., this volume). The difficulties are compounded because it is necessary to obtain vertical profiles of IN, thus requiring airborne instrumentation. The uncertainties in measurements of small ice particles are also quite large. Combined, these two problems represent a large uncertainty in characterizing ice in clouds and in understanding ice formation mechanisms in the atmosphere, which are a blend of primary nucleation through IN and secondary processes through ice multiplication. This area definitely requires further instrument development.

It is often assumed that by measuring the particle size spectrum, the liquid, ice, or TWC can be determined by a simple integration of the spectrum. However, significant sizing errors exist for any measurement, and when the diameters within a size bin of a probe are cubed to calculate volume or mass, the resulting error in the integrated liquid or ice water content becomes large (Baumgardner 1983). Thus, it is recommended that a probe specifically designed for direct measurement of liquid, ice, or TWC, such as a hot wire probe or counterflow virtual impactor, be selected. An icing rate indicator is also very useful to determine whether supercooled liquid was encountered during data collection. If the ramp voltage is measured during the tests, then a rough estimate of LWC can be obtained to provide a check with other instrumentation (Mazin et al. 2001; Cober et al. 2001a).

Imaging probes that measure the shape of the particles are very useful and, in some cases, essential. Many clouds contain both ice and liquid particles (mixed-phase clouds). Cober et al. (2001b) and Korolev et al. (2003) have documented that such clouds occur frequently and discuss some of the associated measurement issues. Summarizing supercooled in-cloud measurements, Isaac et al. (2001) report that 25% of maritime and 49% of continental clouds were characterized as mixed phase. If it is not possible to assess whether ice crystals are present, these particles can be misinterpreted as supercooled large drops, thereby greatly biasing liquid cloud median volume diameter estimates. The hot-wire LWC probes also measure a fraction of the ice particles present, thus overestimating the LWC if the ice mass concentration is high.

There is a need to select instruments that have been evaluated and reported in the open literature. Manufacturers are continuously producing better instruments, but they often do not operate to the specifications provided. Users should look for comparison tests of instrumentation performed in icing wind tunnels (e.g., Strapp et al. 2003). These tests can give evaluations of the effectiveness of the probes and their associated accuracies and limitations over a wide range of environmental conditions. All instrumentation have some weaknesses and strengths. It is important to know what these are before selecting sensors or samplers for a particular application. Instruments and their resulting data should be selected for use only if their accuracies have been documented and demonstrated.

Other concerns that must be considered are:

1. Are the probes adequately de-iced for the temperature and liquid water ranges expected? It is common for newly designed probes not to have adequate de-icing heaters.
2. Will imaging probes and droplet spectrometers fog during rapid descents? Fogging can also occur in climbs into temperature inversions. Fogging and icing signatures (e.g., Brenguier et al. 1993) must be known during the data analysis, and corrections need to be applied.
3. Some probes work well at low aircraft speeds (e.g., typical turboprop speeds) but the electronics are not fast enough for proper operation at high speeds (typical jet aircraft speeds).

Remote-sensing Instruments

The launch of CloudSat (Stephens et al. 2002) and CALIPSO (Winker et al. 2007) in 2006 marked the beginning of a new era in cloud and aerosol spaceborne remote sensing with the start of continuous radar and lidar observations. At the same time, new polarimetric and hyperspectral imagers already in orbit or under preparation provided an improved level of insight into clouds, aerosols, and their impact on weather and climate.

Ground-based remote-sensing observations are often understated for their role in global Earth system remote sensing, yet they excel in temporal sampling resolution, accuracy, and continuity, and provide data that cannot be obtained from satellites (e.g., cloud and aerosol surface radiative forcing). The major drawback of ground-based observations is their limited spatial coverage. Thus, the establishment of networks over the last decade represents an important achievement (Kinne et al., this volume).

Aircraft observations are an indispensable tool for observing clouds. Most satellite applications are tested on aircraft before they are deployed in space, and field measurements are conducted to validate satellite measurements and algorithms on a regular basis; intensive, vertically resolved aircraft observations link the globally extensive ground and spaceborne observations. UAVs are becoming an increasingly important tool for atmospheric research.

Most atmospheric remote sensing is based on the measurement of electromagnetic radiation that has retained information about its interaction with atmospheric constituents through scattering, absorption, and emission. An exception is sodar (sonic detection and ranging), which uses reflection of acoustic waves on atmospheric boundaries and is an important tool for monitoring wind speed and atmospheric stability.

The wavelengths used for atmospheric remote sensing lie within three spectral regions ("windows"), which are characterized by a low opacity of various atmospheric gases within the Earth's atmosphere. This is illustrated in Figure 4.1, where the clear-sky atmospheric opacity is plotted as a function of wavelength and frequency. Also plotted are the normalized Planck curves of solar and terrestrial emission. The first range is called "shortwave" or "solar" window. It is located around the maximum of solar irradiance (490 nm wavelength), originating from blackbody emission of the Sun's photosphere with a temperature of about 5800 K. It comprises the near-ultraviolet, visible, and near-infrared wavelengths and is bounded by stratospheric ozone absorption and molecular scattering at the short wavelength end. The second range ("longwave," "thermal infrared," or "terrestrial" window) is centered at a wavelength of 10 μm (corresponding to the peak in Earth's blackbody emission with a

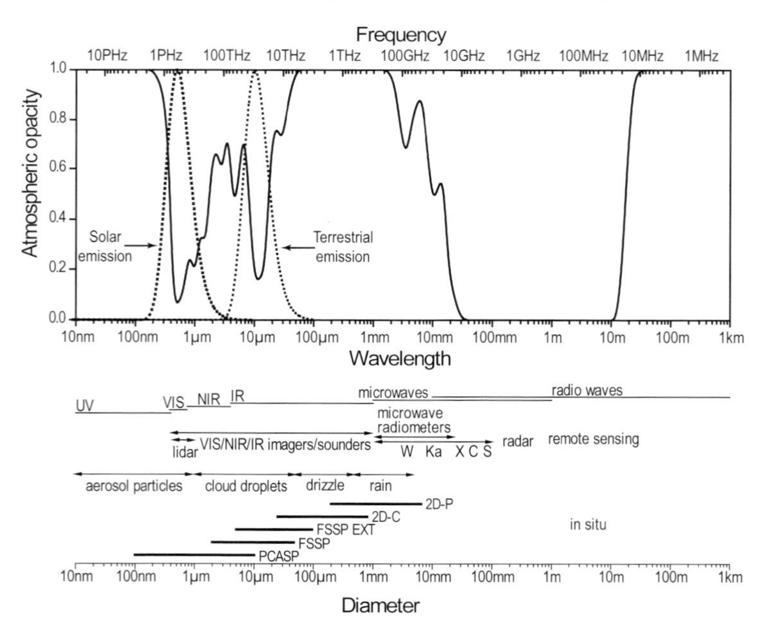

Figure 4.1 Top: atmospheric opacity with normalized solar and terrestrial emission. Bottom: spectral ranges of remote-sensing techniques and size ranges for some *in-situ* instruments.

surface temperature of 290 K). The transition between solar and terrestrial range is often defined at about 4 μm. The terrestrial range extends to 100 μm and includes various carbon dioxide and water vapor absorption features. The third range ("radio window," 1 cm to 30 m wavelength) is used for microwave radiometers and radars. Some remote-sensing techniques and their typical wavelength ranges are shown in Figure 4.1.

If the measured radiation originates from a natural source (solar or terrestrial emission), the technique is called *passive* remote sensing; *active* remote sensing refers to those methods where the radiation originates from the instrument itself. Some examples of passive remote sensing include visible and infrared imagery as well as microwave radiometers. Active techniques include radars, lidars, and ceilometers. Remote sensing can be further distinguished by viewing geometry (*nadir* or *limb*) and by viewing mode (*profilers, imagers*, and *scanners*). Imagers cover large areas with sometimes high horizontal resolution, but they contain only limited information about the vertical structure of the atmosphere. In addition, sensitivity to thin cirrus and aerosols is generally low, especially over bright surfaces. Profilers view the atmosphere on slant paths and thus have high vertical, but very limited horizontal, resolution. They are very sensitive to cirrus, aerosols, and gases because of (a) the large photon path length through the atmosphere and (b) the low surface contributions to the signal. However, the large slant path contributes to strong attenuation and thus the signal saturates at relatively low vertical optical thickness. Some techniques combine nadir and limb viewing using multiple detectors or along/cross-track scanning modes. Spaceborne active instruments use nadir-viewing geometry. Ground-based instruments can be *zenith* viewing (e.g., lidars, sky imagers), *scanning* (e.g., radars), or *tracking* (sun photometers).

Spectrally resolved measurements enable the retrieval of vertical atmospheric structure even for nadir-viewing geometries (AIRS, Chahine et al. 2006; AMSU, Mo 1996), and allow the separation of clouds, aerosols, water vapor, and other gases by means of their spectral signature. On the other hand, to determine the energy budget of clouds and aerosols from space, broadband albedo and emission measurements are required for the solar and terrestrial spectral ranges, respectively.

The polarization state of radiation can be exploited for both passive and active remote sensing. It is useful to detect particle shape and orientation as well as to separate the contributions to top-of-atmosphere radiance from atmosphere and surface (Herman et al. 1997).

Table 4.2 lists cloud and aerosol parameters accessible with remote sensing. The accuracy and sensitivity of a technique to a particular parameter depends on the underlying physics. For example, radars have a high sensitivity to precipitation and can determine its geometrical distribution. This is because the radar reflectivity, given by Rayleigh back-scattering of the emitted signal, is proportional to the sixth moment of the drop size distribution, $\langle D^6 \rangle$, and is therefore heavily weighted by large drops. Microwave radiometers, in contrast,

Table 4.2 Cloud and aerosol parameters, and a technique for retrieval.

	Cloud	Aerosol
Geometry	Cloud top/base height • lidar/ceilometer, NIR, stereo height	Plume/aerosol height • lidar
	Cloud cover • VIS/IR imagers	
	Vertical structure, thickness • radar	Vertical structure, thickness • lidar
Optics	Optical thickness • VIS imagery	Optical thickness, Ångström parameter • sun photometers, imagers, lidar
	Effective radius • VIS/NIR imagery	Single-scattering albedo • scanning sun photometers
	Cloud albedo / radiative forcing • spectral/broadband imagers	Aerosol direct effect / radiative forcing • spectral/broadband imagers
Microphysics	Ice/liquid water path • microwave radiometer/imager	Size distribution • sun photometers, polarimetry
	Crystal shape • polarimetry	Aerosol type (dust, sea salt, ...) • sun photometers, polarimetry
	Precipitation rate; fall speed • microwave radiometer, Doppler radar	
	Thermodynamic phase • VIS/NIR imagery	Mixture state (internal/external) • currently no remote sensing technique

provide a direct measurement of the cloud brightness temperature and emissivity, which can be related to the third moment of the drop size distribution, $\langle D^3 \rangle$, and thus to column-integrated water content (liquid water path, LWP) and precipitation rate. However, little information about the spatial structure can be retrieved. Visible or infrared imagery are heavily weighted by the properties near cloud top. This is the method of choice for determining the cloud radiative forcing because optical thickness, τ, related to $\langle D^2 \rangle$, as well as cloud cover and cloud-top effective drop radius, r_e, can be retrieved directly. Through the simple relationship, LWP = $2/3\ \rho\tau \times r_e$ (where ρ represents water density), LWP can be inferred as well, assuming that the effective radius at cloud top is representative for the whole cloud, which is often not the case because of the vertical cloud structure. Conversely, a cloud column-averaged effective radius can be obtained when optical thickness retrievals are combined with LWP retrievals from a microwave radiometer.

Cloud drop number concentration is not easily accessible through remote sensing; even if a number of moments such as $\langle D^3 \rangle$ and $\langle D^2 \rangle$ are known, assumptions about the *shape* of the drop size distribution are needed to derive its integral, the number concentration. A variety of algorithms exists for retrieving cloud thermodynamic phase (Chylek et al. 2006), but this is particularly difficult when both liquid and ice phase occur. Other parameters that are not well constrained from remote sensing include cloud and aerosol absorption and heating rates. This difficulty is fundamental: spaceborne sensors measure flux on top of the atmosphere; the ground-based counterparts obtain it at the surface; and both are needed simultaneously to determine atmospheric absorption.

Spaceborne aerosol retrievals from imagery suffer from the problem that their radiative signature does not provide sufficient contrast to distinguish it from variability in surface reflectance. In addition, there is currently no method for retrieving aerosol optical thickness in the presence of clouds, because the cloud signal dominates the reflected radiation. Aerosol single-scattering albedo retrievals are even more difficult to obtain, or are possible only for special cases (e.g., from sun glint over water; Kaufman et al. 2002). Ground-based sun photometers are the most reliable source of aerosol optical thickness, but do not work if clouds block the Sun. A promising technique for space- and airborne retrievals is polarimetry. Spaceborne and ground-based lidars provide information about the layering of the aerosols and thin clouds from their backscatter signal; some systems measure extinction profiles directly, and hence optical thickness (column-integrated extinction). From the linear depolarization ratio, information about the particle shape and thermodynamic phase can be deduced: Whereas spherical particles do not change the polarization of the incident radiation, nonspherical particles (ice crystals and some aerosol types) modify the state of polarization. Some information about aerosol particle size can be retrieved from the Ångström parameter, which is a measure of the wavelength dependence of the aerosol optical thickness. It could also be used to distinguish various aerosol types from clouds (whose optical thickness is wavelength-independent) in future hyperspectral satellite observations.

In-situ Instrumentation: Problems and Limitations

Mounting Problems and Flow around and within Probes

A common error when making in-flight observations is the selection of poor mounting locations for probes. All mounting locations should have an engineering assessment of their suitability. This assessment would preferably be performed with model or wind tunnel simulations. Probes should be in the free stream, outside the boundary layer of the aircraft, and not affected by engines, propeller wash, or other probes. Commonly selected poor probe mounting locations are window mounts, top of the fuselage near the cockpit, and wing tips. Good mounting locations can be found underneath the wing and sometimes the belly of the aircraft. King (1984) shows how 100 μm drops can be affected as they flow over the top of a fuselage (Figure 4.2). In this case, close to the skin, the top of the fuselage becomes a shadow zone for drops of this size. Further away from the skin there is a concentration of particle trajectories.

Figure 4.3 shows the top of the fuselage errors for a mounting location on the Canadian Convair 580 during tests of the Nevzorov probe (Isaac et al. 2006). As

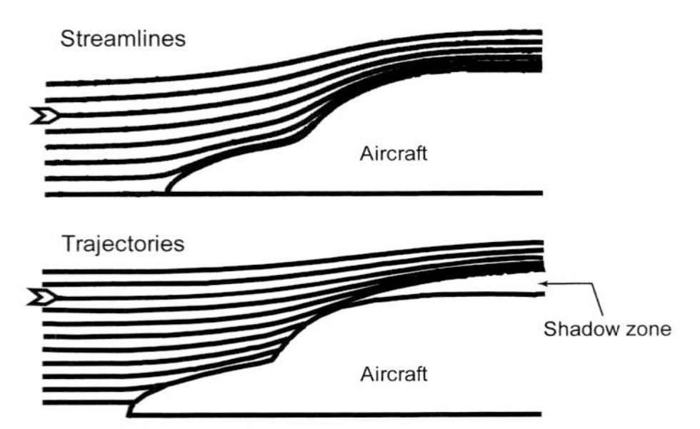

Figure 4.2 Airflow (left to right) over the top of a fuselage showing the streamlines (top) and the trajectories of 100 μm drops traveling over the top of an F27 fuselage traveling at 90 m s^{-1} (bottom) as described by King (1984).

can be seen, there is a shadow zone and the zone of concentration is also delineated for drops of a specified diameter. For the Convair, the shadow zone maximum occurs at a droplet diameter 160 μm. Using more sophisticated models, Twohy and Rogers (1993) have shown that much larger particles, with greater inertia, will cross the streamlines and not be shadowed. It should be noted that the behavior of ice crystals is substantially different from that of spherical water drops. King (1985, 1986) discusses this in more detail. Mounting location errors tend to be highest for the intermediate size particle (100–500 μm).

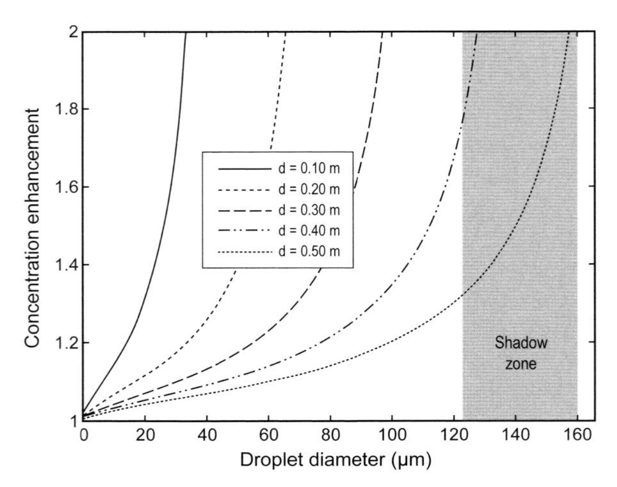

Figure 4.3 Concentration enhancement calculated based on the King (1984) droplet trajectory and airflow model plotted as a function of distance from the fuselage surface of the Canadian Convair 580 aircraft and the droplet diameter (Isaac et al. 2006). Airspeed was assumed to be 100 m s^{-1}. Particles greater than about 120 μm will be in a shadow zone (shaded area). For the Convair, the shadow zone maximum occurs at a droplet diameter 160 μm. Using more sophisticated models, Twohy and Rogers (1993) have shown that much larger particles, with greater inertia, will cross the streamlines and not be shadowed.

Probes themselves can create flow field problems (e.g., Norment 1988). For example, the Cloud Particle Imager (Lawson et al. 1998) has a long flow tube where particles might be disturbed or broken up before sampling. Korolev and Isaac (2005) describe how optical imaging probes can cause the shattering of particles off the probe inlets (see Figure 4.4). Proper software can eliminate many of these shattering problems but users should be aware that it is necessary to make such corrections (e.g., Field et al. 2006).

Defining *In-situ* Probe Sample Volumes

For some probes, it is necessary to calculate the sample volumes accurately. This can be very difficult or relatively simple to do. For the PMS OAP probes, one must recognize that the sample volume can depend on particle size and airspeed in a significant manner (Baumgardner and Korolev 1997). Korolev et al. (1998a) describe how corrections might be made for particle size. Probes (e.g., the PMS FSSP) cannot always count fast enough when particles arrive too quickly. The electronics produce a dead time, which needs to be considered when calculating the sample volume (Brenguier et al. 1994).

In-situ Probe Size Ranges

It is very important to measure particles over the complete size range of interest. This is a problem that can be overlooked. Sometimes, for example, PMS OAP 2-DC probes have been used to determine radar reflectivities. However, it

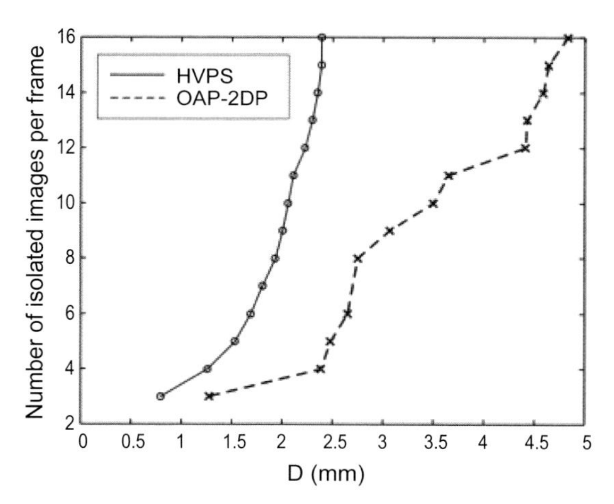

Figure 4.4 Number of isolated images per frame for the PMS 2-DP probe and the SPEC HVPS probe (from Korolev and Isaac, 2005). The value of D was calculated as a size that 97% of all the shattering events for a specified number of isolated images per frame are located at $D > D_{max}$.

is well known that a few large particles can affect the reflectivity because it depends on number concentration multiplied by the diameter to the sixth power. Figure 4.5 shows composite spectra from field projects conducted in southern Ontario, Canada (Isaac et al. 2002). The top, middle, and bottom graphs show the number concentration, mass concentration, and reflectivity concentration, respectively, as a function of TWC. The panels on the left side show the liquid water spectra; those on the right side show the ice water spectra. The basic

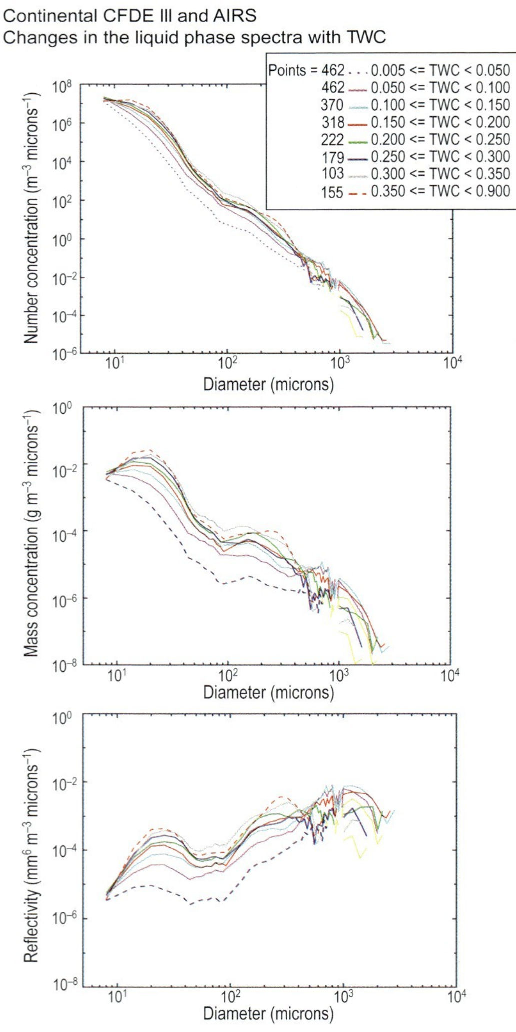

Figure 4.5 Averaged spectra for CFDE III and AIRS sorted by all-liquid and glaciated 30–s values. The "number" of 30-s spectra used for each TWC are shown.

probes used and their nominal size ranges were the PMS FSSP standard range (3–45 μm), the PMS FSSP extended range (5–95 μm), the PMS 2-DC (25–800 μm), and the PMS 2-DP (200–6400 μm). For the 2-DC probe, only particles larger than 100 μm were counted, because of uncertainties in counting and sizing particles from 25–100 μm with this probe. This creates a sizing gap between the FSSP and 2-DC probes, which is visible on many of the plots on the left panel. The analysis techniques, including the phase discrimination method, have been described in detail by Cober et al. (2001b).

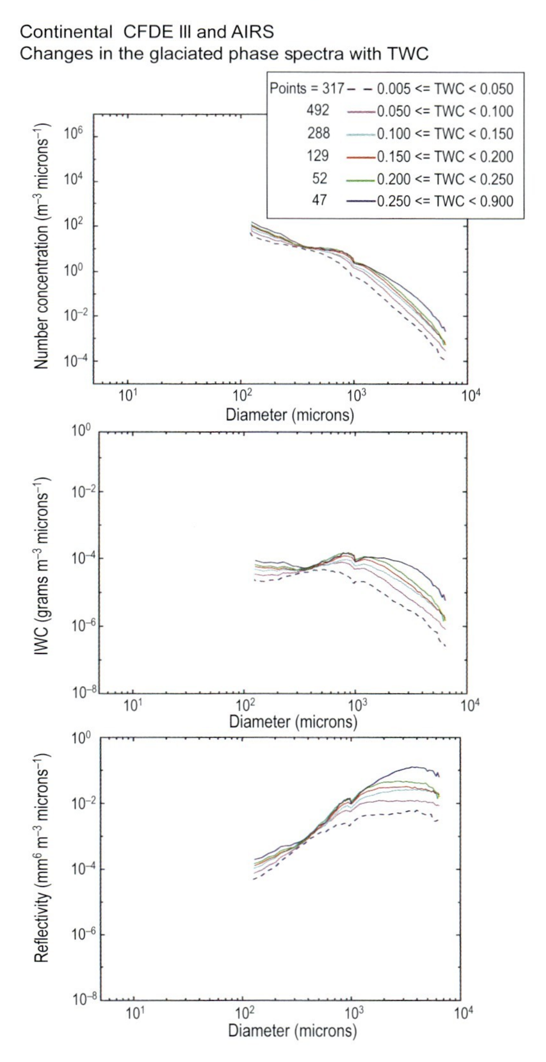

Figure 4.5 (continued)

Figure 4.5 shows that the reflectivity values continue to increase as one goes to increasingly larger sizes. The roll-off at the largest sizes may be a function of the software, and for the ice particle spectra, this may be an indication that the largest sizes were not measured. It should be emphasized that software techniques allow one to measure beyond the maximum size range of the probe, which is 800 μm and 600 μm in the case of PMS 2-DC and 2-DP probes, respectively (see below). However, to do that, one must assume either spherical or symmetrical particles. Obviously, these assumptions can be wrong in the case of ice particles.

In-situ Probe Analysis Software

The analysis of cloud microphysical data often requires sophisticated software. Performing an automated analysis using software is quite difficult. Kingsmill et al. (2004) describe some of the problems. Isaac et al. (2005) provide an example of how different assumptions in the software can lead to different results, as described below.

For a freezing rain case with data collected from the Canadian Freezing Drizzle Experiment III, Table 4.3 illustrates difficulties that can be encountered through the analysis of PMS 2-D imagery (Knollenberg 1981). The analysis was performed using the Cober et al. (2001b) method and a second software package developed through Environment Canada called "2-D Analyzer," which uses the techniques described by Heymsfield and Parrish (1978). Particle diameter is determined from the 2-D imagery by either measuring the dimension in the X direction (direction of flight) or the Y dimension. Diameter can also be computed from the total particle area assuming that the particle is a sphere. Extended area (EXA) is determined using the geometrical reconstruction, after Heymsfield and Parrish (1978), which basically looks at a portion of the drop that is imaged and estimates the diameter of the whole particle. The center-in (CIN) technique must have the center of the particle imaged. The double end element (DEE) technique restricts the analysis to particles that are completely imaged and do not shadow the end elements of the diode array.

Table 4.3 Comparison of various methods of determining LWC during a freezing rain encounter 5.5 min long during CFDE III. H&P: Heymsfield and Parrish (1978).

Analysis technique	Circular particles	Irregular particles	LWC g m^{-3} (125–6400 μm)	LWC g m^{-3} (125–2000 μm)	
DEE	Y	Y	0.262	0.192	2-D Analyzer
CIN	Y	Y	0.28	0.213	2-D Analyzer
CIN	Y	X	0.190 (0.205 with FSSP)	0.190 (0.205 with FSSP)	Cober
EXA	H&P	H&P	0.296	0.184	2-D Analyzer

For both the EXA and CIN techniques, particle geometry must be assumed to determine the probe sample volume accurately. These techniques work well for circular drops but do not apply for irregularly shaped ice particles. The case described in Table 4.3 contained drops between 125–2000 μm in diameter, and ice crystals up to 5000 μm in diameter. The majority of the hydrometeors observed were circular in shape (i.e., drops), and the drop median volume diameter was between 800–1000 μm. A small portion of the mass was in the small droplet size range that would be measured by the PMS FSSP probe (< 100 μm). The 2-D Analyser software provides similar LWC values between the EXA, CIN, and DEE techniques, which should be the case when the majority of the particles images are circular in shape. However, for the range 125–6400 μm, there is a significant disagreement with the LWC from the Cober software. This results from the fact that the 2-D Analyser is interpreting ice crystals with sizes from 2000–6000 μm as drops and computing their associated LWC. The Cober software more accurately segregates the circles (drops) from the non-circles (ice crystals) and hence avoids this problem. When the two software programs are compared over the range 125–2000 μm, the LWC values agree within 10%. This demonstrates that the application of advanced software analysis techniques to determine LWC can be quite erroneous if used blindly. It also provides an example of why 1-D measurements of hydrometeors should not be used to compute LWC, since 1-D instruments cannot separate drops from ice crystals in a mixed-phase environment.

This analysis shows that substantial differences can be obtained when using different analysis techniques. It emphasizes the point that software should be fully understood and used with caution.

In-situ Probe Calibrations

Where possible, calibrations of all probes should be done before and after a field campaign. Many probes can be calibrated in wind tunnels when they are accessible. Strapp and Schemenauer (1982) provided a good illustration of how problems unknown to their users can be detected with tunnel calibrations. They examined 14 Johnson–Williams (J–W) cloud LWC probes with 23 sensor heads from ten research organizations in the National Research Council of Canada's icing tunnel and found:

> ...six of the 14 systems had at least one sensor head with a nonfunctional shell or strut heater on arrival, presumably unknown to its owner. This defect can cause erroneous data at below freezing and above freezing temperatures as observed during these tests. Three systems displayed a large difference in measurement when changing sensor heads. Three more had at least one probe with a strong airspeed dependence. Based on these problems alone, nine of the 14 systems could provide improper measurements if the most unfortunate set of sensor heads were used (Strapp and Schemenauer 1982, p. 106).

More recent tests using the NASA Icing Research Tunnel have been performed by Strapp et al. (2003). Figure 4.6 shows how the King LWC probe compared with the tunnel reference LWC and how the response of this probe rolls off as the median size of the droplets get larger. Beyond median volume diameters of 40 μm, the error grows to over a 20% underestimation. Strapp et al. (2003) also show how optical array probes can measure different concentrations of large drops within the tunnel conditions (Figure 4.7).

Wind tunnels are expensive to operate and are not always available for instrument calibration. It is easier and cheaper to run instrument and data simulators. For example, this could mean putting glass beads or reticles of a known size through the sample area of an optical probe, preferably at speeds close to those experienced during flight. Dye and Baumgardner (1984) describe the calibration of the PMS FSSP using glass beads. Probes should be calibrated

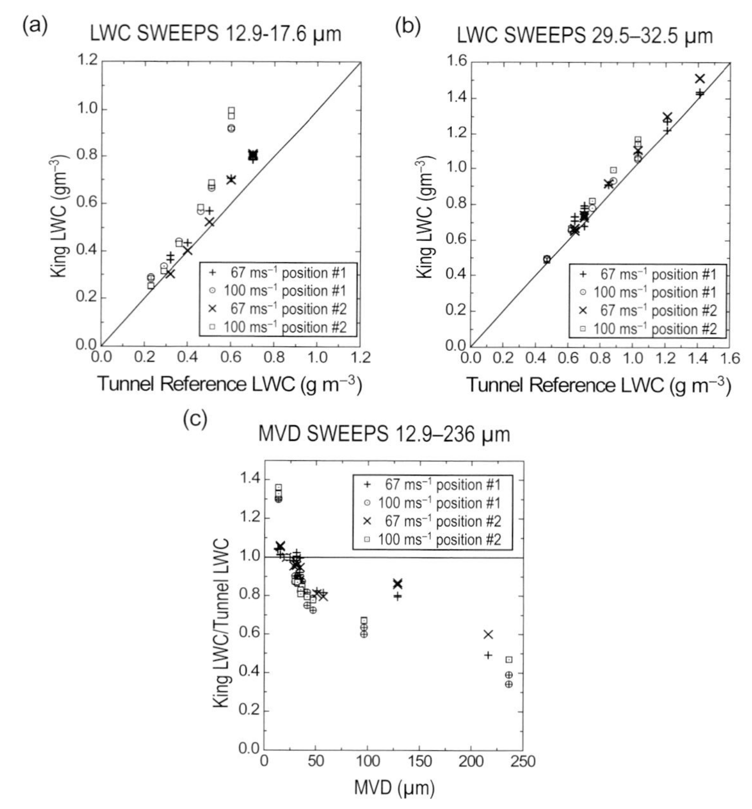

Figure 4.6 PMS King hot-wire probe: (a) comparisons to tunnel reference LWCs for low median volume diameter, MVD, (12.9–17.8 μm); (b) corresponding comparisons for intermediate MVD (29.5–32.5 μm); (c) the ratio of King probe LWC to tunnel LWC as a function of MVD (from Strapp et al. 2003).

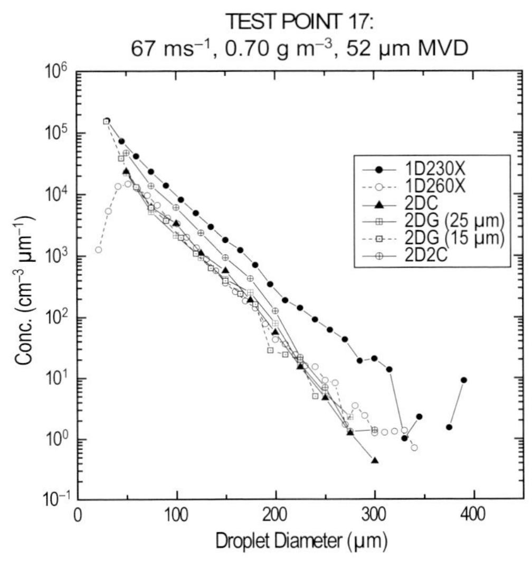

Figure 4.7 Response of various optical array probes run at 67 m s^{-1}, 0.84 g m^{-3} with a median volume diameter (MVD) of 32.9 μm (from Strapp et al. 2003).

at aircraft speeds or, alternatively, the electronic size-response roll-off of the probe with airspeed should be factored into the calibration.

OAPs can also be laboratory tested. Strapp et al. (2001) show how a PMS 2-DC probe's ability to size accurately and its associated depth of field can be assessed in a laboratory setting. Korolev et al. (1998a) provides a theoretical discussion of the problems associated with particle imaging, and calculations of the digitization and out-of-focus imaging effects on size distributions.

Remote-sensing Measurements and Limitations

Visible and Infrared Imagery

Imagers are the "workhorse" sensors for spaceborne cloud and aerosol retrievals, as they provide high global coverage of observations at high spatial resolution. The two different kinds of algorithms are based on (a) reflection of visible light (daytime only) and (b) emission of thermal infrared radiation from cloud top (day and night). Operational retrievals are based on lookup tables, created from radiative transfer models for a given geometry and a set of parameters such as optical thickness, effective radius, cloud-top pressure, and thermodynamic phase. The measured reflected (emitted) radiance is compared to lookup table values, and the corresponding cloud parameters are assigned based upon the modeled values, which approximate most closely the measurements. Emission-based retrievals rely on the temperature contrast between the surface and the (colder) clouds. Therefore, they are capable of detecting high-altitude thin (even subvisible) cirrus for which reflectance-based retrievals fail. Systematic errors in these retrievals arise from vertical and horizontal spatial cloud variability. One example is the albedo bias: because of the nonlinear relationship between optical thickness and cloud albedo, a heterogeneous cloud with the same mean optical thickness as a homogeneous cloud has always a lower albedo than its homogeneous counterpart, which leads to an underestimation of its optical thickness. The effective radius retrievals also become unreliable, especially for broken clouds. The cloud cover derived from imagery depends largely on the instrument spatial resolution (ranging from a few meters to 20 km) and sensitivity of the retrieval to various types of clouds, as well as on cloud optical thickness. Cloud vertical structure is not resolved by nadir-viewing imagers; multi-layer clouds represent a problem especially if the uppermost layer is partly transparent. Ice clouds are particularly difficult for imagery, not only because they can be extremely thin and inhomogeneous but also because of the ice crystal shape; currently, techniques for utilizing multiangle (McFarlane et al. 2005) or polarized (Sun et al. 2006) observations are being explored.

The "classical" AVHRR was used for cloud and aerosol retrievals on a variety of platforms and is still flown on geostationary platforms. Since its first deployment in 1978, both the spatial and spectral resolution of imagers increased: MODIS, flown on NASA's polar orbiting sun-synchronous satellites Terra and Aqua, provides cloud retrievals at 1 km spatial resolution. Aqua is part of the so-called "A-Train," a satellite constellation of various platforms in short sequence with a daily afternoon overpass. MISR (Kahn et al. 2005) was specifically designed to improve aerosol and cloud retrievals by a combination of cameras with different viewing angles onboard Terra. POLDER (Riedi et al. 2001) utilizes the information from polarized reflectance for cloud phase discrimination, crystal shape detection, and aerosol retrievals. Over the last decade, significant progress has been made in the aerosol retrieval from these three platforms, especially over bright land surfaces. However, they are currently only possible under completely cloud-free conditions.

Whereas standard imagers rely on information from a limited number of wavelength bands, *spectral* imagers allow for novel retrieval techniques. In the thermal window, AIRS provides vertical profiles of water vapor, carbon dioxide, and temperature. In the solar spectral range, the capabilities of spectral imaging for combined aerosol–cloud retrievals or the attribution of climate change to various forcing agents are being explored.

Radar

Radar (radio detection and ranging) relies on scattering of microwave radiation by cloud drops and precipitation. The intensity of the back-scattered signal depends on the distance between the radar and the drops, the drop size distribution, and the wavelength of the emitted pulses. In the Rayleigh limit of Mie scattering theory, it is proportional to λ^{-4} and $\langle D^6 \rangle$. Therefore, sensitivity to small cloud drops can only be achieved by using shorter wavelengths. However, at short wavelengths, the signal becomes quickly attenuated as a result of increased scattering and absorption near the edge of the radio window ($\lambda \approx 1$ cm). Hence, the choice of the wavelength depends on the targets: weather radars (5–10 cm wavelength) penetrate non-precipitating clouds whereas cloud radars (1 cm) are sensitive to the smaller-size cloud drops but have shorter range. Most weather radars are scanning and polarized systems, operating in pulse mode. The distance to the cloud is determined from the elapsed time for the pulse's round trip, where the maximum range is determined by the pulse separation. In addition, Doppler radars determine the speed and direction of hydrometeors relative to the radar system. The range in detectable speeds is inversely proportional to the pulse separation time. Signal polarization is used to detect the vertical and horizontal dimension and thus the shape of the scattering object. Radars are used for water as well as for ice clouds. They excel in detecting cloud structure while retrievals of ice and liquid water content profiles are better constrained when combined with data from other instruments.

Radar systems are most commonly deployed at the ground. The first spaceborne radar cloud system was the precipitation radar (f = 13.8 GHz, $\lambda \approx 2$ cm) onboard the TRMM satellite. More recently (2006), the cloud profiling radar (CPR; f = 94 GHz, $\lambda \approx 3$ mm) was launched on CloudSat (Stephens et al. 2002). The frequency, peak power, and dynamic range were chosen to reconcile sufficient sensitivity to small cloud drops and the ability to profile moderately thick clouds. Figure 4.8 shows a CloudSat image from November 9, 2006, as the Canadian Convair 580 was flying underneath during the Canadian CloudSat/CALIPSO Validation Project. A comparison between the satellite radar (W-band) and the aircraft radar (Ka-band) shows some significant differences attributable to wavelength and resolution. However, overall, the CloudSat radar did an excellent job of characterizing the cloud and detecting light precipitation, and it provides coverage over much larger areas than other methods. Figure 4.9 shows the PMS 2-DC ice crystal images from when the aircraft and satellite were observing the same cloud. These diagrams illustrate the general point that *in-situ* aircraft with remote and *in-situ* sensors can be very useful for both testing and validating satellite instrumentation.

Lidar and Ceilometer

Lidar (light detection and ranging) is similar to radar but works with pulses of visible and near-infrared light. Since the wavelength is much shorter than microwave radiation, the back-scattered signal is sensitive to aerosol particles (typical sizes on the order of 0.1–1 μm, compared to precipitation size particles on the order of 0.1–1 mm), but the optical penetration depth is comparatively low ($\tau_{max} \approx 3...4$). The lidar return is determined by the aerosol and molecular extinction profile in a nontrivial manner. The high spectral

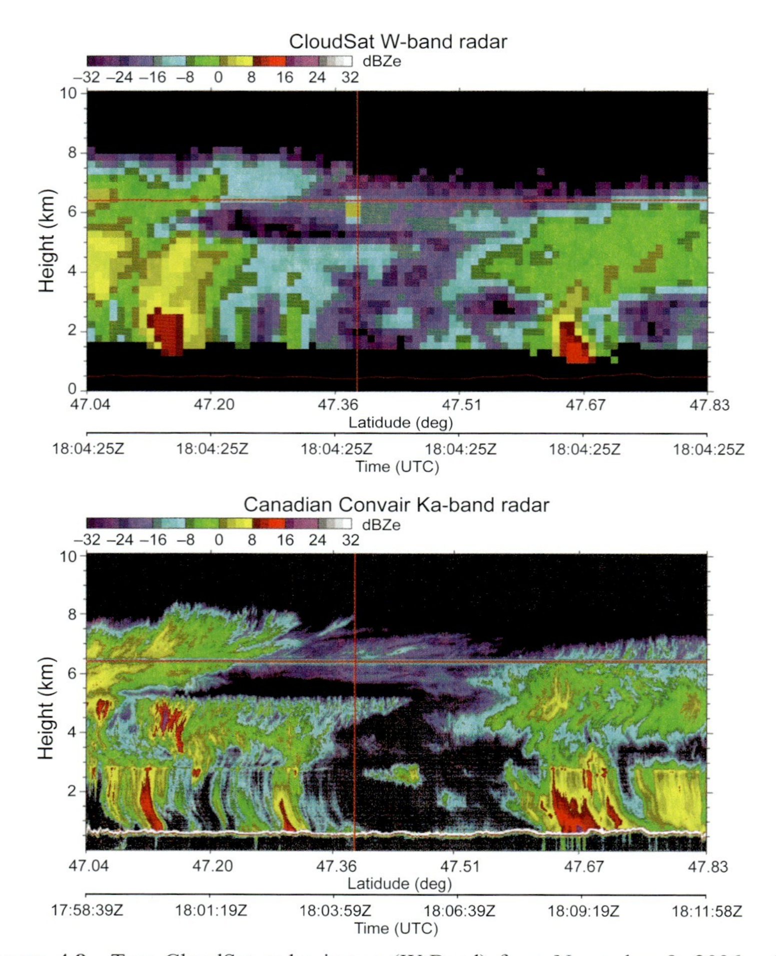

Figure 4.8 Top: CloudSat radar image (W-Band) from November 9, 2006, while passing over southern Ontario at the latitudes and time indicated. The horizontal line across the image shows the level the instrumented Canadian Convair 580 was flying at the time; the vertical line shows the location where it is estimated the two platforms coincide. The bright pixel near the intersection is likely a reflection from the aircraft itself. Bottom: Ka-band radar data from the aircraft with the images matched to correspond with each other. The Convair was flying along the same path as the satellite, but obviously at a slower speed. The Convair radar data is of higher resolution but it matches the properties of the CloudSat image in the main features. (Image selected and produced by Dave Hudak and Peter Rodriguez of Environment Canada.)

resolution lidar (HSRL) separates the molecular and aerosol contribution by detecting the Doppler broadening of the return signal caused by thermal motion of molecules (Shipley et al. 1983). Raman lidars (Ansmann et al. 1990) utilize inelastic molecular scattering of light for determining the concentration of atmospheric gases and aerosol extinction profiles independently. The

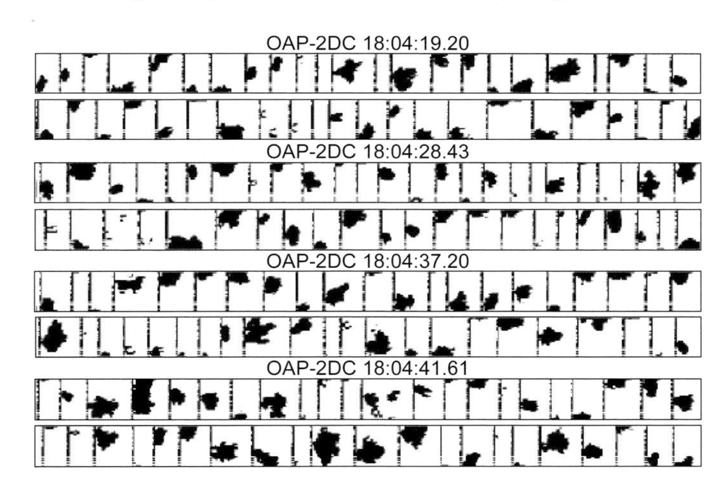

Figure 4.9 PMS 2-DC imagery taken at the time the CloudSat radar was imaging the Convair 580. Vertical bars represent 800 μm. The ice crystals observed were somewhat irregular in shape. (Image produced by Alexei Korolev of Environment Canada.)

depolarization of the lidar return gives information about the shape of the scatterers. A laser ceilometer is a simple type of a lidar where the cloud base is defined at the altitude where the back-scattered signal increases beyond a certain threshold. This threshold may give erroneous cloud base readings for high aerosol optical thickness below the cloud. Optical ceilometers determine cloud base geometrically by measuring the angle of reflectance with a scanning light source and a horizontally separated photocell.

In 2006, CALIOP (Winker et al. 2007) was added to NASA's A-Train onboard CALIPSO, in close coordination with CloudSat and Aqua. It provides vertical profiles of aerosols and thin clouds.

Microwave Radiometers

Passive microwave radiometers detect naturally emitted radiation from clouds and precipitation as well as from the surface, which can be related to LWP and precipitation rate. LWC profiles can be derived in conjunction with radars. The humidity and temperature profile of the atmosphere as well as surface properties must be taken into account for the retrievals. The errors associated with LWP can be as high as 25 g m^{-2} (typical values for thin water clouds range from 10–100 g m^{-2}). The uncertainty attributable to the atmospheric profile can be substantially decreased by the use of dual-frequency radiometers (e.g., Liljegren et al. 2001). These operate at frequencies from 10–30 GHz where the signal is largely determined by precipitation and atmosphere and clouds are virtually transparent. At higher frequencies, the signal is increasingly attenuated, mainly by clouds and water vapor. The addition of high-frequency channels enables the retrieval of the cloud LWP in presence of precipitation, increases the sensitivity to low LWP clouds, and improves the retrievals with respect to the atmospheric profile (Di Michele and Bauer 2006). In particular, the water vapor and temperature profile within clouds can be retrieved.

Vertical and Geographical Variations

Aerosol and cloud microphysical properties vary in vertical distribution, and their properties depend on the airmass where they were formed. Figure 4.10 shows the vertical variation of aerosol particles as a function of back trajectory

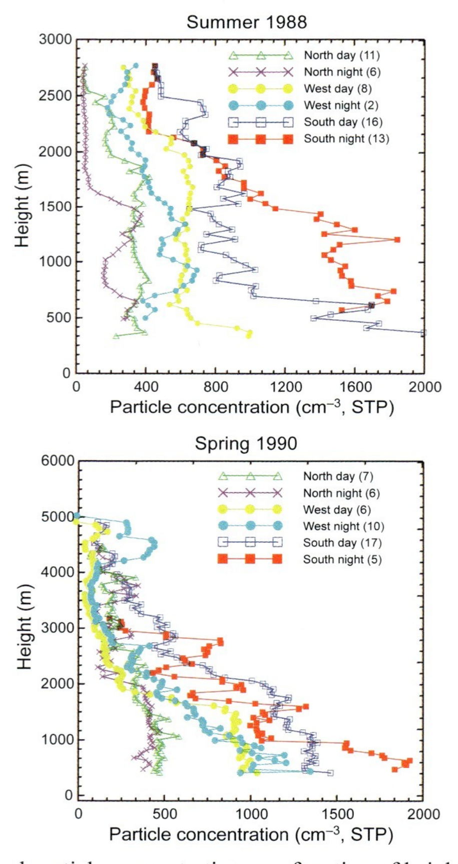

Figure 4.10 Aerosol particle concentration as a function of height for measurements made during 1988 and 1990 field campaigns in southern Ontario (Isaac et al. 1998). A PMS active scattering aerosol spectrometer probe instrument was used in 1988 with a detection size range of 0.17–3 μm. For 1990, a PMS passive cavity aerosol spectrometer probe was used with a size range of 0.12–3 μm.

for a study performed in southern Ontario during the summer of 1988 and the spring of 1990 (Isaac et al. 1998). There is only a small variation with height when airmass back trajectories were from the north, where there were few source regions, whereas there were steep gradients with height when back trajectories were from the south, where the air traveled over heavily populated and industrialized areas. Trajectories from the west showed concentrations between those from the north and south. The use of surface measurements of IN, CCN, or their precursor chemical constituents, without providing some mechanism for dispersing the aerosol particles in the vertical, would provide poor inputs for climate models.

Cloud microphysical properties also depend on geographical location. The distinction between maritime and continental clouds has been well known and documented (see Squires 1958). Isaac et al. (2001) show clear differences between the frequency distributions of droplet concentrations in stratiform clouds in a maritime (CFDE I) and continental (CFDE III and AIRS) environment (see Table 4.4); maritime clouds show lower droplet concentrations and larger drops for the same temperature level.

However, it is unclear whether the clouds in eastern Canada are similar to those in other places of the world. Such a study needs to be done. Korolev et al. (2001) did a statistical analysis of cloud properties by cloud type in the former USSR using a large dataset and showed distinct differences in cloud properties as a function of cloud type. Although the instrumentation and resulting analysis techniques were different, the data are similar to the stratocumulus and stratus measurements of Table 4.4, at least for cloud LWC. Korolev et

Table 4.4 Cloud microphysical summaries, using 30 s averages, for the maritime (CFDE I) and continental cases (CFDE III and AIRS) in terms of static temperature (Ta), droplet number concentration (N_d), total water content (TWC) and median volume diameter (MedVD). Ice crystal concentration is represented as I. For example, 25% of the CFDE I droplet concentrations (N_d) were less than 16 cm^{-3}.

	1%	25%	50%	75%	99%
Maritime	Points = 1154* Ta ≤ 0°C I ≤ 1 l^{-1} TWC ≥ 0.005 g m^{-3}				
Ta (°C)	–20.6	–5.8	–4.1	–2.0	0.0
N_d (cm^{-3})	1	16	52	108	406
TWC (g m^{-3})	0.01	0.07	0.13	0.20	0.47
MedVD (μm)	10	18	24	34	527
Continental	Points = 4759* Ta ≤ 0°C I ≤ 1 L^{-1} TWC ≥ 0.005 g m^{-3}				
Ta (°C)	–24.7	–9.1	–6.2	–3.2	–0.2
N_d (cm^{-3})	2	55	121	233	643
TWC (g m^{-3})	0.01	0.05	0.11	0.21	0.49
MedVD (μm)	10	13	17	22	643

*Liquid and mixed phase, in-icing conditions

al. (2000), Gultepe et al. (2001), Field et al. (2005) and Gayet et al. (2006) all showed similar ice particle (> 100 μm) concentrations around the world, in different cloud types. More coordinated work along this line needs to be performed using similar instrumentation and analysis techniques, so that the measurements can be compared directly.

Scale Effects

Measurements made using aerosol or cloud microphysical probes often examine particle concentrations in volumes less than one cm^{-3} and in most cases less than a few liters. These measurements, however, are used in climate models to describe the average properties of clouds on scales reaching hundreds of kilometers. It is clear that if cloud microphysical data are analyzed over different scales, different answers can be obtained. Gultepe and Isaac (1999, 2007) showed this effect for cloud liquid water, droplet concentration, and parameterizations of cloud cover. Cober et al. (2001c) described this in some detail for cloud LWC and droplet concentration (see Figure 4.11). For climate simulations, especially considering the demanding accuracies required by the sensitivity studies of Slingo (1990) and Rotstayn (1999), these effects are important.

Units of Measurement

One often overlooked problem concerns the units to use when reporting measurements. Modelers tend to use mass units giving the concentrations as a function of a unit mass of air. This has the advantage of being independent of pressure and temperature. Chemists often use such units as well. However, today, most cloud microphysical measurements are reported in volume units. This would

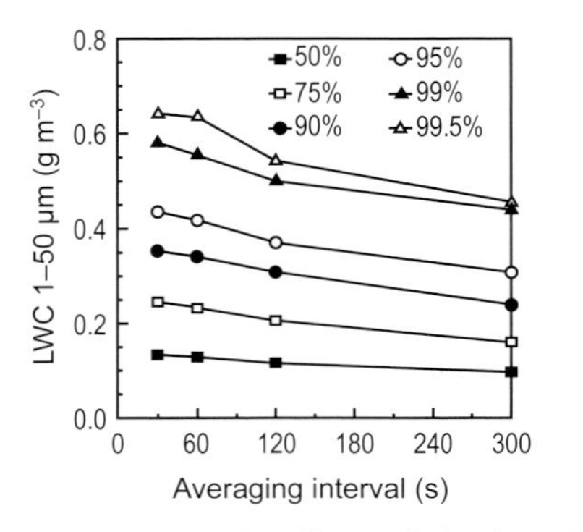

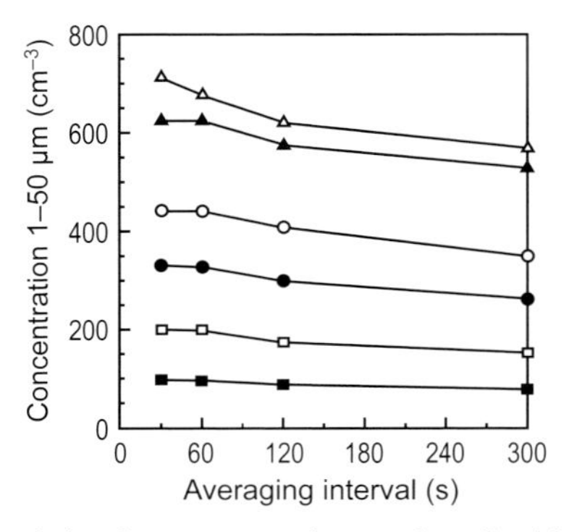

Figure 4.11 Scale effects of cloud LWC and droplet concentration as described by Cober et al. (2001c). 30- and 300-second averaging intervals represent 3 km and 30 km scales, respectively. The plot shows various percentile values of the distribution.

not be a problem if the corresponding pressure and temperatures were also reported. However, this is not done routinely. When analyzing measurements from many different days, the altitude and temperature variations are often ignored and the volume unit concentrations are used for averaging.

Isaac et al. (2004) examined this problem and showed the typical type of "errors" that might result. The data were obtained during four different field projects in eastern Canada, central Canada, and the Arctic. Table 4.5 shows the data analyzed in their original volume units at the measurement level and in volume units that were referenced to a standard temperature and pressure (STP) level of 0°C and 1013 hPa. The STP data presentation is similar to a mass concentration because a simple multiplication factor would convert the numbers to mass units. Presenting the data in this manner, however, allows a direct comparison of the volume and "mass" measurements. It is clearly shown that there can be a factor two between the different methods of presenting the data. The rate of change of LWC and TWC with temperature is less when the data are converted into mass units because measurements at colder temperatures were usually taken at lower pressure altitudes.

Probability Density Functions

Aerosol and cloud microphysical data vary considerably over short distances. It is probably not realistic to represent clouds in climate models with one number for each cloud parameter for grid squares perhaps 10^4 km^2 in size. Table 4.5 shows the distribution for cloud liquid water, cloud total water, droplet number concentration, and ice particle concentration in terms of 25%, 50%, 75%, 95% percentiles and mean values. Gultepe and Isaac (2004) describe the variations in more detail for droplet number concentration and suggest some probability density functions. Modelers are beginning to recognize that they need to consider the variance of important variables. However, the modeling community must present those who are making and analyzing the data some guidance as to what would be most useful.

General Conclusions

In terms of making cloud *in-situ* measurements, there have been many advances in recent years. However, it is difficult to provide suitable measurements for use in climate models and some of the reasons are summarized below:

- Cloud microphysical measurements are collected today without commonly accepted standards for calibration and data analysis. Users are often unaware of the limitations of the calibrations, if any were done, and the problems associated with the data analysis.

Table 4.5 Parameters given here in the original volume units from the measurement level and volume units at standard temperature and pressure (STP) (0°C and 1013 hPa). Data were obtained from 30 s or 3 km averages representing 8596 values for LWC; 15,202 values for TWC; 6297 values for N_d; and 8298 values for N_i.

Liquid Water Content (LWC) vs. Temperature (T)										
T (°C)	LWC ($g\ m^{-3}$)					LWC ($g\ m^{-3}$) at STP				
°C	25%	50%	75%	95%	Mean	25%	50%	75%	95%	Mean
–2	0.04	0.11	0.20	0.35	0.13	0.05	0.13	0.23	0.41	0.16
–6	0.04	0.10	0.18	0.34	0.12	0.05	0.11	0.21	0.40	0.14
–10	0.03	0.10	0.20	0.35	0.13	0.04	0.11	0.25	0.46	0.16
–14	0.02	0.06	0.14	0.41	0.11	0.03	0.08	0.18	0.52	0.15
–18	0.01	0.04	0.07	0.15	0.05	0.02	0.05	0.11	0.24	0.08
–22	0.02	0.04	0.07	0.14	0.05	0.03	0.05	0.11	0.29	0.09
–26	0.01	0.01	0.05	0.15	0.04	0.01	0.02	0.09	0.26	0.08
Total Water Content (TWC) vs. Temperature (T)										
T (°C)	TWC ($g\ m^{-3}$)					TWC ($g\ m^{-3}$) at STP				
	25%	50%	75%	95%	Mean	25%	50%	75%	95%	Mean
–2	0.05	0.12	0.20	0.34	0.14	0.06	0.14	0.25	0.41	0.17
–6	0.04	0.09	0.17	0.32	0.12	0.04	0.11	0.20	0.38	0.14
–10	0.02	0.07	0.15	0.31	0.10	0.03	0.09	0.19	0.41	0.13
–14	0.02	0.04	0.09	0.27	0.08	0.02	0.07	0.14	0.40	0.11
–18	0.01	0.02	0.06	0.15	0.04	0.02	0.04	0.10	0.26	0.08
–22	0.01	0.02	0.04	0.12	0.04	0.02	0.03	0.07	0.23	0.07
–26	0.01	0.01	0.03	0.11	0.03	0.01	0.02	0.05	0.22	0.05

- There is no established data archive for cloud microphysical measurements that should be used for climate simulations.
- The current accuracy of most measurement techniques is at best 15% for cloud LWC and worse for most other variables. Slingo (1990) and Rotstayn (1999) suggest that accuracies better than 5% are required.
- The current methods of measuring IN and ice particle concentrations (at small sizes) are not adequate. For ice particle concentration, the errors are at least a factor of two at the moment. For IN, considering all the potential nucleation mechanisms, all measurements today must be considered as estimates only with unknown errors. These problems demand urgent attention.
- There are no agreed upon formats for providing data to the modeling community in terms of scale of the measurements, probability

Table 4.5 (continued)

Droplet Number Concentration (N_d) vs. Temperature (T)										
T (°C)	N_d (cm^{-3})					N_d (cm^{-3}) at STP				
	25%	50%	75%	95%	Mean	25%	50%	75%	95%	Mean
–2	22	72	151	329	106	25	87	184	371	123
–6	59	121	240	446	165	70	146	286	492	191
–10	43	120	240	568	173	51	143	303	714	214
–14	18	68	142	437	114	27	88	181	520	139
–18	12	28	52	106	45	20	44	70	147	61
–22	14	38	71	87	43	27	63	77	130	59
–26	0	9	38	98	27	0	16	54	142	42

Ice Particle Concentration (N_i) vs. Temperature (T)										
T (°C)	N_i (L^{-1})					N_i (L^{-1}) at STP				
	25%	50%	75%	95%	Mean	25%	50%	75%	95%	Mean
–2	3	8	14	27	10	3	9	18	37	13
–6	3	7	13	26	10	3	9	16	34	12
–10	2	4	9	17	6	2	5	12	26	8
–14	2	5	11	23	8	3	8	18	40	13
–18	2	5	10	20	7	3	9	18	38	13
–22	2	6	9	20	7	3	10	17	36	13
–26	2	5	12	23	8	3	8	23	46	15

density function requirements, or even the unit of measurements (mass versus volume).

From the remote-sensing perspective, current technology has allowed the miniaturization of instruments which could previously not be deployed in space. Advances have been made in energy efficiency, laser technology, cooling, detectors, and onboard calibration. New satellite instrumentation has boosted the understanding of the hydrological cycle, precipitation, and vertical distribution of aerosols and clouds. In addition, data preprocessing, data transfer, retrieval algorithms, and user interfaces (data visualization and download) have improved in the "A-Train era." Challenges in spaceborne applications are limited onboard energy supply, data downlink volume, and the stability of calibration and orbit, as well as aliasing due to limited spatial coverage, which may introduce spurious trends. For the detection of climate signals, long-term observations with high accuracy and stability are required. New projects are underway to tackle these difficulties. At the same time, the continuity of existing capabilities are at risk, such as instruments of the A-Train or solar irradiance measurements in space.

Ground-based remote-sensing observations are particularly useful when organized in networks, such as AERONET, EARLINET, MPL-Net, ARM, or radar networks. They provide detailed measurements where satellite observations have only just begun or are very inaccurate (aerosol optical thickness, single-scattering albedo, vertical distribution and composition, cloud LWC and precipitation). A problem here is the regional nature of such networks. AERONET, for example, is not well represented in Africa and Asia; radar networks are only established in Europe, North America, and parts of Asia. Currently, efforts are underway to extend such networks to obtain better global coverage, and use models and satellites to obtain high-accuracy global datasets of climate-relevant cloud and aerosol parameters. Ground-based networks provide directly the surface radiation budget terms that would otherwise have to be inverted from satellite measurements or estimated from models.

Some general recommendations can be made as follows:

- It would be advantageous if common calibration and data analysis techniques could be established. That way measurements made in different parts of the world by various investigators could easily be compared. Unfortunately, such standards do not exist.
- There is a need to integrate the remote-sensing and *in-situ* measurement communities better so that global or regional datasets can be obtained for use by the modeling community.
- The continuity of existing satellites is currently at risk. It is very important to ensure data continuity in the near future.
- Ground-based remote-sensing networks should be supported and extended because they represent an indispensable tool for measurement of parameters that are not accessible from space. Global datasets require instrument inter-calibration and a common metric for data quality estimates

Acknowledgments

George Isaac would like to thank his Environment Canada colleagues Monika Bailey, Faisal Boudala, Stewart Cober, Dave Hudak, Ismail Gultepe, Alexei Korolev, Peter Liu, Peter Rodriguez, and Walter Strapp, who contributed a great deal to this paper, either through unpublished conference papers or private communications. Peter Pilewskie from the University of Colorado at Boulder provided significant input to the remote sensing section.

References

Ansmann, A., M. Riebesell, and C. Weitkamp. 1990. Measurement of atmospheric aerosol extinction profiles with a Raman lidar. *Opt. Lett.* **15**:746–748.

Baumgardner, D. 1983. An analysis and comparison of five water droplet measuring instruments. *J. Climate Appl. Meteor.* **22**:891–910.

Baumgardner, D., J.-F. Gayet, H. Gerber, A. Korolev, and C. Twohy. 2002. Clouds: *In-situ* measurement techniques. In: Encyclopedia of Atmospheric Scienes, ed. J. R. Holton, J. A. Curry, and J. Pyle, pp. 489–498. London: Academic Press.

Baumgardner, D., and A. Korolev. 1997. Airspeed corrections for optical array probe sample volumes. *J. Atmos. Ocean. Technol.* **14**:1224–1229.

Biter, C. J., J. E. Dye, D. Huffman, and W. D. King. 1987. The drop-size response of the CSIRO liquid water probe. *J. Atmos. Ocean. Technol.* **4**:359–367.

Brenguier, J.-L., D. Baumgardner, and B. Baker. 1994. A review and discussion of processing algorithms for FSSP concentrations. *J. Atmos. Ocean. Technol.* **11**:1409–1414.

Brenguier, J.-L., A. R. Rodi, G. Gordon, and P. Wechsler. 1993. Real-time detection of performance degradation of the forward-scattering spectrometer probe. *J. Atmos. Ocean. Technol.* **10**:27–33

Chahine, M., T. Pagano, H. Aumann et al. 2006. AIRS: Improving weather forecasting and providing new data on greenhouse gases. *Bull. Am. Meteor. Soc.* **87(7)**:911–926.

Chylek, P., S. Robinson, M. K. Dubey et al. 2006. Comparison of near-infrared and thermal infrared cloud phase detections. *J. Geophys. Res.* **111**:D20203. Di Michele, S., and P. Bauer. 2006. Passive microwave radiometer channel selection based on cloud and precipitation information content. *Q. J. Roy. Meteor. Soc.* **132**:1299–1323.

Cober, S. G., G. A. Isaac, and A. V. Korolev. 2001a. Assessing the Rosemount icing detector with *in-situ* measurements. *J. Atmos. Ocean. Technol.* **18**:515–528.

Cober, S. G., G. A. Isaac, A. V. Korolev, and J. W. Strapp. 2001b. Assessing cloud-phase conditions. *J. Appl. Meteor.* **40**:1967–1983.

Cober, S. G., G. A. Isaac, and J. W. Strapp. 2001c. Characterizations of aircraft icing environments that include supercooled large drops. *J. Appl. Meteor.* **40**:1984–2002.

Dye, J. E., and D. Baumgardner. 1984. Evaluation of the forward scattering spectrometer probe. I: Electronic and optical studies. *J. Atmos. Ocean. Technol.* **4**:329–344.

Field, P. R., A. J. Heymsfield, and A. Bansemer. 2006. Shattering and particle interarrival times measured by optical array probes in ice clouds. *J. Atmos. Ocean. Technol.* **23**:1357–1371.

Field, P. R., R. J. Hogan, P. R. A. Brown et al. 2005. Parameterization of ice-particle size distributions for mid-latitude stratiform cloud. *Q. J. Roy. Meteor. Soc.* **131**:1997–2017.

Gayet, J.-F., P. A. Brown, and F. Albers. 1993. A comparison of in-cloud measurements obtained with six PMS 2D-C probes. *J. Atmos. Ocean. Technol.* **10**:180–194.

Gayet, J.-F., V. Shcherbakov, H. Mannstein et al. 2006. Microphysical and optical properties of mid-latitude cirrus clouds observed in the southern hemisphere during INCA. *Q. J. Roy. Meteor. Soc.* **132**:2721–2750.

Gerber, H. 1991. Supersaturation and droplet spectral evolution in fog. *J. Atmos. Sci.* **48**:2569–2588.

Gultepe, I., and G. A. Isaac. 1999. Scale effects on the relationship between cloud droplet and aerosol number concentrations: Observations and models. *J. Climate* **12**:1268–1279.

Gultepe, I., and G. A. Isaac. 2004. Aircraft observations of cloud droplet number concentration: Implications for climate studies. *Q. J. Roy. Meteor. Soc.* **130**:2377–2390.

Gultepe, I., and G. A. Isaac. 2007. Cloud fraction parameterization as a function of mean cloud water content and its variance using *in-situ* observations. *Geophys. Res. Lett.* **34**:L07801.

Gultepe, I., G. A. Isaac, and S. G. Cober. 2001. Ice crystal number concentration versus temperature. *J. Climate* **21**:1281–1302.

Hallett, J. 2003. Measurement in the atmosphere. In: Handbook of Weather, Climate, and Water: Dynamics, Climate, Physical Meteorology, Weather Systems, and Measurements, ed. T. D. Potter and B. R. Colman, pp. 711–720. New York: John Wiley & Sons.

Haman, K. E., S. P. Malinowski, B. D. Struś, R. Busen, and A. Stefko. 2001. Two new types of ultrafast aircraft thermometers. *J. Atmos. Ocean. Technol.* **18**:117–134.

Herman, M., J.-L- Deuze, C. Devaux et al. 1997. Remote sensing of aerosols over land surfaces including polarization measurements and application to POLDER measurements. *J.Geophys. Res.* **102**:17,039–17,049.

Heymsfield, A. J. 2007. On measurement of small ice particles in clouds. *Geophys. Res. Lett.* **34**:L23812.

Heymsfield, A. J., and J. L. Parrish. 1978. A computational technique for increasing the effective sampling volume of the PMS two-dimensional particle size spectrometer. *J. Appl. Meteor.* **17**:1566–1572.

Isaac, G. A., C. M. Banic, W.R. Leaitch et al. 1998. Vertical profiles and horizontal transport of atmospheric aerosols and trace gases over central Ontario. *J. Geophys. Res.* **103**:22,015–22,037.

Isaac, G. A., S. G. Cober, I. Gultepe et al. 2002. Particle spectra in stratiform winter clouds. In: Proc. 11th AMS Conference on Cloud Physics. Ogden, Utah: American Meteorological Society.

Isaac, G. A., S. G. Cober, J. W. Strapp et al. 2001. Recent Canadian research on aircraft in-flight icing. *Canadian Aeronautics & Space J.* **47(3)**:213–221.

Isaac, G. A., S. G. Cober, and J.W. Strapp. 2005. Measuring cloud parameters for in-flight icing certification tests. AIAA 43rd Aerospace Sci. Meeting and Exhibit, Reno, Nevada. AIAA 2005-0857.

Isaac, G. A., I. Gultepe, and S. G. Cober. 2004. Use of mass versus volume units for cloud microphysical parameters. Proc. 14th Intl. Conf. on Clouds and Precipitation, pp. 800–803. Bologna: International Commission on Clouds and Precipitation (IAMAS).

Jensen, J. B., and G. B. Raga. 1993. Calibration of a Lyman-α sensor to measure in-cloud temperature and clear-air temperature. *J. Atmos. Ocean. Technol.* **10**:15–26.

Kahn, R., B. J. Gaitley, J. V. Martonchik et al. 2005. Multiangle Imaging Spectroradiometer (MISR) global aerosol optical depth validation based on 2 years of coincident Aerosol Robotic Network (AERONET) observations. *J. Geophys. Res.* **110(D10)**:D10S04

Kaufman, Y. J., J. V. Martins, L. A. Remer, M. R. Schoeberl, and M. A. Yamasoe. 2002. Satellite retrieval of aerosol absorption over the oceans using sunglint. *Geophys. Res. Lett.* **29**:1928.

King, W. D. 1984. Air flow and particle trajectories around aircraft fuselages. I: Theory. *J. Atmos. Ocean. Technol.* **1**:5–13.

King, W. D. 1985. Air flow and particle trajectories around aircraft fuselages. III: Extensions to particles of arbitrary shape. *J. Atmos. Ocean. Technol.* **4**:539–547.

King, W. D. 1986. Air flow and particle trajectories around aircraft fuselages. IV: Orientation of ice crystals. *J. Atmos. Ocean. Technol.* **3**:433–439.

Kingsmill, D. E., S. E. Yuter, A. J. Heymsfield et al. 2004. TRMM common microphysics products: A tool for evaluating spaceborne precipitation retrieval algorithms. *J. Appl. Meteor.* **43**:1598–1618.

Knollenberg, R. G. 1981. Techniques for probing cloud microstructure. In: Clouds, their formation, Optical properties, and Effects, ed. P. V Hobbs and A. Deepak, pp. 15–91. New York: Academic Press.

Korolev, A., and G. A. Isaac. 2005. Shattering during sampling by OAPs and HVPS. I: Snow particles. *J. Atmos. Ocean. Technol.* **22(5)**:528–542.

Korolev, A. V., and G. A. Isaac. 2006. Relative humidity in liquid, mixed phase, and ice clouds. *J. Atmos. Sci.* **63**:2865–2880.

Korolev, A. V., G. A. Isaac, S. G. Cober, J. W. Strapp, and J. Hallett. 2003. Observations of the microstructure of mixed phase clouds. *Q. J. Roy. Meteor. Soc.* **129**:39–65.

Korolev, A., G. A. Isaac, and J. Hallett. 2000. Ice particle habits in stratiform clouds. *Q. J. Roy. Meteor. Soc.* **126**:2873–2902.

Korolev, A. V., G. A. Isaac, I. P. Mazin, and H. Barker. 2001. Microphysical properties of continental clouds from in-situ measurements. *Q. J. Roy. Meteor. Soc.* **127**:2117–2151.

Korolev, A. V., J. W. Strapp, and G. A. Isaac. 1998a. Evaluation of the accuracy of PMS optical array probes. *J. Atmos. Ocean. Technol.* **15**:708–720.

Korolev, A. V., J. W. Strapp, G. A. Isaac, and E. Emery. 2008. Improved airborne hot-wire measurements of ice water content in clouds. Proc. 15th Intl. Conf. on Clouds and Precipitation, Cancun, Mexico. Intl. Commission on Clouds and Precipitation.

Korolev, A. V., J. W. Strapp, G. A. Isaac, and A. Nevzorov. 1998b. The Nevzorov airborne hot wire LWC/TWC probe: Principles of operation and performance characteristics. *J. Atmos. Ocean. Technol.* **15**:1495–1510

Lance, S., J. Medina, J. N. Smith, and A. Nenes. 2006. Mapping the operation of the DMT continuous flow CCN counter. *Aerosol Sci. and Technol.* **40**:1–13.

Lawson, R. P., B. A. Baker, C. G. Schmitt, and T. L. Jensen. 2001. An overview of microphysical properties of Arctic clouds observed in May and July 1998 during FIRE.ACE. *J. Geophys. Res.* **106**:14,989–15,014.

Lawson, P., A. J. Heymsfield, S. M. Aulenbach, and T. L. Jensen. 1998. Shapes, sizes and light scattering properties of ice crystals in cirrus and a persistent contrail during SUCCESS. *Geophys. Res. Lett.* **25**:1331–1334.

Lawson, R. P., D. O'Connor, P. Zmarzky et al. 2006. The 2D-S (stereo) probe: Design and preliminary tests of a new airborne, high speed, high-resolution particle imaging probe. *J. Atmos. Ocean. Technol.* **23**:1462–1477.

Liljegren, J. C., E. E. Clothiaux, G. G. Mace, S. Kato, and X. Q. Dong. 2001. A new retrieval for cloud liquid water path using a ground-based microwave radiometer and measurements of cloud temperature. *J. Geophys. Res.* **106(D13)**:14,485–14,500.

MacPherson, J. I., and G. A. Isaac. 1977. Turbulent characteristics of some Canadian cumulus clouds. *J. Appl. Meteor.* **16**:81–90.

Mazin, I. P., A. V. Korolev, A. Heymsfield, G. A. Isaac, and S. G. Cober. 2001. Thermodynamics of icing cylinder for measurements of liquid water content in supercooled clouds. *J. Atmos. Ocean. Technol.* **18**:543–558.

McFarlane, S. A., R. T. Marchand, and T. P. Ackerman. 2005. Retrieval of cloud phase and crystal habit from multiangle imaging spectroradiometer (MISR) and moderate resolution imaging spectroradiometer (MODIS) data. *J. Geophys. Res.* **110**:D14201.

Mo, T. 1996 1996. Prelaunch calibration of the advanced microwave sounding unit-A for NOAA-K. *IEEE Trans. Microwave Theory Tech.* **44**:1460–1469.

Norment, H. G. 1988. Three-dimensional trajectory analysis of two drop sizing instruments: PMS OAP and PMS FSSP. *J. Atmos. Ocean. Technol.* **5**:743–756.

Riedi, J., P. Goloub, and R. T. Marchand. 2001. Comparison of POLDER cloud phase retrievals to active remote sensors measurements at the ARM SGP site. *Geophys. Res. Lett.* **28(11)**:2185–2188.

Roberts, G., G. Mauger, O. Hadley, and V. Ramanathan. 2006. North American and Asian aerosols over the eastern Pacific Ocean and their role in regulating cloud condensation nuclei. *J. Geophys. Res.* **111**:D13205.

Rogers, D. C., P. J. DeMott, S. M. Kreidenweis, and Y. Chen. 2001. A continuous-flow diffusion chamber for airborne measurements of ice nuclei. *J. Atmos. Ocean. Technol.* **18**:725–741.

Rose, D., G. P. Frank, U. Dusek et al. 2008. Calibration and measurement uncertainties of a continuous-flow cloud condensation nuclei counter (DMT-CCNC): CCN activation of ammonium sulfate and sodium chloride aerosol particles in theory and experiment. *Atmos. Chem. Phys.* **8**:1153–1179.

Rotstayn, L. D. 1999. Indirect forcing by anthropogenic aerosols: A global climate model calculation of the effeice-radius and cloud-lifetime effects. *J. Geophy. Res.* **104**:9369–9380.

Shipley, S. T., D. H. Tracy, E. W. Eloranta et al. 1983. A high spectral resolution lidar to measure optical scattering properties of atmospheric aerosols. I: Intrumentation and theory. *Appl. Opt.* **23**:3716–3724.

Siebert, H., H. Franke, K. Lehmann, R. Maser et al. 2006. Probing fine-scale dynamics and microphysics of clouds with helicopter-borne measurements. *Bull. Amer. Meteor. Soc.* **87**:1727–1739.

Slingo, A. 1990. Sensitivity of the Earth's radiation budget to changes in low clouds. *Nature* **343**:49–51.

Squires, P. 1958. The microstructure and colloidal stability of warm clouds. II. The causes of variations in microstructure. *Tellus* **10**:262.

Stephens, G. L. et al. 2002. The CloudSat mission and the A-train. *Bull. Am. Meteor. Soc.* **83**:1771–1790.

Strapp, J. W., F. Albers, A. Reuter et al. 2001. Laboratory measurements of the response of a PMS OAP-2DC. *J. Atmos. Ocean. Technol.* **7**:1150–1170.

Strapp, J. W., J. Oldenburg, R. Ide et al. 2003. Wind tunnel measurements of the response of hot-wire liquid water content instruments to large droplets. *J. Atmos. Ocean. Technol.* **6**:791–806.

Strapp, J. W., and R. S. Schemenauer. 1982. Calibrations of Johnson-Williams liquid water content meters in a high-speed icing tunnel. *J. Appl. Meteor.* **1**:98–108.

Sun, W., N. G. Loeb, and P. Yang. 2006. On the retrieval of ice cloud particle shapes from POLDER measurements. *J. Quant. Spect. Radiative Transfer* **101**:435–447.

Twohy, C. H., and D. Rogers. 1993. Airflow and water-drop trajectories at instrument sampling points around the Beechcraft King Air and Lockheed Electra. *J. Atmos. Ocean. Technol.* **4**:566–578.

Winker, D. M., W. H. Hunt, and M. J. McGill. 2007. Initial performance assessment of CALIOP. *Geophys. Res. Lett.* **34**:L19803.

5

Clouds and Precipitation

Extreme Rainfall and Rain from Shallow Clouds

Yukari N. Takayabu[1] and Hirohiko Masunaga[2]

[1]Center for Climate System Research, University of Tokyo, Kashiwa, Chiba, Japan
[2]Hydrospheric Atmospheric Research Center, Nagoya University, Forocho Chikusa-ku, Nagoya, Japan

Abstract

This chapter reviews present knowledge on extreme precipitation and moderate rainfall from low-level clouds. Primary focus is on the statistics of precipitation characteristics rather than on a detailed description of individual case studies. First, observed variability of precipitation from low-level clouds and the existing techniques to separate different microphysical stages from remote-sensing measurements are reviewed. Over the tropical areas of Pacific and Atlantic oceans, the global distribution of shallow rainfall exhibits a "butterfly" pattern. This feature encompasses heavily precipitating regions such as the intertropical, south Pacific, and south Atlantic convergence zones (ITCZ, SPCZ, and SACZ, respectively); the northern hemispheric counterpart of SPCZ and SACZ emerges only when shallow rain is isolated.

The nature of extreme precipitation varies temporally. On a timescale of about a day, extreme precipitation is associated with synoptic-scale disturbances, including a notable example known as tropical plumes or moist conveyer belt, which could give rise to extreme daily precipitation in downstream arid regions. On an hourly timescale, extreme precipitation is caused by mesoscale moisture convergence, which is so intense that it maintains a continuous overturning of saturated air. Satellite observations imply that the global distribution of extreme precipitation shows a systematic difference from the total rainfall map in terms of, for example, the contrast between land and ocean. The distribution of low-level, precipitation-related latent heating associated with warm rain coincides with the butterfly pattern. Its cohabitation and separation with the deep heating suggests that warm rain plays a role in providing a thick layer of moist static energy source to the convection, and that it is also related to the tropical plumes which cause extreme precipitation in the semiarid west coasts of continents.

Precipitation from Low-level Clouds

Low-level clouds are typically non-precipitating, or just lightly precipitating, and are thus unlikely to cause hazardous weather events. Nevertheless, precipitation from low-level clouds is important meteorologically and climatologically for the following reasons: Precipitation from stratocumulus has a direct impact on the dynamic maintenance of its parent cloud (Nicholls 1984). In terms of the so-called second indirect effect of aerosols, an increase in aerosol concentration is hypothesized to suppress cloud droplet growth, leading to a longer cloud lifetime as the cloud drizzles out less efficiently (Albrecht 1989). Precipitation from stratocumulus clouds can thus be a controlling factor of the Earth's radiation budget, although they do not necessarily make a crucial contribution to global water cycle. Tropical and subtropical shallow cumulus yield significant precipitation through warm rain processes. Short and Nakamura (2000) estimated that shallow rainfall accounts for more than 20% of the total rainfall over tropical oceans.

Precipitation from low-level clouds comes generally in the form of warm rain, except in cases where a very low surface temperature does not allow cloud droplets to exist in the liquid phase. The microphysical processes involved in warm rain are relatively well understood (e.g., Mason 1952). Cloud droplets begin to grow in size, initially through vapor condensation to liquid water in supersaturated air. Condensational droplet growth is followed by the drizzling stage, onset by collision and coalescence with distinctly large droplets (20–100 μm) in a turbulent cloud layer. Drizzle turns to warm rain as the droplet coalescence proceeds further and precipitation rate increases.

A number of field campaigns with airborne probes have provided opportunities to study the microphysical properties of clouds in detail. *In-situ* measurements need to be complemented, however, by wide-area observations, such as satellite remote sensing, to obtain robust climatological evidence of large-scale variability. Yet how can we diagnose a microphysical condition from miles away, relying only on information carried by radiation emitted or scattered by cloud particles? To address this concern, let us look at some of the current remote-sensing methodologies used to detect cloud microphysical characteristics and discuss the global climatology of drizzle and shallow rain.

Regional and Global Observations of Drizzle and Warm Rain

Cloud Optical Thickness versus Effective Droplet Radius

Cloud droplets are distributed around 5–10 μm in radius, so that the geometrical optics approximation is valid for cloud droplets at visible wavelengths. Cloud optical thickness, τ_c, measured remotely in the visible (0.65 μm wavelength) spectrum is thus a practical proxy of the second moment of the cloud droplet size distribution integrated over the whole cloud layer:

$$\tau_c \approx 2\pi \int dz \int dr\; n(r) r^2, \tag{5.1}$$

where z, r, and $n(r)$ denote the cloud layer depth, cloud droplet radius, and droplet size distribution, respectively. To translate τ_c into a parameter of meteorological relevance, such as liquid water path (LWP):

$$\text{LWP} \equiv \frac{4}{3}\pi\rho_w \int dz \int dr\; n(r) r^3, \tag{5.2}$$

where ρ_w=1g cm^{-3} is the liquid water density, conversion from observed τ_c to LWP requires knowledge on the relation between the second and third moments of $n(r)$. Combining Equations 5.1 and 5.2:

$$\text{LWP} \equiv \frac{2}{3}\rho_w \tau_c r_e, \tag{5.3}$$

where

$$r_e \equiv \frac{\int dr\; n(r) r^3}{\int dr\; n(r) r^2}, \tag{5.4}$$

and r_e is the effective droplet radius, representing the area-weighted average of cloud droplet radius. Nakajima and King (1990) established a methodology to retrieve cloud optical thickness and effective droplet radius from visible and near-infrared (typically at 3.7 μm) radiances.

The joint use of τ_c and r_e allows different microphysical stages to be separated. LWP increases with r_e and τ_c while cloud droplet number concentration remains nearly constant during the condensational droplet growth. By contrast, the subsequent stage promoted by droplet coalescence is not accompanied by an appreciable increase in LWP. Cloud optical thickness reduces with increasing r_e when LWP is unchanging or is only weakly changing, as inferred from Equation 5.3. The optical thickness versus effective radius diagram constructed from airborne remote sensing (Nakajima and Nakajima 1995) shows a clear contrast between drizzling and non-drizzling clouds as they vary from one area to another. Figure 5.1 shows a pair of sample plots representing both extreme cases. Figure 5.1a depicts clouds that stay at the condensational growth stage, where τ_c and r_e are positively correlated with each other. The correlation turns to negative in Figure 5.1b, implying that droplet collision and coalescence are at work. A very similar evolutionary track in the τ_c-r_e plane is obtained by a bin-microphysical model simulation (Suzuki et al. 2006; see also Nakajima and Schulz, this volume). The effective droplet radius does not reach 15 μm for non-drizzling clouds (Figure 5.1a), whereas r_e can be as large as 20 μm for drizzling clouds (Figure 5.1b). A number of studies, including airborne probe measurements and bin-microphysical model calculations, concluded independently that the transition from the condensational to collisional growth, or the initiation of drizzle, occurs when effective droplet radius reaches about 15 μm (e.g., Gerber 1996; Pinsky and Khain 2002).

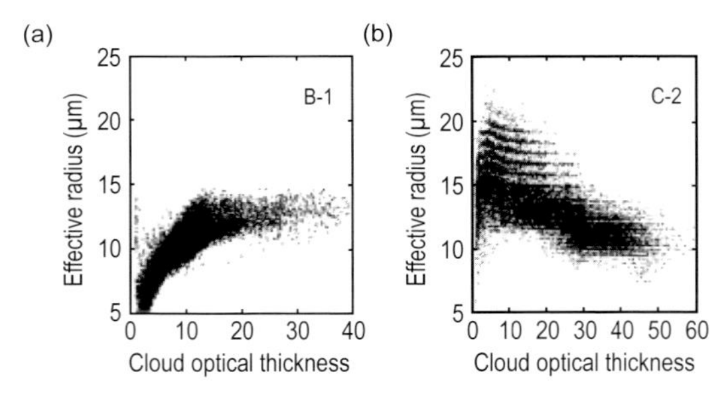

Figure 5.1 Scatterplot of τ_c and r_e for two observed scenes representing different microphysical states (adapted from Nakajima and Nakajima 1995).

Effective Droplet Radius versus Cloud-top Temperature

Within ascending moist air, cloud droplets grow in size so that droplet size increases generally with height inside a cloud layer. The droplet size at cloud top, however, is not uniquely determined for a given cloud thickness. For example, a high concentration of cloud condensation nuclei (CCN) could suppress drizzling for a cloud that otherwise develops deep enough to precipitate. Rosenfeld and Lensky (1998) demonstrated that the cloud-top temperature (T_c) versus an r_e diagram delineates closely the microphysical evolution that occurs in deepening clouds; it relies on the assumption that the local statistics of T_c and r_e are representative of various stages of a developing cloud. Case studies performed by Rosenfeld and Lensky (1998) were successful in identifying the distinct variation of cloud microphysics from one area to another in satellite imagery. Deepening clouds are occasionally not accompanied with significant droplet growth below the level at which droplets freeze (Figure 5.2a, b). In contrast, there are regions where clouds are allowed to reach immediately the drizzling threshold of r_e = 15 µm (Figures 5.2c, d). Rosenfeld and Lensky (1998) hypothesized that this contrast results from the difference in aerosol environment between continental and maritime air. The aerosol influence on clouds, as inferred by Rosenfeld and Lensky (1998) and their subsequent papers, remains controversial (see Ayers and Levin, this volume).

Cloud-top Droplet Radius versus Layer-averaged Droplet Radius

Droplet size increases with height toward the cloud top in non-precipitating clouds; however, the vertical gradient in droplet size should be counteracted by drizzle and rain drops falling toward the cloud base soon after the onset of precipitation. This change in vertical gradient is potentially useful to identify observationally the cloud microphysical stages. The vertical gradient of r_e is, however, not directly observable from the satellite-based schemes introduced above, given that infrared and near-infrared radiation do not penetrate deeply into a cloud layer. In contrast, a liquid cloud can be thoroughly scanned by

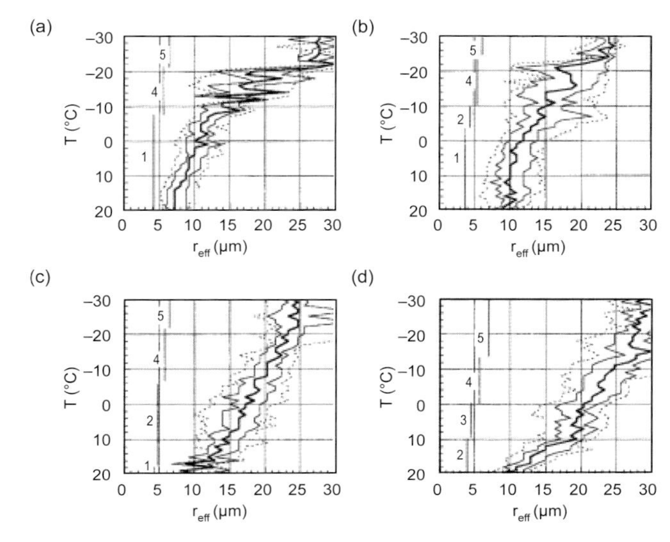

Figure 5.2 Local correlation between T_c and r_e for selected scenes. Different lines are 10th, 25th, 50th, 75th, and 90th percentiles of the r_e for each 1°C interval. Five microphysical stages: (1) condensational growth zone, (2) collisional growth zone, (3) rainout zone, (4) mixed-phase zone, and (5) glaciated zone. Panels (a) and (b) are constructed from continental scenes, and (c) and (d) are from maritime scenes. Adapted from Rosenfeld and Lensky (1998).

microwave radiation, to which cloud water is far less absorptive. Masunaga et al. (2002) proposed a technique to analyze the vertical inhomogeneity in droplet size by the combined use of a visible/infrared imager and a microwave radiometer. In their algorithm, a cloud layer-averaged r_e (or r_e^{ave}) is derived from Equation 5.3 with microwave-retrieved LWP and visible-retrieved τ_c. At the same time, cloud-top r_e (or r_e^{top}) is estimated using the Nakajima and King (1990) method. The partial beam-filling effect in microwave radiometry is corrected using the cloud fraction within a microwave field of view, which is determined from collocated visible radiance. Areas with cloud fractions lower than one-third are excluded from the analysis so that the retrieval noise that results from the uncertainties in surface microwave emission is kept at a minimum. Error analysis and further details in the methodology may be found in Masunaga et al. (2002). Target clouds are limited to those with top temperatures warmer than 273 K.

The global distributions of r_e^{ave} and r_e^{top} are shown in Figure 5.3. The two different versions of effective droplet radius share a common overall pattern, except that the variability of r_e^{ave} is significantly larger in amplitude than r_e^{top}. The effective droplet radii are greatest in the tropical convergence zones, where r_e^{ave} exceeds significantly the drizzling threshold or 15 μm, indicative of the frequent occurrence of warm rain. An intriguing feature is a northern mirror image of the SPCZ, which extends from the equatorial west Pacific

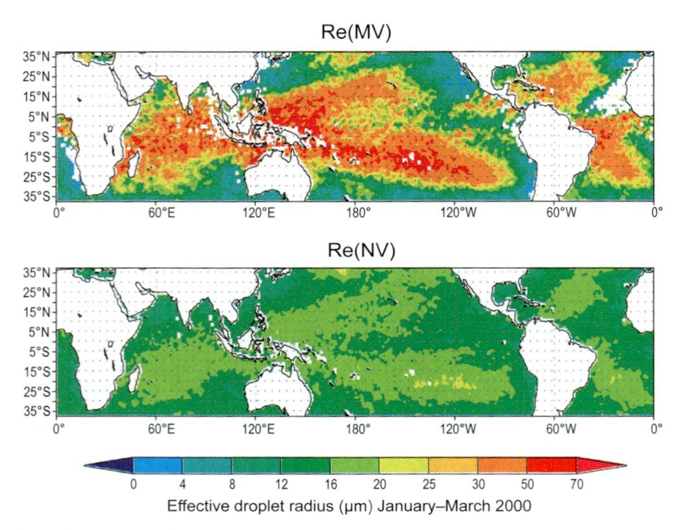

Figure 5.3 The monthly mean global distribution of r_e^{ave} (top) and r_e^{top} (bottom) for January to March, 2000. Adapted from Masunaga et al. (2002).

to the northeast near Hawaii. A similar "butterfly" pattern is observed also in the Atlantic Ocean. A pair of similar butterfly wings is observed in shallow rain climatology (discussed later), whereas the southern wing overwhelms the northern counterpart in magnitude when precipitation from deep clouds is included.

There are areas where effective droplet radius does not reach the drizzling threshold. Particularly notable are subtropical oceans near the western coasts of major continents and the East China Sea, where r_e^{ave} is even smaller than r_e^{top}. The droplet growth in low-level clouds may be suppressed for a number of reasons including high lower-tropospheric stability (Klein and Hartmann 1993) and high aerosol concentration (Twomey 1977; Albrecht 1989). From their analysis of r_e^{ave} and r_e^{top}, Matsui et al. (2004) confirmed that the extent to which the drizzling and warm rain processes proceed is systematically correlated with both lower-tropospheric stability and aerosol concentration. The impacts of static stability and aerosols on low clouds are discussed individually in greater detail elsewhere in this volume (e.g., Bretherton and Hartmann; Feingold and Siebert).

Spaceborne Cloud-profiling Radar

CloudSat, which was launched in 2006, contains W-band (94 GHz) cloud-profiling radar (CPR) and has literally brought a new dimension to our ability to observe clouds from space. As a part of the A-Train constellation, CloudSat is particularly useful when combined with passive sensors on other satellites flying in formation. Stephens and Haynes (2007) have attempted to estimate the droplet coalescence rate based on radar reflectivity observed by CPR together with τ_c and r_e retrieved by MODIS. Suzuki and Stephens (2008) devised a method to separate observationally the condensational growth stage and coalescence stage by a combined use of CPR reflectivity and r_e^{ave} evaluated from the MODIS and AMSR-E. More studies proposing new ideas that exploit the CloudSat and A-Train capabilities are expected in the future.

Global Climatology of Shallow Rain

Shallow cloud tops do not sharply contrast the background surface in satellite infrared imagery. Microwave radiometry is not sensitive either to the depth of a cloud or precipitation layer for a given LWP. It is therefore difficult, in principle, to isolate shallow rain through conventional satellite remote sensing. A breakthrough was brought about by the Tropical Rainfall Measuring Mission (TRMM) precipitation radar (PR), which is able to profile directly the vertical structure of precipitation. Short and Nakamura (2000) analyzed globally PR echo-top height (or storm height) and found that shallow cumulus constitute a distinct peak in the storm height histogram over tropical and subtropical oceans. In this section, we present the global climatology of shallow rain constructed from the 9-year TRMM PR observations: 1998–2006. We analyze the monthly PR near-surface rainfall projected on a half-degree global grid stored in the TRMM 3A25 dataset. Shallow rain is defined as those cases where echo-top heights are substantially lower than the freezing level. According to the TRMM product convention, shallow rain is divided into the two subcategories (isolated and non-isolated), depending on whether or not there is more developed precipitation nearby. As such, non-isolated shallow clouds are presumably a part of organized precipitation systems, such as mesoscale convective systems and tropical cyclones. PR sensitivity (> 17–18 dBZ) to drizzle and light rain (< 1 mm hr^{-1}) is marginal, so that shallow precipitation estimated from PR measurements is attributed mainly to warm rain from (relatively developed) shallow convective clouds. We note that there is so little precipitation from stratocumulus and thus it is not observable in the results below.

In Figure 5.4a, b, monthly rainfall is shown from isolated shallow clouds for February and August, respectively. Isolated shallow clouds precipitate most heavily over subtropical oceans as well as in the ITCZ. The overall global distribution is reminiscent of that for cloud droplet radius (Figure 5.3). The "butterfly" pattern mentioned earlier is again observed in the Pacific and Atlantic oceans. Non-isolated shallow rain (Figures 5.4c, d) contrasts clearly in spatial distribution with the isolated shallow rain. Whereas non-isolated shallow cumulus is, in general, tightly concentrated near the cores of the tropical convergence zones, isolated shallow clouds prefer the fringes of the convergence zones or are outside of the zones, as pointed out by Schumacher and Houze (2003). Non-isolated shallow cumulus can be greater in rain rate than isolated

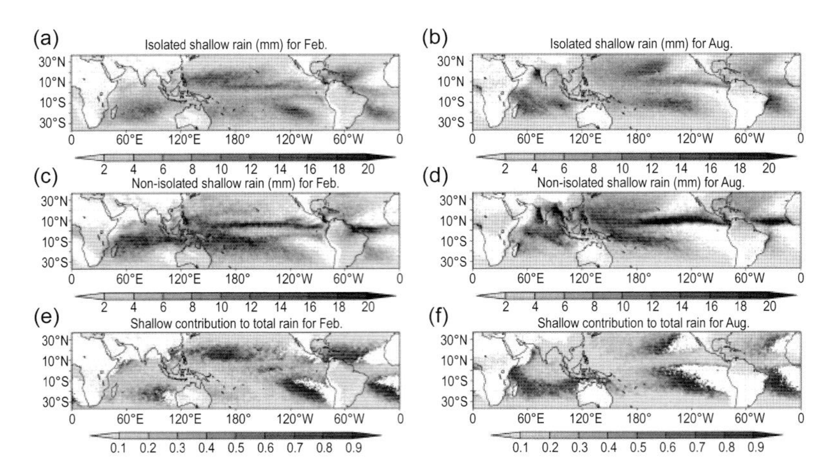

Figure 5.4 The climatology of monthly rainfall from shallow isolated clouds for February (a) and August (b), constructed from the TRMM PR 3A25 dataset. Shallow non-isolated rain is shown in (c) and (d), with the same parameters; (e) and (f) show the ratio of shallow rain (with both isolated and non-isolated) to the total precipitation.

shallow clouds, but the shallow cumulus contribution to total precipitation is only secondary in the tropical convergence zones (Figure 5.4e, f). In contrast, shallow rain explains more than 90% of the total precipitation in several limited regions, including the subtropical oceans, particularly in the winter hemisphere, and at the eastern edge of the SPCZ and SACZ. All of these regions are located under a subsiding branch of the Hadley and Walker circulations, accompanied by a relatively warm sea surface. The strong inversion in such areas hampers deep convection from developing and traps moisture supplied from the underlying ocean, which allows shallow clouds to predominate.

Figure 5.4 shows that there is very little warm rainfall from shallow cumulus over tropical continents compared to tropical oceans, except for the Amazon basin in the wet season (Figure 5.4c). The factors that suppress continental shallow precipitation include (but may not be limited to) the abundance of aerosols acting as CCN and relatively limited moisture supply over land.

Summary

The observed characteristics of large-scale variability in precipitation from low-level clouds have been described. Such observations require a remote-sensing technique to separate drizzling and raining clouds from non-precipitating clouds. Three existing methods were reviewed that utilize (a) cloud optical thickness and effective droplet radius, (b) cloud-top temperature and effective droplet radius, and (c) layer-top and layer-averaged effective droplet radii. Results from these analysis methods are consistent with our knowledge of the low-cloud properties that we have gained from *in-situ* observations in past decades. Satellite observations have shown systematic variability in cloud microphysical status over a wide range of spatial scales beyond the reach of individual field campaigns. Satellite data analysis reveals a large-scale gradient in low-level cloud properties from the deep tropics, where shallow rain occurs frequently, toward the subtropical western coasts of major continents, typical of maritime stratocumulus with little precipitation.

Layer-averaged effective droplet radius is as large as 50 μm or even larger, indicating a frequent occurrence of warm rain in broad areas across the tropical and subtropical oceans. Such areas include not only the ITCZ, SPCZ, and SACZ but also the northern hemisphere counterparts of SPCZ and SACZ. The predominance of shallow rainfall over the northern hemispheric subtropical oceans in February and the southern Indian Ocean in August is confirmed by TRMM PR measurements. Rich moisture supply from warm sea surface and the prevalence of strong trade inversion are presumably major factors that make these regions favorable to shallow rainfall. We do not, however, fully understand why such conditions appear in a quasi-symmetric "butterfly" pattern. In contrast, the tropical convergence zones, which geographically constrain deep convective rainfall, are highly asymmetric about the equator.

As mentioned earlier, precipitation from low-level clouds is not at all hazardous. Shallow rainfall may, however, contribute thermodynamically to extreme precipitation events. Below, we will discuss the possibility that lower tropospheric moistening by shallow clouds could play a role in the development of deeper and more organized convective systems.

Extreme Rainfall

In view of the impact of climate change on human society, changes in extreme rainfall are as important as changes in total rainfall, especially in terms of disaster prevention. It has often been pointed out that while global atmospheric moisture increases with a warming climate (as diagnosed from the Clausius–Clapeyron equation) ~6.5% per Kelvin, projections from climate models indicate that the change in global mean precipitation would be 3.4% per Kelvin. This discrepancy is understood in the context of energy balance, as global mean precipitation is constrained by the atmospheric radiative cooling rate rather than the atmospheric moisture content (e.g., Hartmann and Larson 2002). It is also suggested that while the mean precipitation does not increase at the rate of the atmospheric moisture content, frequency of extreme precipitation would increase more significantly.

Karl et al. (1995) analyzed long-term trends in the daily precipitation station data over the United States, the former Soviet Union (FSU), and China. They found a significant increase in the annual proportion of extreme precipitation category (>50.8 mm d^{-1}) over the U.S. for the period of 1911–1994, although the proportions of moderate (12.7–25.4 mm d^{-1}) and light precipitation categories (2.54–12.7 mm d^{-1}) decreased. A summertime increase in extreme precipitation exceeded 2% of the total summer precipitation; this corresponds to an average of adding one extreme event every two years. They also noted that this increase did not result from an increase in intensity but rather to an increase in the number of days. Over the FSU or China, little systematic change was found in precipitation variability.

In this section, we provide an overview of the future projection of extreme rainfall in climate models. To address current understanding of extreme rainfall, we discuss individual phenomena related to extreme rainfall and emphasize that it cannot be defined without considering temporal or spatial scales.

Extreme Rainfall and Future Change from a Climate Model Perspective

What is "extreme rainfall"? One can easily imagine how the same level of precipitation, 50 mm d^{-1}, can be viewed differently in the wet monsoon region versus in the semiarid subtropics. Certainly, it cannot be defined with one global threshold of rainfall intensity. Thus to address the issue of representing extreme rainfall in the climate models, local values of extreme precipitation (e.g., 20-yr return values of annual extremes of 24-hr precipitation) are often utilized.

Using six climate model experiments, future changes in dynamic and thermodynamic components of mean and extreme precipitation were examined separately by Emori and Brown (2005). For extreme precipitation, the multiyear mean of the fourth largest value was utilized at each grid point. Emori and Brown showed that the global mean change of average precipitation for the 2081–2100 A1B scenario, run from 1981–2000, was 6%, while that for extreme precipitation was as large as 13%. Most of the extreme precipitation change resided in the thermodynamic component, which showed an overall increase even for the areas of modest change or decreasing mean precipitation. They postulated that with a moister atmosphere, the same amounts of mass convergence or divergence results in larger moisture convergence and divergence, and that this brought about larger increases in extreme precipitation.

Kharin et al. (2007) examined extreme precipitation with a 20-yr return period in an ensemble of global coupled climate models as part of the IPCC diagnostic exercise for the Fourth Assessment Report. For future projections of extreme precipitation, multi-model median 20-yr return values of daily precipitation increase with SRES A1B experiment was 6–7% per Kelvin, which was close to the projected Clausius–Clapeyron rate, while the corresponding mean precipitation change was 3.4 % per Kelvin. This result was also consistent with the study by Emori and Brown (2005). Kharin et al. noted that the simulated present-day precipitation extremes were plausible in the extratropics, but in the tropics, large intermodel discrepancies existed. Uncertainties in the tropics were very large, not only in the models but also in the available reanalysis data for comparison, and this abated the confidence in the projected changes in extreme precipitation.

State-of-the-art climate model results suggest strongly that some physical processes associated with extreme precipitation are not well represented in the models, especially over the tropics. Although most model studies on extreme rainfall are based on daily precipitation, we must be mindful that "extremes" vary largely, depending on temporal scales of phenomena as well as the temporal resolution of data. Next we review the observations to examine what phenomena contribute to the extreme precipitation in different temporal resolutions.

Extreme Rainfall from Observational Aspects

Figure 5.5 shows the depth–duration relationships of the world record rainfalls and Japanese record rainfalls, based on precipitation data collected by Jennings (1950) and plotted by Ninomiya and Akiyama (1978, pers. comm.). Depending on the temporal scale, changes in record points indicate that various types of heavy precipitation systems have contributed to rainfall accumulation records on different temporal scales. On a scale of annual to ten days, record precipitation is found in Cherrapunji (northeast India). On a scale of a week to one day, accumulation is greatest in the tropics. From one day to 30 minutes, record levels are found in the subtropics, whereas for less than one hour, record precipitation occur in the midlatitudes and tropics.

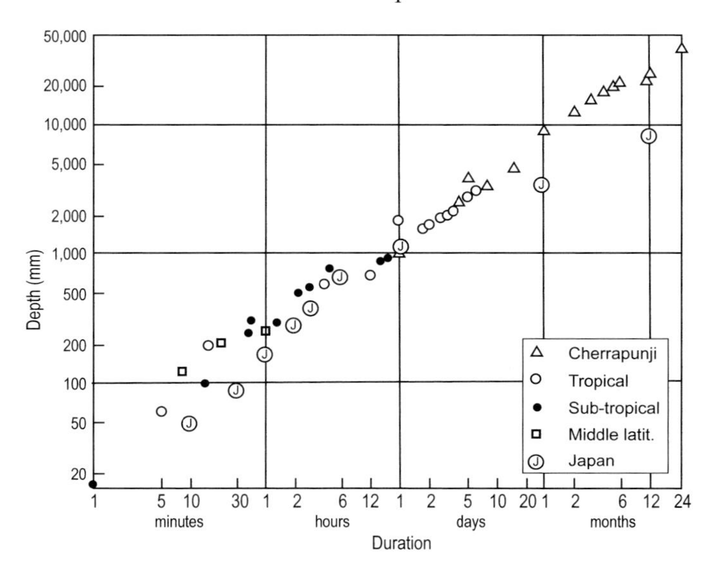

Figure 5.5 Depth–duration relationships of the world record rainfalls and Japanese record rainfalls. Adapted from Ninomiya and Akiyama (1976).

Trenberth (1999) stressed the importance of understanding the change in precipitation that is associated with different types of storms, and recommended that hourly precipitation datasets be collected. Below, we present examples of how various types of storms contribute to extreme precipitation in different regions and on different temporal scales.

Tropical–Extratropical Interactions

Several studies suggest that extreme precipitation occurs over the arid or semiarid regions, along the west coasts of continents, in association with tropical–extratropical interactions. Higgins et al. (2000) utilized 3-day accumulations of precipitation data along the West Coast of the U.S. and found that extreme precipitation events occur during neutral winters just prior to the onset of El Niño. This is a period when intraseasonal activity is large in the tropical Pacific. Jones (2000) examined the daily precipitation in California and showed that frequency of extreme events are more common when tropical convective activity associated with the Madden-Julian Oscillation is high.

Extreme daily precipitation during the winter in the semiarid southwestern U.S., as analyzed by Higgins et al. (2000) and Jones (2000), is related to the enhancements of tropical plumes (i.e., bands of clouds that extend from the deep tropics into subtropical and midlatitudes and last only a few days) (McGuirk et al. 1988; Knippertz and Martin 2007). Figure 5.6 provides an example of the eastern Pacific tropical plume captured by the infrared image from GOES West. Knippertz and Martin (2007) showed that the tropical plume or the "moist conveyer belt" enhances the water vapor fluxes above the planetary boundary layer at around 700 hPa, and causes potential instability in concert with the overriding dry air.

Figure 5.6 An example of the tropical plume over the eastern Pacific, captured in the infrared image of GOES West at 00z January 14, 2007. Obtained from http://www.satmos.meteo.fr/cgi-bin/qkl_sat/quicklook.pl

Tropical plumes are also observed in association with extreme precipitation in arid northwest Africa. Knippertz and Martin (2005) examined three cases of extreme rain over western Africa. For two cases, precipitation exceeded half of the local climatological annual precipitation within just a few days. The other case exceeded the corresponding monthly climatological value. Associated with the deepening of the extratropical trough into the tropics, a tropical plume was observed, which transported moisture in the middle (~600 hPa) level, to produce a potentially unstable stratification in which convection could begin.

Extreme Rain in Japan Associated with the Baiu Front

Japan is located in a region that is subjected to exceptionally heavy rainfall for its latitude. Figure 5.5 shows the rainfall accumulation records for Japan on different timescales. As can be seen, the observed levels of rainfall from several hours to one-day accumulation in Japan are comparable to corresponding world records. This is because extreme rain, which is associated with mesoscale precipitation systems, occurs frequently over Japan, because it is located in a moist subtropical region affected both by subtropical active frontal and depression systems and tropical moist air masses. We note that a similar environmental condition is present in the southeast U.S.

Ninomiya and Akiyama (1978) examined the distributions of rainfall extremes in Japan on different temporal scales and found that the meridional gradient in extreme 10-min accumulated rainfall distribution was small. It was similar to the gradient of the total precipitable water content. They showed that the 10-min rainfall record value corresponded to the maximum precipitable water (~50 mm/10 min) in the atmosphere, or one overturning of the saturated atmosphere. From this point of view, hourly accumulated rainfall represents how many times the moist atmosphere is overturned, or how effective the mesosystems can gather the moisture into the convection. Assuming the typical spatial scale of the extreme rainfall with 50 mm/10 min to be 10^4 km^2, an area of 6×10^4 km^2 needs to be contracted into 10^4 km^2 in one hour to realize the extreme hourly precipitation of 300 mm hr^{-1}. This corresponds to the convergence of about 2×10^{-4} sec^{-1}, which is the typical value of the mesoscale systems. Thus, it is estimated that a continuous overturning for one hour (~300 mm hr^{-1}) is possible with the existence of mesoscale systems. Moreover, distribution of hourly extremes showed a larger meridional gradient compared to the 10-min extremes. This is most likely attributable to the oceanic southerly flow, which prepares a more effective moisture convergence in mesoscale systems, but is not yet limited to particular geographical locations.

Daily extremes, however, were found to be relatively locked into geographic locations. In contrast to hourly rainfall records, the daily extreme value is far from a continuous overturning of the moist atmosphere: There is no ~7200 mm day^{-1} precipitation, and since large-scale disturbances cannot gather moisture as effectively, daily extremes consist of intermittent recovery of mesoscale systems.

Intense Thunderstorms Observed from TRMM

Short-term extreme precipitation is considered to be associated with intense thunderstorms. Employing various kinds of indices obtained from the precipitation features (PFs) database, Zipser et al. (2006) used satellite observations from TRMM to describe the global distribution of extremely intense thunderstorms. Figure 5.7 shows the distributions of extreme (uppermost 0.01% frequency) PFs with different indices. It is indicated that the most extreme PFs are observed over Equatorial Africa, southeastern U.S., and southeast South America. During the months of March–May and June–August, extreme PFs are found in the southeastern area of Africa, northern and eastern areas of India

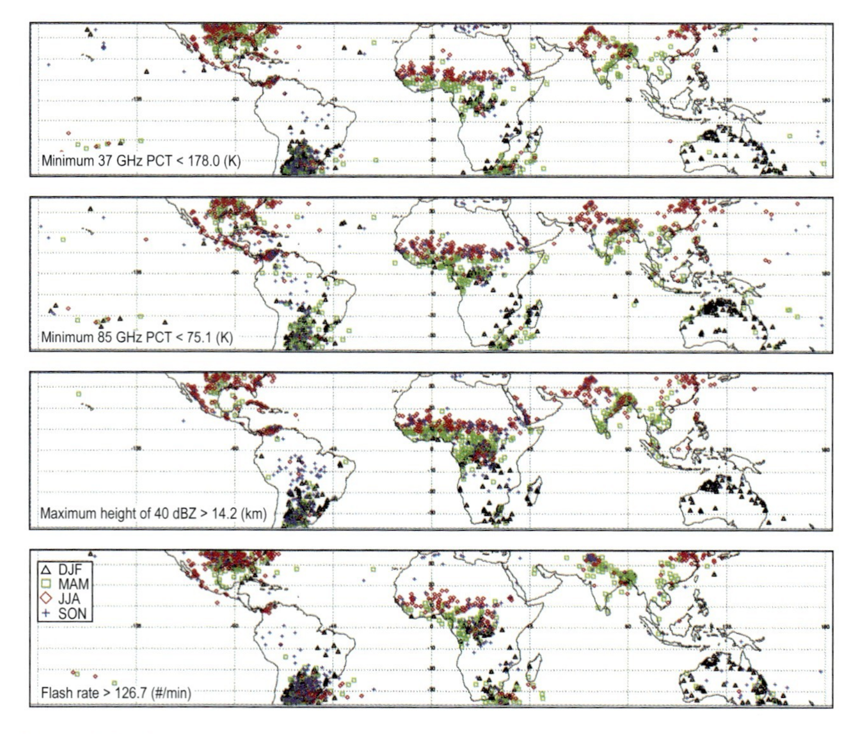

Figure 5.7 Seasonal cycle of the extreme categories with a threshold of 0.01 % occurrences for different parameters obtained from the precipitation feature database. The top two panels are based on minimum precipitation from TRMM microwave observations with the frequency of 37 GHz and 87 GHz, respectively. The third panel is based on the TRMM precipitation radar reflectivity with a threshold of maximum height of 40 dBZ > 14.2 km. The bottom panel is based on the TRMM LIS observations with flash rate > 126.7 min^{-1}. Adapted from Zipser et al. (2006).

and Indochina, and eastern China, whereas from December–February PFs are found in the northern part of Australia. Extremely intense thunderstorms are distributed very differently from total rainfall amount for which regions (e.g., Amazon and the Maritime Continent) would be dominant. These distributions of extreme events correspond instead to the distributions of intense mesoscale convective complexes.

Takayabu (2006) calculated rainyield per flash (RPF) values with the TRMM precipitation radar and lightning imaging sensor (LIS) data. RPF is the total rainfall normalized with lightning flash numbers; thus, it represents the rainfall characteristics irrespective of the rainfall amount. Figure 5.8 presents the 8-year mean distribution of RPF and the average rainfall rate. Regions of most intense thunderstorms are indicated with dark shades of small RPF values. Above all, a strong contrast of RPF between land and ocean is seen, consistent with the observations that most thunderstorms are observed over land. This feature is in contrast with the average rainfall rate, which exhibits rather continuous distribution of tropical rain bands over ocean and over land. The distribution of extremely small RPF values corresponds well to that of extreme PF events in Figure 5.7 over Equatorial Africa, southeastern U.S., southeastern South America, along the periphery of the Tibetan Plateau, and the northern part of Australia. It is also readily observed that regions of largest total annual

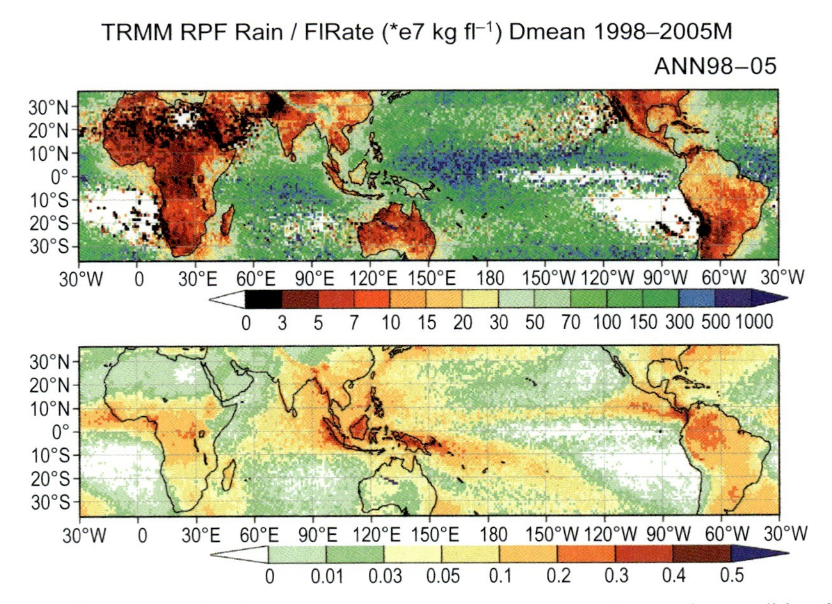

Figure 5.8 Global distributions of 8-yr mean rain yield per flash (a) and unconditional mean rain rate averaged throughout the period (b). Units for the color scales are 10^7 kg fl^{-1} for (a) and mm hr^{-1} for (b). Rain yield per flash averages are obtained by dividing the total precipitation amount by the total flash number for the averaging period.

precipitation, such as Amazonia and the Maritime Continent, are secondary in terms of intensity of the thunderstorms with small RPF values. This is consistent with Zipser et al.'s (2006) conclusions. The overall correspondence of rain characteristics, represented by RPF and PFs, indicate that satellite observations effectively detect extreme rain associated with intense mesoscale convective systems, the distribution of which differs from accumulation extremes which occur over a few days.

Discussion

Thus far, extreme rainfall addressed in climate models has been based primarily on daily precipitation. However, recent satellite observations are able to detect effectively the extreme rainfall associated directly with the mesoscale convective systems. As suggested by prior studies (e.g., Ninomiya and Akiyama 1978; Trenberth 1999; Knippertz and Martin 2005), extreme hourly precipitation corresponds to intense mesoscale systems, whereas extreme daily precipitation is controlled by larger synoptic-scale systems. We must recognize that factors controlling the intensity of synoptic-scale systems and mesoscale systems are entirely different. There has been a discussion with climate model simulations that increasing rate of the extreme rainfall is proportional to the increase of atmospheric moisture. However, while the mesoscale systems' convergence can attain the amount of continuous atmospheric overturning, synoptic-scale convergence cannot maintain continuous overturning. Therefore, when daily extreme precipitation is considered, we must treat both the atmospheric moisture content as well as adequate representations of synoptic-scale systems as essential. Although the assumption may be correct that atmospheric moisture content affects extreme rainfall more directly than the total rainfall amount, quantitative conclusions with daily rainfall extremes require reexamination.

Finally, we wish to discuss the relationship between shallow convection and deep and intense convection. Figure 5.9 shows the distribution of convective heating in terms of Q1–QR, defined by Equation 5.5, at 7.5 km and 2.0 km, respectively, for the December–February season averaged for nine years of TRMM observation. This was calculated with the spectral latent heating (SLH) algorithm introduced by Shige et al. (2004). Q_1 is called as the apparent heating (Yanai et al. 1973),

$$Q_1 \equiv \frac{D\overline{s}}{Dt} = \frac{\partial \overline{s}}{\partial t} + \overline{v} \bullet \nabla \overline{s} + \overline{\omega}\frac{\partial \overline{s}}{\partial p} = Q_R + L(\overline{c} - \overline{e}) - \nabla \bullet \overline{s'v'} - \frac{\partial \overline{s'\omega'}}{\partial p} \tag{5.5}$$

where $s = CpT + gz$, overbars represent large-scale values and primes represent eddy values. L is the latent heat of vaporization, Q_R is the atmospheric radiative heating, c is the condensation rate, and e is the evaporation rate, where Cp is the heat capacity for a constant pressure, T is the temperature, and gz is the geopotential. Note that these two altitudes are typical for cold and warm

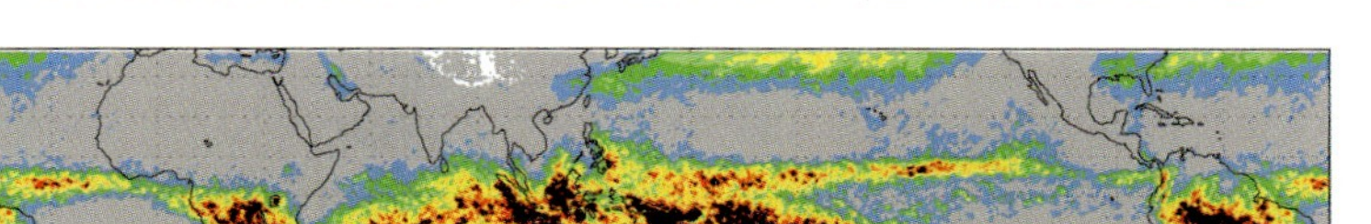

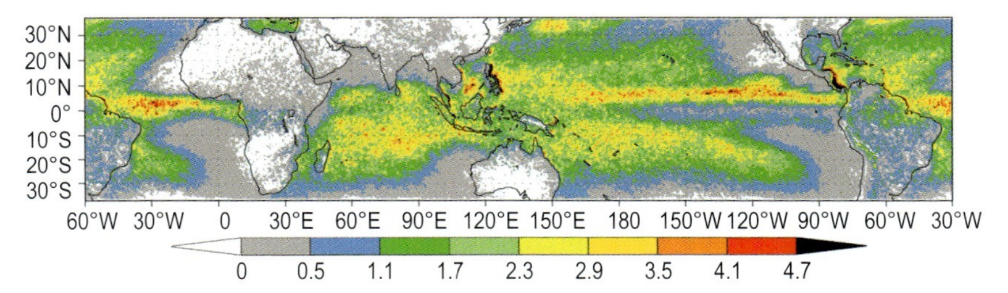

Figure 5.9 Global distributions unconditional mean Q1–QR (apparent heat source minus the radiative heating) averaged for December–February seasons of 1997–2005. Upper panel (a) shows those at the level of 7.5 km and the lower panel (b) is for the 2.0 km. Unit in the gray scale is K d^{-1}.

rainfall, corresponding to two peaks appearing in the average heating profile. From the SLH tables, it is also known that the 2 km heating peak corresponds to the 4–5 km rain top height (not shown).

The distribution of the taller (7.5 km) heating looks similar to the familiar total rainfall distribution. The shallower (2.0 km) heating distribution corresponds to that shown in Figure 5.4 for shallow clouds, and indicates some differences from the distribution of taller heating. There is a more distinct land–sea contrast, and shallow heating is observed almost exclusively over ocean and not much over land, except for over land in the Maritime Continent region. Many studies suggest the importance of preconditioning with the shallow convection to moisten the lower troposphere, prior to the mature convection of oceanic large-scale organized convective systems. Over the Indian Ocean, the western Pacific warm pool, ITCZ, and SPCZ, it is apparent that tall and shallow convective heating coexist. In these regions, shallow convection is considered to prepare the moist lower troposphere for tall convection to start.

There are, however, particular regions where shallow heating is ubiquitous but not the taller heating. These regions are found where the sea surface temperature is relatively high and at the periphery of the subtropical high pressure area latitudes. One region is located in the northern wing of the "butterfly shape" of shallow clouds, extending from the western Pacific warm pool toward the Hawaii Islands (discussed above). A corresponding northern wing without taller heating is also found from the western to central Atlantic. Since the 2.0 km heating peak corresponds to warm rain with 4–5 km rain top height in SLH tables, the lower troposphere is moistened with this shallow convection over these regions, whereas tall convection is suppressed by the subsidence under the influence of subtropical highs. It is interesting to notice that the two tropical plume regions mentioned earlier, over the eastern Pacific and over the Atlantic, extend from the eastern ends of these northern wings to the arid and semiarid continents. It is reported that the axis of moisture transport by tropical plumes is 600–700 hPa and that it originates from moderate convection, which seems to correspond to the warm rain observed here. Tropical plumes bridge lower to middle tropospheric moisture from the warm rain region over the arid and semiarid continents.

In conclusion, these results suggest that a realistic reproduction of synoptic systems as well as a proper representation of shallow convection and its interaction with synoptic-scale systems are indispensable if we wish to reproduce extreme daily precipitation adequately.

References

Albrecht, B. A. 1989. Aerosols, cloud microphysics, and fractional cloudiness. *Science* **245**:1227–1230.

Emori, S., and S. J. Brown. 2005. Dynamic and thermodynamic changes in mean and extreme precipitation under changed climate. *Geophys. Res. Lett.* **32**:L17706.

Gerber, H. 1996. Microphysics of marine stratocumulus clouds with two drizzle modes. *J. Atmos. Sci.* **53**:1649–1662.

Hartmann, D. L., and K. Larson. 2002. An important constraint on tropical cloud–climate feedback. *Geophys. Res. Lett.* **29**:1951.

Higgins, R. W., J.-K. Schemm, W. Shi, and A. Leetmaa. 2000. Extreme precipitation events in the western United States related to tropical forcing. *J. Climate* **13**:793–820.

Jones, C. 2000. Occurrence of extreme precipitation events in California and relationships with the Madden-Julian Oscillation. *J. Climate* **13**:3576–3587.

Karl, T. R., R. W. Knight, and N. Plummer. 1995. Trends in high-frequency climate variability in the twentieth century. *Nature* **377**:217–220.

Kharin, V. V., F. W. Zwiers, X. Zhang, and G. C. Hegerl. 2007. Changes in temperature and precipitation extremes in the IPCC ensemble of global coupled model simulations. *J. Climate* **20**:1419–1444.

Klein, S. A., and D. L. Hartmann. 1993. The seasonal cycle of low stratiform clouds. *J. Climate* **6**:1587–1606.

Knippertz, P., and J. E. Martin. 2005. Tropical plumes and extreme precipitation in subtropical and tropical West Africa. *Q. J. Roy. Meteor. Soc.* **131**:2337–2365.

Knippertz, P., and J. E. Martin. 2007. A Pacific moisture conveyor belt and its relationship to a significant precipitation event in the semiarid southwestern United States. *Wea. Forecast* **22**:125–144.

Mason, B. J. 1952. Production of rain and drizzle by coalescence in stratiform clouds. *Q. J. Roy. Meteor. Soc.* **78**:377–386.

Masunaga, H., T. Y. Nakajima, T. Nakajima et al. 2002. Physical properties of maritime low clouds as retrieved by combined use of tropical rainfall measuring mission microwave imager and visible/infrared scanner: Algorithm. *J. Geophys. Res.* **107(D10)**:4083.

Matsui, T., H. Masunaga, and R. A. Pielke, Sr. 2004. Impact of aerosols and atmospheric thermodynamics on cloud properties within the climate system. *Geophys. Res. Lett.* **31**: L06109.

McGuirk, J. P., A. H. Thompson, and J. R. Schaefer. 1988. An eastern Pacific tropical plume. *Mon. Wea. Rev.* **116**:2505–2521.

Nakajima, T., and M. D. King. 1990. Determination of the optical thickness and effective particle radius of clouds from reflected solar radiation measurements. Part I: Theory. *J. Atmos. Sci.* **47**:1878–1893.

Nakajima, T. Y., and T. Nakajima. 1995. Wide-area determination of cloud microphysical properties from NOAA AVHRR measurements for FIRE and ASTEC regions. *J. Atmos. Sci.* **52**:4043–4059.

Nicholls, S. 1984. The dynamics of stratocumulus: aircraft observations and comparisons with a mixed-layer model. *Q. J. Roy. Metetor. Soc.* **110**:783–820.

Ninomiya, K., and T. Akiyama. 1978. Characteristics of torrential rain in Japan, observed from its intensity and the moisture budgets. Research report No.A-53-4 for the Special Research on Natural Disasters. Tokyo: Mext.

Pinsky, M. B., and A. P. Khain. 2002. Effects of in-cloud nucleation and turbulence on droplet spectrum formation in cumulus clouds. *Q. J. Roy. Meteor. Soc.* **128**:501–534.

Rosenfeld, D., and I. M. Lensky. 1998. Satellite-based insights into precipitation formation processes in continental and maritime convective clouds. *Bull. Amer. Meteor. Soc.* **79**:2457–2476.

Schumacher, C., and R. A. Houze, Jr. 2003. The TRMM precipitation radar's view of shallow, isolated rain. *J. Appl. Meteor.* **42**:1519–1524.

Shige, S., Y. N. Takayabu, W.-K. Tao, and D. E. Johnson. 2004. Spectral retrieval of latent heating profiles from TRMM PR data. Part 1: Development of a model-based algorithm. *J. Appl. Meteor.* **43**:1095–1113.

Short, D. A., and K. Nakamura. 2000. TRMM radar observations of shallow precipitation over the tropical oceans. *J. Climate* **13**:4107–4124.

Stephens, G. L., and J. M. Haynes. 2007. Near global observations of the warm rain coalescence process. *Geophys. Res. Lett.* **34**:L20805.

Suzuki, K., T. Nakajima, T. Y. Nakajima, and A. Khain. 2006. Correlation pattern between effective radius and optical thickness of water clouds simulated by a spectral bin microphysics cloud model. *SOLA* **2**:116–119.

Suzuki, K., and G. L. Stephens. 2008. Global identification of warm cloud microphysical processes with combined use of A-train observations. *Geophys. Res. Lett.* **35**:L08805

Takayabu, Y. N. 2006. Rain-yield per flash calculated from TRMM PR and LIS data and its relationship to the contribution of tall convective rain. *Geophys. Res. Lett.* **33**:L18705.

Trenberth, K. E. 1999. Conceptual framework for changes of extremes of the hydrological cycle with climate change. *Climate Change* **42**:327–339.

Twomey, S. 1977. The influence of pollution on the shortwave albedo of clouds. *J. Atmos. Sci.* **34**:1149–1152.

6

Temporal and Spatial Variability of Clouds and Related Aerosols

Theodore L. Anderson, Rapporteur

Andrew Ackerman, Dennis L. Hartmann, George A. Isaac, Stefan Kinne, Hirohiko Masunaga, Joel R. Norris, Ulrich Pöschl, K. Sebastian Schmidt, Anthony Slingo, and Yukari N. Takayabu

Hygroscopic Properties of Atmospheric Aerosols Appear to be Better Constrained than Other Key Parameters of Warm Cloud Formation

Aerosol–cloud interactions may have played an important role in the observed climate change of the 20th century, and, if so, this may have partly been caused by changes in cloud condensation nuclei (CCN) as a result of industrialization. Sea salt, sulphate, and organics are key components of CCN; the atmospheric sulphate load is known to have changed significantly because of industrialization. The relative change in CCN caused by industrialization depends critically on (a) preindustrial CCN levels, (b) anthropogenic emissions of particles and aerosol precursor gases, (c) atmospheric processing which generates, transforms, and removes aerosol particles, and (d) the ability of particles to act as CCN; that is, their hygroscopicity. Hygroscopicity, as a function of organic mass fraction, has attracted much attention in recent years (e.g., Bates et al. 2006). However, the first three aspects of the problem pose more important uncertainties.

Briefly, preindustrial CCN levels have been estimated almost exclusively using models that remove anthropogenic sources (e.g., Schulz et al. 2006). Andreae (2007), however, has argued that preindustrial CCN levels may have been fundamentally different. Transport models require detailed emission data (i.e., fluxes for all major particle constituents and precursor gases as a function of particle size, injection altitude, and with high temporal and geographical resolution), but such data are presently sparse and uncertain. Atmospheric transformations of aerosols are frequently dominated by in-cloud processing, which can generate particulate mass from precursor gases, remove particles via precipitation, and alter radically the particle size distribution. Currently, models represent all these processes, but observational constraints are again very sparse relative to the full range of atmospheric conditions. As a result, major uncertainties exist in terms of particle concentrations (number and mass), size distribution, and aerosol sources and sinks (primary emissions, secondary formation, dry and wet deposition). These continue to pose obstacles to our understanding of aerosol effects on warm cloud formation in the atmosphere.

Relative to the above issues, the warm-cloud hygroscopic properties of atmospheric aerosols (i.e., the effects of chemical composition on equilibrium water uptake) appear to be fairly well constrained. The effective hygroscopicity of aerosol particles can be efficiently approximated by a single hygroscopicity parameter (κ, Petters and Kreidenweis 2007). With simplifying assumptions, this parameter is easy to calculate from the chemical composition of the particles, and recent laboratory and field measurements suggest that on average it is constrained to fairly narrow value ranges for continental and marine boundary layer aerosols (0.3 ± 0.1 and 0.7 ± 0.2, respectively) (Andreae and Rosenfeld 2008; Kreidenweis et al., this volume; Kinne et al., this volume).

This result is significant because, at present, general circulation models (GCMs) employ a wide range of hygroscopic coefficients for organic particles, and simulations of the 20th-century temperature record are sensitive to this parameter. Thus, it would appear that current understanding of CCN needs to be incorporated as a much tighter constraint (rather than a tuning parameter) in GCMs. At this point, the hygroscopic properties of aerosols do not appear to be a limiting factor for advances in large-scale atmospheric and climate modeling.

This is not to imply that aerosol hygroscopicity is fully understood. For detailed mechanistic studies, CCN closure experiments, and the like, more sophisticated approaches are essential to progress. Improved understanding of the hygroscopic properties of organic substances and organic–inorganic mixtures is needed as well as an improved understanding of kinetic limitations.

Ice Nuclei and Ice Microphysics Still Pose Many Open Questions

Our understanding of ice nucleation (the seemingly tiny fraction of aerosol particles that constitute ice nuclei; the roles of homogeneous and heterogeneous nucleation processes) is currently very poor, as is our knowledge of the "ice-multiplication" processes that generate observed concentrations of small ice crystals. These processes are critical not only for high-latitude precipitation but also for tropical and midlatitude rainfall. Heavy rainfall events generated by mesoscale convective systems, tropical cyclones, and extratropical storms all involve cold rain processes. If we wish to improve our ability to forecast extreme weather and to quantify the global climatology of precipitation, a better understanding of ice microphysics is imperative. Meteorologists have been stymied by these issues for many decades. There are severe limitations in current methods for measuring microphysics of mixed-phase and ice clouds, both in terms of *in-situ* approaches and remote sensing. Even the new satellite techniques (CloudSat and CALIPSO) provide only limited phase information, and only in the vicinity of cloud top.

GCMs are developing parameterization strategies that enable composition to be linked to droplet activation in clouds. In principle, the approach can also be used for ice activation. Unlike droplet activation, however, the links between atmospheric composition and ice formation (nucleation, multiplication) are not well known. Representing interactions between aerosols and ice or mixed-phase clouds in GCMs is problematic without adequate knowledge of ice particle formation.

Little to nothing is known about the collision efficiencies of ice particles, and thus most cloud-resolving bin microphysics models tend to assume that those efficiencies (ice–ice and ice–liquid) are the same as those between water drops, which is certainly a crude simplification that limits the reliability of simulations of mixed-phase and cold clouds with bin microphysics models.

Remote-sensing Limitations

Ice water clouds are far more difficult to characterize from space than liquid water clouds. One level of complexity is added by the shape of ice crystals. In absence of that information, the lookup tables for visible/near-infrared retrievals are usually based on some mixture of crystal shapes. Since the composition of the actual cloud differs from the assumed shape, large errors in both effective radius and optical thickness may be introduced, which then propagate into the ice water path retrieval as well. There is currently no technique available to retrieve crystal shape from satellites. Less problematic is phase discrimination, though this is often merely qualitative and not always consistent between different techniques (e.g., MODIS and POLDER), especially for mixed-phase clouds. With the launch of CloudSat and CALIPSO, it has become possible to validate the cloud-top height products from infrared imagery, and first biases have been identified. However, algorithms for retrievals of liquid water path, crystal size, phase, and the vertical cloud structure remain to be validated with *in-situ* aircraft measurements.

In-situ Measurement Limitations

In-situ probes have persistent problems in ice clouds. This is reflected in the ongoing debate about the existence of small crystals, which have been detected through combinations of (bulk) measurements of ice water content, crystal number, and extinction (e.g., Garrett et al. 2003; Ström et al. 1994). Such small crystals change the warming/cooling properties of cirrus clouds substantially

(Heintzenberg et al. 1995). For crystal-size spectrometers such as the Cloud Aerosol and Precipitation Spectrometer (CAPS), it has been shown in wind tunnel tests that at least part of the small crystals may be attributed to particle breakup ("shattering") on the tips of the probe. A related problem is a major jump in the CAPS-size distributions, which occurs at the transition from the small-size range (forward scattering probe 2–50 μm diameter) to the medium-size range (cloud imaging probe, 50–1500 μm). This jump makes it impossible to derive meaningful properties, such as effective crystal radius from the size distributions, and can most likely be explained not only by shattering alone but through an association with multiple forward scattering peaks from multifaceted ice crystals, which are counted separately by the forward scattering probe but not by the imaging probe. A further practical problem is that there is no agreed upon standard on how to calibrate the probes and process the data and correct for various effects. Often, the probes are not calibrated at all during field experiments. Therefore, it is not unusual that wing-by-wing comparisons of size distributions differ by orders of magnitude in certain size ranges. Bulk probes (e.g., CVI and Nevzorov probes for ice water content and nephelometers and extinctiometers for extinction measurements) help constrain the size distribution obtained from size spectrometers and, when combined, provide an independent measurement for crystal effective radius. However, they may also suffer from shattering.

New techniques, such as Gayet's cloud nephelometer (Gayet et al. 1997) or polarized measurements, enable cloud phase discrimination and can therefore be used in mixed-phase clouds.

Verification of Remote Sensing Using *In-situ* Aircraft

It is essential to verify remote-sensing techniques, both ground- and satellite-based, with *in-situ* aircraft measurements and a clear specification of errors and limitations. For example, satellite retrievals of cloud phase, effective crystal radius, cloud ice water content, and precipitation must be validated taking into account horizontal and vertical inhomogeneities. In the absence of these comparisons, it is rarely possible to attach robust error bars to remotely sensed cloud parameters.

The Confounding Influence of Meteorology

Description of the Problem

Although it is crucial for our understanding of climate change to identify the influences of anthropogenic aerosol and global warming on cloudiness, this is a challenging problem since natural meteorological processes exert such a dominating control over cloud properties and precipitation. For example, in regions of persistent marine stratocumulus, over 80% of the seasonal/geographical variance in cloud cover can be explained by variability in lower tropospheric stability (LTS) (Klein and Hartmann 1993; Wood and Bretherton 2006). Despite the potentially strong susceptibility of marine stratocumulus clouds to aerosol perturbations that result from their low natural droplet concentration and intermediate optical thickness (Platnick and Twomey 1994), on scales larger than ship tracks, it is difficult to identify clearly aerosol effects amid the impacts of variations in LTS, temperature advection, and divergence. Moreover, because changes in aerosol concentration are associated with changes in the large-scale flow (i.e., the meteorology), even perfect observations of coincident clouds and aerosols are insufficient to determine the response of the clouds to aerosols alone (e.g., Mauger and Norris 2007). The extent of this covariance between meteorology and aerosols is poorly understood.

Another reason why large-scale aerosol effects on clouds have not been clearly discerned is that competing mechanisms, each of which is inadequately understood, may be involved. Straightforward physical theory and extensive observations indicate that increased particle number concentration will lead to a higher droplet number concentration at cloud base (e.g., Warner and Twomey 1967; Coakley et al. 1987; Breon et al. 2002), but beyond this, little is known with confidence. For example, hypotheses exist by which aerosols may increase cloud water through suppression of precipitation or decrease cloud water through enhancement of evaporation. These effects will, in turn, change the cloud dynamics (Feingold and Siebert, this volume). Currently, the important research questions are whether (and if so, how) aerosol-induced changes in droplet number affect cloud albedo, cloud cover, and precipitation. For a coupled system, it is both important and difficult to isolate the effects of a single variable.

Strategies for Separating Aerosol and Meteorological Effects on Clouds

Inadequate Methods

Constant liquid water path (LWP). One common approach to "control" for meteorology in efforts to observe the effects of aerosols on cloud albedo in complex systems (i.e., those other than plume studies as typified by ship tracks) is through binning cloud observations by LWP. However, LWP is an emergent property that results from the interplay between meteorological conditions and cloud microphysics, on scales down to that of a cloud element (e.g., a stratocumulus cell). In fact, changing LWP may in some cases be the primary mechanism for aerosol-induced albedo changes. Thus, it is impossible to measure how clouds are being changed by an aerosol perturbation if one eliminates the most important response by design.

Common source region. Another way to distinguish meteorological and aerosol effects on clouds and precipitation in small regions is to segregate the data according to air mass back trajectories from different geographical regions. For example, in eastern Canada, when trajectories are from the south, the air masses come from heavily polluted areas, whereas with northerly trajectories, the air masses pass over areas with fewer aerosol and anthropogenic gas phase source regions. The limitation of this method is that air masses from different regions are usually associated with different meteorological conditions. For instance, air parcels over the ocean with higher aerosol concentration, which are the result of recent contact with a continental region, are likely to be drier and thus have fewer clouds on account of meteorology.

Useful Methods

Small-scale aerosol perturbations. For small-scale effluent plumes (e.g., ship tracks), the assumption of constant meteorology between impacted and unimpacted regions of the cloud appears to be robust in at least some cases (e.g., Radke et al. 1989). Extending the results of such studies to larger scales, however, is not straightforward because the diffusion of plumes and corresponding aerosol effects over larger areas occur at the same timescale as do changes in meteorology.

Meteorological proxies. In addition to identifying source regions by back trajectory, the large-scale meteorological conditions experienced by clouds prior to the time of observation can be obtained from reanalyses and measurements. It is essential to use the history of meteorological forcing rather than data obtained solely from the time of the cloud observation because cloudiness responds to changes in forcing with a time lag. The potential impact of aerosols on cloud properties can be estimated from a large sample of aerosol and cloud observations by comparing average cloud properties between subsets with similar meteorological history but differing aerosol amounts. For example, the study of Mauger and Norris (2007) found that satellite-reported large aerosol optical depth (AOD) was associated with both large low-level cloud fraction and large LTS during the previous three days. Moreover, the difference in cloud fraction between subsamples with large and small AOD, but equivalent LTS, was less than the difference in cloud fraction for subsamples with large and small AOD and unconstrained LTS, thus illustrating the confounding role of LTS in controlling cloud amount. Note that compositing on parameters like LTS only establishes a minimum value for the magnitude of meteorological influence (and an upper limit for aerosol influence), since other unidentified processes may also affect cloudiness. Failure to account adequately for all meteorological effects will lead to an overestimate of the influence of aerosols on cloudiness.

Cloud properties surveyed in diverse aerosol environments. The observation of similar cloud properties across many regions of the world, where diverse aerosol environments are expected, can be used to reject the hypothesis that aerosols are a major influence. For example, the statistics of cloud crystal number concentration were found to be similar in experiments over Canada, the United Kingdom, and the southern hemisphere. Specifically, Korolev et al. (2000), Gultepe et al. (2001), Field et al. (2005), and Gayet et al. (2006) all showed similar ice particle (effective diameter > 100 μm) concentrations around the world for clouds with temperatures warmer than about –35°C. Presumably, these experiments involve very different aerosol environments. Thus, one could infer that the concentration of large ice particles in clouds warmer than about –35°C is not sensitive to aerosol variations. The observation of different cloud properties across regions with diverse aerosol environments is not a sufficient condition for demonstrating an aerosol influence, however, since different regions have different meteorology.

Discussion

What is the appropriate null hypothesis? Many observational studies of aerosol–cloud relationships have mistakenly inferred causation from correlation; that is, if a change in aerosol amount is seen to occur with a change in cloudiness, then aerosols necessarily produced the cloud change. For example, Nakajima and Schulz (this volume) provide an extensive literature review of the apparent sensitivity of cloud properties to aerosols calculated via linear regression with aerosol loading as the sole independent variable. Considering the dominating impact of meteorology on cloudiness, a superior "null" hypothesis would be that aerosols and clouds do not interact, and that observed correlations only arise from their common origins and trajectories in the general circulation. This, however, is too strict a criterion since no predictors of clouds from either observation, forecast, or reanalysis are currently accurate enough to rule out such a "null" hypothesis. Nevertheless, the effects of meteorological factors known to control strongly cloudiness as well as satellite retrieval and other measurement biases must be eliminated before attributing cloud changes to aerosols.

Other confounding aspects of the problem exist. It is possible for clear-sky radiative effects of aerosols to alter directly the meteorology of a region by changing surface and atmospheric radiative absorption (Wendisch et al. 2008). This, in turn, could weaken or strengthen circulations on various scales and consequently change cloudiness. Observational analyses that controlled for meteorology would fail to reveal an indirect effect of this nature. To test such a hypothesis, examination of the sensitivity of atmospheric circulation to an imposed radiative forcing in a global or regional climate model seems to be the best option. One must be careful in interpreting the results, however, since inadequate boundary layer, convection, and cloud parameterizations or inappropriate boundary conditions (e.g., fixed sea surface temperature in long simulations) may cause the model to behave in an unrealistic manner.

Evidence for Large-scale Aerosol Impacts on Cloud Albedo, Cloud Amount, and Precipitation Remain Ambiguous

Evidence for an aerosol influence on clouds at large scales has recently emerged from satellite studies (see Table 17.2 in Nakajima and Schulz, this volume), but these studies have significant limitations. First, there is inadequate consideration of meteorology as an alternative explanation for the reported aerosol–cloud correlation; another concerns the difficulty in distinguishing aerosols from clouds by remote sensing (Martins et al. 2002; Kaufman et al. 2005; Charlson et al. 2007), which could lead to artificial positive correlations between retrieved AOD and cloud fraction in satellite scenes. Studies of aerosol–precipitation relationships suffer from similar shortcomings, and Ayers and Levin (this volume) describe how meteorology is a more plausible explanation for precipitation changes than an aerosol microphysical mechanism is. Moreover, despite decades of research, there is no robust statistical evidence for the efficacy of intentional weather modification through aerosol impacts on cloud microphysics (i.e., "cloud seeding"; see Cotton, this volume). Recent evidence from satellites of cloud perturbations over the remote oceans downwind of fuming volcanoes (Gassó 2008) is suggestive of large-scale aerosol impacts on cloud albedo, yet the absence of data on aerosol concentration means that an aerosol mechanism cannot be conclusively established or quantified.

The potential influence of aerosols on cloud properties can also be assessed by looking for long-term trends in cloudiness in regions that have experienced long-term trends in aerosols. This type of analysis has the benefit of mostly (but not completely) averaging out the confounding effects of synoptic meteorological variability. If aerosols influence cloudiness significantly, and aerosols have strongly increased or decreased in a particular area over several decades, then one would expect to see a physically consistent change in cloudiness over the same time period. For example, the presence of absorbing aerosols in a cloud layer has been hypothesized to decrease cloud fraction because of the absorption of solar radiation and commensurate reduction of humidity (Ackerman et al. 2000). The observation that no long-term reduction of cloud cover has occurred over the Northern Indian Ocean, despite the large increase in absorbing aerosols, indicates that this aerosol mechanism has not actually had a measurable influence on cloud cover in this region (Norris 2001). As another example, an increase in the number concentration of hygroscopic aerosols has been hypothesized to enhance cloud fraction by reducing precipitation loss of cloud water (Albrecht 1989). The observation that cloud cover decreased over Europe during the time period of increasing sulfate emission and increased over Europe during the time period of decreasing sulfate emissions indicates that this aerosol mechanism has not had a measureable influence (Norris and Wild 2007). One observational study by Krüger and Graßl (2002) reported a decrease in cloud albedo over Europe during a time of decreasing sulfate aerosol, but they did not investigate whether this might have instead been the result of meteorology. The North Atlantic Oscillation exhibited a substantial trend during this time period, which could have affected cloudiness over Europe.

A Major Motivation to Study Aerosols and Climate Is to Understand 20th-century Temperature Change

Global mean climate forcing from anthropogenic aerosols is in the process of leveling off and, in strong contrast to forcing from greenhouse gases (GHGs), aerosol forcing is not expected to increase dramatically during the 21st century (Penner et al. 2001). If these projections are correct, then most of the aerosol perturbation to climate that humans will ever impose has already happened.

In global climate simulations of the 20th century, aerosol forcing appears to compensate for differences in climate sensitivity (Kiehl et al. 2007). Better understanding of aerosol forcing has the prospect of improving our knowledge of climate sensitivity and thus our forecasts of future climate change, where the anthropogenic forcing will be increasingly dominated by GHGs.

Anthropogenic aerosols exert regional perturbations with large temporal variations (both increases and decreases). Anthropogenic GHGs, in contrast, exert global-scale perturbations that continuously increase over time. Given these very distinct patterns, it should be possible to separate these two types of perturbation to the climate system. In particular, focusing on regional variations would seem to be a plausible approach for investigating aerosol effects. Although the aerosol is unlikely to exhibit substantial growth in the future, in terms of global mean loading, it will undoubtedly continue to exhibit strong regional signals in areas where it is either growing or diminishing. Nevertheless, the difficulty of separating aerosol from meteorological effects, as previously discussed, must not be overlooked.

Since Some Global Warming Signals Are Now Becoming Evident, We Should Search for Associated Cloud Changes

As radiative forcing by anthropogenic GHGs becomes increasingly strong, we should expect to see specific changes in the Earth's climate system in addition to the rise in global mean temperature. Recent studies have accordingly reported a decline in the summertime Arctic sea ice area (Nghiem et al. 2007), a poleward shift of the Hadley circulation (Fu et al. 2006), a rise in tropopause height (Thomson et al. 2000), and trends toward more precipitation in wet places as well as less precipitation in dry places (Trenberth et al. 2007). Theoretical considerations and global climate models predict that all of these phenomena will occur with global warming, and they illustrate some of the changes we will experience over the coming decades. Considering the great importance of cloud feedbacks to climate sensitivity and our ignorance concerning them, it is essential to investigate the observational record for cloud changes that are likely to be associated with global warming. Although this has been previously difficult because of the large uncertainties in data and substantial natural variability, we expect that cloud signals will become more distinct as greenhouse radiative warming becomes increasingly globally dominant over aerosol radiative cooling. Moreover, distinct regional trend patterns will be more robust than the global average cloud changes. For example, one likely cloud change is a shift of frontal clouds accompanying the poleward retreat of the storm tracks. Cloud simulations in global climate models are unlikely to provide reliable guidance since current models represent cloud processes poorly and inconsistently. Nevertheless, global climate models can provide robust information on particular changes in atmospheric structure and circulation associated with global warming. Reasoning from theory, we can then hypothesize which cloud changes would be physically consistent with these. The identification of such cloud changes in the observed record would be useful for assessing which global climate models are likely to have the most realistic cloud parameterizations for global warming simulations. This, in turn, would help constrain the range of uncertainty in our estimates of cloud feedbacks and climate sensitivity. Although we know many meteorological factors that govern cloudiness (e.g., low-level cloud fraction and lower tropospheric static stability), it is not necessarily clear which factors will be the most influential under global warming. Thus, improved theories are needed.

Clouds interact strongly with mean thermodynamic properties and with organized circulations from the scale of individual convective cells, through mesoscale organization, to synoptic and large scales. These circulations are potentially sensitive to direct effects of GHG and aerosol changes as well as to the changes in temperature and humidity that are expected in association with global warming. Some circulations, such as the mesoscale circulations within convective cloud complexes in the tropics are strongly coupled to cloud microphysical processes (Houze 1982). Changes in the location or intensity of extratropical storms and associated storm tracks, which may occur in response to global warming, could have significant effects upon mean cloud properties and their effect upon climate.

One effect that is widely regarded to be important is the change in dry static stability that will accompany the decrease in temperature lapse rate in response to warming of the tropical sea surface (Knutson and Manabe 1995). The enhanced stability, coupled with radiative cooling rates which are not expected to change much, may decrease the rate of overturning associated with the Hadley and Walker circulations of the tropics. At the same time, the increase in surface saturation humidity associated with the warming will increase the amount of latent energy available in surface air. It is unclear how these two influences will change the structure, frequency, and intensity of tropical convective cloud systems. In addition, the changed large-scale circulation could influence the properties of trade cumulus and stratocumulus clouds, which are known to have a significant influence on the energy balance of Earth. None of these cloud effects are well understood, and it is likely that global climate models do not incorporate them accurately. It is unclear how these macroscale changes will interact with cloud changes associated with changes in aerosol loading, but they are likely to be as important and are not any better understood (cf. Bretherton and Hartmann, this volume).

What data and approaches are available to address cloud macroscale effects? The primary means to address these issues is to gain a complete understanding of the relevant processes, to develop models of these processes, and to test these models rigorously against data. Field programs to address these issues have been mounted in the recent and distant past (e.g., for tropical clouds GATE, TOGA-COARE, KWAJEX). Satellite data are providing a rich dataset which can be used to validate certain aspects of the cloud simulation. For example, MODIS cloud data provide a good representation of the optical properties of convective clouds that are important for the Earth's energy balance, such as the area covered by clouds with particular optical depths and cloud-top temperature pairings. These data can be provided on scales (~5 km) that can be used to test these properties, as they are simulated by a hierarchy of models, including models with high spatial resolution, often called cloud-resolving models. CloudSat and CALIPSO are producing information on the internal vertical structure of these clouds. AIRS is providing global analyses of water vapor and temperature profiles that are coincident with MODIS and CloudSat/CALIPSO data. Reanalysis datasets are providing estimates of global six-hourly analyses of large-scale vertical and horizontal winds that are much more accurate and reliable than previous estimates, even over remote ocean areas. Vigorous testing of emerging models with emerging datasets should lead to progress in understanding and simulating macroscale cloud processes that are critical to climate change.

Existing Cloud, Aerosol, and Radiation Records Need to be Reprocessed, and the Continuity of the Current Observing System Must be Guaranteed

It is very challenging to extract decadal trends from the current observational record for aerosols, clouds, and broadband radiative flux (energy balance). Prospects for improved understanding include (a) reanalysis of existing data, where the possibility exists of correcting known problems; (b) regional-scale analysis, where large secular changes in aerosol emissions and climate change signals are known; and (c) maintaining and expanding high-quality observations into the future.

Since several decades may be needed for human-induced trends in cloudiness and radiation flux to rise above natural variability, it is essential that we improve the satellite record as well as sustain and develop further our present observing system. Currently, the available satellite cloud and radiation datasets with lengthy records, sometimes back to around 1980, are generally not useful in analyzing long-term changes that result from the presence of many inhomogeneities and artifacts. These arise from changes in satellite instrumentation, degradation of sensors, drifts of satellite orbits, and shifts in satellite view angle. Most data originated from weather satellites, which were never intended to be an accurate and stable observing system. Careful processing must therefore be applied to produce a homogeneous record (as is the case for nearly all other climate measurements). Although the amount of resources needed for this task is very small compared to the cost of a satellite, the provision thus far has been inadequate. Most previous corrections and adjustments to satellite records have had an empirical and statistical origin, whereas it would be preferable to develop better physical models for the algorithms. New information gained from recent satellites (e.g., CERES, MODIS, CloudSat, and CALIPSO) will be useful for interpreting the more limited measurements of the older record, and multiple satellite datasets can be brought together to provide a "best guess" of trends and variability. Even if global trends in the reconstructed satellite record suffer from large uncertainties and unknown systematic biases, regional trends are likely to be more robust. The importance of clouds and aerosols in our understanding of the climate provides a strong motivation to overcome the difficulties in getting the best use out of the satellite radiance record.

It is clearly important to make the best use of the existing archive of satellite data on clouds. Reprocessing this data with new or improved algorithms is one approach. Another idea with great potential is to assimilate the satellite radiances into numerical weather prediction (NWP) models to produce reanalyses that include clouds. The existing reanalyses from NWP centers, such as the European Centre for Medium-Range Weather Forecasts (ECMWF) and National Centers for Environmental Protection (NCEP), have created consistent long-term datasets for dynamic and thermodynamic variables. Until recently, NWP centers have avoided the assimilation of cloudy radiances, and the representation of the hydrological cycle has been only partially successful. However, methods for addressing cloud parameters (e.g., radiances) are now being introduced, and this suggests the possibility of using the reanalysis methodology to add clouds to the list of variables and to perform a complete reanalysis of the hydrological cycle. Among other things, this would require knowledge of the error characteristics of the data and a reasonably good simulation of clouds in the model. Although challenging, this objective could deliver datasets with a consistent representation of both clouds and the meteorology, which has thus far been lacking. Currently, data assimilation systems are also being used to provide analyses of the global distribution of aerosols (work in progress at ECMWF). Adding clouds would enable the relationships between the meteorology, clouds, and aerosols to be studied within a consistent framework—a necessary step toward understanding the global impact of aerosols on clouds and climate.

Top-of-atmosphere (TOA) radiation budget for detection of cloud feedback requires the ability to detect global trends on the order of 0.3 W m^{-2} per decade (Loeb et al. 2007). This is extremely challenging and yet critical if we are to improve our knowledge of climate sensitivity. In addition to continuing the satellite programs dedicated to this objective (ERBE/CERES), it makes sense to pursue independent approaches:

1. Monitoring ocean heat content provides a similar capability to detect interannual variations in global energy budget through an entirely independent method.
2. The Earthshine method appears promising, even though initial implementation at a single site may have been flawed. This is a low-cost method such that maintenance of a consistent record over the next three decades (or longer) would not be difficult. Expansion to additional sites (to provide more nearly global coverage) and analysis by other groups would have potentially high value.
3. Whole-earth monitoring from Lagrange point L1 (DSCOVR mission) would provide a continuous, well-calibrated albedo proxy with potential diagnostic value.

Surface sites provide direct measurements of many variables relevant to the aerosol–cloud problem. Variables currently measured include the surface radiation budget (e.g., BSRN, GEBA) and remote sensing of AOD and Ångström coefficient (e.g., AERONET). There are a few sites with a comprehensive range of instruments (e.g., ARM, Cloudnet) that extend the list of variables to include active sensing of the vertical structure of clouds and aerosol layers. Although the distribution of these sites is extremely heterogeneous and is limited to land, they provide a wealth of direct measurements that are used in process studies to detect trends in surface radiation, to support field studies that evaluate the simulation of clouds and aerosols in operational and climate models, and as "ground truth" for satellite retrievals. Compared with new satellites, these sites are relatively inexpensive to establish and maintain. Continuation and indeed expansion of these sites for the coming decades should therefore be a high priority in planning a well-balanced observation system for monitoring aerosols and clouds.

New Strategies Are Required to Observe Key Variables More Accurately for Understanding Perturbed Clouds in the Climate System

Because of the complex and coupled nature of the Earth system, scientific progress requires tight integration of modeling and observational efforts. Both models and observations exist across an extremely wide range of scales, and integration across these different scales is probably the most challenging aspect of developing a coherent scientific strategy in climate research. In addition, there are the usual challenges of developing effective collaborations across various subdisciplinary divides (e.g., *in-situ*/remote sensing, aerosol/cloud, or laboratory/field).

A general strategy for fostering a fully integrated approach would involve (a) modelers framing specific (but realistic) data needs, (b) designing measurement and data archiving protocols that are tailored to modeling needs, (c) getting experimentalists from different groups and communities to combine and synthesize their results intelligently, and (d) getting modelers to actually use the data products. These goals can be facilitated by focusing research efforts on broadly defined domains in a way that attracts the involvement of experimentalists and modelers from many subdisciplines.

In terms of detecting aerosol effects on the global cloud system, some strategic principles include (a) looking for regional/temporal signatures where large aerosol variations are known to exist, (b) developing testable hypotheses, (c) including model validation in experimental design, and (d) combining chemical/physical approaches into one integrated strategy. Aerosols are very variable in time, so it is inherently challenging to compute their integrated effect on clouds and climate. This challenge constitutes a wonderful opportunity for integrating modeling and observational approaches at many scales.

Another strategic challenge is to achieve "critical mass" in field experiments. For example, if we do not get good observations of vertical profile of cloud microphysics, we really have nothing with which to build or constrain a model; that is, there is no way to understand the controlling processes. Putting this in context of large-scale observables and bringing in detailed modeling capability requires a major, coordinated effort.

Strategies for Integrating Modeling/Measurement Activities

One successful strategy is to have modelers in the field during field campaigns. This builds relationships, gives modelers a chance to learn what types of data are realistic, and gives measurement people a chance to become familiar with modeling capabilities and needs. This strategy works well for detailed process models, but a more challenging problem consists of engaging the global climate modeling community. As fully coupled models, GCMs are difficult to run for case studies. Recent projects like VOCALS and AMMA are attempting to integrate modelers working across a full range of temporal and spatial scales. For example, large-scale climate links associated with low clouds affecting ENSO.

Improving GCMs Is Not Straightforward

It is very easy to identify deficiencies in GCMs and yet very hard to correct them. One of the sad aspects of GCMs is that an improved parameterization based on successful field measurements may result in a worse simulation. Why? Implementing any change in a GCM will effect multiple aspects of the model, often degrading model performance in unexpected ways. Every GCM group has wonderful counterintuitive examples of such effects. Thus, implementing a significant change requires that the model be re-tuned to ensure that it continues to simulate successfully known aspects of the climate system. Tuning a GCM, in turn, is a complex, multidimensional optimization problem. Given the importance of GCMs to climate science (not to mention society), it is rather shocking how few people are working on developing and implementing model improvements.

Need for Consistent Global Data

Measurement protocols must be globally consistent to be able to identify geographical differences, trends over time, and to allow valid interpretation of satellite data. Such consistency is facilitated by coordinated monitoring networks (Kinne et al., this volume) that include uniform and rigorously enforced quality control and data analysis procedures. The cost and effort required to establish and maintain such programs is not trivial, although they generally amount to a small fraction of the cost of launching a satellite. There is thus a need for improved financial support for ground-based networks and integrated data analysis.

A major gap in current scientific strategies involves extrapolating from detailed local measurements to consistent global datasets. Effective new strategies will need to include satellites, models spanning the entire range of scales, and mathematical methods to account for varying measurement/retrieval uncertainties as well as spatiotemporal variability. One example is the need for consistent, global aerosol datasets. "The complexity of the aerosol–climate problem implies that no single type of observation or model is sufficient to characterize the current system or to provide the means to predict aerosol impacts in the future with high confidence. Consequently, information must be drawn from multiple observational and theoretical techniques, platforms, and vantage points, and strategies that explicitly plan for the integration and interpretation of the various components need to be designed" (Diner et. al. 2004, p. 1492).

Scaling up the Results of Field Studies to Large Models

Using measurements to impose constraints on models that exist at the same scale is difficult enough. For example, constraining large eddy simulation (LES) models with *in-situ* measurements runs into the difficulty that data are needed simultaneously throughout the domain. It is even more challenging to devise ways of using measurements obtained at one scale to constrain a model at a larger scale. However, this is essential in the development of regional- and global-scale models. At present, there is some effective collaboration between the global climate modeling and measurement communities, a leading example being the GEWEX Cloud System Study (GCSS). However, for the perturbed-cloud problem, there is no framework at present. An obvious suggestion, therefore, would be to establish a GCSS working group on perturbed clouds.

Specific Examples and Proposals

Weekly cycle of aerosols. Look for cloud changes associated with the weekly cycle of industrial aerosol emissions (the "weekend" effect). A weekly cycle of diurnal temperature range has been observed in many of the world's industrialized regions, suggestive of aerosol effects and possibly involving aerosol–cloud interactions (Forster and Solomon 2003; Gong et al. 2006). Research aimed at better elucidating the physical mechanisms would be useful. Advantages of this technique are that it involves short timescales (no need to observe long-term trends) and that it can make use of current observational capabilities.

Regional aerosol studies. Detection of aerosol effects becomes less daunting if we go to the regional scale, where the perturbation is large. Coherent regional patterns should be easier to detect and attribute to causes.

Amazon smoke investigation. The aerosol perturbation associated with dry-season biomass burning over Amazonia is massive, yet highly localized in some cases. The forest provides a uniform environmental context such that there should be minimal meteorological differences between impacted and unimpacted locations. In addition, there are year-to-year variations, since the amount of burning varies with socio-economic factors like the price of soybeans. A problem that must be confronted (in this and other regions) is sorting out the direct thermal effect of aerosol from microphysical effect on clouds.

Recent intensification of industrial emissions in China. Currently, industrial emissions over China constitute a massive aerosol perturbation that has been growing over the past several decades. Several datasets exist for assessing changes in surface radiation and cloudiness in conjunction with the aerosol increases. One suggestion is that aerosol particles are so absorbing that they change deep convection. Investigating this poses an interesting methodological problem. One would want to compare model simulations of the convection with and without the observed aerosol (or with the varying aerosol concentrations observed in the different cases). However, mixed-phase convection is so poorly understood that it is questionable whether model sensitivities to even large aerosol perturbations could be trusted. Investigating aerosol effects on cloud cover would have to take care that cloud cover observations were not directly affected by high aerosol concentrations (Charlson et al. 2007). A more general problem is that the direct thermal effects of the aerosol could alter the meteorological context (e.g., affect the strength of the monsoon circulation or alter convective instability by changing the height of latent heat release). This could lead to cloudiness changes, which need to be separated from aerosol-induced microphysical changes.

Ice formation problem. A sustained laboratory program is needed to identify precisely which particles act as ice nuclei and by which mechanisms. Thereafter it should be possible to design a new generation of *in-situ* airborne devices to identify them. Extensive field programs with airborne *in-situ* sensors for ice nuclei and ice particle size spectra, coupled with ground-based or airborne multi-parameter Doppler cloud radars and lidars sensing the same parameters, should provide a more complete description of the development of frozen precipitation.

Ice formation problem. It is known from field experiments and cloud modeling studies that secondary ice multiplication (e.g., splintering of existing ice crystals or the so-called Hallett–Mossop effect) are important factors in cloud development and for the initiation and intensity distribution of precipitation. Only a few laboratory investigations exist on secondary ice formation. More laboratory experiments are needed to investigate and quantify secondary ice formation, if possible for relevant ice crystal shapes and sizes, under conditions (temperatures, saturations, and turbulence) resembling those encountered in real clouds. Such experiments may be conducted with single droplets and ice crystals or in cloud simulation chambers.

Precipitation. Integrate observed changes in precipitation (Takayabu and Masunaga, this volume) with observed changes in aerosol and other potentially controlling factors.

LES validation for convective clouds. Cumulus updrafts play a central role in the vertical transport of water, momentum, trace chemicals, and heat. Still, there are inadequate observations of the statistical distribution of buoyancy, vertical velocity, and condensate in cumulus updrafts, as a function of cloud depth and aerosol profiles, and surface forcing environment. Such measurements are key constraints on LES and cumulus parameterizations. They are challenging to make from aircraft because of the associated sampling challenges and wetting issues, but shortwave-length radar-based remote-sensing techniques may provide a new opportunity to gather such statistics.

Cloud overlap representation in GCMs. Currently, GCMs predict cloud fraction in each vertical level. Uncertainty as to how to distribute this fractional cloud, both as a function of horizontal resolution and as a function of the distribution of clouds in the other layers of the model, gives model developers considerable freedom in distributing the modeled cloud amount so as to satisfy the radiative constraints at the top of the atmosphere. Experience with different GCMs shows that although they may have large layer-by-layer differences in cloud fraction and cloud amount, suitable choices for the cloud overlap assumption enables each to match the top of the atmosphere radiative constraints. Therefore, providing measurements of cloud overlap would provide further and valuable constraints on the model.

Observational study of cloud lifetime. Satellite measurements with high spatial resolution typically lack temporal resolution. Being able to measure the temporal evolution of clouds as well would provide a valuable constraint on process models of clouds and their environmental interactions. In this regard, for instance, a geostationary observatory that focused a large telescope on particular cloud scenes (to complement field measurements or *in-situ* studies) would be invaluable.

Spectrally resolved radiation budget measurements from space. Stable and accurate observations of changes in the Earth's radiation field over several decades are needed to quantify the forcing and radiative response of the climate system. For the purpose of quantifying trends in the Earth's radiation field, spectrally resolved measurements are especially useful for separation and attribution of the radiative effects from climate forcing and climate response (Goody et al. 1998). The small magnitude of the effects and long integration times required for detection together imply that very stable, absolutely calibrated satellite instruments are required. Several groups are now exploring satellite radiometers designed to detect the radiative forcing, thermal response, and radiative feedbacks in the Earth's climate system. Spectral radiometers can be developed with absolute calibration against traceable standards to insure that trends in the observations are as free as possible of instrumental artifacts (Keith et al 2001; Anderson et al. 2004). Potentially both the infrared and ultraviolet/visible/near-infrared radiation fields could be measured with such instruments. Although the feasibility for detection of infrared GHG forcing has been amply illustrated in modeling and satellite studies (e.g., Haskins et al. 1997), it is important to recognize that the utility of the infrared measurements or detection of longwave cloud feedbacks remains unproven (Leroy et al. 2008). The advantages of the ultraviolet/visible/near-infrared radiation data for detection and estimation of shortwave forcings and feedbacks have not been demonstrated in detail.

Earth observations from Lagrange point L1. Satellites deployed at Lagrange point L1, which are designed to measure the radiation emitted by the sunlight side of the Earth (e.g., the Deep Space Climate Observatory, DSCOVR; Valero et al. 1999), could provide valuable long-term measurements. DSCOVR would include several single-pixel NISTARs with a ground-based calibration chain tied directly to primary national standards. These instruments would measure the total solar, near-infrared, and infrared radiance field emitted by the Earth in the direction of Lagrange point L1. The inherent stability and traceable calibration of these instruments are ideally suited for the detection of secular trends in the Earth's short- and longwave radiation.

Acknowledgments

T. L. Anderson acknowledges support from the U.S. National Science Foundation (grants ATM-0601177 and ATM-0205198).

References

Ackerman, A. S., O. B. Toon, D. E. Stevens et al. 2000. Reduction of tropical cloudiness by soot. *Science* **288**:1042–1047.

Albrecht, B. A. 1989. Aerosols, cloud microphysics, and fractional cloudiness. *Science* **245**:1227–1230.

Anderson, J. G., J. A. Dykema, R. M. Goody, H. Hu, and D. B. Kirk-Davidoff. 2004. Absolute spectrally-resolved radiance, thermal radiance: A benchmark for climate monitoring from space. *J. Quant. Spectrosc. Rad. Trans.* **85**:367–383.

Andreae, M. O. 2007. Aerosols before pollution. *Science* **315**:50–51.

Andreae, M. O., and Rosenfeld, D. 2008. Aerosol-cloud-precipitation interactions. Part 1. The nature and sources of cloud-active aerosols. *Earth Sci. Rev*. **89(1–2)**:13–41.

Bates, T. S., T. L. Anderson, T. Bond et al. 2006. Aerosol direct radiative forcing over the Northwest Atlantic, Northwest Pacific, and North Indian Oceans: Estimates based on *in-situ* chemical and optical measurements and chemical transport modeling. *Atmos. Chem. Phys*. **6**:175–362.

Breon F.-M., D. Tanre and S. Generoso. 2002. Aerosol effect on cloud droplet size monitored from satellite. *Science* **295**:834–838.

Charlson, R. J., A. S. Ackerman, F. A.-M. Bender, T. L. Anderson, and Z. Liu. 2007. On the climate forcing consequences of the albedo continuum between cloudy and clear air. *Tellus* **59**:715–727.

Coakley, J. A., Jr., R. L. Bernstein, and P. A. Durkee. 1987. Effects of ship-track effluents on cloud reflectivity. *Science* **255**:423–430.

Diner, D. J., A. Ackerman, T. P. Anderson et al. 2004. Progressive Aerosol Retrieval and Assimilation Global Observing Network (PARAGON): An integrated approach for characterizing aerosol climatic and environmental interactions. *Bull. Am. Meteor. Soc*. **85(10)**:1491–1501.

Field, P. R., Hogan, R. J., Brown et al. 2005. Parametrization of ice particle size distributions for midlatitude stratiform cloud. *Q. J. Roy. Meteor. Soc.* **131**:1997–2017.

Forster, P. M., and S. Solomon. 2003. Observations of a "weekend effect" in diurnal temperature range. *PNAS* **100**:11,225–11,230.

Fu, Q., C. M. Johanson, J. M. Wallace, and T. Reichler. 2006. Enhanced midlatitude tropospheric warming in satellite measurements. *Science* **312**:1179.

Garrett, T. J., H. Gerber, D. G. Baumgardner, C. H. Twohy, and E. M. Weinstock. 2003. Small, highly reflective ice crystals in low-latitude cirrus. *Geophys. Res. Lett.* **30(21)**:2132.

Gassó, S. 2008. Satellite observations of the impact of weak volcanic activity on marine clouds. *J. Geophys. Res.* **113**:D14S19.

Gayet, J. F., V. Shcherbakov, H. Mannstein et al. 2006. Microphysical and optical properties of midlatitude cirrus clouds observed in the southern hemisphere during INCA. *Q. J. Roy. Meteor. Soc.* **132**:2719–2748.

Gayet, J. F., O. Crépel, J. F. Fournol, and S. Oshchepkov. 1997. A new airborne polar nephelometer for the measurements of optical and microphysical cloud properties. Part I: Theoretical design. *Ann. Geophys.* **15**:451–459.

Gong, D.-Y., D. Guo, and C.-H. Ho. 2006. Weekend effect in diurnal temperature range in China: Opposite signals between winter and summer. *J. Geophys. Res.* **111**:D18113.

Goody, R., J. Anderson, and G. North. 1998. Testing climate models: An approach. *Bull. Amer. Meteor. Soc.* **79**:2541–2549.

Gultepe, I., G. A. Isaac, and S. G. Cober. 2001. Ice crystal number concentration versus temperature. *Int. J. Climatol.* **21**:1281–1302.

Haskins, R. D., R. M. Goody, and L. Chen. 1997. A statistical method for testing a general circulation model with spectrally resolved satellite data. *J. Geophys. Res.* **102**:16,563–16,581.

Heintzenberg, J., Y. Fouquart, A. Heymsfield, J. Ström, and G. Brogniez. 1995. Interactions of radiation and microphysics in cirrus. In: Clouds, Chemistry and Climate. Series I: Global Environmental Change, ed. P. J. Crutzen and V. Ramanathan, pp. 29–54. Berlin: Springer.

Houze, R. A., Jr. 1982. Cloud clusters and large-scale vertical motions in the Tropics. *J. Meteor. Soc. Jpn.* **60**:396–410.

Kaufman, Y. J., L. A. Remer, D. Tanre et al. 2005. A critical examination of the residual cloud contamination and diurnal sampling effects on MODIS estimates of aerosol over ocean. *IEEE Trans.* **43**:2886–2897.

Keith, D. W., and J. G. Anderson. 2001. Accurate spectrally resolved infrared radiance observation from space: Implications for the detection of decade-to-century-scale climatic change. *J. Climate* **14**:979–990.

Kiehl, J. T. 2007. Twentieth century climate model response and climate sensitivity. *Geophys. Res. Lett.* **34**:L22710.

Klein, S. A., and D. L. Hartmann. 1993. The Seasonal Cycle of Low Stratiform Clouds. *J. Climate* **6**:1587–1606.

Knutson, T. R., and S. Manabe. 1995. Time-mean response over the tropical pacific to increased CO_2 in a coupled ocean-atmosphere model. *J. Climate* **8**:2181–2199.

Korolev A. V., G. A. Isaac, and J. Hallett. 2000. Ice particle habits in stratiform clouds. *Q. J. Roy. Meteor. Soc.* **126**:2873–2902.

Krüger, O. and H. Graßl. 2002. The indirect aerosol effect over Europe. *Geophys. Res. Lett.* **29(19)**:1925.

Leroy, S., J. Anderson, J. Dykema, and R. Goody. 2008. Testing climate models using thermal infrared spectra. *J. Climate* **21**:1863–1875.

Loeb, N. G., B. A. Wielicki, W. Su et al. 2007. Multi-instrument comparison of top-of-atmosphere reflected solar radiation. *J. Climate* **20(3)**:575–591.

Martins, J. V., D. Tanré, L. Remer et al. 2002. MODIS cloud screening for remote sensing of aerosols over oceans using spatial variability. *Geophy. Res. Lett.* **29(12)**:8009

Mauger, G. and J. R. Norris. 2007. Meteorological bias in satellite estimates of aerosol–cloud relationships. *Geophys. Res. Lett.* **34**:L16824.

Nghiem, S. V., I. G. Rigor, D. K. Perovich et al. 2007. Rapid reduction of Arctic perennial sea ice. *Geophys. Res. Lett.* **34**:L19504.

Norris, J. R. 2001. Has Northern Indian Ocean cloud cover changed due to increasing anthropogenic aerosol? *Geophys. Res. Lett.* **28**:3271–3274.

Norris, J. R., and M. Wild. 2007. Trends in aerosol radiative effects over Europe inferred from observed cloud cover, solar "dimming," and solar "brightening." *J. Geophys. Res. Atmos.* **112**:D08214.

Penner, J. E., M. Andreae, H. Annegarn et al. 2001. Aerosols, their direct and indirect effects. In: Climate Change 2001: Contribution of Working Group I to the Third Assessment Report of the Intergovernmental Panel on Climate Change, ed. J. T. Houghton, et al., pp. 289–348. New York: Cambridge Univ. Press.

Petters, M. D. and S. M. Kreidenweis. 2007. A single parameter representation of hygroscopic growth and cloud condensation nucleus activity. *Atmos. Chem. Phys.* **7**:1961–1971.

Platnick, S., and S. Twomey. 1994. Determining the susceptibility of cloud albedo to changes in droplet concentrations with the Advanced Very High Resolution Radiometer. *J. Appl. Meteor.* **33**:334–347.

Radke, L. F., J. A. Coakley, and M. D. King. 1989. Direct and remote-sensing observations of the effects of ships on clouds. *Science* **246**:1146–1149.

Schulz, M., C. Textor, S. Kinne et al. 2006. Radiative forcing by aerosols as derived from the AeroCom present-day and pre-industrial simulations *Atmos. Chem. Phys. Discuss.* **6**:5225–5246.

Ström, J., J. Heintzenberg, K. J. Noone et al. 1994. Small crystals in cirrus clouds: Their residue size distribution, cloud water content and related cloud properties. *Atmos. Res.* **32**:125–141.

Thompson, D. W. J., J. M. Wallace, and G. C. Hegerl. 2000. Annular modes in the extratropical circulation. Part II: Trends. *J. Climate* **13**:1018–1036.

Trenberth, K. E., P. D. Jones, P. Ambenje et al. 2007. Observations: Surface and atmospheric climate change. In: Climate Change 2007: The Physical Science Basis. Contribution of Working Group I to the Fourth Assessment Report of the Intergovernmental Panel on Climate Change, ed. S. Solomon, D. Qin, M. Manning et al., pp. 235–336. New York: Cambridge Univ. Press.

Valero, F. P. J. 2006. Keeping the DSCOVR mission alive. *Science* **311**:775

Warner, J., and S. Twomey. 1967. The production of cloud nuclei by cane fires and the effect on cloud droplet concentration. *J.Atmos. Sci.* **24**:704–706.

Wendisch, M., O. Hellmuth, A. Ansmann et al. 2008. Radiative and dynamic effects of absorbing aerosol particles over the Pearl River Delta, China. *Atmos. Environ.* **42(25)**:6405.

Wood, R., and C. S. Bretherton. 2006. On the relationship between stratiform low cloud cover and lower tropospheric stability. *J. Climate* **19**:6425–6432.

7

Laboratory Cloud Simulation

Capabilities and Future Directions

Frank Stratmann[1], Ottmar Möhler[2],
Raymond Shaw[1,3], and Heike Wex[1]

[1]Leibniz Institute for Tropospheric Research, Leipzig, Germany
[2]Institute for Meteorology and Climate Research, Forschungszentrum Karlsruhe, Karlsruhe, Germany
[3]Michigan Technological University, Houghton, MI, U.S.A.

Abstract

Since the atmosphere offers only observation possibilities and does not permit the setting of initial and boundary conditions, laboratory studies are important tools to examine cloud processes under well-defined, repeatable conditions and expand our understanding. An overview is provided of the capabilities and limitations of laboratory studies on physical cloud processes. Aerosol particle hygroscopic growth and activation, droplet dynamic growth, ice nucleation, and droplet–turbulence interactions are addressed and laboratory devices suitable for investigating and simulating these cloud physical processes are presented. These devices range from portable instruments for measuring selected particle and cloud droplet properties, to setups allowing process studies under simulated cloud conditions, to experiments allowing a nearly full-scale cloud simulation. The issue of laboratory-generated particles and their importance in laboratory cloud simulation is discussed. Finally, suggestions are presented for possible future research topics and devices in the field of laboratory cloud simulation. Specific suggestions include the initiation and/or continuation of investigations regarding particle hygroscopic growth and activation, the accommodation coefficients of water vapor on liquid water and ice, aerosol effects on primary ice formation in clouds, aerosol-based parameterizations of cloud ice formation, secondary ice formation/ice multiplication, the production and characterization of particles suitable for cloud simulation experiments, and experiments that combine turbulence and microphysics. The latter topic is important, as it offers the only possible way of simulating and quantifying the possible interactions and feedbacks between the microphysical (activation, growth, freezing) and turbulent transport processes within clouds.

Introduction

Atmospheric clouds constitute a complex system of global importance. Clouds influence climate and are the source of precipitation. Atmospheric clouds occur sporadically in locations that are usually difficult to reach. Therefore, the investigation of atmospheric clouds (i.e., the attempt to understand cloud formation and the dynamic processes within clouds *in situ*) is an ambitious, expensive, and often impossible task. To complicate matters, each cloud is unique; that is, measurements in atmospheric clouds suffer from a lack of reproducibility in terms of their initial and boundary conditions. Together with the complexity of the cloud system itself, this explains why cloud formation and cloud dynamic processes have not yet been quantified and why they are still not well understood.

Experiments are thus needed to allow (a) the isolation, examination, and quantification of individual cloud processes and (b) the simulation of combinations of selected cloud processes. Such investigations must be performed under well-controlled and reproducible initial and boundary conditions. None of these requirements can be fulfilled *in situ* and thus laboratory studies must address the entire range: from selected single cloud-relevant processes to the simulation of clouds on a laboratory scale. To date, laboratory investigations are not, and most likely will never be, able to reproduce all real-world cloud characteristics. In other words, although single characteristics, such as temperature and pressure ranges, timescales, and lifetimes, can be reproduced in the laboratory for selected cloud types, laboratory experiments are incomplete for characteristics such as cloud size, cloud surface area to volume ratio, the absence of any hard surfaces, and the Reynolds number. Nonetheless, laboratory studies are mandatory if we are to increase our understanding of selected cloud processes as well as clouds in general.

In this chapter we address the laboratory investigation of cloud physical processes, excluding, however, cloud chemical processes from our considerations. First we introduce important processes such as hygroscopic growth and activation of aerosol particles, droplet dynamic growth, ice nucleation and droplet freezing, and droplet–turbulence interactions. Thereafter we describe a selection of up-to-date methods, instruments, and facilities for investigating cloud physical processes. In closing we offer suggestions and discuss important future research topics as well as the need for further ideas and facilities.

Cloud Processes Simulated in the Laboratory

Hygroscopic Growth and Activation

The ability of atmospheric aerosol particles to take up water is the primary link to their influence on clouds. For water vapor concentrations below saturation (i.e., relative humidity, RH < 100%), particles may grow hygroscopically, such that when a threshold supersaturation (RH > 100%) is reached, the particles activate to cloud droplets. Both processes—hygroscopic growth and activation—depend strongly on particle properties such as size, structure, and chemical composition (see Kreidenweis et al., this volume).

It is well accepted that hygroscopic growth and activation can be modeled by means of the Köhler equation (Köhler 1923). However, the correct determination of the parameters and coefficients in this equation (e.g., water activity and surface tension), and consequently the consistent theoretical description of both hygroscopic growth and activation, is still a topic of ongoing research. The key is that although hygroscopic growth of solution droplets below 95% RH is controlled by water activity (Raoult term), activation is strongly influenced by the Kelvin term and is both dependent on droplet size and surface tension. This implies that only combined hygroscopic growth and activation measurements are sufficient to characterize fully the particle behavior upon humidification.

Many inorganic and organic substances of atmospheric importance have been examined in the laboratory with respect to their hygroscopic growth and activation. Summarizing the results of these investigations, one may say that (a) in addition to particle size and composition, surface tension exerts a strong influence on the activation process (corresponding sensitivities are given in Wex et al. 2008); (b) the surface tension of an activating droplet might be significantly different from that of water if surface active substances are present, and (c) for most investigated substances, at activation, it seems appropriate to assume ideal behavior of the solution droplet and the surface tension is that of water.

We still lack, however, the means to describe consistently, quantitatively, and efficiently the hygroscopic growth and activation of atmospheric (i.e., internally mixed multicomponent) aerosol particles (internal mixtures of soluble, slightly soluble and insoluble organic and inorganic substances) based on a reasonably small number of physical and chemical particle properties.

Droplet Growth

After activation, cloud droplets grow as a result of further condensation of water molecules on the droplet surface. In the course of this process, water vapor is depleted from and heat is released to the droplets' surrounding. This influences the vapor concentration, the temperature, and consequently the water vapor saturation fields in the vicinity of growing droplets and therefore much of the cloud. Consequently, growth by condensation is one of the key processes that controls cloud properties such as water vapor supersaturation and cloud droplet number and size. One of the main parameters in describing condensational droplet growth is the accommodation coefficient of water vapor on liquid water. This accommodation coefficient has been controversially discussed for a long time. Literature suggests accommodation coefficients between 0.01

and 1.0 (Davidovits et al. 2006); however, we do not understand the reasons for the observed discrepancies.

Another important growth mechanism is droplet collision–coalescence. One driving force behind the collision–coalescence process is the balance of gravitational and fluid drag forces, which causes cloud droplets of different size to fall with different velocities. This differential sedimentation results in collisions between larger and smaller droplets and consequently growth of the larger droplets. However, differential sedimentation is not the only process that causes droplets to collide. Another important mechanism is turbulence-induced droplet motion and it is still not well quantified (discussed further below).

Ice Nucleation and Droplet Freezing

After activation of aerosol particles, the resulting cloud droplets can be further lifted up in the atmosphere, where they can be cooled to temperatures well below 0°C. The spontaneous freezing of pure water, however, is hindered by an energy barrier, mainly because of the surface tension between the new ice germ, formed by stochastic processes, and the surrounding supercooled liquid water phase. Only at temperatures below ca. –35°C are ice germs of sufficient size formed, thereby inducing the freezing of the whole cloud droplet. Homogeneous freezing rates of water droplets measured in two laboratories at simulated cloud conditions were consistent with formulations of classical nucleation theory (DeMott and Rogers 1990; Benz et al. 2005). Recent laboratory measurements of the freezing rates of single water microdroplets demonstrated that the homogeneous freezing process is volume dependent (Duft and Leisner 2004b). These results conflict with the hypothesis that homogeneous freezing is preferentially initiated at the droplet surface, at least for droplets larger than a few micrometers in diameter.

At temperatures between 0° and –35°C, the freezing of supercooled cloud droplets can be induced by solid aerosol particles with specific surface properties, termed ice nuclei (IN). This heterogeneous ice nucleation process is initiated by a particle immersed in the droplet (immersion mode) or in contact with the surface of a droplet (contact mode). If cloud condensation nuclei (CCN) activation happens at sufficiently low temperatures, the same particle can first act as a CCN and then subsequently as an IN (condensation mode). The direct deposition of water vapor to the surface of a solid particle is called deposition-mode ice nucleation. Until now, little has been known about the abundance and nature of heterogeneous IN or the relative importance of the different modes of ice nucleation in mixed-phase clouds. Various ongoing research programs are currently trying to relate the heterogeneous ice nucleation efficiency to specific aerosol properties and to develop aerosol-based formulations for heterogeneous ice nucleation in cloud, weather forecast, and climate models.

At temperatures below –35°C, freezing rates of growing water droplets are sufficiently fast so that the clouds contain only ice particles and no liquid water droplets. These ice clouds (cirrus) are formed either by homogeneous freezing of supercooled water or solution droplets, or by heterogeneous ice nucleation processes (cf. Kärcher and Spichtinger, this volume). The homogeneous freezing threshold of micrometer-sized solution particles increases from a saturation ratio of about 1.4 at –40°C to ca. 1.7 at –80°C. Mineral dust particles are thought to act as very efficient IN at cold temperatures and ice saturation ratios below 1.2 (Möhler, Bunz et al. 2006). Cirrus clouds tend to contain fewer but larger ice crystals if predominantly formed by heterogeneous ice nucleation processes. The number and size of cirrus ice crystals have important consequences for the balance between shortwave cooling and longwave heating of cirrus clouds in the climate system. More experiments and model investigations are needed to assess and quantify the competition between heterogeneous and homogeneous ice formation in cirrus clouds.

Ice Particle Growth and Habits

The ice phase affects the cloud cooling and warming capabilities and, in many cases, initiates precipitation and influences the distribution and intensity of precipitation. The number of primary ice crystals in mixed-phase clouds formed by heterogeneous IN is typically much smaller than the number concentration of droplets. Ice crystals can, however, grow to larger sizes at the expense of the liquid droplets (Bergeron–Findeisen process). Furthermore, the ice phase can be enhanced by secondary processes (e.g., ice splintering, the so-called Hallett–Mossop effect, riming, and ice crystal aggregation) and cause rapid changes to the internal structures of mixed-phase clouds. The quantitative description of these changes requires a better understanding of the basic processes, such as the accommodation of water molecules on ice particles, the rate of ice splinter formation, the collision rates of ice particles with droplets, the aggregation rates of ice crystals, or the influence of turbulence on, for example, condensational growth, aggregation, and riming.

Only a few laboratory experiments have investigated the Hallett–Mossop effect of ice splinter formation upon collisions between supercooled cloud droplets and ice crystals (e.g., Saunders and Hosseini 2001). Most cloud model parameterizations of secondary ice processes in clouds are based on empiric considerations.

Measurements of the mass accommodation coefficient of water vapor on ice are relatively scarce and not entirely consistent with each other. The data tend to support the idea that the accommodation coefficient is a function of temperature, and possibly of supersaturation and crystal size. Recent measurements in a carefully controlled laboratory experiment, involving small crystals suspended in an electrodynamic balance, resulted in an accommodation coefficient of 0.006 ± 0.0015 at a temperature of –50°C (Magee et al. 2006). A very low mass accommodation coefficient would have important implications for cirrus cloud properties and upper tropospheric humidity. Thus, further experiments are needed.

The growth of pristine ice crystals with distinct habits (e.g., columns, plates, needles, or bullet rosettes, or more compact, polycrystalline shape) is controlled mainly by the temperature and supersaturation present during crystal growth. However, it may also depend on the formation mechanism of the ice crystal (see, e.g., Bailey and Hallett 2002). The relation between preferential crystal habit and growth conditions is well established, but little is known about the influence of the ice formation mode on crystal shape. There is also a lack of experimental data for quantifying the effect of crystal habit on crystal growth rates, collision rates, settling velocities, and radiative properties.

As discussed above, the dynamic change of ice number and mass concentration in clouds is still difficult to understand and quantify. Issues requiring further attention include the accommodation coefficient of water vapor on ice and the influence of turbulence or crystal habits on, for example, condensational growth, aggregation, and riming.

Generation and Characterization of Particles Used in Laboratory Cloud Simulations

The properties of the particles used in laboratory cloud research are not the main focus of this chapter. However, since these properties may heavily influence the results obtained during the experiments performed, we provide some thoughts on this topic.

Simulation of aerosol–cloud interactions in the laboratory requires the following:

1. Particles must be generated with well-defined physical and chemical properties (sometimes in large amounts).
2. Particle properties (e.g., by condensation of different vapors) must be tailored to the specific needs of the performed experiment.
3. Particles must be characterized according to their properties of interest with sufficient accuracy.

Some important properties to include in this context are particle diameter, shape, internal structure, volume, mass, surface area, surface properties such as number of active sites, and the chemical particle composition. Here we do not discuss the production or design of particles for a specific experiment; instead we address briefly the characterization of some selected properties.

Characterizing particle properties of interest for a certain cloud experiment represents a true challenge and is sometimes even impossible to achieve. Let us consider the process of heterogeneous ice nucleation in a solution droplet with a solid mineral dust or soot core, to which we add a thin (e.g., 5 nm) organic coating. To characterize the droplet as being frozen, we need to know the volume of the mineral dust core, the amount of organic substance on the core particles, the surface properties of the core particles (e.g., number of active sites), the size of the droplet, and the location of the core within the droplet.

Determination of any of these properties is nontrivial and requires high-end instruments such as differential mobility analyzer (DMA), low pressure impactor, mass spectrometer, and optical particle/droplet spectrometers. The surface properties of the core particle, which is most likely crucial for understanding the ice nucleation process, are currently almost impossible to quantify.

Reliable generation and sufficient characterization of the particles used in laboratory cloud simulations are important issues, and efforts to further our capabilities are needed.

Droplet–Turbulence Interactions

As we attempt to understand and quantify clouds, it is necessary to consider the role of turbulence, including the specific interactions between particles and the turbulence as well as multiscale turbulent transport processes. Droplet growth by condensation and collision–coalescence or aggregation are fundamental Lagrangian processes and depend on the history of individual cloud droplets or ice crystals throughout the cloud. Because cloud particles have a high density compared to air, their trajectories are not the same as that of fluid elements. The thermodynamic fields that they sample and the angles and speeds at which they collide are influenced by the surrounding turbulence. On larger scales, turbulent transport processes cause mixing within a cloud as well as mixing between the cloud and its surroundings. Turbulent transport processes affect strongly the temperature, the water vapor concentration, and saturation distributions within and at the edges of clouds. Consequently, turbulent transport processes influence cloud microphysical processes (e.g., cloud droplet activation and growth) and therefore have effects on cloud microphysical properties (e.g., cloud droplet number, cloud droplet size distribution, and even the small-scale spatial distribution of cloud droplets). To make things even more complex, this is not necessarily a one-way process. Cloud microphysical processes such as condensation and evaporation may have an effect on local turbulence through latent heat effects, resulting in a feedback between cloud microphysics and turbulence. Examples for droplet–turbulence interactions that have been investigated and simulated in laboratory studies are entrainment, inertial particle dynamics, and turbulence-influenced condensational growth and collision–coalescence.

Entrainment, as mentioned above, is one of the main interaction processes between the cloud and its surroundings. It describes the transport of "dry" air, gases, and aerosol particles from the cloud's environment into the cloud by turbulent mixing processes. The opposite process—the transport of cloudy air outside the cloud—is called detrainment. Entrainment processes are thought to be responsible for the broad droplet size distributions observed in real clouds, which is in contrast to results obtained from theoretical calculations. However, the essential physics of the entrainment process are not fully understood: questions remain, for example, about the competing roles of homogeneous and inhomogeneous mixing.

The influence of turbulence on droplet collision–coalescence is one of the key processes that needs to be understood in terms of the growth of droplets within clouds. In a quiescent flow, droplets would collide solely through this differential sedimentation mechanism. However, in a turbulent flow, significant velocity and acceleration components perpendicular to the vertical will enhance the collision probability and thus droplet growth.

These cloud droplet–turbulence interaction processes have been investigated primarily through numerical simulation and field measurements over the last ten years, but controlled laboratory experiments, where specific mechanisms can be isolated and verified, are still needed.

Methods for Simulating Cloud Processes in the Laboratory

Having introduced the main topics of laboratory cloud investigations, let us now turn to some of the important experimental approaches used in these studies. The list of devices discussed below is not meant to be inclusive, but rather is intended to illustrate major techniques currently in use or under development. The laboratory devices presented here range from portable instruments applicable in both the laboratory and the field, to larger-scale chambers for investigating individual cloud processes, and finally to large cloud simulators. We note, however, that currently, and most likely also in the near future, no device is capable of simulating a "real" cloud with all its relevant processes and complexity. As the cloud physical processes described above are investigated with different devices at various scales, they will be classified into the following groups:

1. Devices for measuring selected particle and cloud droplet properties.
2. Devices allowing process studies under simulated cloud conditions.
3. Devices allowing a large-scale cloud simulation.

Devices for Measuring Selected Particle and Cloud Droplet Properties

Here we present instruments that can be used in both laboratory and field studies to determine particle and droplet properties. Specifically, we discuss the humidity tandem differential mobility analyzer (H-TDMA), different CCN and IN counters, and the electrodynamic balance (EDB).

Humidity Tandem Differential Mobility Analyzer

H-TDMA is a tool used widely to simulate and quantify size-segregated hygroscopic growth of airborne particles in the RH range up to 98% (McMurry and Stolzenburg 1989). All H-TDMAs feature the same principle of operation: from an existing dry particle size distribution, a well-defined size fraction is extracted by means of a first DMA. Subsequent to size selection, the aerosol particles are humidified, and the resulting changes in particle size are determined by means of a second DMA. Results are usually given in terms of growth factors (i.e., the ratio of particle/droplet sizes) prior to and after humidification. The main differences between different H-TDMA systems are (a) the humidification method and (b) the manner in which RH and temperature are controlled.

H-TDMAs are excellent tools for simulating and quantifying hygroscopic particle growth in the RH range up to 98%. They have been widely used in both laboratory and field studies. However, they are operated in the RH range where particle growth is controlled mainly by water activity (and not by the Kelvin effect).

Cloud Condensation Nucleus Counters

Particle activation to cloud droplets can be simulated and quantified by means of CCN counters. Activation is achieved by subjecting the aerosol particles to an environment with supersaturated water vapor. Different techniques are available for creating such environments. Adiabatic expansion is one such method, but it usually leads to supersaturations (order of percent to hundreds of percent) larger than those observed in real clouds. Another method involves wetted parallel plates with different temperatures (Radke and Hobbs 1969). Here, the heat and mass transfer processes between the two plates together with the nonlinearity of saturation vapor pressure cause the air–particle–water vapor mixture in the gap between the two plates to become supersaturated, with the supersaturation peaking approximately halfway between the surfaces. Recently, a continuous-flow streamwise thermal-gradient CCN counter has become commercially available (Roberts and Nenes 2005). Here, an aerosol beam is sent through the center of a cylindrical column with wetted walls, such that temperature increases continuously along the column. Supersaturation is achieved because water vapor diffuses faster than heat from the walls to the center of the column.

Measurements using CCN counters have been used to determine the concentration and fraction of activated atmospheric aerosol particles at a given supersaturation, independent of particle size (i.e., the number of aerosol particles activated was compared to the total number concentration). Recently, a DMA has been sometimes used upstream of the CCN counter to obtain information on the size-dependent activation of the examined particles. This allows the retrieval of pairs of values for critical supersaturation and critical diameter for activation.

CCN counters have been widely used in laboratory and field. They are very useful devices for measuring the activation of aerosol particles to droplets. However, they do so under instrument-specific conditions. Timescales and supersaturation profiles inside the instruments differ from those that prevail in the atmosphere, and thus a one-to-one translation from data measured with CCN counters to conditions in atmospheric clouds might not always be feasible.

Ice Nuclei Counters

IN counters are instruments that determine the number concentration of IN in a given particle population at a particular ice supersaturation. They can be viewed as devices to simulate a subset of freezing processes within a cloud. Examples of currently available IN counters are (a) the continuous flow diffusion chamber (CFDC) (Rogers 1988; Rogers et al. 2001), (b) the Zürich ice nucleation chamber (ZINC) (Stetzer et al. 2008), and (c) the fast ice nucleus chamber (FINCH) (Bundke et al. 2007). Both CFDC and ZINC are based on a similar principle but feature different geometries (cylindrical and parallel plate, respectively). In both devices, particles are subjected to an environment supersaturated with respect to ice by means of two "parallel" surfaces that are both coated with ice but feature different temperatures. As a result of heat and mass transfer between the two surfaces and the nonlinearity of the temperature dependence of saturation vapor pressure, the air–particle–water vapor mixture in the gap between the two surfaces is supersaturated with respect to ice, and the supersaturation peaks approximately half way between the surfaces. Depending on the temperature difference between the two surfaces, different supersaturations can be achieved. Since the surface temperatures can be varied, IN number and or ice nucleation probabilities can be determined as functions of ice supersaturation. The number of ice particles is determined by means of an optical particle counter at the outlet of the instrument.

FINCH operates according to a different principle. Here supersaturation is achieved by turbulently mixing warm humid air with cool dry air in a flow tube. Again, ice particles are detected by means of an optical particle counter.

All three instruments have been successfully applied for investigating both atmospheric and laboratory-generated aerosol particles. However, similar to CCN counters, IN counters measure IN numbers under instrument-specific conditions. Timescales and supersaturation profiles inside the instruments differ from those that exist in clouds. Furthermore, IN counters currently lack the possibility to determine ice particle sizes/masses and thus ice particle growth rates. Most of the instruments operate in deposition or condensation nucleation modes, and therefore immersion and contact modes are not measurable.

Electrodynamic Balance

Single aerosol particles and droplets can be levitated and stored in EDBs (for a review, see Davis 1997). Storage time is principally infinite and limited only for practical reasons. EDBs have been used in many fields of aerosol research, including heterogeneous chemistry, deliquescence and efflorescence of salt particles, and optical properties of particles typically larger than a few micrometer in diameter. More recently, EDB devices in temperature-controlled environments have also been used to investigate the freezing of microparticles such as sulfuric acid (Imre et al. 1997) or pure water droplets (Duft and Leisner 2004b).

Several configurations of EDBs have been suggested and used for various applications (Davis 1997). A more sophisticated version of an EDB consists of a central rotationally symmetric torus electrode, with the symmetry axis in vertical orientation and two end cap electrodes on the same symmetry axis. The dimension of an EDB for particle levitation is in the range of 5–10 cm. Optimal storage and levitation results of microdroplets at atmospheric pressure are obtained if the caps and torus are of hyperboloidal shape. An AC voltage with an amplitude on the order of 1 kV and a frequency of a few hundred Hz is applied between the torus electrode and the end caps; this focuses the charged particles in the trap to a small area centered around the symmetry axis. A superimposed DC voltage between the two end caps is adjusted to balance the gravitational force on the particle in the trap.

Mass changes, for example, of evaporating water droplets in the EDB are sensitively monitored with the balance voltage. For spherical droplets, the index of refraction can also be obtained from Mie-scattering features observed optically (e.g., Duft and Leisner 2004a). If the index of refraction is known, the same setup can be used to measure the droplet size accurately. Freezing of droplets is detected by depolarization-sensitive light-scattering techniques. Whereas droplet processes can be studied in EDBs over a wide temperature range, the EDB setup is limited in achieving and controlling high humidities or even supersaturations with respect to ice and water. In addition, possible effects of droplet charge and the applied electrical field should be kept in mind.

Devices Allowing Process Studies under Simulated Cloud Conditions

Let us now turn to those devices that allow the investigation of cloud physical processes at simulated cloud conditions. Again, our discussion is not intended to be complete but features examples for important state-of-the-art-cloud simulation techniques and devices. Specifically, we discuss (in alphabetical order), the aerosol interactions and dynamics in the atmosphere (AIDA) chamber, the Leipzig aerosol cloud interaction simulator (LACIS), the Meteorological Research Institute (MRI) chamber, and the University of Mainz wind tunnel.

Also, we address devices (wind tunnels and chambers) used to study cloud droplet–turbulence interactions.

Aerosol Interaction and Dynamics in the Atmosphere Chamber

Aerosol and cloud processes can be investigated in the AIDA facility of the Forschungszentrum Karlsruhe under a wide range of simulated atmospheric conditions, including temperature, pressure, humidity, as well as trace gas and aerosol constituents (Benz et al. 2005; Möhler, Bunz et al. 2006). Experiments conducted in the large AIDA chamber are complemented by experiments with single levitated microdroplets (see above) as well as the development and application of process models with detailed aerosol and cloud microphysical formulations (Möhler, Field et al. 2006).

At the core of the AIDA facility is a cylindrical aluminum vessel with a volume of 84 m^3. The vessel can be evacuated to pressures below 1 hPa and is located in a thermally insulated container that can be cooled to temperatures as low as –90°C. The cooling system maintains homogeneous temperature conditions for hours to days, with temporal and spatial fluctuations throughout the container below ±0.3°C. Experiments begin at well-defined pressure, temperature, and RH conditions and with a well-specified mix of trace constituents. With respect to ice and/or liquid water, supersaturation is achieved by pumping expansion and corresponding adiabatic cooling of the vessel volume. Thereby, conditions similar to lee wave or convective clouds can be simulated with cooling rates between about 0.1 and 5 K min^{-1}. During a typical cloud expansion simulation run, the pressure in the AIDA vessel is lowered from 1000 to 800 hPa within about 5 to 10 minutes.

A comprehensive set of commercial as well as specially designed and homemade instruments is available at the AIDA chamber for generating various aerosols as well as for measuring physical and chemical particle properties, water vapor and trace gas concentrations, optical particle properties, and ice cloud characteristics. This set includes standard aerosol instrumentation (e.g., condensation nuclei counters and electrical mobility, aerodynamic and optical sizing techniques, Fourier transform infrared, and tunable diode laser spectroscopy) and specially designed scattering, depolarization, and imaging techniques for the detection and characterization of ice particles.

One of the strengths of the AIDA facility is that cloud processes can be investigated under realistic conditions with internal variability and fluctuation of temperature and humidity. A full AIDA cloud simulation cycle includes preparation of aerosol, hygroscopic growth, CCN activation, and, if temperatures are low enough, ice nucleation. Effects of aerosol aging and cloud processing can be investigated in subsequent expansion cycles with the same aerosol. However, the humidity control in the chamber is limited to an accuracy of only a few percent as a result of temperature fluctuations and wall effects.

Leipzig Aerosol Cloud Interaction Simulator

LACIS (Stratmann et al. 2004) is a device to investigate complex phase transition processes, such as particle/droplet hygroscopic growth, activation, ice heterogeneous nucleation, and ice particle growth. LACIS has been designed to simulate accurately cloud processes that take place on short timescales (i.e., up to a minute). Its main focus is on particle size-resolved investigations; it features particle/droplet generation and detection devices capable of producing, detecting, and characterizing monodisperse particle and droplet size distributions. Thermodynamic parameters such as temperature, pressure, RH, critical supersaturation, composition, and concentration of particles/droplets and of chemical composition in the carrier gas can be varied in ranges similar to the lower troposphere.

LACIS consists of a 15 mm diameter laminar flow tube with temperature-controlled walls. The length of the flow tube can vary from 0.5 to 10 m. Residence times up to 60 s and temperatures down to –50°C are feasible. Using high-precision instruments, the thermodynamic conditions in the LACIS flow tube are controlled with extremely high accuracy and reproducibility (temperature about 0.05°C, RH about 0.5%, and critical supersaturation about 0.05%). Particle/droplet size distributions produced in LACIS are determined by means of specially designed optical particle counters mounted along the axis of LACIS. These optical particle counters are capable of counting, sizing, and distinguishing the phase state (e.g., liquid, frozen) of individual droplets inside LACIS.

LACIS may be operated under both sub- and supersaturated conditions with respect to water and ice. Thus it combines the capabilities of simulating hygroscopic growth, dynamic post-activation growth, droplet freezing, and subsequent ice particle growth into one instrument.

LACIS has been used mainly to investigate the hygroscopic growth and activation behavior of different types of aerosol particles. In one recent study, it was applied to connect successfully high RH hygroscopic growth and activation behavior for selected inorganic and organic substances (e.g., Ziese et al. 2008). Recently, initial investigations have been performed on the immersion and deposition freezing of monodisperse dust particles.

Meteorological Research Institute Cloud Chamber

A new dynamic cloud chamber for simulating and investigating IN and CCN processes has been recently developed at the Meteorological Research Institute (MRI) in Tsukuba, Japan (T. Tajiri, pers. comm.). The chamber has a similar design to the Colorado State University cloud chamber (DeMott and Rogers 1990). Adiabatic cloud expansion processes are simulated by synchronously controlling air pressure and wall temperature in the ranges 1013 to 30 hPa and

30 to –100°C, respectively. Adiabatic expansion can be simulated equivalent to updraft speeds of up to 30 m s^{-1}.

The chamber consists of two stainless steel vessels, an outer pressure-controlled chamber, and an inner temperature-controlled chamber. The inner cylindrical vessel has a height of 1.8 m, a diameter of 1 m, and a volume of 1.4 m^3. The walls of the inner vessel are temperature-controlled by a circulating coolant. The particle injection (air supply) port is located at the top of the chamber. A small fan stirs the injected air to achieve homogeneity inside the volume. The chamber is equipped with various instruments for aerosol characterization, CCN measurements, as well as droplet and ice crystal detection.

During cloud formation experiments, the simulated ascents follow dry adiabatic expansion until the air temperature corresponds to the lifting condensation level (LCL). Thereafter, moist adiabatic expansion is simulated. The initial condition of pressure, temperature, dewpoint temperature, and ascent rate must be known to calculate cooling/evacuation rate and LCL. Accurate temperature and pressure controls are made by a combination of feed-forward and three-term PID control methods. The actual temperature and pressure are typically held to within 0.5°C and 0.3 hPa of the command profile.

Among the advantages of the MRI chamber are the wide temperature range (as low as –100°C) and the fact that adiabatic cloud expansion cycles can be simulated with active control of both gas and dew or frost point temperature. The chamber will be used to study cloud formation processes such as CCN activation, ice nucleation, and aspects of artificial cloud seeding. Because of the smaller volume, only a low number of sampling instruments can be operated at the chamber, especially at low expansion rates.

University of Mainz Wind Tunnel

A vertical wind tunnel at the University of Mainz, Germany, is used to investigate and simulate cloud processes such as uptake of trace gases by raindrops, the influences of turbulence on the collisional growth of cloud droplets, and the impaction scavenging of aerosol particles (Pruppacher 1988; Vohl 1989). The wind tunnel allows free suspension of water drops or other hydrometeors at their terminal velocity in its experimental section. The flow inside the wind tunnel is created by two vacuum pumps in the upper horizontal part of the tunnel. The velocity of updraft in the experimental section is controlled by a variable flow sonic nozzle. In the lower horizontal part of the tunnel, there are particle filters and air conditioning units which make it possible to adjust the air humidity and temperature in the experimental section. Trace gases, aerosol particles, or cloud droplets can be introduced into the wind tunnel upstream of the experimental section. Downstream of the experimental section it is possible to take air samples to determine the air temperature and dew point or to measure trace gas concentrations.

Inside the experimental section, the suspended hydrometeors can be observed visually, and after defined residence times, the droplets can be collected and removed from the wind tunnel, fixed in sample bottles, and analyzed by means of ion chromatography to determine their composition.

Thus far, the collisional growth of cloud droplets, impaction scavenging of aerosol particles, and gas uptake by single water drops and/or ice crystals have been studied under laminar and/or turbulent conditions. Droplet immersion and contact freezing processes have also been investigated.

Other Wind Tunnels

We would like to mention two wind tunnels that can be used for measuring interactions between water droplets and turbulence: the wind tunnel at Cornell University Ithaca, New York, U.S.A. (Ayyalasomayajula et al. 2006; Saw et al. 2008, submitted) and the tunnel at the Max Planck Institute for Dynamics and Self Organization in Göttingen, Germany. A key aspect of both tunnels is the ability to reach large turbulence Reynolds numbers, which is necessary when considering turbulence in geophysical flows such as clouds. The Cornell wind tunnel has a cross section of 1 m × 0.9 m, and a length of 20 m, with turbulence driven by an active grid (triangular agitator wings attached to randomly rotating grid bars). It is capable of generating turbulence with a Reynolds number ranging from roughly 10^4 to 10^6. A broad droplet size distribution, with mean diameter approximately 20 μm, is generated by an array of four spray nozzles. The Göttingen wind tunnel, which is currently being assembled in the newly constructed experimental hall and which is expected to be operational by mid-2008, is designed to achieve a Reynolds number of approximately 10^8, the highest Reynolds number ever to be achieved in a standard (i.e., non-superfluid) laboratory flow. The tunnel is 12 m long and the pipe is 1.8 m in diameter. The gas is pressurized to have a density 150 times greater than ambient air. Both tunnels are designed to investigate particle/droplet–turbulence interactions and allow Lagrangian measurements (i.e. tracking of particles/droplets by a high speed camera moving along the side of the tunnel at the mean flow speed).

Warsaw University Cloud Chamber

To investigate interactions between turbulence and cloud droplets at Warsaw University, a glass box of 1.0 m × 1.0 m × 1.8 m (Malinowski et al. 1998; Korczyk et al. 2006) was built. Droplets generated by a commercial ultrasonic humidifier are visualized in a vertical or horizontal cross section through the chamber interior by means of a laser beam formed into the shape of a narrow (~1.2 mm thickness) plane. The light scattered by the illuminated droplets at 90° is detected by photo and video or, more recently, by digital CCD cameras and used to image droplet patterns. Conditions within the chamber are not fully controlled, but are instead monitored. The cloud droplet size spectrum is measured as follows: Droplets are collected on a glass plate (covered with a thin film of paraffin oil) and subsequently imaged by microscope. Images are then processed to determine droplet size and number (i.e., the droplet size distribution). The liquid water content of the cloudy plume is measured with a cotton filter: the increase in mass of the filter after pumping a given volume of cloudy air through it gives the measure of liquid water content. Temperature and humidity profiles within the chamber are monitored using a set of thermocouples and capacitance sensors.

Experiments conducted with this chamber addressed the geometric patterns created by turbulent mixing of the cloud with its environment and the preferential concentrations of cloud droplets in weak turbulence. Currently, a particle imaging velocimetry (PIV) technique is used to retrieve motion of the cloud droplets investigating fine-scale details regarding the turbulent mixing of the cloud with its environment.

Michigan Technological University Cloud Chamber

For studying Lagrangian properties of cloud droplets in turbulence, a laboratory system has been developed at Michigan Technological University in Houghton, MI, U.S.A. (Fugal et al. 2007). The chamber is a cube with speaker-driven jets positioned at each vertex. The jets are randomly forced and interact in the center of the cube, so as to produce approximately homogeneous, isotropic turbulence. Water droplets of a predetermined size are allowed to settle into the chamber from above, after having achieved their terminal fall velocity inside a seeding cylinder. The water droplet positions within the central volume are determined by digitally reconstructing in-line holograms recorded by fast CMOS cameras. The holograms can be recorded at rates up to 6000 frames per second, allowing particles to be tracked in real time. Initial results show clearly the transition between gravitationally and turbulence-dominated Lagrangian velocity and acceleration statistics. For large cloud droplets and small drizzle droplets, velocities that are several times greater than the terminal velocity and accelerations up to ten times the gravitational acceleration are observed as the turbulence intensity is increased.

Devices Allowing a Nearly Full-Scale Cloud Simulation

The last group of devices that we address concerns simulations conducted close to real cloud scale. Two devices—one existing and one currently under discussion—will be presented, both of which feature former mine shafts as cloud simulators. Before discussing the details of such devices, let us consider the rationale behind such huge cloud simulators. Shafts allow the simulation of cloud processes, including the adiabatic expansion of rising air parcels, for reasonably realistic length scales and residence times with hydrometeors being at their respective terminal velocity. Sampling conditions are also relatively easy. Control of the thermodynamic parameters inside the shaft, however, is lower than in usual laboratory experiments. Clouds generated in vertical shafts are thought to permit investigation of central cloud research topics, such as the roles of aerosol particles, turbulence, electrical effects and temperature on cloud history and the formation of precipitation. They also allow experiments on the behavior of complex aerosol (e.g., biomass burning, combustion aerosol) in clouds, the simulation of ice particle interactions in supercooled cloud environments (thought to cause charge separation and therefore lightning), and the investigation of cloud chemical processes under conditions close to those in a real cloud.

Furthermore, vertical shaft clouds exhibit higher Reynolds numbers than can be achieved in laboratory experiments (but still not as high as in the open atmosphere), thus permitting better experimental data to address the question of how microscale turbulence (~ mm to cm length scales) may affect cloud processes such as condensation and collision–coalescence or aggregation. Other possible research topics may include cloud radiative transfer and remote sensing.

Two devices for the generation of shaft clouds are (a) the artificial cloud experimental system (ACES) (e.g., Yamagata et al. 1998, 2004), operated by Hokkaido University in Sapporo, Japan, and (b) the Cloud Physics Facility in South Dakota (Homestake DUSEL), suggested by J. Helsdon.[1] According to our understanding, ACES suffers currently from a lack of financial support and the Cloud Physics Facility at DUSEL has been postponed.

The facility operated by Hokkaido University features an area of 2.5 m × 5 m and a vertical length of 430 m. Updraft velocities can be varied in the range between 0.5 and 2 m s^{-1}, and temperatures range from 13.5 at the bottom to 10.5 at the top of the shaft.

The facility suggested by J. Helsdon would have an area of 3–5 m × 3–5 m, a vertical length of approximately 1 km, an active grid turbulence generator, and *in-situ* measurements of below-, mid-, and cloud-top properties.

One outstanding question is whether shaft clouds permit an accurate amount of adjustment and control of crucial parameters (e.g., temperature, saturation, flow velocity, turbulent intensities, and dissipation rates) to allow the envisioned gain in knowledge and understanding. Furthermore, funding to create and operate such huge devices may be an important constraint.

Future Research and Devices

In the future, we envision that laboratory cloud research will address processes that take place in warm, mixed-phase, and cirrus clouds. These include:

---

[1] http://neutrino.lbl.gov/Homestake/FebWS/presentations/08_Helsdon%20DUSEL%20Homestake%20LOI.pdf

- hygroscopic growth and activation,
- accommodation coefficients of water vapor on liquid water and ice,
- aerosol effects on primary ice formation in clouds,
- aerosol-based parameterizations of cloud ice formation,
- secondary ice formation or ice multiplication,
- generation and characterization of particles used in laboratory cloud simulations,
- specific cloud droplet–turbulence interactions, and
- combining turbulence and microphysics over multiple scales.

Hygroscopic Growth and Activation

There is still a need for understanding and quantifying the effects of slightly soluble substances, droplet solution non-ideality, and partitioning of surface active substances between particle bulk and particle surface, on high RH hygroscopic growth and activation. This holds for both particles that comprise selected single organic components and particles that consist of internal mixtures of soluble, slightly soluble, and insoluble inorganic and organic substances. Closure studies regarding particle hygroscopic growth and activation behavior and derivation of parameters for a consistent description of particle hygroscopic growth and activation are important topics here.

Possible influences of organic (surface-active) substances on the kinetics of hygroscopic growth and activation are another unresolved, and perhaps critical, issue requiring future attention. This is because application of the Köhler theory for modeling or parameterizing hygroscopic growth and activation implies that the droplet is in equilibrium with its surroundings (i.e., kinetic effects influencing droplet growth are neglected). In case this assumption is not justified, usage of the Köhler theory could lead to erroneous predictions in terms of particle/droplet size and critical supersaturation.

The effects of soluble gases (i.e., their uptake into the particle or droplet and effects on particle hygroscopic growth and activation) must be borne in mind. Influences of particle aging and (cloud) processing on hygroscopic growth and activation behavior are also of great interest.

Of special value in this context are measurements that are performed at RHs > 95% and supersaturations up to ca. 0.5%. In this context, instruments such as the H-TDMA, LACIS, EDBs, and CCN counters have been and will continue to be useful tools.

Accommodation Coefficients of Water Vapor on Liquid Water and Ice

In terms of the accommodation coefficient of water vapor on liquid water, new experimental ideas are needed. The accommodation coefficient may be a function of thermodynamic conditions (temperature, pressure, supersaturation), droplet size, or water vapor flux to the droplet surface. Thus, experiments under conditions resembling those of real clouds (including real cloud time-scales) are required.

Our knowledge is also insufficient regarding the accommodation coefficient of water molecules on ice particles, in particular under realistic atmospheric conditions. This is primarily the result of the challenging conditions under which such experiments must be performed. When equipped with suitable detectors to measure ice particle size or (even better) mass, devices such as LACIS, AIDA, CFDC, ZINC, and FINCH could be utilized to determine ice particle growth rates and derive accommodation coefficients. Such measurements, however, require the use of numerical models to evaluate and interpret the resulting data.

Aerosol Effects on Primary Ice Formation in Clouds

At present, little is known about the influences that specific particle properties (e.g., size, chemical composition or surface structure) have on ice heterogeneous nucleation. We lack both a fundamental understanding of the process and factors involved as well as experimental validation of the theoretical tools used to quantify heterogeneous ice formation processes. In particular, experiments that concern the freezing of internally mixed particles consisting of insoluble cores and inorganic and/or organic coatings are needed to identify and quantify the controlling processes and parameters. Laboratory investigations should utilize available or new cloud simulation facilities as well as ice nucleation instruments with sophisticated methods to generate and characterize relevant aerosol particles. In particular, the roles of particle size, surface area, and surface structure and composition (e.g., active surface sites) need to be investigated. Here, AIDA, LACIS, CFDC, ZINC, and FINCH have been and will continue to be useful. In designing experiments, conditions should resemble those of mixed-phase clouds so that insight can be gained, for example, on the relative importance of the different freezing mechanisms.

Aerosol-based Parameterizations of Cloud Ice Formation

A detailed understanding of aerosol-induced ice nucleation processes and their numerical implementation is prerequisite for reliable forecast of clouds, precipitation, and climate change. Homogeneous freezing rates of aerosol particles can be parameterized in numerical models as a function of the temperature, cooling rate, and aerosol parameters. By contrast, aerosol-related parameterizations for heterogeneous ice nucleation processes are more difficult to assess. Most models still describe the abundance of heterogeneous IN only as a function of temperature and humidity. New concepts have only recently been suggested to consider specific aerosol properties for the parameterization of heterogeneous ice nucleation in models. These need to be evaluated for application under variable cloud conditions and with relevant tropospheric aerosol systems. Comparison of results from laboratory cloud simulation experiments is needed to test and improve existing parameterizations, or to develop new ones, because it is difficult to constrain formulations of cloud microphysics in models to field measurements.

Secondary Ice Formation or Ice Multiplication (Splintering)

From field experiments and cloud modeling studies, we know that secondary ice multiplication, such as the splintering of existing ice crystals or the so-called Hallett–Mossop effect, are important factors in cloud development and for the initiation and intensity distribution of precipitation. To date, only a few laboratory investigations have been conducted on secondary ice formation, and thus further experiments are needed to investigate and quantify this, if possible for relevant ice crystal shapes and sizes and under simulated cloud conditions. Such experiments could be conducted with single droplets and ice crystals in EDB setups or in cloud simulation chambers.

The University of Manchester has built a new ice cloud chamber facility suitable for such experiments. The device consists of three large cold rooms arranged vertically above each other on three floors of the building. These are joined by a fall tube (diameter of 1 m and height of 10 m) in which ice cloud properties and, in particular, crystal growth can be studied over timescales much longer than has been possible with smaller chambers. Temperatures in the cold rooms can be controlled to –50°C, and each of the three chambers can be controlled separately. These new facilities will allow previous work on ice particle nucleation, riming, charge transfer, and interaction of radiation with ice particles to be continued and expanded to include new areas such as secondary ice formation and crystal growth.

Generation and Characterization of Particles Used in Laboratory Cloud Simulations

Reliable generation and sufficient characterization of the particles used in laboratory cloud simulations are already important issues in laboratory cloud research. In addition to the identification of particles aimed at increasing our understanding of particle hygroscopic growth and activation behavior, we must ensure that particles are found which are suitable to the investigation and quantification of the different mechanisms involved in ice formation. Both types of investigations imply generation and characterization of multiphase (e.g., insoluble core and liquid mantle), multicomponent particles, well defined with respect to size and chemical composition. In addition, ice formation investigations require well-defined and characterized particle surfaces.

Another interesting topic involves particle/droplet charge. Electrical charges in particles/droplets may influence microphysical processes such as activation, ice nucleation, and droplet and ice particle growth. However, the generation and characterization of particles/droplets larger than one micrometer with well-defined charge levels, as needed when investigation charge influences microphysical processes, still pose a serious problem.

The above requirements can only be fulfilled if both particle generation and characterization techniques are significantly improved. The greatest need lies in the fields of generating and characterizing particle surface properties.

Specific Cloud Droplet–Turbulence Interactions

Investigations of cloud droplet–turbulence interactions (e.g., entrainment, turbulence-influenced condensational and collisional growth, and the spatial and Lagrangian properties of particles in turbulence) are far from complete, and there is a strong need for controlled laboratory experiments to isolate and quantify the mechanisms. Several challenges exist, including the development of experimental facilities that can match all of the relevant dimensionless parameters governing the processes under consideration. For example, when considering the Lagrangian properties of inertial particles in turbulence, it is necessary to match the particle inertia (Stokes number) as well as the particle settling (settling parameter and Froude number). Many laboratory systems tend to produce turbulence with much higher energy dissipation rates than exist in typical clouds, which makes overlap with the correct parameter ranges challenging. Further complicating matters is the difficulty in achieving large, geophysical Reynolds numbers in laboratory systems: any process sensitive to intermittency will be subject to this constraint, and only very large systems, such as the Göttingen wind tunnel or the large mine shafts described earlier, may be able to approach the Reynolds numbers that are expected to exist, for example, in a cumulus cloud.

Combining Turbulence and Microphysics over Multiple Scales

The topics and experiments discussed above address the investigation, simulation, and quantification of single cloud microphysical and turbulent processes. Although vital to our overall understanding of cloud processes, they can only be considered as an initial step in this process. One key topic that has been intensely debated concerns the issue of what really controls cloud properties such as droplet number and size, lifetime and precipitation behavior. An aspect of this problem is the question of homogeneous versus inhomogeneous mixing. Since the early laboratory experiments that initially led to the development of these ideas in the 1980s, essentially no laboratory studies have been dedicated to this important issue. This illustrates a challenge that we face on many issues involving large-scale turbulence: because entrainment and mixing are inherently multiscale processes, it is difficult to achieve such large ranges of spatial and temporal scales in typical laboratory-sized chambers.

To answer or even address these questions, none of the devices and techniques described above appears to be fully suitable. Therefore we suggest that experiments be designed and performed that allow the controlled and well-defined adjustment of both microphysical and turbulence parameters. Only this way can the possible interactions and feedbacks between the microphysical (activation, growth, freezing) and turbulent transport processes within clouds be simulated and quantified. Devices such as small-scale expansion chambers and wind tunnels, or maybe a combination thereof, might be applicable.

Conclusions

In this chapter, we discussed the simulation of clouds in the laboratory and focused on cloud-related topics and effects that have been or could possibly be investigated in laboratory studies. Topics such as aerosol particle hygroscopic growth and activation, droplet dynamic growth, ice nucleation and droplet freezing, and droplet–turbulence interactions were presented, and a number of devices used for laboratory investigation and simulation of relevant cloud processes were discussed. Since cloud physical processes and their simulation involve different scales, we classified cloud investigation/simulation devices into (a) devices that measure selected particle and cloud droplet properties, (b) devices that allow process studies under simulated cloud conditions, and (c) devices that permit a nearly full-scale cloud simulation. However, we emphasize that at the present time, and most likely in the near foreseeable future, no device exists that is capable of simulating a "real" cloud, with all of its relevant processes and complexity.

For the future, we suggest that investigations be continued and/or initiated to address (a) particle hygroscopic growth and activation, (b) the accommodation coefficients of water vapor on liquid water and ice, (c) aerosol effects on primary ice formation in clouds, (d) aerosol-based parameterizations of cloud ice formation, secondary ice formation or ice multiplication, (e) the production and characterization of particles suitable for cloud simulation experiments, and (f) the combination of turbulence and microphysics. We consider the latter to be of particular importance in the simulation and quantification of possible interactions and feedbacks between the microphysical (activation, growth, freezing) and turbulent transport processes within clouds.

Acknowledgments

We acknowledge, in alphabetic order, the very helpful contributions of Karoline Diehl (University of Mainz, Germany), Yasushi Fujiyoshi (Institute for Low Temperature Science, Hokkaido University, Sapporo, Japan), John Helsdon (South Dakota School of Mines & Technology, Rapid City, South Dakota, U.S.A.), Szymon Malinowski (Warsaw University, Poland), Masataka Murakami (Meteorological Research Institute, Tsukuba, Japan), and Zellman Warhaft (Cornell University, Ithaca, New York, U.S.A.).

References

Ayyalasomayajula, S., A. Gylfason, L. R. Collins, E. Bodenschatz, and Z. Warhaft. 2006. Lagrangian measurements of inertial particle accelerations in grid generated wind tunnel turbulence. *Phys. Rev. Lett.* **97**:144507.

Bailey, M. and J. Hallett. 2002. Nucleation effects on the habit of vapour grown ice crystals from –18° to –42°C. *Q. J. Roy. Meteor. Soc.* **128**:1461–1483.

Benz, S., K. Megahed, O. Möhler et al. 2005. T-dependent rate measurements of homogeneous ice nucleation in cloud droplets using a large atmospheric simulation chamber. *J. Photochem. Photobio. A. Chem.* **176**:208–217.

Bundke, U., H. Bingemer, B. Nillius, R. Jaenicke, and T. Wetter. 2008. The FINCH (Fast Ice Nucleus Chamber) counter. Proc. 17th Intl. Conf. on Nucleation and Atmospheric Aerosols. *Atmos. Res.,* in press.

Davidovits, P., C. E. Kolb, L. R. Williams, J. T. Jayne, and D. R. Worsnop. 2006. Mass accommodation and chemical reactions at gas-liquid interfaces. *Chem. Rev.* **106(4)**:1323–1354.

Davis, E. J. 1997. A history of single aerosol particle levitation. *Aerosol Sci. Technol.* **26**:212–254.

DeMott, P. J., and D. C. Rogers. 1990. Freezing nucleation rates of dilute-solution droplets measured between –30°C and –40°C in laboratory simulations of natural clouds. *J. Atmos. Sci.* **47**:1056–1064.

Duft, D., and T. Leisner. 2004a. The index of refraction of supercooled solutions determined by the analysis of optical rainbow scattering from levitated droplets. *Int. J. Mass Spectrom.* **233**:61–65.

Duft, D., and T. Leisner. 2004b. Laboratory evidence for volume-dominated nucleation of ice in supercooled water microdroplets. *Atmos. Chem. Phys.* **4**:1997–2000.

Fugal, J. P., J. Lu, H. Nordsiek, E. W. Saw, and R. A. Shaw. 2007. Lagrangian properties of cloud particles in turbulence obtained by holographic particle tracking. 60th Annual Meeting of the Division of Fluid Dynamics, Abstract JU.00024. Salt Lake City: Amer. Phys. Soc.

Imre, D. G., J. Xu, and A. C. Tridico. 1997. Phase transformations in sulfuric acid aerosols: Implications for stratospheric ozone depletion. *Geophys. Res. Lett.* **24**:69–72.

Köhler, H. 1923. Zur Kondensation des Wasserdampfes in der Atmosphäre, erste Mitteilung. *Geophys. Publ.* **2**:1–15.

Korczyk, P., S. P. Malinowski, and T. A. Kowalewski. 2006. Mixing of cloud and clear air in centimeter scales observed in laboratory by means of particle image velocimetry. *Atmos. Res.* **82**:173–182.

Magee, N., A. M. Moyle, and D. Lamb. 2006. Experimental determination of the deposition coefficient of small cirrus-like ice crystals near –50°C. *Geophys. Res. Lett.* **33**:L17813.

Malinowski, S. P., I. Zawadzki, and P. Banat. 1998. Laboratory observations of cloud-clear air mixing at small scales. *J. Atmos. Ocean. Technol.* **15**:1060–1065.

McMurry, P. H. and M. R. Stolzenburg. 1989. On the sensitivity of particle size to relative humidity for Los Angeles aerosols. *Atmos. Environ.* **23(2)**:497–507.

Möhler, O., H. Bunz, and O. Stetzer. 2006. Homogeneous nucleation rates of nitric acid dihydrate (NAD) at simulated stratospheric conditions. Part II: Modelling. *Atmos. Chem. Phys.* **6**:3035–3047.

Möhler, O., P. R. Field, P. Connolly et al. 2006. Efficiency of the deposition mode ice nucleation on mineral dust particles. *Atmos. Chem. Phys.* **6**:3007–3021.

Pruppacher, H. R. 1988. Auswaschen von atmosphärischen Spurenstoffen durch Wolken und Niederschlag mittels eines vertikalen Windkanals. In: BMFT-Bericht 9/88. Munich: GSF.

Radke, L. F., and P. V. Hobbs. 1969. An automatic cloud condensation nuclei counter. *J. Appl. Meteor.* **8**:105–109.

Roberts, G., and A. Nenes. 2005. A continuous-flow streamwise thermal-gradient CCN chamber for atmospheric measurements. *Aerosol Sci. Technol.* **39**:206–221.

Rogers, D. C. 1988. Development of a continuous flow thermal gradient diffusion chamber for ice nucleation studies. *Atmos. Res.* **22**:149–181.

Rogers, D. C., P. J. DeMott, S. M. Kreidenweis and Y. Chen. 2001. A continuous flow diffusion chamber for airborne measurements of ice nuclei. *J. Atmos. Ocean. Technol.* **18**:725–741.

Saunders, C. P. R., and A. S. Hosseini. 2001. A laboratory study of the effect of velocity on Hallett–Mossop ice crystal multiplication. *Atmos. Res.* **59**:3–14.

Stetzer, O., B. Baschek, F. Luond, and U. Lohmann. 2008. The Zurich Ice Nucleation Chamber (ZINC): A new instrument to investigate atmospheric ice formation. *Aerosol Sci. Technol.* **42**:64–74.

Stratmann, F., A. Kiselev, S. Wurzler et al. 2004. Laboratory studies and numerical simulations of cloud droplet formation under realistic supersaturation conditions. *J. Atmos. Ocean. Technol.* **21**:876–887.

Vohl, O. 1989. Die dynamischen Charakteristika des Mainzer vertikalen Windkanals. Diploma thesis. Institute of Atmospheric Physics, University of Mainz, Germany.

Yamagata, S., S. Baba, N. Murao et al. 1998. Real scale experiment of sulfur dioxide dissolution into cloud droplets generated in artificial cloud experimental system (ACES). *J. Global Environ. Eng.* **4**:53–63.

Yamagata, S., K. Takeshi., Z. Takehiko et al. 2004. Mineral particles in cloud droplets produced in an artificial cloud experimental system (ACES). *Aerosol Sci. Technol.* **38(4)**:293–299.

Wex, H., Stratmann, F., Topping, D., and McFiggans, G. 2008. The Kelvin versus the Raoult term in the Köhler equation. *J. Atmos. Sci.*, in press.

Ziese, M., H. Wex, E. Nilsson et al. 2008. Hygroscopic growth and activation of HULIS particles: experimental data and a new iterative parameterization scheme for complex aerosol particles. *Atmos. Chem. Phys.* **8**:1855–1866.

8

Cloud-controlling Factors

Low Clouds

Bjorn Stevens[1] and Jean-Louis Brenguier[2]

[1]Max Planck Institute for Meteorology, Hamburg, Germany
[2]Meteo-France, CNRM, Toulouse, France

Abstract

The way in which meteorological and aerosol factors conspire to determine the statistics and climatology of layers of shallow (boundary layer) clouds is reviewed, with an emphasis on factors that may be expected to change in a perturbed climate. The paramount role of theory is identified, both in service of advancing our understanding, but also in modeling and attributing specific causes and effects. In particular, it is argued that limits to current understanding of meteorological controls on cloudiness make it difficult, and in many situations perhaps impossible, to attribute changes in cloudiness to aerosol perturbations. Suggestions for advancing our understanding of low cloud-controlling processes are offered; these include renewing our focus on theory, model craftsmanship, and increasing the scope and breadth of observational efforts.

The Idea

The idea that clouds are sensitive to their environment, dramatically illustrated through Figure 8.1, should surprise no one. In introducing Scorer's (1972) cloud atlas, F. H. Ludlam wrote:

> Clouds have an infinite variety of *shapes*, but a limited number of *forms* corresponding to different physical processes in the atmosphere which are responsible for their formation and evolution.

Ludlam continued then to trace the history of the modern cloud classification, beginning with the work of Lamarck (1802) and Howard (1802) and culminating with the 1887 treatise of Abercromby and Hildebrandsson, which presents a cloud classification scheme that is identical, in all essential respects, to that

Figure 8.1 MODIS image from July 5, 2002, showing marine stratocumulus in the vicinity of the Canary Islands off the West African coast.

used by meteorological observers today. It took 85 years to work out the basic *forms* by which we classify clouds; however, the task of relating these forms to specific physical processes (their physical environment if you will)—now some 120 years later—has not yet been completed.

This relating process to form is the focus of this chapter. We concentrate on what one might call a supergenus—*low clouds*—which includes the familiar genera of stratus and cumulus (dating from Howard's classification) and stratocumulus (dating from the 1836 monograph by Kämtz). Our emphasis is on stratocumulus and cumulus. From the point of view of the climate system, the shallowness of the cloud layer (or the property of being "low") is principally manifest as a temperature difference ($T_0 - T_c$) between the surface, T_0, and cloud top, T_c. This temperature difference is related to the physical height of the cloud top; however, it is more fundamental as it helps meter the relative effects of clouds on the effective emissivity of the atmosphere as compared to its albedo. The emissivity (or greenhouse) effect of clouds increases with the difference in temperature between the surface and cloud top, whereas the albedo effect of clouds need not. Because these effects tend to compete with one another, perturbations to the properties and statistics of low clouds affect albedo disproportionately and hence alter more profoundly the net radiative balance of the system as a whole. Contemporary efforts to relate process to the form of low clouds is motivated by this decisive fact.

There are, however, additional, less appreciated motives for studying the physical processes that control the statistics of low clouds. For instance, low clouds are by definition thin, which means the difference between their cloud top and base temperatures is also small. Because of the potential condensation in an adiabatic current spanning the depth of the cloud scales with this temperature difference, the liquid water content in the cloud (even in the absence of precipitation) remains a small fraction of the total water content. We will argue that this removes a constraint from the system that renders the state of shallow clouds more susceptible to environmental perturbations. In plain terms, the development of rain in shallow clouds is not necessary. This is just one way in which they are delicate.

In addition to being decisive and delicate, low clouds are ubiquitous. This property arises from the fundamental asymmetry of moist convection coupled with the need for circulations to conserve mass, which requires ascending air currents to be more vigorous and less spatially extensive than descending currents (Bjerknes 1938). Hence, conditions of gentle subsidence, which favor low clouds, prevail.

Combining their decisiveness with their ubiquity, we estimate that a one percentage-point change in the albedo of low clouds (equivalently a 2–3 percentage-point increase in the average cloud fraction) could result in a 1 W m^{-2} change in the net solar radiation at the top of the atmosphere. Alternatively, such a perturbation could be effected by just a 6 percentage-point increase of the albedo in only the stratocumulus regions, which is on the order of what one might expect from a 0.2 g kg^{-1} moistening of the marine boundary layer, or an increase in ambient droplet concentrations from 75–150 cm^{-3}. In the end, and in light of the perceived delicacies of the system, calculations such as these motivate attempts to answer our phrasing of Ludlam's challenge of relating process to form:

> Quantitatively, what is the relationship between the large-scale environment and the statistics of low clouds?

The answer is paramount to understanding both how clouds may change in the future and, by implication, the climate system.

Our discussion of physical controls on cloudiness is divided in two parts: (a) meteorological controls and (b) the role of the atmospheric aerosol. By meteorology we mean the large-scale dynamic and thermodynamic state that is thought to govern principally cloud macro-structure. The aerosol, by which we mean cloud condensation nuclei (CCN), is principally thought to govern cloud micro-structure. Of course, one is not independent of the other, and a considerable amount of our subsequent discussion will be devoted to the ways in which this fact confounds attempts to meet Ludlam's challenge.

Meteorology

Of the varied forms of moist convection, low clouds are the simplest. They embody fewer processes, and those that are operative typically encompass shorter temporal and smaller spatial scales, making them more amenable to both measurement and simulation. Thus on a phenomenological as well as on a theoretical level, our understanding of shallow clouds is more advanced than that of other cloud forms.

Among the observed relationships between low cloud amount and ambient meteorological conditions, the most compelling are those between seasonal variations in lower tropospheric stability and low-cloud amount in regions where low clouds predominate. Introducing the symbol s to denote the sum of the fluid enthalpy and geopotential, often called the dry-static energy, and adopting a subscript notation so that "s_0" and "s_+" denote values of s at the surface and just above cloud top, respectively, we can define the lower tropospheric stability:

$$\Delta s \equiv s_+ - s_0, \tag{8.1}$$

where Δ is used throughout to denote the difference between a quantity just above cloud top and at the surface. This definition follows most closely what Wood and Bretherton (2006) call the estimated inversion strength (EIS). The observation that cloud incidence tends to increase with Δs goes back to the early stratocumulus studies by Blake (1928) and to some extent Clayton (1896). Recent studies have increasingly shown such relationships to be compelling (Slingo 1987; Klein and Hartmann 1993; Wood and Bretherton 2006), with the latest showing that, if properly constructed, Δs can explain over 80% of interseasonal variance in low-cloud amount. Although deficient because it relates cloud incidence (a non-dimensional quantity) to Δs (a dimensional quantity), these observations are the basis for the long-standing hypothesis that stratocumulus cloud amount will increase in a warmer climate (e.g., Miller 1997). As in a warmer climate, Δs will increase (and does so robustly in general circulation models), as the tropical thermal structure adjusts to a warmer moist-adiabat.

Are there, however, other factors? For instance, surface wind speeds might be thought to play an important role in setting the rate of coastal ocean upwelling, and hence surface temperatures, and thus might provide an important link between cloudiness and the state of the local circulation. To the extent that surface wind speeds are enhanced because of increased cloudiness (and hence large-scale, near-surface cooling), this constitutes a regional-scale feedback process which some suggest might be central to explain features of the current climate (e.g., Philander et al. 1996; Nigam 1997). To the extent that they exist, such feedbacks are likely to compound effects associated with the tendency of greater winds to increase mechanical and biochemical production of the marine aerosol (e.g., Charlson et al. 1987; Latham and Smith 1990).

Free tropospheric humidity may also play a critical role in setting cloudiness. For example, through its critical role in a hypothesized instability process (Deardorff 1980; Randall 1980), Δq is thought to help select among cloud regimes. Well-mixed greenhouse gases, and for that matter, q_+, may also play a role in determining the cloud-top cooling potential of stratocumulus layers, and hence the susceptibility of the cloud layer to decoupling; this, in turn, is thought to be a factor in regulating cloud amount (Turton and Nicholls 1987). During DYCOMS-II, satellite imagery showed that stratocumulus regions break up conspicuously as they drift toward the equator under the influence of increasingly humid mid-free-tropospheric air associated with the North American Monsoon.

Underlying many of the ideas expressed above is the uncertainty about the role of the large-scale divergence of the horizontal wind, $\mathcal{D}$, whose vertical integral is the large-scale vertical motion. The climatology of divergence helps select regions where low clouds predominate, and hence divergence is controlled for when investigating sensitivities between low-cloud amount and other parameters. Climatologically, low clouds are favored in regions of stronger divergence. However, on short time and small spatial scales, such relationships are nontrivial. For instance, all other things being constant, increasing $\mathcal{D}$ lowers cloud top and produces a thinner cloud. In the limit, during periods of offshore flow and strong divergence, stratocumulus is often suppressed as the boundary layer becomes insufficiently deep to support clouds (Weaver and Pearson 1990); if the vertical motion is sufficiently weak, areas of stratiform cloudiness may break down into more cumuliform patterns.

Theory, of course, exists to order the empiricism. To begin, one could simply list the parameters previously discussed and, in doing so, identify what one might call a zeroth-order meteorological vector,

$$m = \{s_0, q_0, \mathcal{V}, \mathcal{D}, s_+, q_+, F_+\}, \tag{8.2}$$

consisting of ostensibly external factors regulating clouds. In addition to the previously introduced variables, $\mathcal{V}$ denotes a surface exchange velocity (proportional to the mean wind) and F_+ denotes the net downwelling longwave radiation above the cloud layer. This vector identifies seven parameters. After multiplying q by L_v, the enthalpy of vaporization, all except $\mathcal{D}$ carry the units of velocity to some power and, because given a surface at saturation and fixed pressure q_0 is given by s_0, m identifies at least a four-dimensional parameter space in which we expect the properties of low clouds to be manifest. The meteorology is neither homogeneous nor stationary; thus, spatial and temporal gradients of m, as well as covariance among its components, will (even in the absence of variations in the aerosol) render the effective dimensionality of the space much larger. This explains why Ludlam's challenge of relating process to form has proved to be much more difficult than Howard's challenge of relating name to form.

For both cumulus and stratocumulus, the theory largely attempts to describe the bulk or integral state of the cloud and subcloud layers (for a review, see Stevens 2006). Because the theory is expressed differently, depending on whether we consider stratocumulus versus shallow cumulus, we discuss each in turn.

Stratocumulus

Lilly's 1968 mixed-layer theory identifies key environmental factors that influence stratocumulus cloud amount (or thickness). In the absence of advective forcing, Stevens (2006) used this theory to derive an expression for the steady-state cloud thickness:

$$h - z_b = h\left\{1 - \frac{\Pi_3}{\alpha_e}\left[\Pi_1 \ln\left(\frac{\Pi_2}{\Pi_2 - \alpha_e \Pi_4}\right) - (1-\alpha_e)(\Pi_2 - \alpha_e)\right]\right\}, \qquad (8.3)$$

where h denotes cloud-top height, z_b cloud-base height, α_e the nondimensional entrainment,

$$\Pi_1 = \frac{s_0 \mathcal{V}}{\Delta F}\left(\frac{R_v T_0}{L_v}\right), \quad \Pi_2 = 1 + \mathcal{V}\frac{\Delta s}{\Delta F}, \quad \Pi_3 = \frac{\mathcal{D}\Delta s}{g\mathcal{V}}, \quad \text{and } \Pi_4 = -\frac{\Delta q}{q_0}. \qquad (8.4)$$

Here g denotes gravity and R_v the gas constant for water vapor. The cloud albedo is an increasing function of its optical thickness, τ, which to a first approximation can be expressed as:

$$\tau = \frac{3}{2}\frac{\mathcal{L}}{r_e}, \qquad (8.5)$$

where $\mathcal{L}$ represents the liquid water path and r_e the effective radius. They are given as:

$$\mathcal{L} \propto (h - z_b)^2 \gamma_l \quad \text{and} \quad r_e \propto \sigma\left[N^{-1}\gamma_l (h - z_b)\right]^{1/3}, \qquad (8.6)$$

where γ_l is taken as the cloud-averaged value of dq_l/dz, N the expected number of activated cloud droplets for an adiabatic cloud, and σ some unspecified measure that encodes effects as a result of the shape of the drop distribution and diabatic (collisions, mixing) effects. Equation 8.6 provides a basis for relating cloud optical properties (or albedo) to cloud geometrical properties. These relationships show that m largely determines the cloud geometric properties, whereas the optical properties depend additionally on the cloud microphysical structure, which in turn may be expected to reflect, at least in part, the properties of the aerosol.[1]

[1] However, especially to the extent clouds precipitate, cloud geometric properties may also be regulated by N.

Relationships such as those above, and the ideas that underlie them, should explain and extend the existing empiricism. There is evidence that the theory is beginning to rise to this challenge. For example, the ideas behind Equation 8.3 can also be used to show that the approach to equilibrium is slow, which means the past matters. This point helps to explain the finding made by Klein et al. (1995) and Pincus et al. (1997) that in a Lagrangian sense, cloud optical properties lag environmental conditions by 12–24 hours, which corresponds roughly to the thermodynamic timescale of the mixed-layer model. The same theory shows that multiple timescales emerge and can lead to differential short- and long-term behavior (e.g., Zhang et al. 2005; Wood 2007), which is also inferred in the observational study by Pincus et al. (1997).

Likewise, the role of microphysical processes identified in Equation 8.6 is the basis for the so-called cloud–water feedback. The idea, first explored by Paltridge (1980), is that γ_l depends on a fixed way to its adiabatic value, which in turn depends on $R_v s_0/(c_p L_v)$, and hence on temperature. The sensitivity of r_e to cloud microphysical alterations through σ has been explored by Chosson et al. (2007). By distributing cloud water according to different mixing scenarios *ex post facto*, they show that the impact of inhomogeneous mixing processes on the droplet size distribution at the top of a stratocumulus layer can affect the optical thickness of the layer more efficiently than the heterogeneous spatial distribution of liquid water (cf. Burnet and Brenguier 2007).

Notwithstanding these initial steps, a full-scale evaluation of the theory has not been attempted for at least two reasons: (a) the theory remains incomplete, or deficient; (b) many of the parameters upon which it is based (most notably q_+, F_+, and $\mathcal{D}$) are difficult to measure. The principal deficiencies in the theory are the lack of a compelling specification of α_e, and the breakdown of a key assumption (well mixedness) under certain conditions. Next, we discuss these two points.

Entrainment

The entrainment rate is expressed non-dimensionally, following Stevens (2006), as $\alpha_e = E\Delta s/\Delta F$, in Equation 8.3. Doing so obscures the fact that it (through the entrainment velocity, E) may depend on m, which means that the full parameter sensitivity of even the equilibrium states of the mixed-layer model remains uncertain.

One idea would be to use large-eddy simulation to estimate α_e, and indeed this has been a strategy adopted by a number of investigators (e.g., Lock and MacVean. 1999; Moeng et al. 1999). This strategy, however, has proved challenging, largely because the fine scale of the entrainment processes makes it difficult to represent with fidelity. This point formed the focus of an intercomparison study by Stevens (2005), who showed that entrainment is sensitive to the detailed numerical formulation of a simulation. As shown in Figure 8.2, even at a relatively fine resolution, the representation of the cloud layer is

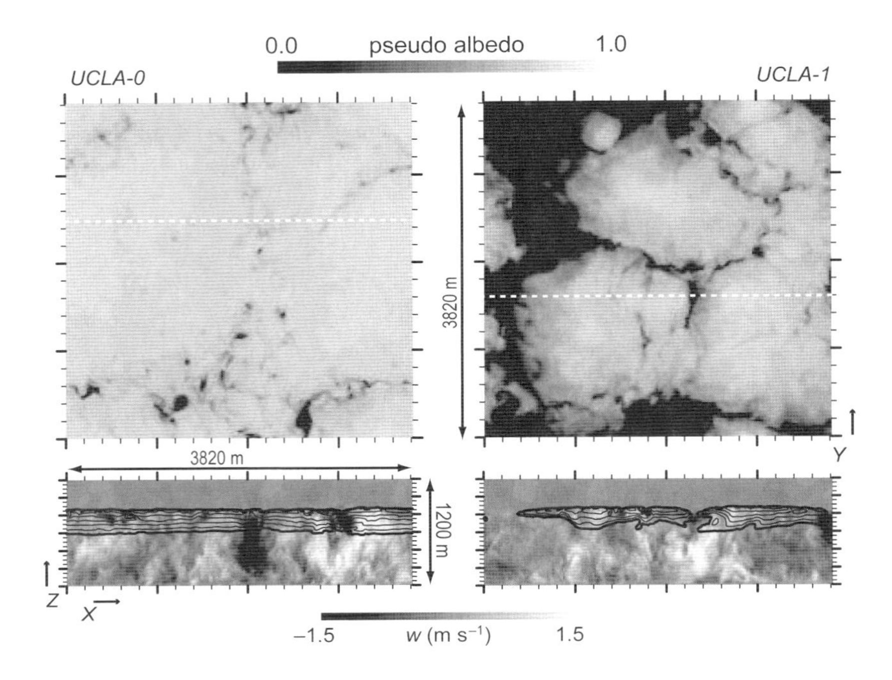

Figure 8.2 Visualization of flow fields from two simulations using the UCLA large-eddy simulation model which differs only in the representation of small-scale (subgrid) mixing. Shown are plan-view images of the albedo estimated from the liquid water path, and cross sections showing vertical velocity (shaded) and cloud water (contoured). Cross-section cuts are indicated by the white dashed line in the plan-view plots. These fields were drawn from simulations wherein $N_x = N_y = 192$ and $\Delta x = \Delta y = 20$ m, $\Delta z = 5$ m in the vicinity of cloud top.

sensitive to the representation of the smallest scales, as manifest in this intercomparison of differences in the representation of the subgrid-scale mixing. The simulations are an example of numerical delicacy. For this particular case, the result on the left (labeled UCLA-0) conforms better to the data. As a point of reference, this simulation was only constructed after it was determined that the default configuration of the model (which produced the image on the right) yielded results that were incompatible with the data. Nonetheless, our ability to begin constraining the models with data in these respects represents a significant step forward and provides an example of the increasingly critical interplay between models, theory, and data.

Decoupling

The mixed-layer limit is believed to become energetically inconsistent in certain regions of parameter space. Roughly speaking, when surface fluxes become significant, relative to the effective cloud-top radiative driving, one can expect a transition to a more cumulus-coupled layer. This transition is often referred to as decoupling because it is associated with increased differentiation (decoupling) between the thermodynamic properties of the cloud and subcloud layer. It was developed as a theory of the stratocumulus to cumulus transition by Bretherton and Wyant (1997), following on ideas developed by Turton and Nicholls (1987) to explain the diurnal cycle. Stevens (2000) used idealized simulations to show that such transitions are sharp, or threshold-like, in parameter space, suggesting the existence of distinct regimes or attractors of the flow. In the decoupled regime one expects significant differentiation between the cloud and subcloud thermodynamic properties, which results in marked reductions in cloudiness (see also Lewellen and Lewellen 2002), as compared to the well-mixed regime.

Regime transitions are thought to underlie the types of numerical delicacies seen in Figure 8.2, wherein small changes in the efficacy of cloud-top mixing may engender profound differences in cloud structure. Because this is an energetic threshold, such a transition may be triggered by a variety of processes (e.g., precipitation). The emergence of such nonlinear behavior implies that the statistics of cloudiness may depend on a richer characterization of the meteorological vector. Thus, for instance, small values of $\mathcal{D}$ imply small values of Π_3 and hence, by Equation 8.3, favor an increase in τ. However, to the extent such states are effectively forbidden, one might expect τ to depend less on m and more on the character of its spatial and temporal fluctuations.

Cumulus

A compelling theoretical framework for cumulus has proved to be more difficult to establish (cf. Albrecht et al. 1979; Bretherton 1993; Bellon and Stevens 2005; Bretherton and Park 2007). In some respects this is surprising, because in certain limits cumulus convection is much simpler than stratocumulus convection. For instance, because radiative fluxes are important primarily to the long-time behavior of boundary layers topped by cumulus convection, they are only indirectly coupled to the turbulent fluxes through the evolution of the mean state and hence surface fluxes. As such it is possible to construct compelling, but purely *transient*, representations of layers of cumulus convection by replacing the surface temperature description with a surface flux prescription. In such a framework one need not make explicit reference to either $\mathcal{D}$ or F_+, which somewhat simplifies the zeroth-order meteorological vector, m (Stevens 2007). To the extent a *stationary* description is desired, such simplifications are not warranted. Simple bulk theories, such as the one layer model of Betts and Ridgway (1989), may be useful. However, even these are complicated by the fact that the radiative cooling is distributed through the layer (rather than concentrated at cloud top as in stratocumulus), so that the fraction of cooling within just the subcloud layer emerges as an additional parameter.

Energetics and Mixing

One of the chief challenges for simple theories of shallow cumulus is the difficulty of reconciling the internal vertical structure with the energetics of the layer. Most research to date focuses on the question of this internal structure; for example, the cloud-base mass flux and its distribution through the cloud layer as constrained by models of lateral mixing (Siebesma 1998; Neggers et al. 2004). Work in this regard suggests that the convective mass flux is well constrained by both the thermodynamic and energetic state of the subcloud layer (Neggers et al. 2007), thereby providing guidance, albeit relatively unexplored, as to how convective cloudiness may change in response to factors like changing surface winds or fluxes, or perhaps changes in the effective saturation pressure of the free troposphere. Recent work also clarifies the mechanism through which the cloud layer deepens and the energetic constraints placed on this process by the subcloud layer (Stevens 2007). In this respect, one important finding is that the cumulus layer deepens primarily as a result of the flux of liquid water into the inversion. This implies that the development of precipitation may help arrest the growth of the cloud layer (Stevens and Seifert 2008). This has implications for cloud feedbacks that exercise microphysical pathways. Clouds that rain less can be expected to deepen more efficiently, but deeper clouds rain more. Similar effects are evident in layers of stratocumulus convection, but are not thought to be so dominant (Stevens et al. 1998; Ackerman et al. 2000; Bretherton et al. 2006).

Convective versus Stratiform Cloudiness

Another important challenge is to determine the respective roles and parameter sensitivity of convective versus stratiform (large-scale) cloudiness. Conceptually, the convective cloudiness can be expected to be proportional to the mass flux and hence constrained by the energetics of the layer; even if processes like the vertical shear of the horizontal wind may yield different cloud distributions for the same mass flux, just as a function of cloud overlap. Stratiform, or large-scale cloudiness in the cloud layer, is less directly related to the convective fluxes and at times may have its own (somewhat more stratocumulus like) dynamics. This complicates the theoretical development and is why most simple models of cumulus layers do not have any predictive skill in determining the amount of cloudiness within the layer (cf. Betts and Ridgway 1989).

Early models (e.g., Slingo 1980) simply tied the large-scale, or stratiform, cloud amount to the relative humidity in a model layer, such that

$$c = f(\mathcal{U}), \tag{8.7}$$

where $\mathcal{U}$ is the large-scale (grid-cell) relative humidity and f is a non-decreasing function between zero and one. Models of the form contained in Equation 8.7 specify implicitly a distribution of the humidity within a grid cell and, through the constancy of the form of f, assume it to be universal. Adaptations of this approach condition the form of f on the cloud regime, for instance by letting it depend on the large-scale vertical velocity or the strength of the cloud-capping inversion (Slingo 1987), or by predicting parameters of a distribution function whose form, and degrees of freedom, are assumed a priori (e.g., Bougeault 1982; Smith 1990; Lewellen and Yoh 1993). Another approach has been to couple the stratiform and convective cloud representations directly (e.g., Sundqvist 1978; Albrecht 1989; Xu and Randall 1996). Doing so introduces different parameter sensitivities, and hence articulates the possibility of different meteorological and microphysical feedbacks. Work on these issues is making increasing use of cloud-resolving or large-eddy simulation models to help decide which strategy is best (e.g., Tompkins 2002), further exemplifying the increasing maturity in the use of models, theory, and data.

The Aerosol

The idea that clouds depend on the atmospheric aerosol, and not just the dynamic and thermodynamic properties of the atmosphere (meteorology), dates back to the earliest cloud studies. In a series of pioneering measurements, Squires and colleagues showed that droplet concentrations in maritime clouds are significantly less than the concentrations characteristic of similar clouds forming in air masses of continental origin (Squires 1956; Squires and Twomey 1966). Squires's (1958) casual observation that shallow marine clouds often rain in less than 30 minutes foreshadowed the mystery of warm rain formation. Subsequent studies, which suggested that anthropogenic perturbations to the aerosol could modulate the propensity of clouds to rain (Warner and Twomey 1967; Warner 1968), stimulated decades of research in weather modification. Twomey's recognition that changes in the aerosol could be important to cloud optical properties (Twomey and Wojciechowski 1969) further motivated research into aerosol–cloud interactions, so that by now they are a major focus of climate research (e.g., Rosenfeld 2006).

Twomey's Calculation

In the absence of other changes, the albedo of clouds with intermediate optical depths (e.g., marine stratocumulus) is especially susceptible to perturbations in cloud drop number concentrations and, by inference, the atmospheric aerosol (Twomey 1974, 1977). Satellite imagery and *in-situ* measurements of ship tracks similar to those shown in Figure 8.3 have provided observational support for these ideas (Radke et al. 1989), as have closure studies on the cloud system scale (e.g., the ACE-2 measurements described by Brenguier et al. 2000). Attempts to incorporate these processes into physically based models have, however, been frustrated by the poor representation of underlying cloud

Figure 8.3 MODIS image from NASA's Aqua satellite showing ship tracks over the Pacific Ocean, just west of British Columbia on January 21, 2008.

processes in existing climate models, as well as a poor understanding of the pathways through which perturbations to the aerosol manifest themselves on the climate system.

Proliferating Pathways

Since Twomey's pioneering studies, the literature has seen a proliferation of hypothesized indirect aerosol–cloud interactions.[2] Perhaps the simplest of this class of effects is that changes in the aerosol, by modifying clear-sky radiative fluxes, may alter the stability of the cloud layer, and hence subsequent cloudiness (e.g., Hansen et al. 1997). Fine-scale modeling studies provide some support for this idea. Ackerman et al. (2000) show that the effect of enhanced absorption of solar radiation by soot in the cloud layer can lead to a reduction in cloudiness to an extent that is not inconsistent with the observational record. Similarly, Feingold et al. (2005) suggest that the reduction in surface fluxes associated with an increase in the aerosol can also reduce cloudiness, thereby counteracting the effects discussed by Twomey. This study, however, begins to hint at the complexity of aerosol–cloud interactions, as the effect of smoke on cloudiness may depend not only on the chemical properties of the aerosol particles (relative role of absorption versus CCN) but also its distribution in the vertical.

Among the more complex ideas are those that connect Squires's interest in aerosol controls on precipitation, with Twomey's focus on cloud optical properties. As an example, Albrecht (1989) argued that changes in the precipitation

[2] This is not to be confused with indirect aerosol–climate interactions, of which the direct aerosol–cloud interaction described by Twomey is a classic example.

rate affected by changes in the aerosol would affect cloud fraction. His hypothesis follows naturally from his cloud model, wherein following Sundqvist (1978), Albrecht couples cloudiness to condensate amount, thereby linking the former to the propensity of clouds to precipitate, as that acts as a sink for the latter. The effects articulated by Albrecht are sometimes called *cloud lifetime effects* because they hinge on the idea that cloudiness depends on condensate amount, which in turn is limited by precipitation. Notwithstanding that most (if not all) work linking cloud lifetime to precipitation actually connects the lifetime of radar echoes to precipitation (e.g., Saunders 1965; Cruz 1973), investigations of these effects using fine-scale models have, just as often, found countervailing processes. For example, Xue and Feingold (2006) and Xue et al. (2008) have shown that the tendency to produce larger drops leads to more cloudiness, as these drops linger. Only when the rain production becomes very efficient does one see a tendency for cloudiness to decrease with more rain, although in such situations secondary effects, such as the formation of cold pools (which are well known to promote the longevity of cloud systems), may counteract this.

Pincus and Baker (1994) introduced a related but conceptually distinct idea, wherein they proposed that increased precipitation could affect the thickness and hence the radiative properties of stratocumulus by altering cloud dynamic processes. Although the details of the mechanism by which this result was manifest in their model (i.e., condensation heating affected a reduction in surface fluxes which in turn reduced the deepening of the layer by entrainment) is probably incorrect, their work was among the first to articulate a cloud–aerosol connection mediated by dynamic (rather than thermodynamic) processes. Their study also merits attention in that it was specifically formulated for layers of stratocumulus, which has been the dominant focus of large-scale model and satellite observational-based attempts to quantify the radiative effects of the aerosol on the climate system as a whole. The observational record is, however, not definitive, and most insight has been developed on the basis of fine-scale modeling studies. As discussed above, large-eddy simulation of stratocumulus-topped layers support the idea that increased precipitation leads to less entrainment (Stevens et al. 1998; Ackerman et al. 2004; Bretherton et al. 2006) and less cloudiness. This is perhaps most dramatically evident in the results of Savic-Jovcic and Stevens (2008), for which precipitating and non-precipitating stratocumulus are represented simply by changing the droplet concentrations from 25 to 200 cm^{-3}. The albedo of these simulated clouds is shown in Figure 8.4 and suggests that precipitation can also have a profound effect on cloud morphology, most likely as a result of the regime transitions discussed above, and not unlike what is observed in pockets of open cells (Stevens et al. 2005). These calculations may also help explain the observed tendency of ship tracks to generate closed-cell cloud forms in regions otherwise consisting of open cells (e.g., on the eastern side of Figure 8.3).

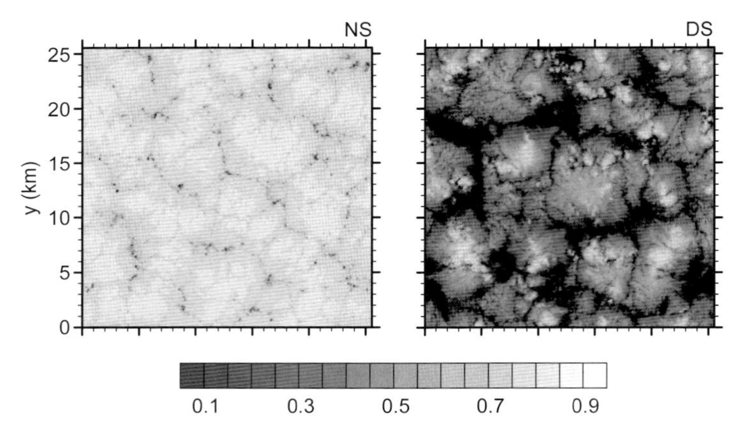

Figure 8.4 Visualization of cloud albedo from large domain simulations with (NS, upper left) and without (DS, upper right), as a function of radiative forcing. Lower simulations show the effect of eliminating the radiative forcing, which we use to represent crudely the effect of solar radiative heating, which to a first approximation can be thought to compensate longwave cooling.

But here again, one finds evidence that the response of the system may be more complex. For instance, Sandu et al. (2008) examined the coupling between aerosol impacts and the diurnal cycle of stratocumulus in simulations for which the radiative and surface fluxes are allowed to respond to the evolution of the cloud layer. Their simulations, albeit on a much smaller domain, suggest that precipitation in the pristine cases reduces the amplitude of the diurnal cycle, partially by inhibiting turbulent mixing at night which then mitigates decoupling during the day (Figure 8.5).

Discussion

From the above review, several themes emerge: (a) attribution demands theory; (b) theory expresses constraints, of which we have too few; and (c) models are necessarily selective in both their empirical and theoretical content.

Attribution Demands Theory

Consider $c(m, a)$ where a introduces a vector describing those aspects of the aerosol affecting cloudiness. It follows that

$$\delta c = \left.\frac{\partial c}{\partial m}\right|_a \delta m + \left.\frac{\partial c}{\partial a}\right|_m \delta a. \tag{8.8}$$

For sake of argument, let us suppose that δc, δm, and δa are all observable;

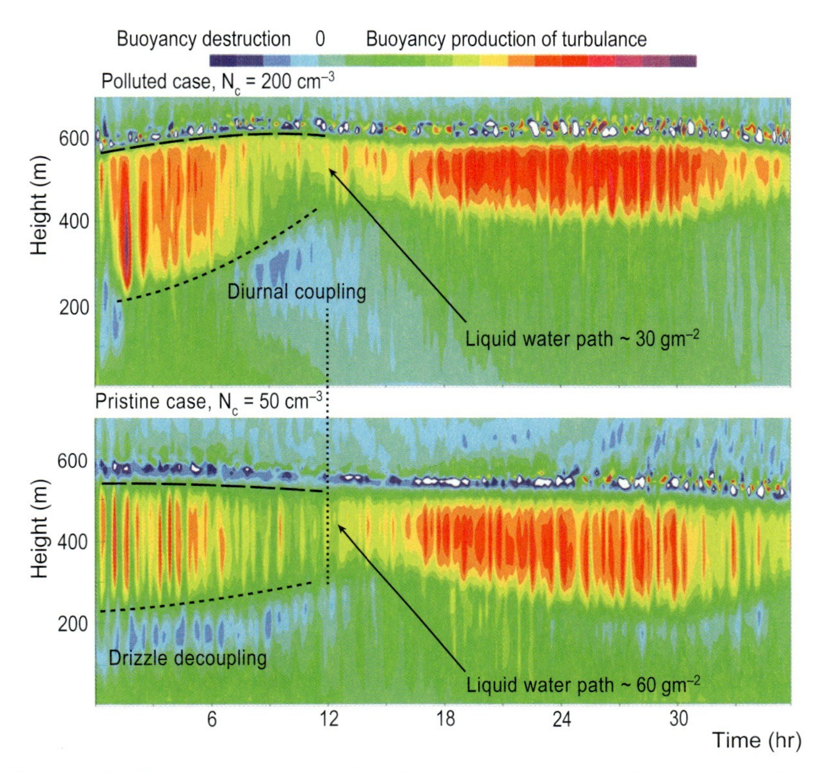

Figure 8.5 Time-height cross section showing the evolution of the buoyancy flux in the stratocumulus-topped boundary layer. The top panel shows a case where droplet concentrations are 220 cm^{-3} and drizzle is weak; the lower panel shows a case with droplet concentrations of 50 cm^{-3} for which drizzle during the night is strong and induces decoupling.

then changes in cloudiness that result from changes in the aerosol (the aerosol effect) can be estimated as:

$$\left.\frac{\partial c}{\partial a}\right|_m = \frac{1}{\delta a}\left[\delta c - \left.\frac{\partial c}{\partial m}\right|_a \delta m\right]. \tag{8.9}$$

Clearly, our ability to estimate such effects is limited by our ability to control for the meteorology. Although this point is well known, what tends to be less well appreciated is the fact that controlling for meteorology does not simply require one to bound δm (which tends to be the approach in the existing, largely flawed, literature) but rather $\left.\frac{\partial c}{\partial m}\right|_a \delta m$. It is precisely the factor $\left.\frac{\partial c}{\partial m}\right|_a$ that gives meaning to the statement that δm is small. In plain terms, the ability to rule out meteorological factors is conditioned on the level of our theoretical understanding, which we have previously argued is primitive.

The difficulty in disentangling meteorological effects on cloudiness from those of the atmospheric aerosol is compounded not only by the poor state of theory, but also because m and a depend on each other in ways that are similarly unclear. Indeed, almost all work that looks for aerosol effects on cloudiness distinct from the pathway articulated by Twomey (1974) explores the idea that $m(m_0, a)$, where m_0 denotes some primitive or initial meteorological state vector whose effects are independent of the aerosol and can presumably be controlled. In this case, one explores the idea that

$$\delta c = \left.\frac{\partial c}{\partial m}\right|_a \left.\frac{\partial m}{\partial m_0}\right|_a \delta m_0 + \left[\left.\frac{\partial c}{\partial a}\right|_m + \left.\frac{\partial c}{\partial m}\right|_a \left.\frac{\partial m}{\partial a}\right|_{m_0} \right] \delta a. \tag{8.10}$$

Likewise, the fact that two air masses have different aerosol properties is almost always an indicator of their differing meteorological histories, which makes it nearly impossible to establish causal relationships between c and a from observations of self-perturbed systems.

Thus, the apparently high susceptibility of cloudiness to the meteorology combined with our poor understanding of exactly how meteorological factors control cloudiness confounds attempts to establish observationally the effect of the atmospheric aerosol on cloudiness.

Theory Expresses Constraints

One little heralded but significant advance in our understanding of the climate system is manifest in the statement "the equilibrium climate sensitivity... is likely to be in the range 2°C to 4.5°C," which is contained in the Fourth Assessment Report of the IPCC (2007). Previous estimates, dating back to the Charney Report (Charney et al. 1979), estimated climate sensitivity between 1.5°C and 4.5°C. The basis for the higher lower-bound (2.0 instead of 1.5) is a better understanding of water vapor feedbacks and reflects largely the acceptance of the argument that small perturbations to the climate system will not affect the distribution of relative humidity. The increasing acceptance of this argument, which dates to Manabe and Wetherald (1967), can be traced to the study by Soden et al. (2005) of upper tropospheric humidity in both satellite observations and general circulation model simulations. Underpinning these results is the idea that the constancy of the distribution of relative humidity in the atmosphere measures the constancy of the atmospheric circulation. This constant circulation hypothesis amounts to saying that the change in the relative humidity, $\delta\mathcal{U}$, is small, equivalently,

$$\frac{\delta\mathcal{U}}{\mathcal{U}} = \varepsilon \ll 1. \tag{8.11}$$

Although Equation 8.11 provides a valuable constraint on water vapor feedbacks, it does little to constrain cloudiness in the lower troposphere. Consider anew the case of stratocumulus initial conditions observed during DYCOMS-II. Above we indicated that a roughly 2% change (from 9.05–9.25 g kg^{-1}) in the boundary layer total-water mixing ratio could increase the cloud liquid water path by more than 35% and the cloud albedo by more than five percentage points. As far as low clouds are concerned, an order ε change in $\mathcal{U}$ can lead to an order one change in cloudiness. More generally if we denote cloud-base height by z_b, then for a shallow well-mixed layer, with surface temperature T_0,

$$z_b \approx z_* \ln\left(\frac{q_0(T_0)}{q}\right), \quad \text{where } z_* = \left(\frac{R_d L_v}{R_v c_p T_0} - 1\right)\frac{g}{R_d T_0} \approx 2000\,\text{m} \quad (8.12)$$

is a moisture scale height. For a cloud layer whose depth $\zeta = h - z_b \approx 200$ m, this implies that

$$\frac{\delta\zeta}{\zeta} = -\frac{z_*}{\zeta}\frac{\delta\mathcal{U}}{\mathcal{U}}, \quad (8.13)$$

where $\mathcal{U}$ denotes here the relative humidity with respect to the saturation humidity at the surface temperature, T_0. If $\delta\mathcal{U}/\mathcal{U}$ is order ε, then $\delta\zeta/\zeta$ is order unity by virtue of the fact that z_*/h is order $1/\varepsilon$. Because, by Equation 8.5, the optical depth is proportional to $\mathcal{L}$ which by Equation 8.6 depends on h^2, the sensitivity of cloud amount to $\mathcal{U}$ is even more pronounced. This makes the point that, at least for stratocumulus, cloud amount is very sensitive to small changes in the humidity within the boundary layer. This may help explain the relative imprecision of simulations of this important cloud regime.

Precipitation, which is central to some of the hypothesized pathways through which the aerosol affects cloudiness, proves even more challenging. Theoretical studies of the droplet collision efficiency (Klett and Davis 1973) suggest that the droplet collection process starts to be significant only when the biggest droplets reach a radius of the order of 20 μm, which corresponds to a mean volume or effective radius of the droplet population slightly greater than 10 μm (for empirical support for this idea, see Gerber 1996; Boers et al. 1998). Increased aerosol particle concentrations, which lead to a decrease of the droplet sizes at a specified liquid water content, can therefore significantly impact precipitation when the cloud depth, and hence the maximum liquid water content, is bounded (e.g., in thin stratocumulus clouds). In deeper convective clouds, however, the effect will only be to raise slightly the level where collection starts to be active. Once the threshold diameter is reached, the collection process is much less sensitive to the number concentration of the cloud droplets (as apparent in bulk parameterizations of the accretion process). Hence near the onset of precipitation, small differences in the depth of the cloud layer can prove decisive in the development of precipitation. Indeed, if cloud-base precipitation scales with h^3, as has been suggested by a variety of field studies (Pawlowska and Brenguier 2003; Van Zanten et al. 2005; Comstock et al. 2005; Geoffroy et al. 2008) we would expect the precipitation rate sensitivity to be a factor of three larger than the relative cloud thickness sensitivity (i.e., $\delta p/p = 3\delta h/h$, where p denotes cloud-base precipitation).

Model Myopia

Given incomplete theoretical descriptions and insufficient empirical constraints, attempts to evaluate the net effect of environmental changes on cloudiness explore necessarily specific and preconceived pathways. Unexpected behavior can only arise on the resolved scales of the simulation, as on the parameterized scales one is limited to the specific preconceptions built into the model, even more so to the extent that parameterizations of distinct processes interact only through changes to the mean state. This weakness, or even flaw, in the approach to exploring the effect of unresolved processes on the climate as a whole might have been foremost in the minds of the IPCC when they approved the following statement (Denman et al. 2007, p. 566):

> The response of the climate system to anthropogenic forcing is expected to be more complex than simple cause and effect relationships would suggest; rather, it could exhibit chaotic behaviour with cascades of effects across the different scales and with the potential for abrupt and perhaps irreversible transitions.

The ultimate clause of this statement suggests that the authors might have had what we call a choleric interpretation in mind; namely, that small changes at the process level can have large consequences that are unanticipated for the system as a whole. Here we prefer to explore a more phlegmatic interpretation and its implications; namely, the idea that large changes at the process level can have small consequences that are unanticipated for the system as a whole.

As a case in point, consider the ideas hypothesized by Albrecht (1989). The standard approach to quantify such effects would be to allow the precipitation efficiency of shallow clouds to depend on the cloud droplet concentration, which in turn is made dependent on the aerosol loading. Then, by comparing simulations with larger and smaller aerosol loadings, one could attempt to quantify the importance of such effects. However, think about the logical structure of such an enterprise: It presumes that an effect on the subgrid scale projects directly onto the resolved scale, which in turn modifies the small scale. If clouds precipitate more readily, will not this information (heating/drying) be most readily evident on the local (cloud) scale, and only gradually work its way through a sequence of scales to the larger scale? Of course, the large-scale model, insofar as its parameterizations are formulated in terms of mean fields (rather than higher-order statistics that incorporate information about the fluctuations and their correlations), is not capable of representing precisely these effects (i.e., those that one would naturally expect to be most predominant). From this perspective, it is not surprising that upon more detailed investigation we find, almost always, that the response of physical systems is much richer and that it is often contrary (e.g., Feingold et al. 1996, 2005; Xue and Feingold 2006; Wood 2007; Xue et al. 2008; Sandu et al. 2008,) to that predicted by the depictions of physical processes studied with the aid of global models or satellite snapshots. Such difficulties are only compounded by bad practice (e.g., the habit of enforcing relationships valid for one scale on entirely other scales or, more specifically, the common practice of applying microphysical concepts designed for single clouds to fields of clouds).

The outcome is that the seduction of using large-scale models to quantify perturbations globally to small-scale physical processes is often greater than the utility; the results can be expected to depend heavily on the (often flawed) conceptual framework underlying particular parameterizations. Moreover, agreement among large-scale models as to the magnitude of a particular effect may just as well represent the poverty of the parameterizations as it does the robustness of an underlying physical principle. In this way, our poor understanding of clouds haunts not only our attempts to estimate aerosol effects on clouds observationally but also numerically.

Outlook

Clearly, these are difficult problems, but science is not afraid of hard problems, particularly important ones. Take, for example, the search for a cure for cancer or the dream of cheap fusion. Both have a long history, and both have seen steady progress. The search for a cancer cure might be a better analogy, because a cloud, like a cancer, finds its expression in many regimes. Given the difficulty of the problem, and the likelihood that progress will be incremental, it appears worthwhile to step back and take stock of which strategies have the best chance of exacting progress.

- *Think globally*: Processes that simply shift the boundary among regimes are likely to scale with the area across which the regime boundary fluctuates. To the extent that such transitions are localized, these types of effects (changing the precipitation efficiency or decoupling boundary), while locally large, may be expected to be relatively small on a global average. Thus priority should be given to identifying and focusing on those changes most likely to affect patterns of cloudiness on the largest possible scales (e.g., lapse-rate feedbacks or the Twomey effect).
- *Look around*, not just up (or down): Shallow cumulus clouds are often difficult to measure with existing satellites, especially on timescales relevant to their physical evolution. *In-situ* measurements are sample limited, and radar measurements tend to suffer from greater absorption (because those radars operate at wavelengths where clouds are reflective), thereby limiting their capacity for scanning. However, better use can be made of scanning cloud radars, especially at remote marine

sites, to yield a larger empirical database elucidating relationships between low clouds and their environment, which can then be the target of theory.

- *Exploit social idiocrasy*: Humanity, through its accidental intervention in natural systems, presents the best opportunity to decouple aerosol perturbations from meteorological ones. Well-known examples are ship tracks, but other less exploited ones are fires, which may or may not be set in Santa Ana conditions, or in regions of biomass burnings. Identifying these opportunities of accidental large-scale intervention may be central to advancing the empiricism.
- *Replace the fetish for quantification with a curiosity for constraints*: By this we mean to suggest that large-scale models might be more fruitfully employed to understand either how the circulations which they do resolve respond to changes made to small-scale physics, or identify robust environmental changes to which clouds should be responding. Such a search for understanding will in the long run better serve the quantitative mandates of our times.
- *Worship hierarchy* (at least model hierarchies): Here the challenge is to identify processes on small scales and work to investigate their impact through the full range of scales, using a hierarchy of models that can be formally related to one another and which produce signatures that can be tested with data. Those arguments explored in this fashion will, in the end, be most compelling.
- *Act locally*: Notwithstanding the primitive state of theory, many large-scale models are not even capable of representing the content of our existing understanding. Therefore, acting to ensure that the physical basis of existing models is sound and able to represent key processes and interactions with fidelity should merit reward.
- *Theory first!* If we can make but only one point, it would be that progress on these and related issues will advance no faster than does the theory. Thus the search for understanding, encapsulating theories of cloud systems capable of explaining the empiricism, should be our priority—not because theory is more important, but rather because it is the crucible in which critical experiments and observations are conducted. Forgetting this compels us to grope forever in the dark.

Acknowledgments

B. Stevens would like to acknowledge UCLA, which supported him during the time when much of this article was prepared. Comments on early drafts of this manuscript by Louise Nuijens, Pier Siebesma, Graham Feingold, Robert Wood, and the editor greatly improved the exposition. We also thank our colleagues for the stimulating discussions during the course of this Forum.

References

Ackerman, A. S., M. P. Kirkpatrick, D. E. Stevens, and O. B. Toon. 2004. The impact of humidity above stratiform clouds on indirect aerosol climate forcing. *Nature* **432**:1014–1017.

Ackerman, A. S., O. B. Toon, D. E. Stevens et al. 2000. Reduction of tropical cloudiness by soot. *Science* **288**:1042–1046.

Albrecht, B. A. 1989. Aerosols, cloud microphysics and fractional cloudiness. *Science* **245**:1227–1230.

Albrecht, B. A., A. K. Betts, W. H. Schubert, and S. K. Cox. 1979. A model of the thermodynamic structure of the trade-wind boundary layer. Part I: Theoretical formulation and sensitivity tests. *J. Atmos. Sci.* **36**:90–98.

Bellon, G., and B. Stevens. 2005. On bulk models of shallow cumulus convection. *J. Atmos. Sci.* **62**:3286–3302.

Betts, A. K., and W. Ridgway. 1989. Climatic equilibrium of the atmospheric convective boundary layer over a tropical ocean. *J. Atmos. Sci.* **46**:2621–2641.

Bjerknes, J. 1938. Saturated-adiabatic ascent of air through dry-adiabatically descending environment. *Q. J. Roy. Meteor. Soc.* **64**:325–330.

Blake, D. 1928. Temperature inversions at San Diego, as deduced from aerographical observations by airplane. *Mon. Wea. Rev.* **56**:221–224.

Boers, R., J. B. Jensen, and P. B. Krummel. 1998. Microphysical and shortwave radiative structure of stratocumulus clouds over the Southern Ocean: Summer results and seasonal differences. *Q. J. Roy. Meteor. Soc.* **124**:151–168.

Bougeault, P. 1982. Cloud ensemble relations based on the gamma probability distribution for the high-order models of the planetary boundary layer. *J. Atmos. Sci.* **39**:2691–2700.

Brenguier, J.-L., P. Y. Chuang, Y. Fouquart et al. 2000. An overview of the ACE-2 CLOUDYCOLUMN closure experiment. *Tellus* **52B**:814–826.

Bretherton, C. 1993. Understanding Albrecht's model of trade cumulus cloud fields. *J. Atmos. Sci.* **50**:2264–2283.

Bretherton, C., P. N. Blossey, and J. Uchida. 2006. Cloud droplet sedimentations, entrainment efficiency and subtropical stratocumulus albedo. *Geophys. Res. Lett.* **34**:L03813.

Bretherton, C., and S. Park. 2007. A new bulk shallow cumulus-topped boundary layer model. *J. Atmos. Sci.* **65**:2174–2193.

Bretherton, C., and M. C. Wyant. 1997. Moisture transport, lower tropospheric stability, and decoupling of cloud-topped boundary layers. *J. Atmos. Sci.* **54**:148–167.

Burnet, F., and J. Brenguier. 2007. Observational study of the entrainment-mixing process in warm convective clouds. *J. Atmos. Sci.* **64**:1995–2011.

Charlson, R. J., J. E. Lovelock, M. O. Andreae, and S. G. Warren. 1987. Oceanic phytoplankton, atmospheric sulphur, cloud albedo and climate. *Nature* **326**: 655–661.

Charney, J., A. Arakawa, D. Baker et al. 1979. Carbon Dioxide and Climate: A Scientific Assessment. Washington, DC: Natl. Research Council.

Chosson, F., J.-L. Brenguier, and L. Schüller. 2007. Entrainment mixing and radiative transfer simulation in boundary layer clouds. *J. Atmos. Sci.* **64**:2670–2682.

Clayton, H. H. 1896. The origin of the stratus-cloud, and some suggested changes in the international methods of cloud-measurement. *Nature* **55**:197–198.

Comstock, K. K., C. S. Bretherton, and S. E. Yuter. 2005. Mesoscale variability and drizzle in southeast Pacific stratocumulus. *J. Atmos. Sci.* **62**:3792–3807.

Cruz, L. A. 1973. Venezuelan rainstorms as seen by radar. *J. Appl. Meteor.* **12**: 119–126.

Deardorff, J. W. 1980. Cloud top entrainment instability. *J. Atmos. Sci.* **37**:131–147.

Denman, K. L., G. Brasseur, A. Chidthaisong et al. 2007. Couplings between changes in the climate system and biogeochemistry. In: Climate Change 2007: The Physical Science Basis. Contribution of Working Group I to the Fourth Assessment Report of the Intergovernmental Panel on Climate Change, ed. Solomon, S., D. Qin, M. Manning et al., pp. 499–588. New York: Cambridge Univ. Press.

Feingold, G., H. Jiang, and J. Y. Harrington. 2005. On smoke suppression of clouds in Amazonia. *Geophys. Res. Lett.* **32(2)**:L02804.

Feingold, G., B. Stevens, W. R. Cotton, and A. S. Frisch. 1996. The relationship between drop in-cloud residence time and drizzle production in numerically simulated stratocumulus clouds. *J. Atmos. Sci.* **53**:1108–1122.

Geoffroy, O., I. Sandu, and J. L. Brenguier. 2008. Relationship between drizzle rate, liquid water path and droplet concentration at the scale of a stratocumulus cloud system. *Atmos. Chem. Phys.* **8**:4641–4654.

Gerber, H. 1996. Microphysics of marine stratocumulus clouds with two drizzle modes. *J. Atmos. Sci.* **53**:1649–1662.

Hansen, J., M. Sato, and R. Ruedy. 1997. Radiative forcing and climate response. *J. Geophys. Res.* **102**:6831–6864.

Howard, L. 1802. On the Modification of Clouds. London: J. Taylor.

IPCC. 2007. Climate Change 2007: The Physical Science Basis. Contribution of Working Group I to the Fourth Assessment Report of the Intergovernmental Panel on Climate Change, ed. S. Solomon, D. Qin, M. Manning et al. New York: Cambridge Univ. Press.

Klein, S. A., and D. L. Hartmann. 1993. The seasonal cycle of low stratiform clouds. *J. Climate* **6**:1587–1606.

Klein, S. A., D. L. Hartmann, and J. R. Norris. 1995. On the relationships among low-cloud structure, sea surface temperature, and atmospheric circulation in the summertime northeast Pacific. *J. Climate* **8**:1140–1155.

Klett, J. D., and M. Davis. 1973. Theoretical collision efficiencies of cloud droplets at small Reynolds numbers. *J. Atmos. Sci.* **30**:107–117.

Lamarck, J.-B. 1802. Annuaire Météorologique, vol. 3, pp. 151–166. Paris: Dentu.

Latham, J., and M. H. Smith. 1990. Effect of global warming on wind-dependent aerosol generation at the ocean surface. *Nature* **347**:372–373.

Lewellen, D. C., and W. Lewellen. 2002. Entrainment and decoupling relations for cloudy boundary layers. *J. Atmos. Sci.* **59**:2966–2986.

Lewellen, W. S., and S. Yoh. 1993. Binormal model of ensemble partial cloudiness. *J. Atmos. Sci.* **50**:1228–1237.

Lilly, D. K. 1968. Models of cloud topped mixed layers under a strong inversion. *Q. J. Roy. Meteor. Soc.* **94**:292–309.

Lock, A. P., and M. K. MacVean. 1999. A parametrization of entrainment driven by surface heating and cloud-top cooling. *Q. J. Roy. Meteor. Soc.* **125**:271–299.

Manabe, S., and R. T. Wetherald. 1967. Thermal equilibrium of the atmosphere with a given distribution of relative humidity. *J. Atmos. Sci.* **24**:241–259.

Miller, R. L. 1997. Tropical thermostats and low cloud cover. *J. Climate* **10**:409–440.

Moeng, C.-H., P. P. Sullivan, and B. Stevens. 1999. Including radiative effects in an entrainment-rate formula for buoyancy driven PBLs. *J. Atmos. Sci.* **56**:1031–1049.

Neggers, R. A., A. P. Siebesma, and G. Lenderink. 2004. An evaluation of mass flux closures for diurnal cycles of shallow cumulus. *Mon. Wea. Rev.* **132**:2525–2538.

Neggers, R. A., B. Stevens, and J. D. Neelin. 2007. Variance scaling in shallow cumulus topped mixed layers. *Q. J. Roy. Meteor. Soc.* 133, 1629–1641.
Nigam, S. 1997. The annual warm to cold phase transition in the eastern equatorial Pacific: Diagnosis of the role of stratus cloud-top cooling. *J. Climate* **10**:2447–2467.
Paltridge, G. W. 1980. Cloud-radiation feedback to climate. *Q. J. Roy. Meteor. Soc.* **106**:895–899.
Pawlowska, H., and J.-L. Brenguier. 2003. An observational study of drizzle formation in stratocumulus clouds for general circulation model (GCM) parameterization. *J. Geophys. Res.* **33**:L19810.
Philander, S. G. H., D. Gu, D. Halpern et al. 1996. Why the ITCZ is mostly north of the equator. *J. Climate* **9**:2958–2972.
Pincus, R., and M. B. Baker. 1994. Effect of precipitation on the albedo susceptibility of marine boundary layer clouds. *Nature* **372**:250–252.
Pincus, R., M. B. Baker, and C. S. Bretherton. 1997. What controls stratocumulus radiative properties? Lagrangian observations of cloud evolution. *J. Atmos. Sci.* **54**: 2215–2236.
Radke, L. F., J. A. Coakley, Jr., and M. D. King. 1989. Direct and remote sensing observations on the effects of ships on clouds. *Science* **246**:1146–1149.
Randall, D. A. 1980. Conditional instability of the first kind upside-down. *J. Atmos. Sci.* **37**:125–130.
Rosenfeld, D. 2006. Aerosols, clouds, and climate. *Science* **312**:1323–1324.
Sandu, I., J.-L. Brenguier, and O. Geoffroy. 2008. Aerosol impacts on the diurnal cycle of marine stratocumulus. *J. Atmos. Sci.* **65(8)**:2705.
Saunders, P. M. 1965. Some characteristics of tropical marine showers. *J. Atmos. Sci.* **22**:167–175.
Savic-Jovcic, V., and B. Stevens. 2008. The structure and mesoscale organization of precipitating stratocumulus. *J. Atmos. Sci.* **65**:1587–1605.
Scorer, R. 1972. Clouds of the World: A Complete Colour Encyclopedia. Melbourne: Lothian Publ.
Siebesma, A. P. 1998. Shallow convection. In: Buoyant Convection in Geophysical Flows, ed. E. J. Plate, E. E. Fedorovich, D. X. Viegas, and J. C. Wyngaard, vol. 513, pp. 441–486. Dordrecht: Kluwer Academic Publ.
Slingo, J. 1980. Cloud parameterization scheme derived from GATE data for use with a numerical model. *Q. J. Roy. Meteor. Soc.* **106**:747–770.
Slingo, J. 1987. The development and verification of a cloud prediction scheme for the ECMWF model. *Q. J. Roy. Meteor. Soc.* **113**:899–927.
Smith, R. N. B. 1990. A scheme for predicting layer clouds and their water content in a GCM. *Q. J. Roy. Meteor. Soc.* **116**:435–460.
Soden, B. J., D. L. Jackson, V. Ramaswamy, M. D. Schwarzkopf, and X. Huang. 2005. The radiative signature of upper tropospheric moistening. *Science* **310**:841–844.
Squires, P. 1956. The microstructure of cumuli in maritime and continental air. *Tellus* **8**:443–444.
Squires, P. 1958. Penetrative downdraughts in cumulli. *Tellus* **10**:381–385.
Squires, P., and S. Twomey. 1966. A comparison of cloud nucleus measurements over central North America and the Caribbean Sea. *J. Atmos. Sci.* **23**:401–404.
Stevens, B. 2000. Cloud transitions and decoupling in shear-free stratocumulus-topped boundary layers. *Geophys. Res. Lett.* **27**:2557–2560.
Stevens, B. 2005. Atmospheric moist convection. *Ann. Rev. Earth Planet. Sci.* **33**: 605–643.

Stevens, B. 2006. Bulk boundary layer concepts for simple models of tropical dynamics. *Theor. Comp. Fluid Dyn.* **20**:379–304.
Stevens, B. 2007. On the growth of layers of non-precipitating cumulus convection. *J. Atmos. Sci.* **64**:2916–2931.
Stevens, B., W. R. Cotton, G. Feingold, and C.-H. Moeng. 1998. Large-eddy simulations of strongly precipitating, shallow, stratocumulus-topped boundary layers. *J. Atmos. Sci.* **55**:3616–3638.
Stevens, B., and A. Seifert. 2008. On the sensitivity of simulations of shallow cumulus convection to their microphysical representation. *J. Meteorol. Soc.*, in press.
Stevens, B., G. Vali, K. Comstock et al. 2005. Pockets of open cells and drizzle in marine stratocumulus. *Bull. Amer. Meteor. Soc.* **86**:51–57.
Sundqvist, H. 1978. A parameterization scheme for non-convective condensation including prediction of cloud water content. *Q. J. Roy. Meteor. Soc.* **104**:677–690.
Tompkins, A. M. 2002. A prognostic parameterization for the subgrid-scale variability of water vapor and clouds in large-scale models and its use to diagnose cloud cover. *J. Atmos. Sci.* **59**:1917–1942.
Turton, J. D., and S. Nicholls. 1987. A study of the diurnal variation of stratocumulus using a multiple mixed layer model. *Q. J. Roy. Meteor. Soc.* **113**:969–1009.
Twomey, S. A. 1974. Pollution and the planetary albedo. *Atmos. Environ.* **8**:1251–1256.
Twomey, S. A. 1977. The influence of pollution on the shortwave albedo of clouds. *J. Atmos. Sci.* **34**:1149–1152.
Twomey, S. A., and T. A.Wojciechowski. 1969. Observations of the geographical variation of cloud nuclei. *J. Atmos. Sci.* **26**:684–688.
Van Zanten, M. C., B. Stevens, G. Vali, and D. H. Lenschow. 2005. Observations of drizzle in nocturnal marine stratocumulus. *J. Atmos. Sci.* **62**:88–106.
Warner, J. 1968. A reduction in rainfall associated with smoke from sugar-cane fires: An inadvertent weather modification. *J. Appl. Meteor.* **7**:704–706.
Warner, J., and S. Twomey. 1967. The production of cloud nuclei by cane fires and the effect on cloud droplet concentrations. *J. Atmos. Sci.* **24**:704–706.
Weaver, C. J., and R. Pearson, Jr. 1990. Entrainment instability and vertical motion as causes of stratocmulus breakup. *Q. J. Roy. Meteor. Soc.* **116**:1359–1388.
Wood, R., 2007. Cancellation of aerosol indirect effects in marine stratocumulus through cloud thinning. *J. Atmos. Sci.* **64**:2657–2669.
Wood, R., and C. S. Bretherton. 2006. On the relationship between stratiform low cloud cover and lower tropospheric stability. *J. Climate* **19**:6425–6432.
Xu, K.-M., and D. A. Randall. 1996. A semi-empirical cloudiness parameterization for use in climate models. *J. Atmos. Sci.* **53**:3084–3102.
Xue, H., and G. Feingold. 2006. Aerosol effects on clouds, precipitation, and the organization of shallow cumulus convection. *J. Atmos. Sci.* **63**:1605–1622.
Xue, H., G. Feingold, and B. Stevens. 2008. Aerosol effects on clouds, precipitation, and the organization of shallow cumulus convection. *J. Atmos. Sci.* **65**:392–406.
Zhang, Y., B. Stevens, and M. Ghil. 2005. On the diurnal cycle and susceptibility to aerosol concentration in a stratocumulus-topped mixed layer. *Q. J. Roy. Meteor. Soc.* **131**:1567–1584.

9

Deep Convective Clouds

Wojciech W. Grabowski[1] and Jon C. Petch[2]

[1]National Center for Atmospheric Research, Boulder, CO, U.S.A.
[2]Met Office, Exeter, U.K.

Abstract

Deep convection plays a key role in the Earth's atmospheric general circulation and is often associated with severe weather. It spans a wide range of spatial scales, from subcentimeter for the cloud microscale to tens and hundreds of kilometers for convective towers and mesoscale convective systems. Deep convection extends, however, far beyond the scale of an individual convective system because its latent heating drives large-scale atmospheric tropical and subtropical circulations, such as the Hadley, Walker, and monsoon circulations. Thus to understand the role of atmospheric deep convection in the climate system, as well as in climate change, key processes across all of these scales must be taken into account. This chapter reviews the relevant aspects of the problem, points out the limitations of current modeling and observational approaches, and suggests areas for future research. It is argued that understanding the role of deep convection in the climate system, as well as in predictions of climate change, requires modeling efforts across all scales, from microscale to global, using a variety of models. Traditional atmospheric general circulation models are not sufficient because representation of deep convection, and how it may change in the perturbed climate, is highly uncertain.

Introduction

Our Earth's climate involves processes across an enormous range of scales, from the planetary down to the subcentimeter level (the latter is referred to as the microscale). Planetary-scale processes are driven by the mean equator-to-pole tropospheric temperature gradient as well as by planetary-scale differences between the oceans and continents. The equator-to-pole temperature gradient results from the radiative imbalance between the low and high latitudes: solar input is larger than thermal loss at low latitudes, with the reverse true at high latitudes. Clouds, and particularly deep convection, play an important role in the meridional energy transport, which compensates for the radiative imbalance between low and high latitudes. In addition, the presence of continents, with their large-scale mountain chains and spatially variable surface characteristics, has an important impact on large-scale atmospheric circulations.

From the perspective of microscale processes, we need to keep in mind that clouds are composed of small water droplets and ice crystals that are suspended in the air (as well as unactivated interstitial particles), and that key climate-related processes take place at the cloud microscale. For instance, scattering, absorption, and emission of solar and thermal infrared radiation by clouds entail interactions between photons and individual cloud droplets and ice crystals. Similarly, microscale processes are involved in the formation of raindrops from much smaller cloud droplets through collision or coalescence. Between the micro- and planetary scale, a plethora of atmospheric processes occur: small-scale turbulence, dry convection, moist shallow and deep convection, mesoscale circulations (waves, mesoscale convective systems, frontal circulations), synoptic-scale weather systems, storm tracks, large-scale land–ocean (monsoon) circulations, and others. All of these processes involve or impact deep convection.

Although deep convection can occur almost anywhere, it is ubiquitous in the tropics and over subtropical and midlatitude continents, particularly during the warm season. The tropics are approximately in a convective-radiative quasi-equilibrium, and deep convection plays a fundamental role in this. Because atmospheric absorption is small, most insolation is absorbed at the surface. At the same time, the troposphere is being constantly cooled and destabilized through thermal infrared emission to space. Solar energy absorbed at the surface is passed on to the atmosphere through surface fluxes, dominated in the tropics by latent heat flux (i.e., surface evaporation). The latent heat is subsequently released in convective clouds, and water returns to the surface as rain. In much of the tropics, latent heating and the surface sensible heat flux offset approximately the radiative cooling of the troposphere.

An important aspect of atmospheric convection, in general, and deep convection, in particular, is its response to the diurnal cycle of solar insolation. Typically, over the tropical oceans, precipitation is the greatest during the early morning hours (Yang and Slingo 2001). This occurs primarily because of the strong radiative cooling of the troposphere at night, although the exact mechanism leading to this peak is still under debate. For weak surface winds, which correspond to a weak mixing in the upper ocean, there is also a lesser secondary peak of convection in the late afternoon that coincides with a weak peak in the sea surface temperature (SST) at this time. Over the summertime and tropical continents, diurnally variable solar energy absorbed at the surface leads to a strong diurnal cycle of surface temperature and surface heat fluxes, and consequently to a strong diurnal cycle of convection. Typically, morning hours are characterized by the development and growth of a well-mixed convective boundary layer as the surface sensible and latent fluxes increase after sunrise. Shallow convective clouds develop as the convective boundary layer deepens, followed by a gradual transition from shallow to deep precipitating convection.

Often, deep convection organizes itself into larger systems, leading to a surface precipitation maximum in the late afternoon or evening. Thereafter, convective dissipation and the development of a stable nighttime boundary layer progresses throughout the night. This classical picture is often modified by the effects of land–sea contrast, orography, and larger-scale organization of convection (e.g., into a mesoscale convective system). For instance, propagating mesoscale systems are responsible for a well-documented summertime late night to early morning maximum of convective intensity over the U.S. Midwest, with the maximum shifting further east as nighttime progresses. These eastward-propagating mesoscale systems develop east of the Rockies (e.g., in eastern Colorado) in the mean vertical shear of the baroclinic zonal flow.

There is a history of various misconceptions and conflicting ideas associated with the clouds-in-climate problem, especially in terms of tropical convection. For example, in his seminal paper, which reviewed issues relating to global warming as they were understood at the end of the 1980s, Lindzen (1990) argued that the increase of convective activity in the tropics attributable to global warming might result in enhanced subsidence, a drier troposphere, and subsequently more thermal radiation escaping to space (because water vapor is the most important greenhouse gas). This would then provide a negative feedback to increase SSTs. However, cloud-resolving model simulations demonstrate that this scenario is unlikely, as warmer SSTs result in more water vapor in the atmosphere with approximately unchanged relative humidity (Tompkins and Craig 1999). Ramanathan and Collins (1991) analyzed changes of tropical cloudiness during the 1987 El Niño and argued that the increase in upper tropospheric anvil clouds, as the SSTs warm, limits the amount of solar radiation reaching the surface and thus prevents any further increase of the SST; this is the thermostat hypothesis. In contrast, Lindzen et al. (2001) argued that satellite observations of tropical western Pacific cloud cover suggest a decrease of cloudiness with increasing SSTs, which would then allow more emission of Earth longwave radiation to space. This infrared iris hypothesis was later challenged by Hartmann and Michelsen (2002) and Rapp et al. (2005). Clearly, the thermostat and iris hypotheses represent conflicting ideas of the response of tropical cloudiness to the increase of the SST, and thus to a warming climate. We wish to point out, however, that these ideas are based on studies concerning the variability of regional climate, and thus extrapolating from them to the problem of global climate change is questionable.

Investigations of cloud changes in the perturbed climate system rely heavily on modeling for several reasons. First, it is difficult to compare characteristics of today's clouds with observations collected thirty or forty years ago because observational techniques (e.g., *in-situ*, remote-sensing, airborne, satellite) have developed tremendously over the last several decades. This problem is similar to that experienced in the analyses of long records of surface temperature measurements, where changes of measurement technology as well as locations at which the measurements were taken make comparison difficult. Second, reliable cloud observations (e.g., from radar or from aircraft) typically provide information on small-scale processes, and it is difficult (if not impossible) to put observed local changes (if they can be detected) into the context of regional or global climate change. New satellite technologies offer hope that small-scale observations from space can be expanded into a regional and global context relevant to climate change (and not just to climate variability as in the thermostat and iris hypotheses). However, such data have to be collected for many years (perhaps decades) before reliable estimates of observed trends can be derived. Finally, and perhaps most importantly, observations can document correlations between various quantities (e.g., enhancement of upper-tropospheric anvil clouds and the increase of the SSTs, as in the thermostat hypothesis), but they cannot establish cause-and-effect relationships (i.e., causality). Establishing causality is needed, however, if we are to achieve an understanding of the interactions (forcings and feedbacks) between various elements of the climate system. Such interactions can only be deduced through numerical modeling using well-designed sensitivity simulations and the appropriate modeling framework.

Although shallow clouds have been demonstrated to be primarily responsible for global climate sensitivity (Bony and Dufresne 2005), deep clouds are a key component of the energy balance of the Earth and may therefore be fundamental to our understanding of climate change as well. A critical point is that deep clouds respond not only to local changes (e.g., resulting from local modifications of surface characteristics, atmospheric stability, and aerosols), but also to changes in large-scale circulation patterns, which provide large-scale forcing for convection. Local effects can be studied with confidence using cloud models and high-resolution (cloud-resolving) mesoscale models. However, the effects of perturbed large-scale circulations are more difficult to assess because deep convection not only responds to these perturbations, it is also responsible for driving them. Thus, deep convection must be considered an integral part of large-scale dynamics—a challenge difficult to meet if the entire range of scales (from an individual convective tower to the global circulation) is to be resolved in a numerical model. There are emerging approaches which address this issue with more confidence. Although discussion in this chapter is divided into separate sections to consider the spatial scales involved, an understanding of the links between the processes across this wide range of scales is vital if we are to assess the effects of deep convective clouds in the perturbed climate system.

The Cloud Scale

A unique aspect of atmospheric moist convection is a strong coupling between cloud-scale dynamics, thermodynamics, and microphysics. Briefly, the latent heating, which results from phase changes in water substance (e.g., the growth of cloud droplets), controls cloud buoyancy; the buoyancy determines the updraft strength; and the updraft strength controls the subsequent growth of cloud droplets and thus latent heating. Deep convection, spanning a wide temperature range (e.g., from ca. 30°C near the surface to ca. –80°C near the tropopause in the tropics), involves both warm rain and ice processes. Formation of cloud particles (water droplets and ice crystals) in the Earth's atmosphere involves small aerosol particles, called cloud condensation nuclei (CCN) or ice-forming nuclei (IN); that is, it involves heterogeneous nucleation (the only exception is the homogeneous freezing of water droplets for ambient temperatures below –40°C). Nucleation of liquid cloud droplets, their growth by diffusion of water vapor, and subsequent formation of drizzle/raindrops through collision–coalescence are relatively well understood at the process level. For ice processes, the situation is significantly more complicated, not only because of the various modes of ice nucleation (contact, condensation freezing, and deposition), but also because of a wide range of ice crystal forms ("habits") and different growth mechanisms, including the diffusion of water vapor, accretion of supercooled water, and aggregation. Despite vigorous research in field and laboratory studies over the last several decades, major gaps still exist in our understanding of ice processes within clouds. This is especially true for ice initiation. For microphysical processes in deep convective clouds, fundamental difficulties arise because the main transport and heating processes occur in strong updrafts that cover only a small area. These processes are difficult to observe *in situ* using instrumented aircraft (especially for severe convection) because of the dangers involved in flying into these regions. It is also difficult, if not impossible, to reproduce the conditions typical of convective updrafts (i.e., rapid changes of temperature and pressure) in a laboratory. Finally, the Reynolds number for laboratory flows is orders of magnitude smaller than in natural clouds.

Small-scale cloud processes can be studied using laboratory and field observations and numerical modeling. Field campaigns, targeting various types of clouds and focusing on small-scale cloud processes as well as the meso- and larger-scale cloud environment, have resulted in a better understanding of clouds and their interactions with the environment. Recognizing that thunderstorms (cumulonimbus clouds) are a serious weather hazard to aviation, the Thunderstorm Project was initiated in Florida and Ohio (U.S.A.) shortly after World War II. For the first time, aircraft, radar, and ground-based observations provided a description of the thunderstorm structure and associated circulations. The most important finding of the project, which still encapsulates most of our current detailed knowledge of deep precipitating convection, was the identification of an individual "cell" as a building block of a cumulonimbus cloud. Each cell proceeds through a life cycle of cloud and precipitation processes in three distinct stages: the cumulus (or development) stage, the mature stage, and the dissipating stage (Byers and Braham 1948).

With improvements of meteorological radars and the emergence of satellite observations, it became apparent in the late 1960s that deep convection is typically organized into "cloud clusters" or "mesoscale convective systems" (i.e., ensembles of cumulonimbus clouds joined in their mature and dissipating stages by extensive upper-tropospheric stratiform anvils, often extending over distances of hundreds of kilometers). These early observations were strongly supported by data collected during the Global Atmospheric Research Program (GARP) Atlantic Tropical Experiment (GATE), an international field campaign conducted over the tropical eastern Atlantic in the summer of 1974 (for a review, see Houze and Betts 1981). Observations in GATE documented how moist convection evolved in response to changes in meteorological conditions (e.g., the phase of a synoptic-scale easterly wave and the presence of lower-tropospheric African easterly jet). The observed mesoscale convective systems consisted of regions of deep convective ascent and surrounding stratiform cloudiness and precipitation. Precipitating anvils featured distinct mesoscale updrafts above the melting level and mesoscale downdrafts beneath it. Convective and mesoscale downdrafts filled the boundary layer with dry and cold (low moist static energy) air, creating distinct convective wakes (often called "cold pools"), which affected the subsequent development of moist convection. Overall, GATE provided strong evidence of the multiscale organization of tropical convection and the role of mesoscale convective systems in the tropical climate.

Over midlatitude continents, deep convection leads often to significant weather hazards: damaging winds, torrential rain, flooding, hail, lightning, and tornadoes. Some of the most damaging convection occurs over subtropical continents, typically in the form of multicell or supercell mesoscale convective systems. Many field campaigns have been dedicated to the study of various aspects of such systems over the past several decades. Pertinent U.S. examples (North America is strongly affected by severe convection) include the National Hail Research Experiment (NHRE), which was conducted over northeastern Colorado from 1972–1976; the Cooperative Convective Precipitation Experiment (CCOPE) conducted in Montana during the spring and summer of 1981; the Oklahoma–Kansas PRE-STORM program during the spring and summer of 1985; and the Bow Echo and MCV (Mesoscale Convective Vortex) Experiment (BAMEX), which was carried out in the spring and summer of 2003 over the central U.S. These field projects provided much of the data that has resulted in our current understanding of continental organized convection and its interaction with the large-scale flow (see Houze 2004).

Model simulations complement observations to develop our understanding of how deep convection works. Non-hydrostatic numerical models with sophisticated microphysical parameterizations, featuring horizontal grid lengths from a few tens of meters to a few kilometers, are key tools for studying deep convection. These are typically referred to as large eddy simulation (LES) models, cloud models (CMs), or cloud system-resolving models (CSRMs), depending on the grid lengths used and the problem under investigation. These models have been used in the past to address various aspects of moist atmospheric dynamics, including deep convection. Typically, CMs have a small domain, high resolution (ca. 100 m), and the most detailed representation of cloud microphysics, often with sizes of cloud and precipitation particles represented using tens or hundreds of size bins. CMs are well suited to study relevant small-scale processes, such as cloud dynamics and microphysics and how these processes may change when the cloud environment is modified. In terms of deep convection, perhaps the best examples are regime changes that occur in response to changes in environment stability (i.e., the convectively available potential energy or CAPE) and ambient wind profile (Thorpe et al. 1982; Weisman and Klemp 1986).

Data collected in field experiments have been used to validate cloud simulations, and they can be used to investigate the effects of climate perturbations on clouds (e.g., to understand how changes in environmental conditions such as environmental stability, shear, or aerosols affect cloud processes). Modification of cloud microphysics through changes of nucleating particles (e.g., from anthropogenic emissions) is commonly referred to as the indirect aerosol effect on climate. This is in contrast to the direct effect, which results from absorption and scattering of radiation by aerosols in cloud-free air. The indirect effect is an uncertain factor in climate change (IPCC 2007), and it is traditionally associated with two further effects: (a) the impact on the size of cloud particles, which changes optical properties of clouds (most importantly, cloud albedo), and (b) the impact on precipitation processes, which can potentially affect abundance, extent, and lifetime of some types of clouds. However, the separation of the indirect effect into these two related aspects is becoming less popular. The indirect effects for warm shallow clouds (such as trade-wind cumulus and subtropical stratocumulus) have been the subject of vigorous theoretical and observational studies. For deep convection, studies on indirect effects are in their infancy, not least because of the uncertainties associated with the modeling of ice processes. For instance, in their simulations of individual deep convective systems, Wang (2005) and Khain et al. (2005) found that a cloud that develops in an environment with a higher concentration of CCN results in stronger convection, higher levels are reached, and more precipitation is produced. It was argued that this occurred through the interaction of warm rain and ice processes, with less effective warm rain processes in high CCN environments resulting in more condensed water arriving at the freezing level and more latent heating aloft because of increased freezing. However, Cui et al. (2006) found the opposite result when modeling convection during CCOPE. In their simulations, increased CCN concentrations led to an inhibition in cloud development and suppression of precipitation. These studies are important because they highlight the key mechanisms that control the life cycle of a single convective cloud. However, as far as climate change is concerned, the extrapolation of single-cloud simulations to a larger-scale response would be wrong.

This is primarily because the models do not consider the feedback between the cloud- and larger-scale processes, as will be discussed in the next section.

Cloud System to Regional Scales

Subsequent field projects, focusing on tropical convection, have strongly reinforced GATE's findings of the importance of the multiscale organization of tropical convection and the role of mesoscale convective systems in the tropical climate. One of the most notable field campaigns was TOGA-COARE, which was conducted from November 1992 to February 1993 over the tropical western Pacific warm pool. In TOGA-COARE, the emphasis was not only on deep convection itself, but also on convection as part of the tropical climate system. Thus, it considered the effects of convection on radiative transfer and upper ocean dynamics, and its role in synoptic and planetary-scale tropical disturbances (e.g., convectively coupled waves, intraseasonal variability). More recently, these ocean-based field experiments have been complemented by a year-long land-based campaign called AMMA. This international and interdisciplinary project took place over Sahel in 2006 and aimed to improve understanding of the West African monsoon and its variability, with an emphasis on daily-to-interannual time scales (Redelsperger et al. 2006).

Data collected in observational campaigns, such as TOGA-COARE and AMMA, can be used to drive model simulations aimed at the understanding of interactions between cloud processes and their larger-scale environment. As mentioned in the previous section, short-term single-cloud simulations are limited because they neglect the feedback between convection (or cloud processes in general) and larger scales. This feedback can occur through the dynamics, microphysics, or radiation on timescales significantly longer than the life cycle of a single cloud. If the profile of convective heating changes when environmental conditions are modified, then this implies different change in the large-scale temperature profile (e.g., different convective response). Such a change affects the large-scale flow and thus the environmental conditions in which subsequent clouds develop. This is the dynamic feedback. Similar arguments can be made for microphysical and radiative feedbacks. For microphysics, a modified amount of precipitation falling from the first cloud affects the large-scale moisture budget, and thus the development of subsequent clouds. For radiation, different cloud water paths or anvil size from the first cloud leads to different radiative heating, and thus the temperature profile in which subsequent clouds develop.

The role of the interactions (forcings and feedbacks) between an ensemble of clouds and all other processes in the system (e.g., radiative transfer, surface fluxes, horizontal advection) is best illustrated using the convective-radiative quasi-equilibrium. In such an equilibrium, radiative cooling of the troposphere (resulting from thermal infrared emission to space which strongly depends on the vertical and horizontal distribution of clouds) is balanced by the total (latent plus sensible) surface heat flux. Depending on surface characteristics (e.g., ocean surface, moist soil, dry soil, with or without vegetation), there will be changes in the partitioning of sensible and latent heat fluxes at the surface just as there will be changes for precipitation. The key point is that the balance between radiative cooling and surface fluxes serves as the externally imposed constraint which the system has to satisfy in equilibrium. So what will happen if, for instance, the characteristics of the atmospheric aerosol are changed? Single-cloud thinking may suggest that there will be less surface precipitation when modified aerosols suppress precipitation development within a single cloud. However, quasi-equilibrium cloud-ensemble thinking implies that surface precipitation cannot change as long as the radiative cooling and thus the surface fluxes remain the same. Then how does the modified system cope with the aerosol change? The answer is that the same amount of surface precipitation is obtained as before, but other elements of the system must respond to maintain this level. For instance, with aerosol particles suppressing formation of precipitation, a slight change of atmospheric stability may allow deeper clouds to develop (or an increase of the contribution of deeper clouds to the entire cloud population), which would then compensate for the microphysical suppression of precipitation that results from modified aerosols (because deeper clouds precipitate more readily than shallow clouds) (for further discussion, see Grabowski 2006). Such reasoning highlights the difference between short-term single-cloud simulations and larger-domain longer-term simulations that feature many clouds at various stages of their life cycle and interact with radiative and surface processes.

Unfortunately, such larger-domain longer-term simulations use typically a model with a somewhat larger grid length (around 1 km) than the single CMs (around 100 m). In addition, they often use computationally more efficient parameterizations of microphysics (far fewer prognostic variables to represent the different cloud species and their size distributions). These models are CSRMs and the simulations carried out with them are generally more relevant for the study of the role of deep convection (or clouds in general) in climate. Examples of such an approach are simulations of the convective-radiative quasi-equilibrium, with relevance to the tropics (Tompkins and Craig 1999) or to the mean Earth's climate (Grabowski 2006). More realistic studies using multi-day cloud ensemble simulations are those driven by observed large-scale conditions (i.e., evolving temperature and moisture advective tendencies referred to as the large-scale forcing, surface characteristics, and the large-scale winds) using the methodology developed by Grabowski et al. (1996) and Xu and Randall (1996). Such simulations mimic the response of convection to large-scale atmospheric processes over an area comparable to a single grid-box of a traditional atmospheric general circulation model (AGCM). Together with the observations, such simulations can be applied to improve parameterizations of cloud processes in AGCMs (see Randall et al. 2003 and references therein).

Like many of the modeling studies within meteorology (and in other disciplines involving multiscale processes), CSRM simulation places multiple demands on computer resources. Key areas that compete for computer time are:

- Three-dimensional dynamics: Although 2-D (slab-symmetric height-distance geometry) versions of CSRMs can be and are commonly run, many scale interactions associated with cloud systems involve 3-D processes and may behave differently in 2-D and 3-D models.
- Domain size: Since a simulation needs to capture both the cloud systems and the associated compensation of descending cloud-free air, a large computational domain is needed (several hundreds of kilometers squared) for organized convection.
- Vertical and horizontal resolution: Individual convective clouds (including shallow convection) are the building blocks of a convective system, and to resolve these requires an appropriate resolution that typically is significantly higher than afforded in CSRM simulations. With low spatial resolution, uncertain subgrid-scale parameterizations (e.g., for entrainment or unresolved microphysical processes) become important (or even critical).
- Total length of the integration: CSRMs are typically run for periods of days or even weeks to represent life cycles of cloud systems and/or to capture the response of a cloud system to evolving large-scale conditions.
- Model complexity: Representation of physical processes (e.g., microphysics, aerosols, radiative transfer, boundary layer, land surface, and ocean) can vary substantially. For instance, for cloud microphysics, single-moment bulk schemes and detailed (bin) microphysics are contrasting choices, with two-moment schemes providing a sensible compromise (Phillips et al. 2007; Morrison and Grabowski 2008).

Thus, CSRM-based experiments must be carefully designed to address the specific question being addressed. Even then, the final choice for any experiment is usually a compromise, with less-than-ideal choices for some aspects. For example, many CSRM simulations use a grid length of 1 km or larger, or are run in 2-D, both of which have been shown to be poor in representing the development of deep convection (Petch 2006). In addition, CSRMs often use a simple single-moment bulk microphysical scheme, which is computationally fast but unable to represent the role of aerosols in a physically realistic way compared to schemes typically used in the single-cloud studies.

If they are well designed, CSRM simulations can be used to assess the impact of various "climate perturbations" on deep convection. For instance, observed large-scale conditions from field observations (GATE, TOGA-COARE, or AMMA) can be used with modified surface characteristics, modified atmospheric composition (e.g., increased concentrations of greenhouse gases), or different atmospheric aerosols to investigate the response of the cloud field to such modifications. Such simulations permit an understanding of relevant feedbacks in the system. Examples of relevant questions include:

- How does deep convection change in response to increases in the tropical SST? (For CSRM simulations, see Tompkins and Craig 1999.)
- Given the surface characteristics and large-scale conditions (i.e., the large-scale forcing), how does deep convection respond to the increase in the concentration of carbon dioxide in the column? We are unaware of any such simulations performed thus far.
- Given the large-scale forcing, how will the convective boundary layer and deep convection respond to a perturbed land surface? If drier land-surface conditions are assumed, surface latent heat flux will be initially reduced at the expense of the sensible flux (thus resulting in a deeper boundary layer), and more solar radiation will be reflected by the surface (drier soil has typically higher albedo; longwave emissivity will change as well). Will these changes result in a positive or negative feedback as far as surface characteristics are concerned? In other words, will clouds and convection increase the soil moisture in return or will they dry out the soil even further? How will the response depend on the large-scale forcing (e.g., the horizontal moisture advection)? To our knowledge, no such simulations have yet been performed.
- When more CCN and/or IN exist in the column (e.g., because of anthropogenic emissions), how will an ensemble of clouds respond? (See discussion in Grabowski 2006.)
- How is deep convection impacted by stronger radiative heating in the clouds (e.g., because of a smaller effective radius of the ice clouds or presence of absorbing CCN and/or IN)?

As mentioned, one weakness of the CSRM approach is that one typically needs to compromise on resolution and/or the complexity of representation of relevant processes (e.g., cloud microphysics), as compared to a single CM. Another weakness is that by applying the prescribed effects of large-scale dynamics (the large-scale forcing), one cannot investigate how changes in deep convection feed back on the regional and global scales. Since it is clear that large-scale conditions (e.g., the strength of large-scale forcing and its temporal variability) play a vital role in the evolution of deep convection, a critical question is: Do we understand how large-scale conditions may change in the perturbed climate, and how does the response of convection feed back onto them? We argue below that the answer is difficult since contemporary climate models provide little confidence as far as the coupling between deep convection and large-scale atmospheric dynamics is concerned.

Global Effects

Deep convection is the engine of large-scale tropical circulations, such as the Hadley and Walker circulations. Paleoclimatic data suggest that the tropical SSTs have changed no more than a few degrees over the last 100 million years, with the largest SST changes occurring at mid- and high latitudes. If such a paradigm holds true for the current climate change, one might anticipate most of the effects outside the tropics. However, even a modest SST increase (e.g., 1°C) implies a significant change of the depth of the tropical troposphere (for CSRM simulations, see Tompkins and Craig 1999; for an analysis of observed trends and those simulated in global climate models, see Santer et al. 2003). Together with the change in mean temperature outside the tropics, this might affect the extent of the Hadley cell, and indeed there are hints that this is actually happening (Seidel et al. 2008). This is an important aspect because descending branches of the Hadley circulation between 15° and 30° latitude determine the location of the driest places on Earth. Moreover, a larger increase of the mid- and high-latitude temperatures compared to the tropics will reduce the equator-to-pole temperature gradient, and subsequently affect the dynamics of baroclinic eddies responsible for much of the weather in midlatitudes (Bengtsson et al. 2006). Differences in the anthropogenic forcing between the northern and southern hemispheres (e.g., through both direct and indirect aerosol effects) may lead to further complex responses of the climate system. The same may apply to regional changes driven by modifications of the SST, land use, and atmospheric aerosol. Pertinent examples of regional changes in precipitation and cloudiness, which have received significant attention in recent years, include a strong drying trend between the 1950s and 1980s over the Sahel (Held et al. 2005), changes in the precipitation pattern in south eastern China (Cheng et al. 2005), and changes in Australian rainfall and cloudiness (Rotstayn et al. 2007). Since these changes are mostly over subtropical summertime continents, most of the change is associated with convective precipitation, such as mesoscale convective systems over Sahel and monsoon precipitation over southeastern China. It follows that the coupling between convective and large-scale dynamics as well as the role of surface, radiative, and microphysical processes play a major role in these changes.

In many respects, contemporary AGCMs (i.e., the atmospheric components of climate models like those used by IPCC 2007), are poorly suited to address quantitatively the coupling between cloud processes and large-scale dynamics. First, AGCMs typically include simple representations of cloud-scale and mesoscale dynamics, including deep convection. In particular, the organization of moist convection into mesoscale convective systems and their impact on the large-scale horizontal momentum budget is seldom considered in traditional AGCMs. It follows that the representation of the coupling between convective and larger-scale processes in traditional AGCMs is problematic. As far as tropical climate is concerned, this problem is best illustrated by the unrealistic behavior of equatorially trapped waves (Kelvin waves in particular), which typically propagate too quickly in AGCMs and resemble dry waves rather than convectively coupled waves. Related to this, AGCMs are poor representations of intraseasonal variability (i.e., the Madden-Julian Oscillation) and monsoonal flows. For deep convection over warm-season continents, traditional AGCMs tend to misrepresent the phase of the diurnal cycle of surface precipitation, with convective precipitation in the model developing too early during the day. Another common problem is the frequency and intensity of convective precipitation, with AGCMs typically precipitating too often and with low intensity. These differences are significant because frequency and intensity of convective precipitation is an important factor in the prediction of climate change. For instance, less frequent and more intense precipitation events (with the same time-mean precipitation) imply more potential for flash floods and severe weather. If contemporary AGCMs cannot predict the variability of the current climate and poorly couple the (parameterized) cloud-scale processes to the larger scales, how can we trust them to estimate future climate accurately?

A fundamental drawback of contemporary AGCMs, directly related to their coarse spatial resolution, is the fact that the effects of parameterized small-scale and mesoscale processes (e.g., clouds and cloud systems) are immediately felt by large-scale dynamics. In nature, small-scale and mesoscale dynamics are the first to respond, and only the residual imbalances are available to drive large-scale circulations. For instance, the effect of the size of cloud particles on radiative fluxes operates at cloud-scale level, not at scales resolved by an AGCM. Similarly, differences in convective heating (e.g., from different mesoscale organization or aerosol effects) are first felt by small-scale and mesoscale processes. In a contemporary AGCM, such differences affect immediately large-scale circulations, since the dynamic response is possible only at large scales because of a lack of small-scale and mesoscale dynamics. It follows that changes in large-scale circulations, which result from modifications of small-scale processes predicted by contemporary AGCMs, need to be treated with caution. This leads us to question seriously studies that investigate the aerosol impact on clouds by simply altering the cloud scheme in an AGCM. Although the results may be interpreted as the changes immediately effecting large-scale circulations, responses on the cloud scale and mesoscale are completely neglected.

The above discussion suggests that we need to address the coupling between deep convection and large-scale dynamics in AGCMs to understand better the role of clouds in a perturbed climate. One way to address this is to use an AGCM that resolves cloud-scale processes, a cloud-resolving non-hydrostatic AGCM. Such a model has been developed at Japan's Frontier Research Center for Global Change, the home of the Earth Simulator (Miura et al. 2007; Collins and Satoh, this volume). Unfortunately, the cloud-resolving AGCM is about six orders of magnitude more computationally demanding than a traditional AGCM and thus cannot be applied to extended climate simulations within the near future. However, with well-focused research using a model like this, perhaps on a large regional scale rather than the entire globe, we can improve our understanding of the response of the larger-scale to cloud-scale perturbations, and this understanding can be used to improve traditional AGCMs. We stress, however, that even though this high-resolution global model is referred to as "cloud resolving" and does not apply any convective parameterization, its grid length is around 3 km, so it resolves scales of only about 10 km at best. Therefore, individual clouds (even large deep convective cells) are not truly resolved; the more appropriate term to describe such a model is "convection-permitting," not "cloud-resolving." The implications of forcing individual convective cells to act on scales larger than is physically reasonable is not yet fully understood, and we recommend that this be investigated as part of the growing area of research using large domain CSRMs.

Another modeling approach developed recently to bridge the gap between large-scale dynamics and convective processes is the cloud-resolving convection parameterization, or "super-parameterization," proposed by Grabowski and Smolarkiewicz (1999). The idea is to use a CSRM (typically, a 2-D slab-symmetric model) in each column of a large-scale model to represent explicitly small- and mesoscale processes. This approach is only 2–3 orders of magnitude more expensive than current climate models, but is still cost prohibitive for long climate simulations. However, as with a fully cloud-resolving AGCM, this approach could serve as a very useful tool to improve our understanding of the response of the large-scale processes to cloud-scale perturbations. As with global CSRMs, well-designed experiments can enhance our understanding of the role of deep convection in a perturbed climate and improve traditional AGCMs.

Khairoutdinov et al. (2007) applied the same approach to the NCAR's Community Climate System Model and referred to this approach as the multiscale modeling framework (MMF). MMF improved the representation of convectively coupled equatorially trapped synoptic-scale waves and statistics of convective precipitation over summertime continents, but it introduced other biases previously not seen in traditional AGCMs. A limitation of this approach, which is far from understood, is that the convective and mesoscale processes are not truly coupled to large-scale dynamics, although they are better coupled to each other.

Other, more radical, approaches are also being considered to allow modeling of the broad range of scales linking convection to the large scale. One is to rescale the governing non-hydrostatic equations (in particular, the vertical momentum equation) so that the scale gap between the large-scale and convective circulations is reduced. This allows 3-D moist convective motions to be simulated using horizontal grid lengths of tens of kilometers (Kuang et al. 2005; Garner et al. 2007). Another approach, more similar to MMF, is to use very different grid lengths in the two horizontal directions. This permits convection to be explicitly represented but reduces the total requirement for grid points. Shutts (2006) used this technique specifically to study the upscale effects of deep convection. Such approaches may provide a valuable alternative to a cloud-resolving AGCM and MMF, especially for climate change simulations. However, more research using these approaches is needed to understand fully the implications of the compromises made to constrain the computational expense.

Summary

Moist precipitating convection plays a crucial role in the Earth's atmosphere, and understanding how it may change as a result of anthropogenic climate forcing is critical. Deep convection spans a wide range of spatial scales, from the cloud microscale (where growth of individual drops and ice crystals takes place) to the scale of a mesoscale convective system (i.e., scales from the sub-centimeter to tens and hundreds of kilometers). However, moist convection extends far beyond the scale of an individual cloud or cloud system because convective heating associated with precipitating convection drives atmospheric large-scale circulations, such as Hadley and Walker circulation in the tropics and monsoon circulations in the subtropics. In midlatitudes, deep convection over summertime continents plays an important role in the hydrological cycle by providing a significant fraction of the annual surface precipitation. It is also associated with severe weather events, such as flash floods, damaging winds, hail, lightning, and tornadoes. How the intensity and frequency of these events change as the climate evolves has significant implications for a large fraction of the Earth's population. Deep convection over continents responds to changes in land-surface characteristics (e.g., changes in land use, deforestation, irrigation), surface insolation (e.g., from direct aerosol effects), and environmental conditions (e.g., CAPE and shear). Although changes of shallow clouds have been shown to dominate global climate sensitivity, modifications of deep convective clouds are of fundamental importance for regional climate change in the subtropics and midlatitudes. Areas such as the Sahel, the Indian subcontinent, and southeastern China depend on summertime convective precipitation for their freshwater supply.

Numerical modeling is crucial to our understanding of the role of deep convection in the context of the climate change. The development and testing of numerical models relies on field and laboratory observations. Today's understanding of cloud-scale processes (such as cloud dynamics and microphysics), and how these may change when environmental conditions evolve, stems from the combination of laboratory studies, field experiments, and numerical modeling on a wide range of scales. Unfortunately, it is difficult to extrapolate this knowledge to the climate change problem. Single-cloud thinking (e.g., the suppression or enhancement of precipitation from a single cloud when nucleating aerosols are changed) is of limited use as far as the climate problem is concerned. The cloud-ensemble thinking (i.e., when one considers interactions between clouds, radiation, mesoscale circulations, and surface processes) can tell us much more about cloud systems in a perturbed climate.

As deep convective systems both impact and are impacted by larger-scale circulations, it follows that a global or near global-scale model is needed to study in depth the role of deep convection in a perturbed climate. Although traditional AGCMs can produce many realistic aspects of the current climate, they do not allow changes related to the cloud scale (such as a change to CCN) to act on the cloud-scale dynamics, thermodynamics, and microphysics. Instead, changes impact directly onto the large-scale circulation, and this is unrealistic. Therefore, cloud-resolving large-scale models (e.g., cloud-resolving AGCMs) are needed to study the coupling between convective processes and the large-scale circulations with confidence. Such a model has been developed in Japan (see Collins and Satoh, this volume), and several other climate modeling centers are actively pursuing the development of additional models. We note, however, a major limitation to these models: at present (and probably for the next few decades) these models as well as other, less extensive versions (e.g., the super-parameterization or use of anisotropic grids) cannot be used for the multidecadal experiments that are typically conducted with the more traditional AGCMs. In addition, the current grid lengths employed by these models (typically around 3 km) do not adequately capture cloud-scale processes.

Given the limitations of all the modeling frameworks, how can the role of deep convection in a perturbed climate be studied? We believe that it is imperative to run well-designed experiments using all the available modeling tools, constrained by observations where appropriate and possible. Experiments with each framework (single CMs, CSRMs, global CSRMs, and AGCMs) can be designed to complement each other, and the results would be instrumental in broadening our understanding of the role that deep convection plays in a perturbed climate. In particular, CMs and CSRMs could be used to develop and validate more robust parameterizations of cloud processes in AGCMs. Although each model has its own limitations, a more comprehensive picture could be obtained by combining our understanding gained from the range of models available.

Currently, researchers who investigate climate change work predominately with traditional AGCMs (see IPCC 2007) and add increasingly complex processes to the models. Although this work should continue, we believe that the limitations of such a framework are significant, especially when considering the role of deep convection. Therefore, we recommend highly that there should be a shift of emphasis within the climate research community to use the wider range of available tools. By following all approaches, we should eventually be able to gain a better understanding of the interaction between deep precipitating convection and large-scale processes, and this would enable more confident predictions of climate change. Such efforts, perhaps coordinated at the international level by the World Meteorological Organization or the United Nations, should be part of any future assessment made by the Intergovernmental Panel on Climate Change.

Acknowledgments

Dr. Leon Rotstayn suggested relevant references for the discussion in the section concerning large-scale processes. Comments on an early draft on this manuscript by Met Office's Andy Brown and NCAR's Mitch Moncrieff, as well as comments from participants of the Forum and two anonymous reviewers, are acknowledged. NCAR is sponsored by the National Science Foundation.

References

Bengtsson, L., K. I. Hodges, and E. Roeckner. 2006. Storm tracks and climate change. *J. Climate* **19**:3518–3543.

Bony, S., and J.-L. Dufresne. 2005. Marine boundary layer clouds at the heart of tropical cloud feedback uncertainties in climate models. *Geophys. Res. Lett.* **32**:L20806.

Byers, H. R., and R. R. Braham, Jr. 1948. Thunderstorm structure and circulation. *J. Meteor.* **5**:71–86.

Cheng, Y., U. Lohmann, J. Zhang et al. 2005. Contribution of changes in sea surface temperature and aerosol loading to the decreasing precipitation trend in southern China. *J. Climate* **18**:1381–1390.

Cui, Z., K. S. Carslaw, Y. Yin, and S. Davies. 2006. A numerical study of aerosol effects on the dynamics and microphysics of a deep convective cloud in a continental environment. *J. Geophys. Res.* **111**:D05201.

Garner, S., D. M. Frierson, I. M. Held, O. Pauluis, and G. K. Vallis. 2007. Resolving convection in a global hypohydrostatic model. *J. Atmos. Sci.* **64**:2061–2075.

Grabowski, W. W. 2006. Indirect impact of atmospheric aerosols in idealized simulations of convective-radiative quasi-equilibrium. *J. Climate* **19**:4664–4682.

Grabowski, W. W., and P. K. Smolarkiewicz. 1999. CRCP: A cloud resolving convection parameterization for modeling the tropical convecting atmosphere. *Physica D* **133**:171–178.

Grabowski, W. W., X. Wu, and M. W. Moncrieff. 1996. Cloud resolving modeling of tropical cloud systems during Phase III of GATE. Part I: Two-dimensional experiments. *J. Atmos. Sci.* **53**:3684–3709.

Hartmann, D. L., and M. L. Michelsen. 2002. No evidence for iris. *Bull. Amer. Meteor. Soc.* **83(2)**:249–254.

Held, I. M., T. L. Delworth, J. Lu, K. L. Findell, and T. R. Knutson. 2005. Simulation of Sahel drought in the 20th and 21st centuries. *PNAS* **102**:17,891–17,896.

Houze, R. A. 2004. Mesoscale convective system. *Rev. Geophys.* **42**:RG4003.

Houze, R. A., and A. K. Betts. 1981. Convection in GATE. *Rev. Geophys. Space Phys.* **19**:541–575.

IPCC. 2007. Climate Change 2007: The Physical Science Basis. Contribution of Working Group I to the Fourth Assessment Report of the Intergovernmental Panel on Climate Change, ed. S. Solomon, D. Qin, M. Manning et al. New York: Cambridge Univ. Press.

Khain, A., D. Rosenfeld, and A. Pokrovsky. 2005. Aerosol impact on the dynamics and microphysics of deep convective clouds. *Q. J. Roy. Meteor. Soc.* **131**:2639–2663.

Khairoutdinov, M., C. A. DeMott, and D. A. Randall. 2007. Evaluation of the simulated interannual and subseasonal variability in an AMIP-style simulation using the CSU Multiscale Modeling Framework. *J. Climate* **21**:413–431.

Kuang, Z., P. N. Blossey, and C. S. Bretherton. 2005. A new approach for 3D cloud-resolving simulations of large-scale atmospheric circulation. *Geophys. Res. Lett.* **32**:L02809.

Lindzen, R. S. 1990. Some coolness concerning global warming. *Bull. Amer. Meteor. Soc.* **71**:288–299.

Lindzen, R. S., M.-D. Chou, and A. Y. Hou. 2001. Does the Earth have an adaptive iris? *Bull. Amer. Meteor. Soc.* **82**:417–432.

Miura, H., M. Satoh, H. Tomita et al. 2007. A short-duration global cloud-resolving simulation with a realistic land and sea distribution. *Geophys. Res. Lett.* **34**:L02804.

Morrison, H., and W. W. Grabowski. 2008. Modeling supersaturation and subgrid-scale mixing with two-moment bulk warm microphysics. *J. Atmos. Sci.* **65**:792–812.

Petch, J. C. 2006. Sensitivity studies of developing convection in a cloud-resolving model. *Q. J. Roy. Meteor. Soc.* **132**:345–358.

Phillips, V., L. J. Donner, and S. T. Garner. 2007. Nucleation processes in deep convection simulated by a cloud-system-resolving model with double-moment bulk microphysics. *J. Atmos. Sci.* **64**:738–761.

Ramanathan, V., and W. Collins. 1991. Thermodynamic regulation of ocean warming by cirrus clouds deduced from observations of the 1987 El Niño. *Nature* **351**:27–32.

Randall, D., S. Krueger, C. Bretherton et al. 2003. Confronting models with data: The GEWEX cloud systems study. *Bull. Amer. Meteor. Soc.* **84**:455–469.

Rapp, A. D., C. Kummerow, W. Berg, and B. Griffith. 2005. An evaluation of the proposed mechanism of the adaptive infrared iris hypothesis using TRMM VIRS and PR measurements. *J. Climate* **18**:4185–4194.

Redelsperger, J.-L., C. Thorncroft, A. Diedhiou et al. 2006. African Monsoon Multidisciplinary Analysis (AMMA): An international research project and field campaign. *Bull. Amer. Meteor. Soc.* **87**:1739–1746.

Rotstayn, L. D., W. Cai, M. R. Dix et al. 2007. Have Australian rainfall and cloudiness increased due to the remote effects of Asian anthropogenic aerosols? *J. Geophys. Res.* **112**:D09202.

Santer, B. D., M. F. Wehner, T. Wigley et al. 2003. Contributions of anthropogenic and natural forcing to recent tropopause height changes. *Science* **301**:479–483.

Seidel, D. J., Q. Fu, W. J. Randel, and T. J. Richler. 2008. Widening of the tropical belt in the changing climate. *Nature Geosci.* **1**:21–24.

Shutts, G. J. 2006. Upscale effects in simulations of tropical convection on an equatorial beta-plane. *Dyn. Atmos. Oceans* **42**:30–58.

Thorpe, A. J., M. J. Miller, and M. W. Moncrieff. 1982. Two-dimensional convection in non-constant shear: A model of mid-latitude squall lines. *Q. J. Roy. Meteor. Soc.* **108**:739–762.

Tompkins, A. M., and G. C. Craig. 1999. Sensitivity of tropical convection to sea surface temperature in the absence of large-scale flow. *J. Climate* **12**:462–476.

Wang, C. 2005. A modeling study of the response of tropical deep convection to the increase of cloud condensation nuclei concentration. 1. Dynamics and microphysics. *J. Geophys. Res.* **110**:D21211.

Weisman, M. L., and J. B. Klemp. 1986. Characteristics of isolated convective storms. In: Mesoscale Meteorology and Forecasting, ed. P. S. Ray, pp. 331–358. Boston: Am. Meteorological Soc.
Xu, K.-M., and D. A. Randall. 1996. Explicit simulation of cumulus ensembles with the GATE Phase III data: Comparison with observations. *J. Atmos. Sci.* **53**:3710–3736.
Yang, G.-Y., and J. Slingo. 2001. The diurnal cycle in the tropics. *Mon. Wea. Rev.* **129**:784–801.

10

Large-scale Controls on Cloudiness

Christopher S. Bretherton and Dennis L. Hartmann

Department of Atmospheric Sciences, University of Washington, Seattle, WA, U.S.A.

Abstract

The climatological distribution of clouds is tightly coupled to large-scale circulation. Net cloud radiative forcing is mainly the result of boundary layer clouds in large-scale subsidence. Deep convective cloud systems exert long- and shortwave cloud forcing that nearly cancel out each other. The extent of this cancellation depends strongly on the vertical motion profile, suggesting that if the cancellation is not coincidental, dynamic feedbacks probably play a role in its maintenance. Low cloud radiative forcing is tied to how cold the surface is compared to the free troposphere. It is an open question how this correlation should be represented in a way that generalizes to a perturbed climate. Simple empirical representations of deep and low cloud forcing are shown to provide strong feedbacks on an idealized model of a tropical overturning circulation. Global weather and climate models, however, still have profound difficulties in accurately representing the cloud response to large-scale forcings.

Introduction

Clouds result from the condensation of water vapor. Since cloud particles generally sediment and tend to evaporate in subsiding air, clouds are most frequently observed in regions where the motion has been upward; this can both act to induce supersaturation as a result of adiabatic cooling and to lift condensed water higher in the atmosphere. However, regions of time-mean subsidence are not necessarily cloud-free, because vertical motion is organized on many scales: from turbulent eddies to moist convection to transient storm systems. In addition, the microphysical diversity and complexity of clouds makes their response to a given pattern of vertical motion challenging to model. Thus atmospheric general circulation models (GCMs) struggle to simulate skillfully the relationships between cloudiness and its large-scale controls in the current climate and how this will change in climates perturbed by changed greenhouse gas or aerosol concentrations.

In this chapter, we relate some observed aspects of cloud climatology to their dynamic controls. We describe the impact of empirically deduced radiative feedbacks from deep convective clouds and from boundary layer clouds in an idealized model of the tropical Walker circulation. On this basis, we examine the physical nature of the deep and low cloud radiative forcing (CRF) parameterizations, and whether they might generalize to other climates. We conclude with a discussion of the modeling challenges and biases in simulating the relation between cloudiness and large-scale circulations.

Large-scale Climatology of Cloudiness

The planetary-scale organization of cloudiness can easily be seen by comparing global maps of cloud properties with global maps of time-mean vertical motion. Figure 10.1 shows the annual mean distribution of total cloud amount from the International Satellite Cloud Climatology Project (ISCCP) (Rossow and Schiffer 1999). The biggest cloudiness maxima are over the extratropical oceans, where baroclinic eddies drive substantial vertical motion over the moist surface of the ocean. Also evident is a banded structure in which the upward motion near the equator enhances cloud fraction; the downward motion in the subtropics generally suppresses cloudiness, except at the eastern margins of the subtropical oceans where it combines with low sea surface temperature (SST) to form the ideal environment for marine stratocumulus clouds, which are trapped under an inversion that is supported by large-scale downward

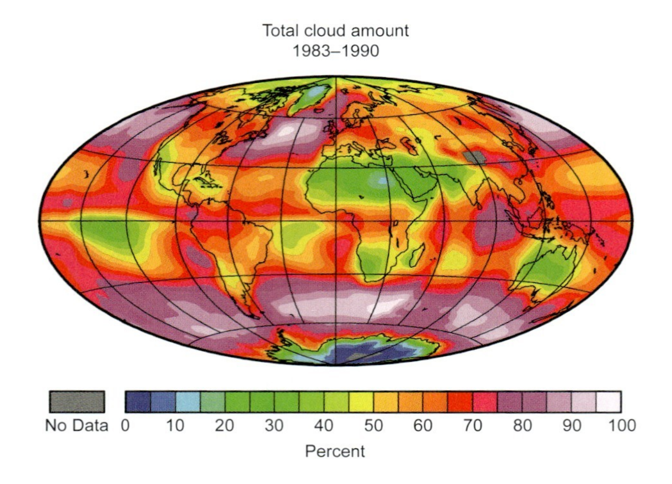

Figure 10.1 Annual mean total cloud fraction from ISCCP.

motion. The contribution of satellite-visible low-topped marine boundary layer clouds to the total cloudiness (Figure 10.2) maximizes in these regions and over the extratropical oceans, especially during summertime (Klein and Hartmann 1993). Extensive stratus clouds and fog form in regions of warm advection, while stratocumulus and shallow cumulus clouds are ubiquitous in the downward mean motion and cold advection behind cold fronts associated with midlatitude storms.

A cursory examination of the spatial organization of clouds shows a strong relationship to the general circulation of the atmosphere. Less immediately apparent is the strong role of radiative processes in controlling these distributions. Deep convection drives strong latent heating and large-scale upward motion. By mass continuity, an equal amount of downward mass flux must exist somewhere away from the convection. Energy conservation requires that globally averaged heating of the atmosphere through latent heat release must be balanced by radiative cooling of the atmosphere (aside from a small residual resulting from upward surface sensible heat flux).

These arguments show that latent heating rate and thereby global-scale precipitation and evaporation are constrained by the radiative cooling rate of the atmosphere rather than by surface humidity. These two quantities respond quite differently to climate perturbations, even though most of the radiative cooling comes from emission from water vapor and clouds. This can lead to important large-scale constraints on convective cloud structures (Hartmann and Larson 2002), as well as on the sensitivity of precipitation and evaporation rates to global mean surface temperature (Held and Soden 2006). Hartmann and Larson (2002) argued that the balance between clear-sky radiative cooling and convective heating suggests that the tops of anvil clouds in the tropics should

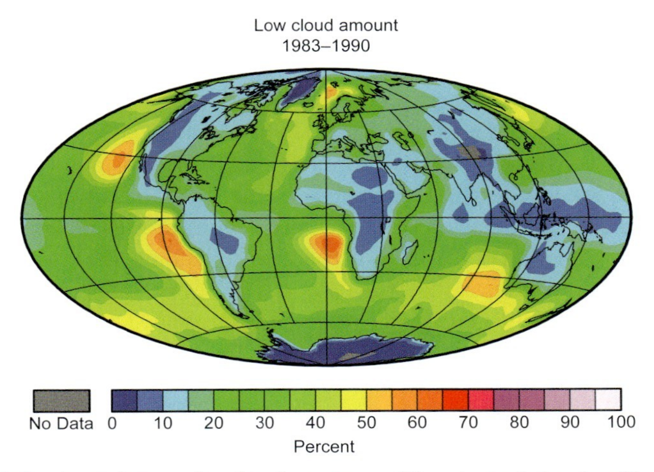

Figure 10.2 Annual mean fractional coverage of low-topped clouds with cloud-top pressures greater than 680 hPa from ISCCP.

remain at a constant temperature under climate change. In addition, radiative cooling (and hence the atmospheric circulation and clouds) can be affected immediately by greenhouse gas perturbations, even before they have induced a change in surface temperature and humidity (Gregory and Webb 2008).

Net Radiative Effect of Cloudiness

One way to measure the importance of clouds is to assess their impact on the radiation balance at the top of the atmosphere. Clouds affect both solar and terrestrial radiation transmission. The effect on the top-of-atmosphere (TOA) radiation balance depends on the radiative properties of the cloud particles and the morphology of the cloud. To a first order, the reduction of net upward longwave radiative flux attributable to clouds, or longwave cloud radiative forcing (LWCRF), depends on the cloud-top temperature at TOA. It is usually positive and largest for clouds with high, cold tops. The shortwave cloud radiative forcing (SWCRF) is usually negative at TOA because clouds reflect sunlight. SWCRF depends on the liquid and ice water path and, secondarily, on cloud particle size and habit. The sum of LWCRF and SWCRF at TOA is the net CRF (NCRF). The left panel of Figure 10.3 shows that the NCRF of an idealized tropical cloud layer increases as its cloud-top temperature cools and decreases as its optical depth increases. High clouds in the tropics can have either strong warming or cooling effects on the TOA radiation budget, depending on the optical depth of the clouds. Low-topped clouds always have a net radiative cooling effect.

The right panel of Figure 10.3 shows the fractional coverage of clouds in the East Pacific with visible optical depths in logarithmic categories and in 5°C intervals of cloud-top temperature as measured by MODIS. The volume under this histogram is approximately equal to the total cloud fraction. Large amounts of cloud are present with both positive and negative NCRF, and the net effect of tropical convective cloud systems is fairly small. Later, we will return to this observation (namely, that short- and longwave CRF of tropical oceanic deep convection cancel surprisingly well) and look at its dynamic implications to ask whether this is a mere coincidence of the current climate.

In this region, boundary layer clouds with tops at high temperatures are also common, and the probability distribution function (PDF) of their optical depth integrates to have a strong negative NCRF. The net negative global effect of clouds on the Earth's energy balance can be attributed primarily to marine boundary layer clouds.

Large-scale Vertical Velocity Profiles and Cloud Properties

In the tropics, a tight relationship exists between the vertical structure of deep convective cloud ensembles, their radiative effects, and the vertical structure

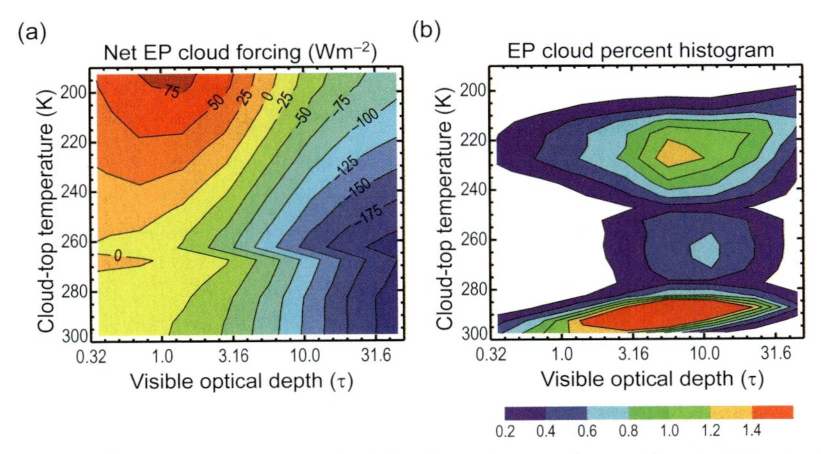

Figure 10.3 (a) Net effect of tropical clouds on the top-of-atmosphere radiation budget as a function of cloud-top temperature and cloud optical depth. (b) MODIS-derived histogram of East Pacific (EP) ITCZ cloud cover (in units of percent cloud fraction per bin) binned jointly by visible optical depths (in logarithmic categories) and cloud-top temperature (in 5°C intervals) (Kubar et al. 2007).

of the vertical velocity field. Climatological profiles of vertical pressure velocity, ω, show "top heavy" upward motion in the West Pacific warm pool and "bottom heavy" upward motion in the East Pacific ITCZ (Back and Bretherton 2006; Yuan and Hartmann, submitted), as shown in Figure 10.4a, b. Back and Bretherton (submitted) have correlated this to the lower SST and stronger meridional SST gradients in the East Pacific. Figure 10.4 also shows that different reanalyses have a large spread in the amplitude of the 850 hPa upward motion in the East Pacific ITCZ, even though all have similar vertical motion profiles. This illustrates that there are not enough observations to prevent large uncertainties in reanalysis-derived vertical motion in the tropics. Presumably, this reflects differences in the moist physical parameterizations (especially for deep convection) between the forecast models that are used for creating the reanalyses. On daily timescales, this problem becomes even more severe and needs to be kept in mind when correlating observed cloud properties with vertical motion.

In the tropical free troposphere, the adiabatic cooling attributable to vertical motion closely counterbalances diabatic heating, which in rainy regions derives primarily from latent heating and secondarily from radiation. Bottom-heavy upward motion implies bottom-heavy latent heating, which requires more shallow cumuli and less deep convection. Indeed, on average, cloud ensembles in the East Pacific ITCZ have less LWCRF per (negative) unit of SWCRF, and a stronger negative effect on the radiation balance, than cloud ensembles in the West Pacific ITCZ (Kubar et al. 2007), as shown in Figure 10.4c. This demonstrates how important it is, when interpreting cloud observations, to consider

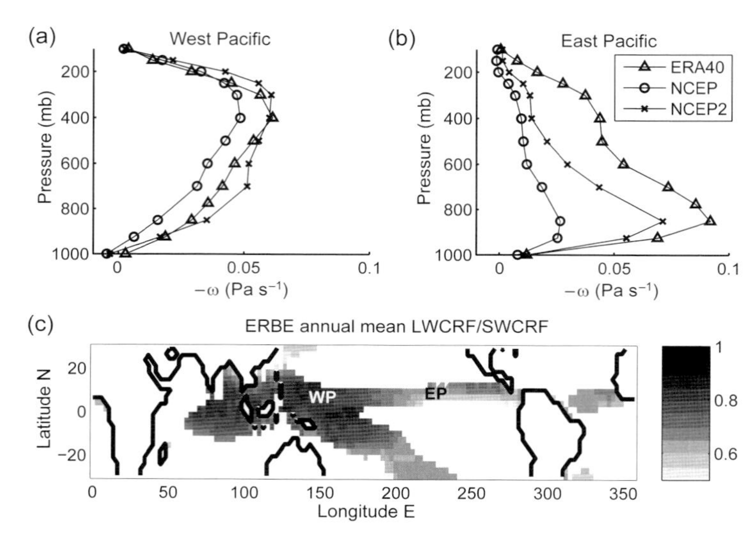

Figure 10.4 Annual mean vertical motion profiles in (a) West Pacific (WP) and (b) East Pacific (EP) based on three reanalyses. ERBE-deduced annual mean LWCRF/SWCRF ratio is shown in (c); WP and EP are marked. Panels (a) and (b) were adapted from Fig. 3 in Back and Bretherton (2005).

carefully the dynamic regime from which they are taken. When analyzed in this way, observations provide a tighter constraint on GCM simulations of clouds and their radiative forcing.

Observations show clearly that the large-scale velocity field, the general circulation, and the associated convection and cloud properties are closely linked. Large-scale dynamic processes and the constraints that the radiation budget imposes on clouds and circulation are important. Changes in the large-scale circulation that accompany climate change are likely to feed back strongly on cloud properties. Expected climate responses to global warming, such as a weakened Walker circulation in the tropics (Vecchi et al. 2006), an expanded Hadley cell (Seager et al. 2007), and northward shifting of storm tracks (Yin 2005), are likely to produce significant cloud feedbacks. In addition, clouds will both respond and feed back on the distribution of SST within the tropics as it changes under global warming or in response to aerosols, with a significant influence on global CRF. The response of convective and marine boundary layer clouds is likely to be affected by changes in SST gradients in the tropics, and this may have a significant effect on the TOA radiation budget (Barsugli et al. 2006; Zhu et al. 2007).

Cloud-radiative Feedbacks on Tropical Circulations

In the preceding section, we documented how vertical motion controls clouds on large space and timescales. Here, we discuss radiative feedbacks of clouds on the large-scale circulation and the associated surface temperature gradients. Our focus is on the tropical oceans, where SST and rainfall biases in coupled GCMs (e.g., insufficiently cool SSTs under the eastern subtropical oceans, spurious eastern Pacific double ITCZ) have often been attributed to cloud feedbacks.

Peters and Bretherton (2005) showed the effect of cloud feedbacks on an idealized model of the tropical Walker circulation above a slab ocean. The tropical troposphere was represented with a single mode of vertical motion following Neelin and Zeng (2000) and included simplified feedbacks between conditional instability, convection, clouds, water vapor, and surface/TOA radiation. They modeled an east–west slice along the equatorial Pacific. The circulation was forced by removing heat from the eastern part of the slab ocean (representing equatorial upwelling) but not from the west. Clearly this type of model cannot represent the differences between bottom-heavy and top-heavy vertical motion profiles and shallower versus deeper precipitating cumuli that are seen over the tropical oceans (e.g., Takayabu and Masunaga, this volume). However, because it is simple, it can illuminate some of the fundamental mechanisms by which clouds radiatively influence tropical general circulation.

Peters–Bretherton Representation of Cloud Radiative Effects

Two cloud-radiative feedbacks are considered in the Peters–Bretherton (2005) model. For consistency, we will follow their sign conventions, even though other conventions might be more natural for the discussion here. The atmospheric CRF, $-R^{cld}$, is defined as the radiative heating of the atmospheric column attributable to the presence of cloud. The surface CRF, S^{cld}, is defined as the radiative heating of the surface as a result of the presence of cloud. They used empirical fits to satellite observations to relate these radiative forcings to large-scale predictors. These fits are of independent interest, as they encapsulate the primary radiative effects of low-latitude oceanic clouds on the real tropical atmosphere–ocean system.

Peters and Bretherton distinguished between deep convective clouds in rainy regions and boundary layer clouds in dry regions, which have rather different radiative effects. They found that on monthly timescales and synoptic space scales, deep convective cloud radiative effects scale with precipitation, whereas boundary layer CRFs scale with lower tropospheric stability (LTS).

Figure 10.5a, b shows the satellite-derived atmospheric radiative heating, $-R^{cld}$, and surface cloud radiative cooling, $-S^{cld}$ (from ISCCP-FD; Zhang et al. 2004) versus monthly precipitation, P (from Xie and Arkin 1997). All of these fields have retrieval uncertainties discussed in the references above, but are thought to be accurate to 10–20% at most locations. Each point is a climatological monthly average over a $2.5° \times 2.5°$ grid box over the tropical oceans (20°S–20°N); L is the latent heat of vaporization. To the right of the vertical

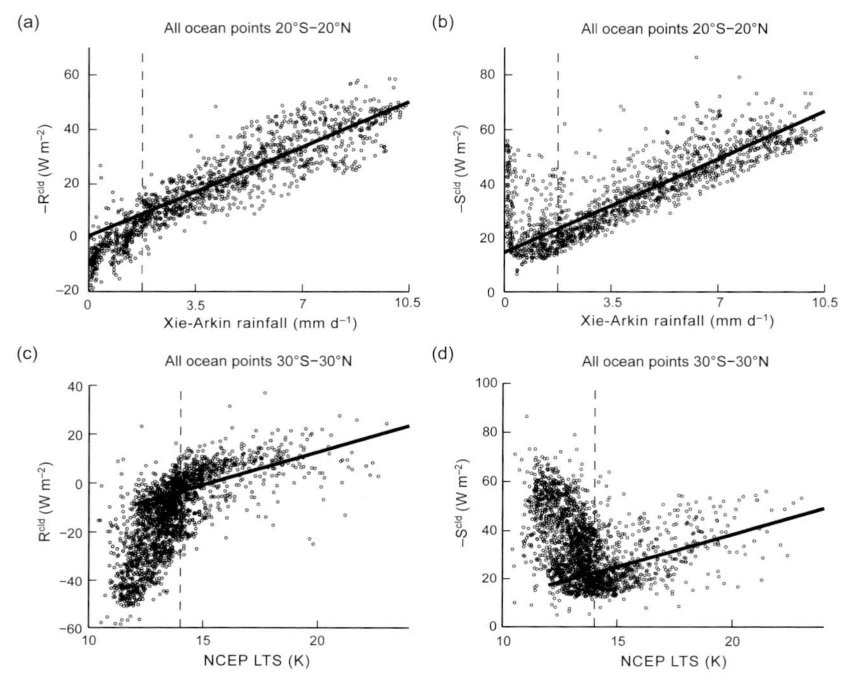

Figure 10.5 Scatterplots (a) and (b) represent low-latitude ocean gridpoints of monthly mean satellite-derived cloud-induced atmospheric column radiative heating, $-R^{cld}$, and surface radiative cooling, $-S^{cld}$, versus rainfall. Fit lines are superposed. The same is shown in (c) and (d) but for lower tropospheric stability (LTS), with fit lines superposed. In (c), R^{cld} is plotted instead of $-R^{cld}$ because boundary layer clouds tend to produce atmospheric radiative cooling at high LTS. In each plot, the fit is best to the right of the dashed line, which marks an approximate climatological threshold between deep convection and boundary layer clouds. Adapted from Peters and Bretherton (2005).

dashed line, in regions of significant deep convective rainfall, the data are well fit by the linear relationships:

$$-R^{cld} = rLP, \tag{10.1a}$$

$$-S^{cld} = 17 \text{ W m}^{-2} + rLP, \tag{10.1b}$$

where $r = 0.17$ and $P > 1.8$ mm d^{-1}. These can be interpreted as anvil cirrus-induced greenhouse heating of the atmosphere and shading of the sea surface from deep convection, both increasing at the same rate with rainfall. Together, Equations 10.1a and 10.1b imply that the net TOA CRF in rainy regions of the tropical oceans is $S^{cld} - R^{cld} \approx -17$ W m^{-2}, independent of the amount of deep convection as measured by rainfall rate, a result which we will return to below. Equation 10.1a also implies that the cloud-induced atmospheric radiative heating in deep convective regions is a constant and non-negligible fraction r of the latent heating LP of the atmospheric column. These fits are tropics-wide composites. Local deviations from these fits are caused by the vertical structure (top heaviness) of the cloud cover, effects of wind shear on upper-level clouds, and other factors.

Lower tropospheric stability, LTS = θ(700 hPa) – θ(1000 hPa), is a good predictor of boundary layer cloud amount. Klein and Hartmann (1993) documented that stratus cloud amount is correlated to LTS, both seasonally and geographically. LTS is also a good predictor of both atmospheric and surface cloud radiative cooling over the low-latitude oceans, as seen in Figure 10.5c, d. In Figure 10.5c, the vertical axis is R^{cld}, whereas in Figure 10.5a it was $-R^{cld}$. Peters and Bretherton (2005) made this choice because in contrast to deep convective clouds, boundary layer clouds cool the atmosphere ($R^{cld} > 0$).

Figure 10.5c, d shows two regimes separated by a threshold lower tropospheric stability, $\text{LTS}_d = 14$ K (dashed line). The regime $\text{LTS} < \text{LTS}_d$ left of the dashed line corresponds to deep convection, which we have already considered. Over the warmest oceans, LTS is the lowest and rainfall is the highest; thus Figure 10.5c, d shows another view of the same behavior depicted in Figure 10.5a, b. Where $\text{LTS} > \text{LTS}_d$, right of the dashed line, deep convection is rare and boundary layer clouds dominate. Here, the following empirical linear fits apply:

$$R^{cld} = 3 + \sigma(\text{LTS} - \text{LTS}_d), \tag{10.2a}$$

$$-S^{cld} = 22 + \sigma(\text{LTS} - \text{LTS}_d), \tag{10.2b}$$

where $\sigma = 3$ W m^{-2} K^{-1}. The ocean experiences net cloud-induced cooling ($-S^{cld} > 0$) because of cloud shading, partly compensated by a cloud-induced increase in downwelling longwave radiation. The atmosphere experiences net cooling ($R^{cld} > 0$) as a result of cloud enhancement of longwave radiative cooling. Both the ocean and atmospheric cloud radiative cooling increase at about the same rate with LTS.

Response of the Peters–Bretherton Model to Cloud Radiative Feedbacks

The Peters–Bretherton model exhibits important model sensitivity to both deep convective and boundary layer cloud radiative feedbacks. Figure 10.6 compares steady-state solutions with no cloud radiative feedbacks, radiative feedbacks attributable only to the deep convective clouds, and radiative feedbacks as a result of both deep convective and low boundary layer clouds. The left edge of the domain is the warm pool; the right edge is the cold pool, where energy is being withdrawn more quickly from the ocean mixed layer. Over the warm pool, the model forms a region of ascent, deep convection, and rainfall separated by a sharp edge from a region of mean subsidence and no precipitation over the cold pool. The deep convective cloud feedback has no impact on the pattern of mean vertical motion (left panel) or rainfall, but flattens the SST gradients over the warm, rainy regions (center panel). The boundary layer cloud

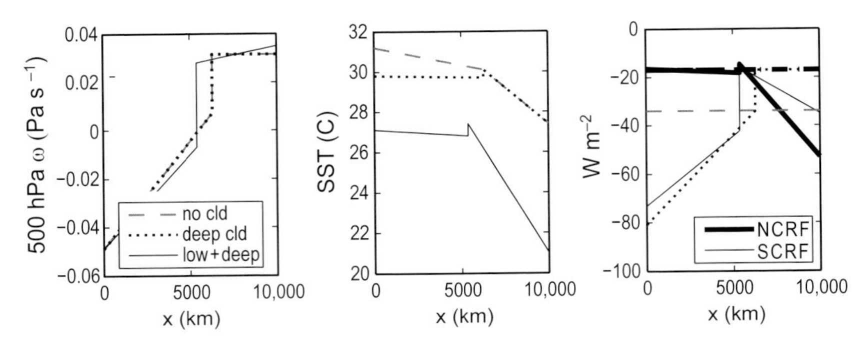

Figure 10.6 Steady state 500 hPa ω (left), sea surface temperature (SST) (center), and net and surface cloud radiative forcing (CRF) (right) in the Peters–Bretherton model, forced by heat removal from the ocean mixed layer that linearly increases from 0 at the left edge to 80 W m^{-2} at the right edge. Dashed line: no cloud feedbacks (horizontally uniform CRF); chain-dash: only deep cloud feedbacks; solid line: low and deep cloud feedbacks. Rainfall (mm d^{-1}) is approximately $160(0.03-\omega_{500})$.

feedback significantly cools SST over the East Pacific, further enhancing the low cloud cover there, intensifying the Walker circulation, and narrowing and slightly cooling the rainy region.

In both cases, the clouds feed back on the atmospheric circulation by their net TOA radiative effect (NCRF, right panel). The circulation diverts energy out of regions of atmospheric ascent into regions of descent. Extra local TOA radiative cooling must be compensated by more advective energy import, which requires stronger subsidence. Hence, the TOA radiative energy loss that results from boundary layer clouds over the cool SST regions strengthens the subsidence there. Deep convective clouds have no TOA radiative effect, hence no effect on the circulation in this model.

The SST adjusts to balance the surface energy budget. More deep convection reduces surface insolation (a negative surface cloud radiative forcing, or SCRF, visible in the right panel). This cirrus anvil shading effect is a powerful negative feedback on further SST rise. Ramanathan and Collins (1991) suggested that this feedback would keep the warmest tropical SSTs fixed close to 30°C even in highly perturbed climates, keeping tropical climate sensitivity very weak. Waliser and Graham (1993) and others pointed out the correct interpretation; namely, that the shading feedback keeps the warmest SSTs (now 30°C) from substantially exceeding the threshold SST for deep convection (now 26°C). Given a forcing such as CO_2 doubling, the threshold SST adjusts as much as is needed to bring the global energy budget into balance.

In summary, in a simple model, horizontal gradients in TOA cloud radiative forcing act as atmospheric circulation feedbacks, and cloud shading helps regulate SST. We believe that these ideas can also serve as useful guides for thinking about the real atmosphere.

Generality of Cloud Relationships versus the Large Scale

The two empirical relationships shown in Equations 10.1 and 10.2 between clouds and their large-scale environment are interesting features of the current climate. Can they also be generalized to perturbed (e.g., ice age or greenhouse) climates? If so, this would provide powerful constraints on low-latitude climate sensitivity. Because we cannot observe other climates, this question must be addressed either through physical arguments that transcend the current climate or through detailed and trustworthy process models that reproduce these relationships. In neither case is the answer yet conclusive, but the physical arguments are worth reviewing.

Balance of LWCRF and SWCRF in Tropical Oceanic Deep Convection

If LWCRF and SWCRF stayed roughly balanced over marine deep convection regions in a changed climate, then any tropical-mean NCRF change could not come from these regions. There is not enough deep convective area or boundary layer cloud over tropical land masses to affect tropics-wide NCRF to a large extent. Thus tropical NCRF changes would have to come mainly from boundary layer cloud regimes.

This idea is consistent with intercomparisons of IPCC AR4-coupled global climate models. Bony and DuFresne (2005) analyzed simulations with 15 atmospheric models using specified historical SSTs, aggregating the model cloud response using tropical (30°S–30°N) vertical velocity binning. They found that the interannual variability in binned NCRF was caused primarily by subsidence regions, and that different GCMs produced substantially different levels of interannual variability in those regions. They also analyzed the NCRF change associated with equilibrium CO_2 doubling in coupled versions of these models. Again, the change was larger and showed larger intermodel differences in subsidence regions. They concluded that boundary layer clouds, especially trade cumulus regimes which cover much of the subtropical oceans, are the principal cause of tropical NCRF changes. However, cloud formation from deep convection in GCMs relies on a cascade of uncertain parameterizations, so model consensus is an unsatisfying and perhaps unreliable substitute for a physical argument.

Even in the current climate, the balance of LWCRF and SWCRF holds only when we aggregate in time and space. As pointed out earlier, the ratio of LWCRF to SWCRF is affected by dynamics, because the vertical distribution of clouds is tightly linked to the mean vertical motion profile. Thus, it is only appropriate to talk of a balance after averaging across the vertical velocity regimes seen over the tropical rainfall belts. Over land, deep-convective SWCRF tends to exceed LWCRF, perhaps because of the diurnal cycle or perhaps for microphysical reasons.

Kiehl (1994) suggested that the balance is a coincidence of the current climate that stems from the typical optical thickness and cloud-top temperature of cirrus anvils. If so, there is no reason to think the changes of LWCRF and SWCRF would track each other in a future climate. Hartmann et al. (2001) suggested that it may be maintained dynamically by the relative inefficiency of horizontal energy exchange between atmospheric columns in deep convective regions, and hence would hold in other climates. Further study of these two hypotheses seems warranted, given their different implications for the role of deep convection in climate sensitivity.

This is most feasible within the model world. The Hartmann et al. (2001) hypothesis should apply just as well to a coupled atmosphere–ocean model as to the real planet. Thus, it would imply that in the climatology of coupled GCMs, LWCRF and SWCRF might be individually biased, but that these biases should compensate. In particular, one might expect LWCRF and SWCRF to compensate better in the coupled climate than in a specified-SST GCM simulation, which does not support the Hartmann et al. mechanism. This would be an interesting test to perform on a suite of coupled models.

A second test would be to look at compensation of LWCRF and SWCRF in cloud-resolving models (CRMs) run to radiative-convective equilibrium in different climates. Tompkins and Craig (1999) performed such simulations in doubly-periodic domains above SSTs of 298, 300 and 302 K. They found approximate compensation between LWCRF and SWCRF in all simulations. Since this setup did not include any horizontal variations, it excludes the Hartmann et al. mechanism, yet reproduces the SWCRF-LWCRF compensation. Bretherton (2007) discussed a CRM analogue to the idealized Walker circulation model discussed in the previous section. In that simulation, LWCRF was approximately 70% as large as SWCRF throughout the simulated region of deep convection (P. Blossey, pers. comm.), so there was no compensation of the two cloud forcings even in a domain-averaged sense. The difference between the ratio of LWCRF/SWCRF in this simulation and those of Tompkins and Craig suggests that this ratio is strongly affected by CRM microphysical parameterizations. Both of these CRM studies favor a Kiehl-type argument, but further CRM study of the control of LWCRF/SWCRF in tropical oceanic deep convection is warranted.

Boundary Layer Cloud and Lower Tropospheric Stability

Lower tropospheric stability is a good predictor of boundary layer cloud impacts on atmospheric and surface radiation balance, as encapsulated in Equation 10.2. Can this relation be quantitatively extended to other climates? If so, this would provide an important step toward understanding low cloud feedbacks on climate sensitivity.

Physically, one can argue that this relation reflects how the typical thermodynamic profile of the lower troposphere changes between stratocumulus and shallow cumulus cloud regimes. Marine stratocumulus clouds tend to have a low, strong capping inversion. The strong inversion inhibits entrainment of dry air from above, and the low inversion top allows a well-mixed turbulent cloud-topped layer to persist. The strong inversion and thick layer of stable stratification between this inversion and 700 hPa both contribute to large LTS. If the inversion is weaker, entrainment tends to deepen the boundary layer, which then decouples into a cumuliform structure with less cloud just below the capping inversion. The weaker, higher inversion leads to smaller S. This argument would apply to other climates, but not necessarily with the same constants as in Equation 10.2.

In a warmer climate, LTS tends to increase across the tropics. The free tropospheric stratification, which is determined mainly by deep convection over the warmest parts of the tropics, roughly follows a moist adiabat. In a warmer climate, the moist adiabatic $d\theta/dz$ increases, contributing to an increase in LTS. Does this then increase low cloud amount and act as a negative climate feedback? GCM simulations by Medeiros et al. (2008) show that if tropical SSTs are uniformly warmed by 2 K, then low cloud cover is largely unaltered despite increased LTS across the entire tropics. One could represent this in Equation 10.2 by a shift toward higher LTS_d in the warmer climate.

Wood and Bretherton (2006) proposed a variant on LTS, called Estimated Inversion Strength (EIS), which removes the effect of free-tropospheric stratification changes that track a moist adiabat. EIS correlates very well with LTS across the low-latitude oceans, but is a better predictor than LTS of stratus cloud fraction over the midlatitude oceans. In these regions, the free-tropospheric stratification tracks a cooler adiabat than in the tropics. This suggests that EIS might also be a "climate-invariant" low cloud predictor, which might be applicable to other climates with perturbed moist adiabats. This hypothesis could be tested using carefully designed sensitivity studies with large-eddy simulations of boundary layer clouds.

Modeling Clouds in Large-scale Circulations

Cloud feedbacks on large-scale circulations are a challenge for GCMs. They involve complex interactions between resolved-scale fields and a suite of parameterized moist processes, including microphysics, cloud fraction, cumulus convection, boundary layer turbulence, surface fluxes, and radiation. In low latitudes, almost all clouds are intimately associated with some form of convection and turbulence. Even in extratropical cyclones, the ice and liquid water paths in the main region of large-scale ascent will be sensitive to microphysical parameterization uncertainties; there will certainly be turbulence-driven boundary layer clouds (convective behind the cold front, possibly shear-driven in the warm sector) and possibly also embedded deep convection. Designing a single parameterization to respond correctly to the full range of large-scale conditions encountered around the globe is challenging; to design and improve a system of tightly coupled parameterizations that work together as intended is even more so. Furthermore, many clouds are thin, and hence poorly resolved by the vertical grid of a typical GCM. The parameterization of cloud–aerosol interactions adds another layer to the challenge. Hence, results derived from GCMs about cloud–aerosol interaction and cloud feedbacks should be critically analyzed and traced to plausible and testable physical mechanisms before they are taken too seriously.

Global models that resolve the cloud processes at much higher spatial resolution may help with these challenges. In a GCM, in which grid spacing typically exceeds 100 km, there is often enormous horizontal cloud heterogeneity within grid cells. The moist physical parameterizations must be aggregated to the grid-cell scale at which they interact using assumptions about this heterogeneity. In practice, the heterogeneity is rarely treated in a fully consistent manner across all parameterizations, leading to such non-sequiturs as rain with no cloud.

It is much more appealing if the cloud physical parameterizations and dynamics can interact at the scale of individual eddies, within which horizontal heterogeneity is not as severe. This explains the success of large-eddy simulation (using grid resolutions of 100 m or less to simulated boundary layer cloud systems) and cloud-resolving modeling (using resolution of 1–5 km to simulate deep convective cloud systems). Global climate CRMs are in their infancy, but they show promise and are already being applied to study cloud-radiative responses on global scales (see Collins and Satoh, this volume). Global and regional numerical weather prediction models are being run without deep cumulus parameterizations at similar or better resolutions. Super-parameterization (Grabowski and Petch, this volume) is a shortcut to a global CRM: a small CRM is placed at each grid column of a GCM, and the CRMs interact through their averaged effects on the GCM grid. As with a global CRM, the cloud processes in a super-parameterized GCM interact on the large-eddy scale. Although super-parameterization involves roughly 100-fold more computations than a regular GCM, it is much cheaper than a full global CRM of the same resolution. A potential advantage of super-parameterization over a global CRM is that the CRM resolution in a super-parameterization can be made much finer or even customized to the geographic location or weather regime, making global cloud-resolving simulations of boundary layer clouds and their cloud and aerosol feedbacks possible. However, current super-parameterizations use 4 km horizontal resolution, which is inadequate for this purpose. None of these global high-resolution simulation methods resolve all cloud processes, and all involve uncertain microphysical and other parameterizations (see Grabowski and Petch, this volume). Hence there is no guarantee that they could represent clouds or their interactions with large-scale circulations much better than a conventional GCM; there are still significant biases in their simulated climate and cloud climatology (e.g., Khairoutdinov and Randall 2005). Thus we return to observational tests of how well models of all types simulate different cloud regimes and their relation to large-scale dynamics.

Biases in the PDFs of the vertical distribution and thickness of clouds are of particular relevance to modeling cloud–aerosol interaction and cloud feedbacks, because simulated clouds with biased vertical structure can be expected to show a distorted response to changes in aerosol, temperature, or vertical motion. Zhang et al. (2005) and Wyant et al. (2006a, b) documented such biases by applying an "ISCCP simulator" to model output; this partitions clouds by their cloud-top pressure and cloud optical depth to mimic a similar ISCCP satellite-derived dataset.

Zhang et al. (2005) segregated results into latitude belts to examine different cloud regimes. Wyant et al. (2006a) used "Bony-binning," in which monthly-mean ω_{500} is used to sort low-latitude clouds into dynamic regimes. Mean ascent ($\omega_{500} < 0$) favors deep convection, and mean subsidence ($\omega_{500} > 0$) favors boundary layer cloud. Figure 10.7 (from Wyant et al. 2006b) applies this approach to compare ISCCP-like cloud statistics from two conventional GCMs and a super-parameterized GCM with satellite observations. All three models simulate too little optically thin cloud, except at the tropopause, and too much optically thick cloud at all levels. For tropical deep convective regimes, con-

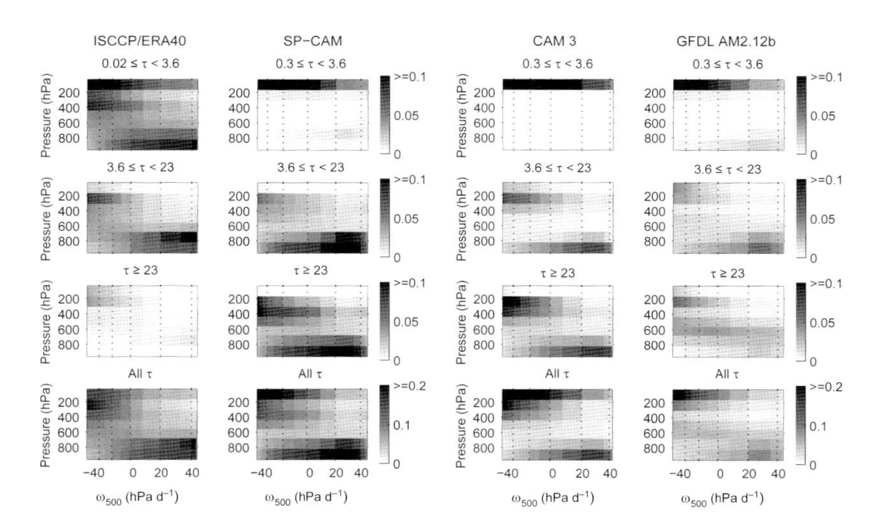

Figure 10.7 Probability distribution of cloud-top pressure conditional on monthly mean ω_{500} based on all grid-column months in 30°S–30°N. Rows show partitioning into thin, medium, and thick optical depth categories. Bottom row includes all cloud optical depths, τ. Left column: ISCCP observations. Other columns show model-derived ISCCP simulator results from super-parameterized CAM-SP and the CAM3 and AM2 GCMs. ω_{500} scale stretched to show frequency of occurrence. Adapted from Fig. S1 in Wyant et al. (2006b).

ventional GCMs also underestimate mid-topped cloud in ascent regimes; the super-parameterized GCM reduces this bias.

Another well-known bias of most GCMs is that the diurnal cycle of cumulus convection is advanced by several hours compared to observations (e.g., Yang and Slingo 2001). Because cumulus convection releases latent heat, this results in continental-scale feedbacks with vertical motion over the tropics. Again, super-parameterization reduces this bias (Khairoutdinov et al. 2005).

We have emphasized the limits of our ability to model accurately the interaction of clouds and dynamics in the current climate. These limits relate closely to our lack of fundamental understanding of the empirical controls on tropical deep and low cloud forcing. New tools, such as cloud-resolving global modeling, and new observations, such as the NASA A-Train suite as well as new efforts to mine the historical data record, may lead to progress if cleverly applied, but no pat answer is in sight. Therefore, we must tread carefully and test comprehensively as we add processes to global climate models, such as aerosols and chemistry, which interact closely with clouds.

Acknowledgments

The authors acknowledge support from NASA grant NNG05GA19G and NSF grant ATM-0336703 for research presented herein. We would like to thank Matt Wyant and Larissa Back for redrafting figures for use in this paper, and Jost Heintzenberg, Anthony Illingworth, Jon Petch, Yukari Takayabu, Rob Wood, and an anonymous reviewer for helpful comments.

References

Back, L. E., and C. S. Bretherton. 2006. Geographic variability in the export of moist static energy and vertical motion profiles in the tropical Pacific. *Geophys. Res. Lett.* **33**:L17810.

Barsugli, J. J., S. I. Shin, and P. D. Sardeshmukh. 2006. Sensitivity of global warming to the pattern of tropical ocean warming. *Climate Dyn.* **27**:483–492.

Bony, S., and J.-L. Dufresne. 2005. Marine boundary layer clouds at the heart of cloud feedback uncertainties in climate models. *Geophys. Res. Lett.* **32**:L20806.

Bretherton, C. S. 2007. Challenges in numerical modeling of tropical circulations. In: The Global Circulation of the Atmosphere, ed. T. Schneider and A. H. Sobel, pp. 302–330. Princeton: Princeton Univ. Press.

Gregory, J., and M. J. Webb. 2008. Tropospheric adjustment induces a cloud component in CO_2 forcing. *J. Climate* **21**:58–71.

Hartmann, D. L., and K. Larson. 2002. An important constraint on tropical cloud–climate feedback. *Geophys. Res Lett.* **29(20)**1951.

Hartmann, D. L., L. Moy, and Q. Fu. 2001. Tropical convection and the energy balance at the top of the atmosphere. *J. Climate* **14**:4495–4511.

Held, I. M., and B. J. Soden. 2006. Robust responses of the hydrological cycle to global warming. *J. Climate* **19**:5686–5699.

Khairoutdinov, M. F., C. DeMott, and D. A. Randall. 2005. Simulation of the atmospheric general circulation using a cloud-resolving model as a super-parameterization of physical processes. *J. Atmos. Sci.* **62**:2136–2154.

Kiehl, J. T. 1994. On the observed near cancellation between longwave and shortwave cloud forcing in tropical regions. *J. Climate* **7**:559–565.

Klein, S. A., and D. L. Hartmann. 1993. The seasonal cycle of low stratiform clouds. *J. Climate* **6**:1587–1606.

Kubar, T. L., D. L. Hartmann, and R. Wood. 2007. Radiative and convective driving of tropical high clouds. *J. Climate* **20**:5510–5526.

Medeiros, B. P., B. Stevens, I. M. Held et al. 2008. Aquaplanets, climate sensitivity and low clouds. *J. Climate,* in press.

Neelin, J. D., and N. Zeng. 2000. A quasi-equilibrium tropical circulation model formulation. *J. Atmos. Sci.* **57**:1741–1766.

Peters, M. E., and C. S. Bretherton. 2005. A simplified model of the Walker circulation with an interactive ocean mixed layer and cloud-radiative feedbacks. *J. Climate* **18**:4216–4234.

Ramanathan, V., and W. Collins. 1991. Thermodynamic regulation of ocean warming by cirrus clouds deduced from observations of the 1987 El Niño. *Nature* **351**:27–32.

Rossow, W. B., and R. A. Schiffer. 1999. Advances in understanding clouds from ISCCP. *Bull. Amer. Meteor. Soc.* **80**:2261–2287.

Seager, R., M. F. Ting, I. Held et al. 2007. Model projections of an imminent transition to a more arid climate in southwestern North America. *Science* **316**:1181–1184.

Tompkins, A. M., and G. C. Craig. 1999. Sensitivity of tropical convection to sea surface temperature in the absence of large-scale flow. *J. Climate* **12**:462–476.

Vecchi, G. A., B. J. Soden, A. T. Wittenberg et al. 2006. Weakening of tropical Pacific atmospheric circulation due to anthropogenic forcing. *Nature* **441**:73–76.

Waliser, D. E., and N. E. Graham. 1993. Convective cloud systems and warm-pool sea surface temperatures: Coupled interactions and self-regulation. *J. Geophys. Res.* **98**:12881–12893.

Wood, R., and C. S. Bretherton. 2006. On the relationship between stratiform low cloud cover and lower tropospheric stability. *J. Climate* **19**:6425–6432.

Wyant, M. C., C. S. Bretherton, J. T. Bacmeister et al. 2006a. A comparison of tropical cloud properties and responses in GCMs using mid-tropospheric vertical velocity. *Climate Dyn.* **27**:261–279.

Wyant, M. C., M. Khairoutdinov, and C. S. Bretherton. 2006b. Climate sensitivity and cloud response of a GCM with a superparameterization. *Geophys. Res. Lett.* **33**:L06714.

Xie, P., and P. A. Arkin. 1997. Global precipitation: A 17-year monthly analysis based on gauge observations, satellite estimates, and numerical model outputs. *Bull. Amer. Meteor. Soc.* **78**:2539–2558.

Yang, G.-Y., and J. Slingo. 2001. The diurnal cycle in the tropics. *Mon. Wea. Rev.* **129**:784–801.

Yin, J. H. 2005. A consistent poleward shift of the storm tracks in simulations of 21st century climate. *Geophys. Res. Lett.* **32**:L18701.

Zhang, M. H., W. Y. Lin, and S. A. Klein et al. 2005. Comparing clouds and their seasonal variations in 10 atmospheric general circulation models with satellite measurements. *J. Geophys. Res.* **110**:D15S02.

Zhang, Y. C., W. B. Rossow, A. A. Lacis, V. Oinas, and M.I. Mishchenko. 2004. Calculation of radiative fluxes from the surface to top of atmosphere based on

ISCCP and other global datasets: Refinements of the radiative transfer model and the input data. *J. Geophys. Res.* **109**:D19105.

Zhu, P., J. J. Hack, J. T. Kiehl, and C. S. Bretherton. 2007. Climate sensitivity of tropical and subtropical marine low cloud amount to ENSO and global warming due to doubled CO_2. *J. Geophys. Res.* **112**:D17108.

11

Cloud-controlling Factors of Cirrus

Bernd Kärcher[1] and Peter Spichtinger[2]

[1]Deutsches Zentrum für Luft- und Raumfahrt, Institut für Physik der Atmosphäre, Oberpfaffenhofen, Germany
[2]ETH Zurich, Institute for Atmosheric and Climate Science, Zurich, Switzerland

Abstract

Factors controlling cirrus clouds comprise small- and large-scale atmospheric dynamics, ice nucleation behavior of natural and anthropogenic particles, and interaction with terrestrial and solar radiation. Current understanding of these factors is summarized. Key uncertainties in this active area of research are addressed, and future developments aimed at reducing these uncertainties are outlined.

Introduction

Various definitions of cirrus clouds exist in the scientific literature (Lynch et al. 2002). In this chapter, we define cirrus as clouds that primarily inhabit the global upper troposphere and tropopause region and which lack a liquid water phase; that is, they are entirely composed of ice crystals. This is equivalent to a temperature criterion constraining the ice formation process, as liquid water droplets freeze spontaneously below ~235 K, depending on their size. Upper tropospheric cirrus ice crystals may subsequently sediment and populate lower atmospheric regions, which we assume to be cloud-free to avoid overlap with other (e.g., mixed-phase) cloud types. Consequently, to form and persist, cirrus require relative humidities over ice (RHI) that exceed 100%; hence cirrus develop predominantly in ice-supersaturated regions. Low-level ice clouds and ice fogs of the Arctic and Antarctic (including Diamond Dust) are not considered cirrus. In addition, high-latitude ice clouds appear well within the lower stratosphere during the polar winter, and these are known as type II polar stratospheric clouds.

The water budget in the climatically sensitive region of the upper troposphere and the radiation balance at the top of the atmosphere are greatly influenced by the amount of condensate in the form of ice and its vertical distribution. Cirrus cloud coverage is substantial according to satellite climatologies (Wang et al. 1996; Rossow and Schiffer 1999) and affects the atmosphere and climate globally, because cirrus trap longwave radiation effectively and can be strong reflectors of incident solar radiation (Stephens et al. 1990).

In addition to their climatic relevance, cirrus clouds are unique among the Earth's plethora of cloud types for two reasons. First, the prevalence of rather high ice nucleation thresholds (tens of percent supersaturation with respect to ice) and long growth and sublimation timescales (owing to the low-temperature environment) result in regions of persistent supersaturation inside and outside of cirrus, as well as a lack of a clear distinction between primary ice (nucleating and growing by vapor diffusion) and secondary (aggregating and precipitating) ice. Second, because ice crystals may survive substantial subsaturations, the transition between clear and cloudy air is rather continuous; this contrasts with other liquid-phase clouds, where cloud and water-saturated regions are almost identical. Sedimentation of ice crystals constitutes an important factor in determining cirrus morphology and development. Finally, cirrus evolve often in a rather stable thermodynamic environment, contrary to most other tropospheric clouds types.

Most of the uncertainty surrounding climate change prediction using general circulation models (GCMs) arises from interactions and feedbacks between dynamic, microphysical, and radiative processes affecting cirrus (Zhang et al. 2005). Model climates are sensitive to even small changes in cirrus coverage (Lohmann and Kärcher 2002) or ice microphysics (see Jakob in Lynch et al. 2002). Optically thin cirrus in the tropical tropopause layer dehydrate the air entering the stratosphere, affecting water vapor and hence ozone concentrations there (Holton and Gettelman 2001). As cirrus ice crystals might scavenge fine aerosol particles and soluble trace gases, ice crystal sedimentation may lead to a vertical redistribution of these substances. Cirrus particles trap or adsorb chemically active trace gases, such as nitric acid (Kärcher and Voigt 2006), and initiate heterogeneous halogen chemistry (Thornton et al. 2007), both of which affect ozone. Settling ice crystals trigger glaciation of warm clouds (e.g., altocumulus), changing their precipitation efficiency by producing deep ice-cloud layers and modulating the hydrological cycle (Herzegh and Hobbs 1980).

It is difficult to define cirrus based on atmospheric measurements. Lidar instruments are capable of detecting cirrus with optical depths as low as 10^{-4} at wavelengths in the visible part of the radiation spectrum, whereas many satellite sensors require visible optical depths of ~0.1 for cirrus to be detected via their forcing in the terrestrial spectrum. Usually, optical spectrometers have acceptable sampling statistics only when measuring ice particle concentrations *in situ* above ~0.1 cm^{-3}, although counterflow virtual impactors can be designed to detect concentrations as low as 0.005 cm^{-3}. The size range 1–10 μm may be populated with both haze droplets and small ice crystals and is particularly difficult to study experimentally (Kärcher and Solomon 1999).

Meteorology textbooks present a host of cirrus cloud types based primarily on their visual appearance. Here, we adopt a simpler categorization by covering the basic types of stratiform cirrus: cirrus formed *in situ* by ice nucleation processes; anvil cirrus arising from deep convective outflow (cumulonimbus clouds); and contrail cirrus that result from jet aircraft operations. *In-situ* cirrus, in which ice nucleates both on supercooled liquid aerosol particles and a special class of heterogeneous ice nuclei (IN), are ubiquitous at all latitudes. Anvil cirrus are widespread in the tropics; along with upper level *in-situ* cirrus, they affect climate most strongly. Contrail cirrus are the only single, purely anthropogenic cloud type that affect midlatitude radiative forcing on regional scales, and are arguably the most obvious human influence on the Earth's climate. They have been invoked to explain a sudden increase in high cloud cover in the U.S.A. at the beginning of the jet era (Liou et al. 1990). Anvil and contrail cirrus have distinct sources that are not primarily connected to *in-situ* ice nucleation processes. The special class of frontal cirrus (cirrostratus) may exhibit features that are characteristic of both anvil and *in-situ* cirrus. For instance, ice condensate can be lofted up in mixed-phase frontal clouds or it may nucleate on aerosol particles in upper cloud levels.

We begin by summarizing the current understanding of cloud-controlling factors of cirrus. We will not provide a complete review, as virtually all aspects of cirrus have been treated elsewhere (Lynch et al. 2002). Instead, our goal here is to supplement the existing body of knowledge with more recent work appropriate to the goals of this Forum. We will focus on what is less well known about cirrus and outline directions for future research to consider.

Current Understanding

Dynamic Controls

The regional appearance of contrail clusters without natural cirrus cloudiness and the large horizontal extent of tropical anvils imply that the upper troposphere is often supersaturated with respect to ice. Three cooling mechanisms control cirrus formation and development by increasing the relative humidity past ice saturation: (a) adiabatic vertical air motions acting on a wide range of spatial and temporal scales; (b) turbulent mixing of air parcels caused by dynamic instabilities or wind shear; and (c) diabatic effects arising from longwave heating or shortwave cooling. The latter two mechanisms evolve on timescales that are probably too long to cause cloud formation, but which affect the life cycle of cirrus after formation. Once cirrus has formed, diabatic processes may couple back to internal cloud dynamics.

Adiabatic vertical air motion is caused by a range of gravity waves spanning spatial and temporal scales from 1–1000 km and from Brunt-Väisälä periods to seasonal timescales, respectively. As is true for all clouds, cirrus represent a multiscale system; motions and processes that transpire on different scales interfere with each other. To estimate the consequences for cirrus evolution, it is imperative that we have an accurate prediction of air temperature changes (cooling rates, $\omega = |dT/dt|$) and corresponding changes in RHI. Variability in water vapor caused by advective or diffusive transport is another factor affecting cirrus, but it is considered of secondary importance for cirrus formation (Kärcher and Ström 2003). This is different to lower tropospheric clouds, where fluctuations of total water appear to be more important than temperature variability (Tompkins 2003).

Aircraft measurements suggest typical horizontal extensions of ice-supersaturated regions of ~150 km in the midlatitude upper troposphere (occasionally few 1000 km) (Gierens and Spichtinger 2000). Ice supersaturation occurs in vertical layers 0.6–1.4 km (occasionally 3–5 km) thick at mid- and high latitudes (Downing and Radke 1990; Spichtinger et al. 2003a; Treffeisen et al. 2007), but the variability in layer thicknesses is large. Ice-supersaturated regions are colder and/or moister than the surrounding air masses (Spichtinger et al. 2003b). Regions of ice supersaturation coincide with areas where the mean RHI as well as its variability is high (Gettelman et al. 2006a). For instance, an ice-supersaturated region extending 1000–2000 km horizontally and up to 3 km vertically and lasting for more than 24 h was observed to form near the outflow region of a warm conveyor belt (Spichtinger et al. 2005b).

Cirrus clouds coexist with regions where RHI > 100%. This is obvious when comparing regions with a high frequency of supersaturation with the spatial distribution of cirrus, both exhibiting similar patterns. Sometimes cirrus develop fallstreaks (*virga*) that may have a larger vertical extent than the associated supersaturated layers, because large sedimenting ice crystals may survive in subsaturated air for a long time (Pruppacher and Klett 1997). The similarity between large-scale patterns of high relative humidity and cirrus occurrence implies that synoptic cold pools define the overall thermodynamic conditions in which *in-situ* cirrus formation and evolution takes place (Newell et al. 1996). Connections between radiatively important microphysical properties of optically thin cirrus and atmospheric humidity, large-scale updraft velocity, and other meteorological parameters have been illuminated by satellite retrievals (Stubenrauch et al. 2004).

Observed cirrus properties cannot, however, be reproduced in simulations in which ice formation is forced only with synoptic cooling rates ($\omega \approx 1$ K/h) (Kärcher and Ström 2003). Instead, cirrus cloud microphysical properties are controlled to a large part by mesoscale variability in ω (Ström et al. 1997; Haag and Kärcher 2004). Mesoscale (1–100 km) temperature fluctuations arise from mesoscale gravity waves; these waves could be excited by convection (Bretherton and Smolarkiewicz 1989), stratified flows over orography

(Smith 1979), geostrophic adjustment (Plougonven et al. 2003), or baroclinic instability (Wang and Zhang 2007). Evidence for the generation of cirrus by Kelvin-Helmholtz wave breaking and internal gravity waves is available (Marsham and Dobbie 2005; Spichtinger et al. 2005a). Cirrus occur more frequently over mountainous regions (Dean et al. 2005). Away from these source areas, a persistent background of such fluctuations driven by high-frequency gravity waves is observed at all latitudes and seasons in cirrus levels (Gary 2006), typically leading to mean $\omega \approx 10$ K/h. Turbulence is associated with all kinds of these waves.

Anvil and contrail cirrus have dynamic origins that differ from cirrus formed *in situ*. Convective anvil cirrus materialize through detrainment of frozen (or rapidly freezing liquid) condensate from deep cumulus updrafts. Anvils develop into stratiform cirrus in favorable conditions, becoming a significant source of upper tropospheric cloud ice at midlatitudes (Tiedtke 1993). The initial ice crystal number density and mass are largely determined by the evolution of the freezing drop size distribution during upward transport (in turn linked to the updraft speed) and by the rate of entrainment of drier ambient air. Given the variability in tropical storm sizes (~100–1000 km) and intensities of mesoscale convective systems, anvils exhibit a wide range of optical properties and lifetimes (see DelGenio in Lynch et al. 2002). The regions with the highest frequencies of deep convection are collocated with regions where thin tropical tropopause cirrus is prevalent (Dessler et al. 2006). Large updraft speeds in the tops of overshooting convective plumes enable the freezing of entrained liquid particles and lead to the formation of thin cirrus well above the main convective detrainment level that contain copious small (radius ~10 μm) ice crystals (Jensen and Ackerman 2006). As mentioned above, ice in frontal cirrus may either form *in situ* (in upper levels) or originate from the upward transport of frozen condensate.

Atmospheric conditions enabling the formation of jet contrails can be derived by analyzing the bulk moisture and heat budgets in cooling aircraft plumes. Plume cooling is brought about by rapid (~0.1 s) isobaric mixing of an initially free, hot turbulent jet with colder ambient air, leading to transient water-supersaturated states. The maximum temperature and minimum relative humidity at which contrails form are determined by ambient temperature, pressure and RHI, specific heat of jet fuel combustion, emission index of water vapor, and the overall aircraft propulsion efficiency (Schumann 1996). Thus contrail formation is easier to predict than *in-situ* cirrus formation, because contrail occurrence is related to local temperature, not to ω. While contrail formation occurs along airplane flight paths, the development of the initially line-shaped clouds into persistent, extended contrail cirrus decks is controlled by weather patterns and the multiscale dynamic processes determining atmospheric ice supersaturation.

Turbulence is often found near the tropopause (Worthington 1999) and is thus an essential component for the evolution of all three types of cirrus clouds.

Turbulence in cirrus is linked to both the dynamic state of the background flow (mainly atmospheric stability) and the microphysical and radiative processes that occur within clouds (see Quante and Starr in Lynch et al. 2002). The complex spatial structure of cirrus results from interactions between localized turbulence, diabatic heating, vertical mixing, and particle sedimentation (Jensen et al. 1998; Whiteway et al. 2004).

Aerosol Effects

Aerosol particles serve as precursors to ice crystals in cirrus clouds that form *in situ*. Homogeneous freezing of liquid particles and heterogeneous ice nucleation at solid surfaces of mixed-phase or insoluble particles contribute to a variable proportion to ice initiation in cirrus. Secondary ice multiplication (i.e., rime splintering, fragmentation during evaporation, or collisions between crystals) is important for the glaciation of supercooled water clouds and is much less relevant in understanding ice formation in cirrus (Cantrell and Heymsfield 2005). Direct ice nucleation from aerosol particles is not the primary determinant for the ice content in anvil and contrail cirrus, but may contribute to their ice budgets at some point during their life cycle.

In cirrus levels, supercooled aqueous particles contain mainly dissolved sulfuric acid (H_2SO_4) and organics (Murphy et al. 1998) and perhaps ammonium, although this is difficult to detect *in situ*. A laboratory analysis of 18 different aqueous solutions (including salts and organics) in the form of emulsion drops (radii ~1–10 μm) provided unequivocal evidence that their equilibrium freezing temperatures are independent of the chemical nature of the solutes (Koop 2004). Critical relative humidities describing the onset of homogeneous freezing have been inferred as a function of temperature and have been used widely in models. Soluble organics tend to lower the freezing rates relative to pure aqueous H_2SO_4 owing to their lower hygroscopicity and possibly a reduced accommodation coefficient of water molecules; both result in less water uptake (Kärcher and Koop 2005) and cause organic-rich particles to remain preferentially unfrozen (Cziczo et al. 2004a). Independent measurements in an aerosol chamber as well as in the field have confirmed the critical RHI (~150–170%) (Möhler et al. 2003; Haag et al. 2003; Hoyle et al. 2005), also for ammonium sulfate particles, which have been controversially discussed (Abbatt et al. 2006).

The temporal development of RHI in an adiabatically rising air parcel is controlled by cooling of air (increasing RHI) and depositional growth of ice particles (reducing RHI) that nucleate once the homogeneous freezing threshold is passed. The total number of ice crystals forming in such a cooling event and their initial size is determined by the nucleation, τ_n, and growth, τ_g, timescales (Kärcher and Lohmann 2002). The former is inversely proportional to ω and depends on fundamental properties of the nucleation rate coefficient; the latter is inversely proportional to the ice saturation vapor pressure. If $\tau_n > \tau_g$, initial growth is faster than freezing, and the ice crystals lose their memory about their initial size. In this fast growth regime, realized over much of the midlatitudes, the number of ice crystals formed is very insensitive to details of the freezing aerosol size distribution but rises as $\omega^{3/2}$, rendering the exact knowledge of the cooling rate very important. This is contrary to the activation of cloud condensation nuclei in warm clouds, where the dependence of cloud drop number density on ω is weaker. Background mesoscale temperature fluctuations in conjunction with homogeneous freezing are capable of explaining the high number densities (~0.1–10 cm^{-3}) of small ice crystals routinely measured in cirrus forming *in situ* at all latitudes (Kärcher and Ström 2003; Jensen and Pfister 2004; Hoyle et al. 2005).

At low temperatures, changes in liquid particle number and size can cause a relatively weak indirect effect on total ice crystal concentration (changes of either sign within a factor of ~2) (Kärcher and Lohmann 2002). According to climate model simulations that include homogeneous freezing and tuned subgrid-scale components of the vertical wind speed, the eruption of Mt. Pinatubo caused no significant effect on cirrus radiative forcing consistent with ISCCP data analyses; however, a noticeable radiative impact of cirrus changes, after very intense volcanic eruptions, cannot be ruled out (Lohmann et al. 2003). Changes in ice crystal properties brought about by enhanced SO_2 emissions in China and Southeast Asia and subsequent enhanced aerosol nucleation may impact future stratospheric humidity by lifting more ice crystals of smaller size into the tropical lower stratosphere (Notholt et al. 2005). More significant aerosol indirect effects are possible if efficient IN and liquid particles compete during cirrus formation (DeMott et al. 1997).

Heterogeneous ice nucleation involves several distinct modes of action (nucleation modes) and requires particles that exhibit particular surface features supporting ice formation (see DeMott in Lynch et al. 2002). In cirrus levels, the predominant nucleation modes are immersion freezing and perhaps deposition nucleation. Immersion nuclei are enclosed within a liquid particle and their surfaces trigger ice formation in the surrounding liquid; deposition nuclei have dry surfaces at which ice nucleates directly from the vapor phase. Immersion nuclei are likely abundant and constitute a subset of the insoluble particle fraction. Efficient deposition nuclei that derive from the Earth's surface are probably rare in the upper troposphere, because they act most likely as cloud nuclei and are removed from the atmosphere before they reach cirrus levels. Surface crystallization of supercooled water has been proposed as a process competing with bulk homogeneous freezing (Tabazadeh et al. 2002); however, its relevance for ice nucleation from aerosol particles in cirrus conditions has not been demonstrated. Ice nuclei relevant to cirrus formation include mineral dust, crystallized salts in mixed-phase particles, and perhaps black carbon soot and sea salt. For a given IN type, chemical composition and size are important factors in the determination of IN activity. Preactivation may occur when a particle once formed ice, experiences subsaturation, and later forms ice at lower RHI than the unactivated particle in otherwise similar conditions.

Although soluble sulfates and organics dominate, at times, the composition of free tropospheric particles at northern latitudes, the fraction of IN is typically enriched in mineral dust and fly ash, metallic components, and, to a much smaller degree, carbon, potassium, and other trace species (DeMott et al. 2003). Convective storms have the potential to transport gases and particles, including sea salt or mineral dust, from the boundary layer up to the upper troposphere (Cziczo et al. 2004b). It is unlikely, however, that significant amounts of very potent IN deriving from surface sources, such as pollen (Diehl et al. 2001) or Saharan dust (Sassen et al. 2003), are able to reach frequently cirrus levels away from deep convection. Most of those ice precursors are likely to be removed by precipitation scavenging in the lower troposphere. In terms of the efficiency of black carbon (soot) particles for cirrus ice formation, available laboratory and field information is inconclusive (Kärcher et al. 2007).

At a given cooling rate, deposition of water vapor on ice crystals formed from IN at low supersaturations reduces or halts the increase in RHI in a rising air parcel, leading to fewer homogeneously frozen particles relative to pure homogeneous cloud formation. Hence, adding IN to a liquid particle population reduces the total number of nucleated ice crystals. This mechanism was coined the "negative Twomey effect" (Kärcher and Lohmann 2003) in association with the traditional Twomey effect in warm clouds, where the addition of cloud condensation nuclei enhances the cloud droplet number density. The negative Twomey effect can lead to reductions of the total ice crystal concentration by up to a factor of 10. According to model simulations, it causes lower cirrus albedo as a result of increased effective radii and decreased ice water contents, as well as nonlinear changes in cirrus occurrence, optical extinction, and subvisible cloud fraction (Haag and Kärcher 2004). The exact magnitude of these changes depends on the ratio between the cooling rate and rate of increase in ice mass attributable to water uptake on heterogeneously nucleated ice particles. If sufficient IN are available and the cooling rates are slow enough to activate only a fraction of these IN (and none homogeneously), more ice crystals can form, compared to a case without IN, but this effect is much weaker than the negative Twomey effect (Kärcher et al. 2006).

One satellite case study of polluted aerosol and ice cloud properties over the Indian Ocean lends support for the existence of the negative Twomey effect in the upper troposphere (Chylek et al. 2006). Extensive lidar studies in the same region, however, yield inconclusive results (Seifert et al. 2007). *In-situ* measurements provide direct evidence for an impact on cirrus via heterogeneous nucleation (Haag et al. 2003; DeMott et al. 2003), suggesting background IN concentrations of $\sim$10–30 l^{-1} at northern midlatitudes. In the majority of cases, the presence of IN did not prevent homogeneous freezing from occurring, but presumably modified cirrus properties. However, observed differences in cirrus properties cannot easily be traced back to IN (Gayet et al. 2004), partly because of the difficulty of separating aerosol effects from effects caused by variability in dynamic forcing.

Observations of the response of deep convective clouds to aerosol particles and soluble trace gases, in general, and to anvil properties, in particular, are very rare (Cziczo et al. 2004b). Only a limited number of numerical studies have addressed this issue using models that couple transport and entrainment in deep convection with detailed cloud microphysics. To describe anvil cirrus properly, the entire cloud evolution must be known, including warm and mixed-phase precipitation and dynamic-microphysical-radiative feedbacks. In general, ice phase processes respond very sensitively to even small changes in the cloud environment (e.g., continental vs. maritime, thermal stability, moisture vertical profile), assume aerosol properties, and represent multiple pathways that tie liquid and ice phase processes together. As a result, local concentrations of ice particles and IN are only poorly correlated.

Some numerical studies of continental deep convection suggest the production of more but smaller anvil ice crystals with increasing aerosol input, either from surface sources or through entrainment of mid-tropospheric particles (Khain et al. 2004; Fridlind et al. 2004). By contrast, Cui et al. (2006) report the opposite behavior and point to differences in convective strength (whether or not cloud tops reach homogeneous freezing levels) and in assumed ranges of aerosol concentrations as possible explanations. In model studies of intense tropical thunderstorms, Connolly et al. (2006) suggest a strong influence of aerosols via the ice phase on cloud microphysics and dynamics; other studies hint at IN not acting as key players in determining anvil cirrus radiative effects (Carrió et al. 2007). Regardless, ice processes responsible for driving the response of cloud evolution to aerosol in such simulations are highly uncertain, and measurements of small ice crystals in deep convective clouds, which many numerical studies use as a guide, are fraught with uncertainties.

Jet engines produce copious plume particles that serve as contrail IN (Kärcher 1999): soot particles formed during fuel combustion, ultrafine particles composed of H_2SO_4, and organics formed in the plume mainly on emitted chemiions (charged molecular clusters) before ice saturation is reached. Available observations are consistent with activation of $\sim 10^4$–10^5 cm^{-3} of plume aerosols into small (some 0.1 μm) water droplets that freeze immediately. A surplus of plume particles and the strong dynamic control of contrail formation driven by large plume cooling rates ($\sim$1000 K s^{-1}) cause the properties of nascent contrails to be rather insensitive to details of the ice nucleation process.

Radiative Effects

Cirrus clouds have a varying effect on the radiation budget (Stephens and Webster 1981). Optical properties of ice crystals are determined by the refractive index of bulk ice as well as crystal size, shape, and orientation relative to incident light. Shape and orientation render the determination of radiative forcing by cirrus significantly more complicated than that of warm clouds composed of spherical water droplets. For sufficiently low optical depth, absorption of infrared radiation and re-emission at colder temperatures dominates scattering of solar radiation, in which case cirrus warm. As optical depth becomes larger, the solar albedo effect increases, leading to a net cooling. The albedo increase is larger for smaller effective ice crystal radii. This picture does not change significantly when supersaturation is allowed inside thin midlatitude cirrus (Fusina et al. 2007). The radiative response of cirrus is further influenced by their coverage, altitude and thickness, spatial inhomogeneity, and numerous other factors, including ice crystal size distribution, solar zenith angle, surface albedo, and presence of clouds and water vapor column.

Thin cirrus cool the surface and exert a net warming within and at the top of the atmosphere; optically thick ice clouds still warm the atmosphere on the whole but cool the surface and the upper atmosphere (Chen et al. 2000). Generally, the net radiative forcing by cirrus results from a difference of two large numbers and is therefore difficult to measure or calculate accurately. Hence, the transition between net heating and net cooling in a cirrus cloud layer is difficult to predict. This problem is aggravated by the fact that the spatial structure on the cloud scale can significantly affect the cirrus radiative response attributable to 3-D effects (horizontal photon transport) (Zhong et al. 2008).

Whereas *in-situ* cirrus visible optical depths rarely exceed ~5 (thin, high cloud according to the ISCCP classification), those values range from ~10–50 for anvil cirrus. Line-shaped contrails have mean optical depths ~0.1–0.5, similar to thin cirrus. Detailed numerical studies revealed that radiation can have dramatic effects on cirrus causing significant differences in cloud inhomogeneity and lifetime compared to studies ignoring radiative effects (Dobbie and Jonas 2001). Solar radiation was found to be important in layers with optical depth in the range of 1–2, whereas infrared radiation was more important in thicker layers (> 2). Thin cirrus with optical depth < 1were found to be only marginally affected by radiation. Depending on the initial cloud layer stability, radiation interactions can produce a more vigorous dynamic evolution, including cellular enhancements at the scale of the layer thickness, and can increase ice water path and lifetime. Anvil cirrus or midlatitude cirrus layers in summer (over warm surfaces) can be maintained through turbulence generated by radiative heating of lower cloud parts (see Quante and Starr in Lynch et al. 2002). Buoyancy lifts and cools the cloud layer and serves to maintain it against the dissipating effects of heating (Köhler 1999). Radiative cooling at cloud tops may create supersaturations high enough to initiate ice nucleation. Similar radiative stabilization may also operate in some contrail cirrus evolving in supersaturated air according to numerical simulations (Jensen et al. 1998).

Cirrus clouds are strong absorbers in the infrared spectral region; hence, from brightness temperature differences, they can be detected day and night using remote-sensing techniques. Remote sensing works well over the ocean because the signal to noise ratio is large enough to determine high clouds relatively accurately. Over land, detection is more difficult and uncertainties are higher. Often, various remote-sensing methods differ in quantitative results about cloud frequencies of occurrence or cloud amount and properties, depending on the type of observation (e.g., nadir vs. limb sounding) and underlying retrieval methods (Stephens and Kummerow 2007). High frequencies of occurrence are found in the tropics over Amazonia, central Africa, and the Indonesian warm pool (Rossow and Schiffer 1999). The Intertropical Convergence Zone (ITCZ) is also characterized by high cirrus cloud amounts. Storm tracks mark high probabilities to find cirrus in extratropical latitudes.

Life Cycle

Remote sensing has been applied to study the life cycle of clouds or cloud systems in a few cases, and only little is known in this respect about cirrus. Tropical convection or midlatitude synoptic weather systems in combination with radiative effects are presumably responsible for seasonal and diurnal variability in the amount of cirrus. Satellite measurements have revealed pronounced diurnal cycles of cirrus over land (maximum occurrence during the late afternoon), in particular in the tropics and at midlatitudes in summer (Wylie and Woolf 2002). The single maximum indicates the building of convective systems and the spreading of anvils from them. Diurnal cycles have been found to be small or absent during winter over the continental U.S.A. In the western tropical Atlantic ITCZ, dual maxima (morning and late afternoon) have been found, the causes of which are not well understood. Observed seasonal variations in the tropics and subtropics are tied to cycles of convective activity and movement of the ITCZ (Stubenrauch et al. 2006). The seasonal cycle in cirrus occurrence at midlatitudes is weaker and more closely linked to synoptic-scale dynamics. Here, the highest probability of occurrence occurs in spring and autumn, whereas the lowest probability occurs in summer, and is more pronounced over the ocean than over land.

Estimating the life cycle of cirrus clouds from satellite data is possible, but difficult. Safer conclusions can be reached by employing modeling tools in support of the remote sensing, such as Lagrangian trajectory calculations (Luo and Rossow 2004). The outflow of convective systems is first transformed into cirrostratus and subsequently into thin cirrus. The timescales of these processes have been estimated as ~6–12 h for the transition to cirrostratus and ~1 d for the transition to thin cirrus. The cirrus amount has an e-folding time of ~5 d; the mean lifetime of the convectively driven cirrus is ~10 h. A comparable study addressing the life cycle of midlatitude cirrus does not exist.

Virtually nothing is known about the life cycle of contrail cirrus, which is essential for an assessment of the aviation climate impact. Available *in-situ* information is scarce and covers only contrail ages up to ~30–60 min, occasionally longer when tracking individual contrails in remotely sensed data. These measurements do not allow representative statistics. The contrail life cycle is not yet represented in global models.

Key Uncertainties

Supersaturation

The SPARC Water Vapor Assessment Report (Kley et al. 2000) discussed different techniques to measure water vapor and relative humidity in the atmosphere (balloon-borne sondes, aircraft instruments, remote sensing). Systematic and sometimes significant differences in measured RHI values have been found by comparing various instruments, particularly in low-pressure and low-temperature conditions. Satellite observations are employed to infer relative humidity without having been designed to measure RHI (Spichtinger et al. 2003b; Gettelman et al. 2006b), so these measurements are very uncertain. More comparisons between different *in-situ* instruments have been carried out in a number of recent field campaigns. At temperatures above ~210 K, different *in-situ* techniques yield similar RHI values, although uncertainty ranges are still substantial (~5–10% standard deviation in absolute terms). Discrepancies are large and unresolved at lower temperatures (Peter et al. 2006). Some data taken in extremely cold tropical conditions (180–185 K) suggest differences in measured water vapor mixing ratios by a factor of two, yielding RHI-values much above the thresholds for homogeneous ice formation (Jensen et al. 2005).

Current models employed to simulate the deposition of H_2O molecules on ice surfaces rely on very simple concepts of bulk diffusional growth of ice particles with idealized shapes (e.g., spheroids). Surface effects that may limit water uptake are lumped into a bulk deposition coefficient, α, for H_2O molecules, impinging on a homogeneous ice surface, and the rate of uptake is proportional to the supersaturation (i.e., the H_2O saturation vapor pressure over ice). Model simulations using the saturation vapor pressure of hexagonal ice and $\alpha=0.1$–1 appear to be consistent with observations of homogeneous freezing (Möhler et al. 2003; Haag et al. 2003) and cirrus formation in the field (Kärcher and Ström 2003; Hoyle et al. 2005) above ~210 K. In colder conditions, the situation reverses. Cubic ice with a larger vapor pressure may nucleate, depending on particle composition (Murray and Bertram 2007), before it transforms into hexagonal ice on the timescale of hours (Murphy 2003). The deposition coefficient may decrease with decreasing temperature or increasing supersaturation (Wood et al. 2001), in some cases perhaps by blockage of active H_2O adsorption sites via co-adsorbed HNO_3, which stabilizes in the form of the nitric acid trihydrate (step pinning) (Gao et al. 2004). Growth laws, which ignore physical processes that determine the nucleation of growth steps on realistic ice crystal facets, may no longer be applicable at low temperatures, low pressures, and small ice crystal sizes. The mobility of H_2O molecules in supercooled liquid particles may decrease at very low temperatures because of an increased viscosity of the solution, suppressing ice germ formation. All of these issues retard ice formation and may explain, at least qualitatively, observations of ice supersaturations above the currently expected homogeneous freezing levels. Water activity of liquid aerosol particles needs to be known to compute homogeneous freezing nucleation rate coefficients (Koop 2004). In turn, activity depends on the vapor pressure of supercooled liquid water, for which available expressions contain large uncertainties up to some tens of percent at temperatures well below the spontaneous freezing point of pure water (Murphy and Koop 2005). Detailed studies addressing these issues are urgently required.

Vertical Velocities

Owing to the strong dynamic control of cirrus formation, measuring and predicting vertical velocities over a range of spatial scales is of paramount importance for the simulation of cirrus clouds. The mesoscale appears to be especially relevant to ice nucleation in cirrus (Kärcher and Ström 2003). However, mesoscale processes (horizontal scales of several 10 km) are neither captured by cloud-resolving models (~100 m) nor by global models (~100 km) and must therefore be prescribed or parameterized. (Below, we return to this issue in the context of new developments regarding cloud-system-resolving models.) For instance, a subgrid-scale component of the vertical velocity tuned in proportion to the turbulent kinetic energy is used to drive the homogeneous nucleation parameterization in the climate model ECHAM (Lohmann and Kärcher 2002).

In many general circulation models (GCMs), parameterizations for the gravity wave drag of mountain waves are used for a better representation of the stratospheric circulation (Lott and Miller 1997; Scinocca and McFarlane 2000). Attempts to use such parameterizations for the representation of orographic cirrus in GCMs have been reported only recently (Dean et al. 2007; Joos et al. 2008). A consistent approach to parameterize vertical velocity (i.e., temperature changes) and water vapor fluctuations from gravity waves and other sources for use in GCM cloud microphysical schemes has not yet been developed.

To make matters worse, vertical wind speeds are notoriously difficult to measure *in situ*. Evaluation of aircraft probes determining vertical velocities contain large biases (10–30 cm s^{-1}) when evaluated along short flight segments (Bögel and Baumann 1991). Statistics of such data comprising many flight hours are more reliable but do not yield local information on cooling rates. Cooling rates inferred from measured air parcel displacements relative to isentropic trajectories using the microwave temperature profiler are relatively accurate. The statistic of such mesoscale temperature fluctuations provided by Gary (2006) covers large portions of the northern hemisphere, lower stratosphere, and tropopause region but does not contain signals arising from lee waves.

Ice Particle Measurements

Measurements of particle concentrations with optical particle spectrometers that have shrouds or inlets (e.g., the FSSP and the CAS) could be affected by shattering of large ice crystals (Gardiner and Hallett 1985; Field et al. 2006; McFarquhar et al. 2007). Shattering leads to unreasonably large ice particle number concentrations and to artificial broadening of inferred size distributions. This is highly relevant, as small ice crystals have been estimated to dominate cirrus optical extinction. Shattering seems to be a serious issue for several cloud probes in the presence of a sufficient (yet relatively small) number of large (> 100 μm) ice crystals. Quantifying the effects of shattering in previous field measurements requires careful analyses. Measurements in anvil cirrus appear to be particularly prone to errors. A recent study suggests that conclusions from some previous studies of the radiative effect of small ice particles require reevaluation (Heymsfield 2007).

Accurate determination of ice crystal shape is very important for the calculation of parameters crucially affecting the overall radiative effects of cirrus. These parameters include ice particle density, surface areas, and effective (optically active) radius. Although cloud imaging probes can be used to infer ice crystal habits with reasonable accuracy (Lawson et al. 2006), measuring habits of small (<50 μm) ice particles is still challenging (Baumgardner et al. 2005). Determination of crystal habit is a prerequisite to the accurate determination of optical properties and terminal fall speeds of large ice particles. Using a static diffusion chamber, laboratory studies have provided a classification of ice crystal habits, among other properties, at ice supersaturations and air pressures comparable to those in the atmosphere in the temperature range 203–253 K (Bailey and Hallett 2004). Columnar shapes appear to be prevalent below 233 K. With increasing supersaturation and decreasing temperature, bullet rosettes, needles, and plates appear with higher frequencies. These findings are largely consistent with field observations. Differences in shapes bring about changes in growth rates of up to ± 50% under identical growth conditions. The shape-dependence of mass growth rates affects the competition for available condensate and is often not adequately considered in models.

In the laboratory studies, temperature appears to be the primary factor controlling ice crystal habits, followed by supersaturation and pressure (via H_2O diffusivity) as secondary factors. However, sedimentation, updrafts within cloud, and inhomogeneities in temperature and RHI fields might at times result in a poor correlation between habits and temperatures observed in real cirrus clouds (Field and Heymsfield 2003). A spread in ice nucleation thresholds further adds to the complex spatial distribution of ice crystal habits detected *in situ*. Therefore, field measurements are difficult to interpret without the help of models for which the laboratory findings provide a useful guide.

Ice Initiation and Growth

A great number of atmospheric measurements of high relative humidities close to the homogeneous ice nucleation limits, the prevalence of mesoscale temperature fluctuations generating high local cooling rates, reliable laboratory measurements of homogeneous freezing rate coefficients (enabling sound representation of ice formation in models), and the frequent observation of numerous small ice particles strongly suggest that homogeneous freezing is a ubiquitous pathway for cirrus formation (DeMott et al. 2003; Haag et al. 2003; Law et al. 2005). The predominance of homogeneous freezing does not rule out possible effects of IN on cirrus properties, as IN number concentrations appear to be low enough to modify cirrus properties instead of frequently preventing homogeneous freezing from occurring. However, geographical distribution, size distribution and chemical composition, and ice nucleation modes of IN at cirrus levels are not well known. Although it may constitute an insignificant portion of the total aerosol, IN may exert a significant control on the microphysical structure of cirrus. Since the competition between liquid particles and IN for available water during nucleation depends sensitively on vertical air motion variability, uncertainties in the latter have strong impacts on predicted indirect IN effects on cirrus.

Mineral dust particles can act as efficient IN over a wide range of temperature conditions. In the accumulation mode size range, they may survive lower tropospheric cloud processing and reach the upper troposphere. Presumably, those residual dust particles might exhibit specific seasonal and geographical patterns, owing to the variable source strength (dust storms), which have not yet been systematically explored. There is mixed evidence for their ice nucleation behavior at mid-tropospheric temperatures (Sassen et al. 2003; Ansmann et al. 2008). While being transported to cirrus altitudes, their ice-nucleating ability may further change because of ongoing chemical processing and coagulation with background particles. Using surrogates of dust, laboratory studies have shown that the IN activity may decrease with decreasing size (in the range 50–200 nm), depending on the chemical nature of a surface coating (Archeluta et al. 2005; Knopf and Koop 2006). Onset RHI values of ice nucleation span ranges from near ice saturation up to values required for liquid particle freezing.

Ice formation by soot particles remains very poorly understood (Kärcher et al. 2007). Some soot particles act as IN, as they have been found as residuals in ice crystals formed on upper tropospheric aerosol samples (Chen et al. 1998). The ice nucleation behavior of soot particles in various nucleation modes seems to depend very sensitively on size and surface characteristics, supersaturation, and temperature. As for mineral dust, ice nucleation often occurs over a range of RHI, which can be attributed to individual particle characteristics in the samples. Many laboratory studies have not quantified the nucleation properties on a single particle basis, rendering interpretation of those results difficult. Further, atmospheric aging processes are not well quantified, and associated uncertainties propagate directly into global model predictions of upper tropospheric soot abundance (Hendricks et al. 2004). Thus, studies have often used soot samples of unknown relevance to atmospheric soot.

Heterogeneous effects on ice nucleation may also occur in partially soluble particles in which crystalline inorganic or organic phases (in particular oxalic acid) can form (Zobrist et al. 2006; Abbatt et al. 2006). Ice formation in mixed-phase particles may compete with deliquescence of crystalline solids at low supersaturations, bringing about more complicated pathways to cirrus formation than by fully insoluble IN alone (Colberg et al. 2003). Such effects have not yet been detected *in situ*.

Once ice crystals evaporate and release insoluble core particles upon which they have nucleated, it is possible that the cores facilitate ice formation in a subsequent nucleation event by lowering the nucleation threshold to form ice again (preactivation). This mechanism is poorly understood experimentally and theoretically. Preactivation has been identified as being potentially important for dust (Knopf and Koop 2006). The IN behavior of soot particles may change once they are released from evaporating ice particles, because they may change their size and shape (i.e., they fractionate into smaller pieces or aggregate into larger clusters). Short-lived or persistent contrails may act as preactivating agents for exhaust soot particles.

Even if the nucleation puzzle was solved and the evolution of ice crystal habits as a function of ambient conditions were known, incomplete knowledge about the physical nature of ice surface kinetics strongly limits atmospheric applications (Wood et al. 2001). Once H_2O molecules impinge onto flat facets (terraces) at the ice surface, through diffusion from the gas phase, they migrate toward ledges (growth sites) and may become incorporated into the lattice with a certain probability before desorption. Ice growth is promoted either by screw dislocations (a continuous ledge source) at low local supersaturation, or by two-dimensional nucleation (at local ledge sources) requiring higher local supersaturation. The number and spacing of ledges in turn are functions of the ambient supersaturation and crystal-specific features. Models capable of predicting ledge properties for cirrus ice crystals are not available.

Indirect Aerosol Effects

Most previous cloud-resolving simulations (e.g., those performed in the working group 2 of GEWEX, which focused on cirrus cloud processes) did not study aerosol effects in cirrus in detail. Instead, parameterizations of IN activity originally developed for mixed-phase cloud conditions have been employed and extrapolated to lower temperatures; sometimes studies did not include homogeneous freezing from supercooled aerosols and mostly neglected small-scale dynamic variability. Details of how IN could modify cirrus properties and frequency of occurrence in realistic dynamic conditions and with aerosol forcing supported by observations were first suggested by near-global, domain-filling trajectory simulations of midlatitude cirrus, using a parcel model with parameterized ice particle sedimentation (Haag and Kärcher 2004). More recently, the negative Twomey effect has been studied in large-eddy simulations (LES) on the cloud scale (Spichtinger and Gierens 2008a). These simulations indicate that under some conditions, the initial macroscale evolution of cirrus may be markedly affected by IN, while overall radiative properties are still dominated by liquid particle freezing. Cirrus formation and ice particle sedimentation are found to feed back to the aerosol distribution by nucleation scavenging and vertical redistribution. Large ice crystals may survive substantial distances when falling through subsaturated air. When interacting with liquid water clouds, they initiate the Bergeron–Findeisen process and change precipitation rates.

Because of the wide range of dynamic, thermodynamic, and aerosol-related parameters, indirect aerosol effects on cirrus cannot be properly quantified, and final conclusions on the importance of IN for cirrus radiative forcing cannot yet be provided. Modification of lower-level clouds by cirrus has not been studied in detail.

Radiative Forcing

Radiative transfer simulations with prescribed cirrus microphysical properties have illustrated the fact that for a given ice water path, the net cloud forcing is basically determined by the effective ice crystal radius. Decreasing the latter causes the instantaneous radiative forcing to change sign from warming to cooling. High numbers of small ice crystals are especially prevalent in young contrail cirrus and some anvil cirrus. The effective radius at which the transition to cooling occurs depends strongly on the assumed ice crystal habit (Zhang et al. 1999). The crucial impact of habit on cloud solar and thermal infrared radiative response has also been demonstrated in field measurements (Wendisch et al. 2005, 2007). In addition, ice particle habits affect the remote sensing of cirrus. Much uncertainty is introduced in these issues as a result of the poor understanding of the factors controlling ice particle shapes.

Cirrus radiative forcing is known to depend on horizontal inhomogeneity on scales unresolved by global models (< 100 km) (Fu et al. 2000; Carlin et al. 2002), affecting the mean climate state. Including effects of spatial inhomogeneity in global model simulations may change radiative forcings by tens of W m^{-2} (Gu and Liou 2006). For experimental investigations of the impact of these inhomogeneities, new measurement techniques have been developed that combine nadir observations and in-cloud radiation measurements (Schmidt et al. 2007), besides lidar instruments.

Cirrus radiative forcing is also very sensitive to the vertical inhomogeneity (i.e., the vertical distribution of cloud ice on scales of a few 100 m). Small ice modes originate from homogeneous freezing and are predominantly observed in upper cloud layers, where supersaturation levels are highest (Miloshevich and Heymsfield 1997). Large ice modes are typically observed in lower cloud regions and are fed by sedimenting and aggregating ice crystals. Although understanding of ice particle aggregation as an isolated process has advanced (Field and Heymsfield 2003; Westbrook et al. 2004), a rigorous treatment of factors affecting aggregation in the presence of diffusional growth, sedimentation, and small-scale turbulence has not been performed and dedicated observations are rare (Westbrook et al. 2007). Early heterogeneous nucleation of few IN may also contribute to large ice modes. Interaction of turbulence, nucleation, and growth may lead to a more complex layering of ice crystal sizes according to an LES model (Spichtinger and Gierens 2008b). Tails at low values of probability distribution functions (PDFs) of ice water content (Haag and Kärcher 2004) and optical depth (Immler and Schrems 2002) may well be affected by sedimentation and aggregation processes, as well as by IN and by low cooling rates. The impact of sedimentation has been highlighted by Jakob (in Lynch et al. 2002) using a numerical weather prediction model. He showed that cirrus radiative forcing may be subject to changes of some 10 W m^{-2} given uncertainties in ice particle terminal fall velocities in the range 0.1–2 m s^{-1}.

The understanding of the global net radiative response of various cirrus cloud types is still uncertain. There are crucial uncertainties concerning the extinction component attributable to small ice particles and the host of factors controlling anvil microphysical properties in mesoscale convective systems, among other issues. In addition, the annual and global mean picture of cirrus cloud radiative forcing derived from the ISCCP D2 data contains substantial simplifications in treating the vertical layering of cloud, the radiative transfer in cirrus, and in assumptions about nighttime radiative fluxes (Chen et al. 2000).

Cirrus Representation in GCMs

In the ECMWF-integrated forecast system, a simple method of representing ice supersaturation before cirrus formation via homogeneous freezing has been implemented into the operational scheme, which uses a prognostic cloud fraction variable (Tompkins et al. 2007). This step has led to a significant improvement of the relative humidity fields at the tropopause, and to corresponding increases of the net incoming solar and outgoing longwave radiation of ~2 W m^{-2} that almost cancel each other globally. The updated scheme still relies, however, on bulk mass microphysics with a temperature-dependent water/ice phase partitioning and uses saturation adjustment after cirrus formation (i.e., it does not allow in-cloud supersaturation to occur).

A different approach has been taken in the ECHAM climate model. An improved prognostic two-moment (ice particle number and mass) microphysics package is currently being implemented that parameterizes homogeneous freezing in competition with heterogeneous ice nucleation on the subgrid scale; cloud ice derived from different aerosol sources (liquid and IN) are tracked separately for the first time in a GCM framework (Kärcher et al. 2006). This enhanced microphysical complexity allows more interactions between dynamics and aerosols during cirrus formation to be studied on a global scale. However, the cloud fraction is still diagnosed based on resolved relative humidity and predicts an overcast grid box already at saturation. Thus, it cannot handle the supersaturation now allowed in cirrus conditions. This basic inconsistency is also present in alternative approaches to represent ice supersaturation and aerosol–ice interactions in GCMs with diagnostic cloud fraction (Gettelman and Kinnison 2007; Liu et al. 2007).

A central assumption common to virtually all GCM cloud schemes is that cloud particles form as soon as saturated conditions are surpassed. Likewise, cloud particles are not allowed to exist in subsaturated conditions. These are fair assumptions for low-level tropospheric clouds that form and exist close to water saturation. Hence, statistical cloud schemes based on PDFs of total water (and perhaps other variables) allow the cloud fraction and cloud water content to be diagnosed, once the moments of the underlying PDFs are known. These central assumptions do not hold in the case of pure ice clouds, indicating that current parameterization concepts need to be fundamentally modified (Kärcher and Burkhardt 2008), in particular the use of cloud schemes that diagnose cloud fraction based on resolved relative humidity. Ultimately, GCM parameterizations that provide a consistent description between cloud fraction, ice supersaturation, and ice nucleation form a sound basis for subgrid parameterizations of cirrus cloud inhomogeneity and overlap.

Validation of GCMs

When evaluating the water budget in climate models, the emphasis is usually on warm clouds. Comprehensive efforts to validate GCM predictions of ice water content, effective ice crystal sizes, and other crucial cirrus properties have not yet been reported. Current GCM results cannot be accurately constrained by available satellite data (Zhang et al. 2005). However, the microphysics of ice clouds is often used for tuning the models (see Jakob in Lynch et al. 2002), particularly sedimentation rates that control simulated ice water content. Consequently, there are indications that optical properties of natural cirrus are not represented well, at least in some climate models (Lohmann et al. 2007). Accurate global model validation is seriously complicated by lack of climatologically representative data (e.g., from airborne measurements), uncertainties in defining what has actually been detected (e.g., detection efficiencies of cloud in remote sensing), and uncertainties in measurements (e.g., frequencies and absolute values of supersaturation). Few concepts exist to describe systematic strategies for a process-oriented validation of GCMs (Eyring et al. 2005).

Therefore, the proper validation of ice microphysical processes and simulated cirrus properties remains an elusive goal.

Climate Change

Climate change may bring about changes in the global distribution of temperature, relative humidity, and cooling rates attributable to greenhouse gas emissions. The distribution and efficacy of ice-nucleating aerosols may also change. These changes could significantly modify cirrus properties (Kärcher and Ström 2003). Any effort to ascribe cause to changes or trends of cirrus cloud properties requires a careful evaluation of the underlying processes.

While the atmosphere can hold more moisture at higher temperatures, it is likely that relative humidity remains largely unaffected (Held and Soden 2000), but possible changes in upper tropospheric relative humidity cannot be excluded. The tropopause height rises as a result of tropospheric warming and simultaneous stratospheric cooling (Santer et al. 2003). Peak clear-sky supersaturations and cirrus cloud tops occur frequently, very close to the tropopause. A change in the vertical extension of the troposphere may imply a change in the vertical distribution of relative humidity and hence modify cirrus occurrence and properties. If cirrus form at warmer/colder temperatures (i.e., different distances relative to the future tropopause height), then the number of ice crystals will decrease/increase (Kärcher and Lohmann 2002) and affect the radiative response of cirrus (Fusina et al. 2007).

Changes in the meridional temperature gradient affect the atmospheric circulation. In a warmer climate, storm tracks might be shifted toward higher latitudes, and midlatitude weather systems might become weaker (Yin 2005; Bengtsson et al. 2006). This may cause changes in mesoscale variability in vertical velocities, which have been identified as the key controlling factors of cirrus formation, besides temperature. Finally, properties of IN from natural and anthropogenic sources might continue to change (Bigg 1990). The indirect aerosol effect and changes in dynamic forcing patterns may alter cirrus cloud properties by similar amounts (Haag and Kärcher 2004). Thus, differentiating between natural and anthropogenic causes of possible cirrus changes represents a great challenge.

It is difficult to predict if and how properties of anvil cirrus might change in the future. It seems plausible, however, that the frequency of occurrence of anvils changes in proportion to the number of deep convective events in the tropics. As air traffic continues to grow at a substantial rate and the atmosphere is not saturated with respect to contrail occurrence, contrail cirrus will be a more frequent phenomenon in decades to come. A larger occurrence of contrail cirrus might tap more condensable ambient water by growth and sedimentation. This may cause the atmosphere not to reach, or to reach later, the thresholds for formation of natural cirrus. As for *in-situ* cirrus, the behavior of anvil and contrail cirrus in a future climate cannot be predicted with confidence unless the processes controlling their formation and evolution are understood. This includes the development of a predictive capability for feedbacks between microphysics, radiation, and dynamics across various spatial and temporal scales, including potential experimental clarification.

The Way Ahead

Measurement Techniques

The design of airborne measurements that have the potential to unravel the relative roles of homogeneous freezing and heterogeneous ice nucleation (i.e., the indirect effect of IN on cirrus clouds) is very challenging. An atmospheric closure experiment would require a combination of at least high-resolution measurements of aerosol composition and ice nucleation activity, vertical air motion, relative humidity, and cirrus ice particle size distribution in the range 1–1000 μm. Cloud ice properties could be predicted directly from dynamic and aerosol information, ideally in combination with numerical modeling.

Phase-dependent particle sampling in conjunction with particle spectrometers has recently improved our capabilities of cold cloud analyses (Mertes et al. 2007). Progress has recently been made to reduce uncertainties attributable to shattering; the conditions in which contributions from small particles are significant have been identified and methods have been proposed to correct available data (Field et al. 2006; Heymsfield 2007). The use of different measurement techniques on the same measuring platform leads to a better estimate of the impact of shattering (Gayet et al. 2006). Wind tunnel studies could also be helpful to characterize the influence of shattering or to design inlets with reduced impact. A better knowledge of ice particle size distributions in combination with habit distributions will enable improved cirrus cloud radiation modules for use in GCMs to be developed.

There is an urgent need for improved real-time *in-situ* measurement capabilities of ice nucleation from upper tropospheric particles (Stetzer et al. 2007). Chemical composition of IN is best measured in conjunction with mass spectrometry (Cziczo et al. 2004a). Fast IN sensors for cirrus applications are becoming available, as revealed by a recent workshop[1] on ice nucleation measuring systems. New ways of combining different instruments should be fully exploited, for example, through the combination of counterflow virtual impactors, continuous flow diffusion chambers and aerosol mass spectrometry (Cziczo et al. 2003).

Laboratory experiments are well suited to develop and test airborne instrumentation, but may also be employed to study inherent differences in methods for measuring particles and humidity in the atmosphere and in the

[1] http://lamar.colostate.edu/~pdemott/IN/INWorkshop 2007.htm/

laboratory. Laboratory-based studies should be performed to examine details of fundamental processes of ice formation (e.g., what makes certain particles better IN than others). In this regard, direct sampling of particles of atmospheric relevance would be most fruitful, as techniques for generating aerosol particles that closely resemble their atmospheric counterparts are limited. Finally, evolutionary studies under controlled initial conditions and thermodynamic histories enable the detailed investigation of processes related to ice formation and growth and ice crystal optical properties.

High-quality fields of relative humidity require very accurate measurements of temperature and water vapor mixing ratio. The current suite of satellite sensors is not fully adequate to study supersaturation in the upper troposphere, despite the valuable insights gained from recent evaluations of microwave limb sounder (MLS) and AIRS measurements. Higher resolution—both vertically and horizontally—and better cloud clearing are required for these observations. This heightens the current role of *in-situ* measurements. The scientific community is making headway in understanding pending discrepancies between various *in-situ* measurement techniques that often rely heavily on extensive laboratory calibrations (Peter et al. 2008). Dedicated campaigns aimed at an intercomparison of different measurement techniques and improved theoretical concepts of small ice crystal growth at low temperatures and pressures will support these efforts. Radiosondes underestimate upper tropospheric relative humidity even when corrections are applied that enable the detection of supersaturation (Spichtinger et al. 2003a). Recent developments of low-cost yet accurate frost-point hygrometers (Verver et al. 2006) could improve routine measurements of ice supersaturation in cirrus conditions. The value of radiosonde measurements can be increased by including information about clouds, e.g., by modifying aerosol backscatter-sondes. This would also yield more information on the vertical structure of cirrus clouds on a routine basis.

An improved estimate of vertical air motion variability is necessary to detect any aerosol impact on cirrus properties. As a prerequisite to drive physically based cirrus parameterizations, it is important to obtain cloud-scale vertical velocity PDFs (Kärcher and Lohmann 2002). In a GCM framework, those might be derived from gravity wave drag parameterizations, and be validated with MTF data and a systematic evaluation using advanced LES cirrus models. Mesoscale gravity wave structure could also be explored using high-resolution GPS retrievals of upper tropospheric temperatures.

The potential of new airborne instrumentation can be fully exploited by means of two new research aircraft, HALO[2] (in Germany) and HIAPER[3] (in the U.S.A.), which have altitude, range, and endurance capabilities well suited to perform critical atmospheric research. The research communities have made great efforts to transform business jets, which serve as host aircraft,

[2] http://www.halo.dlr.de

[3] http://www.hiaper.ucar.edu

into cutting-edge observational platforms, which will operate mainly in the upper troposphere and lower stratosphere region, probing horizontal scales ~0.1–1000 km. Both HIAPER and HALO will deploy comprehensive sets of instrumentation for a wide variety of scientific applications, including research into cirrus clouds and their radiative and chemical effects, to meeting most of the research needs discussed here over the next decades.

After the first flights of a backscatter lidar (LITE, in 1996) and a precipitation radar (TRMM, in 1997) in space, a new era of satellite observations has emerged with the completion, in 2006, of the NASA/CNES A-Train (afternoon) constellation of five low orbit satellites flying in formation. The CALIPSO lidar and the CloudSat cloud-profiling radar provide unprecedented global surveys of cloud properties and high-resolution vertical profiles (at resolutions below ~100 m for lidar and ~500 m for radar) with spatial and temporal coverage needed to evaluate parameterizations of clouds and cloud-related processes in global models (Chepfer et al. 2008). Besides advancing our understanding of moist (especially convective) processes in the climate system, ice clouds and upper tropospheric humidity may become a strong focus of research using these tools. However, it is also clear that much remains to be done to render data from these new sources useful for the development of GCMs (Sassen and Wang 2008).

Modeling Techniques

The use of bulk microphysics in cirrus cloud-resolving models has a long tradition (Starr and Cox 1985). Most existing schemes only prognose the cloud ice mass and are therefore computationally efficient enough to be included in a multidimensional model framework (in particular, in LES models). However, surprisingly few three-dimensional LES simulations of cirrus have been reported to date. LES models and GCMs have included, only recently, two-moment schemes (based on ice particle mass and number concentrations), but only very few of them allow for coupling with a detailed aerosol model for an advanced treatment of ice nucleation. Such schemes remain very valuable in either simulating case studies based on observations or in conducting sensitivity studies in terms of the wide range of parameters that control cirrus formation and evolution (Spichtinger and Gierens 2008b). In one- or two-moment schemes, microphysical processes and radiative properties must be heavily parameterized as a function of ice water content (and ice particle number density). To arrive at more realistic simulations that enable a better control of underlying processes, more size-resolved aerosol and ice microphysics modules in cirrus-resolving models are required (Jensen et al. 1994).

An alternative approach is currently being developed at DLR-IPA, in which the EULAG LES model[4] is coupled with a sophisticated aerosol–ice–radiation

---

4 http://www.mmm.ucar.edu/eulag

package to obtain a benchmark LES cirrus model. This package tracks liquid background particles and a number of IN types on a moving size grid and computes ice nucleation rates for each aerosol particle type and size. Ice particle growth, sedimentation, aggregation, light absorption, trace gas uptake, evaporation, and particle core return are simulated using a Lagrangian tracking method for a large number of simulation particles (clusters carrying a number of real ice crystals). The detailed location and history of individual ice particle clusters (i.e., nucleation core, supersaturation) and physical properties (i.e., shape, vapor pressure) can be tracked as a function of time. Such simulations may have the potential to set new standards for analyzing field observations and improving GCM parameterizations.

The use of new concepts for multiscale modeling of atmospheric flows for detailed cirrus studies appears to be a promising research avenue. Adaptive grid methods, new closure models for turbulence, modified approaches for nudging and nesting, as well as the prescription of spatially and/or temporally variable environmental conditions could be introduced in cloud modeling efforts. Such concepts are being developed and applied to cirrus within the priority program Multiscale Modeling in Fluid Dynamics and Meteorology[5] funded by the German Research Foundation. Until a better understanding of ice crystal growth has emerged from new theories or models that combine faceted crystal growth with shape and surface kinetic effects, process modeling must rely on the simplified concepts of diffusional growth in which these effects are treated empirically by parameters such as deposition coefficients and capacitance factors.

Another new development concerns the design of 1–5 km scale-resolving models, commonly referred to as cloud system-resolving or cloud ensemble models (Tompkins and Craig 1998; Grabowski and Moncrieff 2001). Those can be run over wide domains (1000 km to global) and allow interactions of individual clouds with the large-scale circulation to be studied. They have not yet been applied to study cirrus in detail. As mentioned before, ice nucleation in cirrus and the temperature fluctuations that drive the formation process are linked to the mesoscale (1–100 km). On similar scales, inhomogeneities in the three-dimensional structure of cirrus have been found to be important for determining their radiative effects, including cloud overlap issues (Hogan and Kew 2005). Therefore, attempts to employ cloud system-resolving models to study the interactive chain of dynamic forcing, ice nucleation, and radiation appear to be very promising. As a first step, existing parameterization schemes for homogeneous ice nucleation and initial growth (Kärcher and Lohmann 2002) could be directly applied in such models to facilitate the improvement of their cirrus microphysics. Within this context, the use of very high-resolution numerical weather prediction models would also be helpful.

---

[5] http://emm.mi.fu-berlin.de/DFG-MetStroem/

Despite recent progress, cirrus clouds are still insufficiently represented in global models. A number of GCMs have been coupled to sophisticated aerosol modules (Jacobson 2001; Adams and Seinfeld 2002; Stier et al. 2005; Lauer et al. 2005), but the coupling of these modules to the GCM cloud schemes, and to cirrus schemes in particular, is in its infancy. The introduction of separate prognostic ice variables and a physically based parameterization of cirrus formation enable the simulation of ice supersaturation on the resolved scales (Lohmann and Kärcher 2002). Realistic dynamic forcing of ice nucleation requires the inclusion of temperature fluctuations induced by gravity waves from orography (Dean et al. 2007; Joos et al. 2008) and other sources. On the one hand, aerosol modules should allow the prediction of background aerosols and IN and need to be coupled to suitable ice nucleation parameterizations that allow for competition between different ice-nucleating particle types (Kärcher et al. 2006; Liu et al. 2007). On the other, the cirrus cloud schemes should enable the prediction of a number of ice tracers for each type of ice-forming aerosols, and the GCM radiation modules ideally make use of this information to compute the radiative responses of the individual ice types. This includes predictions of accurate ice crystal size information. Efforts are underway to arrive at a consistent description of cirrus cloud fraction, dynamic forcing, ice microphysics, and ice supersaturation in GCMs (Kärcher and Burkhardt 2008). Recent progress in representing contrail cirrus as a distinct cirrus cloud type, with coverage and optical properties different from natural cirrus, provides valuable guidance (Burkhardt et al. 2008).

Concluding Remark

Cirrus cloud research is intimately connected with important societal issues, such as climate change, weather prediction, and aircraft operations. The direct impact of cirrus clouds on climate is based on their substantial radiative effects. The partial control of stratospheric water vapor concentrations and the possible impact on tropospheric ozone chemistry provide further examples of how cirrus cloud physics is closely tied to climate change issues. Improved weather prediction, as a result of advanced representation of cirrus in forecast models, may prove very valuable in furthering the quality of medium-range forecasts. To allow sustainable growth of aviation, it is imperative for the climate impact of contrail cirrus clouds to be known accurately so that appropriate mitigation strategies can be developed.

Acknowledgments

P. S. gratefully acknowledges partial support by the European Commission through the Marie Curie Fellowship "Impact of mesoscale dynamics and aerosols on the life cycle of cirrus clouds."

References

Abbatt, J. P. D., S. Benz, D. J. Cziczo et al. 2006. Solid ammonium sulfate aerosols as ice nuclei: A pathway for cirrus cloud formation. *Science* **313**:1770–1773.

Adams, P. J., and J. H. Seinfeld. 2002. Predicting global aerosol size distributions in general circulation models. *J. Geophys. Res.* **107**:4370.

Ansmann, A., M. Tesche, D. Althausen et al. 2008. Influence of Saharan dust on cloud glaciation in southern Morocco during the Saharan Mineral Dust Experiment. *J. Geophys. Res.* **113**:D04210.

Archeluta, C. M., P. J. DeMott, and S. M. Kreidenweis. 2005. Ice nucleation by surrogates for atmospheric mineral dust and mineral dust/sulfate particles at cirrus temperatures. *Atmos. Chem. Phys.* **5**:2617–2634.

Bailey, M., and J. Hallett. 2004. Growth rates and habits of ice crystals between –20°C and –70°C. *J. Atmos. Sci.* **61**:514–544.

Baumgardner, D., H. Chepfer, G. B. Raga, and G. L. Kok. 2005. The shapes of very small cirrus particles derived from *in situ* measurements. *Geophys. Res. Lett.* **32**:L01806.

Bengtsson, L., K. I. Hodges, and E. Roeckner. 2006. Storm tracks and climate change. *J. Climate* **19**:3518–3543.

Bigg, E. K. 1990. Measurement of concentrations of natural ice nuclei. *Atmos. Res.* **25**:397–408.

Bögel, W., and R. Baumann. 1991. Test and calibration of the DLR Falcon wind measuring system by maneuvers. *J. Atmos. Ocean. Technol.* **8**:5–18.

Bretherton, C. S., and P. K. Smolarkiewicz. 1989. Gravity waves, compensating subsidence and detrainment around cumulus clouds. *J. Atmos. Sci.* **46**:740–759.

Burkhardt, U., B. Kärcher, M. Ponater, K. Gierens, and A. Gettelman. 2008. Contrail cirrus supporting areas in model and observations. *Geophys. Res. Lett.* **35**:L16808.

Cantrell, W., and A. J. Heymsfield. 2005. Production of ice in tropospheric clouds: A review. *Bull. Amer. Meteor. Soc.* **86**:795–807.

Carlin, B., Q. Fu, U. Lohmann, G. G. Mace, K. Sassen, and J. M. Comstock. 2002. High cloud horizontal inhomogeneity and solar albedo bias. *J. Climate* **15**:2321–2339.

Carrió, G. G., S. C. van den Heever, and W. R. Cotton. 2007. Impacts of nucleating aerosol on anvil-cirrus clouds: A modeling study. *Atmos. Res.* **84**:111–131.

Chen, T., W. B. Rossow, and Y. Zhang. 2000. Radiative effects of cloud-type variations. *J. Climate* **13**:264–286.

Chepfer, H., S. Bony, D. Winker et al. 2008. Use of CALIPSO lidar observations to evaluate the cloudiness simulated by a climate model. *Geophys. Res. Lett.* **35**:L15704.

Chylek, P., M. K. Dubey, U. Lohmann et al. 2006. Aerosol indirect effect over the Indian Ocean. *Geophys. Res. Lett.* **33**:L06806.

Colberg, C. A., B. P. Luo, H. Wernli, T. Koop, and Th. Peter. 2003. A novel model to predict the physical state of atmospheric $H_2SO_4/NH_3/H_2O$ aerosol particles. *Atmos. Chem. Phys.* **3**:909–924.

Connolly, P. J., T. W. Choularton, M. W. Gallagher et al. 2006. Cloud-resolving simulations of intense tropical Hector thunderstorms: Implications for aerosol–cloud interactions. *Q. J. Roy. Meteor. Soc.* **132**:3079–3106.

Cui, Z., K. S. Carslaw, Y. Yin, and S. Davies. 2006. A numerical study of aerosol effects on the dynamics and microphysics of a deep convective cloud in a continental environment. *J. Geophys. Res.* **111**:D05201.

Cziczo, D. J., P. J. DeMott, C. Brock et al. 2003. A method for single particle mass spectrometry of ice nuclei. *Aerosol Sci. Technol.* **37**:460–470.

Cziczo, D. J., P. J. DeMott, S. D. Brooks et al. 2004a. Observations of organic species and atmospheric ice formation. *Geophys. Res. Lett.* **31**:L12116.

Cziczo, D. J., D. M. Murphy, P. K. Hudson, and D. S. Thomson. 2004b. Single particle measurements of the chemical composition of cirrus ice residue during CRYSTAL–FACE. *J. Geophys. Res.* **109**:D04201.

Dean, S. M., J. Flowerdew, B. N. Lawrence, and S. D. Eckermann. 2007. Parameterisation of orographic cloud dynamics in a GCM. *Climate Dyn.* **28**:581–597.

Dean, S. M., B. N. Lawrence, R. G. Grainger, and D. N. Heuff. 2005. Orographic cloud in a GCM: the missing cirrus. *Climate Dyn.* **24**:771–780.

DeMott, P. J., D. J. Cziczo, A. J. Prenni et al. 2003. Measurements of the concentration and composition of nuclei for cirrus formation. *PNAS* **100**:14,655–14,660.

DeMott, P. J., D. C. Rogers, and S. M. Kreidenweis. 1997. The susceptibility of ice formation in upper tropospheric clouds to insoluble aerosol components. *J. Geophys. Res.* **102**:19,575–19,584.

Dessler, A. E., S. P. Palm, W. D. Hart, and J. D. Spinhirne. 2006. Tropopause-level thin cirrus coverage revealed by ICESat/Geoscience Laser Altimeter System. *J. Geophys. Res.* **111**:D08203.

Diehl, K., C. Quick, S. Matthias-Maser, S. K. Mitra, and R. Jaenicke. 2001. The ice-nucleating ability of pollen. Part I: Laboratory studies in deposition and condensation freezing modes. *Atmos. Res.* **58**:75–87.

Dobbie, S., and P. R. Jonas. 2001. Radiative influences on the structure and lifetime of cirrus clouds. *Q. J. Roy. Meteor. Soc.* **127**:2663–2682.

Dowling, D. R., and L. F. Radke. 1990. A summary of the physical properties of cirrus clouds. *J. Appl. Meteorol.* **29**:970–978.

Eyring, V., N. R. P. Harris, M. Rex et al. 2005. A strategy for process-oriented validation of coupled chemistry–climate models. *Bull. Amer. Meteor. Soc.* **86**:1117–1133.

Field, P. R., and A. J. Heymsfield. 2003. Aggregation and scaling of ice crystal size distributions. *J. Atmos. Sci.* **60**:544–560.

Field, P. R., A. J. Heymsfield, and A. Bansemer. 2006. Shattering and particle interarrival times measured by optical array probes in ice clouds. *J. Atmos. Oceanic Technol.* **23**:1357–1371.

Fridlind, A. M., A. S. Ackerman, E. J. Jensen et al. 2004. Evidence for the predominance of mid-tropospheric aerosols as subtropical anvil cloud nuclei. *Science* **304**:718–722.

Fu, Q., B. Carlin, and G. Mace. 2000. Cirrus horizontal inhomogeneity and OLR bias. *Geophys. Res. Lett.* **27**:3341–3344.

Fusina, F., P. Spichtinger, and U. Lohmann. 2007. Impact of ice supersaturated regions and thin cirrus on radiation in the midlatitudes. *J. Geophys. Res.* **112**:D24S14.

Gao, R. S., P. J. Popp, D. W. Fahey et al. 2004. Evidence that ambient nitric acid increases relative humidity in low-temperature cirrus clouds. *Science* **303**:516–520.

Gardiner, B. A., and J. Hallett. 1985. Degradation of in-cloud Forward Scattering Spectrometer Probe measurements in the presence of ice particles. *J. Atmos. Oceanic Technol.* **2**:171–180.

Gary, B. L. 2006. Mesoscale temperature fluctuations in the stratosphere. *Atmos. Chem. Phys.* **6**:4577–4589.

Gayet, J.-F., J. Ovarlez, V. Shcherbakov et al. 2004. Cirrus cloud microphysical and optical properties at southern and northern midlatitudes during the INCA experiment. *J. Geophys. Res.* **109**:D20206.

Gayet J.-F., V. Shcherbakov, H. Mannstein et al. 2006. Microphysical and optical properties of midlatitude cirrus clouds observed in the southern hemisphere during INCA. *Q. J. Roy. Meteor. Soc.* **132**:2719–2748.

Gettelman, A., W. D. Collins, E. J. Fetzer et al. 2006b. Climatology of upper-tropospheric relative humidity from the Atmospheric Infrared Sounder and implications for climate. *J. Climate* **19**:6104–6121.

Gettelman, A., E. J. Fetzer, A. Eldering, and F. W. Irion. 2006a. The global distribution of supersaturation in the upper troposphere from the Atmospheric Infrared Sounder. *J. Climate* **19**:6089–6103.

Gettelman, A., and D. E. Kinnison. 2007. The global impact of supersaturation in a coupled chemistry–climate model. *Atmos. Chem. Phys.* **7**:1629–1643.

Gierens, K., and P. Spichtinger. 2000. On the size distribution of ice-supersaturated regions in the upper troposphere and lowermost stratosphere. *Ann. Geophys.* **18**:499–504.

Grabowski, W. W., and M. W. Moncrieff. 2001. Large-scale organization of tropical convection in two-dimensional explicit numerical simulations. *Q. J. Roy. Meteor. Soc.* **127**:445–468.

Gu, Y., and K. N. Liou. 2006. Cirrus cloud horizontal and vertical inhomogeneity effects in a GCM. *Meteorol. Atmos. Phys.* **91**:223–235.

Haag, W., and B. Kärcher. 2004. The impact of aerosols and gravity waves on cirrus clouds at midlatitudes. *J. Geophys. Res.* **109**:D12202.

Haag, W., B. Kärcher, J. Ström et al. 2003. Freezing thresholds and cirrus cloud formation mechanisms inferred from *in situ* measurements of relative humidity. *Atmos. Chem. Phys.* **3**:1791–1806.

Held, I. M., and B. J. Soden. 2000. Water vapor feedback and global warming. *Ann. Rev. Energy Environ.* **25**:441–475.

Hendricks, J., B. Kärcher, A. Döpelheuer et al. 2004. Simulating the global atmospheric black carbon cycle: A revisit to the contribution of aircraft emissions. *Atmos. Chem. Phys.* **4**:2521–2541.

Herzegh, P. H., and P. V. Hobbs. 1980. The mesoscale and microscale structure and organization of clouds and precipitation in midlatitude cyclones. II: Warm frontal clouds. *J. Atmos. Sci.* **37**:597–611.

Heymsfield, A. J. 2007. On measurements of small ice particles in clouds. *Geophys. Res. Lett.* **34**:L23812.

Hogan, R. J., and S. F. Kew. 2005. A 3-D stochastic cloud model for investigating the radiative properties of inhomogeneous cirrus clouds. *Q. J. Roy. Meteor. Soc.* **131**:2585–2608.

Holton, J. R., and A. Gettelman. 2001. Horizontal transport and the dehydration of the stratosphere. *Geophys. Res. Lett.* **28**:2799–2802.

Hoyle, C. R., B. P. Luo, and Th. Peter. 2005. The origin of high ice crystal number densities in cirrus clouds. *J. Atmos. Sci.* **62**:2568–2579.

Immler, F., and O. Schrems. 2002. LIDAR measurements of cirrus clouds in the northern and southern midlatitudes during INCA (55°N, 53°S): A comparative study. *Geophys. Res. Lett.* **29**:L015077.

Jacobson, M. Z. 2001. Global direct radiative forcing due to multicomponent anthropogenic and natural aerosols. *J. Geophys. Res.* **106**:1551–1568.

Jensen, E. J., and A. S. Ackerman. 2006. Homogeneous aerosol freezing in the tops of high-altitude tropical cumulonimbus clouds. *Geophys. Res. Lett.* **33**:L08802.

Jensen, E. J., A. S. Ackerman, D. E. Stevens, O. B. Toon, and P. Minnis. 1998. Spreading and growth of contrails in a sheared environment. *J. Geophys. Res.* **103**:13,557–13,567.

Jensen, E. J., and L. Pfister. 2004. Transport and freeze-drying in the tropical tropopause layer. *J. Geophys. Res.* **109**:D02207.

Jensen, E. J., J. B. Smith, L. Pfister et al. 2005. Ice supersaturations exceeding 100% at the cold tropical tropopause: Implications for cirrus formation and dehydration. *Atmos. Chem. Phys.* **5**:851–862.

Jensen, E. J., O. B. Toon, D. L. Westphal, S. Kinne, and A. J. Heymsfield. 1994. Microphysical modeling of cirrus. 1. Comparison with 1986 FIRE IFO measurements. *J. Geophys. Res.* **99**:10,421–10,442.

Joos, H., P. Spichtinger, U. Lohmann, J.-F. Gayet, and A. Minikin. 2008. Orographic cirrus in the global climate model ECHAM5. *J. Geophys. Res.* **113**:D18205.

Kärcher, B. 1999. Aviation-produced aerosols and contrails. *Surv. Geophys.* **20** :113–167.

Kärcher, B., and U. Burkhardt. 2008. A Cirrus cloud scheme for general circulation models. *Q. J. Roy. Meteor. Soc.* **134**:1439–1461.

Kärcher, B., J. Hendricks, and U. Lohmann. 2006. Physically based parameterization of cirrus cloud formation for use in global atmospheric models. *J. Geophys. Res.* **111**:D01205.

Kärcher, B., and T. Koop. 2005. The role of organic aerosols in homogeneous ice formation. *Atmos. Chem. Phys.* **5**:703–714.

Kärcher, B., and U. Lohmann. 2002. A Parameterization of cirrus cloud formation: Homogeneous freezing including effects of aerosol size. *J. Geophys. Res.* **107**:4698.

Kärcher, B., O. Möhler, P. J. DeMott, S. Pechtl, and F. Yu. 2007. Insights into the role of soot aerosols in cirrus cloud formation. *Atmos. Chem. Phys.* **7**:4203–4227.

Kärcher, B., and J. Ström. 2003. The roles of dynamical variability and aerosols in cirrus cloud formation. *Atmos. Chem. Phys.* **3**:823–838.

Kärcher, B., and C. Voigt. 2006. Formation of nitric acid/water ice particles in cirrus clouds. *Geophys. Res. Lett.* **33**:L08806.

Khain, A., A. Pokrovsky, M. Pinsky, A. Seifert, and V. T. J. Phillips. 2004. Simulation of effects of atmospheric aerosols on deep turbulent convective clouds by using a spectral microphysics mixed-phase cumulus cloud model. Part 1: Model description and possible applications. *J. Atmos. Sci.* **61**:2963–2982.

Kley, D., J. M. Russell III, and C. Phillips. 2000. SPARC Assessment of upper tropospheric and stratospheric water vapor. WCRP-113, WMO/TD-No.1043, SPARC Report No.2, pp. 312.

Knopf, D. A., and T. Koop. 2006. Heterogeneous nucleation of ice on surrogates of mineral dust. *J. Geophys. Res.* **111**:D12201.

Köhler, M. 1999. Explicit prediction of ice clouds in general circulation models. PhD thesis, 167 pp. University of California at Los Angeles.

Koop, T. 2004. Homogeneous nucleation in water and aqueous systems. *Z. Phys. Chem.* **218**:1231–1258.

Lauer, A., J. Hendricks, I. Ackermann et al. 2005. Simulating aerosol microphysics with the ECHAM/MADE GCM. Part I: Model description and comparison with observations. *Atmos. Chem. Phys.* **5**:3251–3276.

Law, K., L. Pan, H. Wernli et al. 2005. Processes governing the chemical composition of the extra-tropical UTLS. *IGAC Newsletter* **32**:2–22.

Lawson, R. P., D. O'Connor, P. Zmarzly et al. 2006. The 2D-S (stereo) probe: Design and preliminary tests of a new airborne, high speed, high-resolution particle imaging probe. *J. Atmos. Oceanic Technol.* **23**:1462–1477.

Liou, K. N., S. C. Ou, and G. Koenig. 1990: An investigation of the climatic effect of contrail cirrus. In: Air Traffic and the Environment: Background, Tendencies and Potential Global Atmospheric Effects, ed. U. Schumann, pp 154–169. Lecture Notes in Engineering. Berlin: Springer.

Liu, X., J. E. Penner, S. J. Ghan, and M. Wang. 2007. Inclusion of ice microphysics in the NCAR Community Atmospheric Model Version 3 (CAM3). *J. Climate* **20**:4526–4547.

Lohmann, U., and B. Kärcher. 2002. First interactive simulations of cirrus clouds formed by homogeneous freezing in the ECHAM GCM. *J. Geophys. Res.* **107**:4105.

Lohmann, U., B. Kärcher, and C. Timmreck. 2003. Impact of the Mt. Pinatubo eruption on cirrus clouds formed by homogeneous freezing in the ECHAM GCM. *J. Geophys. Res.* **108**:4568.

Lohmann, U., P. Stier, C. Hoose et al. 2007. Cloud microphysics and aerosol indirect effects in the global climate model ECHAM5-HAM. *Atmos. Chem. Phys.* **7**:3425–3446.

Lott, F., and M. Miller. 1997. A new subgrid-scale orographic parameterization: Its formulation and testing. *Q. J. Roy. Meteor. Soc.* **123**:101–127.

Luo, Z. Z., and W. B. Rossow. 2004. Characterizing tropical cirrus life cycle, evolution, and interaction with upper-tropospheric water vapor using Lagrangian trajectory analysis of satellite observations. *J. Climate* **17**:4541–4563.

Lynch, D. K., K. Sassen, D. O'C. Starr, and G. Stephens, ed. 2002. Cirrus. New York: Oxford Univ. Press.

Marsham, J., and S. Dobbie. 2005. The effects of wind shear on cirrus: A large-eddy model and radar case-study. *Q. J. Roy. Meteor. Soc.* **131**:2937–2955.

McFarquhar, G. M., J. Um, M. Freer, D. Baumgardner, G. L. Kok, and G. Mace. 2007. Importance of small ice crystals to cirrus properties: Observations from the Tropical Warm Pool International Cloud Experiment (TWP-ICE). *Geophys. Res. Lett.* **34**:L13803.

Mertes, S., B. Verheggen, S. Walter et al. 2007. Counterflow virtual impactor based collection of small ice particles in mixed-phase clouds for the physico-chemical characterisation of tropospheric ice nuclei: Sampler description and first case study. *Aerosol Sci. Technol.* **41**: 848–864.

Miloshevich, L. M., and A. J. Heymsfield. 1997. A balloon-borne continuous cloud particle replicator for measuring vertical profiles of cloud microphysical properties: Instrument design, performance, and collection efficiency analysis. *J. Atmos. Oceanic Technol.* **14**:753–768.

Möhler, O., O. Stetzer, S. Schaefers et al. 2003. Experimental investigations of homogeneous freezing of sulphuric acid particles in the aerosol chamber AIDA. *Atmos. Chem. Phys.* **3**:211–223.

Murray, B. J., and A. K. Bertram. 2007. Strong dependence of cubic ice formation on droplet ammonium to sulfate ratio. *Geophys. Res. Lett.* **34**:L16810.

Murphy, D. M. 2003. Dehydration in cold clouds is enhanced by a transition from cubic to hexagonal ice. *Geophys. Res. Lett.* **30**:2230.

Murphy, D., and T. Koop. 2005. Review of the vapour pressure of ice and supercooled water for atmospheric applications. *Q. J. Roy. Meteor. Soc.* **131**:1539–1565.

Murphy, D. M., D. S. Thomson, and M. J Mahoney. 1998. *In situ* mesurements of organics, meteoritic material, mercury, and other elements in aerosols at 5 to 19 kilometers. *Science* **282**:1664–1669.

Newell, R. E., Y. Zhu, E. V. Browell et al. 1996. Upper tropospheric water vapor and cirrus: Comparison of DC-8 observations, preliminary UARS microwave limb sounder measurements and meteorological analyses. *J. Geophys. Res.* **101**:1931–1941.

Notholt, J., B. P. Luo, S. Fueglistaler et al. 2005. Influence of tropospheric SO_2 emissions on particle formation and the stratospheric humidity. *Geophys. Res. Lett.* **32**:L07810.

Peter, Th., M. Krämer, and O. Möhler. 2008. Upper tropospheric humidity: A Report on an international workshop. *SPARC Newsletter* **30**:9–15.

Peter, Th., C. Marcolli, P. Spichtinger, T. Corti, M.B. Baker, and T. Koop. 2006. When dry air is too humid. *Science* **314**:1399–1402.

Plougonven, R., H. Teitelbaum, and V. Zeitlin. 2003. Inertia gravity wave generation by the tropospheric midlatitude jet as given by the Fronts and Atlantic Storm-Track Experiment radio soundings. *J. Geophys. Res.* **108**:4686.

Pruppacher, H. R., and J. D. Klett. 1997. Microphysics of Clouds and Precipitation, 2nd ed. Dordrecht: Kluwer Acad. Publ.

Rossow, W. B., and R. A. Schiffer. 1999. Advances in understanding clouds from ISCCP. *Bull. Amer. Meteor. Soc.* **80**:2261–2287.

Santer, B. D., F. Wehner, T. M. L. Wigley et al. 2003. Contributions of anthropogenic and natural forcing to recent tropopause height changes. *Science* **301**:479–483.

Sassen, K., P. J. DeMott, J. M. Prospero, and M. R. Poellot. 2003. Saharan dust storms and indirect aerosol effects on clouds: CRYSTAL-FACE results. *Geophys. Res. Lett.* **30**:1633.

Sassen, K., and Z. Wang. 2008. Classifying clouds around the globe with the CloudSat radar: 1-year of results. *Geophys. Res. Lett.* **35**:L04805.

Schmidt, K.S., P. Pilewskie, S. Platnick, G. Wind, P. Yang, and M. Wendisch. 2007. Comparing irradiance fields derived from Moderate Resolution Imaging Spectroradiometer airborne simulator cirrus cloud retrievals with solar spectral flux radiometer measurements. *J. Geophys. Res.* **112**:D24206.

Schumann, U. 1996. On conditions for contrail formation from aircraft exhausts. *Meteor. Z.* **5**:4–23.

Scinocca, J. F., and N. A. McFarlane. 2000. The parameterization of drag induced by stratified flow over anisotropic orography. *Q. J. Roy. Meteor. Soc.* **126**:2353–2393.

Seifert, P., A. Ansmann, D. Müller et al. 2007. Cirrus optical properties observed with lidar, radiosonde, and satellite over the tropical Indian Ocean during the aerosol-polluted northeast and clean maritime southwest monsoon. *J. Geophys. Res.* **112**:D17205.

Smith, R. B. 1979. The influence of mountains on the atmosphere. *Adv. Geophys.* **21**:187–230.

Spichtinger, P., and K. Gierens. 2008a. Modelling of cirrus clouds. Part 2: Competition of different nucleation mechanisms. *Atmos. Chem. Phys. Discuss.* **8**:9061–9098.

Spichtinger, P., and K. Gierens. 2008b. Modelling of cirrus clouds. Part 1: Model description and validation. *Atmos. Chem. Phys. Discuss.* **8**:601–686.

Spichtinger, P., K. Gierens, and A. Dörnbrack. 2005a. Formation of ice supersaturation by mesoscale gravity waves. *Atmos. Chem. Phys.* **5**:1243–1255.

Spichtinger, P., K. Gierens, U. Leiterer, and H. Dier. 2003a. Ice supersaturation in the tropopause region over Lindenberg, Germany. *Meteor. Z.* **12**:143–156.

Spichtinger, P., K. Gierens, and W. Read. 2003b. The global distribution of ice-supersaturated regions as seen by the microwave limb sounder. *Q. J. Roy. Meteor. Soc.* **129**:3391–3410.

Spichtinger, P., K. Gierens, and H. Wernli. 2005b. A case study on the formation and evolution of ice supersaturation in the vicinity of a warm conveyor belt's outflow region. *Atmos. Chem. Phys.* **5**:973–987.

Starr, D.'O.C., and S. K. Cox. 1985. Cirrus clouds. Part I: A cirrus cloud model. *J. Atmos. Sci.* **42**:2663–2681.

Stephens, G. L., and C. D. Kummerow. 2007. The remote sensing of clouds and precipitation from space: A review. *J. Atmos. Sci.* **64**:3742–3765.

Stephens, G. L., S.-C. Tsay, P. W. Stackhouse Jr., and P. J. Flatau. 1990. The relevance of the microphysical and radiative properties of cirrus clouds to climate and climatic feedback. *J. Atmos. Sci.* **47**:1742–1753.

Stephens, G. L., and P. J. Webster. 1981. Clouds and climate: Sensitivity of simple systems. *J. Atmos. Sci.* **38**:235–247.

Stetzer, O., B. Baschek, F. Lüönd, and U. Lohmann. 2007. The Zurich Ice Nucleation Chamber (ZINC) – A new instrument to investigate atmospheric ice formation. *Aerosol Sci. Technol.* **42**:64–74.

Stier, P., J. Feichter, S. Kinne et al. 2005 . The aerosol–climate model ECHAM5-HAM. *Atmos. Chem. Phys.* **5**:1125–1156.

Ström, J., B. Strauss, T. Anderson et al. 1997. *In situ* observations of the microphysical properties of young cirrus clouds. *J. Atmos. Sci.* **54**:2542–2553.

Stubenrauch, C. J., A. Chedin, G. Rädel, N. A. Scott, and S. Serrar. 2006. Cloud properties and their seasonal and diurnal variability from TOVS path-B. *J. Climate* **19**:5531–5553.

Stubenrauch, C. J., F. Eddounia, and G. Rädel. 2004. Correlations between microphysical properties of large-scale semi-transparent cirrus and the state of the atmosphere. *Atmos. Res.* **72**:403–423.

Tabazadeh, A., Y. S. Djikaev, and H. Reiss. 2002. Surface crystallization of supercooled water in clouds. *PNAS* **99**:15,873–15,878.

Thornton, B. F., D. W. Toohey, A. F. Tuck et al. 2007. Chlorine activation near the midlatitude tropopause. *J. Geophys. Res.* **112**:D18306.

Tiedtke, M. 1993. Representation of clouds in large-scale models. *Mon. Wea. Rev.* **121**:3040–3061.

Tompkins, A. M. 2003. Impact of temperature and total water variability on cloud cover assessed using aircraft data. *Q. J. Roy. Meteor. Soc.* **129**:2151–2170.

Tompkins, A. M., and G. C. Craig. 1998. Radiative-convective equilibrium in a three-dimensional cloud-ensemble model. *Q. J. Roy. Meteor. Soc.* **124**:2073–2097.

Tompkins, A., K. Gierens, and G. Rädel. 2007. Ice supersaturation in the ECMWF Integrated Forecast System, *Q. J. Roy. Meteor. Soc.* **133**:53–63.

Treffeisen, R., R. Krejci, J. Ström et al. 2007. Humidity observations in the Arctic troposphere over Ny-Ålesund, Svalbard, based on 15 years of radiosonde data. *Atmos. Chem. Phys.* **7**:2721–2732.

Verver, G., M. Fujiwara, P. Dolmans et al. 2006. Performance of the Vaisala RS80A/H and RS90 Humicap Sensors and the Meteolabor "Snow White" Chilled-Mirror Hygrometer in Paramaribo, Suriname. *J. Atmos. Oceanic Technol.* **23**:1506–1518.

Wang, P. H., M. P. McCormick, P. Minnis, G. S. Kent, and K. M. Skeens. 1996. A 6-year climatology of cloud occurrence frequency from SAGE II observations (1985–1990). *J. Geophys. Res.* **101**:29,407–29,429.

Wang, S., and F. Zhang. 2007. Sensitivity of mesoscale gravity waves to the baroclinicity of jet-front systems. *Mon. Wea. Rev.* **135**:670–688.

Wendisch, M., P. Pilewski, J. Pommier et al. 2005. Impact of cirrus crystal shape on solar spectral irradiance: A case study for subtropical cirrus. *J. Geophys. Res.* **110**:D03202.

Wendisch, M., P. Yang, and P. Pilewski. 2007. Effects of ice crystal habit on thermal infrared radiative properties and forcing of cirrus. *J. Geophys. Res.* **112**:D08201.

Westbrook, C. D., R. C. Ball, P. R. Field, and A. J. Heymsfield. 2004. Universality in snowflake aggregation. *Geophys. Res. Lett.* **31**:L15104.

Westbrook, C. D., R. J. Hogan, A. J. Illingworth, and E. J. O'Connor. 2007. Theory and observations of ice particle evolution in cirrus using Doppler radar: Evidence for aggregation. *Geophys. Res. Lett.* **34**:L02824.

Whiteway, J., C. Cook, M. Gallagher et al. 2004. Anatomy of cirrus clouds: Results from the Emerald airborne campaigns. *Geophys. Res. Lett.* **31**:L24102.

Wood, S. E., M. B. Baker, and D. Calhoun. 2001. New model for the vapor growth of hexagonal ice crystals in the atmosphere. *J. Geophys. Res.* **106**:4845–4870.

Worthington, R. M. 1999. Tropopausal turbulence caused by the breaking of mountain waves. *J. Atmos. Sol. Terr. Phys.* **59**:1543–1547.

Yin, J. H. 2005. A consistent poleward shift of the storm tracks in simulations of 21st century climate. *Geophys. Res. Lett.* **32**:L18701.

Zhang, M. H., W. Y. Lin, S. A. Klein et al. 2005. Comparing clouds and their seasonal variations in 10 atmospheric general circulation models with satellite measurements. *J. Geophys. Res.* **110**:D15S02.

Zhong, W., R. J. Hogan, and J. D. Haigh. 2008. Three-dimensional radiative transfer in midlatitude cirrus clouds. *Q. J. Roy. Meteor. Soc.* **134**:199–215.

Zobrist, B., C. Marcolli, T. Koop et al. 2006. Oxalic acid as a heterogeneous ice nucleus in the upper troposphere and its indirect aerosol effect. *Atmos. Chem. Phys.* **6**:3115–3129.

12

Cloud-controlling Factors

A. Pier Siebesma, Rapporteur

Jean-Louis Brenguier, Christopher S. Bretherton, Wojciech W. Grabowski, Jost Heintzenberg, Bernd Kärcher, Katrin Lehmann, Jon C. Petch, Peter Spichtinger, Bjorn Stevens, and Frank Stratmann

Introduction

To address the multitude of issues that contribute to a thorough understanding of clouds, science needs to consider and incorporate a wide variety of biological, physical, and chemical processes that are embedded in flows. A host of processes emerge from the categories of atmospheric dynamics—kinetics, diffusion, and turbulence—and span scales from the millimeter, or smaller, scale to the planetary dimension (see Figure 12.1). The development of a simulation system, or theory, that would incorporate the full range of all relevant cloud-controlling factors does not appear to us to be foreseeable, and it is unlikely that we will ever have a reference model that incorporates all relevant processes and their interactions.

Nonetheless, a variety of models or theories continue to be developed in an attempt to describe specific sets of processes over a subset of relevant scales. These models and theories, as well as the connections between them, form an informal network or hierarchy. We refer to this network as being informal because the connections between models cannot be rigorously justified; we term it hierarchical in the sense that the flow of information through the hierarchy tends to have an aggregative character, wherein one attempts to represent the statistical behavior of clouds on small scales in larger-scale models.

The informal nature of the network raises two important issues: How can formalism (i.e., the connections) be increased among processes and models at different scales? How should the network be justified in the context of each problem one wishes to solve, in the present case perturbed clouds.

By "perturbed clouds," we mean clouds that are perturbed either by anthropogenic aerosols, which initiate such perturbations at the smallest spatial

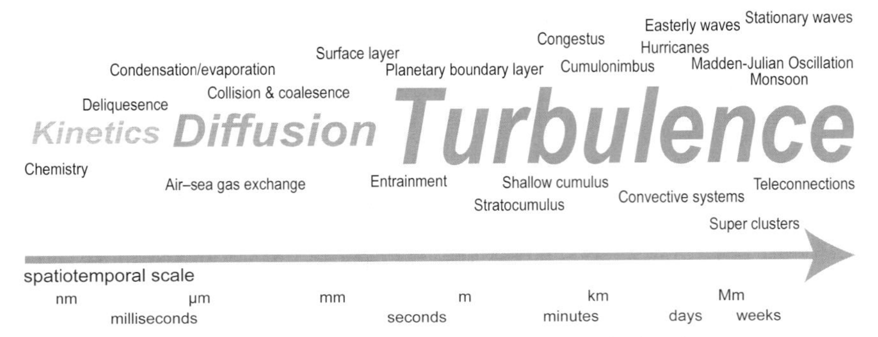

Figure 12.1 Depiction of the continuum of relevant cloud-related processes across the full range of spatiotemporal scales, showing the underlying categorical behavior (gray text) from which processes emerge.

scales (i.e., the microscale), or through global warming attributable to additional greenhouse gases. These perturbed clouds include possibly all tropospheric clouds, including contrails which are a direct consequence of anthropogenic influences, and may include polar stratospheric clouds as well.

What are the practical considerations of this network? For individual models that can encompass at most three orders of magnitude of spatial (horizontal) scales, we need to rethink carefully how to use these in the network. On the smallest scale, direct numerical simulations (DNS) are used, ranging from the millimeter scale to the 1-meter scale. Large eddy simulations (LES) resolve partly the turbulence and the individual clouds up to a scale of 100 km but need to parameterize microphysics and small-scale turbulence. At the next level, cloud-resolving models (CRMs) address cloud clusters up to scales of 1000 km, but these require additional parameterized turbulence and clouds for more coarse CRMs in the boundary layer (see Grabowski and Petch, this volume). On the largest scale, general circulation models (GCMs) are employed. They have the obvious advantage of not requiring lateral boundary conditions; however, as they operate at resolutions of typically 100 km, they fail to represent explicitly most of the cloud-controlling factors and essential cloud processes. Perhaps surprisingly, the effects of perturbed clouds in our climate system have been primarily studied by using these GCMs. One important theme that emerged from our discussions is the need to develop a more optimal use of this hierarchical network of models to answer questions on the representation of clouds in our climate system. This is a difficult task, as we do not yet understand completely how the various interactions propagate across the many scales.

Similar arguments hold for experimental research, which ranges from small-scale laboratory experiments that address microphysical issues to large-scale field experiments and global satellite observations.

Connecting Scales in Global Climate Models

Traditional Pathways and Shortcomings

GCMs have been the major modeling tool used to address the issue of perturbed clouds, despite the fact that many relevant cloud processes are not explicitly resolved by them. In our opinion, such models are useful for specific aspects that they resolve (i.e., the large-scale dynamics), but care must be taken when interpreting the impact of cloud-scale physical, chemical, and dynamic processes in these models, as they are represented imperfectly. Figure 12.2 illustrates how the various scales are connected in most of the state-of-the-art GCMs today: The box on the right represents all of the resolved processes on a mean grid box state (scales of typically 100–500 km); on the left, boxes depict the unresolved processes relevant for clouds, including turbulent processes, moist convection, cloud microphysics, and radiation. Traditionally, impacts of these processes are modeled in a statistical manner in terms of the grid box-averaged fields: temperature, T, specific humidity, q_v, and the velocity, $\mathbf{v}$ (u, v, w). The impact of these statistical models, or parameterizations, are directly communicated with the grid box mean state, as indicated in Figure 12.2.

With this approach, we face at least four categories of errors and uncertainties. The first relates to our inherent lack of knowledge of certain physical processes. A prime example is our lack of understanding of microphysics for mixed-phase and, in part, ice clouds; they are challenging to simulate irrespective of resolution. For these types of processes, further experimental research is especially required (discussed in detail below). In addition to new experimental research, we require new theoretical concepts (e.g., a consistent framework for explicit ice supersaturation in GCMs).

The second category of errors relates to processes that we understand at some level of aggregation, but for which there is no clear way to parameterize these in GCMs, mainly because of the nonlinear character of these processes. Examples include radiative flux transfer functions $R(q_l, q_v, T)$, which describe the vertical radiative fluxes as a function of the atmospheric conditions, and

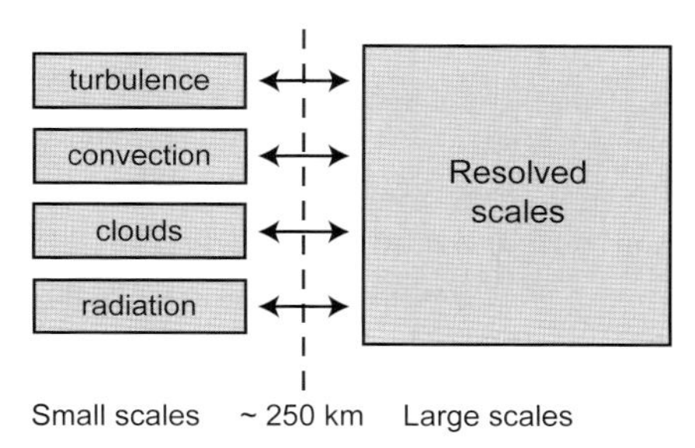

Figure 12.2 Depiction of the interaction between resolved and parameterized unresolved cloud-related processes (convection, turbulence, clouds and radiation) in present-day climate models.

to a certain extent warm-phase microphysical processes. In terms of the latter, consider a cloud-scale adjustment scheme for cloud liquid water, q_l, and cloud fraction, c, in which it is assumed that condensation occurs if supersaturation occurs and cloud droplets coalesce into rain drops at a rate A, where for the sake of argument we take A following Kessler (1969):

$$\begin{aligned} c = H(q_t - q_s),\ q_l = (q_t - q_s)H(q_t - q_s), \\ A = K(q_l - q_{cr})H(q_l - q_{cr}), \end{aligned} \tag{12.1}$$

where H is the Heaviside step function, q_s is the saturation specific humidity, $q_t = q_v + q_l$ is the total water-specific humidity, K an inverse timescale, and q_{cr} a critical threshold of the liquid water below which there is no conversion to rain. In a GCM, these processes must be represented by averaging over a family of clouds; however, typically, the above relationships are simply implemented in terms of grid-scale properties, which causes biased errors because the underlying relationships are nonlinear:

$$\begin{aligned} \overline{c(q_t, q_s)} \geq c(\overline{q_t}, \overline{q_s}), \\ \overline{q_l(q_t, q_s)} \geq q_l(\overline{q_t}, \overline{q_s}), \\ \overline{A(q_l)} \geq A(\overline{q_l}), \end{aligned} \tag{12.2}$$

where overbars denote grid box averages. In a similar fashion, applying the grid box-averaged liquid water in radiative transfer calculations leads to a systematic overestimation of the albedo α (i.e., $\overline{\alpha(q_l)} \leq \alpha(\overline{q_l})$, the so-called plane-parallel cloud albedo bias). For calculations capable of representing the scales at which the physical relationships are valid (in our example, the cloud scale), these biases are negligible, which partly explains the success, if not authority, of fine-scale calculations.

The third category of errors relates to the fact that current practice largely requires various subgrid processes to interact through modifications to the mean state, as depicted in Figure 12.2. In reality, however, virtually all relevant subgrid processes displayed in Figure 12.2 are related to the same joint probability density function (PDF) of vertical velocity, w, q_t, and the liquid water (or ice) potential temperature θ_l. Vertical transport through turbulence, unresolved mesoscale gravity waves, and convection creates subgrid variability in temperature, moisture, and vertical velocity. This subgrid variability determines cloud amount and cloud condensate. It is also these subgrid updraft velocities that influence the cloud particle number concentrations through the cloud condensation nuclei (CCN), droplet freezing, or ice nuclei (IN) activation (ice-forming nuclei constitute the subset of aerosol particles on which ice forms). Conversely, cloud amounts influence the vertical stability and therewith the vertical transport of moisture and heat. Finally, the subgrid variability of cloud amounts influences radiative heating profiles. Thus, in reality, all of the subgrid processes interact with each other, and this is not usually taken into account by GCMs, which leads to biases and inconsistencies.

A last category of errors is associated with the mean field representation of subgrid processes, whereby there is a one-to-one correspondence between the subgrid processes and the local (in space and time) resolved state. Although such a quasi-equilibrium assumption might be defendable in situations in which the grid box size is much larger than the typical size of the subgrid process of interest, or timescales are long compared to the timescale of the process being represented, the assumption breaks down if the spatiotemporal size of a grid box becomes comparable to the size of the subgrid process of interest. In that case, resolution is still too coarse to resolve the process of interest; however, the grid box is also too small to allow a statistical parameterization that assumes quasi-equilibrium. An interesting example is the convective mass flux, indicating convective activity within a grid box, which can be reasonably represented by a single value for grid box sizes of several hundred kilometers and a range of possible values for a given mean state if the grid box has a typical size of tens of kilometers. This observation motivates a *stochastic* approach rather than a *mean field* representation.

New Pathways

The most direct, but also computationally most demanding alternative is to use global CRMs (see Figure 12.3a). With this class of models, it is now possible to achieve a resolution of 1 to ~5 km for time integrations of a month (Collins and Satoh, this volume). This has the obvious advantage of allowing processes beyond this scale to be represented, in principle, in a more realistic, consistent way. High computational costs, however, prohibit long (years to many decades) or repeated simulations.

A computationally less demanding, but still challenging alternative involves cloud-resolving convective parameterization or super-parameterization (see Grabowski and Petch, this volume), where a two-dimensional CRM is embedded in each grid box of a conventional GCM (see Figure 12.3b). Unlike traditional nesting techniques, this approach uses the fine-scale model to parameterize the fine-scale physics rather than span the space distended by the grid box itself. The computation advantage over the previous approach rests on the degree to which the fine-scale model leaves out some of these intermediate scales. This has the advantage of allowing cloud processes to interact on the more appropriate scale of a CRM grid point, while global circulation feedbacks are still being simulated. The disadvantage is that mesoscale processes are often poorly represented as the intermediate scales are missing. With current computational resources, super-parameterized GCMs are becoming practical for five- to ten-year integrations, thus providing an alternative framework to explore the global effect of greenhouse gas and aerosol perturbations on clouds.

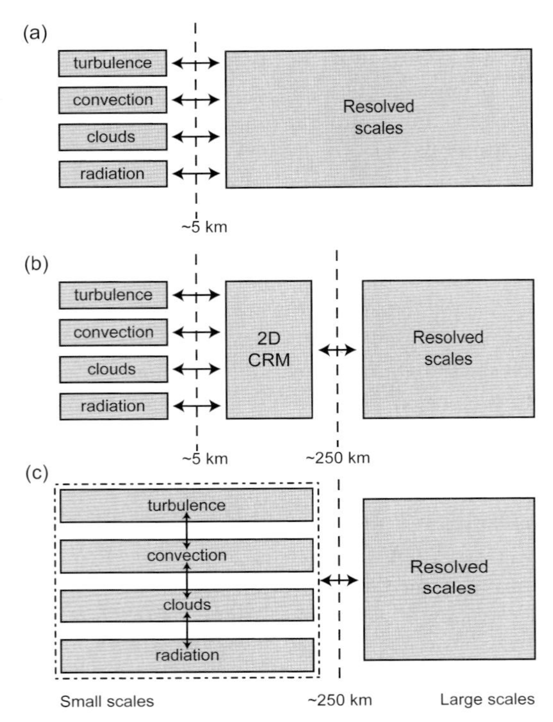

Figure 12.3 New pathways to improve the representation of cloud-related processes for the next generation of climate models: (a) global CRMs, (b) cloud-resolving convective parameterization or super-parameterization and (c) interactive parameterizations that communicate with each other directly rather than through the mean resolved state.

For both methods, clouds must be adequately resolved to be realistic. A 4-km horizontal resolution (as currently used) is still too coarse to resolve deep convection properly, let alone to represent boundary layer clouds, which are believed to be the most sensitive cloud types for cloud–climate feedbacks. Therefore, boundary clouds, turbulence, and microphysics require parameterization, and biases associated with the poor resolution of deep convection should be expected. Hence, we need to invest computational resources in even higher resolution studies (not forgetting the all important role of vertical resolution) to gain understanding of the implications of running these global CRMs and super-parametrizations at these coarse resolutions. Limited area tests should be conducted to compare runs using 4-km grid lengths with reference runs using mesh sizes well below 100 m for different cloud regimes. An analysis of such simulations should focus additionally on the representation of convection and cloud processes.

A third alternative is to develop more consistent and sophisticated parameterizations for clouds, turbulence, convection, and microphysics. If the joint PDF $P(q_t,\theta_l, w)$ of the vertical velocity w, the total water specific humidity, q_t, and the liquid (ice) water potential temperature, θ_l, for a given grid box could be estimated, a consistent treatment of convective and turbulent transport, cloud properties and their interaction with radiation can be inferred. Aerosol and microphysical processing could also be addressed in this framework. Various different pathways for these PDF-based schemes can be explored in which extra subgrid information is obtained through extra prognostic equations of higher-order moments or of cloud variables, through multiple updraft equations as well as through the use of an assumed shape of the PDF. This approach is proving successful for cloudy boundary layer processes, as it addresses what we referred to above as the second (nonlinear processes that can generate biased errors) and third category (a consistent treatment of the relevant processes) errors. However, it is computationally demanding because of the large number of prognostic equations and the short timestep needed for numerical stability, and it requires numerous closure assumptions.

Finally, stochastic parameterizations that provide a framework to account for deviations from the mean field prediction may be necessary, if not convenient, especially for GCMs that use resolutions finer than 100 km. In such a stochastic approach, the constraint of having only a one to one correspondence between the mean state and the subgrid response is relaxed. As a result, different subgrid-scale responses can occur that are selected stochastically, leading in general to an increased variability of the climate system.

Use of Process Models to Improve Cloud Representation in Climate Models

For any of the pathways described above, we must strive to improve our understanding of the cloud processes that remain unresolved in GCMs over the coming decades. By evaluating GCMs with global observational datasets, weak points in the representation of certain cloud regimes can be identified. Subsequently, by making use of theory and fine-scale flow solvers (such as LES and CRMs) guided by field studies, more insight can be gained into the cloud processes that require parameterizations in GCMs. Intercomparisons of these process models can advance theoretical understanding, thereby promoting the development of new parameterizations which can be further tested in a single column model environment or through a numerical weather prediction (NWP) model. As a final step, the parameterizations can be evaluated in a full GCM using global observations to assess improvements in our ability to simulate the current climate system. This top-down/bottom-up approach has proven to be a slow but successful way of making use of the network of models. In particular, it has led to demonstrable progress in our representation of top-entrainment in stratocumulus, lateral entrainment and detrainment in shallow cumulus, and closures for convection schemes. By carefully comparing observations and the fine-scale modeling frameworks (LES, CRMs), these models (and best practices in their use) have improved substantially over the last ten years. This strategy has also been effective in formulating key questions, which in turn serve to direct field and laboratory studies. For example, despite high resolutions of up to 5 m, LES intercomparison studies show great uncertainty in cloud-top entrainment rates in stratocumulus, a crucial dynamic process for this cloud type. As a result, the DYCOMS-II field study (Stevens and Brenguier, this volume) was designed, and it successfully narrowed the uncertainty in cloud-top entrainment rates for several well-observed cases. Similarly, LES results inspired a recent field study, RICO (Rauber et al. 2007), and an ongoing field study, VOCALS[1], designed to quantify the role of precipitation and its interaction to aerosol and microphysical processes in trade-wind cumulus and stratocumulus, respectively.

Finally, these process models have been useful to study and define processes that are not well represented in GCMs. Prime examples are the dependence of tropospheric humidity on moist convection, the transition from shallow to deep convection in relation to the diurnal cycle, and the formation of ice in stratiform cirrus by homogeneous freezing of supercooled aerosols.

Connecting Scales Using Alternative Modeling Techniques

Accepting the limitations of GCMs to study perturbed clouds, let us now proceed to discuss a number of alternatives.

Numerical Weather Prediction

The NWP model is a useful tool in the evaluation of (perturbed) clouds. Since most of the cloud processes, in which we are interested, operate on short timescales, we are able to view the impact of current cloud parameterizations on forecasts. It is therefore useful to run climate models in a NWP mode, as many climate biases are visible within two days. The performance of cloud parameterizations can be critically evaluated using ground-based profiling stations, such as those from ARM and Cloudnet (Illingworth and Bony, this volume). Furthermore, aerosol impacts may be testable in forecast models.

There is a large potential for further exploitation of reanalysis methods, as at present they are not sufficiently constrained by cloud and boundary layer properties. Introducing constraints to reflect these quantities in a better manner would advance the use of a wider variety of data as analysis products are developed. This could be very powerful in helping us to untangle meteorological from aerosol effects. It must be recognized, however, that even if an aerosol is not explicitly included in the analysis, its effects on cloud or thermal structure may be indirectly evident.

[1] http://www.eol.ucar.edu/projects/vocals/

Large Eddy Simulation Models, Cloud-resolving Models, and Super-parameterized Single Column Models

We have already mentioned that this class of models offers a way to improve the representation of unresolved processes in climate models. However, one can also imagine ways of using these models more directly to study perturbed clouds. To this end, it would be helpful to have "model problems" that exhibit aspects of the global cloud response of a type of model in a conceptually simpler framework. One potentially useful framework is to understand the response of clouds in an appropriately chosen "single column" framework, in which the effect of the large scale is encapsulated in specified advective forcings of temperature, moisture, and surface boundary conditions.

In reality, at any given location, these forcings constantly change as weather systems move through, thus generating a distribution of clouds in the column. In other words, at any location, the climatological cloud distribution is built up through patterns of weather variability. Thus, to utilize a single column framework to understand the cloud processes, it may be necessary for the clouds in this column to be simulated for months or years with realistic covariability in the forcing to build up a climatologically realistic response. This approach must be repeated in the perturbed climate (with a correspondingly different time series of forcings) to understand how the cloud response changes. To predict the cloud properties and their sensitivity to climate perturbations, the forcings can either be applied to a single column model or, preferably, a small-domain CRM. A similar framework can be used to study clouds perturbed by anthropogenic aerosols.

Model Simplification

Over the last decade, the climate modeling community has tended to increase the complexity of models by adding more processes or components (e.g., aerosols, chemistry, carbon-cycle, vegetation) as well as, but to a lesser extent, by increasing the model resolution. It appears that this tendency has not resulted in a better understanding of cloud–climate feedbacks, as the uncertainty in climate sensitivity has not decreased with time. In fact, the range of global warming estimates for the end of the 21st century, has increased with the inclusion of carbon-cycle feedbacks in coupled ocean–atmosphere models. Although this reflects, in part, the manner in which scientific progress is made, it is nonetheless disappointing, because the aim of climate modeling is not only to provide global warming estimates or global cloud feedback estimates but to learn how the climate works and responds to external perturbations.

A better understanding of the physical mechanisms that underlie the cloud–climate feedbacks produced by climate models would enable us to design a strategy to evaluate these feedbacks using observations. By simplifying models (rather than making them more complex) and by conducting idealized experiments, we should be able to identify key critical processes, to arrange them hierarchically, and to test resultant ideas or theories. Interpretation frameworks may then be proposed to help us understand the physics of cloud–climate feedbacks and intermodel differences in cloud–climate feedbacks.

With GCMs, this could be done, for instance, by simplifying the large-scale boundary conditions (e.g., aquaplanet versions of the models), by reducing the dimensionality of the system (e.g., 2-D or 1-D model versions derived from the 3-D model), or even by removing some processes (e.g., by replacing complex microphysical schemes with simpler ones). Models of intermediate complexity might also be implemented. Simple conceptual models (e.g., 2-box models) may be viewed as the ultimate step in this simplification process. We need, however, to approach these conceptual models with a sense of caution, as there is the risk of inadvertently neglecting interactions or feedbacks, which could result in biased behavior. For instance, Albrecht's conceptual model hypothesised how an increase of nucleating aerosol particles could suppress the development of precipitation in boundary clouds and increase cloud fraction (Albrecht 1989). Although this mechanism appears to be relevant in some circumstances, recent fine-scale modeling studies suggest a plethora of circumstances for which the relationship between cloudiness and precipitation is exactly the opposite (see Stevens and Brenguier; Brenguier and Wood, both this volume).

In general, the extent to which simplified models can be useful in reproducing and interpreting complex model results remains to be tested and quantified through comparison with process-resolving fine-scale models. Simplified models are useful in conceptualizing the essential processes, and therefore they can provide guidance if these processes are realistically represented in complex GCMs.

We strongly recommend that simplified climate models and/or experiments be developed with the aim of better identifying and *understanding* the key processes that control cloud–climate feedbacks. Ultimately, our trust in the results of a climate model is leveraged against this understanding. An ideal situation would be for each global climate model to be associated with a suite of simpler or more idealized model versions to support the analysis and understanding of the results obtained. This would, in a sense, formalize the hierarchy.

Global Climate Models: What Level of Complexity Is Necessary, and What Drives Complexity?

Does higher complexity provide better (more reliable) estimates of climate sensitivity? In climate models, the complexity of all modules (e.g., radiation, diffusion, dynamics, cloud microphysics, and convection) should be more or less balanced. It is obviously not an easy task to qualify what "more or less" means. However, one should ask whether a given GCM can cope with the complexity of a proposed new parameterization. It is, for instance, not helpful to incorporate a full-fledged spectral microphysics parameterization when the GCM is unable to predict the cloud-scale updraft velocities.

The level of complexity required in a GCM will depend on the question being posed. For example, even though contrail cirrus contribute only a small amount to the global-mean radiative forcing, they need to be studied in order to quantify the significant regional effects already in place or their potential effects in a future climate. This task requires that the contrail cirrus microphysical and radiative properties need to be investigated within GCMs at the same level of sophistication that exists in natural cirrus. Similarly, the surface tension of organics should not matter for the present-day simulation of clouds; however, if one is interested in the past or future perturbation of organic aerosols and their effects on clouds and climate, it may well be important.

Because GCMs are used for different applications, and since people prefer to utilize only one model to address all possible questions of interest, it appears that this path toward increasing complexity will continue, as new processes or submodules are added. To increase our understanding of climate, however, it may be more beneficial to reduce complexity and employ a suite of different models in an effort to understand which processes are necessary to reproduce a specific phenomenon.

Tuning and Metrics of Global Climate Models

Models used to project behavior of clouds in the perturbed climate system are generally optimized by adjusting uncertain parameters according to a mixture of developed intuition and empirical experiences. In effect, top-of-the-atmosphere radiative fluxes are adjusted by tuning cloud optical properties. The task becomes more difficult as model complexity increases. The use of formal nonlinear optimization methods offers a way to systematize these procedures and possibly identify solutions to improve desired model metrics. If more complete observational datasets become available on both cloud properties and their radiative properties, this will put strong constraints on the freedom in the tuning of the cloud properties. We would expect that such observational constraints will decrease the large disagreement in cloud properties in existing GCMs and improve the physical basis for representing the climate system.

Fundamental Problems in Cloud Processes

We have discussed problems and errors of the representation of cloud processes that are primarily related to a lack of resolution. These problems would theoretically disappear if only we had enough computational power to resolve them, given that we fully understood the physical mechanisms involved. However, an increase in resolution requires most of the (scale-dependent) parameterizations in GCMs to be revised. These revisions of code physics may turn out to be very challenging. On a more fundamental level, it must be recognized that we are also coping with processes for which we do not even know the principal processes, let alone their governing equations. These type of problems are predominantly related to microphysics, in general, and further to the ice phase in cold and mixed-phase clouds, in particular.

In addition, hygroscopic growth and activation behavior of atmospheric aerosol particles are not yet fully understood. In this context, possible kinetic effects deserve mention: key words here are surface active substances and their influences on water uptake ("accommodation coefficient") during droplet activation and growth processes (see Stratmann et al., and Kreidenweis et al., both this volume). Further experiments are thus needed to understand and quantify the effects of (a) slightly soluble, surface active substances and soluble gases on high relative humidity hygroscopic growth and activation (i.e., CCN closure experiments), and (b) the influence of kinetic limitations during droplet activation and growth processes.

Ice processes are significantly more complicated than those associated with warm rain. Below, we highlight several fundamental issues associated with the ice phase in mixed-phase clouds (in particular deep convection) and in cold cirrus clouds (at temperatures below the spontaneous freezing point of supercooled water droplets).

Mixed-phase Clouds

Mixed-phase clouds are loosely defined as clouds which contain liquid and ice in close proximity, so that the liquid and ice particles interact microphysically. They are common in subfreezing conditions (below 0°C) and are associated with a wide variety of systems, including deep convection, fronts, and orographic systems. A critical aspect of mixed-phase clouds is their colloidal instability; that is, ice tends to grow through vapor deposition at the expense of liquid water because saturation vapor pressure is lower with respect to ice than liquid (the Bergeron–Findeisen mechanism). Liquid water is generated through upward motion within the cloud. Thus, the existence of liquid water in mixed-phase clouds results from a balance between its evaporation via the Bergeron–Findeisen mechanism and its generation via vertical motion. The accretion of liquid droplets onto snow (riming) may also play an important role in the depletion of liquid, especially if high concentrations of large ice crystals are present. The amount of liquid water in mixed-phase clouds and its lifetime is important to the climate because small liquid droplets have a much larger impact on the cloud radiative forcing relative to the larger ice crystals. In mixed-phase clouds, supercooled liquid water has also been important for weather forecasting because of its role in aircraft icing.

Assuming saturation with respect to water, the supersaturation with respect to ice, and thus the strength of the Bergeron–Findeisen mechanism, increases with decreasing temperature. The dominant microphysical controls on the Bergeron–Findeisen mechanism (as well as depletion of liquid water through riming) are the concentration, size, and shape of the ice crystals. These parameters determine how quickly water vapor can be taken up by the ice. However, vapor depositional growth of ice particles remains uncertain from a basic physical standpoint; additional laboratory measurements and theoretical understanding are needed to better constrain this process (Stratmann et al., this volume). Studies have suggested that larger updraft speeds are required to sustain liquid water in mixed-phase clouds as the concentration and/or size of the ice particles increase. For example, in weakly forced mixed-phase stratus, which are endemic to the Arctic and have relatively weak updrafts, models have suggested strong sensitivity of the lifetime of these clouds to small increases in crystal concentration; in the presence of stronger updrafts (e.g., in deep convection), however, the existence of liquid may be less sensitive to crystal concentration. Concentrations of small ice crystals have been difficult to observe in the past, although new techniques and instrumentation have led to better observational constraints (Kärcher and Spichtinger, this volume). Nonetheless, small ice crystals remain difficult to measure.

The concentration of crystals in mixed-phase clouds may be largely dependent on the concentration of IN and, under some conditions, secondary ice formation processes (i.e., ice multiplication), such as the production of ice splinters during riming via the Hallet–Mossop mechanism. Still, ice nucleation processes remain poorly understood, and concentrations of ice crystals often far exceed concentrations of IN, even in conditions that do not support any known ice multiplication processes. Thus, additional laboratory studies are required to expand our understanding of ice particle formation, along with field-deployable instruments to study ice initiation processes in real clouds, which cannot be measured by current methods (Stratmann et al., this volume).

Cold Cirrus Clouds

Cold cirrus clouds are composed of ice crystals.The interplay between dynamic and aerosol impacts on stratiform (non-convective) cirrus is less complex than for mixed-phase clouds. The dynamic forcing is driven by small-scale vertical wind variability (gravity waves, turbulence) on scales of tens of kilometers. Those regions of small-scale variability are embedded in larger-scale ice-supersaturated regions, where cirrus formation takes place *in situ* through ice nucleation. These meteorological factors led to the development of a relatively simple mechanistic understanding of the ice formation process in rising air parcels. In anvil clouds and the lower parts of frontal cirrus, ice is formed within a mixed-phase cloud environment and transported aloft. In such cases, *in-situ* nucleation can occur after most of the preexisting ice mass has been removed by sedimentation.

The relative role of meteorological factors and aerosol impact on cirrus properties and their variability (e.g., PDF of ice crystal concentration) has been addressed by Kärcher and Spichtinger (this volume), based on *in-situ* data and emphasizing the key role of meteorological factors. The balance in a rising air parcel between increasing supersaturation attributable to cooling and decreasing supersaturation by forming ice condensate implies a predominantly dynamic control of cirrus formation, with the total initial ice crystal concentration being a strong function of cooling rates or vertical velocities. Heterogeneous IN may modulate cirrus formation in regions with relatively small cooling rates. Surprisingly few LES or CRM studies are available that address the links between dynamic and aerosol control of cirrus.

Several field measurements in the tropical and midlatitude upper troposphere have suggested a ubiquitous background of mesoscale (scale of tens of km) temperature fluctuations, leading to typical mean cooling rates of the order 10 K h^{-1}. The origin of these background fluctuations is not well known, but is thought to be generated by gravity waves caused by flow over terrain, and amplifying with height. A predictive capability of geographical and seasonal dependences of this type of forcing is lacking. In such a dynamic environment, the effect of an IN population on cirrus properties depends primarily on the IN number concentration, the ice nucleation threshold (as determined by size, chemical composition and other factors), and the local cooling rate. Theory shows that the likely effect of adding IN to the ubiquitous liquid aerosol background is to reduce actually the total number of ice crystals formed in most cases, keeping other factors fixed. Numerical simulations show that this effect should be robust (though smaller in magnitude) when variability in the dynamic forcing conditions consistent with observations is explained. Adding IN tends to weaken the homogeneous freezing process, leading to slightly larger effective ice crystal sizes and less bright cirrus, and perhaps more significant changes in cloud frequency of occurrence. However, although these theoretical and numerical results are entirely plausible, no closure field studies are available to demonstrate that these processes are actually at work in nature. The current limitations in measuring relative humidity in upper tropospheric conditions accurately (e.g., to within a few percent in absolute terms) limit our ability to discriminate between ice nucleation behavior of various IN in the field. After ice initiation, large ice crystals are generated by sedimentation and aggregation processes. The largest ice particle dimensions that have been observed rarely exceed a few millimeters, and precipitation formation (graupel, hail, snow) commences only after ice crystals have fallen into rather warm cloud layers at low altitudes. The longevity of stratiform cirrus permits radiative heating to feed back to cloud dynamics: absorption of thermal emission by large ice crystals may induce internal convective instability (in particular in a less stable thermodynamic cloud environment), prolonging cirrus lifetime by additional cooling and possibly triggering turbulence-induced ice nucleation.

Further Issues for Ice Clouds

In addition to the issues noted above, there are additional characteristic differences between ice and warm clouds that are common to all ice-containing clouds. These include the coupling between the cloud phase and the moisture field, radiative/microphysical effects of particle shape, and the lack of observational data to constrain cloud models.

In general, clouds that contain liquid water droplets are always close to water saturation. By contrast, if ice forms in a warm cloud at very low concentrations, it can turn into a mixed-phase cloud with the ice phase growing by the diffusion of water vapor and accretion of supercooled water. If the ice concentration is very high, growth occurs primarily by vapor diffusion. This difference is crucially important for local cloud characteristics and aerosol effects, but they also affect significantly the bulk measures in mixed-phase clouds, such as the latent heating or precipitation fallout. Low-temperature ice clouds are often far away from equilibrium; that is, they are less strongly coupled to the ice-supersaturated moisture field because of relatively long supersaturation relaxation timescales. The evolution of their local characteristics depends more sensitively on the history of individual air parcels. Therefore, relative humidity-based, diagnostic GCM cloud schemes fail generally to capture the observed behavior of relative humidity and cirrus in the upper troposphere.

The various forms, shapes ("habits"), and surface characteristics of ice crystals compared to spherical cloud droplets affect scattering and absorption of shortwave radiation and thus the albedo of an ice cloud. Inhomogeneities in the cloud structure affect radiative properties as well, rendering radiative transfer calculations as very challenging. For the same reason, retrieval algorithms used to interpret ground- or satellite-based remote-sensing data of ice clouds are far more uncertain than for low-level water clouds. Moreover, the sedimentation velocities of ice crystals vary significantly, depending on the particle mass and shape (e.g., the habit for pristine crystals and the degree of riming or aggregation for snowflakes), and affect the simulation of cirrus cloud life cycles. Finally, ice initiation has been intensively studied in the laboratory, but it is only poorly understood on the theoretical level. There is a substantial lack of atmospheric IN measurements in all (particularly in the highest and coldest) regions of the troposphere; based on discussion at a recent workshop[2] on IN instruments, this situation will hopefully improve in the near future (Stratmann et al., this volume).

All of the above issues render the treatment of the ice phase in process models (LES, CRM models) difficult, let alone their representations in models that rely on cloud parameterizations (e.g., mesoscale models and especially in GCMs). Therefore, validation studies of LES and CRM simulations (e.g.,

[2] http://lamar.colostate.edu/~pdemott/IN/INWorkshop2007.htm, Intl. Workshop on Comparing Ice Nucleation Measuring Systems.

using the GCSS strategy) are an important step. How to transfer the knowledge from small-scale models into GCMs via physically based parameterizations poses, however, another challenge.

The difficulty in providing a robust, quantitative answer to how aerosols affect the distribution of ice in cirrus lies in the fact that our understanding of how small changes in formation conditions (i.e., initial ice crystal size distribution) affect cloud ice distribution is very limited. The further development of clouds is strongly tied to dynamic forcing and the evolution of the moisture field. It is unclear how long an initial perturbation of a cloud attributable to IN may last, given the potentially long lifetime of cirrus. We also do not know whether changes in aerosols significantly alter the net cloud radiative forcing. Exploring this parameter space poses a challenge for emerging LES studies of cirrus, including detailed aerosol and ice microphysics. On the GCM level, we are beginning to explore this issue as tools become available to parameterize the relevant subgrid-scale processes and their mutual interactions. Closure measurements that are successful in disentangling aerosol and dynamic effects on cirrus may be possible in the near future given the substantial progress that is currently being made in airborne instrumentation.

Entrainment

To measure entrainment of stratocumulus-topped boundary layers, tracer measurements are employed, which enable an estimate of the entrainment velocity. Depending on the variability of the tracers within and above the boundary layer, they can provide very accurate estimates of this entrainment velocity, particularly when performed with multiple tracers. However, they tell us nothing about the physical mechanism by which air is entrained into the boundary layer, and this information is critical if we wish to relate our parameterizations to physical principles and processes (e.g., sedimentation). Methods using near-stationary platforms, such as ACTOS, and laboratory studies under well-defined turbulence conditions (Stratmann et al., this volume), could provide valuable information about the details of the entrainment interface and nature of the entrainment process itself, and thus merit support. Such methods could be similarly useful for constraining the entrainment interface along cloud boundaries.

Overlap and Heterogeneity Assumptions

Currently, GCMs predict cloud fraction in each vertical level. Uncertainty on how to distribute this fractional cloud, both as a function of horizontal resolution and as a function of the distribution of clouds in the other layers of the model, gives model developers considerable freedom in distributing the modeled cloud amount so as to satisfy the radiative constraints at the top of the atmosphere. Experience with different GCMs show that while there may be large layer-by-layer differences in cloud fraction and cloud amount, suitable choices for the cloud overlap assumption enables each to match the top-of-the-atmosphere radiative constraints. Thus, providing measurements of cloud overlap along with cloud horizontal inhomogeneity (e.g., as from the Cloudnet and ARM remote-sensing sites) would provide further and valuable constraints on the models. In addition, such an approach should encourage representations of clouds that encapsulate the entirety of the vertical column, rather than processes on a grid by grid-level basis.

Cloud Lifetime

Satellite measurements with high spatial resolution lack typically temporal resolution. Being able to measure the temporal evolution or even life cycle of clouds would provide a valuable constraint on process models of clouds and their environmental interactions. For instance, a geostationary observatory, such as Meteosat Second Generation, could provide invaluable inputs. Going a step further, one could imagine and promote the deployment of a geostationary observatory that could steer and focus a large telescope on particular cloud scenes (to complement field measurements or *in-situ* studies).

Observational Strategies and Proposals

Observational Strategies to Quantify Aerosol-perturbed Clouds

To a first order, the main cloud-controlling factor is relative humidity, which combines two atmospheric state parameters: specific humidity, q_v, and temperature, T. Clouds form if supersaturation occurs in a clear-sky region, i.e., if $q_v > q_s(T)$. Since both temperature and specific humidity vary strongly in both time and space, as a result of meteorology, the performance of the existing observation, forecast, and reanalysis systems is far below what is necessary to discriminate an aerosol impact from changes attributable to the meteorology (see Stevens and Brenguier, this volume). This holds both for the cloud-controlling factors and for the cloud macrophysical properties that are supposed to be impacted by aerosol particles (radiative properties and precipitation at the surface), and is valid for all cloud types. This makes it extremely hard to demonstrate the effect of aerosols on clouds on an observational basis (see Kärcher and Spichtinger, Stevens and Brenguier, and Brenguier and Wood, all this volume).

In attempting to detect an aerosol impact on cloud macrophysical properties (e.g., radiative properties or precipitation), field studies face the same obstacle that weather modification has experienced for over fifty years. Indeed, the sensitivity of these macrophysical properties to meteorology is so high, and our capability of measuring cloud-controlling factors so limited, that the number of case studies is generally too small to reduce statistically the variability caused by cloud-controlling factors below the level of expected aerosol impacts. One way to optimize the chances of detecting an aerosol signal in the meterological noise is to choose cases in which the perturbation of the aerosol particle is large, the variability of the meteorology small, and the covariance between the two negligible. Fires, particularly those that result from human activity, offer a chance to examine the effects of large-scale changes in the atmospheric aerosol independent of changes in the meteorological environment. Indeed, fires are preferable under certain meteorological conditions. Whether or not fires actually develop, however, often depends on mere chance. For instance, during periods of Santa Ana Winds (offshore flow over the southwestern United States and northwestern Mexico) fires are common, but they are invariably triggered by chance human activity. By sampling aerosol and cloud properties, well downstream of the fire regions, during periods of the Santa Anna Winds, with and without fires, it may be possible to develop a dataset within which meteorological and aerosol effects are independent. The utility of such an approach could first be explored using the satellite record; then, if warranted, it could be realized by targeted deployments of suitably instrumented aircraft over several fire seasons. We recognize that even this may be difficult to achieve. Apart from the rigor of obtaining sufficient samples, one must sample well downwind of the fires because conditions that favor such winds do not favor clouds. Doing so, however, yields measurements in a more disperse aerosol, thereby lessening its impact. Perhaps more importantly, such a strategy allows sufficient time for the development of meteorological biases because of the direct effect of the smoke on radiative fluxes in the evolving cloud-free air.

Another clear example of anthropogenic aerosols that perturb clouds is the existence of ship tracks in stratocumulus. These were the subject of the 1994 MAST experiment (Durkee et al. 2000) as well as numerous remote-sensing studies. These studies suggest that when cloud droplet concentrations exceed 100 cm^{-3}, the cloud in the ship track has a similar or lower liquid water content than that of the ship track itself, suggesting a negative secondary indirect effect on climate (Platnick et al. 2000). When cloud droplet concentrations are much lower, however, the ship track tends to be much thicker than the surrounding stratocumulus. There are almost no realistic modeling studies of ship tracks that quantitatively reproduce these results, including the possibly large effects of circulations between the perturbed cloud in the ship track and the unperturbed surrounding clouds. Ship track measurements provide a good opportunity to test whether LES models can simulate the response of a stratocumulus cloud to a large aerosol perturbation, but as yet this opportunity has been underutilized.

Finally, one can also follow the opposite approach: a situation could be chosen where aerosol properties are similar but cloud-controlling factors vary significantly. Such situations occur every day, at all locations, over the diurnal cycle, with a significant and precisely reproducible change in sun radiative forcing. Over such a period and if air masses are sufficiently homogeneous in terms of aerosol properties, this reverse approach might actually be more successful in detecting an aerosol impact on cloud macrophysical properties than doing so directly. The chance of success of the two approaches should be evaluated following the same methodology as recommended by the WMO in the design of weather modification experiments (NRC 2003). Since aerosols have a diurnal cycle attributable to photochemistry, vertical mixing, and diurnal variation of human and natural aerosol/chemical sources, concern was expressed about the applicability of this approach.

Strategies Using Satellite Measurements

Satellites provide wonderful global observations of cloud properties, but their ability to provide sensible inputs on cause and effect relationships is often questioned. It is interesting to note that this same criticism is not applied to ground remote-sensing or *in-situ* observations from instrumented aircraft, although they too provide just snapshots of various physical parameters. There are two differences which may explain such contrasting views:

1. Time evolution: Observations of clouds help demonstrate, for instance, that clouds are responsible for generating precipitation and not the opposite, because time series show that clouds come before precipitation. This simplistic example shows that time evolution is a crucial ingredient when establishing cause and effect relationships. Present cloud satellite instruments, mainly polar orbiting, are not suited for short timescale evolution experiments, but the increasing performance of geostationary satellites, in terms of cloud observations and time resolution, opens new opportunities.
2. Spatial distributions: Most satellite observations, based on passive sensors, are two-dimensional and cannot document the vertical distribution of the components. For instance, statistics on horizontal correlations of aerosol and cloud liquid water path cannot discriminate aerosols located higher than cloud layers, hence limited to radiative interactions, from those entering cloud base that can control cloud microphysics. The availability of the A-Train observations (Anderson et al. 2005), some of which are vertically resolved, opens new opportunities.

Lacking these two capabilities, satellite observations leave numerous possible explanations open for the observed correlations between aerosol and cloud properties, in terms of cause and effect relationships (cf. Nakajima and Schulz, this volume).

Experimental Strategies for Cloud Ensembles

The last decade has witnessed a number of field studies (DYCOMS-II, RICO, EPIC-2001) designed to measure processes related to the cloud ensemble. Whereas earlier attempts to make such measurements often required many platforms (BOMEX, ATEX, GATE), the emergence of a wide variety of satellite products as well as dropsondes and aircraft-based remote sensing make it possible for a single, long-endurance aircraft instrumented for *in-situ* measurements of dynamics, radiation, gas and aerosol chemistry, and microphysics both to constrain important remaining aspects of the cloud environment and to measure the statistical properties of clouds in this environment. These studies appear most promising for understanding the behavior of cloud ensembles, which is critical in advancing parameterizations as statistical representations of fields of clouds. Combining such measurements with long-term space and ground-based remote sensing offers further opportunity for progress. One measurement that is, however, conspicuous in its absence is that of cloud fraction, such as might be measured by a highly sensitive scanning cloud (K-band) radar, or a spaceborne observatory capable of being focused on a given experimental area (for discussion on the satellite techniques CloudSat/CALIPSO, see Anderson et al., this volume).

Conclusions

Processes that impact clouds span scales from the molecular to the planetary level, a fact that makes their description in a single reference model unimaginable. This implies that large-scale models will continue to hinge on parametric (or statistical) representations of cloud processes on at least some scale for the foreseeable future. As we endeavor to represent present-day and perturbed climates, we must recognize that this introduces a source of uncertainty at best, and likely bias. The origins of uncertainty in our representation of clouds in present-day GCMs are:

1. a fundamental lack of knowledge of some key cloud processes (e.g., upper tropospheric ice supersaturation and role of ice nucleating particles in mixed-phase and cold ice clouds),
2. a lack of knowledge about how to represent the aggregate properties (statistics) of processes that are well understood at their native scale,
3. a lack of interaction among subgrid processes,
4. a breakdown of quasi-equilibrium assumptions at the intermediate scales.

While some of these issues may play less of a role when increasing the resolution of GCMs (category 2 and 3), others (category 1) remain or might even play a larger role (category 4). We propose a number of new pathways for

GCMs, including global GCRMs and cloud-resolving convective (super) parameterization (Grabowski and Petch, this volume). However, even with such approaches, it will remain important to have consistent and interactive treatments of the remaining unresolved scales, such as boundary-layer cloud processes. When moving to higher resolutions, stochastic parameterizations might become more relevant as quasi-equilibrium assumptions break down. We have discussed the more fundamental problems (category 1) pertaining to cloud microphysics (e.g., primary and secondary ice formation, droplet/ice particle growth kinetics, entrainment, cloud overlap) and have made recommendations for both observational studies (laboratory and field experiments) and high-resolution process models.

GCMs should not be used as the sole instrument to quantify the effect of perturbed clouds in our climate system. Instead, a better use of the full hierarchy of models could improve our understanding of the key processes that control cloud–climate feedbacks. For instance, process-resolving, fine-scale models can be used under perturbed conditions to understand and quantify the cloud dynamic response to specified perturbations. Simplified frameworks (aquaplanets, 2-D models, column models with large-scale feedbacks) are useful for conceptualizing these physical cloud processes and provide guidance of how they could be operating in complex GCMs. They can also advance our understanding upon which our confidence in our climate predictions ultimately rests.

Finally, we have proposed appropriate experiments to study perturbed clouds by aerosols. As it is extremely hard to detect an aerosol signal in the meteorological noise, we suggest that cases be chosen in which the perturbation of an aerosol is large compared to the meteorological variability and the covariance between the two is neglible. Proposed candidates are fires triggered by human activity and perturbed stratocumulus clouds in ship tracks. In addition, we recommend that the reverse approach be considered; namely, aerosol properties remain similar but meteorological factors vary significantly (e.g., the diurnal cycle).

References

Albrecht, B. A. 1989. Aerosols, cloud microphysics, and fractional cloudiness. *Science* **245**:1227–1230.

Anderson, T. L., R. J. Charlson, N. Bellouin et al. 2005. An "A-train" strategy for quantifying direct climate forcing by anthropogenic aerosols. *Bull. Amer. Meteor. Soc.* **86**:1795–1809.

Durkee, P. A., K. J. Noone, and R. T. Bluth. 2000. The Monterey area ship track experiment. *J. Atmos. Sci.* **57**:2523–2553.

Kessler, E. 1969. On the Distribution and Continuity of Water Substance in Atmospheric Circulations. Meteorol. Monogr., vol. 32. Boston: Amer. Meteorol. Soc.

NRC. 2003. Critical Issues in Weather Modification Research. Board on Atmospheric Sciences and Climate, Division on Earth and Life Studies. Washington, DC: National Academies Press.

Platnick, S., P. A. Durkee, K. Nielsen et al. 2000. The role of background cloud microphysics in the radiative formation of ship tracks. *J. Atmos. Sci.* **57**:2607–2624.

Rauber, R. M., B. Stevens, H. T. Ochs et al. 2007. Rain in shallow cumulus over the ocean: The RICO campaign. *Bull. Amer. Meteor. Soc.* **88**:1912–1928.

13

Cloud Particle Precursors

Sonia M. Kreidenweis[1], Markus D. Petters[1],
and Patrick Y. Chuang[2]

[1]Department of Atmospheric Science, Colorado State
University, Ft. Collins, CO, U.S.A.

[2]Department of Earth and Planetary Science, University
of California, Santa Cruz, CA, U.S.A.

Introduction

In this chapter, we examine the physicochemical properties of atmospheric particles that determine which subsets of the ambient aerosol can serve as cloud particle precursors for both warm (liquid) and cold (mixed-phase and ice) clouds. We recognize that not only do the particle properties play a role in determining which particles participate in cloud formation, so do environmental variables such as the air mass thermodynamic history and vertical motions. For a discussion of this larger context, which determines the activation behavior of populations of particles and the resulting influences on the microphysical and radiative properties of clouds, the reader is referred to Feingold and Siebert as well as Kärcher and Spichtinger (this volume). Our focus here is on the atmospheric aerosol itself, linking various particle sources to their contributions to cloud particle precursor populations. To make these connections, we begin with a brief review of the current status of understanding of the relationships between aerosol properties and cloud condensation nuclei (CCN), and between aerosol properties and ice-forming nuclei, including identification of the sensitivities to the key variables needed for predictive closure. Thereafter we discuss how changes in aerosol sources induced by anthropogenic activities in the past as well as in the future might influence the cloud particle precursor properties, and highlight the major uncertainties that remain to be addressed.

Relevant Properties: Warm Cloud Particle Precursors

The well-known equation that expresses equilibrium between the environmental relative humidity with respect to liquid water, RH_w, and the vapor pressure of water over a spherical, aqueous solution droplet of diameter D is:

$$RH_w(D) = a_w \exp\left(\frac{4\sigma_{s/a} M_w}{D\rho_w RT}\right), \tag{13.1}$$

where RH_w has been expressed as a fraction, a_w is the water activity of the solution, $\sigma_{s/a}$ is the surface tension of the particle/air interface, R is the ideal gas constant, T is temperature, and the partial molar volume of water in solution has been approximated as the ratio of the molecular weight of water, M_w, to its density, ρ_w. For a choice of dry diameter, D_{dry}, known or assumed relationships between composition, drop density, and water activity can be used in Equation 13.1 to compute $RH_w(D)$. This curve has a maximum which is identified as the critical supersaturation needed for droplet activation, s_c ($s = RH_w - 1$).

Taking Equation 13.1 as the governing equation, it is possible to list the relevant particle-related variables for cloud drop formation. For individual particles, these are surface tension, appearing in the exponential term; temperature, which also appears in the exponential term explicitly and implicitly in water activity and density; particle composition, which affects both D, in the exponential term, and water activity; and particle size (expressed here as D_{dry}). Equation 13.1 assumes that the wet particle is spherical, and sphericity is generally assumed for both the wet and dry particles to compute volumes and masses from diameters. This assumption, however, may not be valid for atmospheric particles, and particle shape should also be included as a variable. An aqueous solution and corresponding water activity may not exist over the full range of assumed particle water contents. Thus, particle phase state and constituent solubilities must also be considered. Finally, for a *population* of particles, these variables and their distribution across the population must be important, and thus the mixing state (i.e., particle-to-particle variations in composition) is also a key property.

The classic Köhler equation is obtained from Equation 13.1 by first making an assumption about the form of the water activity—composition relationship—the most common of which is:

$$a_w^{-1} = 1 + \nu\Phi \frac{n_s}{n_w} = 1 + \nu\Phi \frac{\rho_s M_w}{\rho_w M_s} \frac{V_s}{V_w}, \tag{13.2}$$

where n_s is the moles of solute, n_w the moles of water, $\nu\Phi$ accounts for solute dissociation and solution nonidealities, ρ_s is the density of dry solute, ρ_w is the density of water, V_s is the volume of dry solute, and V_w the volume of water. This definition can be carried through an analysis to derive an expression for the critical supersaturation for a single solute, fully dissolved particle of size D_{dry}:

$$\text{(a)}\quad \ln RH_{w,c} \cong s_c = \left(\frac{4A^3}{27B}\right)^{1/2},$$

$$\text{(b)}\quad A = \frac{4M_w\sigma_{s/a}}{\rho_w RT}, \tag{13.3}$$

$$\text{(c)}\quad B = \nu\Phi\frac{\rho_s M_w}{\rho_w M_s}D_{\text{dry}}^3.$$

Because of the assumptions used to derive them, Equation 13.3 must be modified to treat multiple dry aerosol components, including insoluble species.

A number of recent studies have examined the sensitivity of predicted critical supersaturations to the various particle-related parameters. McFiggans et al. (2006) provide a comprehensive review of the literature regarding the effects of physical and chemical aerosol properties on cloud drop formation, including observational attempts at aerosol–CCN closure. With respect to properties controlling a particle's ability to serve as a CCN, their discussion highlights the poorly characterized CCN activity of the organic aerosol fraction and the role of surface tension and film-forming compounds in modifying activation behavior. They note that although surfactants are present in cloud water samples, partitioning of surfactants between the bulk and the surface of a droplet will modify calculated surface tension depressions. McFiggans et al. (2006) cite a need for better understanding of whether such partitioning effects are significant in atmospheric droplets. Although they list numerous factors that can affect the critical supersaturation at which a particle activates, McFiggans et al. did not rank the various factors in relative importance. In this chapter, we present an analysis aimed at developing such a ranking.

To simplify the examination of the relative role of dry particle composition, we propose an alternate form of Equation 13.2 (Petters and Kreidenweis 2007; see also Rissler et al. 2006):

$$a_w^{-1} = 1 + \kappa\frac{V_s}{V_w}. \tag{13.4}$$

When used in Equation 13.1 to predict critical supersaturation as a function of dry diameter, lines of constant κ describe the $s_c - D_{\text{dry}}$ relationship expected for a particle of fixed composition (Figure 13.1). Summarizing both theoretical calculations and available observations, Petters and Kreidenweis (2007) provided a table of values of κ for many compounds, including inorganic species commonly found in atmospheric aerosols such as sodium chloride and ammonium sulfate, as well as values representative of the behavior of complex aerosols, such as secondary organic aerosol (SOA) generated in the ozonolysis of several different precursors. Figure 13.1 shows typical ranges of κ for inorganic species, organic aerosols, and several other measured aerosol types. Petters and Kreidenweis (2007) demonstrated that the CCN activities of

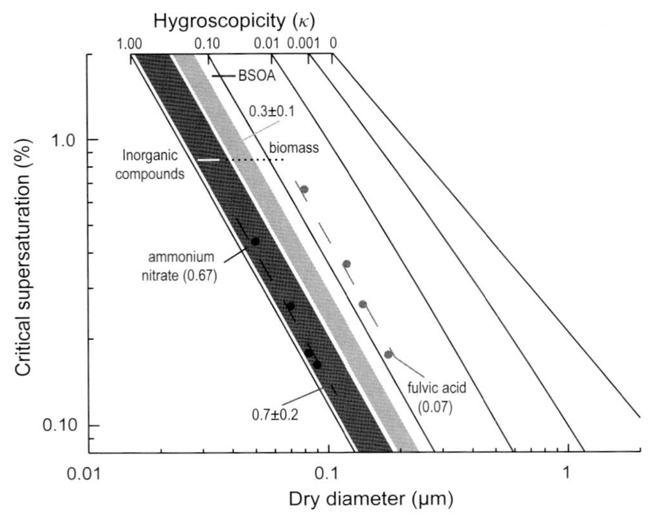

Figure 13.1 Relationship between critical supersaturation and dry diameter, adapted from Petters and Kreidenweis (2007). Fulvic acid and ammonium sulfate points are from Svenningsson et al. (2006) and demonstrate that experimental data organize approximately along lines of constant hygroscopicity. Shading indicates the hygroscopicity ranges for continental (light gray, 0.3±0.1) and marine (dark gray, 0.7±0.2) aerosols suggested by Andreae and Rosenfeld (2008). Also shown are ranges in hygroscopicity for biomass burning aerosols dominated by organic compounds (dotted line) and having substantial mass fractions of inorganic compounds (white line) (unpublished), and biogenic secondary organic aerosols (BSOA; Prenni et al. 2007b).

complex particle types, such as SOA, follow, to a good approximation, lines of constant κ, greatly simplifying the calculation of the critical supersaturation; in particular, the intractable problem of establishing the complete speciation of the organic aerosol is avoided.

In Equation 13.4, κ can represent a mixture of components by summing the volume of fraction-weighted component κ values, including components with $\kappa = 0$, that contribute to the initial dry volume but not to any subsequent water uptake. However, as written, Equation 13.4 does not take particle phase state into account. In other words, water uptake is assumed to be continuous over the full range $0 < a_w \leq 1$, unless some additional constraint is imposed. For compounds that are not infinitely soluble in water, a stable solution is formed only for water contents (and thus solution activities) larger than the solubility limit; this constraint is manifested as deliquescence of a particle at a size-dependent RH_w. For some sparingly soluble compounds and submicrometer particle sizes, deliquescence may occur only at water supersaturations (Hori et al. 2003; Bilde and Svenningsson 2004), leading to $s_c - D_{\text{dry}}$ relationships which do not follow the κ isolines (Kreidenweis et al. 2006). In the general case, for multiple components each having individual solubilities, the κ framework can be extended to account for gradual dissolution (Petters and Kreidenweis 2008).

In fitting experimental data to Equations 13.1 and 13.4 to obtain a value of κ, an assumption is required for the particle/air surface tension, $\sigma_{s/a}$. It is possible to attempt to fit both κ and $\sigma_{s/a}$ using a single dataset, but these cannot be fitted uniquely. An assumption fixing the value of $\sigma_{s/a}$ is needed to fit humidified tandem differential mobility analyzer (H-TDMA) data or s_c–D_{dry} data, but derived values of κ are more sensitive to the assumption in the latter case. The literature contains a number of recent observational and laboratory studies which suggest that some constituents of the atmospheric aerosol depress surface tension and thus enhance CCN activity (e.g., Facchini et al. 1999), although it has also been argued that the overall effect in mixtures and at the typical dilutions experienced at activation are not large enough to play a significant role (e.g., Sorjamaa et al. 2004; Topping et al. 2007). Petters and Kreidenweis (2007) suggest that the surface tension of pure water at 298.15 K be used for purposes of fitting and applying κ values. For surface-active compounds, this results in a derived value of κ that is higher than would be determined from bulk thermodynamic data, reflecting the enhancement of CCN activity attributable to surface tension depression. When applied in mixture calculations, the fitted κ value captures in part the effective contribution of the surface-active component to the lowering of the mixture surface tension.

Sensitivity of Liquid Cloud Particle Formation to Particle Properties

We now estimate the sensitivity of individual particle critical supersaturation to relevant particle properties. Although κ isolines are not linear and parallel over the entire domain, for an infinitely soluble dry particle with $\kappa > 0.2$, the analog of Equation 13.3c can be derived, where $B = \kappa D^3_{\text{dry}}$, leading to:

$$s_c^2 \propto \frac{\sigma_{s/a}^3}{\kappa D_{\text{dry}}^3 T^3}. \tag{13.5}$$

We note that Equations 13.3a and 13.5 are inaccurate for small κ, breaking down completely for $\kappa = 0$, and also do not apply at subsaturated conditions (Petters and Kreidenweis 2007). From Equation 13.5, we can deduce the influence on critical supersaturation of variations in surface tension, composition, dry diameter, and temperature. For example, at fixed supersaturation,

$$\frac{d\kappa}{\kappa} \approx -3\frac{dD_{\text{dry}}}{D_{\text{dry}}}. \tag{13.6}$$

In other words, a 10% decrease in diameter must be compensated by a 30% increase in hygroscopicity to maintain constant critical supersaturation. This demonstrates that critical supersaturation is three times more sensitive to particle size variations than to composition variations. Similar considerations lead to the other entries in Table 13.1a.

Table 13.1a Matrix of the intrinsic sensitivities of the properties governing CCN activity. Each entry denotes the value of the partial derivative of the numerator (columns) with respect to the denominator (rows), where the numerator and denominators are the natural logarithm of the indicated parameter. The table can be used to assess the relative importance on critical supersaturation of particle size (expressed as dry particle diameter D_{dry}), composition (expressed as hygroscopicity parameter κ), temperature (T) and surface tension ($\sigma_{s/a}$). For example, for liquid drop nucleation, $\partial \ln \kappa / \partial \ln D_{dry} = -3$, indicating that the same critical supersaturation can be maintained for a particle if a 10% decrease in D_{dry} is compensated by a 30% increase in κ.

	Liquid drop nucleation			
	$\boldsymbol{D_{dry}}$	$\boldsymbol{\kappa}$	$\boldsymbol{T}$ †	$\boldsymbol{\sigma_{s/a}}$
$\boldsymbol{D_{dry}}$	1	–3	–1	1
$\boldsymbol{\kappa}$	–1/3	1	–1/3	1/3
$\boldsymbol{T}$	–1	–3	1	1
$\boldsymbol{\sigma_{s/a}}$	1	3	1	1

† Sensitivity to temperature was evaluated assuming that κ is independent of temperature and that the surface tension of pure water is constant. In reality, surface tension increases with decreasing temperature, introducing some cross-correlation into the sensitivity table.

The intrinsic sensitivities in Table 13.1a must be combined with typical variations in each variable in order to rank their relative importance. Let us consider now excursions in each parameter of Equation 13.5 and list the selected ranges, $d \ln X_i$ (see Table 13.1b). For diameter, we consider both a large variability in submicrometer size, 0.01–1 μm, and a smaller excursion of 0.08–0.2 μm, intended to capture typical variations in the median diameter of the accumulation mode. Andreae and Rosenfeld (2008) summarize many published estimates of aerosol composition and hygroscopicity in terms of equivalent κ values. They suggest that continental particles, with more significant mass fractions of non- and less-hygroscopic components, can be well represented by $\kappa = 0.3 \pm 0.1$ and marine particles by $\kappa = 0.7 \pm 0.2$. We note that close

Table 13.1b Range in parameters considered for estimation of sensitivities (see Tables 13.1c and 13.3b). acc: accumulation mode; mar: marine; cont: continental; dil: dilute.

	$d \ln D_{dry}$	$d \ln \kappa$	$d \ln T$	$d \ln \sigma_{s/a}$
Range of D_{dry}, all sizes 0.01–1 μm	4.6			
Range of D_{dry}, acc 0.08–0.2 μm	0.92			
Range of κ, all 0.01–1.2		4.8		
Range of κ, mar 0.2–0.4		0.69		
Range of κ, cont 0.5–0.9		0.59		
Range of T 240–303 K / 210–240 K			0.23/0.13	
Range of $\sigma_{s/a}$, all 0.025–0.075 N m^{-1}				1.1
Range of $\sigma_{s/a}$, dil 0.06–0.075 N m^{-1}				0.22

to sources, on urban and smaller scales and in the boundary layer, the mixing processes (such as coagulation and condensation), which serve to homogenize composition as well as to move the aerosol particles toward a single mode in the accumulation size range, have not had sufficient time to operate. Thus particles observed at such scales may exhibit larger variability in κ values; that is, the aerosols may contain multiple modes with possible κ values $1.2>\kappa>0.01$, suggesting a maximum range of $d \ln \kappa \sim 4.8$. For liquid drop formation, the relevant temperature range is 240–303 K. Finally, we consider variations in surface tension which span the wide range of values for atmospheric aerosols that have been speculated in the literature (approximately 0.025 N m^{-1}, up to the value for water, 0.075 N m^{-1}) and a smaller range, 0.06 N m^{-1} to the pure water value, which is likely more representative of surface tensions for dilute solutions in particles near the point of activation.

Table 13.1c combines the intrinsic sensitivities with the range estimates to produce an estimate of the sensitivities between two parameters, X_i and X_j:

$$\text{sensitivity} = \left(\frac{d \ln X_i}{d \ln X_j}\right)_{\text{intrinsic}} \frac{\left(d \ln X_j\right)_{\text{excursion}}}{\left(d \ln X_i\right)_{\text{excursion}}}. \tag{13.7}$$

From Table 13.1c, it is clear that T, κ, and $\sigma_{s/a}$ have comparable influences on CCN activity, if excursions in T, κ, and $\sigma_{s/a}$ are limited to smaller ranges, as appropriate for well-mixed marine or continental aerosols exhibiting minimal surface tension suppression. If small temperature changes of a few degrees, such as might be associated with climate change, are considered, then the relative influence of T on critical supersaturation is very small. Whether sensitivity of critical supersaturation is higher for κ or D_{dry} depends to a large part on the ranges of variations one wishes to consider reasonable for the various parameters. The sensitivities of s_c to variations in diameter are almost

Table 13.1c Relative sensitivities, of column with respect to row variables, limiting allowed variations to those observed in the atmosphere, as shown in Table 13.1b. To obtain the values shown, the excursions in Table 13.1b have been multiplied by the absolute values of the intrinsic sensitivities, as given in Table 13.1a (see text for definition). The temperature range 240–303 K has been applied, appropriate for liquid drop nucleation.

	Liquid drop nucleation					
	D_{dry}, all	**D_{dry}, acc**	**κ, all**	**κ, mar**	**κ, cont**	**T**
κ, all	2.9	0.58				
κ, mar	20	4.0				
κ, cont	23	4.7				
T	20	4.0	7.0	1.0	0.86	
$\sigma_{s/a}$, all	4.2	0.84	1.5	0.21	0.18	0.21
$\sigma_{s/a}$, dil	21	4.2	7.3	1.01	0.89	1.1

always the largest, except for the case of accumulation-mode particle sizes with large variations in κ, which might occur for an externally mixed aerosol. This observation highlights the importance of aerosol mixing state, which can result in large particle-to-particle excursions in κ within a size range, exerting a large influence on variability of critical supersaturations in the accumulation mode aerosol.

In addition to the direct role of temperature in Equation 13.5 evaluated above, temperature variations also modify particle thermodynamic properties, including hygroscopicity, surface tension, and phase state. It is generally assumed that variations in solution water activity (and hence in κ) with temperature are small, and that a_w–composition relationships obtained at room temperature can be extrapolated to all tropospherically relevant temperatures. For individual inorganic compounds, calculations which include temperature variations show that κ is within a few percent of that computed when ignoring the temperature effects (Clegg et al. 1998), as demonstrated in Figure 13.2. However, the thermodynamic data on which the calculations are based is sparse, and the validity of this assumption is unknown for more complex particle types including those dominated by organic constituents.

The initial phase state of a particle can significantly raise the critical supersaturation required for activation, if supersaturated conditions are required to initiate dissolution (e.g., Bilde and Svenningsson 2004). However, many atmospheric particles are mixtures of a number of constituents, and thermodynamic arguments suggest they can retain water to very low RH (Marcolli et al. 2004), which implies that in the atmosphere, solid-to-liquid phase transitions only infrequently influence CCN activity. In Figure 13.3 we explore the

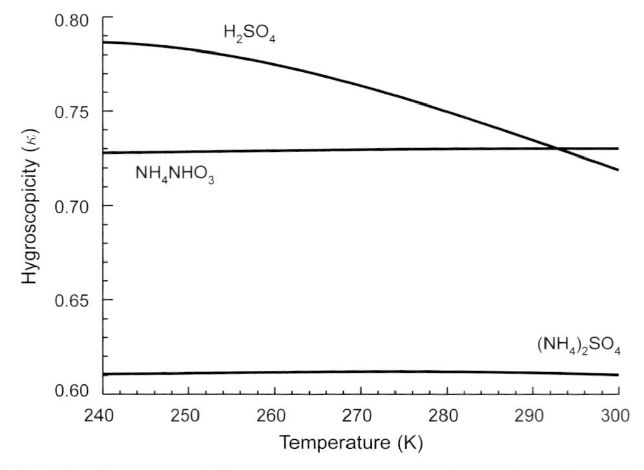

Figure 13.2 The hygroscopicity parameter computed as a function of temperature from the water activities predicted by the Aerosol Inorganics Model (AIM), Model II (Clegg et al. 1998), evaluated at a_w=0.9985, corresponding approximately to the dilution at CCN activation.

effect of a related property, solubility (C_i), on predicted CCN activity. Here solubility is expressed as the volume of compound per unit volume of water in the saturated solution. For $C_i > 10^{-1}$, the initial particle or component phase state has no impact on CCN activity. For $C_i < 5 \times 10^{-3}$, the particle or component may be treated as effectively insoluble. In the intermediate regime, the effects are more complex and lead to enhanced sensitivity of CCN activity to D_{dry} (i.e., $s_c - D_{\text{dry}}$ relationships that do not follow κ-isolines; see Petters and Kreidenweis 2008). We note, however, that this highly sensitive regime applies to a limited region of solubilities, and only a few studies (Hori et al. 2003; Bilde and Svenningsson 2004) have observed solubility-limited behavior for laboratory-generated, single-component organic proxy aerosols. Thus, despite the sensitivity of CCN activity to solubility, we argue that in practice, it may be sufficient to know if a constituent or entire particle falls into the effectively insoluble or effectively infinitely soluble regime, and assign only moderate importance to this parameter.

We close this section with a brief discussion of the broader picture; namely, factors that influence the activation of a population of particles to cloud drops (see also Feingold and Siebert, this volume). In some cases in clean regions

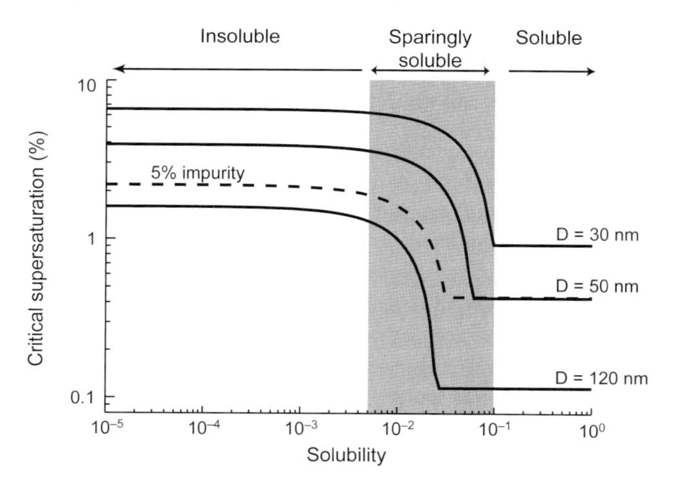

Figure 13.3 Variation of critical supersaturation with solubility C_l (volume of compound per unit volume of water in the saturated solution) for particles having assumed $\kappa_l = 0.6$ and dry diameters of 30, 50, and 120 nm (solid lines). The dashed line represents a 50 nm particle of mixed composition, with 95% by volume of a substance with $\kappa_l = 0.6$ and 5% of an infinitely soluble substance with $\kappa_2 = 0.3$. The plot delineates three solubility regimes that impact CCN activity in the size range $30 < D_{\text{dry}} < 500$ nm: (1) $C_l < \sim 5 \times 10^{-3}$, where solubility of that compound is negligible, regardless of the value of κ, and the compound may be treated as insoluble ($\kappa = 0$). (2) $C_l > \sim 10^{-1}$, where solubility is large enough so the compound may be treated as infinitely soluble. (3) $\sim 5 \times 10^{-3} < C_l < \sim 10^{-1}$, where solubility can strongly impact CCN activity, and the compound's CCN behavior, either as a pure compound or in a mixture, must be treated as discussed in Petters and Kreidenweis (2008).

and/or in the presence of strong updrafts, virtually all of the available particles may be nucleated to drops, and any variations in Table 13.2 parameters are much less important than total aerosol number concentrations in influencing total drop number concentrations. If excess numbers of particles are available, then drop number concentrations also depend on the spectrum of the individual particles' critical supersaturations as well as on the competition in the forming cloud between generation and consumption of water supersaturation, which determines the peak supersaturation achieved in each cloud element. For a single choice of input aerosol properties, Figure 13.4 demonstrates the effects of variations in surface tension on the spectrum of critical supersaturations, hygroscopicity, temperature, particle number median diameter, geometric standard deviation of the particle population, and total particle number concentration. The complexity of the response of critical supersaturation to

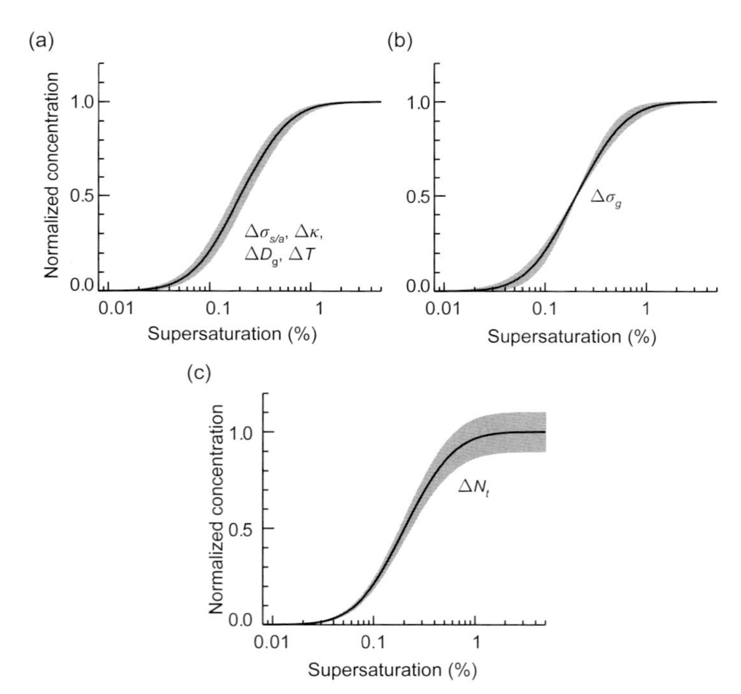

Figure 13.4 Variation of CCN activation spectrum, expressed as the normalized number concentration of particles active as a function of the supersaturation (abscissa). The calculations were performed for a log-normal aerosol distribution having a geometric number mean diameter of 100 nm and geometric standard deviation of 1.8, with the following assumed properties and environmental variables: $\kappa=0.3$; $\sigma_{s/a}=0.065$ N m^{-1}; and T=280 K. In (a), the shaded region indicates the effect of a ±10% excursion in surface tension (response is opposite in sign to the others in this panel), temperature, or median diameter, or a ±30% excursion in κ. Panels (b) and (c) indicate the effects of ±10% excursions in geometric standard deviation and total aerosol number concentration, respectively.

various perturbations for this single case gives an indication of the difficulty in generalizing sensitivities. Nevertheless, it is immediately obvious from Figure 13.4 that at high supersaturations, where almost all particles are able to activate, there is sensitivity only to the total number concentration of particles. At intermediate supersaturations, all of the other variables can modify the number concentration active at a selected supersaturation, with approximately similar influences on changes in number concentration for similar percentage excursions (except for κ, which is three times less influential, as discussed above), as noted already in Table 13.1a for surface tension, diameter, and temperature. Finally, we wish to highlight two additional, important points. First, when cloud dynamics are taken into account, sensitivities of drop number concentrations to all of the variables considered here are lower than the sensitivities computed for the response of CCN concentrations to those variables. The lower sensitivity arises because the number concentration of activated droplets is tied to the peak supersaturation achieved in an updraft, which changes in response to changes in the input aerosol. Second, particle dissolution kinetics (Asa-Awuku et al. 2008; Taraniuk et al. 2007) and reduction of mass accommodation coefficients, and thus drop growth rates (Chuang et al. 1997; Nenes et al. 2001), can also alter peak supersaturations. Thus, it is not possible to use Figure 13.4 to evaluate directly the effect of a parameter perturbation on drop number concentration. These issues are discussed in detail by McFiggans et al. (2006), who also summarize many recent modeling and observational studies aimed at CCN–drop number concentration closure. Even more complexity exists at scales larger than a single updraft, where factors controlling cloud formation and maintenance are not well understood and may prove to be of more importance to the global radiation balance than perturbations induced by changes in the atmospheric aerosol (for further discussion on cloud-controlling factors, see Bretherton and Hartmann; Stevens and Brenguier; Grabowski and Petch; Kärcher and Spichtinger; and Feingold and Siebert, all this volume).

Relevant Properties: Ice Cloud Particle Precursors

In discussing the particle properties relevant to ice formation in the atmosphere, the first distinction that must be made is with respect to freezing mechanism. Homogeneous freezing (i.e., nucleation of an ice crystal from solution and subsequent freezing of the particle) is important at temperatures colder than about –38°C and can occur either in supercooled, dilute droplets or in aqueous solution droplets. Heterogeneous freezing can proceed via several processes, all requiring the presence of a solid surface with properties that catalyze ice nucleation either from the gas phase or in a cloud drop. Added complications are that ice crystal nucleation is strongly temperature dependent and, unlike cloud drop formation, stochastic (i.e., there is a probability per unit time for nucleation to take place in a given volume; Vali 2008). Nevertheless, the onset conditions for nucleation in *individual* particles can be identified with characteristic supersaturations and temperatures: since nucleation rates vary so dramatically across small changes in these variables, they can often be approximated as step functions.

Heterogeneous Freezing

Heterogeneous ice nuclei (IN) can play important roles in determining cloud particle phase state, for supercooled clouds between 0°C and –38°C. It is the onset of freezing that determines when such clouds transition into a colloidally unstable regime. With the appearance of ice crystals in clouds, precipitation and lightning processes become greatly accelerated. This highlights the importance of onset freezing temperatures associated with the nuclei, in addition to their number concentrations. Some bacterial IN can nucleate ice at temperatures as warm as –2°C, whereas observed onset temperatures for dust IN range from –8° to –25°C. Typical number concentrations of IN in the atmosphere range from ~0.001 to 0.01 cm^{-3}, although values up to 1 cm^{-3} have been observed in dust plumes (DeMott et al. 2003b). Their very low number concentrations, compared with that of the total ambient aerosol, which even in extremely clean conditions of 10 cm^{-3} is four orders of magnitude larger, present unique measurement and modeling challenges.

For $T > -38$°C, most ice nucleation mechanisms require the presence of a formed, or forming, cloud droplet. The droplets are usually sufficiently dilute to ignore the presence of a dissolved solute. Effective heterogeneous freezing nuclei often have a component with a crystallographic near-match to the structure of hexagonal ice (Vonnegut 1947); however, crystal polarity (Gavish et al. 1992) also plays a role. Ice germs are thought to form on surface irregularities, or active sites (Vali 1994; Marcolli et al. 2007). It appears that the presence of an aqueous solution can deactivate sites for germ formation for heterogeneous nucleation of haze at cirrus temperatures (Möhler et al. 2005; Koehler 2007; Zobrist et al. 2008), perhaps implying that coatings of hygroscopic solutes may reduce ice nucleation efficiency of dusts in dilute cloud droplets also. In the atmosphere, however, hygroscopic coatings may be necessary to activate the particle into a cloud droplet and thus a necessary condition for immersion freezing to proceed (Vali 1994; Andreae and Rosenfeld 2008). Mineral dusts, some strains of bacteria, and certain organic compounds have been found to initiate ice in laboratory studies at temperatures warmer than –20°C. Ice crystal residuals collected in field study sampling of clouds in this temperature regime have not been investigated to date because the ice crystals must be separated reliably from the cloud droplet population. Recently developed ice counterflow virtual impactors (Mertes et al. 2007) may fill this gap. Electron microscopy analyses of residual particles from ice crystals formed inside continuous flow diffusion cloud chambers (DeMott et al.

2003a) at warmer temperature regimes will also provide new information on composition. Those studies that have characterized composition of ice-forming particles in the atmosphere, mostly limited to ice crystal residuals collected at cirrus temperatures, show salts, crustal materials, carbonaceous particles, and bacteria (Jayaweera and Flanagan 1982; Heintzenberg et al. 1996; DeMott et al. 2003a; Twohy and Poellot 2005; Targino et al. 2006). Little, however, is known about the specific sources, their seasonality, and their strength, the combination of which apparently drive the global IN budget.

Homogeneous Freezing

Cloud droplets freeze spontaneously when cooled below ~–38°C (Jeffery and Austin 1997). In the presence of dissolved solutes, the freezing point is depressed, and metastable liquid phases can exist even at much colder temperatures. Nevertheless, ice germ nucleation may occur in submicrometer solution particles at relative humidities subsaturated with respect to liquid water. The most widely applied parameterization for computing homogeneous freezing nucleation rates in such aqueous haze particles is that of Koop et al. (2000), who reviewed a large body of experimental data and showed that freezing temperature was associated with a unique solution water activity. Koop et al. provide empirical fits for computing nucleation rates, if the solution water activity–composition relationship is known; variations in these relationships and in dry particle densities among compounds lead to small changes in the computed aqueous particle volume that can affect nucleation rates, which are based on the volume of solution. Thus, particle hygroscopicity and surface tension of the solution, as discussed above with respect to CCN activity, are again relevant parameters.

Sensitivity of Ice Cloud Particle Formation to Particle Properties

We now estimate the sensitivity of ice crystal formation to relevant particle properties (see also Kärcher and Spichtinger, this volume, who consider dynamic controls on cirrus formation in addition to aerosol influences). Figure 13.5 demonstrates the relationships between homogeneous freezing relative humidity (FRH), dry diameter, temperature, and hygroscopicity, as calculated from the Koop et al. (2000) parameterization. As expected, the lowest RH_w onset is associated with the coldest temperature (i.e., the largest supercooling). Large dry particle size and particle hygroscopicity lead to more favorable conditions for freezing because the available droplet volumes are larger, because of the action of the Kelvin effect, which increases solution concentration in equilibrium at the specified RH_w for smaller particles. The results in Figure 13.5 have been used to estimate numerically sensitivities to D_{dry}, κ, T, and $\sigma_{s/a}$, which are displayed in Table 13.2a. Freezing rates are, by far, most sensitive to

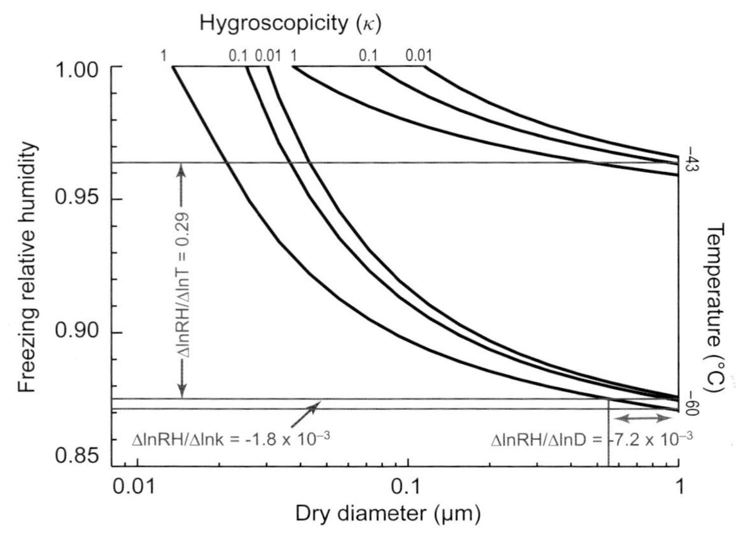

Figure 13.5 Calculated relative humidity, using the parameterization of Koop et al. (2000), for 0.1% of particles to freeze homogeneously within 5 s at two different temperatures (–43° and –60°C) and three different hygroscopicities ($\kappa = 1, 0.1, 0.01$). Gray lines demonstrate evaluation of the sensitivity of freezing relative humidity to T, D_{dry}, and κ at $D_{dry} = 1.0$ μm.

temperature. Weighting by the typical excursions in $d\ln x$, as discussed above for warm clouds (Table 13.1b), we find the sensitivities presented in Table 13.2b. We have modified the unique matrix elements shown, from those selected for presentation in Table 13.1c, to more clearly demonstrate the strong sensitivity to temperature above all other variables. In general, the next most important variable is D_{dry}, and sensitivity to κ and to surface tension is minimal.

Because of the low temperatures and sometimes low relative humidities in the mid to upper troposphere, particle phase state may be a more important factor in influencing conditions for onset of homogeneous freezing than it is in influencing warm cloud formation: for some particle types and sizes, it is possible that no aqueous phase exists until water supersaturation conditions. However, the deliquescence relative humidities for ammonium sulfate (Onasch et al. 1999), some organic compounds (Brooks et al. 2002), and some mixed particle types (Brooks et al. 2002; Marcolli et al. 2004; Parsons et al. 2004) have been examined at low temperatures in the laboratory, and results suggest thus far that in most cases freezing processes would not be inhibited because of high deliquescence RH_w. Further, since ice cloud formation can occur for such a low number concentration of particles, even if only a small fraction of available particles are deliquesced, these will be sufficient to initiate ice formation. We have therefore estimated that homogeneous freezing nucleation is only somewhat sensitive to initial phase state of individual

Table 13.2a As in Table 13.1a, but for homogeneous freezing nucleation. The intrinsic sensitivities have been evaluated numerically, using the formulae recommended by Koop et al. (2000).

	Homogeneous freezing nucleation			
	D_{dry}	κ	$T^{\dagger}$	$\sigma_{s/a}$
D_{dry}	1	~–4	0.024	~2
κ	~–1/4	1	0.006	~1/2
T	42	156	1	–83
$\sigma_{s/a}$	~1/2	~2	–0.012	1

† Sensitivity to temperature was evaluated assuming that κ is independent of temperature and that the surface tension of pure water is constant. In reality, surface tension increases with decreasing temperature, introducing some cross-correlation into the sensitivity table.

particles. Before homogeneous freezing can set in, a particle must take up water to dilute its content until a minimal level of temperature-dependent water activity is reached. Consequently, we estimate some sensitivity to the composition mixing state, which determines the distribution of hygroscopic material and thus of κ.

All of the sensitivities of heterogeneous freezing onset conditions to particle properties were estimated using qualitative arguments, since no definitive theoretical treatments exist to date. It is reasonable to assume that surface properties, which control the number of active sites, are the most important variable, and that D_{dry} assumes importance because active sites may scale with surface area (Vali 1994; Marcolli et al. 2007). It is expected that the probability of ice germ formation on an active site immersed inside a cloud droplet increases exponentially with decreasing temperature (e.g., Vali 1994 and references therein); surprisingly, however, some atmospheric observations do not follow

Table 13.2b Sensitivities, of column variable with respect to row variable, limiting allowed variations to those observed in the atmosphere, as shown in Table 13.1b. To obtain the values shown, the excursions in Table 13.1b have been multiplied by the absolute values of the intrinsic sensitivities, as given in Table 13.3a (see text for definition). The temperature range 210–240 K has been applied, appropriate for homogeneous freezing.

	Homogeneous freezing nucleation					
	κ, all	**κ, mar**	**κ, cont**	**T**	**$\sigma_{s/a}$, all**	**$\sigma_{s/a}$, dil**
D_{dry}, all	0.26	0.04	0.03	1.18	0.12	0.02
D_{dry}, acc	1.30	0.19	0.16	5.89	0.60	0.12
κ, all				4.51	0.46	0.09
κ, mar				31.40	3.19	0.64
κ, cont				36.72	3.73	0.75
T					0.10	0.02

this trend (Gultepe et al. 2001; Prenni et al. 2007a). In fact, deposition nuclei at cirrus temperatures also freeze at constant ice supersaturation with no sensitivity of freezing onset RH_i to temperature for $T<-40^{\circ}C$ (Koehler et al. 2007). Nevertheless T is likely to be important in heterogeneous freezing processes; thus we have assigned a moderate sensitivity to this variable. Coatings of soluble material degrade freezing activity for deposition nuclei under cirrus conditions (Möhler et al. 2005; Koehler 2007), so we expect there to be strong sensitivity to initial phase state and composition mixing state, and some sensitivity to solubility; the exact value of hygroscopicity of the soluble material, in contrast, is not so important. The importance of phase state is also highlighted by recent observations that the water soluble compounds ammonium sulfate and oxalic acid can nucleate ice via deposition if they are in their crystallized form (Abbatt et al. 2006; Zobrist et al. 2006).

Future Changes in Cloud Particle Precursor Types, Properties, and Concentrations

For purposes of understanding the effects of past and future changes in aerosol populations on cloud particle precursor concentrations, we separate the problem into two parts: factors that affect particle properties, such as hygroscopicity, and additional factors that affect particle residence times, although clearly these are related (e.g., through effects of hygroscopicity on precipitation scavenging). As a basis for discussing the impacts of past and future changes in aerosol burden and composition on CCN number concentrations, we use the differences in preindustrial and present-day aerosols as simulated by Tsigaridis et al. (2006). Those authors found a nonlinear response of global annual average aerosol burdens (expressed in Tg) to emissions (expressed in Tg y^{-1}): while emissions of inorganic aerosol precursors increased by factors of 3–6, burdens of corresponding aerosol species increased by factors of only 2.6–2.9. Dust and sea salt were kept constant for all simulations, and represented nearly the same fractions of total aerosol burden in both cases (see their Figure 13.3). The main differences are that increased contributions of inorganic aerosol components in present-day aerosols create "more acidic and hygroscopic aerosols," with significantly higher associated aerosol water contents. Their model does not simulate aerosol mixing state, but since many secondary inorganic aerosol species partition to the particle phase via condensation, it is implied that more internal mixing occurs in the present day, as indicated by significantly shorter simulated times for primary organic aerosol and black carbon to transform from hydrophobic to hydrophilic. Table 13.3 shows the modeled preindustrial and present-day relative mass fractions of constituents other than dust and sea salt. We choose representative κ values for each component (e.g., $\kappa_{SOA}=0.1$; $\kappa_{POA}=\kappa_{BC}=0$; all other $\kappa=0.7$) and compute the volume-weighted mixture κ, assuming that the aerosol particles are internally mixed (at some distance

Table 13.3 Present and preindustrial mass fractions (excluding contributions from dust and sea salt) for secondary organic aerosol (SOA), primary organic aerosol (POA), black carbon (BC), sulfate, ammonium, nitrate, and methanesulfonic acid (MSA), as presented by Tsigaridis et al. (2006). Overall κ values were calculated by weighting the assumed component hygroscopicities with the modeled volume fractions; we assumed that the mass and volume fractions are identical.

	κ	Present	Preindustrial
SOA	0.1	0.31	0.46
POA	0.0	0.24	0.19
BC	0.0	0.04	0.02
Sulfate	0.7	0.13	0.09
Ammonium	0.7	0.13	0.09
Nitrate	0.7	0.12	0.09
MSA	0.7	0.02	0.25
	overall κ	**0.32**	**0.28**

from the source). We find that the difference in mixture κ between the two time periods is less than ~15%. Of greater consequence would be the persistence of unmixed aerosols farther from source regions in preindustrial times, because of lower condensation rates for sulfate, nitrate, and ammonium. However, if less condensation onto primary organic aerosol (POA) and black carbon (BC) produced fewer internally mixed particles, the results would be two populations: one, containing POA and BC, with smaller κ than the computed average, and one, containing only secondary species, characterized by a larger κ. Unless the particle size in the two modes varied significantly from that of the completely mixed aerosol, the net effect on CCN spectra would be small. In summary, past and future changes in composition alone are probably insufficient to have a large impact on CCN activity.

From the sensitivities estimated in Table 13.1c, changes in lifetime that might affect mean particle size (and total number concentrations) are probably more important influences on CCN populations than changes in overall composition. Indeed, we have shown here that a small change in particle size has as much impact on CCN activity as a much larger change in that particle's composition. Tsigaridis et al. (2006) used present-day meteorology in their preindustrial simulations, so that computed changes in the aerosol were entirely driven by the applied changes in emissions. If past and future climates are associated with significant changes in spatial and temporal patterns of precipitation, the main removal mechanism for aerosols, particle lifetimes will change from their present-day values. Similarly, if future climates are associated with more frequent and persistent high-pressure systems over regions having significant particle precursor emissions, we can expect enhanced condensation and growth of the accumulation mode. An unresolved question is whether the increases in global aerosol mass burden from the preindustrial era to the present day, and any changes in meteorology, were associated with similar increases in global aerosol number concentrations, so that mean aerosol size was not greatly affected. Since coagulation processes tend to reduce aerosol number concentrations rapidly, and process ambient submicrometer size distributions toward a single, accumulation-mode peak, we speculate that number and mass increases were not proportional, and that the mean diameter of the accumulation mode (again, at some distance from primary sources) is probably larger in the present day than in preindustrial times. For the extreme assumption that preindustrial number concentrations were similar to those in the present day, the modeled increase in the atmospheric burden of non-sea salt, non-dust aerosol over this period reported by Tsigaridis et al. (2006), 2.7 to 4.8 Tg, would result in a ~20% increase in mean particle size, equivalent in its effects on CCN activity to a ~60% increase in κ, much larger than the effect of varying chemical composition alone. Assuming that homogeneous ice nucleation follows the predictions of Koop et al. (2000), neither the projected changes in hygroscopicity nor in mean accumulation-mode particle size have the potential to affect homogeneous freezing processes substantially. In contrast, however, any changes in dust and biological particle emissions strength, transport patterns, and lifetimes could play a strong role in frequency and altitude of ice cloud formation. Although Tsigaridis et al. (2006) assumed dust emissions were unchanged, other studies have considered the impacts of land use change, prolonged drought, and other climate variables. To date there is no consensus on the past or future changes in dust global cycles, and emissions estimates of biological particles active in cloud formation are virtually unknown for even the present day. Both are needed to develop estimates of future aerosol–cloud precursor relationships.

Status of "Closure" and Recommended Research Directions

Warm Cloud Formation

CCN closure studies evaluate the consistency between measured CCN number concentrations at a certain supersaturation and predictions based on measured particle size distributions and chemical composition or hygroscopicity. McFiggans et al. (2006) summarized several closure studies which show that predicted CCN number concentrations, while generally within ~20% of observations, are almost always biased high compared with direct measurements. This bias, with exception of the Vestin et al. (2007) result, is also evident in more recent studies (Broekhuizen et al. 2006; Mochida et al. 2006; Medina et al. 2007; Stroud et al. 2007). Although the closure studies are often within experimental uncertainties, the consistent overprediction is troubling and not easily explained. Many of the suggested modifications to Köhler theory, such as surface tension depression effects by aerosol organic components, lead to even greater predicted CCN concentrations. It is important to point out that Equation 13.1 is an equilibrium expression, and this has motivated some investigators to search for non-equilibrium effects, related to particle growth kinetics, that may reduce the closure bias. Reduced growth rates may prevent particles expected to activate under equilibrium conditions from forming supermicrometer droplets and being detected as CCN in instruments with finite growth times. At least two mechanisms have been identified to explain such a phenomenon: (a) it has been hypothesized that organic films are ubiquitous on atmospheric particles (Gill et al. 1983); (b) if the dry soluble mass in a particle is unable to dissolve during the particle's exposure to supersaturated conditions, then the s_c required for activation is increased over the equilibrium value. Results from laboratory experiments have been interpreted as exhibiting the latter effect (e.g., Hegg et al. 2001).

In the laboratory, the presence of a film at the gas/liquid interface has been shown to reduce the rate of condensational growth (e.g., Chuang 2003, and references therein). Mass accommodation coefficients as low as 10^{-5} have been measured for a single component film. However, in the atmosphere, where the organic fraction comprises numerous species, it is unclear to what extent reductions in mass accommodation occur. Recent measurements of the growth rate of atmospheric particles by Ruehl et al. (2008) have found that particles that grow only one-half and one-tenth as quickly as ammonium sulfate particles were commonly found at four sites ranging from rural to urban, representing up to 80% and 20% of all particles, respectively. Some recent studies address also the possibility that dissolution kinetics may play a role in droplet activation in the atmosphere (Asa-Awuku et al. 2008; Taraniuk et al. 2007).

A significant problem with the interpretation of CCN data is introduced by the need to calibrate the supersaturation in the CCN instruments, since calculated supersaturations, based on temperature and vapor gradients, do not agree with inferred supersaturations, based on ammonium sulfate test aerosol and Köhler theory. This disagreement is found in both static (Snider et al. 2006) and continuous flow (Rose et al. 2008) CCN instruments. Further, no universally accepted relationship between critical supersaturation and dry particle diameter exists for ammonium sulfate, the typical calibration aerosol, and assumed κ values for calibrations ranging from 0.45 to 0.7. Thus, reported $s_c - D_{\mathrm{dry}}$ relationships for even the same particle types may vary between studies using different calibration assumptions: the $0.45 \leq \kappa \leq 0.7$ range represents up to a 50% bias in the $s_c - D_{\mathrm{dry}}$ relationship that is added to the measurement uncertainties. To achieve equivalence of datasets, it may be better to adopt the suggestion of Roberts et al. (2006) to report all D_{dry} measurements at a particular supersaturation relative to the D_{dry} assumed for ammonium sulfate at that supersaturation, although this method still leaves uncertain the precise relationship between particle size and actual critical supersaturation.

Accurate measurements of particle size distributions and CCN number concentrations are prerequisites for achieving closure. Additionally, closure studies must often rely, at least in part, on assumptions about aerosol mixing state, solution surface tension, particle shape, density, or refractive index, particle phase state during sizing, organic component hygroscopicity, insoluble fractions, and mass accommodation coefficient, which can be difficult to constrain well. Thus, even successful closure does not necessarily indicate predictive power: it should be viewed as a test for consistency within a reasonable parameter space. In several studies, consistency could not be demonstrated, even by this less stringent standard, suggesting biases in the sizing, number concentration, and CCN measurements themselves.

As discussed above, cloud droplet activation is sensitive to the dry volume-equivalent diameter, which must be inferred from optical, aerodynamic, or mobility-based measurement systems. Basic sizing and number concentration errors stemming from uncertainties in instrument operation parameters (e.g. flow rates, electric fields, laser strength, binning, counting) are usually on the order of 10–20%, and are worse for particles larger than about 500 nm, which are present in relatively low number concentrations; however, these largest particles generally activate at the lowest s_c, and are thus very important contributors to the CCN budget at low supersaturations. Even insoluble, large particles, when coated with small amounts of hygroscopic material, may readily serve as CCN at low supersaturations. This is seen from Figure 13.1, which shows that a 500 nm particle activates at a supersaturation of 0.1% of its hygroscopicity $\kappa > 0.01$. This value of κ corresponds to ammonium sulfate comprising just 1.7% of the volume of an otherwise nonhygroscopic particle. Conversion uncertainties from operationally defined size to volume-equivalent size are often 10% and sometimes as large as 100%, adding to the uncertainties. In contrast, overall particle hygroscopicity can usually be measured within ~30% (Petters and Kreidenweis 2007) using either H-TDMA or size-resolved CCN measurements. Although these techniques are difficult to apply to particles larger than 500 nm, the s_c of those particles is generally low enough that accurate characterization of their number concentrations is likely a more important issue than determination of their hygroscopicity in evaluating their role in CCN budgets. The many unidentified, semivolatile organic species present in tropospheric particulate matter are often cited as problematic in attempting to characterize completely the chemical composition of particulate matter, and are implicated as primary causes of the inability to model accurately observed CCN spectra from measurements of particle properties (e.g., McFiggans et al. 2006). However, we believe this situation is not so dire, because thus far experience has shown that variations in the hygroscopicity of typical ambient organic aerosols are not so large. Further, in mixtures, if the volume fraction of very hygroscopic (generally inorganic) material dominates, then the κ of those constituents dominates the volume-weighted κ for the particle, and the volume fraction of the organic aerosol is more important than the assumptions regarding its hygroscopicity, for $0.1 > \kappa_{org} > 0$.

Both solubility and composition mixing state have been adjusted within reasonable bounds in some CCN closure studies to achieve better agreement with CCN measurements, but again this is more of a consistency check than an indicator of predictive power. As we discussed above, solubility is important, but probably more as an indicator of whether an aerosol constituent expresses its actual κ value during activation, or whether it behaves as a nearly insoluble particle. We are presently unaware of any evidence indicating that sparingly soluble components influence CCN activity to the extent that atmospheric particles containing them exhibit the anticipated highly size-dependent critical supersaturations. However, solubility is a strong function of temperature, and most studies are performed near 298 K, much warmer than is relevant to most of the troposphere. There is a need to extend CCN closure studies, including characterization of particle properties, to colder temperatures.

In recent studies, Gasparini et al. (2006) and Vestin et al. (2007) combined measured size distributions and size-resolved hygroscopicity, using an H-TDMA system, to predict CCN spectra, and compared the results with direct CCN observations. An advantage of using the H-TDMA to determine approximate κ characterizing a size bin is that external mixing within that bin can be resolved and included in the model. These studies represent some of the first published attempts at this type of closure, and the encouraging results indicate that future studies should make use of size-resolved hygroscopicity and mixing state, evaluated at high time resolution, to remove some of the persistent uncertainties plaguing such closure. The findings support our contention that measured size distributions and number concentrations are often a significant source of error, and especially that techniques for accurately measuring (and predicting) particle size and number concentration for $D_{dry} > 500$ nm are needed to increase confidence in aerosol–CCN closure.

Models of aerosol–cloud interactions seek to use predicted aerosol components, and modeled or assumed size distributions, to construct CCN spectra for input to cloud microphysics modules. The studies by Gasparini et al. (2006) and Vestin et al. (2007) did not have direct measurements of chemical composition, but used H-TDMA data to model each aerosol size bin as composed of spherical, homogeneously mixed particles with an effective hygroscopicity. The success of this approach suggests that exact chemical composition is not important and that overall hygroscopicity can be captured by assuming either infinitely soluble or completely insoluble components. However, the method applied in those studies did not demonstrate the ability to use chemical composition data to predict κ and subsequently CCN spectra. Single-particle mass spectrometry is presently the only online method by which individual particle composition and mixing state can be measured. However, microscope-based single particle analyses are also useful to help define mixing state, and they have greatly advanced in recent years with the development of the environmental microscope (e.g., Ebert et al. 2002).

In summary, current measurement capabilities and uncertainties suggest that aerosol–CCN closure can only be achieved within ~50% uncertainty in number concentrations. Most studies are well inside this bound, although the consistent overprediction bias may point to systematic errors which may be amenable to elimination through further study. For evaluation of aerosol indirect effects, uncertainty in liquid water path and cloud size may be more influential than this uncertainty in aerosol–CCN relationships. Nevertheless, recent work suggests that improved aerosol–CCN closure can be realized if size-resolved hygroscopicity and mixing state are measured. A remaining need, however, is to demonstrate closure between size-resolved chemical composition and size-resolved hygroscopicity. For this purpose, less focus is needed on resolving small differences in hygroscopic behavior between the myriad aerosol chemical components, and the role of solubility limitations is also of secondary importance; assignment of appropriate κ values to a limited set of measured composition variables is probably sufficient. Finally, we note that the community seems to be coming to consensus on these points: Andreae and Rosenfeld (2008) also state that the most important characteristics defining a particle's CCN activity are size and fraction of soluble (i.e., highly hygroscopic) matter. McFiggans et al. (2006) argue that size-resolved composition is important in achieving aerosol–CCN closure, primarily in the range 40 nm $< d_g <$ 200 nm, together with mixing state of the components in individual particles. Finally, there is a need to develop tested methods that can be used to partition aerosol mass into size and hygroscopicity (including mixing-state) bins in order to compute CCN concentrations for modeling aerosol–cloud interactions. Most likely, models will need to simulate a mixing of components in various size classes and cannot treat the atmospheric aerosol as a purely external or purely internal mixture.

Ice Cloud Formation

Most of the same variables and their uncertainties apply to ice crystal formation, as in the above discussion for CCN. Direct measurements of suspended (i.e., not subject to influences from a substrate on which they are deposited) number concentrations of particles activated to ice crystals are in their infancy, and we are aware of no studies that have attempted to reconcile measured ice crystal number concentrations produced by heterogeneous nucleation with measured aerosol populations and properties, although such relationships are urgently needed by modelers. A number of such relationships have been suggested in the literature, but most are not constrained by observational data, and thus remain to be validated. The role of biologically derived particles in ice formation has been demonstrated in the past, but no standardized techniques exist at the moment to isolate them from the ambient aerosol and permit them to be counted and characterized. Thus, it has not been possible to date to quantify their contributions to ambient IN concentrations.

As with CCN instruments, supersaturations achieved in instruments designed to measure ice formation onset conditions can be computed from other measured parameters, but it would be helpful to have a means of independently verifying the supersaturation, as is done with known salt calibration aerosols for CCN counters. Recent work has used the predictions of the Koop et al. (2000) homogeneous freezing parameterization, applied to ammonium sulfate aerosols, to check for consistency in nucleation onset conditions, but even this relatively simple attempt at closure has been surprisingly difficult. It is not known if the results thus far point to errors in instrument supersaturations or to the failure of the parameterization as applied to ammonium sulfate in some temperature ranges. Major, multi-year and multi-investigator instrument intercomparison efforts, aimed at all freezing mechanisms, are currently underway, in recognition of the need to advance the current state of measurement technology as well as to obtain much-needed data for aerosol–cloud interaction modeling.

Measurement campaigns are necessary to increase the database of vertically resolved IN concentrations, extended to identify the particle types contributing to the population as a function of temperature and relative humidity. Perhaps equally important, from the standpoint of improving modeling of aerosol–cloud interactions and future climate, measurements are needed of the fluxes of these particle types to the atmosphere. This represents a significant measurement challenge, particularly for the critically important class of IN represented by biological particles, which can be expected to vary not only with vegetation type, microbial populations, and season, but also with climate variables such as water availability. Present-day techniques for DNA identification are more sensitive to the presence of a wider array of bioaerosols than could be detected in the past using cultivation-dependent methods (Fierer et al. 2008), but associated techniques for determining which of these biological particles are active as IN remain to be developed.

Importance of Spatial Scales

Finally, we revisit the discussion of the sensitivity of cloud particle formation to the particle properties discussed earlier. The discussion to this point has focused on the governing equation and on closure with point measurements; that is, on relatively small scales, as most of the atmospheric measurements were also conducted at time resolutions short enough to represent small spatial scales (except for multi-hour chemical composition measurements). Our contention is that at small spatial scales, uncertainties in aerosol size-resolved compositional mixing state, including the problem of measuring chemical composition and hygroscopicity on a single particle basis, dominate closure uncertainties, and have the ability to measure sizes and number concentrations of larger particles also contributing to uncertainty. In contrast, at larger scales, such as are represented in climate models, the sensitivity to D_{dry} assumes a greater importance in the overall uncertainty: an average size distribution for such a large spatial scale might produce an averaged CCN spectrum that bears little resemblance to activation spectra that are relevant at cloud scales. Furthermore, predicted aerosol size distributions in models depend strongly on processes that are not well represented, including cloud chemistry, coagulation and wet deposition, leading to potentially large errors in mean D_{dry}. Although hygroscopicity variations over the same spatial scale may occur, as we discussed earlier, mean κ values for aerosol populations distant from source regions do not seem to vary that significantly, and mixing state may also be less important in well-aged aerosols. Our final point, then, is that the relative sensitivities for cloud drop formation—and homogeneous freezing—as listed in Tables 13.1c and 13.2b, may vary with spatial scale and with the particular application.

Acknowledgments

The authors wish to acknowledge support from the Ernst Strüngmann Forum for their participation in this activity. SMK acknowledges support from NOAA's Climate Goal under NOAA Award Number GC05-021. MDP acknowledges support from NASA under NNG06GF00G.

References

Abbatt, J. P. D., S. Benz, D. J. Cziczo et al. 2006. Solid ammonium sulfate aerosols as ice nuclei: A pathway for cirrus cloud formation. *Science* **313(5794)**:1770–1773.

Andreae, M. O., and D. Rosenfeld. 2008. Aerosol–cloud–precipitation interactions. Part 1: The nature and sources of cloud active aerosols. *Earth Sci. Rev.* **89(1–2)**:13–41.

Asa-Awuku, A., A. P. Sullivan, C. J. Hennigan, R. J. Weber, and A. Nenes. 2008. Investigation of molar volume and surfactant characteristics of water-soluble organic compounds in biomass burning aerosol. *Atmos. Chem. Phys.* **8**:799–812.

Bilde, M., and B. Svenningsson. 2004. CCN activation of slightly soluble organics: The importance of small amounts of inorganic salt and particle phase. *Tellus B* **56(2)**:128–134.

Broekhuizen, K., R. Y. W. Chang, W. R. Leaitch, S. M. Li, and J. P. D. Abbatt. 2006. Closure between measured and modeled cloud condensation nuclei using size-resolved aerosol compositions in downtown Toronto. *Atmos. Chem. Phys.* **6**:2513–2524.

Brooks, S. D., M. E. Wise, M. Cushing, and M. A. Tolbert. 2002. Deliquescence behavior of organic/ammonium sulfate aerosol. *Geophys. Res. Lett.* **29(19)**:1917.

Chuang, P. Y. 2003. Measurement of the time scale of hygroscopic growth for atmospheric aerosols. *J. Geophys. Res. Atmos.* **108(D9)**:4282.

Chuang, P. Y., R. J. Charlson, J. H. Seinfeld. 1997. Kinetic limitations on droplet formation in clouds. *Nature* **390**:594–596.

Clegg, S. L., P. Brimblecombe, and A. S. Wexler. 1998. Thermodynamic model of the system H^+-NH_4^+-Na^+-SO_4^{2-}-NO_3^--Cl^--H_2O at 298.15 K. *J. Phys. Chem.* **102(12)**:2155–2171.

DeMott, P. J., D. J. Cziczo, A. J. Prenni et al. 2003a. Measurements of the concentration and composition of nuclei for cirrus formation. *PNAS* **100(25)**:14,655–14,660.
DeMott, P. J., K. Sassen, M. R. Poellot et al. 2003b. African dust aerosols as atmospheric ice nuclei. *Geophys. Res. Lett.* **30(14)**:1732.
Facchini, M. C., M. Mircea, S. Fuzzi, and R. J. Charlson. 1999. Cloud albedo enhancement by surface-active organic solutes in growing droplets. *Nature* **401(6750)**:257–259.
Fierer, N., Z. Liu, M. Rodriguez-Hernandez et al. 2008. Short-term temporal variability in airborne bacterial and fungal populations. *Appl. Environ. Microbiol.* **74**:200–207.
Gasparini, R., D. R. Collins, E. Andrews et al. 2006. Coupling aerosol size distributions and size-resolved hygroscopicity to predict humidity-dependent optical properties and cloud condensation nuclei spectra. *J. Geophys. Res. Atmos.* **111(D5)**:D05S13.
Gavish, M., J. L. Wang, M. Eisenstein, M. Lahav and L. Leiserowitz. 1992. The role of crystal polarity in alpha-amino-acid crystals for induced nucleation of ice. *Science* **256(5058)**:815–818.
Gill, P. S., T. E. Graedel, and C. J. Weschler. 1983. Organic films on atmospheric aerosol-particles, fog droplets, cloud droplets, raindrops, and snowflakes. *Rev. Geophys.* **21(4)**:903–920.
Gultepe, I., G. A. Isaac, and S. G. Cober. 2001. Ice crystal number concentration versus temperature for climate studies. *Int. J. Climatol.* **21(10)**:1281–1302.
Hegg, D. A., S. Gao, W. Hoppel et al. 2001. Laboratory studies of the efficiency of selected organic aerosols as CCN. *Atmos. Res.* **58(3)**:155–166.
Heintzenberg, J., K. Okada, and J. Strom. 1996. On the composition of non-volatile material in upper tropospheric aerosols and cirrus crystals. *Atmos. Res.* **41(1)**:81–88.
Hori, M., S. Ohta, N. Murao, and S. Yamagata. 2003. Activation capability of water soluble organic substances as CCN. *J. Aerosol. Sci.* **34(4)**:419–448.
Jayaweera, K., and P. Flanagan. 1982. Investigations on Biogenic Ice Nuclei in the Arctic Atmosphere. *Geophys. Res. Lett.* **9(1)**:94–97.
Jeffery, C. A., and P. H. Austin. 1997. Homogeneous nucleation of supercooled water: Results from a new equation of state. *J. Geophys. Res.* **102(D21)**:25,269–25,279.
Koehler, K. 2007. The impact of natural dust aerosol on warm and cold cloud formation. Ph.D. dissertation. Fort Collins: Dept. of Atmospheric Science, Colorado State Univ.
Koehler, K., S. M. Kreidenweis, P. J. DeMott, A. J. Prenni, and M. D. Petters. 2007. Potential impact of Owens (dry) Lake dust on warm and cold cloud formation. *J. Geophys. Res.* **112(D12)**:D12210.
Koop, T., B. P. Luo, A. Tsias, and T. Peter. 2000. Water activity as the determinant for homogeneous ice nucleation in aqueous solutions. *Nature* **406(6796)**:611–614.
Kreidenweis, S. M., M. D. Petters, and P. J. DeMott. 2006. Deliquescence-controlled activation of organic aerosols. *Geophys. Res. Lett.* **33(6)**:L06801.
Marcolli, C., S. Gedamke, T. Peter, and B. Zobrist. 2007. Efficiency of immersion mode ice nucleation on surrogates of mineral dust. *Atmos. Chem. Phys.* **7(19)**:5081–5091.
Marcolli, C., B. P. Luo, and T. Peter. 2004. Mixing of the organic aerosol fractions: Liquids as the thermodynamically stable phases. *J. Phys. Chem. A.* **108(12)**:2216–2224.
McFiggans, G., P. Artaxo, U. Baltensperger et al. 2006. The effect of physical and chemical aerosol properties on warm cloud droplet activation. *Atmos. Chem. Phys.* **6**:2593–2649.
Medina, J., A. Nenes, R. E. P. Sotiropoulou et al. 2007. Cloud condensation nuclei closure during the International Consortium for Atmospheric Research on Transport

and Transformation 2004 campaign: Effects of size-resolved composition. *J. Geophys. Res.* **112(D10)**:D10S31.

Mertes, S., B. Verheggen, S. Walter et al. 2007. Counterflow virtual impact or based collection of small ice particles in mixed-phase clouds for the physico-chemical characterization of tropospheric ice nuclei: Sampler description and first case study. *Aerosol. Sci. Technol.* **41(9)**:848–864.

Mochida, M., M. Kuwata, T. Miyakawa et al. 2006. Relationship between hygroscopicity and cloud condensation nuclei activity for urban aerosols in Tokyo. *J. Geophys. Res.* **111(D23)**:D23204.

Möhler, O., C. Linke, H. Saathoff et al. 2005. Ice nucleation on flame soot aerosol of different organic carbon content. *Meteorol. Z.* **14(4)**:477–484.

Nenes, A., S. Ghan, H. Abdul-Razzak, P. Y. Chuang, and J. H. Seinfeld. 2001. Kinetic limitations on cloud droplet formation and impact on cloud albedo. *Tellus* **53**:133–149.

Onasch, T. B., R. L. Siefert, S. D. Brooks et al. 1999. Infrared spectroscopic study of the deliquescence and efflorescence of ammonium sulfate aerosol as a function of temperature. *J. Geophys. Res. Atmos.* **104**:21,317–21,326.

Parsons, M. T., D. A. Knopf, and A. K. Bertram. 2004. Deliquescence and crystallization of ammonium sulfate particles internally mixed with water-soluble organic compounds. *J. Phys. Chem. A.* **108(52)**:11,600–11,608.

Petters, M. D., and S. M. Kreidenweis. 2007. A single parameter representation of aerosol hygroscopicity and cloud condensation nucleus activity. *Atmos. Chem. Phys.* **7**:1961–1971.

Petters, M. D., and S. M. Kreidenweis. 2008. A single parameter representation of hygroscopic growth and cloud condensation nucleus activity. Part 2: Including solubility. *Atmos. Chem. Phys. Discuss.* **8**:5939–5955.

Prenni, A. J., J. Y. Harrington, M. Tjernstrom et al. 2007a. Can ice-nucleating aerosols affect arctic seasonal climate? *Bull. Amer. Meteor. Soc.* **88(4)**:541–550.

Prenni, A. J., M. D. Petters, P. J. DeMott et al. 2007b. Cloud drop activation of secondary organic aerosol. *J. Geophys. Res. Atmos.* **112**:D10223.

Rissler, J., A. Vestin, E. Swietlicki et al. 2006. Size distribution and hygroscopic properties of aerosol particles from dry-season biomass burning in Amazonia. *Atmos. Chem. Phys.* **6**:471–491.

Roberts, G., G. Mauger, O. Hadley, and V. Ramanathan. 2006. North American and Asian aerosols over the eastern Pacific Ocean and their role in regulating cloud condensation nuclei. *J. Geophys. Res.* **111(D13)**:D13205.

Rose, D., S. S. Gunthe, E. Mikhailov et al. 2008. Calibration and measurement uncertainties of a continuous-flow cloud condensation nuclei counter (DMT-CCNC): CCN activation of ammonium sulfate and sodium chloride aerosol particles in theory and experiment. *Atmos. Chem. Phys.* **8**:1153–1179.

Ruehl, C. R., P. Y. Chuang, and A. Nenes. 2008. How quickly do cloud droplets form on atmospheric particles? *Atmos. Chem. Phys.* **8**:1043–1055.

Snider, J. R., M. D. Petters, P. Wechsler, and P. S. K. Liu. 2006. Supersaturation in the Wyoming CCN instrument. *J. Atmos. Ocean. Technol.* **23(10)**:1323–1339.

Sorjamaa, R., B. Svenningsson, T. Raatikainen et al. 2004. The role of surfactants in Kohler theory reconsidered. *Atmos. Chem. Phys.* **4**:2107–2117.

Stroud, C. A., A. Nenes, J. L. Jimenez et al. 2007. Cloud activating properties of aerosol observed during CELTIC. *J. Atmos. Sci.* **64(2)**:441–459.

Svenningsson, B., J. Rissler, E. Swietlicki et al. 2006. Hygroscopic growth and critical supersaturations for mixed aerosol particles of inorganic and organic compounds of atmospheric relevance. *Atmos. Chem. Phys.* **6**:1937–1952.

Taraniuk, I., E. R. Graber, A. Kostinski, and Y. Rudich. 2008. Surfactant properties of atmospheric and model humic-like substances (HULIS). *Geophys. Res. Lett.* **34**:L16807.

Targino, A. C., R. Krejci, K. J. Noone, and P. Glantz. 2006. Single particle analysis of ice crystal residuals observed in orographic wave clouds over Scandinavia during INTACC experiment. *Atmos. Chem. Phys.* **6**:1977–1990.

Topping, D. O., G. B. McFiggans, G. Kiss et al. 2007. Surface tensions of multi-component mixed inorganic/organic aqueous systems of atmospheric significance: Measurements, model predictions and importance for cloud activation predictions. *Atmos. Chem. Phys.* **7**:2371–2398.

Tsigaridis, K., M. Krol, F. J. Dentener et al. 2006. Change in global aerosol composition since preindustrial times. *Atmos. Chem. Phys.* **6**:5143–5162.

Twohy, C. H., and M. R. Poellot. 2005. Chemical characteristics of ice residual nuclei in anvil cirrus clouds: Evidence for homogeneous and heterogeneous ice formation. *Atmos. Chem. Phys.* **5**:2289–2297.

Vali, G. 1994. Freezing rate due to heterogeneous nucleation. *J. Atmos. Sci.* **51(13)**:1843–1856.

Vali, G. 2008. Repeatability and randomness in heterogeneous freezing nucleation. *Atmos. Chem. Phys. Discuss.* **8**:4059–4097.

Vestin, A., J. Rissler, E. Swietlicki, G. P. Frank, and M. O. Andreae. 2007. Cloud-nucleating properties of the Amazonian biomass burning aerosol: Cloud condensation nuclei measurements and modeling. *J. Geophys. Res.* **112(D14)**:D14201.

Vonnegut, B. 1947. The nucleation of ice formation by silver iodide. *J. Appl. Phys.* **18(7)**:593–595.

Zobrist, B., C. Marcolli, T. Koop et al. 2006. Oxalic acid as a heterogeneous ice nucleus in the upper troposphere and its indirect aerosol effect. *Atmos. Chem. Phys.* **6**:3115–3129.

Zobrist, B., C. Marcolli, T. Peter, and T. Koop. 2008. Heterogeneous ice nucleation in aqueous solutions: The role of water activity. *J. Phys. Chem.* **112**:3965–3975.

14

Cloud–Aerosol Interactions from the Micro to the Cloud Scale

Graham Feingold[1] and Holger Siebert[2]

[1]NOAA Earth System Research Laboratory, Boulder, CO, U.S.A.
[2]Leibniz Institute for Tropospheric Research, Leipzig, Germany

Abstract

The effect of suspended particles (aerosol) on clouds and precipitation from the micro to the cloud scale has been well studied through laboratory, *in-situ*, and remote-sensing data but many uncertainties remain. In particular, there is scant observational evidence of aerosol effects on surface precipitation. Clouds and precipitation modify the amount of aerosol through both physical and chemical processes so that a three-way interactive feedback between aerosol, cloud microphysics, and cloud dynamics must be considered. The fundamental cloud microphysical properties are driven by dynamics; vertical motions and mixing processes between the cloud and its environment determine the concentration of cloud water, a key parameter for both climate and precipitation. However, aerosol particles can significantly affect the microphysics and dynamics of clouds by changing the size distribution of drops, their ability to grow to raindrops, their rates of evaporation, and their mixing with the environment. The physical system is strongly coupled and attempting to separate aerosol effects has only been done using some simple constructs, some of which will be shown to be of dubious utility. Both observations and modeling suggest that not only the magnitude, but perhaps also the sign of these effects, depends on the larger-scale meteorological context in which aerosol–cloud interactions are embedded. Some alternate approaches are considered, as we explore the possibility of self-regulation processes that may act to limit the range over which aerosol significantly affects clouds.

Historical Perspective

Historical evidence of the effects of aerosol on atmospheric processes abounds, as seen in ancient texts all the way through to modern times. Common human experience tells us that less sunlight reaches the surface under cloudy skies and that aerosol laden skies (e.g., from biomass burning or volcanic eruptions) produce hazy skies and colorful sunsets. In 1783, Benjamin Franklin experienced a particularly cool summer in Europe and attributed it to "fog" associated with a volcanic eruption in Iceland, earlier in the year, that visibly dimmed the sunlight. Evidence that aerosol particles are complicit in cloud droplet formation came in the late 19th century, when P.-J. Coulier (1875), J. Aitken (1880), and C. T. R. Wilson (1897) showed that aerosol particles are necessary for cloud droplet formation. Subsequent work by Aitken (1923) and Köhler (1936) strengthened further the link between aerosol and cloud droplet formation. Houghton (1938) discussed the importance of collision and coalescence of cloud droplets for rain formation and even recognized the role of giant hygroscopic nuclei, a subject that is of great interest to this day. Howell (1949) developed a numerical model of the growth of a population of droplets in a rising parcel of air and laid the foundations of similar tools used today. Analytical solutions to these equations (Twomey 1959) are still in common use. Thus, although the early literature did not focus on aerosol as a means of *perturbing* cloud microphysical processes, the link between aerosol and clouds has existed in the scientific literature for over a century.

Pollution and Climate Change

A more direct link between pollution particles and climate change was raised by Twomey (1974), who considered the fact that the positive correlation between aerosol number concentration, N_a, and drop number concentration, N_d, would result in smaller droplets and more reflective clouds *ceteris paribus*; that is, all else, particularly the amount of condensed water, being equal. This constitutes a shortwave radiative forcing since low-level clouds radiate in the longwave at temperatures close to the surface temperature. The concept is a hypothetical one since it is known that feedbacks also result in modifications to liquid water. It is applied in global climate models by modifying the reflectance of fixed cloud fields. The recently released IPCC (2007) report still regards this "cloud albedo effect" as the single biggest unknown in climate change predictions. It is also of note that although many aspects of the effect of aerosol particles on clouds are well-known, the global radiative (and therefore climate) implications are still poorly constrained. For example, most satellite remote-sensing studies quantify the change in cloud reflectance, or cloud drop size, in response to an aerosol perturbation without stratifying for similar meteorological conditions (e.g., cloud liquid water path, LWP). Thus "*ceteris paribus*" is often ignored, resulting in considerable ambiguity: Are clouds more reflective because of higher N_d and smaller drops, or because of higher liquid water content (LWC), or some combination of the two?

Pollution and Precipitation

One of the oldest problems in cloud microphysics is the inability to describe quantitatively broad drop size distributions that are observed (compared to the narrow, modeled size distributions that assume adiabatic growth) and the onset of precipitation. Broadening has been attributed to entrainment-mixing, aerosol effects (such as giant aerosol), and turbulence effects on collision–coalescence. The first experimental evidence of a link between pollution and precipitation processes known to us appeared in a paper by Gunn and Phillips (1957). They performed experiments in a giant cloud chamber and observed that when clean air was drawn into the chamber, large, precipitation-sized drops were produced; polluted air produced smaller drops and no precipitation. These observations were confirmed in natural clouds a decade later by Warner (1968), but subsequent analysis by Warner yielded inconclusive results. The experimental evidence for this link is still a subject of enormous debate, particularly for deep convective cloud systems (e.g., Levin and Cotton 2007; see also Cotton, this volume; Ayers and Levin, this volume).

The implications for climate change of aerosol effects on precipitation were considered by Albrecht (1989), who used a numerical model of a cloudy boundary layer to show that an increase in N_d results in clouds with less collision–coalescence, less precipitation, and therefore increased LWP and higher albedo. This constitutes a positive feedback to the albedo effect since Twomey (1974) had considered the radiative impacts for static clouds of the same LWP. In addition, Albrecht (1989) suggested that a suppressed precipitation process might increase the lifetime of a cloud. In the intervening decades, the precipitation suppression hypothesis has become known as the "Albrecht effect," or "lifetime effect." It has also come under increasing scrutiny in recent years as observational evidence for this phenomenon (*in-situ*, satellite remote sensing) has been pursued. Various aspects of this linkage between aerosol, precipitation, and climate had been qualitatively confirmed (e.g., the suppression of precipitation by aerosol in warm rain), but to date there is still considerable debate over the effects of aerosol on cloud fraction, cloud LWP, and cloud lifetime, and the quantitative relationship between aerosol and surface precipitation.

Aerosol Effects on Clouds: Deconstructing the Constructs

The Albedo Effect

There is very little argument about the existence of aerosol perturbations to cloud albedo, provided that clouds are stratified by similar amounts of condensed water. The classic example is of ship tracks, which manifest themselves as bright linear features in a cloudy background associated with aerosol particles emitted from ship stacks.

The primary discussion centers around the *degree* of this effect; that is, the extent to which pollution contributes to the particle number concentration at diameters greater than ~0.05 μm, and the extent to which N_d increases with increasing aerosol.

Parameters Affecting Activation: Number and Size versus Composition

There is some controversy over the relative importance of various aerosol parameters (number, size, composition) in determining N_d. Whereas inorganic sulfate particles were assumed to be the dominant aerosol type vis-à-vis cloud microphysics, research in the last decade has pointed to the prevalence of organics (both primary and secondary), and much work has addressed their influence on clouds relative to inorganics. It has been suggested that the surface tension-reducing properties of some organics may significantly enhance the number of activated droplets by reducing the Kelvin (curvature) effect. However, modeling attempts, which consider composition by taking into account a variety of (sometimes competing) factors (e.g., molecular weight, surface tension, van't Hoff factor, and solubility), have shown that the effects of composition are relatively small ($\leq 15\%$) compared to aerosol parameters such as N_a and size, or dynamical parameters such as updraft velocity (Ervens et al. 2005). These findings echo the words of Houghton (1938), who upon examining the form of the Köhler curves stated: "This would seem to indicate that hygroscopic nuclei are not much more effective than neutral [insoluble] nuclei of the same size." A number of exceptions to this may exist: the presence of film-forming compounds could act as a barrier to droplet growth. Composition, as manifested in an external mixture of both hydrophopic and hygroscopic particles, is also of great importance in determining N_d. This, and other composition issues, are discussed in much more depth by Kreidenweis et al. (this volume).

It is important to note that the fact that aerosol composition may significantly influence the equilibrium diameter of a particle does not equate to a concomitant effect on the number of activated drops. The system, comprising a population of particles growing in an updraft, is self-regulating to a degree. Composition effects that reduce (increase) condensation are compensated for by an increase (decrease) in available vapor for activation of smaller particles, and/or growth of existing particles. Therefore, calculations of N_d based on equilibrium calculations, rather than dynamic calculations, must always overestimate the importance of composition for N_d.

Perhaps the single, greatest uncertainty in droplet activation and the growth of a population of drops is that of the mass accommodation, α_c, defined as the fraction of water vapor molecules arriving at a droplet that condenses. Values ranging from 0.03 to 1 have been measured experimentally and proposed based on theoretical arguments. Some of the experimental variability may be associated with the purity of the water droplets, as well as temperature and pressure. The effects of α_c are significant; larger values (order 0.3–1) allow growing haze droplets to grow much more efficiently, thus reducing the supersaturation and N_d significantly. Conversely, smaller values (on the order of 0.05) allow the supersaturation to build up, resulting in higher N_d. The effect is considerably larger than those associated with identified composition effects. Drop number concentration closure experiments suggest $\alpha_c \sim 0.06$ (Conant et al. 2004), but more direct measurements show that α_c may vary by more than a magnitude depending on location and time (Ruehl et al. 2008).

We would be remiss in not mentioning at least one case where an increase in N_a has been hypothesized to result in a *decrease* in N_d. Large, or giant particles (> a few μm in size), and in sufficient concentrations (order 1 cm^{-3}), have been shown in modeling studies to suppress the development of supersaturation and prevent the activation of smaller particles (O'Dowd et al. 1999). Again we stress the importance of calculating N_d as a dynamic process. Had one performed equilibrium calculations of N_d in the presence of giant nuclei, N_d would simply have increased by the number concentration of giant nuclei, rather than decreased.[1]

Finally, in the case of ice nucleation unlike that for water droplets, insoluble particles tend to be better ice nuclei than soluble particles. It is noteworthy that most parameterizations pose ice crystal concentration, N_i, as a function of temperature and sometimes ice supersaturation. The absence of explicit representation of the total aerosol population in which these ice nuclei reside may at least partially explain the very large dynamic range of observed N_i when compared to predictions from said parameterizations.

Albedo Susceptibility

An understanding of the increase in N_d for an incremental increase in aerosol is an essential component of the shortwave albedo effect but does not constitute an understanding of the radiative response. For the simple case of clouds, which exhibit a linear increase in LWC with increasing altitude (adiabatic or subadiabatic), cloud optical depth τ_c can be shown to scale with $N_d^{1/3}$ $LWP^{5/6}$ (e.g., Boers and Mitchell 1994). Cloud optical depth is therefore two-and-a-half times more sensitive to LWP than to N_d. For thin clouds, cloud albedo is linearly dependent on τ_c and exhibits therefore similar sensitivities. This points to the importance of stratification by LWP if a pure aerosol effect on cloud optical depth is sought. Alternatively, in a dynamic system with variable LWP, the extent of variability in LWP must be considered alongside the variability in the aerosol (and N_d) perturbation. A useful approach has been to consider the cloud susceptibility S_0 (Platnick and Twomey 1994), defined as the change in cloud albedo A for an incremental increase in N_d. For a plane-parallel cloud, at constant LWP and fixed distribution breadth (constant "micro"),

---

[1] Giant nuclei, in concentrations as low as 10^{-3} cm^{-3}, also expedite the formation of rain (Houghton 1938), as will be discussed later.

$$S_0 = \left.\frac{dA}{dN_d}\right|_{\text{micro}} = \frac{A(1-A)}{3N_d}, \tag{14.1}$$

or

$$S_0' = \left.\frac{d\ln A}{d\ln N_d}\right|_{\text{micro}} = \frac{(1-A)}{3}. \tag{14.2}$$

S_0' decreases monotonically with A and has the advantage of being independent of N_d.[2] Relaxing the assumptions of constant LWP and breadth yields

$$S_0'' = S_0'\left[1 + \frac{5}{2}\frac{d\ln \text{LWP}}{d\ln N_d} + \frac{d\ln k}{d\ln N_d}\right], \tag{14.3}$$

where k is a parameter inversely proportional to drop distribution breadth, and the double prime implies that microphysical assumptions have been relaxed. We note the strong dependence of susceptibility on LWP changes. When spectral broadening (smaller k) is associated with increasing N_d (because of competition for water vapor in the relatively polluted, condensation-dominated regime), S'' is diminished, whereas when broadening is associated with a reduction in N_d (the cleaner, coalescence-dominated regime), S'' is enhanced.

A complete description of the radiative response of a cloud system to an aerosol perturbation requires 3-D radiative transfer modeling based on a description of both microphysical properties (drop size distributions) and macroscale properties, such as the spatial distribution of cloud optical depth, LWP, cloud fraction, cloud morphology, and distances between clouds. Moreover, the regions of aerosol haze surrounding clouds can also have a substantial radiative impact (Koren et al. 2007; Charlson et al. 2007). Zuidema et al. (2008) have demonstrated that cloud microphysical and macrophysical responses to perturbations in aerosol can work in unison or counter one another to produce either stronger or weaker albedo response to a change in aerosol.

Higher-order Aerosol Perturbations to Clouds

The effect of aerosol on clouds without stratification by LWP is a far more complex construct because it addresses an evolving system which includes multiple feedbacks over the life cycle of a cloud. Parts of this construct, such as the aerosol suppression of warm rain, are fairly well established, at least from a qualitative point of view. Other aspects, such as the increase in LWP, cloud fraction, and cloud lifetime are highly uncertain and either completely lack observational support (e.g., cloud lifetime) or are plagued by measurement difficulties, artifacts (e.g., satellite measurement of aerosol effects on cloud

[2] These susceptibility calculations are posed in terms of N_d perturbations. Similar relationships could be developed in terms of N_a perturbations by linking N_d to N_a and other parameters; a nontrivial problem.

fraction), and the fact that observed correlations are not necessarily indicative of causal relations. In the case of cold clouds, there is even greater uncertainty because of significantly increased complexity and a dearth of measurements, or measurement difficulties.

Unlike the case of the albedo effect, there is no underlying normalization by LWP in the "lifetime effect" because it attempts to address the LWP response. Separation of dynamical influences on LWP and other cloud properties, as opposed to aerosol effects, is therefore the primary challenge that has eluded us for decades. The fact that so many coupled processes act in unison suggests that new approaches need to be considered.

Increase in Cloud Fraction and Liquid Water Path

A central tenet of the "lifetime effect" is that both LWP and cloud fraction increase in response to an increase in (scattering) aerosol attributable to suppression of collision–coalescence and precipitation. (The special case of absorbing aerosol will be discussed separately below.) The ACE-2 field campaign illustrated the pitfall of assuming that polluted clouds generate clouds with greater LWP. There, pollution events were associated with drier and warmer air (Brenguier et al. 2003) and, consequently, lower LWP, which was most likely further reduced by free tropospheric entrainment. Thus, it is not surprising that satellite studies have produced mixed messages. These studies are correlative in nature, and causality is much harder to establish. Satellite remote sensing is also not without measurement challenges: aerosol and cloud measurements are not collocated; the closer one gets to a cloud, the more difficult it is to distinguish cloud from hydrated aerosol ("cloud contamination"). Some of these problems are alleviated by surface and/or aircraft-based remote sensing.

Modeling studies have established causal links between both negative and positive responses of cloud LWP to aerosol but we await direct observational support. We will dwell on this point since it is of central importance. When dry air is entrained into positively buoyant clouds, evaporative cooling can cause buoyancy reversal. Negatively buoyant parcels produce more turbulence kinetic energy, which enhances entrainment and mixing. In the case of stratocumulus clouds, this process has been hypothesized to break up clouds (cloud-top entrainment instability).

Some recent modeling studies have begun to challenge the generality of the assumption that LWP increases with increasing aerosol. Wang et al. (2003) used large eddy simulations (LES) of stratocumulus clouds to show that LWP decreased in response to an aerosol perturbation for idealized non-precipitating clouds. This decrease was linked to the fact that polluted clouds have smaller drops and thus a shorter evaporative timescale than clean clouds. The result is a stronger entrainment rate and a decrease in cloud water. We wish to highlight the fact that bulk microphysical schemes assume instantaneous condensation/evaporation and therefore (a) cannot simulate this feedback and

(b) overestimate entrainment rates. This same result was also demonstrated for shallow cumulus (Xue and Feingold 2006). Ackerman et al. (2004) simulated precipitating and non-precipitating stratocumulus clouds for a variety of environmental soundings and reinforced the importance of evaporation rates at cloud top. They showed that only under very low N_d or humid conditions above cloud top did an increase in aerosol result in an increase in LWP. Under more polluted conditions and/or drier free tropospheric air, entrainment dominated and cloud LWP decreased with increasing aerosol (thus opposing the Twomey effect).

Can Aerosol Influence the Behavior of Small-scale Mixing in Clouds?

Above, we discussed the fact that aerosol can influence cloud dynamics via modification to the condensation/evaporation timescale (Wang et al. 2003). The tightly coupled nature of the cloud droplet/dynamical system at these small scales (order centimeters to meters) requires some discussion of mixing mechanisms. When a cloudy air parcel mixes with ambient subsaturated air, two bounding scenarios have been considered:

1. Homogeneous mixing: where turbulent mixing is much faster than the rate at which droplets can react to their new environment. Under these conditions all droplets experience approximately the same thermodynamic conditions of the newly mixed air parcel followed by evaporation.
2. Inhomogeneous mixing: where turbulent mixing is comparably slow and the boundary regions between entrained and cloudy air exist long enough to ensure that the droplets in the interfacial regions evaporate while other regions remain unaffected by mixing.

These mixing scenarios have different influences on droplet size distributions. For homogeneous mixing, the mean drop radius, $\bar{r}$, is shifted to smaller sizes as all drops experience the same subsaturation. For inhomogeneous mixing, $\bar{r}$ is maintained but N_d is reduced by the number of evaporated droplets. The reduction in N_d means that the mixed air parcel may subsequently experience higher supersaturations compared to parcels that have not experienced mixing, and produce larger droplets than those associated with adiabatic growth. Entrainment of unactivated nuclei is more likely to result in activation of smaller particles. Both processes suggest that inhomogeneous mixing will generate broader droplet spectra than those resulting from homogeneous mixing.

The relevant number for examining these processes is the Damköhler number, D_a, defined as the ratio between the timescale typical for the mixing process τ_m, and a timescale for the reaction of the droplets to their new (subsaturated) thermodynamic environment τ_e. For a given air parcel with typical dimension l and energy dissipation rate ε describing the turbulence intensity, the time τ_m for complete mixing of this parcel is given by classical turbulence theory, $\tau_m = (l^2/\varepsilon)^{1/3}$. Although it is quite straightforward to estimate ε from cloud measurements, it is not clear what the governing length scales, l, are. In fact, l will likely change during the course of a cloud's life cycle.

Definition of an appropriate timescale for reaction (evaporation) is also not straightforward. A typical timescale can be defined as the evaporation time τ_e of a droplet with mean radius $\bar{r}$ in an environment with ambient subsaturation $S - 1$ and can be derived from $r\frac{dr}{dt} \propto \frac{S-1}{r}$ to yield $\tau_e \propto \frac{\bar{r}^2}{|S-1|}$. However, as evaporation proceeds, the humidity increases, and if saturation is reached before all droplets have evaporated, then a phase relaxation timescale $\tau_p \propto (\int rn(r)dr)^{-1} \sim (N_d\bar{r})^{-1}$ is more appropriate. It has been proposed that one use the minimum value of τ_e and τ_p to avoid unrealistically large values of τ_p in strongly subsaturated regions or very large values of τ_e close to saturation. Although τ_p and τ_e appear to have very different dependences on $\bar{r}$($\bar{r}^{-1}$ and $\bar{r}^{-2}$, respectively), calculation of τ_e and τ_p for drop populations with the same LWC results in similar values for both (since $N_d \sim \bar{r}^{-3}$ for constant LWC).

A decrease in the evaporation timescale will lead to a shift in the mixing process towards the inhomogeneous extreme (*ceteris paribus*) with attendant reduction in N_d, broadening of the droplet size distribution, and an increased chance of producing larger droplets. This behavior tends to reduce the albedo response to an increase in aerosol number concentration. Furthermore, the production of a few large droplets can initiate a more effective coalescence process which could lead to the onset of precipitation and counter the conventional higher-order effects as posed by Albrecht (1989).

The suggested linkage between τ_e and τ_m raises the question of how D_a (= τ_m/τ_e) might respond to an aerosol perturbation. This is represented schematically in Figure 14.1. An aerosol perturbation reduces τ_e. The shorter τ_e will likely generate stronger mixing as a result of a more efficient evaporation process. How this affects τ_m and D_a will depend on how ε and/or the characteristic size of the entrained blob of air l are affected by evaporation (vertical arrows). Large eddy simulations of non-precipitating cumulus (Xue and Feingold 2006; Jiang et al. 2006) have shown that turbulence kinetic energy increases in response to increasing aerosol because enhanced evaporation rates generate more negative buoyancy and stronger horizontal buoyancy gradients, which generates stronger vorticity. This suggests a shorter τ_m (downward arrow) and modulation of increases in D_a resulting from reduced τ_e.

Increase in Cloud Lifetime

There is no observational support for the claim that an increase in aerosol increases cloud lifetime. In stratiform clouds, such as stratocumulus, it is impractical to define a cloud lifetime since the existence of a cloud for many hours does not mean that individual parcels of cloudy air exist for many hours; rather,

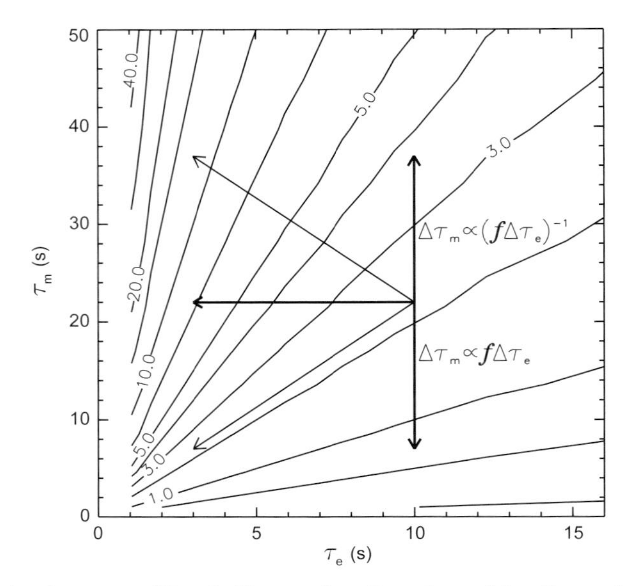

Figure 14.1 Contours of Damköhler number D_a and possible effects of aerosol on D_a. Calculations are based on modified gamma drop size distributions for a range of N_d and fixed breadth parameter. An aerosol perturbation of 300 cm^{-3} versus 50 cm^{-3} results in a reduction in τ_e (left arrow). The end point in D_a space (thin solid arrows) depends on whether τ_m increases (up arrow) or decreases (down arrow) as a result of the perturbation.

these parcels cycle in and out of clouds, and cloud elements constantly form and decay.

In the case of convective cumulus clouds, it is more straightforward to calculate the lifetime of a cloud cell. Radar has been used to track the lifetime of precipitating cells; however, this is very different from the lifetime of the cloud. Anecdotal evidence of clouds in very clean conditions precipitating almost as soon as they form has been noted, but we are unaware of any systematic observations of aerosol effects on cloud lifetime. Modeling work has shown no effect of aerosol perturbations on the lifetime of populations of cumulus clouds (Jiang et al. 2006). There are also indications that aerosol effects may depend on the size of the perturbed clouds and that small clouds may respond differently from large clouds.

The "lifetime" concept is one that climate modelers have sustained because it is relatively easy to simulate. In a climate model, study of the "lifetime effect" is equivalent to changing the rate at which cloud water converts to rain and has little to do with cloud lifetime. It does, however, artificially force water to rain out earlier or later on timescales on the order of half an hour (the typical timestep of a global climate model). Since this is a significant fraction of the lifetime of a real cloud, lifetime studies in climate models are ill-posed because they attempt to simulate many unresolved processes.

The "Semi-direct" Effect

The so-called "semi-direct" effect is a poignant example of how aerosol might affect clouds "indirectly"[3] by influencing cloud dynamics over a range of spatial scales.[4] Light-absorbing aerosol generate local heating, and their presence at the appropriate concentrations may modify the atmospheric stability and suppress vertical motion and cloud formation (Hansen et al. 1997). However, in general, absorbing aerosol layers may either stabilize or destabilize the boundary layer depending on where the aerosol is located, so that this factor alone may not explain observed reductions in cloudiness.

We expand our definition of the semi-direct effect to include the fact that the presence of aerosol, particularly light-absorbing aerosol, is very effective at reducing the downwelling solar radiation at the surface. Therefore, balance in the net surface radiation requires that the upwelling surface sensible and latent heat fluxes are reduced. Over land, where surface heating is a primary driver for convective clouds, this can have a significant impact on cloud fraction and cloud depth. This process links aerosol–cloud interactions to land surface type, another large-scale control that cannot be ignored. Together these mechanisms provide a means for aerosol to reduce the cloud fraction of surface-forced clouds (Koren et al. 2004; Feingold et al. 2005).

Cloud Effects on Aerosol

Clouds are not only affected by aerosol, they also exert significant effects on aerosol. These effects, sometimes lumped into a category of "cloud processing of aerosol," comprise a number of different processes such as *washout* (the removal of aerosol by rain falling to the surface), *convective redistribution* (the vertical transport of aerosol by clouds), *coalescence processing* (modification in the number and size of aerosol particles resulting from repeated drop coalescence events), *chemical processing* (the formation of nonvolatile mass attributable to aqueous chemical reactions, discussed briefly below), and *new particle formation around clouds* (addressed in somewhat greater detail). Since most cloud droplets evaporate and removal by precipitation is a relatively rare occurrence, aerosol processing by clouds is a major factor in the aerosol particle life history.

In terms of *chemical processing*, more than half of global sulfate is produced in clouds via aqueous reactions (e.g., Scott and Hobbs 1967); clouds may also be a significant source of secondary organic aerosol (Ervens et al. 2008). Aqueous production of aerosol is manifested in a bimodal size distribution: the first (smaller) mode comprises unactivated aerosol particles whereas

---

[3] We use "indirectly" here to convey that we are not referring to one of the classical aerosol indirect effects.

[4] Other examples include the self-organization of cloud structures at the mesoscale.

the second is made up of activated particles, upon which additional nonvolatile mass has formed. The implications range from significant increases in light scattering to modification in activation in subsequent cloud cycles.

New particle formation events can produce number concentrations that exceed background concentrations by orders of magnitude (Hegg et al. 1990; Perry et al. 1994; Clarke et al. 1998). These freshly formed particles can grow by condensation to Aitken mode particles (20 nm < diameter < 100 nm) over the course of many hours, and thus serve as cloud condensation nuclei (CCN) or influence the radiative properties directly.

Most explanations of increased ultrafine number concentrations are based on binary homogeneous nucleation of sulfuric acid (H_2SO_4) and water vapor (H_2O); however, ternary nucleation ($H_2SO_4 - NH_3 - H_2O$) or H_2SO_4 ion clusters may also play a role. For all of these mechanisms, an elevated concentration of sulfuric acid is essential, and the high relative humidity in the vicinity of clouds promotes further particle formation. The sulfuric acid can be produced by oxidation of SO_2 with photochemically produced OH, the latter requiring high actinic fluxes. The sulfuric acid vapor has to be supersaturated before condensation can occur, so low temperatures and high humidity are ideal. In addition, clean (vis-à-vis aerosol) conditions are favored, otherwise gases will simply condense onto existing particles.

Cloud edges with high actinic fluxes, high relative humidity, and relatively low temperatures, combined with convection and therefore vertical transport of precursor gases, are thought to be an ideal environment for new particle formation. However, many details of these processes are still not well understood; discrepancies between observed and modeled nucleation rates exist, with models typically underestimating these rates significantly. Many open issues remain, including:

1. Ammonia is usually not measured in clouds and thus ternary nucleation is poorly understood.
2. What possible influence do ion clusters have in the vicinity of clouds?
3. What is the exact location of the nucleation process? Most airborne particle measurements have a spatial resolution of only 50–100 m. This is of the same order of magnitude as the detrainment zone itself, which has been identified as one favored region for new-particle formation.
4. Drop and/or ice particle shattering on inlets of fast-flying aircraft may explain the observed high number concentrations of small particles (Weber et al. 1998).

Reconstructing the Constructs and Possible Ways Forward

A Proposed New Construct for Warm Clouds

Current constructs of aerosol effects on clouds are useful provided that they can be related to observations that prove causality. We propose here a simple construct that may assist in addressing aerosol effects on precipitation. Consider an analog to the albedo susceptibility S_0', which we will term "*precipitation susceptibility*," R_0', defined as:

$$R_0' = -\frac{d \ln R}{d \ln N_d}. \tag{14.4}$$

R_0' expresses the relative change in precipitation R for a relative increase in N_d in warm clouds. The minus sign is applied so that a positive R_0' will reflect the conventional wisdom that for warm rain, R and N_d are negatively correlated. (Cold clouds, and the ensemble of results from a population of interacting clouds, may respond quite differently; e.g., Orville and Chen 1982.) To begin, we have performed some simple calculations with a cloud parcel model that includes warm microphysical processes (activation, condensation, collision–coalescence), solved using size-resolved methods, to test the usefulness of this approach. The parcel rises at constant velocity and neglects entrainment. It is run for a range of LWP ($50 < \text{LWP} < 1100$ g m^{-2}) and CCN concentrations (25 cm^{-3} < CCN < 500 cm^{-3}). Results suggest that R_0' is a function of LWP (Figure 14.2). There appear to be three regimes:

1. A low LWP regime where R_0' is small, primarily because low LWP limits the cloud's ability to generate precipitation.
2. A mid-LWP regime where R_0' increases steadily with increasing LWP. Here the precipitation process is not limited by LWP and addition of N_d tends to suppress R.
3. A high LWP regime where R_0' begins to decrease, most likely because there is sufficient condensed water for clouds to precipitate regardless of N_d.

These results should be considered qualitative and subjected to further scrutiny, given the limitations of the model (primarily the neglect of sedimentation and entrainment)—limitations that become increasingly severe as LWP increases. However, results from an ensemble of 500 trajectories from two LES simulations of stratocumulus show agreement in the low LWP regime (horizontal line in Figure 14.2).[5] More importantly, remarkable new results from CloudSat radar analyses (M. D. Lebsock and G. L. Stephens, pers. comm.) show that Figure 14.2 is qualitatively reasonable. Their data show that deeper,

[5] Our analysis of 500 trajectories derived from LES of ASTEX and FIRE cases (LWP ~ 100–200 g m^{-2}) yields a maximum rainrate of $R \sim \text{LWP}^{1.6}/N_d^{0.67}$, i.e., $R_0' = 0.67$.

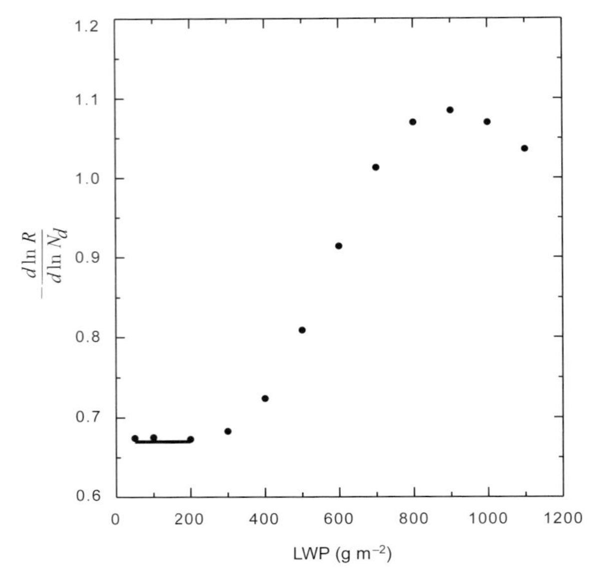

Figure 14.2 Precipitation susceptibility as a function of the liquid water path (LWP). Note the existence of three distinct regimes: (a) low R'_0 at low LWP; (b) steadily increasing R'_0 at intermediate LWP; (c) decrease in R'_0 at high LWP. The solid horizontal line at $R'_0 = 0.67$ is derived from analysis of large eddy simulations trajectories for ASTEX and FIRE-I stratocumulus modeling studies.

warm clouds tend to exhibit a stronger reduction in rain with increasing aerosol than do shallow clouds.

This seems to be a construct worth pursuing, regardless of the functional form of Equation 14.4, since it provides guidance as to where efforts might be invested in exploring the sensitivity of precipitation to aerosol perturbations. Other parameters which should be considered include aerosol size distribution parameters and perhaps dynamic parameters such as turbulence kinetic energy and/or cloud-top heights (to the extent that the latter are not reflected in LWP). In cold convective clouds, one should include total condensate and perhaps even separate terms for ice and water content.

Equation 14.4 is amenable to testing with both *in-situ* and remote-sensing (satellite, as described above, and surface-based) observations. Its logarithmic form should alleviate requirements of accurate precipitation and drop concentration measurements. From aircraft, ACE-2 and DYCOMS-II field experiments have already derived expressions for cloud base precipitation as a function of cloud depth H and N_d of the form $R \sim H^3/N_d$ (e.g., van Zanten et al. 2005) or $R \sim \text{LWP}^{1.5}/N_d$ (for linearly stratified clouds), consistent with $R'_0 = 1$. Moreover, these relationships have been tied to scavenging rates (Wood 2006), providing further motivation for their study. From space, microwave radiometers and radar provide measures of precipitation, as well as LWP, and retrievals of column-integrated N_d can be derived under various assumptions. From the surface, R, LWP, and N_d can be derived using similar remote-sensing techniques.

Self-Regulation Mechanisms and Non-Monotonic Behavior in Aerosol–Warm Cloud Systems

The previous discussion was peppered with examples of multiple feedbacks, some of which work in unison and some that counter one another. We outline a number of these below and consider the implications. Note that these effects are somewhat artificially separated in this listing and that more complex interactions are also possible.

1. Aerosol effects on LWP and cloud fraction: An increase in aerosol simultaneously acts to slow collision–coalescence and to enhance evaporation and reduce drop fall velocities. Xue et al. (2008) consider the possibility of non-monotonic responses of LWP to increasing aerosol perturbations in cumulus and stratocumulus clouds and the existence of two regimes: (a) at low aerosol concentrations, aerosol and LWP are positively correlated; (b) at high aerosol concentrations, aerosol and LWP are negatively correlated. The suggestion is that there is a limit to the degree of LWP buildup resulting from an aerosol perturbation. This may have some support in the satellite study of Han et al. (2002), who showed that ~1/3 of cases followed the first regime, ~1/3 followed the second regime, and ~1/3 showed no clear response.
2. Albedo susceptibility: Associated with the above regimes are corresponding albedo susceptibility responses. Clouds with low A have high S_0'; an increase in aerosol (and associated increase in N_d) produces monotonic increases in A but decreases in S_0'. However, if one relaxes the assumption of constant LWP (Equation 14.3), it is clear that macroscale cloud properties may significantly affect susceptibility. Although albedo is likely to respond monotonically with increasing aerosol, S_0'' may not, because of LWP and cloud fraction responses (Zuidema et al. 2008).[6]
3. The "semi-direct" effect: Under relatively clean conditions, an increase in aerosol containing an absorbing component will generate brighter clouds because many of these particles are hygroscopic enough to act as CCN. With further increases in aerosol, however, the absorption will be strong enough to suppress surface fluxes and stabilize the atmosphere, with concomitant reduction in τ_c, LWP, and cloud fraction (Jiang and Feingold 2006).

[6] Note that to the extent that the R and LWP responses to N_d are related, there may be an interesting link between R_0' and S_0''.

4. Deep convective clouds: In our modeling studies (S. Tessendorf and G. Feingold, pers. comm.), we have found non-monotonic responses of precipitation to aerosol in single deep convective, mixed-phase clouds. Warm clouds tend to have an active warm-rain process that facilitates the production of surface precipitation. On the other hand, polluted clouds enjoy the benefits of a higher freezing level and invigoration due to latent heat of freezing, resulting in an increase in the mass of condensed or frozen hydrometeors, which assists in generating precipitation via accretion processes. Clouds that are affected by moderate levels of pollution seem not to enjoy either of these benefits and may precipitate less than clean or polluted clouds.
5. Aerosol effects on clouds in the longwave: Thin clouds in the wintertime high latitudes experience an aerosol effect in the longwave that has not yet been discussed. Clouds with LWP <25 g m^{-2} act as gray bodies with emissivity $\varepsilon<1$ (e.g., Garrett and Zhao 2006). An aerosol perturbation to these clouds will increase the downwelling longwave radiation and warm the surface. An analog to the shortwave albedo susceptibility developed by Garrett ($S_{LW}=-(1-\varepsilon)\ \ln(1-\varepsilon)/3N_d$) shows that in the longwave, the maximum susceptibility is at $\varepsilon = 0.6$ and low N_d. Thicker clouds act as black bodies ($\varepsilon=1$) whereas thinner clouds are simply too thin to have much of an effect.

These examples are all suggestive of subtleties that preclude simple assessment of aerosol perturbations to clouds as expressed by the albedo and lifetime effects. They also point to the possibility that in terms of both radiative forcing and precipitation, neither the cleanest nor the most polluted clouds are particularly susceptible to perturbations, but that some intermediate regime may be where efforts should be expended. This "intermediate regime" will depend on the problem at hand and the meteorological context, so that we hesitate to venture further. An interesting case in point is that of giant nuclei, which are known to expedite the collision–coalescence process. Their effect has been shown to be negligible in clean clouds which already have an active warm-rain process, progressively more significant (in a relative sense) in polluted clouds, but most important (in an absolute sense) in moderately polluted conditions.

It is important to distinguish between absolute effects and relative measures such as susceptibility (either albedo or precipitation). For example, the fact that albedo susceptibility is low does not imply that cloud radiative forcing is low, rather that aerosol perturbations will have little impact. One illustration of this is the stark difference in reflectance between relatively polluted, non-precipitating stratocumulus (high albedo, low S_0'), and cleaner, precipitating stratocumulus (low albedo, high S_0') which sometimes manifest themselves as closed cell convection and pockets of (precipitating) open cells, respectively (e.g., Xue et al. 2008).

Statistical Methods

The complexity of the aerosol–cloud–precipitation system is such that it may render the reductionist approach impractical. Statistical methods may need to be employed alongside *in-situ* and remote observations and numerical modeling. These would assess the primary factors driving aerosol effects on radiative forcing and precipitation using methodologies such as principal components analysis, factor separation, data assimilation, and others.

Summary

In conclusion, we provide a synthesis of the implications for aerosol effects on clouds and precipitation and offer a new framework in which the effects of aerosol on precipitation can be considered.

1. *Aerosol number concentration and size:* The available body of evidence suggests a dominant role for aerosol number concentration and size in determining drop concentrations in warm clouds. Composition is of secondary importance, except perhaps in very polluted conditions and for weak updrafts. Composition effects that are important include the mixing state of the particles (e.g., an externally mixed, hydrophobic mode) and the mass accommodation coefficient relevant to atmospheric conditions.

   For ice clouds, it is worth noting that characteristics of the total aerosol population, of which ice nuclei are a subset, do not appear in common ice nucleation parameterizations. This may explain the very wide range of predicted ice crystal concentrations.
2. *The coupled aerosol–cloud–dynamical system: Self regulation in response to aerosol perturbations?* Clouds respond to aerosol in a manner that is strongly dependent on meteorological context. Clouds are dynamic entities that are influenced by meteorological factors which control convection. Aerosol particles, a necessary ingredient for cloud droplet formation, are not the primary driver for clouds. Rather, they can modify cloud microphysics in sometimes subtle ways to generate feedbacks that may amplify or dampen their influence. These feedbacks have the potential to make aerosol effects on clouds of major climatic and hydrological importance. The extent to which this is true is the challenge of this community.

   The simple constructs, now decades old, referred to as the "albedo" and "lifetime" effects, have been shown to be of somewhat limited use. This is especially true of the lifetime effect, which requires separation of meteorological and aerosol influences in a tightly coupled system. Even for a given meteorological context, the response of a cloud to an aerosol perturbation may not be monotonic. Based on modeling studies, we suggest the possible existence of two regimes: a cleaner regime where aerosol suppresses precipitation and results in increased cloudiness, and a more polluted regime where aerosol reduces cloud fraction because of more efficient evaporation. A similar non-monotonic response has been suggested for cloud response to absorbing aerosol: under clean conditions, clouds become brighter with increasing aerosol, but above some threshold aerosol absorption dominates, convection is suppressed, and cloud optical depth is reduced.

We also raise the possibility that aerosol may influence the nature of entrainment-mixing through effects on evaporation timescales and feedbacks of the aforementioned to the timescales of mixing (Figure 14.1). As yet, no firm conclusions have been drawn.

3. *A proposed new construct for investigating aerosol effects on precipitation*: We propose investigating the utility of the "precipitation susceptibility,"

$$R_0' = \frac{-d \ln R}{d \ln N_d}, \tag{14.5}$$

where R is rainrate, an analog of the albedo susceptibility. Some very simple calculations suggest that clouds in some intermediate LWP regime exhibit increasing susceptiblity to aerosol perturbations (Figure 14.2). There may be a maximum in R_0' at high LWP, beyond which clouds tend to precipitate regardless. There is new observational support for this hypothesis from spaceborne cloud radar. It is proposed that we identify parameters other than LWP that affect R_0' through statistical analyses of large observational datasets (satellite and surface remote sensing, *in-situ*) and model outputs. More generally, an increased use of statistical analyses of this kind may be required alongside the reductionist approach to expedite progress in the field.

Acknowledgments

The authors wish to acknowledge the support from the Ernst Strüngmann Forum. GF acknowledges support from NOAA's Climate Goal.

References

Ackerman, A. S., M. P. Kirkpatrick, D. E. Stevens, and O. B. Toon. 2004. The impact of humidity above stratiform clouds on indirect aerosol climate forcing. *Nature* **432**:1014–1017.

Aitken, J. 1880. On dust, fogs, and clouds. *Proc. Roy. Soc. Edinb.* **9**:14–18.

Aitken, J. 1923. Collected scientific papers. London: Cambridge Univ. Press.

Albrecht, B. A. 1989. Aerosols, cloud microphysics, and fractional cloudiness. *Science* **245**:1227–1230.

Boers, R., and R. M. Mitchell. 1994. Absorption feedback in stratocumulus clouds: Influence on cloud top albedo. *Tellus A* **46**:229–241.

Brenguier, J.-L., H. Pawlowska, and L. Schuller. 2003. Cloud microphysical and radiative properties for parameterization and satellite monitoring of the indirect effect of aerosol on climate. *J. Geophys. Res.* **108**:8632.

Charlson, R. J., A. S. Ackerman, F. A.-M. Bender, T. L. Anderson, and Z. Liu. 2007. On the climate forcing consequences of the albedo continuum between cloudy and clear air. *Tellus B* **59**:715–727.

Clarke, A. D., J. L. Varner, F. Eisele et al. 1998. Particle production in the remote marine atmosphere: Cloud outflow and subsidence during ACE 1. *J. Geophys. Res.* **103(D13)**:16,397–16,409.

Conant, W. C., T. M. VanReken, T. A. Rissman et al. 2004. Aerosol, cloud drop concentration closure in warm cumulus. *J. Geophys. Res.* **109(D13)**:D13204.

Coulier, P. J. 1875. Note sur une nouvelle propriete de l'air, J. de Pharmacie et de Chirnie, Paris, Ser. 4, **22(1)**:165–172 and **22(2)**:254–255.

Ervens, B., A. G. Carlton, B. J. Turpin et al. 2008. Secondary organic aerosol yields from cloud-processing of isoprene oxidation products. *Geophys. Res. Lett.* **35**:L02816.

Ervens, B., G. Feingold, and S. M. Kreidenweis. 2005. The influence of water-soluble organic carbon on cloud drop number concentration. *J. Geophys. Res.* **110**:D18211.

Feingold, G., H. Jiang, and J. Y. Harrington. 2005. On smoke suppression of clouds in Amazonia. *Geophys. Res. Lett.* **32(2)**:L02804.

Garrett, T. J., and C. Zhao. 2006. Increased Arctic cloud longwave emissivity associated with pollution from mid-latitudes. *Nature* **440**:787–789.

Gunn, R. and B. B. Phillips. 1957. An experimental investigation of the effect of air pollution on the initiation of rain. *J. Meteor.* **14**:272–280.

Han, Q., W. B. Rossow, J. Zeng, and R. Welch. 2002. Three different behaviors of liquid water path of water clouds in aerosol–cloud interactions. *J. Atmos. Sci.* **59**:726–735.

Hansen, J., M. Sato, and R. Ruedy. 1997. Radiative forcing and climate response, *J. Geophys. Res.* **102**:2663–2668.

Hegg, D. A., L. F. Radke, and P. V. Hobbs. 1990. Particle production associated with marine clouds. *J. Geophys. Res.* **95**:13,917–13,926.

Houghton, H. G. 1938. Problems connected with the condensation and precipitation processes in the atmosphere. *Bull. Amer. Meteor. Soc.* **19**:152–159.

Howell, W. E. 1949. The growth of cloud drops in uniformly cooled air. *J. Meteor.* **6**:134–149.

IPCC. 2007. Climate Change 2007: Summary for Policymakers. The Physical Science Basis. Contribution of Working Group I to the Fourth Assessment Report of the Intergovernmental Panel on Climate Change, edited by S. Solomon, D. Qin, M. Manning et al. New York: Cambridge Univ. Press.

Jiang, H., and G. Feingold. 2006. Effect of aerosol on warm convective clouds: Aerosol–cloud–surface flux feedbacks in a new coupled large eddy model. *J. Geophys. Res.* **111**:D01202.

Jiang, H., H. Xue, A. Teller, G. Feingold, and Z. Levin. 2006. Aerosol effects on the lifetime of shallow cumulus. *Geophys. Res. Lett.* **33**:L14806.

Köhler, H. 1936. The nucleus in and the growth of hygroscopic droplets. *Trans. Faraday Soc.* **32**:1152–1161.

Koren, I., Y. J. Kaufman, L. A. Remer, and J. V. Martins. 2004. Measurement of the effect of Amazon smoke on inhibition of cloud formation. *Science* **303**:1342–1345.

Koren, I., L. A. Remer, Y. J. Kaufman, Y. Rudich, and J. V. Martins. 2007. On the twilight zone between clouds and aerosols. *Geophys. Res. Lett.* **34**:L08805.

Levin, Z., and W. R. Cotton. 2007. Aerosol pollution impact on precipitation: A scientific review. WMO/IUGG Intl. Aerosol Precipitation Science Assessment Group (IAPSAG), pp 485. Geneva: WMO.

O'Dowd, C. D., J. A. Lowe, M. H. Smith, and A. D. Kaye. 1999. The relative importance of non-sea-salt sulphate and sea-salt aerosol to the marine cloud condensation nuclei population: An improved multi-component aerosol–cloud droplet parametrization. *Q. J. Roy. Meteor. Soc.* **125**:1295–1313.

Orville, H., and J.-M. Chen. 1982. Effects of cloud seeding, latent heat of fusion, and condensate loading on cloud dynamics and precipitation evolution: a numerical study. *J. Atmos. Sci.* **39**:2807–2827.

Perry, K. D., and P. V. Hobbs. 1994. Further evidence for particle nucleation in clear air adjacent to marine cumulus clouds. *J. Geophys. Res.* **99(D11)**:22,803–22,818.

Platnick, S., and S. Twomey. 1994. Determining the susceptibility of cloud albedo to changes in droplet concentration with the Advanced Very High Resolution Radiometer. *J. Appl. Meteor.* **33**:334–347.

Ruehl, C. R., P. Y. Chuang, and A. Nenes. 2008. How quickly do cloud droplets form on atmospheric particles? *Atmos. Chem. Phys.* **8**:1043–1055.

Scott, W. D., and P. V. Hobbs. 1967. The formation of sulfate in water droplets. *J. Atmos. Sci.* **24**:54–57.

Twomey, S. 1959. The nuclei of natural cloud formation. Part II: The supersaturation in natural clouds and the variation of cloud droplet concentration. *Geofis. Pura Appl.* **43**:243–249.

Twomey, S. 1974. Pollution and the planetary albedo. *Atmos. Environ.* **8**:1251–1256.

Van Zanten, M. C., B. Stevens, G. Vali, and D. H. Lenschow. 2005. Observations of drizzle in nocturnal marine stratocumulus. *J. Atmos. Sci.* **62**:88–106.

Wang, S., Q. Wang, and G. Feingold. 2003. Turbulence, condensation and liquid water transport in numerically simulated nonprecipitating stratocumulus clouds. *J. Atmos. Sci.* **60**:262–278.

Warner, J. 1968. A reduction in rainfall associated with smoke from sugar-cane fires: An inadvertent weather modification? *J. Appl. Meteor.* **7**:247–251.

Weber, R. J., A. D. Clarke, M. Litchy et al. 1998. Spurious aerosol measurements when sampling from aircraft in the vicinity of clouds. *J. Geophys. Res.* **103(D21)**:28,337–28,346.

Wilson, C. T. R. 1987. Condensation of water vapour in the presence of dust-free air and other gases. *Phil. Trans. Roy. Soc. Lond. A* **189**:265–307.

Wood, R. 2006. The rate of loss of cloud droplets by coalescence in warm clouds. *J.Geophys. Res.* **111**:D21205.

Xue, H., and G. Feingold. 2006. Large eddy simulations of trade wind cumuli: investigation of aerosol indirect effects. *J. Atmos. Sci.* **63**:1605–1622.

Xue, H., G. Feingold, and B. Stevens. 2008. Aerosol effects on clouds, precipitation, and the organization of shallow cumulus convection. *J. Atmos. Sci.* **65**:392–406.

Zuidema, P., H. Xue, and G. Feingold. 2008. Shortwave radiative impacts from aerosol effects on marine shallow cumuli. *J. Atmos. Sci.* **65(6)**:1979.

15

Weather and Climate Engineering

William R. Cotton

Colorado State University, Department of Atmospheric Science, Fort Collins, CO, U.S.A

Abstract

In this chapter I present an overview of the concepts and status of the science of weather and climate engineering. I begin by discussing the concepts of seeding clouds through glaciogenic and hygroscopic seeding. I review the status of research on these concepts for increasing rainfall, decreasing hail damage, and reducing hurricane intensity. Thereafter I present an overview of the concepts for climate engineering to counter greenhouse warming. These include seeding in the stratosphere with sulfate-producing gases and aerosols, and carbonaceous aerosols. I also consider hygroscopic seeding of marine stratocumulus boundary layer clouds to enhance their albedo and cause a cooling effect. Also considered is seeding mid-level stratus clouds to enhance their albedo during the day and increasing outgoing longwave radiation during the night time. Cirrus clouds present a major obstacle to climate modification owing to their widespread global coverage and their tendency to warm the surface, thus reinforcing greenhouse warming. Speculations on the seeding of carbonaceous aerosols to clear cirrus through a semi-direct effect are presented. Most of the proposed concepts require a great deal of research to quantify their impacts and potential adverse consequences.

I include a long list of the reasons as to why we should *not* apply climate engineering. Despite these, I anticipate that if we find ourselves in a true climate crisis, politicians will call for climate engineering measures in an attempt to alter adverse climate trends. If this should ever be the case, let us be sure that we do so with the most advanced level of knowledge of the climate system and the full consequences of our actions.

Introduction

I have been tasked with the job of providing a position paper on weather modification and geoengineering. I have titled this paper "Weather and Climate Engineering" and, in doing so, have discarded the normally used term "weather modification" so that it relates better to climate engineering. Moreover, I have chosen the term "climate engineering" as I am going to focus mainly on hypotheses for engineering changes in the Earth's albedo or longwave radiation rather than discuss policies for reducing carbon emissions, sequestration of carbon, and so forth. At the outset I wish to state that this is written from the perspective of a scientist who is naturally skeptical about the many claims of how humans can influence weather and climate. This philosophy of what I will call "healthy skepticism" grew out of my graduate training in weather modification research, where there were many claims of great success in modifying the weather. After over fifty years of research, there is still no strong physical and statistical evidence that these early claims were ever realized.

I have carried that skepticism into the area of climate change where there seems to be a consensus among the scientific community (IPCC 2007) that human production of CO_2 is causing a global warming trend. I do not deny that the evidence is very strong that we are in a period of global climate warming and that adding CO_2 to the atmosphere will contribute to warming. However, I remain skeptical that current global warming trends are due solely to human causes and that other causes of natural climate variability are not the major contributing factors. This perspective on weather modification and climate change is discussed by Cotton and Pielke (2007).

Thus it is with this philosophical view that I venture into the domain of weather and climate engineering. I do so in spite of my belief that we should not be further "mucking" up our environment but also recognize that we may at some point in the future have to resort to modifying weather and climate in order to provide sufficient water resources and a climate capable of sustaining the population of humans on this planet.

Weather Engineering

Over the last fifty years, weather engineering or weather modification has mainly focused on cloud seeding. There have been a few hypotheses advanced to modify weather or regional climate that do not involve cloud seeding: Anthes (1984) hypothesized that trees and crops could be planted in mesoscale patches to provide optimum forcing for regional weather circulations and convective precipitation; Black and Tarmy (1963) suggested using waste asphalt to make strips of low-albedo surfaces to enhance mesoscale circulations and rainfall. By and large, however, most activity in weather engineering has been associated with cloud seeding.

During this time, deliberate cloud seeding has been pursued, with the goal of increasing precipitation through the injection of specific types of particles into clouds. Efforts to understand the processes involved have led to a significant body of knowledge about clouds and about the effects of the seeding aerosol. A number of projects focused on the statistical evaluation of whether a seeding effect can be distinguished in the presence of considerable "natural variability."

In this chapter, I review briefly the fundamental concepts of cloud seeding. It is not my intent to provide a complete assessment of the current status of cloud seeding research. For this, I direct the reader to more comprehensive weather modification assessments in NRC (2003), Cotton and Pielke (2007), Silverman (2001, 2003), Garstang et al. (2005), and Levin and Cotton (2008).

Deliberate cloud seeding experiments can be divided into two broad categories: *glaciogenic* and *hygroscopic* seeding. Glaciogenic seeding occurs when ice-producing materials (e.g., dry ice or solid CO_2, silver iodide, liquid propane) are injected into a supercooled cloud to stimulate precipitation via the ice particle mechanism. The underlying hypothesis for glaciogenic seeding is that there is commonly a deficiency of natural ice nuclei and therefore insufficient ice particles for the cloud to produce precipitation as efficiently as it would in the absence of seeding.

Hygroscopic seeding, by contrast, was generally used in the past to enhance rain from warm clouds (for a review of early hygroscopic seeding research, see Cotton 1982). More recently, however, it has been applied to mixed-phase clouds as well. The goal of hygroscopic seeding is to increase the concentration of *collector drops* that can grow efficiently into raindrops by collecting smaller droplets and by enhancing the formation of frozen raindrops and graupel particles. This is done by injecting into a cloud (generally at cloud base) large or giant hygroscopic particles (e.g., salt powders) that can grow rapidly through the condensation of water vapor to produce collector drops.

Glaciogenic Cloud Seeding

Glaciogenic cloud seeding can be subdivided into *static cloud* and *dynamic cloud* seeding. Static cloud seeding refers to the use of glaciogenic materials to modify the microstructures of supercooled clouds and precipitation. Many hundreds of such experiments have been conducted over the past fifty years. Some have been operational cloud seeding experiments (many of which are still being carried out around the world), which rarely provide sufficient information to form a decision as to whether or not they have modified either clouds or precipitation. Others have been well-designed scientific experiments that have yielded extensive measurements and modeling studies to permit an assessment of whether the artificial seeding modified cloud structures and, if the seeding was randomized, what effects the seeding had on precipitation. Although there still is some debate about what constitutes firm "proof" (see NRC 2003; Garstang et al. 2005) that seeding affects precipitation, it is generally required that *both* strong physical evidence of appropriate modifications to cloud structures and highly significant statistical evidence be obtained.

Glaciogenic Seeding of Cumulus Clouds

The static seeding concept has been applied to supercooled cumulus clouds and tested in a variety of regions. Two landmark experiments (Israeli I and Israeli II), carried out in Israel, have been described in peer-reviewed literature. These experiments were carried out by researchers at the Hebrew University of Jerusalem, hereafter the experimenters. These two experiments were the foundation for the general view that under appropriate conditions, cloud seeding increases precipitation (e.g., NRC 1973; Sax et al. 1975; Tukey et al. 1978a, b; Simpson 1979; Dennis 1980; Mason 1980, 1982; Kerr 1982; Silverman 1986; Braham 1986; Cotton 1986a, b; Cotton and Pielke 1992; 2007; Young 1993).

Nonetheless, a reanalysis of those experiments by Rangno and Hobbs (1993) suggests that the appearance of seeding-caused increases in rainfall in the Israel I experiment was due to "lucky draws" or a Type I statistical error. Furthermore, they argue that during Israel II naturally heavy rainfall over a wide region encompassing the north target area gave the appearance that seeding caused increases in rainfall over the north target area. At the same time, lower natural rainfall in the region encompassing the south target area gave the appearance that seeding decreased rainfall over that target area. This speculation, however, could not explain the positive effect when the north target area was evaluated against the north upwind control area. Details of this controversy can be found in the March 1997 issue of the *Journal of Applied Meteorology* (Rosenfeld 1997; Rangno and Hobbs 1997a; Dennis and Orville 1997; Rangno and Hobbs 1997b; Woodley 1997; Rangno and Hobbs 1997c; Ben-Zvi 1997; Rangno and Hobbs 1997d). Some of these responses clarified issues; others left many questions unanswered.

Another noteworthy experiment was carried out in the high plains of the United States (High Plains Experiment or HIPLEX-1; Smith et al. 1984). Analysis of HIPLEX-1 revealed the important result that after just five minutes, there was no statistically significant difference in the precipitation between seeded and non-seeded clouds (Mielke et al. 1984). Cooper and Lawson (1984) found that while high ice crystal concentrations were produced in the clouds by seeding, the cloud droplet region, where the crystals formed, evaporated too quickly for the incipient artificially produced ice crystals to grow to appreciable sizes. Instead, they formed low-density, unrimed aggregates that had the water equivalent of only drizzle drops, which were too small to reach the ground before evaporating. Schemenauer and Tsonis (1985) affirmed the findings of Cooper and Lawson in a reanalysis of the HIPLEX data, emphasizing their own earlier findings (Isaac et al. 1982) that cloud lifetimes were too short in the HIPLEX domain for seeding to have been effective in the clouds targeted for seeding (i.e., those with tops warmer than –12°C). Although the experiment failed to demonstrate statistically all the hypothesized steps, the problems could be traced to the physical short lifetimes of the clouds (Cooper and Lawson 1984; Schemenauer and Tsonis 1985).

Glaciogenic Seeding Winter Orographic Clouds

The static mode of cloud seeding has also been applied to orographic clouds. Precipitation enhancement of orographic clouds by cloud seeding has several advantages over cumulus clouds. The clouds are persistent features that produce precipitation even in the absence of large-scale meteorological disturbances. Much of the precipitation is spatially confined to high mountainous regions, thus making it easier to set up dense ground-based seeding and observational networks. Moreover, the "natural variability" of orographic clouds is less than cumulus clouds, thus making it easier to identify a "cause and effect."

The landmark randomized cloud seeding experiments conducted at Climax, near Fremont Pass, Colorado (referred to as *Climax I* and *Climax II*), reported by Grant and Mielke, suggested increases in precipitation of 50% and more on favorable days (e.g., Grant and Mielke 1967; Mielke et al. 1970, 1971). These results were widely viewed as demonstrating the efficacy of cloud seeding (e.g., NRC 1973; Sax et al. 1975; Tukey et al. 1978a, b), even by those most skeptical of cloud seeding claims (e.g., Mason 1980, 1982).

Nonetheless, Hobbs and Rangno (1979; Rangno and Hobbs 1987, 1993) questioned both the randomization techniques and quality of data collected during those experiments and concluded that the Climax II experiment failed to confirm that precipitation can be increased by cloud seeding in the Colorado Rockies. Even so, in their reanalysis, Rangno and Hobbs (1993) did show that precipitation increased by about 10% in the combined Climax I and II experiments.

Two other randomized orographic cloud seeding experiments, the Lake Almanor Experiment (Mooney and Lunn 1969) and the Bridger Range Experiment (BRE) as reported by Super and Heimbach (1983) and Super (1986), suggested positive results. These particular experiments, however, used high elevation AgI generators, which increase the chance that the AgI plumes get into the supercooled clouds. Moreover, both experiments provided physical measurements that support the statistical results (Super and Heimbach 1983, 1988).

Finally, Ryan and King (1997) reviewed over 14 cloud seeding experiments covering much of southeastern, western, and central Australia as well as the island of Tasmania. They concluded that static seeding over the plains of Australia is not effective. They argue that for orographic stratiform clouds, there is strong statistical evidence that cloud seeding increased rainfall, perhaps by as much as 30% over Tasmania, when cloud top temperatures are between –10° and –12°C in southwesterly airflow. The evidence that cloud seeding had similar effects in orographic clouds over the mainland of southeastern Australia is much weaker. Note that the Tasmanian experiment had both strong statistical and physical measurement components and thus meets or at least comes close to meeting the NRC (2003) criteria for scientific "proof." Benefit/cost analysis of the Tasmanian experiments suggests that seeding has a gain of about 13/1. This is viewed as a real gain to hydrologic energy production.

In summary, the "static" mode of cloud seeding has been shown to cause the expected alterations in cloud microstructure, including increased concentrations of ice crystals, reductions of supercooled liquid water content, and more rapid production of precipitation elements in both cumuli (Isaac et al. 1982; Cooper and Lawson 1984) and orographic clouds (Reynolds 1988; Super and Boe 1988; Super et al. 1988; Super and Heimbach 1988; Reynolds and Dennis 1986). The documentation of increases in precipitation on the ground due to static seeding of cumuli, however, has been far more elusive with the Israeli experiment (Gagin and Neumann 1981), providing the strongest evidence that static seeding of cold-based, continental cumuli can cause significant increases of precipitation on the ground. The evidence that orographic clouds can cause significant increases in snowpack is far more compelling, particularly in the more continental and cold-based orographic clouds (Mielke et al. 1981; Super and Heimbach 1988).

The most challenging obstacles to evaluating cloud seeding experiments to enhance precipitation are perhaps the inherent "natural variability" of precipitation in space and time as well as the inability to increase precipitation amounts to better than ~10%. The latter puts great demands on measuring accuracy and the duration of the experiments. The fact that the evidence is stronger that glaciogenic seeding increases precipitation in orographic clouds than cumulus clouds is probably a consequence of their lower "natural variability." This is also consistent with the findings of Levin and Cotton (2008), who show that the strongest evidence that aerosol pollution reduces precipitation is from orographic clouds. Again, the fact that orographic clouds exhibit a lower level of "natural variability" may be a major contributing factor.

Dynamic Glaciogenic Seeding

Thus far we have considered only static seeding, in which the principal thrust is to modify the microstructures of clouds generally for the purpose of enhancing precipitation. There is, however, another glaciogenic seeding hypothesis in which the cloud-scale dynamics of a cloud is enhanced by stimulating buoyancy and upward motions of air. This is referred to as *dynamic cloud seeding*. In principal, this can be done by glaciating convective clouds so that large quantities of latent heat are released by the freezing of copious liquid water to invigorate updrafts in the cloud. This can be particularly effective if, prior to seeding, the tops of the clouds are restricted by a shallow stable layer produced by a temperature inversion. In this case, the sudden release of a large quantity of latent heat might provide enough buoyancy to push the top of the cloud through the stable layer and into a region where the air is naturally unstable. The cloud might then rise to much greater heights than it would have done naturally. To some extent the distinction between static seeding and dynamic seeding is rather artificial. As has been shown in recent modeling studies of the effects of aerosol pollution on stratocumuli, small cumuli, and cumulonimbi, any modification of precipitation from these clouds contributes to nonlinear dynamical responses in the clouds and cloud systems (Levin and Cotton 2008).

In a series of randomized experiments carried out in Florida in 1968 and between 1970–1973 (called the Florida Area Cumulus Experiment or FACE), it was found that precipitation (measured by radar) from isolated cumulus clouds ~5 km in diameter, which were artificially seeded to induce explosive growth, was about twice that of the unseeded control clouds (e.g., Simpson and Woodley 1975; Woodley et al. 1982). The seeded clouds rained more than the control clouds since they were bigger and lasted longer, rather than because their rainfall rates were significantly greater.

In FACE II, the attempt was made to confirm and replicate the results of FACE I by going the additional step of specifying the manner in which clouds would respond to seeding based on what was understood to have been the response in FACE I. Although there were several suggestions of seeding effects on some clouds and some days (e.g., Woodley et al. 1983), the overall experiment officially failed to confirm the results of FACE I (Flueck et al. 1981; Nickerson 1979, 1981). In essence, the experiment succumbed to the high "natural variability" of these storms.

In recent years the dynamic seeding strategy has been applied in Thailand and West Texas. Results from exploratory dynamic seeding experiments over West Texas have been reported by Rosenfeld and Woodley (1989, 1993). Analysis of the seeding of 183 convective cells suggests that seeding increased the maximum height of the clouds by 7%, the areas of the cells by 43%, the durations by 36%, and the rain volumes of the cells by 130%. The results are encouraging, but such small increases in vertical development of the clouds are hardly consistent with earlier exploratory seeding experiments.

As a result of their experience in Texas, Rosenfeld and Woodley (1993) proposed an altered conceptual model of dynamic seeding in which explosive vertical development of seeded clouds is not emphasized. As pointed out by Silverman (2001), however, application of the revised hypothesis in Thailand (Woodley et al. 2003a, b) indicated rainfall enhancement, but the results did not reach statistical significance. Moreover, the enhanced downdraft presumably produced by it did not appear to be delayed (Woodley et al. 1999b).

In summary, the concept of dynamic seeding is a physically plausible hypothesis that offers the opportunity to increase rainfall by much larger amounts than simply enhancing the precipitation efficiency of a cloud. It is a much more complex hypothesis, however, requiring greater quantitative understanding of the behavior of cumulus clouds and their interaction with each other, with the boundary layer, and with larger-scale weather systems.

Hygroscopic Cloud Seeding

Hygroscopic seeding was mainly used in the past in warm clouds, where no ice is present. More recently, this type of seeding method has been tried in mixed-phase clouds. The aim in seeding warm clouds is to enhance drop growth by coalescence, thus improving the efficiency of rainfall formation. However, seeding mixed-phase clouds seems to affect both drop growth and ice formation, probably through the efficient formation of graupel particles. As the modeling studies of Cotton (1972), Murray and Koenig (1972), and Scott and Hobbs (1977) suggest, a cloud composed of larger supercooled drops is likely to glaciate much faster than one composed primarily of small drops. Appropriately sized salt particles, water droplets from sprays of either water or saline solution (Bowen 1952; Biswas and Dennis 1971; Cotton 1982; Murty et al. 2000; Silverman and Sukarnjanasat 2000), and hygroscopic flares (Mather et al. 1997; WMO 2000) have been used for seeding. Recent enthusiasm for the concept was motivated by Mather's (1991) study, which demonstrated the effects of paper-mill effluent on precipitation. Statistical results, observations, and modeling results provided some evidence that precipitation may be enhanced under certain conditions and with optimal seed drop size spectra (Farley and Chen 1975; Rokicki and Young 1978; Young 1996; Reisin et al. 1996; Yin et al. 2000a, b). Seeding experiments using hygroscopic flare particles provided statistical support for rainfall increases due to seeding based on single cloud analyses (Mather et al. 1997; Bigg 1997; WMO 2000; Silverman 2003). Model simulations suggest that the increase in rainfall amounts stems from the increase in graupel numbers and masses, which are generated by the increased concentrations of large drops (Yin et al. 2000a, b). Such increases could generate more rain, but it is not clear how these procedures can affect the clouds for such a length of time, as some of the measurements suggest (e.g., Silverman 2003).

In both South Africa (Mather et al. 1997) and Mexico (WMO 2000), hygroscopic flares have been applied to mixed-phase convective cloud systems in limited physical and statistical experiments. Aircraft microphysical measurements were made to verify some of the processes involved. Radar-measured 30 dBZ volumes, produced by the convective complexes, were tracked by automated software, and various storm and track properties were calculated. These two sets of experiments produced remarkably similar results in terms of the difference in radar-estimated rainfall between the seeded and non-seeded groups. The South African data have been reevaluated independently by Bigg (1997) and Silverman (2000); both concluded that there is statistically significant evidence of an increase in *radar-estimated* rainfall from seeded convective cloud systems.

Mather et al. (1997), Bigg (1997), and Silverman (2000) all allude to apparent dynamic effects of seeding clouds, manifest in the seeded cloud systems, being longer-lived. It was speculated that the relation between (a) the amount of precipitation (negative buoyancy due to the weight of condensed water) and evaporation, (b) the characteristics of the downdraft that is generated, and (c) the downdraft and the storm organization, evolution, and lifetime determines the dynamic effect of seeding on rainfall. Another factor not mentioned is the possible consequences of altered raindrop size distributions. If seeding shifts the raindrop size distribution to smaller raindrops, then greater sub-cloud evaporation would ensue, which would alter cold-pool dynamic effects. If seeding shifts the raindrop spectrum to larger drops, the opposite response would be expected (Yin et al. 2001).

It appears that continental convective storms are remarkably sensitive to changes in the cloud condensation nuclei (CCN) ingested at cloud base. For example, both the South African and Mexican experiments with hygroscopic flares show very strong signals in terms of increased storm lifetime in seeded storms, increases in reflectivity aloft, and increases in storm densities. Thus, these hygroscopic flare seeding experiments suggest that it is possible, under appropriate conditions, to produce large differences in cloud properties by injection of hygroscopic particles into cloud bases.

In its assessment of weather modification research, the NRC (2003) concluded that the South African and Mexican experiments have demonstrated responses in clouds to treatment in accordance with understanding of the chain of physical reactions leading to precipitation. However, since the analyzed statistical results are for radar-defined floating targets, they still do not prove that rainfall can be increased by hygroscopic seeding on the ground for specific watersheds. Moreover, since seeding may alter the size spectrum of raindrops, which alters the radar return, uncertainties exist in the evaluation of actual rain amounts for seeded versus non-seeded floating targets. Finally, since the main response to seeding found in the South African, Mexican, and Thailand experiments is delayed in time for as much as 1 to 6 hours, following the cessation of seeding, we lack a clear understanding of the actual processes that could lead to such a physical response.

Thus while the results of the hygroscopic seeding experiments are quite promising, they still do not constitute a "proof" that hygroscopic seeding can enhance rainfall on the ground over an extended area. The areas affected by cloud seeding have not yet been characterized. In after-the-fact analyses, several rain enhancement projects have reported evidence for physical effects outside the area or timing originally designated as the target, or beyond the time interval when seeding effects were anticipated. For example, in recent large particle hygroscopic and glaciogenic seeding trials involving warm-base convective clouds in Thailand and Texas, increases in rain were reported 3 to 12 hours after seeding was conducted (Woodley et al. 1999), well beyond the time that direct effects of seeding were expected and possibly outside the target area.

Other Cloud Seeding Applications

Cloud seeding strategies have also been applied to hail suppression and reduction in hurricane intensity. Following many years of research, scientific field confirmation of the concepts of hail suppression has been largely unsuccessful (Atlas 1977; Federer et al. 1986). Only long-term statistical analyses of non-randomized, operational programs have provided convincing evidence, suggesting that seeding can significantly reduce hail frequency (Mesinger and Mesinger 1992; Smith et al. 1997; Eklund et al. 1999; Rudolf et al. 1994; Dessens 1998). An advantage of evaluating an operational program is that often one can work with long-period records (e.g., forty years in Yugoslavia), whereas randomized research programs cannot typically get funding for more than five years or so. The disadvantage is that one cannot totally eliminate concerns about "natural variability" in the climate (see comments in NRC 2003).

Likewise, attempts to modify hurricanes (Simpson et al. 1963; Simpson and Malkus 1964; Gentry 1974; Sheets 1981) failed to confirm the proposed hypotheses. In a sense, Stormfury succumbed to the very large "natural variability" of hurricanes, including a period of very low hurricane frequency from the mid-1960s to the mid-1980s (Sheets 1981). As a result, interest and government funding for hurricane modification plummeted. Although there have been recent simplified modeling studies that suggest that seeding hurricanes with pollution-sized aerosols may weaken storm intensity (Rosenfeld et al. 2007; Cotton et al. 2007), these concepts need to be developed more quantitatively.

Implications of Cloud Seeding Research to Climate Engineering

The scientific community has established a set of criteria to determine when there is "proof" that seeding has enhanced precipitation (NRC 2003; Garstang et al. 2005): *both* strong physical evidence of appropriate modifications to cloud structures and highly significant statistical evidence are required. Likewise, for firm "proof" that climate engineering is affecting climate, or even that CO_2 is modifying climate, both strong physical evidence of appropriate modifications to climate and significant statistical evidence should be required.

A lesson from evaluating cloud seeding experiments is that "natural variability" of clouds and precipitation can be quite large and can thus inhibit conclusive evaluation of even the best-designed statistical experiments. The same can be said for evaluating the effects of climate engineering or whether human-produced greenhouse gases are altering climate. If the signal is not strong, then evaluating whether human activity has produced some observed effect (cause and effect) requires much longer time records than is available for most, if not all, data sets. To do so, we have to resort to "proxy" data sets, which results in uncertainties in calibrations, inconsistencies between older data estimates and more recent measurements, large noise in the data, and inadequate coverage of sampling of the selected control variables. Thus, at present, we do not have an adequate measure of the "natural variability" of climate, which makes venturing into climate engineering hazardous indeed.

Climate Engineering

The climate system is far more complicated than many scientists lead us to believe. There is considerable evidence that the planet Earth is warming. Furthermore, the concentrations of CO_2 are also increasing at alarming rates. The question is: Are these cause-and-effects or is the planet warming for other reasons? The so-called "hockey stick" paper of Mann et al. (1998) provides the strongest evidence that the current period of global warming is unprecedented over the last 1000 years or so. Their analysis is based on proxy data that includes ice cores, tree rings, marine sediments, and historical sources from Europe and Asia. These data are therefore evidence for warming in the northern hemisphere; proxy and historical data for the southern hemisphere are very sparse. This paper, however, has been criticized by a number of scientists (e.g., McIntyre and McKitrick 2003; von Storch et al. 2006) as having major problems in the statistical treatment of the data. The National Research Council (2006) reported on an independent evaluation of Mann et al.'s conclusion and on the use of proxy data, and concluded that the last few decades of the 20th century were warmer than any period in the last 400 years. They stated that the conclusion reached by Mann et al.—that it was warmer than any period in the last 1000 years—is plausible but that there is less confidence that the warming was unprecedented for periods prior to 1600, owing to fewer proxies at fewer locations available prior to 1600. They noted that none of the reconstructions indicated that it was warmer during the Medieval Climate Optimum than during the end of the 20th century. That there were regions of the northern hemisphere that were warmer during the Medieval Climate Optimum can be seen from reconstructions of surface temperatures for the Sargasso Sea (Robinson et al. 2007). These data suggest that the warming period that we are experiencing has been going on for over 300 years (i.e., since the end of the Little Ice Age) and that the Medieval Climate Optimum period 1000 years ago was much warmer. There is also circumstantial evidence that the climate in Greenland, for example, was much warmer during the Medieval Climate Optimum period as the glaciers were much reduced in coverage and the seas were more open to navigation. Considering the scarcity of data, I find it difficult to conclude that we know enough about the "natural variability" of climate over the last 1000 years to state that this recent period of warming is unprecedented.

CO_2 is clearly a major absorber of longwave radiation and therefore contributes to so-called greenhouse gases. We need, however, to keep in mind that CO_2 is not the major greenhouse gas; water vapor has that distinction. Thus, much of the greenhouse warming in models is due to feedbacks that involve higher concentrations of water vapor in the atmosphere, which then contributes to most of the greenhouse warming. Clouds are very important absorbers of longwave radiation as well as the albedo of our planet. Low clouds tend to enhance the Earth's albedo (a cooling affect) while having little influence on the longwave radiation budget because their temperatures are close to that of the Earth's surface. On the other hand, high clouds tend to absorb more longwave radiation while (with the exception of optically thick tropical anvil-cirrus clouds) reflecting small amounts of shortwave radiation; therefore they serve as greenhouse warmers. Because models depend on rather crude parameterizations of clouds, it is still uncertain how clouds respond to a warming planet and to the enhanced water vapor content of the atmosphere. Are there more high clouds versus low clouds in a warming planet? How does cloud variability change with latitude? Increased cloud cover at high latitudes contributes to a warming trend in the Arctic since the annually averaged surface energy budget at high latitudes is dominated by longwave radiation.

While greenhouse gases, especially water vapor, are a major contributor to the habitability of planet Earth, is the variability of these gases the dominant contributor to climate change?

What are some of the other competing processes that change the forcing of our climate system? These are reviewed in Cotton and Pielke (2007) and include the following:

- changes in solar luminosity and orbital parameters,
- changes in surface properties,
- natural and human-induced changes in aerosols and dust,
- differential temporal responses to external forcing by the atmosphere and oceans.

Changes in Earth orbital parameters, the so-called Milankovitch cycle (Imbrie and Imbrie 1979; Berger 1982), are believed to be responsible for the onset of ice ages. It is unable, however, to explain the current warming trend, as it predicts we will be moving into an ice age in the next 5000 years. Although there is evidence of a small variation in the sun's irradiance, the amount of variability is too small to account for recent climate variations, let alone any over the last 1000 years. While many studies have suggested statistical correlations between varying solar parameters and the Earth's climate (i.e., Svensmark and Friis-Christensen 1997), the physical causes of those correlations are for the most part not well founded (Sun and Bradley 2002). Nonetheless, this does not mean that some unknown amplification process related to solar parameters could be contributing to the current warming trend. It remains as part of the uncertainty in climate prediction.

Variations in land-surface properties affect the planetary albedo and alter the surface energy budget such that the Bowen ratio can be changed. Human activity contributes to changes in surface properties through agricultural land use and urbanization. Moreover, changes in land use and vegetation respond to climate changes in a nonlinear way, thus altering both the planetary albedo and the surface energy budget. Although changes in land-surface properties are a significant contributor to the planetary energy budget, they do not probably rank as high as greenhouse warming (IPCC 2007). Nonetheless, the IPCC's estimates are based only on changes in albedo and do not include changes in sensible and latent heat fluxes which should make changes in global climate by land-use changes larger than estimated by IPCC.

Cotton and Pielke (2007) devoted an entire chapter to human-induced changes in aerosols. The chapter considers both the direct and semi-direct effects of aerosols and dust as well as indirect effects that alter the Earth's albedo and hydrologic budget through alterations in cloud properties. Large uncertainties exist in estimating the consequences of aerosols on climate largely because of the fact that a major contributor is related to cloud processes, which are poorly represented in GCMs. Nonetheless, it is generally believed that human-induced changes in aerosols contribute to a net cooling in the climate system, which offsets greenhouse warming by roughly one-third that of greenhouse gas warming (IPCC 2007), or to what is sometimes referred to as "global dimming." Some GCM simulations of greenhouse warming and direct and indirect aerosol effects (Liepert et al. 2004) show that the indirect and direct cooling effects of aerosols reduce surface latent and sensible heat transfer and, as a result, act to dry the atmosphere and thereby substantially weaken greenhouse gas warming. Since greenhouse warming causes a moistening of the atmosphere, and aerosol direct and indirect cooling counteracts that, the potential influence of aerosols on global climate could be far more significant than previously thought.

A major "wild card" in the climate system is naturally produced aerosols and specifically aerosols in the lower stratosphere induced by volcanic activity. Until recently, I had thought that volcanic activity was purely random. A series of papers by Reid Bryson and colleagues (Bryson and Goodman 1980a, b; Bryson 1982, 1989; Goodman 1984) suggests otherwise. These papers suggest that volcanic activity is modulated by the Sun–Moon–Earth tidal variations. Under this scenario, periods of global warming, such as we are now experiencing, can be attributed to periods of very low volcanic activity as seen between 1920 and 1940 (Robock 1979) and the Medieval Climate Optimum period. Periods of extensive cooling, like the Little Ice Age, were periods of maximum alignment of the Sun–Moon–Earth tidal forcing, which contributed to very active episodes of volcanic activity and global cooling. The consequence of this is that forecasts of global greenhouse gas warming are at the mercy of climate variability due to volcanic activity. Periods of greater than normal volcanic activity could completely override or mask the forcing by greenhouse gases. Is it possible that the current warming period is due to a period of below normal volcanic activity?

Finally, the atmosphere and ocean have very different timescales of response to external forcing: the atmospheric timescale is on the order of months, the ocean mixed layer is on the order of ten years, and the deep ocean is 100 years.

Thus, the current climate is being influenced by changes in external forcing that occurred as long as 100 years ago. This mismatch between ocean and atmosphere response to external forcing is a major contributor to "natural variability" of the climate system.

Based on this overview of cloud seeding, I will now address climate engineering. I will only focus on climate engineering as it pertains to engineering changes in global albedo and top-of-atmosphere longwave radiation emission by aerosols and cloud modification. I will not go into the broader context of geoengineering, which includes such things as capturing and disposing of CO_2 from flue gas streams, increasing net CO_2 uptake in the terrestrial biosphere, increasing net CO_2 uptake in the oceans, carbon sequestration, alternate energies, or even changing the albedo of oceans and land surfaces.

Emulating Volcanoes

Volcanoes are a major "wild card" in the climate system. A major volcanic eruption distributes large quantities of dust and debris into the upper troposphere and lower stratosphere. More importantly, it introduces large quantities of SO_2 into the lower stratosphere where it is subjected to slow gas-to-particle production, particularly the formation of sulfuric acid drops. These highly soluble drops scatter solar radiation, thus reducing the amount of sunlight that reaches the surface. A single major eruption can produce a reduction in solar radiation that can last for anything up to two years and can result in residual heat loss in the ocean-mixed layer for as long as ten years.

The idea of introducing sulfate aerosols into the stratosphere dates back to Budyko (1974) and Dyson and Marland (1979), and has received recent prominence by the Nobel Laureate, Paul Crutzen (2006). The idea is to burn S_2 or H_2S carried into the stratosphere by balloons, artillery guns, or rockets to produce SO_2. Crutzen suggests that to enhance residence time, and thereby minimize the mass required, the gases should be introduced in the upward stratospheric circulation branch in the tropics where slow gas-to-particle conversion can take place. He estimates that 1.9 Tg S would be required to offset 1.4 W m^{-2} warming by CO_2, and that this can be achieved by continuous deployment of about 1–2 Tg S per year for a total cost of $25–50 billion per year. To compensate for a doubling of CO_2 (estimated 4 W m^{-2} warming), Crutzen estimates that 5.4 Tg S per year are needed with corresponding cost increases. As with volcanoes, we can expect the sky to be whitened and that red sunsets and sunrises will prevail. One adverse consequence of SO_2 seeding the stratosphere is that stratospheric ozone would be reduced. Crutzen noted that El Chichón introduced 3–5 Tg S in the stratosphere and reduced ozone by 16% at 20 km altitude, whereas Mt. Pinatubo introduced 10 Tg S, which contributed to a 2.5% reduction in column ozone loss. I imagine that this could translate into rates of increased incidence of skin cancer by higher UV radiation amounts.

Another option noted by Crutzen would be to release soot particles in the lower stratosphere by burning fossil fuels. Like the nuclear winter hypothesis (see review by Cotton and Pielke 2007), the resulting soot particles would absorb solar radiation, which would deplete solar radiation reaching the surface but also warm the stratosphere. This warming could have undesirable consequences in terms of changes in stratospheric circulations and ozone depletion. It would be less costly to deliver as only 1.7% of the mass of sulfur would be needed to produce the same cooling effect.

Rather than manufacture scattering or absorbing aerosols *in situ*, it has also been suggested that mirrors can be introduced in space. The National Academy of Sciences proposed deployment of something like 55,000 mirrors with a surface area of 100 km^2 into the Earth's orbit (NAS 1992). These mirrors would deplete about 2% of solar radiation, but each mirror would also create a shadow in the shape of an eclipse. A greater number of smaller mirrors have also been proposed, but these would create a flickering of sunlight. In addition to the installation costs, removing the mirrors, if undesirable responses developed, would be difficult and expensive.

Early (1989) has also proposed the introduction of a solar shield at the Sun–Earth Lagrange point (1.5×10^6 km from Earth). This could reflect 2% of solar radiation reaching Earth, but would cost in the trillion dollar range to install. In the event of adverse responses, removal would be easier and less costly than a large number of mirrors.

When we consider such options, we must keep in mind that altering solar radiation is quite different from changing longwave radiation. CO_2 traps longwave radiation both day and night, whereas reducing sunlight affects only daytime radiation and, on the annual average, most strongly in equatorial regions and in summer seasons at high latitudes (Govindasamy and Cadeira 2000). Govindasamy and Cadeira performed GCM simulations that emulated the above proposed changes in solar radiation by lowering the solar constant by 1.8%, which would offset a doubling of the CO_2 scenario. As anticipated, the climate-engineered simulations produced the most cooling in the tropics but reduced the amplitude of the diurnal cycle over land by only 0.1K. Govindasamy and Cadeira found that the amplitude of the seasonal cycle was greater in the climate-engineered simulation than in the double CO_2 simulation at high latitudes because there was more sea ice in the climate-engineered simulation. Sea ice tends to insulate the warmer underlying waters from the overlying air, resulting in colder winters and amplification of the seasonal cycle. Little change in the model's hydrological cycle was noted in either their double CO_2 simulation or the climate-engineered simulation.

Back to Cloud Seeding

Purposeful cloud seeding has been mainly designed to increase precipitation or modify storms. Could cloud seeding strategies, however, be designed either to increase cloud albedo or the amount of longwave radiation escaping to space? As reviewed by Levin and Cotton (2008), there is plenty of evidence to show that pollution aerosols are inadvertently modifying cloud radiative properties and precipitation. One problem in implementation is that any aerosols introduced into the troposphere, particularly the lower troposphere, must be done almost continuously as the residence times are so short, generally on the order of a few days or at most a few weeks. This limitation makes these options much more costly than those proposed for the stratosphere or space unless this is done on the basis of business as usual; that is, these changes would be associated with normal industrial operations, ship operations, or aircraft operations. One motivation for Crutzen's (2006) paper is that as we become successful in cleaning up industrial aerosol pollution, which current estimates suggest causes a significant cooling effect that offsets greenhouse warming, the amplitude of global warming will become greater. Moreover, much of the estimated cooling by aerosol pollution may be a result of indirect effects associated with clouds. Hence, we should consider how to pollute clouds without producing disastrous adverse consequences. Admittedly this may be impossible; however, let us speculate how cloud seeding could be applied to climate engineering.

More Ship Tracks!

The strongest evidence that we have that pollution aerosols increase cloud albedo comes from ship tracks (Figure 15.1). In fact, Porch et al. (1990) referred to them as the *Rosetta Stone* connecting changes in aerosols over the oceans and cloud albedo effects on climate.

Measurements show that ship tracks contain higher droplet concentrations, smaller droplet sizes, and higher liquid water contents than surrounding clouds (Radke et al. 1989). The tracks are often as long as 300 km or more and about 9 km wide (Durkee et al. 2000). They typically form in relatively shallow boundary layers between 300–750 m deep. They do not form in boundary layers deeper than 800 m (Durkee et al. 2000).

Figure 15.1 A number of ship tracks (i.e., clouds formed from the exhaust of ships' smokestacks) can be seen north and west of the smoke plume. Image courtesy of the SeaWiFS Project, NASA GSFC, and ORBIMAGE.

It Is Therefore Hypothesized That We Should Produce More Ship Tracks

The regions most susceptible to those changes are oceanic subtropical high pressure regions. One could redesign ship routes (with economic incentives) for high sulfur-containing coal-burning ships to sail along the windward regions of subtropical highs. These could be supplemented with additional "*albedo enhancer*" ships to sail back and forth along the windward side of marine stratocumulus cloud layers in the vulnerable regions. Research is needed to estimate the number of supplemental ships and economic incentive costs to achieve a desirable increase in global-averaged albedo. I expect the costs would be prohibitive. There is modeling evidence that not all clouds respond to increasing aerosol pollution with an increase in albedo (Jiang et al. 2002; Ackerman et al. 2004; Lu and Seinfeld 2005). Thus the science of cloud responses to aerosols must be advanced before this hypothesis could be implemented as a strategy. Of course, there are adverse consequences of purposely polluting clouds that we must consider, including acid drizzle, but at least the regions affected would be well offshore, away from most human activity.

The idea behind hygroscopic seeding of marine stratocumulus clouds is not new, as Latham (1990, 2002) proposed generating seawater drops around 1 μm in size near the ocean surface to enhance droplet concentrations. A spray of seawater drops would be produced either by high volume atomizers or blowing air through porous pipes to produce air bubbles that would rise to the sea surface and burst, much like a natural wave action produces the bubbles. The former technique has the advantage that one can be more certain that the salt particles thus produced would have an optimum size for competing with natural CCN and thereby increase droplet concentrations once the particles are lofted into clouds in the marine turbulent boundary layer. The advantage of this technique is that raw materials would be free and non-polluting. However, the production and movement of a large number of generating floats or derricks would be very costly indeed. Latham claims that the power requirements for their operation could be supplied by solar, wave action, or even wind power. He actually proposes the development of sailing ships based on the Magnus effect, where spinning towers would not only develop the aerodynamic lift to propel the ships but would also drive the sea-spray generators (Latham et al. 2008). Figure 15.2 presents an artist's concept of such Magnus force-based sailing ships designed for sea-spray generation. Rough estimates of the climatic effects of deploying a large ensemble of such ships to produce sea-spray over a large area have been made with a GCM. The GCM, however, does not consider possible negative dynamic responses, such as enhanced entrainment or those that will result from alterations in drizzle, which cloud-resolving simulations have suggested (Jiang et al. 2002; Ackerman et al. 2004; Lu and Seinfeld 2005). Thus, the GCM estimates probably err on the side of yielding a greater cooling effect than can be achieved in reality.

Figure 15.2 Artist's concept of a Magnus-effect sailing ship for sea-spray generation. From Latham et al. 2008; used with permission from the artist, John MacNeill.

Overall, the approach to climate engineering using hygroscopic seeding concepts is worth examining more fully with models and limited field experiments. I must admit to being skeptical that one could implement such a strategy on a near continuous basis over large enough areas to counter greenhouse warming significantly.

Mid-level Stratus Seeding

Mid-level stratus clouds, also called altostratus, are ubiquitous throughout large regions of the middle latitudes. A typical elevation of these clouds is about 3 km above sea level, and during the cold seasons many of these clouds are supercooled. Normally, mid-level stratus are thought to play a neutral role in the Earth's radiation budget since they reflect about as much solar radiation as they absorb longwave radiation. However, this near radiative balance might be upset by worldwide selective cloud seeding. For instance, consider nonfreezing stratus clouds: One can imagine a systematic seeding of these clouds by day with pollution aerosols (small hygroscopic particles) to increase their albedo and by night seed with giant CCN or conventional hygroscopic seeding materials to cause them to rain-out, thereby making them more transparent to longwave radiation. This would shift their contribution to the global radiative balance to a net cooling effect. A similar strategy could be followed for supercooled stratus. In that case, one could again seed with pollution aerosols during the daytime to increase their albedo but at night seed with glaciogenic seeding materials, such as AgI. It has been shown a number of times that seeding supercooled stratus will reduce the total condensate path of those clouds, thus making

Figure 15.3 This racetrack pattern, approximately 20 miles long, was produced by dropping crushed dry ice from an airplane. The safety pin-like loop at the near end of the pattern resulted when the dry ice dispenser was inadvertently left running as the airplane began climbing to attain altitude from which to photograph results. From Havens et al. (1978). Photo, taken in 1948, used with permission from Dr. Vincent Schaefer.

them more transparent to longwave radiation. Figure 15.3 shows a classic example of clearing supercooled stratus by seeding with glaciogenic materials.

How could this be done globally in a cost-effective manner? Some industries with tall stacks could have their affluent doped with the appropriate aerosol. Use of commuter aircraft with their jet fuels doped with aerosol generators is another possibility. Also, the use of unmanned aerial vehicles or blimps for aerosol dispersal could be considered. Potential adverse consequences, however, are likely, including impacts on precipitation, local cold temperature extremes (which would also impact fossil fuel demands), and the hydrological cycle.

Overall, this approach to countering greenhouse gas warming is more costly and less feasible than hygroscopic seeding of marine stratocumuli.

Seeding Cirrus Clouds or Making More Contrails

On an annual average, clouds cover between 60–65% of the Earth (Rossow and Dueñas 2004) and much of that cloud cover consists of middle and high clouds. It is thought that cirrus clouds contribute globally to a warming of the atmosphere due to their contribution to downward transfer of longwave radiation. In other words, they act as a greenhouse agent. Human activity is already modifying cirrus clouds through the production of aircraft contrails. Kuhn (1970) found that contrails depleted solar radiation and increased downward longwave radiation, but during the daytime their shortwave influence dominates and they contribute to a net surface cooling. Kuhn (1970) calculated that if contrails persist over 24 h, their net effect would be cooling. Others have concluded that they lead to surface warming (Liou et al. 1991; Schumann 1994), but Sassen (1997) notes that the sign of the climatic impact of contrails is dependent upon particle size. Global estimates of the effects of contrails are that they contribute to a net warming (Minnis et al. 2004).

It has even been proposed seeding in clear air in the upper troposphere to produce artificial cirrus, which would warm the surface enough to reduce cold-season heating demands (Detwiler and Cho 1982). Thus the prospects for seeding cirrus to contribute to global surface cooling do not seem to be very good.

The only approach that might be feasible is to perform wide-area seeding with soot particles, which would absorb solar radiation and warm cirrus layers enough to dissipate perhaps cirrus clouds. This strategy would be similar to that proposed by Watts (1997) and Crutzen (2006) for implementation in the stratosphere. As noted by Crutzen (2006), only 1.7% of the mass of sulfur is needed to produce a similar magnitude of surface cooling. Application at cirrus levels in the upper troposphere would have the double benefit of absorbing solar radiation, thus contributing to surface cooling and dissipating cirrus clouds which would increase outgoing longwave radiation. Of course, the soot that becomes attached to ice crystals would reduce the albedo of cirrus, thus countering the longwave warming effect to some degree. In addition, there is evidence that soot particles can act as ice nuclei, thus contributing to greater concentrations of ice crystals by heterogeneous nucleation but possibly reduced crystal production by homogeneous nucleation (DeMott et al. 1994; Kärcher et al. 2007). Thus it would be best to engineer carbonaceous aerosol to be ineffective as ice nuclei.

The possible adverse consequences of such a procedure can only be conjectured at this time but are most likely to impact the hydrological cycle. Complex chemical, cloud-resolving, and global models are required to evaluate the feasibility of this approach and to estimate possible adverse consequences. The feasibility of this approach in terms of implementation strategies is probably comparable to seeding sulfates in the lower stratosphere. The costs would be similar to Crutzen's estimates for stratospheric seeding.

Summary and Recommendations

In this chapter, I have provided an overview of both weather engineering (cloud seeding) and climate engineering. I have shown that there are a number of lessons learned from cloud seeding evaluation, such as that *both* strong physical evidence of appropriate modifications to the climate system and highly significant statistical evidence are required. This will be quite challenging, as I find it hard to imagine that randomized statistical experiments can be designed and implemented for long enough time periods to isolate the modification signals from the background "natural variability" of the climate system.

As I mentioned above, if we as a scientific community require the same standards of "proof" imposed on the weather modification community for evaluating cloud seeding hypotheses as for evaluating whether human-produced greenhouse gases are changing climate (which I think we should), we are a long way from being able to say that CO_2 is altering climate. Likewise, for firm "proof" that climate engineering is affecting climate, the required levels of physical model evaluations and statistical evaluations will be extremely challenging. What is needed, first of all, is a demonstrated climate model forecast skill that is large enough to be able to extricate the climate modification signal from the "natural variability" or "noise" of the climate system. Once this predictive skill is achieved, then there is the opportunity to apply advanced statistical methods that use model–output statistics and observed response variables that can confirm the hypothesis. Moreover, this climate forecast model should be able to identify and quantify unexpected undesirable consequences of climate engineering.

Alan Robock (2008) recently wrote a paper entitled "Twenty Reasons Why Geogenineering May be a Bad Idea." He notes that one possible response to climate engineering to mitigate greenhouse gas warming is that precipitation is likely to be modified both globally and regionally. Some countries may find themselves in a drought in response to climate engineering. Many of the cloud-related climate engineering hypotheses are likely to impact the hydrological cycle, especially those hypotheses associated with modification of mid- and high-level clouds. Other reasons listed by Robock (2008) were:

- Continued ocean acidification.
- Ozone depletion.
- Effects on the biosphere.
- Enhanced acid precipitation.
- Effects on cirrus clouds (reference to S seeding in the stratosphere).
- Whitening of the sky (reference to S seeding in the stratosphere).
- Less solar radiation for solar power, especially for those requiring direct solar radiation.
- Rapid warming when it stops.
- How rapidly could effects be stopped?
- Environmental impacts of aerosol injection.
- Human error.
- Unexpected consequences.
- Schemes perceived to work will lessen the incentive to mitigate greenhouse gas emissions.
- Use of the technology for military purposes.
- Commercial control of technology.
- Violates current treaty.

- Would be tremendously expensive.
- Even if it works, whose hand will be on the thermostat? How could the world agree on the optimum climate?
- Who has the moral right to advertently modify the global climate?

In regard to unexpected consequences, I have already stated that I do not believe that we understand all the factors that effect climate variability nor have we demonstrated a climate forecast skill to merit implementing a climate warming mitigation strategy. What if I am right that volcanic activity is a major "wild card" in the climate system? Now, suppose we implement one of the climate engineering concepts outlined above to cool the planet in opposition to greenhouse warming. If successful, this cooling will lead to ocean responses on timescales of decades to perhaps a century. In the meantime, suppose we find ourselves in the midst of a period of enhanced volcanic activity. The cooling trend following volcanic activity combined with our "engineered" cooling trend could drive us into a Little Ice Age or worse. I expect the consequences of that would be far worse than global warming.

Despite these concerns, climate engineering is an issue that cannot be ignored. The gravity of this situation is emphasized by the fact that any of the aerosol- or cloud-related climate engineering schemes would have to be continued for centuries because of the long atmospheric residence times of the atmospheric greenhouse gases. I recommend that major international initiatives be planned throughout the world, using the most advanced models in the design of specific climate engineering projects. Before climate engineering can, however, be implemented, fundamental research must first advance our *quantitative* understanding of the climate system, of climate variability, the scientific possibilities of climate engineering, technical requirements, social impacts, and requisite political structures needed. Climate engineering should be considered a "last gasp" measure to prevent catastrophic consequences of a changing climate.

Another lesson learned from cloud seeding, which I have not mentioned previously, is that cloud seeding is often called upon by politicians to demonstrate *that they are doing something* during periods of drought and major water shortages or following major catastrophes. This has occurred despite the lack of strong scientific evidence that cloud seeding actually works. I refer to this as the use of *political placebos*. If we find ourselves in a true climate crisis, I anticipate that politicians will call for climate engineering measures to alter the adverse climate trends. If this should ever be the case, let us be sure that we act based on the most advanced level of knowledge of the climate system as well as of the full consequences of our actions.

Acknowledgments

I would like to acknowledge Brenda Thompson for her assistance with this paper. This material is based upon work supported by the National Science Foundation under Grant No. ATM-0526600. Any opinions, findings, and conclusions or recommendations expressed in this material are those of the author and do not necessarily reflect the views of the National Science Foundation.

References

Ackerman, A. S., M. P. Kirkpatrick, D. E. Stevens, and O. B. Toon. 2004. The impact of humidity above stratiform clouds on indirect aerosol climate forcing. *Nature* **432**:1014–1017.

Anthes, R. A. 1984. Enhancement of convective precipitation by mesoscale variations in vegetative covering in semiarid regions. *J. Climate Appl. Meteor.* **23**:541–554.

Atlas, D. 1977. The paradox of hail suppression. *Science* **195**:139–145.

Ben-Zvi, A. 1997. Comments on "A new look at the Israeli cloud seeding experiments." *J. Appl. Meteor.* **3**:255–256.

Berger, W. H. 1982. Climate steps in ocean history-lessons from the Pleistocene. In: Climate in Earth History, ed. W. H. Berger and J. C. Crowell, Panel Co-Chairmen, pp. 43–54. Washington, DC: National Academy Press.

Bigg, E. K. 1997. An independent evaluation of a South African hygroscopic cloud seeding experiment, 1991-1995. *Atmos. Res.* **43**:111–127.

Biswas, K. R., and A. S. Dennis. 1971. Formation of rain shower by salt seeding. *J. Appl. Meteor.* **10**:780–784.

Black, J. F., and B. H. Tarmy. 1963. The use of asphalt coating to increase rainfall. *J. Appl. Meteor.* **2**:557–564.

Bowen, E. G. 1952. A new method of stimulating convective clouds to produce rain and hail. *Q. J. Roy. Meteor. Soc.* **78**:37–45.

Braham, R. R., Jr. 1986. Rainfall enhancement: A scientific challenge. In. Rainfall Enhancement—A Scientific Challenge, vol. 21, pp. 1–5. AMS Meteor. Monogr. Boston: Amer. Meteor. Soc.

Bryson, R. A. 1982. Volcans et climat. *La Rescherche* **13(135)**:844–853.

Bryson, R. A. 1989. Late quaternary volcanic modulation of Milankovitch climate forcing. *Theor. Appl. Climatol.* **39**:115–125.

Bryson, R. A., and B. M. Goodman. 1980a. The climatic effect of explosive volcanic activity—Analysis of the historical data. In: Proc. Symp. on Atmospheric Effects and Potential Climatic Impact of the 1980 Eruptions of Mount St. Helens. Nov. 18–19, 1980. Washington, DC: NASA Conf. Publ. 2240.

Bryson, R. A., and B. M. Goodman. 1980b. Volcanic activity and climatic changes. *Science* **27**:1041–1044.

Budyko, M. I. 1974. The method of climate modification. *Meteor. Hydrol.* **2**:91–97.

Cooper, W. A., and R. P. Lawson. 1984. Physical interpretation of results from the HIPLEX-1 experiment. *J. Clim. Appl. Meteor.* **23**:523–540.

Cotton, W. R. 1972. Numerical simulation of precipitation development in supercooled cumuli, Part I. *Mon. Wea. Rev.* **11**:757–763.

Cotton, W. R. 1982. Modification of precipitation from warm clouds—A review. *Bull. Amer. Meteor. Soc.* **63**:146–160.

Cotton, W. R. 1986a. Testing, implementation and evolution of seeding concepts—A review. *Meteor. Monogr.* **21**:63–70.

Cotton, W. R. 1986b. Testing, implementation, and evolution of seeding concepts—A review. In: Precipitation Enhancement—A Scientific Challenge, ed. R. R. Braham, Jr. *Meteor. Monogr.* **43**:139–149.

Cotton, W. R., and R. A. Pielke 1992. Human Impacts on Weather and Climate. Ft. Collins, CO: ASTeR Press.

Cotton, W. R., and R. A. Pielke, Sr. 2007. Human Impacts on Weather and Climate. Cambridge Univ. Press.

Cotton, W. R., H. Zhang, G. M. McFarquhar, and S. M. Saleeby. 2007. Should we consider polluting hurricanes to reduce their intensity? *J. Weather Modif.* **39**:70–73.

Crutzen, P. J. 2006. Albedo enhancement by stratospheric sulfur injections: A contribution to resolve a policy dilemma? *Climate Change* **77**:211–219.

DeMott, P. J., M. P. Meyers, W. R. Cotton. 1994. Parameterization and impact of ice initiation processes relevant to numerical model simulations of cirrus clouds. *J. Atmos. Sci.* **41**:77–90.

Dennis, A. S. 1980. Weather Modification by Cloud Seeding. New York: Academic Press.

Dennis, A. S., and H. D. Orville. 1997. Comments on "A new look at the Israeli cloud seeding experiments." *J. Appl. Meteor.* **36**:277–278.

Dessens, J. 1998. A physical evaluation of a hail suppression project with silver iodide ground burners in southwestern France. *J. Appl. Meteor.* **37**:1588–1599.

Detwiler, A., and H. Cho. 1982. Reduction of residential heating and cooling requirements possible through atmospheric seeding with ice-forming nuclei. *J. Appl. Meteor.* **21**:1045–1047.

Durkee, P. A., R. E. Chartier, A. Brown et al. 2000. Composite ship track characteristics. *J. Atmos. Sci.* **57**:2542–2553.

Dyson, F. J., and G. Marland. 1979. Technical fixes for the climatic effects of CO_2. In: Workshop on the Global Effects of Carbon Dioxide from Fossil Fuels, Miami Beach, FL, March 7–11, 1977, ed. W. P. Elliott and L. Machta, pp. 111–118. U.S. DOE, CONF-770385, UC-11.

Early, J. T. 1989. Space-based solar screen to offset the greenhouse effect. *J. Brit. Interplanetary Soc.* **42**:567–569.

Eklund, D. L., D. S. Jawa, and T. K. Rajala. 1999. Evaluation of the western Kansas weather modification program. *J. Wea. Mod.* **31**:91–101.

Farley, R. D., and C. S. Chen. 1975. A detailed microphysical simulation of hygroscopic seeding on the warm process. *J. Appl. Meteor.* **14**:718–733.

Federer, B., A. Waldvogel, W. Schmid et al. 1986. Main results of Grossversuch IV. *J. Climate Appl. Meteor.* **25**:917–957.

Flueck, J. A., W. L. Woodley, R. W. Burpee, and D. O. Stram. 1981. Comments on "FACE rainfall results: Seeding effect or natural variability?" *J. Appl. Meteor.* **20**:98–107.

Gagin, A., and J. Neumann. 1981. The second Israeli randomized cloud seeding experiment: Evaluation of results. *J. Appl. Meteor.* **20**:1301–1311.

Garstang, M., R. Bruintjes, R. Serafin et al. 2005. Weather modification: Finding common ground. *Bull. Amer. Meteor. Soc.* **86**:647–655.

Gentry, R. C. 1974. Hurricane modification. In: Climate and Weather Modification, ed. W. N. Hess, pp. 497–521. New York: Wiley.

Goodman, B. M. 1984. The climatic impact of volcanic activity. Ph.D. Thesis, Dept. of Meteorology, Univ. of Wisconsin-Madison.

Govindasamy, B., and K. Cadeira. 2000. Geoengineering Earth's radiation balance to mitigate CO_2-induced climate change. *Geophys. Res. Lett.* **27**:2141–2144.

Grant, L. O., and P. W. Mielke, Jr. 1967. A randomized cloud seeding experiment at Climax, Colorado 1960–1965. Proc. 5th Berkeley Symp. on Mathematical Statistics and Probability, vol. 5, pp. 115–131. Berkeley: Univ. of California Press.

Havens, B. S., J. E. Jiusto, and B. Vonnegut. 1978. Early history of cloud seeding. Project Cirrus Fund, SUNY-ES/328, Albany: Langmuir Lab.

Hobbs, P. V., and A. L. Rangno. 1979. Comments on the Climax randomized cloud seeding experiments. *J. Appl. Meteor.* **18**:1233–1237.

Imbrie, J., and K. P. Imbrie. 1979. Ice Ages: Solving the Mystery. Short Hills, NJ: Enslow Publ.

IPCC. 2007. Climate Change 2007: The Physical Science Basis. Contribution of Working Group I to the Fourth Assessment Report of the Intergovernmental Panel on Climate Change, ed. S. Solomon, D. Qin, M. Manning et al. New York: Cambridge Univ. Press.

Isaac, G. A., J. W. Strapp, and R. S. Schemenaur. 1982. Summer cumulus cloud seeding experiments near Yellowknife and Thunder Bay, Canada. *J. Appl. Meteor.* **21**:1266–1285.

Jiang, H., G. Feingold, and W. R. Cotton. 2002. Simulations of aerosol-cloud-dynamical feedbacks resulting from entrainment of aerosol into the marine boundary layer during the Atlantic Stratocumulus Transition Experiment. *J. Geophys. Res.* **107**:4813.

Kärcher, B., O. Möhler, P. J. DeMott, S. Pechtl, and F. Yu. 2007. Insights into the role of soot aerosols in cirrus cloud formation. *Atmos. Res.* **7**:4203–4227.

Kerr, R. A. 1982. Cloud seeding: One success in 35 years. *Science* **217**:519–522.

Kuhn, P. M. 1970. Airborne observations of contrail effects on the thermal radiation budget. *J. Atmos. Sci.* **27**:937–942.

Latham, J. 1990. Control global warming? *Nature* **347**:339–340.

Latham, J., 2002. Amelioration of global warming by controlled enhancement of the albedo and longevity of low-level maritime clouds. *Atmos. Sci. Lett.* **3**:52–58.

Latham, J., P. Rasch, C.-C. Chen et al. 2008. Global temperature stabilization via controlled albedo enhancement of low-level maritime clouds. *Phil. Trans. Roy. Soc.* **336**:3969–3987.

Levin, Z., and W. R. Cotton. 2008. Aerosol pollution impact on precipitation: A scientific review. Report from the WMO/IUGG Intl. Aerosol Precipitation Science Assessment Group (IAPSAG). Geneva: WMO.

Liepert, B. G., J. Feichter, U. Lohmann, and E. Roeckner. 2004. Can aerosols spin down the watercycle in a warmer and moister world? *Geophys. Res. Lett.* **31**:L06207.

Liou, K. N., J. L. Lee, S. C. Ou, A. Fu, and Y. Takano. 1991. Ice cloud microphysics, radiative transfer and large-scale cloud processes. *Meteor. Atmos. Phys.* **46**:41–50.

Lu, M.-L., and J. H. Seinfeld. 2005. Study of the aerosol indirect effect by LES of marine stratocumulus. *J. Atmos. Sci.* **62**:3909–3932.

Mann, M. E., R. S. , and M. K. Hughes. 1998. Global-scale temperature patterns and climate forcing over the past six centuries. *Nature* **392**:779–787.

Mason, B. J. 1980. A review of 3 long-term cloud-seeding experiments. *Meteor. Mag.* **109**:335–344.

Mason, B. J. 1982. Personal reflections on 35 years of cloud seeding. *Contemp. Phys.* **23**:311–327.

Mather, G. K. 1991. Coalescence enhancement in large multicell storms caused by the emissions from a Kraft paper mill, *J. Appl. Meteor.* **30**:1134–1146.

Mather, G. K., D. E. Terblanche, F. E. Steffens, and L. Fletcher. 1997. Results of the South African cloud seeding experiments using hygroscopic flares. *J. Appl. Meteor.* **36**:1433–1447.

Matveev, L. T. 1984. Cloud Dynamic. Dordrecht: Reidel Publ.

McIntyre, S., and R. McKitrick. 2003. Corrections to the Mann et al. (1998) proxy data base and the Northern Hemispheric average temperature series. *Energy Environ.* **14**:751–771.

Mesinger, F., and N. Mesinger. 1992. Has hail suppression in Eastern Yugoslavia led to a reduction in the frequency of hail? *J. Appl. Meteor.* **31**:104–111.

Mielke, P. W., Jr., L. O. Grant, and C. F. Chappell. 1970. Elevation and spatial variation effects of wintertime orographic cloud seeding. *J. Appl. Meteor.* **9**:476–488.

Mielke, P. W., Jr., L. O. Grant, and C. F. Chappell. 1971. An independent replication of the Climax wintertime orographic cloud seeding experiment. *J. Appl. Meteor.* **10**:1198–1212.

Mielke, P. W., Jr., G. W. Brier, L. O. Grant, G. J. Mulvey, and P. N. Rosenweig. 1981. A statistical reanalysis of the replicated Climax I and II wintertime orographic cloud seeding experiments. *J. Appl. Meteor.* **20**:643–659.

Mielke, P. W., Jr., K. Berry, A. S. Dennis, P. L. Smith, J. R. Miller, Jr., and B. A. Silverman. 1984. HIPLEX-1: Statistical evaluation. *J. Appl. Meteor.* **23**:513–522.

Minnis, P., J. K. Ayers, R. Palikonda, and D. Phan. 2004. Contrails, cirrus trends, and climate. *J. Climate* **17**:1671–1685.

Mooney, M. L. and G. W. Lunn. 1969. The area of maximum effect resulting from the Lake Almanor randomized cloud seeding experiment. *J. Appl. Meteor.* **8(1)**:68–74.

Murray, F. W., and L. R. Koenig. 1972. Numerical experiments on the relation between microphysics and dynamics in cumulus convection. *Mon. Wea. Rev.* **100**:717–732.

Murty, A. S. R., et al. 2000. Eleven-year warm cloud seeding experiment in Maharashtra state, India. *J. Weather Modif.* **32**:10–20.

NAS (National Academy of Sciences; National Academy of Engineering, and Institute of Medicine). 1992. Policy Implications of Greenhouse Warming: Mitigation, Adaptation, and the Science Base. Washington, DC: National Academy Press.

Nickerson, E. C. 1979. FACE rainfall results: Seeding effect or natural variability? *J. Appl. Meteor.* **18**:1097–1105.

Nickerson, E. C. 1981. Reply: The FACE-1 seeding effect revisited. *J. Appl. Meteor.* **20**:108–114.

NRC (National Research Council). 1973. Weather and Climate Modification: Progress and Problems. Washington, DC: GPO.

NRC. 2003. Critical Issues in Weather Modification Research. Board on Atmospheric Sciences and Climate, Division on Earth and Life Studies. Washington, DC: National Academy Press.

NRC. 2006. Surface Temperature Reconstructions for the Last 2000 Years. Board on Atmospheric Sciences and Climate, Division on Earth and Life Studies. Washington, DC: National Academy Press.

Porch, W. M., C.-Y. J. Kao, and R. G. Kelley, Jr. 1990. Ship trails and ship induced cloud dynamics. *Atmos. Environ.* **24AI**:1051–1059.

Radke, L. F., J. A. Coakley, Jr., and M. D. King. 1989. Direct and remote sensing observations of the effects of ship tracks on clouds. *Science* **246**:1146–1149.

Rangno, A. L., and P. V. Hobbs. 1987. A re-evaluation of the Climax cloud seeding experiments using NOAA published data. *J. Clim. Appl. Meteor.* **26**:757–762.

Rangno, A. L., and P. V. Hobbs. 1993. Further analyses of the Climax cloud-seeding experiments. *J. Appl. Meteor.* **32**:1837–1847.

Rangno, A. L., and P. V. Hobbs. 1997a. Reply to Woodley. *J. Appl. Meteor.* **36**: 253–254.

Rangno, A. L., and P. V. Hobbs. 1997b. Reply to Ben-Zvi. *J. Appl. Meteor.* **36**: 257–259.

Rangno, A. L., and P. V. Hobbs. 1997c. Reply to Rosenfeld. *J. Appl. Meteor.* **36**: 272–276.

Rangno, A. L., and P. V. Hobbs. 1997d. Reply to Dennis and Orville. *J. Appl. Meteor.* **36**:279–279.

Reisin, T., S. Tzivion, and Z. Levin. 1996. Seeding convective clouds with ice nuclei or hygroscopic particles: A numerical study using a model with detailed microphysics. *J. Appl. Meteor.* **35**:1416–1434.

Reynolds, D. W. 1988. A report on winter snowpack augmentation. *Bull. Amer. Meteor. Soc.* **69**:1290–1300.

Reynolds, D. W., and A. S. Dennis. 1986. A review of the Sierra Cooperative Pilot Project. *Bull. Amer. Meteor. Soc.* **67**:513–523.

Robinson, A. B., N. E. Robinson, and W. Soon. 2007. Environmental effects of increased atmospheric carbon dioxide. *J. Amer. Phys. Surg.* **12**:79–90.

Robock, A. 1979. The “Little Ice Age”: Northern Hemispheric average observations and model calculations. *Science* **206**:1402–1404.

Robock, A. 2008. Twenty reasons why geogenineering may be a bad idea. *Bull. Atom. Sci.* **64(2)**:14–18

Rokicki, M. L., and K. C. Young. 1978. The initiation of precipitation in updrafts. *J. Appl. Meteor.* **17**:745–754.

Rosenfeld, D. 1997. Comment on “A new look at the Israeli cloud seeding experiments.” *J. Appl. Meteor.* **36**:260–271.

Rosenfeld, D., A. Khain, B. Lynn, and W. L. Woodley. 2007. Simulation of hurricane response to suppression of warm rain by sub-micron aerosols. *Atmos. Chem. Phys.* **7**:3411–3424.

Rosenfeld, D., and W. L. Woodley. 1989. Effects of cloud seeding in west Texas. *J. Appl. Meteor.* **28**:1050–1080.

Rosenfeld, D., and W. L. Woodley. 1993. Effects of cloud seeding in west Texas: Additional results and new insights. *J. Appl. Meteor.* **32**:1848–1866.

Rudolf, B., H. Hauschild, W. Rueth, and U. Schneider. 1994. Terrestrial precipitation analysis: Operational method and required density of point measurements. In: Global Precipitations and Climate Change, ed. M. Desbois and G. Desalmond, pp. 173–186. NATO ASI Series I, 26. New York: Springer.

Ryan, B. F., and W. D. King. 1997. A critical review of the Australian experience in cloud seeding. *Bull. Amer. Meteor. Soc.* **78**:239–354.

Sassen, K. 1997. Contrail-cirrus and their potential for regional climate change. *Bull. Amer. Meteor. Soc.* **78**:1885–1903.

Sax, R. I., S. A. Changnon, L. O. Grant et al. 1975. Weather modification: Where are we now and where are we going? An editorial overview. *J. Appl. Meteor.* **14**:652–672.

Schemenauer, R. S., and A. A. Tsonis. 1985. Comments on “Physical Interpretation of Results from the HIPLEX-1 Experiment.” *J. Appl. Meteor.* **24**:1269–1274.

Schumann, U. 1994. On the effect of emissions from aircraft engines on the state of the atmosphere. *Ann. Geophysicae* **12**:365–384.

Scott, B. C., and P. V. Hobbs, 1977. A theoretical study of the evolution of mixed-phase cumulus clouds. *J. Atmos. Sci.* **34**:812–826.

Sheets, R. C. 1981. Tropical cyclone modification: The Project STORMFURY Hypothesis. Miami, Florida, NOAA Technical Report ERL 414-AOML30.

Silverman, B. A. 1986. Static mode seeding of summer cumuli: A review. In: Rainfall Enhancement: A Scientific Challenge. AMS Meteor. Monogr., vol. 21, pp. 7–24, Boston: Amer. Meteor. Soc.

Silverman, B. A. 2000. An independent statistical reevaluation of the South African hygroscopic flare seeding experiment. *J. Appl. Meteor.* **39**:1373–1378.

Silverman, B. A. 2001. A critical assessment of glaciogenic seeding of convective clouds for rain enhancement. *Bull. Amer. Meteor. Soc.* **82**:903–924.

Silverman, B. A. 2003. A critical assessment of hygroscopic seeding of convective clouds for rainfall enhancement. *Bull. Amer. Meteor. Soc.* **84**:1219–1230.

Silverman, B. A., and W. Sukarnjanasat. 2000. Results of the Thailand warm-cloud hygroscopic particle seeding experiment. *J. Appl. Meteor.* **39**:1160–1175.

Simpson, J. S. 1979. Comment on "Field experimentation in weather modification." *J. Amer. Statist. Assoc.* **74**:95–97.

Simpson, R. H., and J. S. Malkus 1964. Experiments in hurricane modification. *Sci. Amer*. **211**:27–37.

Simpson, J., and W. L. Woodley. 1975. Florida Area Cumulus Experiments 1970–1973 rainfall results. *Appl. Meteor.* **14**:734–744.

Simpson, R. H., M. R. Ahrens, and R. D. Decker. 1963. A cloud seeding experiment in Hurricane Esther, 1961. National Hurricane Res. Proj. Rep. No. 60. Washington, DC: U.S. Dept. of Commerce, Weather Bureau.

Smith, P. L., A. S. Dennis, B. A. Silverman et al. 1984. HIPLEX-1: Experimental design and response variables. *J. Clim. Appl. Meteor.* **23**:497–512.

Smith, P. L., R. Johnson, D. L. Priegnitz, B. A. Boe, and P. J. Mielke, Jr. 1997. An exploratory analysis of crop hail insurance data for evidence of cloud seeding effects in North Dakota. *J. Appl. Meteor*. **36**:463–73.

Super, A. B. 1986. Further exploratory analysis of the Bridger Range Winter Cloud Seeding Experiment. *J. Appl. Meteor.* **25(12)**:1926–1933.

Sun, B., and R. S. Bradley. 2002. Solar influences on cosmic rays and cloud formation: A reassessment. *J. Geophys. Res*. **107(D14)**:4211.

Super, A. B., and B. A. Boe. 1988. Microphysical effects of wintertime cloud seeding with silver iodide over the Rocky Mountains. Part III: Observations over the Grand Mesa, Colorado. *J. Appl. Meteor*. **27**:1166–1182.

Super, A. B., B. A. Boe, and E. W. Holroyd. 1988. Microphysical effects of wintertime cloud seeding with silver iodide over the Rocky Mountains. Part I: Experimental design and instrumentation. *J. Appl. Meteor*. **27**:1145–1151.

Super, A. B., and J. A. Heimbach. 1983. Evaluation of the Bridger Range Winter Cloud Seeding Experiment Using Control Gages. *J. Appl. Meteor*. **22(12)**:1989–2011.

Super, A. B., and J. A. Heimbach. 1988. Microphysical effects of wintertime cloud seeding with silver iodide over the Rocky Mountains. Part II: Observations over the Bridger Range, Montana. *J. Appl. Meteor.* **27**:1152–1165.

Svensmark, H. and E. Friis-Christensen 1997. Variation of cosmic ray flux and global cloud coverage—a missing link in solar-climate relationships. *J Atmos-Terrest. Physics* **59**:1225–1232.

Tukey, J. W., L. V. Jones, and D. R. Brillinger. 1978a. The Management of Weather Resources, vol. 1, Proposals for a National Policy and Program, Report of the Statistical Task Force to the Weather Modification Advisory Board. Washington, DC: GPO.

Tukey, J. W., D. R. Brillinger, and L. V. Jones. 1978b. Report of the Statistical Task Force to the Weather Modification Advisory Board, vol. 2. Washington, DC: GPO.

von Storch, H., E. Zorita, J. M. Jones, F. Gonzalez-Rouco, S. F. B. Tett. 2006. Response to comment on "Reconstructing past climate from noisy data." *Science* **312**:529.

Watts, R. G., 1997. Engineering response to global climate change: Planning a research and development agenda, pp. 379–427. Boca Raton, FL: CRC Press.

WMO-World Meteorological Organization. 2000. Report of the WMO Workshop on Hygroscopic Seeding, WMP Report No. 35, WMO/TD No. 1006. Geneva: WMO.

Woodley, W. L. 1997. Comments on "A new look at the Israeli cloud seeding experiments." *J. Appl. Meteor.* **36**:250–252.

Woodley, W. L., B. Jordan, A. Barnston et al. 1982. Rainfall results of the Florida Area Cumulus Experiment, 1970–1976. *J. Appl. Meteor.* **21**:139–164.

Woodley, W. L., A. Barnston, J. A. Flueck, and R. Biondini. 1983. The Florida Area Cumulus Experiment's second phase (FACE-2). Part II: replicated and confirmatory analyses. *J. Appl. Meteor.* **22**:1529–1540.

Woodley, W. L., D. Rosenfeld, W. Sukarnjanaset et al. 1999. The Thailand cold-cloud seeding experiment: 2. Results of the statistical evaluation. Proc. Seventh WMO Conf. on Weather Modification,Chaing Mai, Thailand, WMO, pp. 25–38.

Woodley W. L., D. Rosenfeld, and B. A. Silverman. 2003a. Results of on-top glaciogenic cloud seeding in Thailand. Part I: The demonstration experiment. *J. Appl. Meteor.* **42**:920–938.

Woodley W. L., D. Rosenfeld, and B. A. Silverman. 2003b. Results of on-top glaciogenic cloud seeding in Thailand. Part II: Exploratory analyses. *J. Appl. Meteor.* **42**:939–951.

Yin, Y., Z. Levin, T. G. Reisin, and S. Tzivion. 2000a. The effects of giant condensation nuclei on the development of precipitation in convective clouds: A numerical study. *Atmos. Res.* **53**:91–116.

Yin, Y., Z. Levin, T. G. Reisin and S. Tzivion. 2000b. Seeding convective clouds with hygroscopic flares: Numerical simulations using a cloud model with detailed microphysics. *J. Appl. Meteor.* **39**:1460–1472.

Yin, Y., Z. Levin, T. G. Reisin, and S. Tzivion. 2001. On the response of radar-derived properties to hygroscopic flare seeding. *J. Appl. Meteor.* **40**:1654–1661.

Young, K. C. 1993. Microphysical Processes in Clouds, pp. 335–336. New York: Oxford Univ. Press.

Young, K. C. 1996. Weather modification: A theoretician's viewpoint. *Bull. Amer. Meteor. Soc.* **77**:2701–2710.

16

Air Pollution and Precipitation

Greg Ayers[1] and Zev Levin[2]

[1]CSIRO Marine and Atmospheric Research, Aspendale, Australia
[2]Department of Geophysics and Planetary Science,
Tel Aviv University, Ramat Aviv, Israel

Abstract

The physical hypothesis that air pollution in the form of small particles should lead to less efficient formation of precipitation has been established for several decades and is widely considered to be scientifically sound. As yet, however, no convincing proof exists that such a microphysical control of precipitation efficiency has been the prime cause of rainfall reduction in any area of the globe. There is a need for new experimental designs to test this hypothesis in a holistic way, taking into account all possible confounding influences on rainfall trends in a climate that is clearly not stationary in the face of global warming and natural decadal variability.

Introduction

Precipitation (including rain, snow, and hail) is the primary mechanism for transporting water from the atmosphere back to the Earth's surface. Precipitation amounts and temporal/geographical distribution are affected primarily by atmospheric dynamics; however, precipitation is also influenced by cloud microphysical processes associated with aerosol properties, which are primarily responsible for cloud drop and ice crystal formation. Changes in precipitation regimes and the frequency of extreme weather events (e.g., floods, droughts, severe ice/snow storms, monsoon fluctuations, and hurricanes) are of great importance to life on our planet. Thus, a plausible hypothesis is that by influencing the amounts, chemical composition, and distribution of natural and anthropogenic aerosols, changes in precipitation of significance to local communities may possibly occur. However, quantitative testing of that hypothesis has proved difficult.

Much of the work that was conducted over the years addressed the issues of the effects of aerosols on clouds and was motivated by the desire to understand (so as to predict) natural precipitation formation better as well as to underpin the novel idea of artificial weather modification, following Schaefer's pioneering work on freezing in supercooled water clouds (Schaefer 1946). The fundamental scientific understanding of cloud microphysical processes achieved in those years was summarized sequentially by Fletcher (1962), Mason (1971), and Pruppacher and Klett (1978). Based on many measurements and models, a general consensus emerged that, everything else being equal, the addition of more cloud condensation nuclei (CCN) to a cloud leads to the formation of smaller and more numerous cloud drops. It was also observed that the addition of giant CCN to clouds can lead to the formation of a few larger cloud drops (e.g., Mather 1991) and wider drop size distributions. Furthermore, recent work in shallow orographic clouds has revealed that riming efficiency in polluted clouds is smaller, leading to smaller snow crystals (Borys et al. 2000, 2003). All these observations and related modeling studies suggest as a sound physical hypothesis that, other things being equal, the consequence of particulate pollution on clouds should be a reduction in precipitation.

Unfortunately, the connection between aerosol concentration and the amount of precipitation on the ground is not yet clear. This is partly because feedbacks between the microphysical, radiative, and dynamical processes must exist and can sometimes lead to enhancement or to suppression of precipitation via atmospheric dynamic rather than cloud microphysical effects of aerosol. Recent examples in support of this proposition on the global scale may be found in the work of Rotstayn and Lohmann (2002), Rotstayn (2007), and Rotstayn et al. (2007). Similarly, warming due to greenhouse gases (GHG) is expected to increase atmospheric water vapor. Global circulation models coupled with mixed-ocean layer models show that increased aerosol concentrations lead to more clouds and thus to reduced solar radiation that reaches the surface. This reduces sensible and latent heat fluxes from the surface and thus reduces precipitation (Liepert et al. 2004). In other words, the effects of aerosols on precipitation could occur through their effects on the radiation in addition to their microphysical effects on clouds. Over the years there have been numerous attempts to shed more light on this connection, but the results vary widely between increases in rainfall, decreases in rain amounts, and no connection at all.

The World Meteorological Organization (WMO) and International Union of Geodesy and Geophysics (IUGG) recognized the importance of this issue and formed a group in 2004 to assess the knowledge and suggest directions for future research. The final report from this group was unable to reach an unambiguous conclusion that a systematic reduction in precipitation is the demonstrated result of particle pollution that enhances CCN levels (Levin and Cotton 2008).

In this chapter, we summarize in three parts some of the main points related to the effects of aerosol–cloud precipitation interactions. We do not intend to give an exhaustive review of the literature, but rather to illustrate the complexities involved in this issue. We begin with a discussion of the issues by separating the potential effects of aerosols on different types of clouds. This is because the effects could be significantly different depending on whether these are convective or orographic, or whether the affected clouds are downwind of urban centers. Using Australia as an example, we explore thereafter the complexities of the evidence and arguments about trends in precipitation and their causal attribution. Finally, we offer some remarks about the manner in which the scientific uncertainty on this issue has been reflected in the community.

The Effects of Aerosols on Clouds

Warm clouds are those that contain no ice. Measurements have shown that an increase in CCN from natural or anthropogenic sources increases cloud drop concentrations and reduces cloud drop size (leading to the first indirect effect or the "Twomey" effect). These ideas have been confirmed by many *in-situ* measurements following the early work by Squires (1958a, b), Twomey and Squires (1959), and Warner and Twomey (1967), for example by analysis of ship tracks using satellite images (Coakley et al. 1987; Durkee et al. 2000, 2001).

Rosenfeld and Lensky (1998) developed a method of using satellite images to estimate the effective radius of cloud particles near cloud tops as a function of temperature (a surrogate for cloud top altitude). Their analysis suggested that an increase in aerosol optical depth corresponds to slower growth of the cloud drops, because of the increase in their number concentrations and decrease in their effective radius. Slow growth in the warm clouds may lead to elimination or the suppression of precipitation development. Radar echoes from the TRMM satellite in combination with the estimate of the effective radius at cloud top were interpreted as showing that the development of precipitation in these clouds diminishes, although the number of such documented cases has been small and is controversial (Ayers 2005; Rosenfeld et al. 2006).

Airborne and ground measurements of clouds and precipitation in the Amazon region (Andreae et al. 2004) showed that clean continental clouds with relatively low numbers of aerosol particles, behaved similarly to marine clouds; namely, growth by coalescence occurs early and rapidly as the cloud develops. On the other hand, clouds that developed in the smoky atmospheres grew deeper with precipitation particles growing higher up in the clouds. Andreae et al. (2004) argued that the slow growth of the drops led to ice formation higher up in the clouds and to enhanced updrafts attributable to an increase of condensed water and resultant increased release of latent heat. Such clouds sometimes led to hail and lightning formation. These conditions were even more pronounced when the clouds developed just above the fire region (named pyroclouds). In such cases, input of heat and large numbers of smoke particles led to invigoration of the cloud, resulting in taller clouds and larger cloud particles. There is no evidence, however, that these clouds produced more or less rain amounts on the ground.

Aerosol Impact on Rainfall on the Ground

Convective Clouds

Warner and Twomey (1967) and Warner (1968) summarized the potential effects of sugarcane smoke on rainfall by looking at multidecadal rainfall records from stations up- and downwind of these prolific anthropogenic aerosol sources. Despite the expectation that there would be a direct correlation between increased pollution and rain suppression, they could not conclusively see one (Warner 1971). In his 1971 paper, Warner stated, "It would be surprising if the microphysics of a cloud played no part in determining its rainfall, but we must await further results if this is to be adequately demonstrated."

Rain enhancement of up to 30% from warm clouds downwind of paper mills in the state of Washington was reported by Hobbs et al. (1970). Analyzing the same case through the use of a one-dimensional numerical model, Hindman et al. (1977) concluded that the emitted giant CCN from the paper mill could not by themselves account for the observed large increase in rainfall, and that the total effects of heat, water vapor, and CCN from the paper mill in combination may be responsible for the increased precipitation.

It is valuable to note that Mather (1991; Mather et al. 1997) reported an increase in radar echo from mixed-phase convective clouds affected by particles emitted from paper mills in South Africa. His observations led to a large field experiment for rain enhancement using hygroscopic particles.

Through the use of MODIS and TRMM satellite data, Lin et al. (2006) analyzed the effects of forest fires on precipitation in the dry season in the Amazon region. They reported an increase in cloud heights and in precipitation with increases in aerosol optical depth. The increased cloud height led to enhanced growth of ice crystals, which culminated in heavier precipitation. However, despite the good correlation between these variables, they could not unequivocally establish causal links between aerosols and the observed changes in cloud height or with precipitation increases. The role of enhanced convection attributable to the heat from the fires and/or from absorption of solar radiation by the smoke itself could not be excluded.

There have been many numerical simulations of single convective clouds, suggesting that increase in CCN concentrations leads to reduced precipitation. However, the input of only a few giant CCN (a few per litter) is enough to enhance precipitation (e.g., Yin et al. 2000; Teller and Levin 2006). Teller and Levin (2006) showed that increasing the giant CCN from 1–10 per cm^{-3} can compensate for a decrease in precipitation amounts due to increased CCN from 200–400 cm^{-3}. Simulating a 3-D case of Crystal Face storm, in which dust particles were present, Van der Heever and Cotton (2006) demonstrated that giant CCN could affect the rain intensity early in the development of the storm by permitting higher supercooled water contents to be lifted to higher levels and freeze. Several hours later the enhanced CCN simulations produced less precipitation because the cold pools, which were associated with the evaporation of the falling hydrometeors, decoupled themselves from the sea breeze convergence zone. The overall effect is, therefore, not straightforward to allow predictions with certainty.

Effects of Urban Pollution on Rainfall

Extensive studies were conducted to explain the seemingly anomalous behavior of the precipitation around La Porte, Indiana, which is located downwind of Chicago. Local records suggest an upward shift in warm season rainfall, thunderstorms, and hail from the late 1930s to about 1965. The puzzling thing about this case is the fact that the anomaly appeared and then disappeared. In reviewing the observations, Changnon (1980) concluded that the microphysical effects must have played a role but without some dynamic or meteorological influence, this effect could not have occurred.

In the 1970s, a large field experiment (METROMEX) was conducted around St. Louis, Missouri, motivated by the examination of historical data, which revealed summer precipitation increases in the immediate downwind area of the city (Figure 16.1). The records show increases in rainfall (10–17%), moderate rain days (11–23%), heavy rainstorms (80%), thunderstorms (21%), and hail storms (30%) (Changnon et al. 1971). In his summary of METROMEX, Braham (1974) reported that the CCN production from the city was about 10^4 cm^{-2} s^{-1} (i.e., much higher than was present in the surrounding rural areas), accompanied by an increase in cloud drop concentrations and a decrease in drop

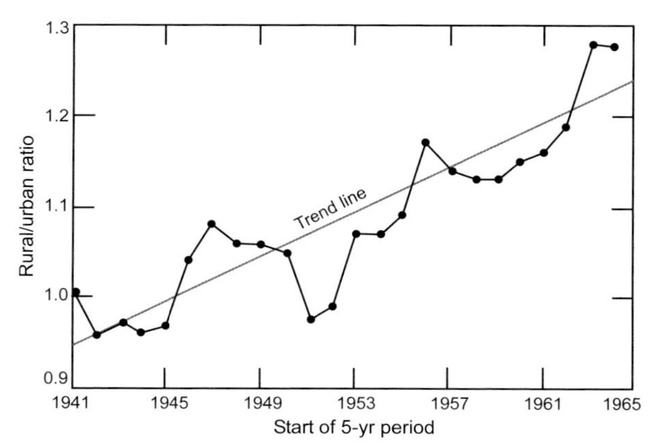

Figure 16.1 Five-year moving averages and time trend of Centerville (downwind of St. Louis) summer rainfall, 1941–1968 (after Changnon et al. 1971).

size. However, radar echoes from the large droplets in these clouds usually occurred lower in the atmosphere than their counterparts in the rural surroundings. This seems to contradict our physical understanding of cloud growth; however, Braham concluded that one way to explain the observations was to assume that the urban area also emitted giant CCN, which were not detected by the CCN sampling methods in use, but which could be responsible for the increased precipitation.

More recently, Van der Heever and Cotton (2007) simulated the effects of pollution on precipitation during the passage of a storm in the St. Louis area. The thermodynamic data that they used were from a recent specific day in which ordinary thunderstorms were prevalent, whereas aerosol input data were based on averages obtained during the METROMEX experiment. The simulation was carried out using two main cases: one with the city of St. Louis without pollution and the other with the pollution containing both small CCN and giant CCN. The model results show the expected temporal and spatial distribution of the rain changes as a result of the effects of pollution (Figure 16.2). At the beginning of the storm, polluted clouds produced much heavier precipitation; however, as the storm progressed, the difference between the integrated amounts of rain from the beginning of the storm of the polluted minus the clean case diminished. After 1.5 hour, the integrated rain amount over the whole area was higher in the clean case. This work demonstrates the complexity of the interaction between aerosols and precipitation. Part of the complexity appears attributable to the fact that the initial rain cleans the atmosphere of pollution, thus reducing the effects that pollution will have on further rainfall. In addition, downdrafts produced by the precipitation enhance the development of neighboring clouds, thus increasing the integrated rain amounts over the whole

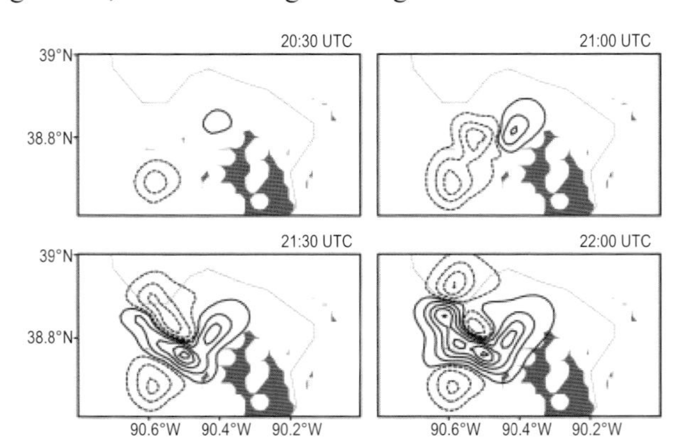

Figure 16.2 Model results showing accumulated surface precipitation from clean-polluted clouds around the city of St. Louis, Missouri. Solid lines represent pollution suppressing precipitation; Dash lines represent the opposite. Contour interval is 5 mm starting from 1 mm. Note the changes in spatial and temporal distribution of rain (after Van Den Heever and Cotton 2007).

area. Van der Heever and Cotton (2007) concluded that the effect of pollution on precipitation in an urban setting depends strongly on the background aerosol concentrations. Adding more pollution to an already polluted atmosphere has very little effect on precipitation amounts. They further indicated that effects of land use (e.g., soil moisture, release of sensible and latent heat fluxes, modification of wind stress and more) play a dominant role in those precipitation anomalies. A key factor, as far as precipitation response is concerned, is whether the secondary convection (forced by cold pools) remains coupled to the urban land-use driven mesoscale circulation or not.

Using four years of observation data obtained from the NASA Earth Observing System (EOS) MODIS, *in-situ* AERONET, and *in-situ* EPA PM2.5 data for one mid-latitude city (New York) and one subtropical city (Houston), Jin et al. (2005) analyzed diurnal, weekly, seasonal, and interannual variations of urban aerosols with an emphasis on summer months. Their research reveals that spatial and temporal urban aerosol optical depth varies dynamically as a result of various parallel factors: human activity, land-cover changes, cloud–aerosol interactions, and chemical processes. Diurnal, seasonal, and interannual variations of aerosol optical depths were examined and found to be largely affected by weather conditions; however, the optical depth peaks often during the rush hours in the morning and evening. Analysis of monthly mean aerosol optical thickness and rainfall did not show strong relationships between aerosol and rainfall in a climatological sense.

In this analysis, virtually no seasonality was observed for rainfall over Houston and New York City, suggesting that aerosols affect rainfall less than larger-scale processes (e.g., land use, urban heat island effects). Around Houston, the TRMM satellite-based accumulated rainfall data show that the maxima in monthly mean rainfall occurred in October 2000, May 2001, and September 2002. This is consistent with the transition between seasons in this region. In general, New York's rainfall had less month-to-month variation than Houston, with a maximum slightly above 200 mm/month in October 2002. The effective radius for water clouds was lower in New York City than in Houston, suggesting that there were either more aerosols in New York City than in Houston, or thinner clouds. The lack of direct relationship between rainfall and urban aerosol optical thickness implies that urban rainfall anomalies are not fully related to changes in aerosol. This observation is consistent with the earlier conclusions from METROMEX (Ackerman et al. 1978).

As can be seen from the above discussion, in spite of many measurements, there is no conclusive evidence that aerosol pollution from urban regions affects precipitation in a consistent, systematic, or repeatable manner.

Precipitation from Orographic Clouds

Borys et al. (2000, 2003) provide some evidence that pollution can delay precipitation in winter orographic clouds in the Rocky Mountains. Their analysis shows that pollution increases the concentration of CCN, and thus cloud drops, leading to the formation of smaller cloud drops. The reduced drop size leads to less efficient riming and therefore to smaller ice crystals (Figure 16.3), smaller fall velocities, and less snowfall.

Givati and Rosenfeld (2004) analyzed about 100 years of records of orographically forced convective precipitation in regions located downwind of pollution sources and compared them to precipitation in regions unaffected by these sources. In their study, they documented the precipitation trends in the orographic enhancement factor, R_o, which is defined as the ratio between precipitation over the hill with respect to the upwind, lowland precipitation amount (Givati and Rosenfeld 2004). Two geographical areas were chosen for this study: California and Israel. The topography in both regions is similar, although the mountains in Israel are much lower than the Sierra Nevada. Their statistical results for both locations show that downwind of pollution sources, on the upslope of mountains and mountain tops, orographic precipitation is reduced by ~20% and ~7%, respectively. It was hypothesized that this decrease is attributable to an increase in droplet concentrations and a decrease in droplet size. Farther downwind on the lee side of mountains, the amount of precipitation increased by ~14%. Givati and Rosenfeld postulated that this increase is because smaller cloud particles take a longer time to grow, allowing the winds aloft to carry them over the mountain top (see earlier study of similar effects, produced by deliberate over-seeding with ice-producing particles, by Hobbs 1975a, b). However, they hypothesized that the integrated rainfall amount over the whole mountain range was reduced by (an assumed but not substantiated) progressively increased pollution over the years. Subsequent studies show similar decreasing trends in R_o over a few western states in the U.S. (Rosenfeld and Givati 2006; Griffith et al. 2005) and the eastern slopes of the Colorado Rockies (Jirak and Cotton 2006). Givati and Rosenfeld argue that although absolute precipitation amounts and R_o are affected by fluctuations in the atmospheric circulation patterns, such as those associated with the Pacific

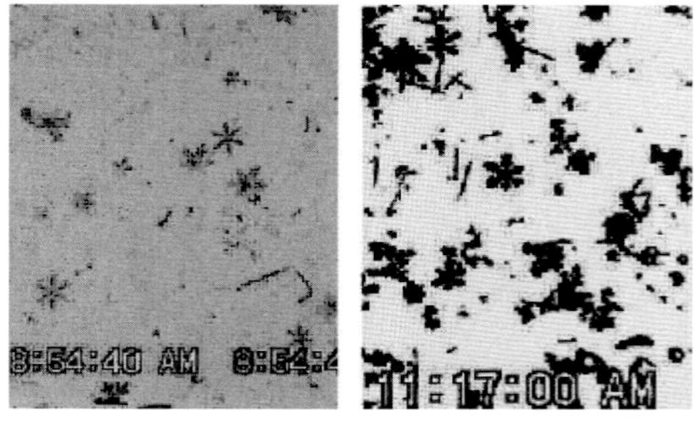

Figure 16.3 Light riming of ice crystals in clouds affected by pollution (left) compared to heavier riming in non-polluted clouds (right). (From Borys et al. 2003, by permission of the American Geophysical Union.)

Decadal Oscillation and the Southern Oscillation Index, these cannot explain the observed spatial variational trends in R_o.

Recently, Alpert et al. (2008) re-analyzed the data from Israel using the orographic ratio method, taking the ratio between the stations on the Samaria Mountain and the stations located upwind along the seashore, as well as stations on the mountains in the western Galilee and the seashore near the city of Haifa. Their results show effects opposite to those reported by Givati and Rosenfeld (2004, 2005); namely, the orographic ratio actually increased (see Figure 16.4) over the years. Alpert et al. concluded that at least in Israel, factors other than aerosol pollution dominate the precipitation amount in orographic clouds. They showed that by calculating R_o for all the stations on the mountain against the stations along the coast (and not stations located in and downwind of the urban centers, as was done by Givati and Rosenfeld), there are more cases in which R_o *increased* rather than decreased over the years (Figure 16.5). It can be clearly seen that the majority of pairs show an increase in orographic ratio between the Samaria Mountains and the seashore, in contrast to the conclusion of Givati and Rosenfeld (2004). Alpert et al (2008) argued further that the orographic ratio is not an appropriate method to estimate the effect of pollution on rainfall. This is because the ratio can decrease not only by decreasing the numerator but also by increasing the denominator. To make things worse, many of the stations upwind of the mountains used by Givati and Rosenfeld (2004, 2005) were located in or downwind of urban pollution sources, where the rain actually increased over the years. Thus, these stations were affected not only by pollution but also by other, probably much more important, factors such as urban effects (e.g., urban heat island, changes in frictional velocity,

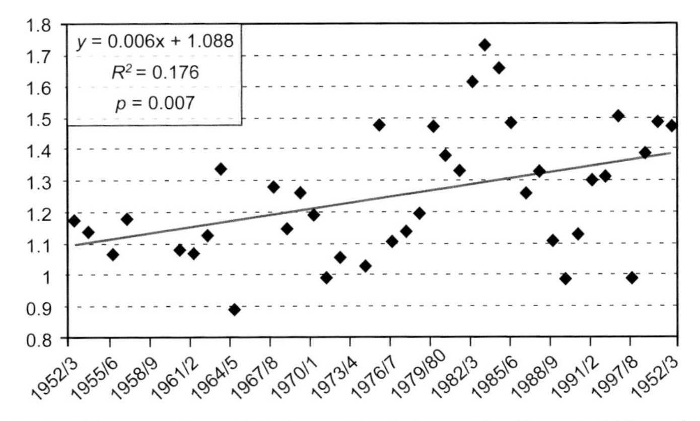

Figure 16.4 The annual precipitation ratios between the Samaria hills and the central coast clusters, for the period 1952–1998, are plotted along with the best-fit line. The dates on the abscissa represent the winter season (November to April), which is the rainy period in Israel. Note the significant increasing trend of the orographic ratio ($r = 0.42$, $p = 0.007$) in contrast to the results of Givati and Rosenfeld (2004). After Alpert et al. (2007).

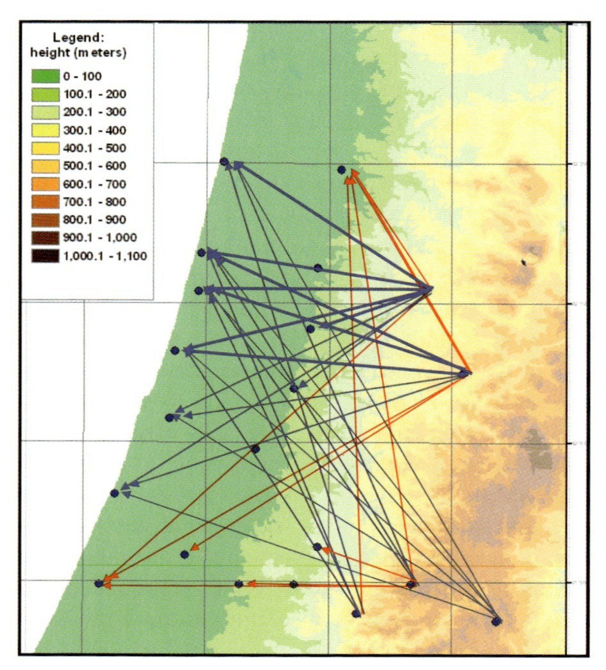

Figure 16.5 The orographic ratio (R_o) in central Israel between mountain stations and those along the shore and inland. White lines indicate an increase in R_o over the past fifty years and black lines indicate the opposite. The thicker the lines, the larger R_o is. Note that most R_o indicate an increase over the years, in contrast to the report by Givati and Rosenfeld (2004). Figure based on Alpert et al. (2008).

land-use change). Such urban effects on increased downwind precipitation have been found by many other investigators in many other locations (e.g., Braham 1974; Landsberg 1981; Goldreich 2003; Goldreich and Manes 1979).

In a more recent publication, another version of the orographic analysis was presented on hypothesized suppression of precipitation attributable to pollution in China (Rosenfeld et al. 2007). This paper relies on data from very few rain gauge stations near two cities and one station on the mountain top. Rosenfeld et al. (2007) reached strong conclusions about the suppression of precipitation by pollution, by making a connection between visibility and rainfall, where visibility is used in the paper as an indicator of pollution levels. However, their results show (Figure 4 in Rosenfeld et al. 2007) that under the same low visibility conditions, the amount of precipitation in the urban area increased (not suppressed) more than over the mountain. Thus, in contrast to the conclusion by the authors, using the orographic ratio R_o, it is impossible to separate increases attributable to the urban effects (including pollution in the urban area) from the factors affecting precipitation over the mountain. This reinforces one of the points made by Alpert et al (2008), and mentioned above: care must be taken in the choice of the stations used for evaluating the orographic ratio. It is clear that inherent variability in the climate system and changes in local population distribution will always yield particular regional (or rain gauge) patterns that, depending on choice of rain gauge, can yield diverse signals. Moreover, simple empirical correlation of surrogate variables based on historical time series cannot take into account the evident non-stationary nature of climate known to be associated with anthropogenic global warming as well as natural decadal variability. New and better statistical methods need to be developed to separate the different factors affecting precipitation.

In summary, although pollution does affect clouds, the hypothesized effects on precipitation are not yet clearly understood. In fact, even the effects of clouds on aerosols (e.g., scavenging of aerosols during the rainy period) have hardly been considered in these correlation studies. It seems quite probable that other factors, such as synoptic-scale processes, urban effects, and mesoscale dynamic factors, dominate the precipitation amounts over the microphysical processes, while long-term trends cannot be divorced from changing weather patterns driven by global warming and other long-term drivers of climate change. We will delve into this complexity by closely looking at what we know about the effects of pollution on precipitation in Australia, as but one possible case study, and then try to reflect more broadly about the potential effects elsewhere.

Complexity at Regional Level: Australia

Australia provides a useful example of the complexities inherent in seeking evidence of an aerosol-induced impact on precipitation. As a case study, it has singular advantages: it is an isolated continent effectively immune from long-range transport of particle pollution from elsewhere; it is the driest continent, and thus has a long history of scientific interest and effort in atmospheric and cloud physics research relevant to the question at hand; it is relatively unpopulated (approximately 21 million in 2006) comparable with its land area (ca. 7,600,000 km^2); and the majority of its population (~15 million, or almost two thirds) inhabit just eight cities (the State and Territory capitals), leading to strong contrasts in air composition between the isolated city air sheds and the vast, clean regions in between. Thus experiments contrasting "clean" and "polluted" conditions, un-confounded by long-range transboundary air pollution, are possible in Australia, which would be impossible in most countries of the northern hemisphere (Bigg and Turvey 1978).

Pollution

From the 1950s through the 1980s, Twomey and Bigg spent considerable effort trying to understand the characteristics and climatology of CCN, condensation (or Aitken) nuclei (CN), and ice nuclei (IN) in Australia, and more generally. Of particular note, with respect to CCN, was Twomey's systematic and comprehensive investigation of variations in cloud-active aerosols at the midlatitude site of Robertson, New South Wales. This site was located in a rural region well over 100 km to the southwest of the mega-metropolitan region covered by Sydney; it was due west of the heavily industrialized steel-making city of Wollongong, due east of the rural township of Bowral, and southeast of the twin towns of Mittagong and Moss Vale. Twomey et al. (1978) showed that air masses arriving at Robertson after passage over the major cities Sydney or Wollongong had average CCN levels that exceeded 800 cm^{-3}, air masses passing over the smaller towns of Mittagong and Moss Vale had concentrations averaging 400–800 cm^{-3}, air from Bowral typically contained 300 cm^{-3}, and air masses from other, relatively unpopulated regions exhibited CCN levels of only ~150 cm^{-3} (Figure 16.6). Clearly anthropogenic activities enhanced CCN levels severalfold over the natural, continental background levels, with plumes of cloud-active aerosols wandering over the Australian land mass as dictated by location of the urban areas and meteorology.

Bigg and Turvey (1978) reported on the results of an ambitious airborne program from 1974–1977 that mapped the distribution of CN across the whole continent as a function of altitude from the surface to about 6000 m in altitude. Their work reinforced and amplified Twomey's findings: median CN concentration over the continent in the well-mixed and relatively unpolluted boundary layer was very low by world standards at 680 cm^{-3} (very consistent with Twomey's CCN level of ~150 cm^{-3} for "clean air," given that CCN are a subset of CN). In contrast, CN plumes from population centers were greatly enhanced in concentration, with Bigg and Turvey (1978) stating:

> Particle production by one of the smaller metropolitan areas at 4×10^{19} s^{-1} was shown to exceed the estimates from all natural sources, indicating that particles resulting from human activities may often dominate the continental aerosol over much of Australia.

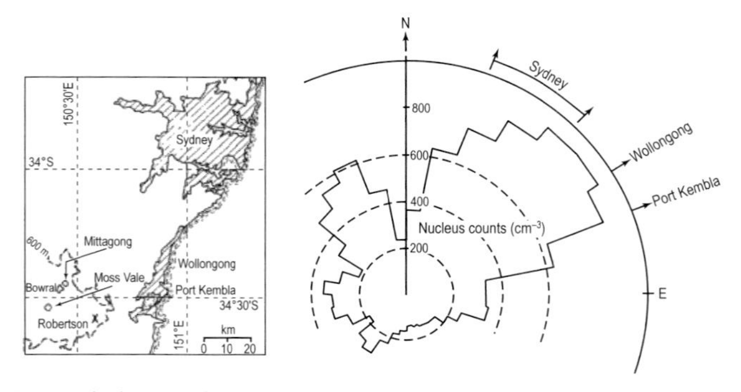

Figure 16.6 Wind rose of CCN concentrations for air masses arriving at Robertson, NSW, from 1968–1973. Reproduced from Twomey et al. (1978).

Bigg extended his measurements to evaluate total particle emission flux over Australia by developing a methodology in which an aircraft made continuous CN measurements between the surface and the boundary layer inversion at fixed distances downwind of the source. Convolved with wind speed measurements also taken by the aircraft as a function of altitude, the downwind plume CN cross section yielded an instantaneous transport rate through the sampled slice of plume, or an instantaneous "emission flux" measured at the given downwind distance. This methodology was then applied to the isolated Mt. Isa copper and lead smelter complex in tropical western Queensland as the experimental source. The repeated airborne flux measurements confirmed yet again that anthropogenic aerosol plumes exhibited greatly enhanced CN and CCN concentrations over natural, continental levels, and also traveled great distances (> 600 km). Ayers et al. (1979) and Bigg et al. (1978) provide examples of plume cross sections measured, plus derived flux estimates that decrease out to 200 km from Mt. Isa but increase thereafter as gas-to-particle conversion of the SO_2 in the smelter plume created new particles by homogeneous nucleation. Subsequently, Carras and Williams (1981) tracked the Mt. Isa plume to an astonishing 1800 km downwind, at which point it exited the Australian continent over the Indian Ocean. The importance of anthropogenic activities as a source of CCN is underscored by the estimate of Ayers et al. (1979) that Mt. Isa alone had a CCN emission flux equivalent to ~0.1% of the total, natural, global emission flux for CCN active at 0.5% supersaturation.

The same methodology was applied to a program of particle emission flux measurements from a representative number of urban centers across the country. Figure 16.7 shows CN flux as a function of population downwind of a range of towns and cities (Ayers et al. 1982). A theoretical analysis by Manton and Ayers (1982) confirmed the physical plausibility of the essentially linear

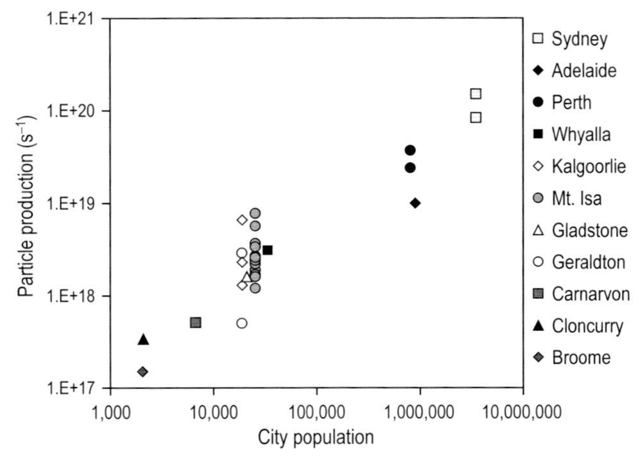

Figure 16.7 Condensation nuclei flux as a function of population downwind of a number of Australian towns and cities (Ayers et al. 1982).

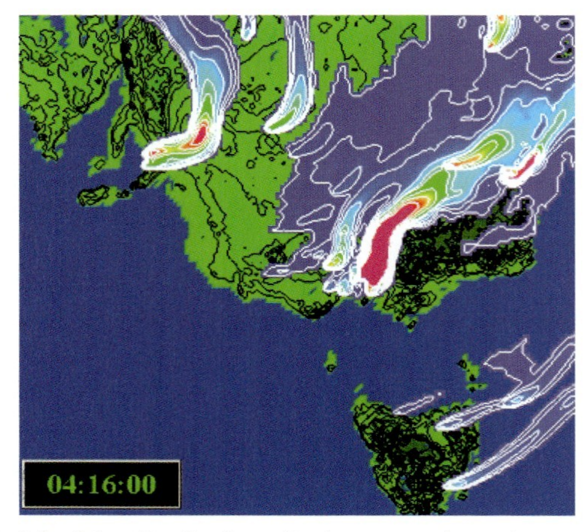

Figure 16.8 Model of the distribution of primary condensation nuclei concentration at 500 m over SE Australia on October 21, 1998, at 16.00 local time. Source function used was the relationship shown in Figure 16.6 (Ayers et al. 1982) convolved with a gridded population density database available at 1 km resolution. The chemical transport model used for this simulation was TAPM run at 12 km resolution.

relationship over more than three orders of magnitude evident from Figure 16.7: the flux–population relationship is 8×10^{13} particles per person per second. Use of this figure plus gridded population density, as a surrogate explanatory variable for primary CN emissions from human activities in Australia, enables modeling of emissions from towns of all sizes in any part of the country, as illustrated in Figure 16.8. Extrapolation to the whole continent based on total population led immediately to the conclusion that total anthropogenic particle production over Australia, at $\sim 10^{21}$ s^{-1}, exceeded the natural particle emission rate by more than an order of magnitude (Bigg and Turvey 1978).

Aerosol and Cloud Properties

The role of variability in CN and CCN concentrations as a determinant of variability in cloud microphysical structure at the condensation level has been studied in a number of Australian experiments over both the continent and ocean. Here we focus only on continental examples; for completeness, we note that additional work paralleling Bigg's marine CN studies addressed the marine environment upwind of Australia. One marine example is the work of Boers et al. (1994, 1998) and Boers and Krummel (1998), which couple data on long-term, systematic seasonal variations in CCN measured at Cape Grim (determined by the seasonal cycle in dimethylsulfide emissions from the Southern Ocean) with airborne measurements in the Southern Ocean Cloud Experiment (SOCEX) on marine stratocumulus properties upwind of Cape Grim. Figure 16.9 shows the remarkable agreement achieved between average

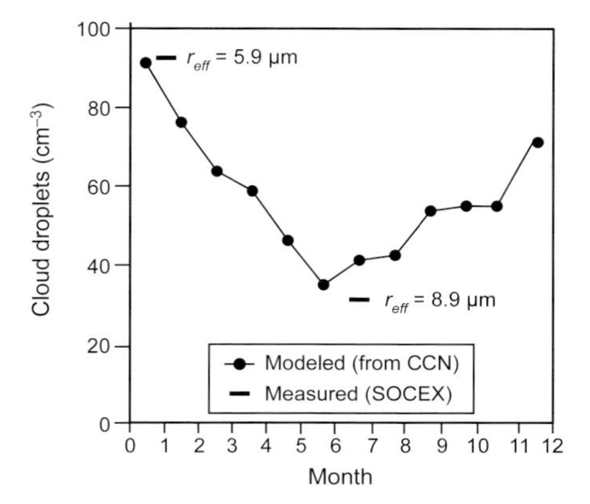

Figure 16.9 Measured (SOCEX) summer (January) versus winter (June) average droplet concentrations and effective radii near cloud base in marine stratocumulus clouds upwind of Cape Grim, Tasmania, compared with seasonal cycles modeled using an explicit cloud model and the observed monthly mean CCN concentrations measured at Cape Grim (Boers et al. 1994, 1998).

summer and winter cloud droplet concentrations just above cloud base and droplet concentrations predicted by a cloud model with explicit CCN-droplet size dependence. Here, we consider three continental examples in detail.

Effects of Smoke from Sugar Cane Fires

Warner and Twomey (1967) presented subcloud CCN and in-cloud droplet concentration data from just above the condensation level taken in airborne surveys up- and downwind of massive sugar cane fires in NE Queensland (removal of foliage by burning prior to harvest was at that time a normal practice). Their data has been replotted in Figure 16.10, demonstrating a clear

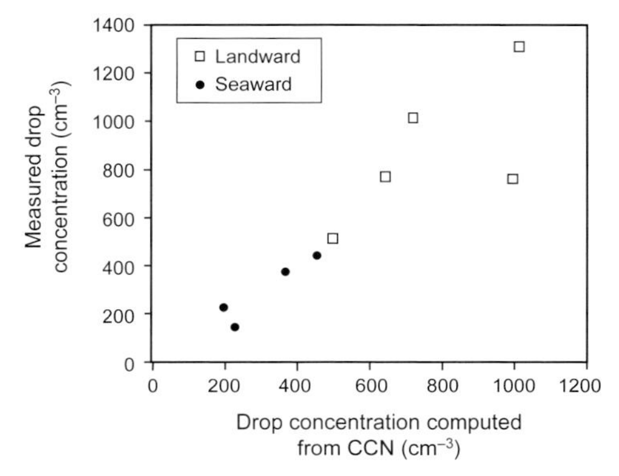

Figure 16.10 Relationship measured between subcloud CCN concentrations and cloud droplet concentration near cloud base upwind and downwind of sugar cane fires in NE Queensland (Warner and Twomey 1967).

relationship between CCN numbers below and cloud droplet concentrations above the cloud base.

Effects of Mt. Isa Copper Smelter Plume

Comparable results were obtained in studies conducted 56 km downwind of Mt. Isa in which cloud droplet size distributions were determined just above the condensation level in fair-weather cumulus clouds growing in air masses inside and outside the sulfate aerosol plume from the Mt. Isa smelter complex. Figure 16.11, redrawn from Ayers (1981), depicts cloud droplet spectra measured in two sets of clouds on March 5 and 6, 1977. Clearly evident is an elevation in cloud droplet number comparable to that noted by Warner and Twomey in smoke-affected clouds downwind of cane fires, plus the concomitant reduction in cloud droplet size that must accompany the concentration increase, assuming that everything else (e.g., liquid water content, atmospheric temperature and humidity profiles, and synoptic conditions) is the same.

Effects of the Adelaide Plume

Application by Rosenfeld (2000) of the Rosenfeld and Lensky (1998) algorithms to a remotely sensed cloud field recorded over Adelaide on August 12, 1997, showed unmistakable evidence of "pollution tracks" in clouds downwind of Adelaide and nearby rural cities. These pollution tracks showed up as broadening regions of high-concentration/low-effective radius droplets in cloud. This pattern, observed using remote-sensing technology, confirmed the evolution of spreading CN plumes demonstrated downwind of Adelaide and other Australian cities by Bigg and Turvey (1978), Ayers et al. (1979, 1982), and Manton and Ayers (1982). The explicit connection between the city as a

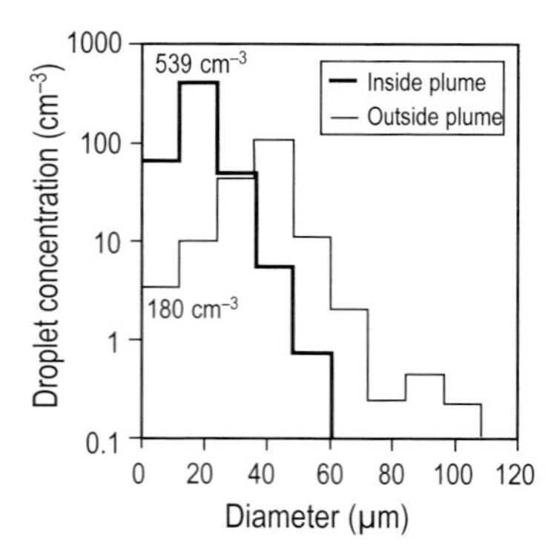

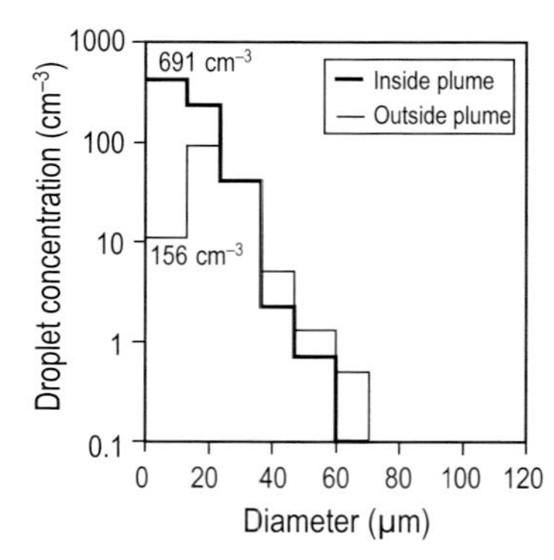

Figure 16.11 Droplet distributions measured close to the cloud base in fair weather cumulus clouds near Mt. Isa growing in air inside or outside the plume from the Mt. Isa copper smelter (Ayers 1981). Top: March 5, 1977; bottom: March 6, 1977.

prolific aerosol source and a resultant increase in cloud droplet numbers or decrease in cloud droplet size provided a modern reconfirmation of the earlier works of Warner, Twomey, and Ayers in NE Queensland and at Mt. Isa.

Precipitation Suppression

The previous sections provide compelling evidence that even over a relatively unpopulated continent, such as Australia, anthropogenic aerosols dominate the aerosol number concentration, and that where these aerosols are mixed up into cloud, they affect cloud microphysical properties in the anticipated ways: increasing cloud droplet number concentration and decreasing droplet effective radius. Although this clearly has the potential to affect atmospheric radiative transfer via an effect on cloud albedo, as foreshadowed by Twomey (1977), it is not the so-called "first indirect effect" (IPCC 2001) that is the focus here, but rather the subsequent effect, if any, on formation of precipitation.

It has been known for some time that a broad droplet distribution with a significant population of cloud droplets (radii > 15 μm) is needed for precipitation to occur efficiently in warm clouds (Fletcher 1962), as has the hypothesis that polluted clouds having small droplets and narrow droplet size distribution might be less efficient in production of rain. However, only recently has "rainfall suppression by pollution" been claimed to have been *demonstrated* (Rosenfeld 2000). From experience, we know that no matter how clear the effects of aerosols are on cloud microphysical properties, it has proved extremely difficult to test the subsequent hypothesis that microphysical changes significantly alter precipitation. This comment stems from the 50-year history of an inability by the scientific community to demonstrate unequivocally that well-designed, intentional weather modification experiments based on aerosol dispersal in cloud can demonstrate altered precipitation patterns at scales above that of single clouds. It is worth emphasizing the need for caution on this issue by restating a key introductory sentence from the U.S. National Academies report (BASC 2003): "The committee concludes that there is still no convincing scientific proof of the efficacy of intentional weather modification efforts."

Effects of Smoke from Sugar Cane Fires

The demonstration by Warner and Twomey (1967) that biomass burning aerosols from cane fires dramatically altered downwind cloud properties provided Warner with a natural laboratory in which to seek evidence of concomitant reductions in rainfall associated with aerosol pollution. The experimental design was relatively straightforward. Warner (1968) presented an analysis of 60 years of rainfall records for stations up- and downwind of the cane-growing regions of NE Queensland, and sought a decrease in downwind rainfall with time that mirrored the history of expanding cane production. Warner produced the cautious conclusion that such a signal appeared to be evident in the data, but noted that "the possibility that other factors caused the particular climatic changes observed cannot be eliminated."

Based on this promising result, Warner (1971) went on to conduct a more comprehensive analysis, which was presented at the 1971 International Conference on Weather Modification in Canberra. However, this more complete analysis led him to revise the earlier conclusion: "What has been said above may explain why it was not possible to detect an association between increasing cane production and rainfall. Nevertheless, we must conclude that the present study gives no support to the idea that association found between cane fires and rainfall at Bundaberg was due to inhibition of the coalescence process by a reduction in average cloud droplet size." A null result.

Effects of the Adelaide Plume

In contrast to Warner's carefully argued null result, Rosenfeld (2000) claimed positive evidence for long-term suppression of rain and snow formation through anthropogenic aerosols downwind of Adelaide. A careful reading of Rosenfeld's paper, however, suggests a need for caution. His conclusion is not based on a representative sample, since it is based on the detailed analysis of only a single satellite image, which is an impossibly small "instantaneous" sample from which to reach such a substantial conclusion concerning long-term spatial and temporal rainfall patterns in SE Australia. Moreover, there was an absence of supporting evidence to validate any of the (a) remotely sensed cloud properties, (b) pattern of aerosol pollutants across the remotely sensed cloud fields, (c) spatial distribution of liquid water content in the clouds, and (d) pattern of rainfall at the ground. Each of these factors was considered in detail by Ayers (2005), who concluded: "In the case of rainfall suppression over Australia presented by Rosenfeld (2000), the analysis presented above identifies major doubts about the conclusions reached; indeed the conclusions in that work are invalid based on the evidence available. This is the same position reached by Warner (1971) almost two decades earlier." A rebuttal of this conclusion was subsequently published by Rosenfeld et al. (2006). However the rebuttal fails to address either the problems of detail or the two fundamental flaws evident in Rosenfeld (2000): neither (a) the experimental design used in that work nor (b) the specific day chosen for study is capable of providing a scientifically valid test of the rainfall suppression hypothesis.

Precipitation in Australia

Trends

Rainfall trends over the last century across Australia have been analyzed by a number of people (e.g., Suppiah and Hennessy 1998; Hennessy et al. 1999; Smith 2004). For all-Australian rainfall, Smith (2004) identified a positive long-term trend over the century, though differences in sign of the trend and with respect to season, along with high variability, emerge when the data are broken down to regional scale (Suppiah and Hennessy 1998; Hennessy et al. 1999; Smith 2004). Significantly, a generalized Australia-wide long-term suppression of rainfall caused by increasing population and increasing pollution emissions over the decades is not supported by these data: it would be illogical to argue rainfall suppression on average when long-term rainfall on average has increased. This is also the case when the southeast region discussed by Rosenfeld (2000) is considered. Figure 16.12 plots the average rainfall time series for SE Australia, which reveals a positive trend up until 1995, followed by a sharp reduction thereafter. This stands in sharp contrast to Rosenfeld's qualitative claim that rainfall decreased over decades in this region since pollutant emissions have grown slowly with population. The more so, since the decline, post-1996, corresponds with the implementation of a National Environmental Protection Measure for Air (the Air NEPM) in 1998, which resulted in a stabilization of urban air quality in the decade thereafter, during which rainfall sharply declined. Evidently the recent rainfall decline cannot be ascribed a priori to a concomitant increase in level of pollution. It is equally evident that simple correlations between rainfall and assumed or implied levels of "pollution" do not provide a particularly robust test of the rainfall suppression hypothesis. This is because the inherent variability in the climate system will *always* yield particular regional (or rain gauge) patterns that, depending on choice of rain gauge, can be interpreted in a variety of ways (see the discussion above relating to the effects of pollution on orographic precipitation in Israel). We conclude that in the absence of comprehensive time series data constraining long-term changes

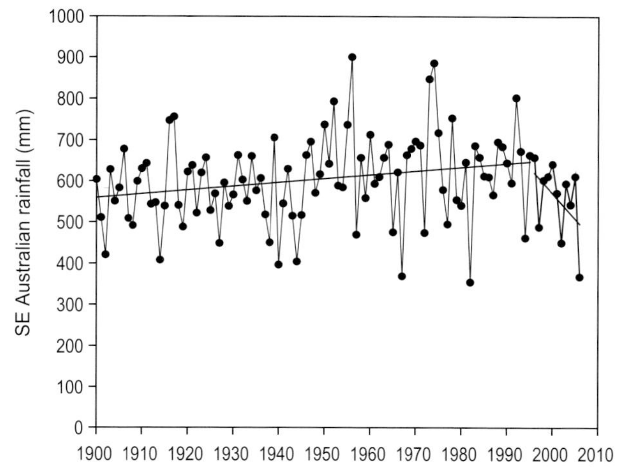

Figure 16.12 Long-term average rainfall for southeastern Australia. Separate trend lines are shown for 1900–1995 and 1996–2006 (data source: Australian Bureau of Meteorology).

in microphysical processes and the CCN environment, it is not possible to ascribe rainfall trends as being microphysically driven.

Moreover, since it is clear that climate variation over the latter part of last century in particular has led to changes in atmospheric and oceanic circulation, explanations for rainfall trends must include dynamic as well as microphysical explanations (i.e., caused by the non-stationary nature of the climate system), a proposition that we examine next.

As a lead-in to that discussion, Figure 16.13 depicts pictorially the continental rainfall trends from 1970–2006. It is evident that there are major patterns of change at work across the continent, which do not show any broad-scale qualitative relationship with population center (and hence pollutant) distribution. The decade-long national drying trend from 1996, evident in Figure 16.13, is also visible in the continental picture across the southwestern, southern, and eastern parts of the continent, as in the long-term trend to increased tropical rainfall in NW Western Australia.

Explanations of Rainfall Trends

Space limitations preclude the presentation of more than two regional examples. For the first example, we continue the discussion of rainfall trends in SE Australia, where Rosenfeld (2000) supported his claim of rainfall suppression through air pollution by alluding to a "decreasing trend in snow cover in the Snowy Mountains in the period 1897–1991" along with "decreases" (although these were acknowledged as statistically insignificant) in "snow, winter temperature and total winter rainfall." Nicholls (2000) analyzed, in detail, the average rainfall in the Snowy Mountains district (District 71), which had shown

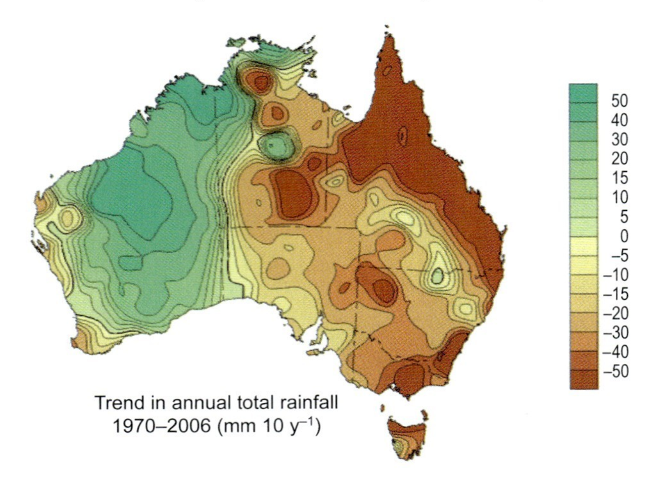

Figure 16.13 Rainfall trends across Australia, 1970–2006, in mm per decade (data source: Australian Bureau of Meteorology).

a strong long-term decrease in precipitation during the period 1913–1992, and which at face value appeared to support Rosenfeld's suggestion. Nicholls showed, however, that the strong decline was artificial, that it was produced by the closure of a high altitude/high rainfall-reporting site in 1954, which meant that data from only one high rainfall site, rather than two, were contributing to the district average, leading to an artificial decline in the average over the century. His reconstruction of a composite high-elevation record showed no significant decline. This example highlights one of the major challenges in time series studies: it is imperative to ensure that time series data are of consistent quality over a decades-to-century time span.

On the separate issue of a strong decreasing trend over time, evident from the snow depths in the Snowy Mountains upon an initial observation in October, Nicholls (2005) produced another elegant analysis of the relevant data to demonstrate that the trend between 1962 and 2002 was not principally related to changes in precipitation, but rather to significant regional warming over the last half-century, which is best linked to effects of climate variation (Figure 16.14). Snow depth declined primarily because of melting (i.e., temperature), not because of a strong decrease in precipitation.

A third observation concerning rain-bearing systems in SE Australia is that only very small changes in dynamic forcing are sufficient to yield variations in precipitation. Manton (1979) demonstrated this susceptibility by detecting the orographic influence on rainfall from individual weather systems using pluviograph data from eight sites, each of which had a range in elevation of only 200 m. Manton's hypothesis was that small spatial variations in dynamic forcing

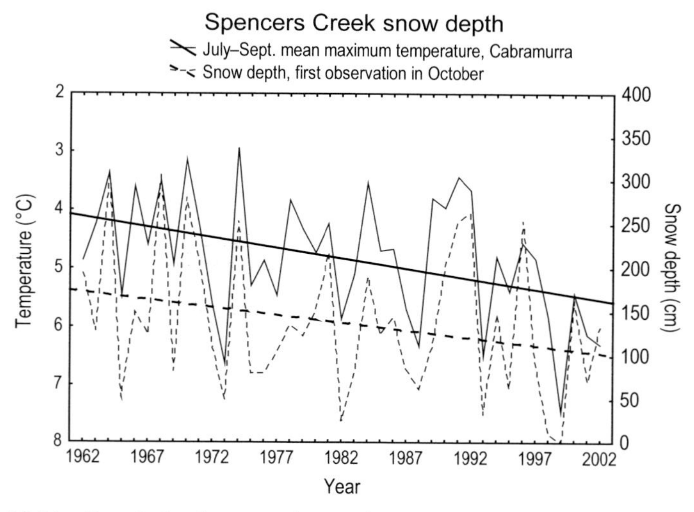

Figure 16.14 Correlation between time series trend and variability in snow depth in the Snowy Mountains, SE Australia, with time series and trend in maximum air temperature. After Nicholls (2005).

may induce systematic variations in rainfall from a cloud system which would otherwise be statistically homogenous. His confirmation of this hypothesis can be expected to have temporal as well as spatial implications: analysis of time series trends in precipitation over multiple decades, when evidence of an aerosol microphysical effect was sought, will need to account for the confounding causality associated instead with trends in dynamic forcing (e.g., greenhouse-induced shifts in regional dynamics, changes surface albedo, roughness, moisture availability from land-use change).

For the second example, we move from southeast to southwest Australia, where a secular decline in precipitation, on the order of 20%, has been under way since the mid-1970s. Figure 16.15 shows the associated reductions of inflow to water storages for the capital city of Perth; in February 2007, this led the State government to commission a seawater desalination plant to supply 45 gigaliters per year of potable drinking water to the city's residents.

Although it has been claimed in the Australian news media (e.g., ABC News, January 9, 2007) that the Perth rainfall decline was predominately caused by the cloud–microphysical consequences of air pollution, we are unaware of any scientific analyses or publications that support this assertion. To the contrary, extensive analyses undertaken by the multiagency Indian Ocean Climate Initiative (IOCI 2002) have identified a range of other potential contributors to the decline that are coincident with large-scale changes in atmospheric dynamics and circulation patterns. One significant argument favoring controls on rainfall in this region being large-scale is the observation that there is a very strong correlation between rainfall, at all timescales, and mean sea-level pressure (IOCI 2002). More detailed analyses show influences including changes in Indian Ocean temperature and circulation, potential effects of stratospheric ozone depletion on the Southern Annular Mode, large-scale greenhouse forcing (concluded to provide 50% of the declining precipitation signal), and land-use change (e.g., Lyons 2002; Pitman et al. 2004; Cai et al. 2003, 2005, 2007; Cai and Cowan 2006; Timball and Arblaster 2006; Timbal et al. 2006).

In addition, it is important to note that multicentury global climate model simulations suggested that the 30-year drying trend has also a finite probability of occurrence as part of natural, long-term climate variability. When analyzing any time series trend, it is therefore important to include natural variability as one component of an explanatory hypothesis. An overall analysis of the change in wintertime atmospheric states by Frederiksen and Frederiksen (2007), which used NCEP/NCAR reanalyses for the periods 1949–1968 and 1974–1994, led them to conclude that from 1975–1994 the primary process involved in rainfall reduction in SW Western Australia was a reduction of intensity of cyclogenesis and southward deflection of some storms. A poleward shift in storm tracks would appear to be consistent with the observed freshening of intermediate waters of the Southern Ocean (between 55°S and 65°S), interpreted by Wong et al. (1999) as an increase in precipitation over this oceanic region. In addition, Frederiksen and Frederiksen (2007) point out that the observed changes

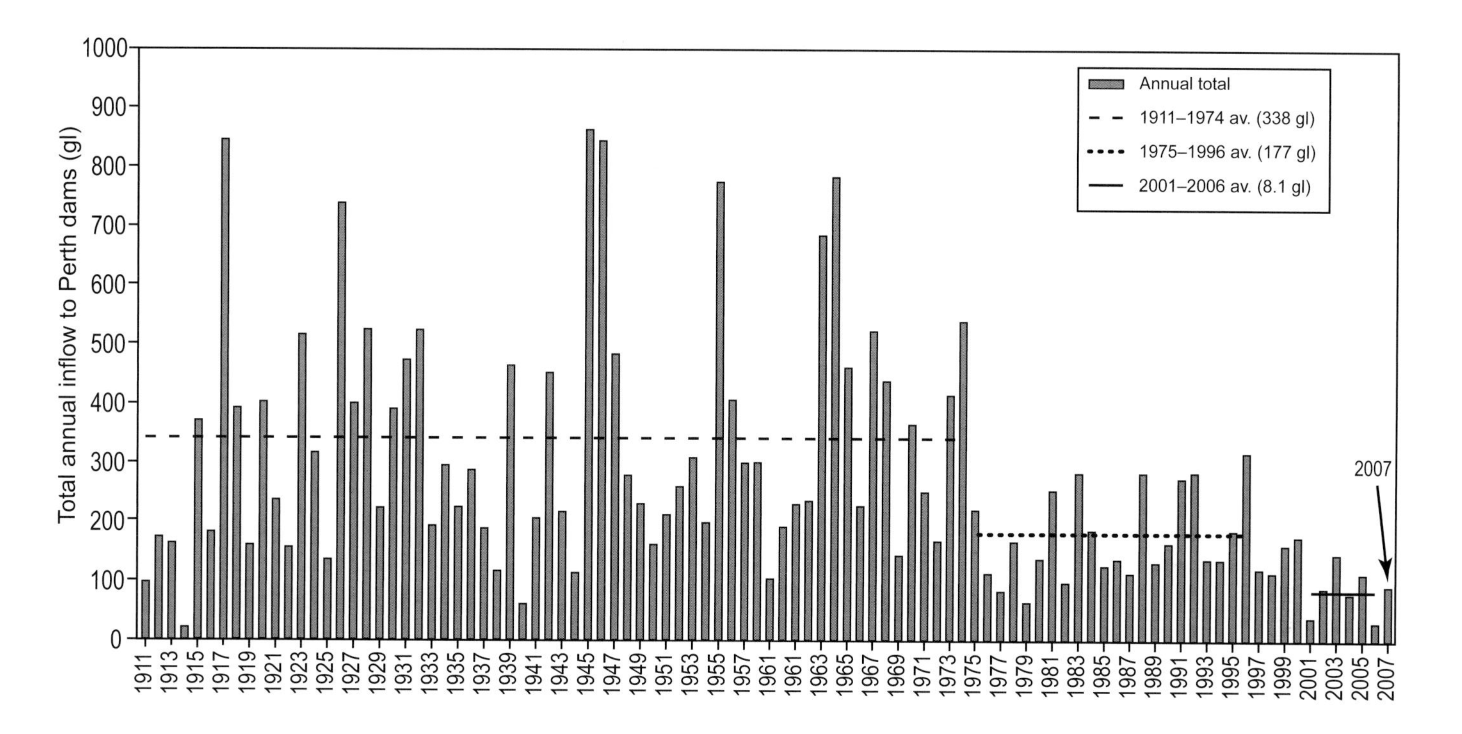

Figure 16.15 Inflow to Perth water storages since 1911, showing a decline over the two decades 1975–1995, and another decline in the decade since, consistent with the general drying trend across Australia shown in Figures 16.13 and 16.14.

in the climate of the southern Hemisphere, including reduction in equator-to-pole tropospheric temperature gradient and consequent reduction in zonal flow, are consistent with the effects of anthropogenic greenhouse forcing. However, they show appropriate restraint in stating that at this stage such a conclusion has not yet been established.

A second evaluation, this time of temporal variability in dominant synoptic types influencing SW Western Australia (Hope et al. 2006), confirmed that the frequency of occurrence of troughs associated with wet conditions had declined in the region while the frequency of synoptic categories associated with high pressure and dry conditions over the continent had increased, consistent with the findings of Frederiksen and Frederiksen (2007). A cogent summary of all these influences and a comprehensive list of relevant publications set within the context of overall detection and attribution of Australian climate change has been provided by Nicholls (2006). Notably, in this very balanced evaluation, is the acknowledgment that the Indian Ocean Climate Initiative (IOCI 2002) had considered increased local aerosol pollution as one possible causative agent for decreasing rainfall trends, but concluded that it could not be more than a secondary (unlikely) contributor in the face of evident dynamical signals. The key findings from the IOCI have been recently summarized by Bates et al. (2007), and Rotstayn (2007) provides an additional argument on the need to understand dynamical effects of pollutant aerosol.

Synthesis

Four key conclusions can be drawn from our Australian case study. First, it has been demonstrated beyond doubt that even for this relatively unpopulated country (on average fewer that 3 people per km^2), anthropogenic emissions of CN and CCN at aggregate level dominate over natural emissions. Given the concentration of a majority of the population in fewer than ten large cities, aerosol plumes containing high concentrations of CN and CCN separated by regions of lower, or "background" CN and CCN concentration have been well known since the late 1960s. These observations are in agreement with many other studies that have been conducted in other parts of the world over the last few decades, which show increases in CN and CCN concentrations downwind of populated centers.

Second, given this situation of identifiable spatial gradients in CN and CCN levels, it has been demonstrated unequivocally that clouds forming in more polluted air contain, near the cloud base, higher concentrations of smaller droplets than clouds forming in less polluted air. Again this observation is not unique and can be generalized to other regions of the world, as has been demonstrated in many publications (e.g., Garrett and Hobbs 1995).

Third, where temporal and spatial trends in precipitation have been analyzed in Australia, changes in atmospheric dynamic processes offered the best explanations, rather than changed cloud microphysics, all other things being equal. This, too, is in agreement with other studies, such as METROMEX, which showed that although cloud microphysical properties have changed as a result of aerosol pollution, precipitation had actually increased.

Fourth, however, we must acknowledge that the work done to date is incapable of definitively ruling out microphysical forcing on precipitation efficiency as a part of the story in some circumstances (e.g., the presence of giant CCN as a way to increase precipitation development even under polluted environments). Although the observations suggest that the effect on total precipitation on the ground attributable to modification in cloud microphysics is relatively small, no experiments have been, in our opinion, yet designed and carried out anywhere that consider and link, in an integrated way, process studies to demonstrate causality, with time-series studies to define precipitation trends, while covering concurrently both microphysical and dynamic forcing on precipitation processes (i.e., all confounders at once) and their interactions. Better statistical methods need to be developed and applied to separate the relative contributions of the different processes and their synergistic effects. A recommendation along this line was recently outlined in the WMO/IUGG report (Levin and Cotton 2008). We suggest that this challenge (i.e., experimental design) will need to be met before a definitive scientific view can be achieved on the otherwise plausible hypothesis that anthropogenic aerosol emissions have the potential to affect precipitation through microphysical pathways.

The Human Dimensions: Pollution, Precipitation, and Polarization

One of the key features of the long history of scientific endeavor on artificial weather modification is the polarization of opinion that has resulted not only within scientific circles but more widely in the general community as well as within the ranks of the news media, legislators, and policy-makers. As scientists we need to be careful how we communicate our results to the general public, governments, and policy-makers, not declaring a hypothesis as proof, until it is proven. This point is germane to the way the issue of pollution suppressing rainfall has played out in the public domain in Australia, Israel, and other countries. This has parallels with the issue of artificial rain enhancement, so we again mention the headline finding of the U.S. National Academies report: "The committee concludes that there is still no convincing scientific proof of the efficacy of intentional weather modification efforts" (BASC 2003). It also recommended against conducting weather modification operations at this time, stating that this was premature until more research had clarified scientific issues that (despite evident promise) still remained unresolved. This scientific view was not well received in the operational cloud-seeding community, or by some of its supporters in the scientific community. Further consequences of this polarity in views are evident, for example, in the WMO Commission for Atmospheric Sciences. During its fourteenth session in Cape Town in 2006, the draft version of the WMO Statement on Weather Modification and the WMO Guidelines for Planning of Weather Modification Activities, written by the WMO Expert Working Group on Cloud Physics and Chemistry and Weather Modification Research, was not approved. Modifications, which included more positive statements about weather modification, had to be inserted before approval was reached. We observe that the polarity evident in that case has spilled over to, and become linked with, the issue of inadvertent weather modification by pollution (specifically, the hypothesis of widespread, aerosol-induced reduction in precipitation).

Once again, the situation in Australia provides a clear example of experiences that have repeated elsewhere around the world: complex, polarized interactions unfolded across the scientific, commercial, news media, and public policy domains, primarily as a result of strong arguments by one or two individuals who claimed that rainfall suppression by pollution is proven, even down to the level of gigaliters of water involved (e.g., Figure 10 in Ayers 2005). Another specific illustration of this is the issue of rainfall decline in SW Western Australia, which was explored in detail in the previous section. A particular difficulty in such cases is the confusing media storyline: the media report that aerosol pollution suppresses rainfall, but that additional aerosols, in the form of cloud-seeding materials, can fix the problem.

A common feature in press reports that link aerosol pollution and cloud seeding is the lack of acknowledgment that the science of cloud seeding itself contains significant uncertainty. For example, in Australia there has been limited public acknowledgment of the comprehensive scientific review carried out by the U.S. National Academies of Science quoted earlier, or of a recent Australian analysis (Fletcher 2002), which concluded that both precipitation suppression by pollution and cloud seeding as an antidote should not be seen as scientific certainties. Similarly, there has been little public exposure of results from the more recent independent statistical review of historical cloud seeding in Israel, where conclusions were also very muted. Three relevant quotes from the Abstract of the independent review into cloud seeding in Israel are (Kessler et al. 2006):

> Based on the review of previous studies and the results obtained in this study, it can be generally concluded that the effectiveness of cloud seeding on the rainfall quantities has not been proven beyond any doubt.

> Therefore, it can be concluded that cloud seeding has some positive impact on rainfall enhancement. Subdividing the rainfall intensities into ranges, a positive effect was mostly found within the middle range of 3–15 mm/day. In the upper range (higher than 15 mm/day), the seeding effectiveness is diminishing and even becomes negative. This suggests that cloud seeding at a time of heavy storms has little impact with possible negative effects.

> No supporting evidence was found for the thesis of Givati and Rosenfeld (2005) regarding the decline in the orographic precipitations due to the increase of air pollution.

As noted earlier, Alpert et al. (2008) lend additional support to the latter conclusion, when they demonstrated that rain suppression, attributable to pollution in Israel, cannot be identified from the available data. This contrasts the conclusions of Givati and Rosenfeld (2004, 2005).

Therefore, in our opinion, the coupling of the two issues to produce a single proposition that (a) pollution suppresses rainfall and (b) cloud seeding can fix the problem does not, as yet, have a sound scientific basis.

It is our hope that dialog resulting from this Forum will temper the external debate, as in the past it has often been quite polarized, which has only served to obscure the issues in a most unhelpful way. We hope that the science community will provide agreed upon advice, acknowledging caveats, even though the science is not settled. To assist in this, we point to a publication that illuminates the sorts of issues inherent in achieving such a result: Polacheck (2006) provides a careful deconstruction of the problems inherent in a situation where a high profile journal and media exposure run ahead of the scientific community's consensus in an area where policy implications abound—a significant parallel to our situation involving pollutant effects upon rainfall. Another, by Surowiecki (2004), argues convincingly that the consensus of a community ("a crowd"), as long as it contains independence and diversity, will inevitably come to better conclusions than those made separately by individual experts, no matter how expert any particular guru is. This is a useful perspective with which to view media commentary stating that pollution suppresses precipitation.

References

ABC News. 2007. WA air pollution contributing to lack of rain, research says. Jan 9, 2007.

Ackerman, B., S. A. Changnon, G. L. Dzurisin et al. 1978. Summary of METROMEX. Causes of Precipitation Anomalies, vol. 2., pp. 395. Champaign: Illinois State Water Survey.

Alpert, P., N. Halfon, and Z. Levin. 2008. Does pollution really suppress precipitation in Israel? *J. Appl. Meteor. Climatol.* **47(4)**:933–943.

Andreae, M. O., D. Rosenfeld, P. Artaxo et al. 2004. Smoking rain clouds over the Amazon. *Science* **303**:1337–1342.

Ayers, G. P. 1981. Natural aerosols, anthropogenic aerosols and cloud properties over Australia. In: Proc. 7th Intl. Clean Air Conference, pp. 375–385. Adelaide: Ann Arbor Science.

Ayers, G. P. 2005. Air pollution and climate change: Has air pollution suppressed rainfall over Australia? *Clean Air Environ. Qual.* **39**:51–57.

Ayers, G. P., E. K. Bigg, and D. E. Turvey. 1979. Aitken particle and cloud condensation nucleus fluxes in the plume from an isolated industrial source. *J. Appl. Meteor.* **18**:449–459.

Ayers, G. P., E. K. Bigg, D. E. Turvey, and M. J. Manton. 1982. Urban influences on condensation nuclei over a continent. *Atmos. Environ.* **16**:951–954.

Bates, B. C., P. Hope, B. F. Ryan, and I. Smith. 2007. Key findings from the Indian Ocean Climate Initiative and their impact on policy development in Australia. *Climate Change* **89(3–4)**:339.

BASC. 2003. Critical Issues in Weather Modification Research. U.S. Natl. Academies Board on Atmospheric Sciences and Climate. Washington, DC: National Academies Press.

Bigg, E. K., G. P. Ayers, and D. E. Turvey. 1978. Measurements of the dispersion of a smoke plume at large distances from the source. *Atmos. Environ.* **12**:1815–1818.

Bigg, E. K., and D. E. Turvey. 1978. Sources of natural atmospheric particles over Australia. *Atmos. Environ.* **12**:1643–1655.

Boers, R., G. P. Ayers, and J. L. Gras. 1994. Coherence between seasonal variation in satellite-derived cloud optical depth and boundary layer CCN concentrations at a mid-latitude Southern Hemisphere station. *Tellus* **46B**:123–131.

Boers, R., J. B. Jensen, and P. B. Krummel. 1998. Microphysical and short-wave radiative structure of stratocumulus clouds over the Southern Ocean: summer results and seasonal differences. *Q. J. Roy. Meteor. Soc.* **124(545A)**:151–168.

Boers, R., and P. B. Krummel. 1998. Microphysical properties of boundary layer clouds over the southern ocean during ACE 1. *J. Geophsy. Res.* **103(D13)**:16,651–16,663.

Borys, R. D., D. H. Lowenthal, S. A. Cohn, and W. O. J. Brown. 2003. Mountaintop and radar measurements of anthropogenic aerosol effects on snow growth and snow rate. *Geophys. Res. Lett.* **30(10)**:1538.

Borys, R. D., D. H. Lowenthal, and D. L. Mitchel. 2000. The relationship among cloud physics, chemistry and precipitation rate in cold mountain clouds. *Atmos. Environ.* **34**:2593–2602.

Braham, R. R., Jr. 1974. Cloud Physics of urban weather modification: A preliminary report. *Bull. Amer. Meteor. Soc.* **55**:100–106.

Cai, W. J., and T. D. Cowan. 2006. SAM and regional rainfall in IPCC AR4 models: Can anthropogenic forcing account for southwest Western Australian winter rainfall reduction? *Geophys. Res. Lett.* **33**:L24708.

Cai, W., T. Cowan, M. Dix et al. 2007. Anthropogenic aerosol forcing and the structure of temperature trends in the southern Indian Ocean. *Geophys. Res. Lett.* **34**:L14611.

Cai, W., G. Shi, and Y. Li. 2005. Multidecadal fluctuation sin winter rainfall over southwest Western Australia simulated in the CSIRO Mark 3 coupled model. *Geophys. Res. Lett.* **32**:L12701.

Cai, W., P. Whetton, and D. J. Karoly. 2003. The response of the Antarctic Oscillation to increasing and stabilized atmospheric CO_2. *J. Climate* **16**:1525–1538.

Carras, J. N., and D. J. Williams. 1981. The long-range dispersion of pollution from an isolated source. *Atmos. Environ.* **15**:2207–2215.

Changnon, S. A., Jr. 1980. More on the LaPorte anomaly: A review. *Bull. Amer. Meteor. Soc.* **61**:702.

Changnon, S. A., Jr., F. A. Huff, and R. G. Semonin. 1971. METROMEX: an investigation of inadvertent weather modification. *Bull. Amer. Meteor. Soc.* **52**:958–968.

Coakley, J. A., Jr., R. L. Bernstein, and P. A. Durkee. 1987. Effects of ship track effluents on cloud reflectivity. *Science* **255**:423–430.

Durkee, P. A., R. E. Chartier, A. Brown et al. 2000. Composite ship track characteristics. *J. Atmos. Sci.* **57**:2542–2553.

Durkee, P. A., K. J. Noone, R. J. Ferek et al. 2001. The impact of ship-produced aerosols on the microstructure and albedo of warm marine stratocumulus clouds: A test of MAST hypotheses 1i and 1ii. *J. Atmos. Sci.* **57**:2554–2569.

Fletcher, N. H. 1962. The Physics of Rainclouds. Cambridge: Cambridge Univ. Press.

Fletcher, N. H. 2002. Advice on the impact of pollution on rainfall and the potential benefits of cloud seeding. A report to the Secretary Department of the Environment and Heritage, 5 Dec. 2002. [http://www.environment.gov.au/water/publications/urban/cloud-seeding.html]

Frederiksen, J. S., and C. S. Frederiksen. 2007. Inter-decadal changes in Southern Hemisphere winter storm track modes. *Tellus A* **59(5)**:599–617.

Garrett, T. J., and P. V. Hobbs. 1995. Long-range transport of continental aerosols over the Atlantic Ocean and their effects on cloud droplet size distributions. *J. Atmos. Sci.* **52**:2977–2984.

Givati, A., and D. Rosenfeld. 2004. Quantifying precipitation suppression due to air pollution. *J. Appl. Meteor.* **43**:1038–1056.

Givati, A., and D. Rosenfeld. 2005. Separation between cloud-seeding and air-pollution effects. *J. Appl. Meteor.* **44**:1298–1315.

Goldreich, Y. 2003. The Climate of Israel, Observations, Research and Applications. New York: Cluwer Academic Press.

Goldreich, Y., and A. Manes. 1979. Urban effects on precipitation patterns in the greater Tel-Aviv area. *Arch. Meteor. Geophys. Biokl.* **27B**:213–224

Griffith, D. A., M. E. Solak, and D. P. Yorty. 2005. Is air pollution impacting winter orographic precipitation in Utah? Weather modification association. *J. Weather Modif.* **37**:14–20.

Hennessy, K. J., R. Suppiah, and C.M. Page. 1999. Australian rainfall changes, 1910–1995. *Aust. Meteor. Mag.* **48**:1–13.

Hindman, E. E., II, P. V. Hobbs, and L. F. Radke. 1977. Cloud condensation nucleus size distributions and their effects on cloud droplet size distributions. *J. Atmos. Sci.* **34**:951–955.

Hobbs, P. V. 1975a. The nature of winter clouds and precipitation in the Cascade Mountains and their modification by artificial seeding. Part I: Natural conditions. *J. Appl. Meteor.* **14**:783–804.

Hobbs, P. V. 1975b. The nature of winter clouds and precipitation in the Cascade Mountains and their modification by artificial seeding. Part III: Case studies of the effects of seeding. *J. Appl. Meteor.* **14**:819–858.

Hobbs, P. V., L. F. Radke, and E. Shumway. 1970. Cloud condensation nuclei from industrial sources and their apparent effect on precipitation in Washington State. *J. Atmos. Sci.* **27**:81–89.

Hope, P. K., W. Drosdowsky, and N. Nicholls. 2006. Shifts in the synoptic systems influencing southwest Western Australia. *Climate Dyn.* **26**:751–764.

IOCI. 2002. Climate variability and change in southwestern Western Australia. Perth: IOCI.

IPCC. 2001. IPCC Third Assessment Report: Climate Change 2001, ed. R. T. Watson et al. Cambridge: Cambridge Univ. Press.

Jin, M., J. M. Shepherd, and M. D. King. 2005. Urban aerosols and their variations with clouds and rainfall: A case study for New York and Houston. *J. Geophys. Res.* **110**:D10S20.

Jirak, I. L., and W. R. Cotton. 2006. Effect of air pollution on precipitation along the front range of the Rocky Mountains. *J. Appl. Meteor. Climate* **45**:236–246.

Kessler, A., A. Cohen, and D. Sharon. 2006. Assessment of the effect of operational cloud seeding in northern Israel. Abstract (in English) of the full report entitled: Analysis of the Cloud Seeding Effectiveness in Northern Israel, report to the Israeli Water Authority. Haifa: Environmental and Water Resources Eng.

Landsberg, H. H. 1981. The Urban Climate. New York: Academic Press.

Levin, Z., and W. R. Cotton. 2008. Aerosol Pollution Impact on Precipitation: A Scientific Review. WMO and IUGG Report. Heidelberg: Springer.

Lin, J. C., T. Matsui, R. A. Pielke, Sr., and C. Kummerow. 2006. Effects of biomass-burning-derived aerosols on precipitation and clouds in the Amazon Basin: A satellite-based empirical study. *J. Geophys. Res.* **111**:D19204.

Lyons, T. J. 2002. Clouds prefer native vegetation. *Meteor. Atmos. Phys.* **80**:131–140.

Manton, M. J. 1979. Modification of rainfall by topographical variations. *Contr. Atmos. Phys.* **52**:58–66.

Manton, M. J., and G. P. Ayers. 1982. On the number concentration of aerosols in towns. *Boundary Layer Meteorol.* **22**:171–181.

Mason, B. J. 1971. The Physics of Clouds. Oxford: Oxford Univ. Press.

Mather, G. K. 1991. Coalescence enhancement in large multicell storms caused by the emissions from a Kraft paper mill. *J. Appl. Meteor.* **30**:1134–1146.

Nicholls, N. 2000. An artificial trend in district average rainfall in the Snowy Mountains. *Aust. Meteor. Mag.* **49**:255–258.

Nicolls, N. 2005. Climate variability, climate change and the Australian snow season. *Aust. Meteor. Mag.* **54**:177–185.

Nicholls, N. 2006. Detecting and attributing Australian climate change: A review. *Aust. Meteor. Mag.* **55**:199–211.

Pitman, A. J., G. T. Narisma, R. Pielke, and N. J. Holbrook. 2004. The impact of land cover change on climate of south west Western Australia. *J. Geophys. Res.* **109**:D18109.

Polacheck, T. 2006. Tuna longline catch rates in the Indian Ocean: Did industrial fishing result in a 90% rapid decline in the abundance of large predatory species? *Marine Policy* **30**:470–482.

Pruppacher, H. R., and J. D. Klett. 1978. Microphysics of Clouds and Precipitation. Dordrecht: Reidel.

Rosenfeld, D. 2000. Suppression of rain and snow by urban and industrial air pollution. *Science* **287**:1793–1796.

Rosenfeld, D., J. Dai, X. Yu et al. 2007. Inverse relations between amounts of air pollution and orographic precipitation. *Science* **315**:1396–1398.

Rosenfeld, D., and A. Givati. 2006. Evidence of orographic precipitation suppression by air pollution induced aerosols in the western U.S. *J. Appl. Meteor. Climatol.* **45**:893–911.

Rosenfeld, D., and I. M. Lensky. 1998. Satellite-based insights into precipitation formation processes in continental and maritime convective clouds. *Bull. Amer. Meteor. Soc.* **79**:2457–2476.

Rosenfeld, D., I. M. Lensky, J. Peterson, and A. Gingis. 2006. Potential impacts of air pollution aerosols on precipitation in Australia. *Clean Air Environ. Qual.* **40**:43–49.

Rotstayn, L. D. 2007. Possible impacts of pollution aerosols on Australian climate and weather: A short review. *Clean Air Environ. Qual.* **41**:26–32.

Rotstayn, L. D., W. J. Cai, M. R. Dix et al. 2007. Have Australian rainfall and cloudiness increased due to the remote effects of Asian anthropogenic aerosols? *J. Geophys. Res.* **112(D9)**:D09202.

Rotstayn, L. D., and U. Lohmann. 2002. Tropical rainfall trends and the indirect aerosol effect. *J. Climate* **15**:2103–2116.

Schaefer, V. J. 1946. The production of ice crystals in a cloud of supercooled water droplets. *Science* **104**:457–459.

Smith, I. N. 2004. An assessment of recent trends in Australian rainfall. *Aust. Meteor. Mag.* **53**:163–173.

Squires, P. 1958a. The microstructure and colloidal stability of warm clouds. Part I: The relation between structure and stability. *Tellus* **10**:256–261.

Squires, P. 1958b. The microstructure and colloidal stability of warm clouds. Part II: The causes of the variations in microstructure. *Tellus* **10**:262–271.

Suppiah, R., and K. J. Hennessy. 1998. Trends in total rainfall, heavy rain events and number of dry days in Australia, 1910-1990. *Int. J. Climatol.* **18**:1141–1164.

Surowiecki, J. 2004. The Wisdom of Crowds. London: Abacus.

Timball, B., and J. Arblaster. 2006. Land cover change as an additional forcing to explain rainfall decline in the south west of Australia. *Geophys. Res. Lett.* **33**:L07717.

Timball, B., J. Arblaster, and S. Power. 2006. Attribution of the late 20th century rainfall decline in south west Australia *J. Climate* **19**:2046–2062.

Twomey, S. A. 1977. The influence of pollution on the shortwave albedo of clouds. *J. Atmos. Sci.* **34**:1149–1152.

Twomey, S., K. A. Davidson, and K. J. Seton. 1978. Results of five years' observations of cloud nucleus concentration at Robertson, New South Wales. *J. Atmos. Sci.* **35**:650–656.

Twomey, S., and P. Squires. 1959. The influence of cloud nucleus population on the microstructure and stability of convective clouds. *Tellus* **11**:408–411.

van den Heever, S., and W. R. Cotton. 2007. Urban aerosol impacts on downwind convective storms. *J. Appl. Meteorol.* **46**:828–850.

Warner, J. 1968. A reduction in rainfall associated with smoke from sugar-cane fires: An inadvertent weather modification. *J. Appl. Meteorol.* **7**:247–251.

Warner, J. 1971. Smoke from sugar-cane fires and rainfall. In: Proc. Intl. Conf. on Weather Modification, pp. 191–192. Canberra: Amer. Meteor. Soc.

Warner, J., and S. Twomey. 1967. The production of cloud nuclei by cane fires and the effect on cloud droplet concentration. *J.Atmos. Sci.* **24**:704–706.

Wong, A. P. S., N. L. Bindoff, and J. A. Church. 1999. Large-scale freshening of intermediate waters in the Pacific and Indian oceans. *Nature* **400**:440–443.

17

What Do We Know about Large-scale Changes of Aerosols, Clouds, and the Radiation Budget?

Teruyuki Nakajima[1] and Michael Schulz[2]

[1] Center for Climate System Research, University of Tokyo, Kashiwa, Chiba, Japan
[2] Laboratoire des Sciences du Climat et de l'Environnement, CEA/CNRS, Gif-sur-Yvette, France

Abstract

In this chapter we examine how aerosol and cloud fields undergo perturbations by anthropogenic activities. Recent surface observations and satellite remote sensing have detected signatures of large-scale changes in the atmospheric aerosol amounts, associated changes in the cloud fraction, and microphysical structures on a global scale. Models can simulate these signatures fairly well, but problems still remain. Fields of anthropogenic aerosol optical depth (AOD) from several atmospheric models have been found to be consistent with the spatial pattern obtained from satellite data. Further studies are needed to differentiate between natural and anthropogenic aerosols and to interpret observed temporal and regional trends in aerosol parameters. The strength of the cloud–aerosol interaction can be characterized by the regression of AOD or aerosol index (AI) on cloud droplet number (N_c). From recent studies, the corresponding slopes $d\log(N_c)/d\log(A)$ vary between 0.19 and 0.7. Further work is needed to see whether such variability is the result of methodological problems or differences in cloud environments for which the studies have established the cloud–aerosol relation.

Introduction

Our understanding and, hence, modeling of the aerosol and cloud system is mired by large uncertainties. As we attempt to characterize the extent to which clouds are perturbed by humankind in modern times, we cannot afford to ignore any process, even those that are still unknown. It is imperative for us to accumulate evidence of the changing atmospheric particle environments.

In this chapter, we present the signature and features of changing aerosols, clouds, and radiation fields, because these provide evidence of alterations in the clouds and radiation budget by aerosols on the global scale. Thereafter we discuss changes attributable to natural and anthropogenic causes, as quantified by satellite remote sensing and surface measurements, which have greatly improved over the last two decades.

We begin by reviewing studies that report on the anthropogenic perturbation of the aerosol system, which may, in turn, perturb clouds. At the present time, our understanding of aerosol–cloud interactions is not acute enough to extrapolate an aerosol perturbation into measurable cloud characteristics. Based on the evidence of the perturbations in the aerosol system, we propose a framework for studies that have quantified the relationship between aerosol parameters and cloud properties (e.g., cloud optical thickness, cloud droplet radius, or liquid water path) to be subsequently reviewed. Other Earth system components (e.g., greenhouse gas levels or altered land surface attributable to land use), though important, are not addressed, as these fall outside the focus of our mandate. We wish to emphasize, however, that they cannot be ignored, as they may also contribute to the perturbation of the cloud system.

Large-scale Changes of Aerosols

Anthropogenic trace substances can be currently found in precipitation everywhere around the globe as a result of the long-range transport of aerosols. In a strict sense, it is impossible nowadays to study a pristine natural atmosphere. However, if we are to identify or quantify the degree of anthropogenic perturbation, we must have an understanding of pristine conditions. Several methods have been used to estimate and characterize the anthropogenic fraction of aerosols:

- chemical analysis of ice core data,
- analysis of observed recent trends of aerosol parameters,
- characterization of differences in clean and polluted areas of aerosol composition,
- process understanding of natural aerosol emissions,
- reconstruction of the evolution of anthropogenic emissions,
- transport modeling of preindustrial and present-day conditions.

Ultimately, characterization needs to translate the uncertainty of anthropogenic aerosol perturbation into the uncertainty of any particular process involving aerosols and inducing a cloud perturbation. Some findings that indicate an anthropogenic perturbation of atmospheric aerosols may not be directly relevant to the cloud system (e.g., sulfate deposition on the high plateau of

Greenland). However, they may still provide an estimate of the extent of the aerosol perturbation, which could ultimately be related to parameters directly relevant for the cloud system.

In this section we address the following questions:

- What is the history of anthropogenic perturbations to the aerosol content of the atmosphere, and how does it compare to present-day conditions? Is a historical versus current view of perturbations consistent with our knowledge of anthropogenic emissions?
- What is the current regional distribution of anthropogenic aerosols?
- How can we translate anthropogenic perturbations of aerosol parameters (e.g., deposition, mass concentration, optical depth and emissions) into perturbations of parameters relevant for clouds (e.g., cloud condensation nuclei (CCN) concentrations and radiative budgets)?

Anthropogenic Perturbation of the Aerosol Content in the Atmosphere

Glaciers provide an invaluable record of past precipitation. Deep ice cores reveal information on the preindustrial concentration levels of key aerosol components. However, the anthropogenic enhancement of these concentrations visible in the most recent snow deposits, compared to the preindustrial levels, can only be regarded as minimal estimates of aerosol perturbation levels in the atmosphere, as most glaciers are located far from primary anthropogenic source regions. Thus, such estimates can only reflect a diluted state of the perturbed atmosphere. Actual levels for specific regions or for a global average will most likely be higher.

Sulfate concentrations in ice cores suggest that aerosol concentrations increased dramatically as of 1900 and continued at an elevated level until ca. 1950, after which they rose steeply until 1980. As of 1980, levels began to decline, coincident with the introduction of sulfur dioxide abatement schemes in power plants. In 1980, an approximate fourfold increase in sulfate concentrations, compared to preindustrial levels, was found in Greenland and the Canadian Arctic (Legrand et al. 1997; Goto-Azuma and Koerner 2001). In the Alps, ice cores show larger increases, attributable to their proximity to European industrial sources. In the Alps, Preunkert et al. (2001) observed that sulfate concentrations in ice increased from preindustrial levels of 80 ng g^{-1} to 860 ng g^{-1} in 1980, followed by a decrease to 600 ng g^{-1} in the 1990s; that is, sulfate was enhanced anthropogenically by a factor of ten. Goto-Azuma and Koerner (2001) found that nitrate levels in the Arctic stayed at preindustrial levels until 1950, after which it began to climb up until the 1980s, reflecting an anthropogenic enhancement by a factor of 2–3. Black carbon trends in Greenland seem to demonstrate that a maximum level (a fivefold increase over the levels from 1870) was reached by 1920. Air pollution control measures reduced these levels by 1950 (McConnell et al. 2007), decoupling the black carbon trend in Greenland from that of sulfate after 1940. Compared to the early 19th century, it appears that current black carbon concentrations were enhanced only by a factor of two (McConnell et al. 2007). The correlation analysis by McConnell et al. suggests that fires from conifer-rich boreal forests were, and still are, responsible for these levels at this site. Several other studies support the general trends of sulfate, nitrate, and black carbon in ice cores, either through a direct comparison to emission scenarios or through transport modeling studies (e.g., Boucher and Pham 2002; Preunkert et al. 2001).

Trends established for the last decades indicate that the modern-day perturbation of anthropogenic aerosols is changing. From AVHRR retrievals, Mishchenko et al. (2007b) derived a global decrease (only over the ocean) of total AOD, from 0.14 (1988–1990) to 0.11 (2003–2005). If this trend was caused solely by anthropogenic aerosol changes, then this would represent a relatively large increase as compared to model estimates for the year 2000: oceanic anthropogenic AOD of 0.022 (Schulz et al. 2006). If both models and observations are correct, then by the end of the 1980s, anthropogenic AOD should have been as high as 0.05, exerting a direct aerosol forcing effect two times greater than today. However, total global aerosol emissions have not been reported to have decreased by a factor of two since 1980. Natural aerosol trends and climate-driven changes in columnar aerosol burdens may have influenced these AOD trends, but it is questionable whether we understand past emission trends well enough, or whether we have misjudged the capability of the satellite sensors to detect trends or the models to derive an anthropogenic AOD. It is worth noting that a recent decrease in tropospheric AOD may have contributed to the modern-day increases in measurements for surface solar fluxes. Several sites of the BSRN network show increases after 1990 in downward solar radiation of 5–10 W m^{-2}, whereas a decline of the same order of magnitude in surface solar fluxes was observed between 1960 and 1990 (Wild et al. 2005; Liepert and Tegen 2002).

Since around 1970, systematic measurements of *in-situ* aerosol properties have been performed at surface sites (e.g., aerosol extinction, absorption, aerosol composition, concentration and wet deposition), and several authors have attempted to analyze the trends. For the Arctic, Quinn et al. (2007) report that in Alert, Alaska, from 1982–2004, peak concentrations of particulate sulfate decreased from 0.8–0.3 μg S m^{-3}. Peak black carbon levels at Alert decreased only slightly, from 250 ng m^{-3} in 1990 to 150 ng m^{-3} in 2000, whereas absorption measurements in Barrow showed almost no trend during 1990–2005. Scattering is reported to have decreased from 20 mm^{-1} in 1980 to a minimum of 10 mm^{-1} in 1995, and since then it is reported to be increasing again. Based on a model analysis of observed concentrations, Heidam et al. (2004) reports that particulate sulfate has decreased by 30% in Northern Greenland between 1990 and 2000.

Such trends are sometimes related to regional changes in emissions. A recent example is the increase in East Asian emissions, which has most prominently been detected by satellites to reveal increases in NO_2 (Richter et al. 2005) and AOD (Mishchenko et al. 2007a). Modern regional aerosol trends are potentially important, because they allow cloud–aerosol interactions to be tested using modern instrumentation.

Regional Distribution of Anthropogenic Aerosols and Forcing Efficiency

Two arguments support the idea that a better understanding of anthropogenic AOD would help our understanding of perturbed clouds. First, global coverage of satellite-derived AOD and its interpretation in terms of natural or anthropogenic origin are crucial if we are to understand regional differences, for example, in cloud perturbation. Second, a considerable number of studies use AOD–cloud–parameter relationships to understand aerosol-induced cloud perturbations; thus, the anthropogenic contribution to total AOD is of primary interest. Ice core data and long-term, intensive *in-situ* observations are sparse, yet they need to be assimilated to derive a large-scale assessment of the anthropogenic perturbation by large-scale models, surface networks, or remote sensing. In other words, different types of information need to be integrated if we are to derive global anthropogenic AOD fields.

Spaceborne sensors cannot easily distinguish between anthropogenic and natural aerosols. However, retrievals based on multi-wavelength sensors, such as the MODIS instrument, make use of the dependence of light scattering on particle size. Since most anthropogenic aerosol particles of relevance are caused by combustion processes, they often dominate the fine-mode fraction of the aerosol particles. A fine-mode AOD fraction can be estimated from the wavelength-dependent aerosol extinction signal. From this notably dimethylsulfide-derived sulfate, fine fractions of dust and sea salt as well as natural biomass burning must be subtracted before the fine-mode AOD signal can be interpreted as anthropogenic. Kaufman et al. (2005) suggest classifying the MODIS observations according to dominant aerosol types. In doing so, the average fractions of additional aerosols can be estimated and global maps of anthropogenic AOD constructed. These global maps can be compared to model estimates of the anthropogenic AOD, which are based on the difference of a simulation of preindustrial and present-day conditions. To demonstrate, we compare in Figure 17.1 the MODIS-derived anthropogenic AOD field and the corresponding modeled field for the month of July, using LMDzT–INCA (see description in Kinne et al. 2006 and Textor et al. 2006). The general patterns of high anthropogenic AOD values close to the industrialized coasts as well as to South Africa show up in both the satellite data analysis and model simulation.

A broader comparison is desirable to document our understanding of anthropogenic AOD. In addition, model data should be filtered for the presence of MODIS observations to obtain comparable global averages. This can be done by using the AeroCom project aerosol model intercomparison results, as described by Schulz et al. (2006), and six years of MODIS observations. In

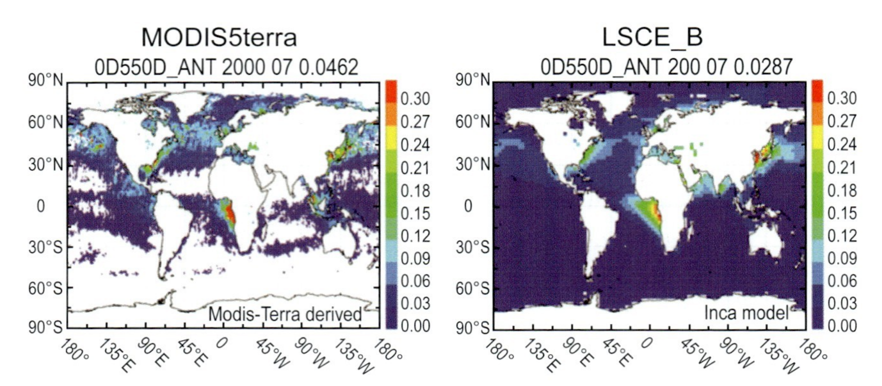

Figure 17.1 Anthropogenic aerosol optical depth as derived from MODIS fine-mode AOD and total AOD using the calculation proposed by Kaufman et al. 2005 (left figure) for the month of July, 2000. INCA anthropogenic AOD is derived from two simulations using AeroCom present-day and preindustrial emissions.

Figure 17.2, a range of multiannual observations with MODIS is established for each month and shown together with the individual model-derived seasonal cycles. The figure indicates considerable interannual variability in the MODIS-derived anthropogenic AOD. A maximum is observable in early summer, and values seem to be slightly higher in late autumn than in winter and early spring. However, the observed seasonal cycle is small, yet seasonal amplitude is the same size as the interannual variability itself. Model-derived anthropogenic AOD is, on average, below the MODIS-derived values. Since all models used

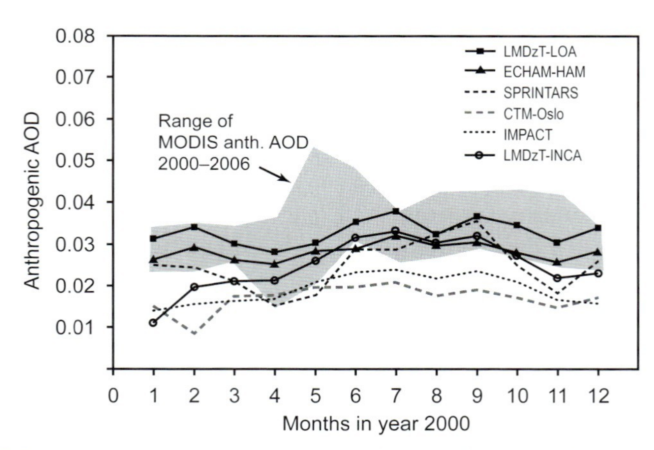

Figure 17.2 Ocean-only anthropogenic aerosol optical depth simulated by AeroCom models and derived from a MODIS collection of five fine-mode and total AOD (after Kaufman et al. 2005). The light gray band shows the data range from six years of MODIS data. Model-simulated AOD fields are filtered for presence of MODIS 2001 observations. The models used meteorological fields for the year 2000 and AeroCom emissions from year 2000 and 1750.

the same seasonal cycle of emissions, a particular observation can be made to explain consistently low values in April/May and higher values in late boreal summer. Note, however, that the monthly AOD evolution is not highly correlated among models. If we assume that the general transport conditions were rather similar if not identical, because reanalyzed meteorological fields from the year 2000 were used in all models, then we would find variance in the lifetime of anthropogenic aerosols and, hence, removal processes between models.

Altogether, we can conclude that two independent methods show rather similar results with respect to an estimate of the anthropogenic AOD. Satellite-derived estimates of the anthropogenic AOD are slightly higher. Interannual variability, as derived from six years of MODIS observations and intermodel diversity, are of similar magnitude. The monthly anthropogenic AOD, as derived from the two methods, is found in a band ranging from 0.01 to 0.05, which may be interpreted as an estimate of the uncertainty range of this parameter. Table 17.1 summarizes differences in AOD between land and ocean. Both satellite-derived estimates and model results suggest that anthropogenic AOD over land is on a global average three times higher than over ocean.

In addition to regional differences in anthropogenic AOD, the sensitivity of the radiative balance or the cloud system to AOD may also vary regionally. It would thus be useful to study this sensitivity in greater detail. As a first step, we investigate a forcing efficiency, which is defined as the ratio of total direct aerosol radiative forcing (RF) over anthropogenic AOD. By looking at forcing efficiencies, one may characterize the way aerosols influence directly the radiative balance of the Earth. One must eliminate variation of the RF attributable to

Table 17.1 AeroCom model simulation results of anthropogenic aerosol optical depth (AOD), top-of-the-atmosphere radiative forcing (RF), and forcing efficiency per unit optical depth (NRF). Grid-box area-weighted means are given over land and ocean in clear-sky conditions between 60°S and 60°N. Observation-based values reported by Yu et al. (2006).

Model	Ocean			Land		
	AOD	RF ($W\ m^{-2}$)	NRF ($W\ m^{-2}\ \tau^{-1}$)	AOD	RF ($W\ m^{-2}$)	NRF ($W\ m^{-2}\ \tau^{-1}$)
UMI	0.019	–0.68	–38	0.057	–1.33	–24
UIO_CTM	0.018	–0.69	–45	0.058	–1.64	–28
LOA	0.033	–0.67	–16	0.093	–1.47	–14
LSCE	0.025	–0.89	–34	0.064	–1.35	–21
MPI.HAM	0.023	–0.57	–26	0.075	–1.10	–24
GISS	0.011	–0.33	–38	0.026	–0.42	–17
SPRINTARS	0.026	–0.32	–11	0.081	–0.63	–14
AeroCom mean	0.022	–0.59	–29	0.065	–1.14	–20
Yu et al. (2006)	0.031	–1.10	–37	0.088	–1.80	–20

simple variation of AOD. A spatial variation in forcing efficiency indicates that aerosols in various regions perturb the radiative budget differently. Table 17.1 reports how the clear-sky forcing efficiency may differ between land and ocean regions (Schulz et al. 2006; Yu et al. 2006). It appears that the forcing efficiency is almost twice as negative over the ocean. Several factors influence the forcing efficiency: particle absorption and hemispheric backscatter, surface albedo, vertical distribution of aerosols and of the atmospheric temperature. The more negative forcing efficiency values, observed over the ocean, are attributable to higher land than ocean surface albedo and more absorbing anthropogenic aerosols over land. The comparison to the observationally based estimate reveals that both the anthropogenic AOD and the clear-sky forcing efficiency are smaller in the models. Thus, both factors contribute to a smaller direct aerosol forcing value derived from the AeroCom models.

Finally, we suggest that there are only very few regions in which the direct radiative effect of aerosols is considerably perturbed by anthropogenic aerosols. The compilation of modeled forcing fields by Schulz et al. (2006) can be used to see where the major contributions to the total direct aerosol forcing can be expected. This may guide efforts to improve our understanding of the aerosol forcing through measurements. Table 17.2 reports the area fraction in which significantly negative clear-sky forcing with values less than –2 Wm^{-2} can be found in the models. The contribution of the forcing from that area to the total direct aerosol forcing is then computed for each model. As a result, 45–65% of the clear-sky direct aerosol forcing is expected to originate from only 10–15% of the Earth's surface.

Anthropogenic CCN Enhancements and Aerosol Forcing

Anthropogenic perturbations of the various aerosol parameters can impact the cloud system through different processes. Of direct relevance is that part of the aerosol particle spectrum which can act as CCN. In addition, any change in the

Table 17.2 Global area fraction A in which clear-sky direct aerosol forcing (CS–RF) is more negative than –2 W m^{-2} in six AeroCom models and contribution of clear-sky direct aerosol forcing in this area to total clear-sky forcing by aerosols.

	Area fraction A, where CS–RF < –2 Wm^{-2}	Area fraction A, where CS–RF < –2 Wm^{-2}
LOA	14%	64%
LSCE	14%	54%
UMI	10%	44%
UIO-CTM	11%	60%
MPI-HAM	13%	59%
SPRINTARS	9%	55%

temperature profile within the troposphere resulting from direct RF will perturb cloud formation. For both CCN concentrations and RF, similar problems are encountered as we attempt to investigate their anthropogenic perturbation and role in the perturbation of clouds:

- No direct observations exist in the preindustrial or pristine atmosphere.
- Current observations are difficult and sparse, or have just recently been initiated.
- Anthropogenic and natural components cannot strictly be observed independently in the ambient atmosphere.
- The relationship between observable parameters (e.g., particle concentration, composition, size distribution, optical depth, phase function and CCN concentration) and the direct/indirect radiative effect of the aerosol depends on the meteorological and chemical environment.
- Both natural and anthropogenic components need to be known to assess properly the potential of CCN and RF to perturb clouds.

Deriving or estimating pristine CCN concentrations requires a conceptual model or a full aerosol–chemistry model, verified with caution against remote, clean locations of the globe. Recently, Andreae (2007) suggested that continental and oceanic CCN concentrations may have been similar, which would contrast current conditions, where polluted regions show typical enhancements of 50–300% over remote regions. A better understanding of the anthropogenic CCN concentrations is needed and, when available, will most likely have considerable impact on the modeling of the aerosol indirect effect. It is worth noting that the majority of climate models assume a rather simple relationship of anthropogenic sulfate aerosol mass and cloud properties (Boucher and Lohmann 1995); they neglect any variation in natural aerosols or other aerosol components that would indirectly alter the relationship between sulfate mass and CCN.

A slightly more straightforward assessment can be made for direct RF by aerosols. Multicomponent aerosol models are positioned to become state-of-the-art tools for the estimation of global aerosol RF. Within AeroCom, nine different global models with detailed aerosol modules have independently produced instantaneous direct RF as a result of anthropogenic aerosols (Schulz et al. 2006). Figure 17.3 shows the major components needed to characterize direct and semi-direct aerosol forcing as maps, illustrating the spatial location of such direct aerosol forcing. For the semi-direct effect, Figure 17.3d displays the solar atmospheric forcing (the difference between top of the atmosphere and surface forcing), which heats the atmospheric column and has thus the potential to influence the formation of clouds. This forcing is estimated at +0.82 ± 0.17 W m^{-2}. The local annual average maxima of this atmospheric forcing exceed +5 W m^{-2}, confirming the regional character of aerosol impacts on climate. Such regional warming of the atmospheric column attributable to

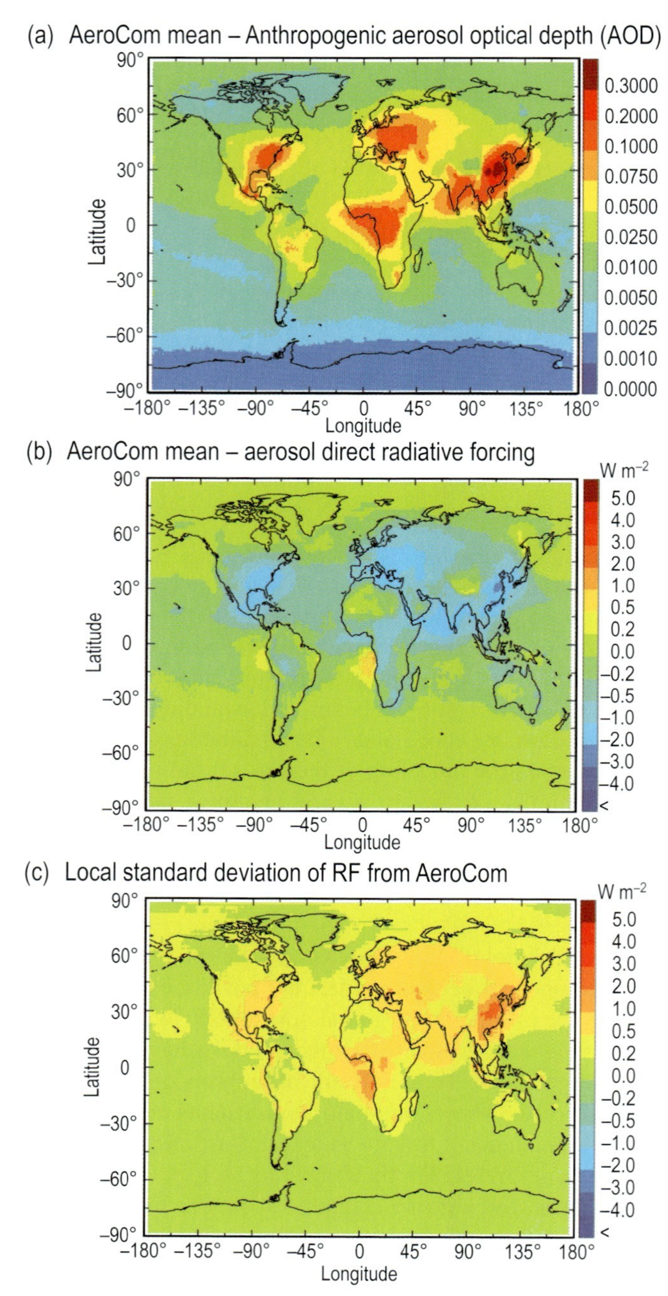

Figure 17.3 Mean annual fields derived from the regridded AeroCom model simulations of (a) anthropogenic aerosol optical depth; (b) total direct aerosol radiative forcing; (c) local standard deviation of radiative forcing (RF) from nine models.

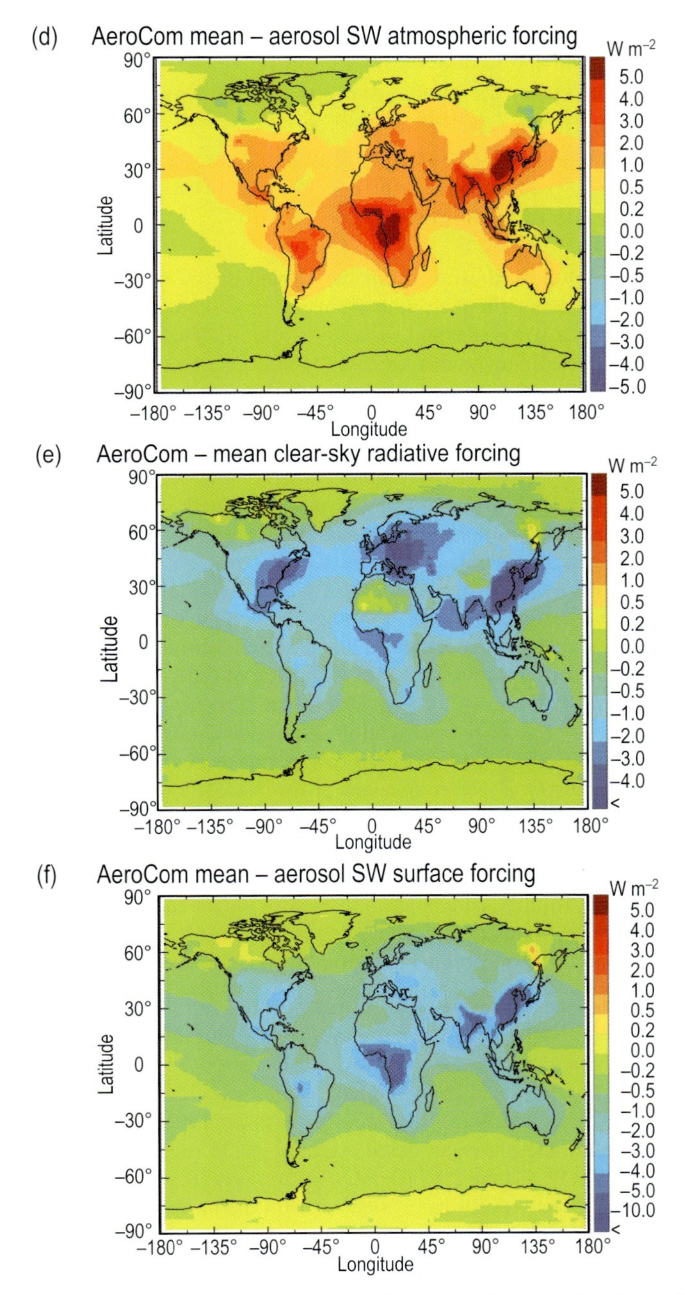

Figure 17.3 (continued) (d) Atmospheric shortwave forcing of column by aerosols; (e) clear-sky aerosol RF; (f) aerosol surface forcing. See values in Table 5 of Schulz et al. (2006).

solar absorption by soot has been measured, for example, in the outflow of the Indian subcontinent (Ramanathan et al. 2007). In terms of the amplitude of the hydrological cycle, the annual average surface forcing is estimated to be $-1.02 \pm 0.23 \text{Wm}^{-2}$.

In Working Group I's contribution to the IPCC Fourth Assessment Report, Forster et al. (2007) summarized current knowledge on global estimates of the direct RF at the top of the atmosphere:

> A central model-derived estimate for the aerosol direct RF is based here on a compilation of recent simulation results using multi-component global aerosol models...This is a robust method for several reasons. The complexity of multi-component aerosol simulations captures nonlinear effects. Combining model results removes part of the errors in individual model formulations. As shown by Textor et al. (2006), the model-specific treatment of transport and removal processes is partly responsible for the correlated dispersion of the different aerosol components. A less dispersive model with smaller burdens necessarily has fewer scattering and absorbing aerosols interacting with the radiation field. An error in accounting for cloud cover would affect the all-sky RF from all aerosol components. Such errors result in correlated RF efficiencies for major aerosol components within a given model. Directly combining total aerosol RF results gives a more realistic aerosol RF uncertainty estimate. The AeroCom compilation suggests significant differences in the modeled local and regional composition of the aerosol..., but an overall reproduction of the total [AOD] variability can be performed (Kinne et al. 2006). The scatter in model performance suggests that currently no preference or weighting of individual model results can be used (Kinne et al. 2006). The aerosol RF taken together from several models is more robust than an analysis per component or by just one model. The mean estimate...of the total aerosol direct RF is -0.2 W m^{-2}, with a standard deviation of ± 0.2 W m^{-2}. This is a low-end estimate for both the aerosol RF and uncertainty because nitrate (estimated as -0.1 W m^{-2}...) and anthropogenic mineral dust (estimated as -0.1 W m^{-2}...) are missing in most of the model simulations. Adding their contribution yields an overall model-derived aerosol direct RF of -0.4 W m^{-2} (90% confidence interval derived from 20 model results: 0 to -0.8 W m^{-2}).
>
> Three satellite-based measurement estimates of the aerosol direct RF have become available, which all suggest a more negative aerosol RF than the model studies...Bellouin et al. (2005) computed a TOA aerosol RF of -0.8 ± 0.1 W m^{-2}. Chung et al. (2005), based upon similarly extensive calculations, estimated the value to be -0.35 ± 0.25 W m^{-2}, and Yu et al. (2006) estimated it to be -0.5 ± 0.33 W m^{-2}. A central measurement-based estimate would suggest an aerosol direct RF of -0.55 W m^{-2}. Figure 2.13 [of the IPCC report] shows the observationally based aerosol direct RF estimates together with the model estimates published since the TAR [Third Assessment Report] and the AeroCom model results.

Note that the measurement-based estimate is based on very few estimates and should be understood as the result of an expert assessment rather than the result of a statistic analysis:

> The discrepancy between measurements and models is also apparent in oceanic clear-sky conditions where the measurement-based estimate of the combined aerosol direct radiative effect (DRE) including natural aerosols is considered unbiased. In these areas, models underestimate the negative aerosol DRE by 20 to 40% (Yu et al. 2006). The anthropogenic fraction of [AOD] is similar between model and measurement based studies. Kaufman et al. (2005) used satellite-observed fine-mode [AOD] to estimate the anthropogenic [AOD]. Correcting for fine-mode [AOD] contributions from dust and sea salt, they found 21% of the total [AOD] to be anthropogenic, while the [model compilation] suggests that 29% of [AOD] is anthropogenic. Finally, cloud contamination of satellite products, aerosol absorption above clouds, not accounted for in some of the measurement-based estimates, and the complex assumptions about aerosol properties in both methods can contribute to the present discrepancy and increase uncertainty in aerosol RF. (Forster et al. 2007, p. 169, 171)

Considering that the observational and model-based forcing estimates represent both valid methods, the IPCC assessed that an intermediate direct aerosol RF estimate with a large uncertainty (–0.5 [–0.9 to –0.1] W m^{-2}) reflects our current knowledge, containing most available RF estimates. The forcing chart expresses satisfactorily the important role that aerosols play as the major greenhouse gas opponent in constituting the total anthropogenic forcing of current climate (for further discussion, see Haywood and Schulz 2007).

Cloud Field Change and Aerosol Effects

Evidence of perturbed clouds on the global scale are more difficult to pinpoint from observations because spatial and temporal variations of the cloud field are large and complex. The effective droplet radius (CDR) of water clouds has been used frequently to study the signature of perturbed clouds because the spatial distribution of CDR is relatively homogeneous, even when clouds show a large variability in the cloud optical thickness (COT). Large-scale signatures, such as enhanced COT of ship trail clouds, and reduced CDR around the continents have been observed from satellites (Coakley et al. 1987; Radke et al. 1989; Nakajima et al. 1991; Han et al. 1994; Kawamoto et al. 2001). Kawamoto and Nakajima (2003) reported distinct seasonal changes of CDR, depending on location. However, an observed long-term change of CDR could not be clearly linked to either an interannual change in cloud perturbation or a shift in the satellite's orbit. These signatures can be compared with the sulfate emission increase in the populated regions to establish a link between human activity and satellite-derived regional signatures (Chameides et al. 2002; Kawamoto et al. 2004, 2006). Comparison of satellite-observed aerosol and cloud parameters is a more direct approach to investigate the mechanism of the phenomena. This approach has become more useful since satellite-derived aerosol and cloud products were constructed from various satellite imagers: AVHRR, POLDER,

MODIS, and GLI (King et al. 1999). We note that detailed interpretation using models is indispensable. The preceding discussion shows that the global distribution of total AOD at a wavelength of 550 nm (AOD) can be observed and modeled within an accuracy of 0.05.

Figure 17.4 shows global distributions of the CDR in water clouds derived from a MODIS and MIROC+SPRINTARS global climate simulation (Takemura et al. 2006). Here the area of large aerosol loading corresponds well to the area of reduced CDR, with values as small as 8 μm, both in the satellite and model results. The observed aerosol plumes are much larger in the North Pacific and smaller in the northern Indian Ocean than simulated

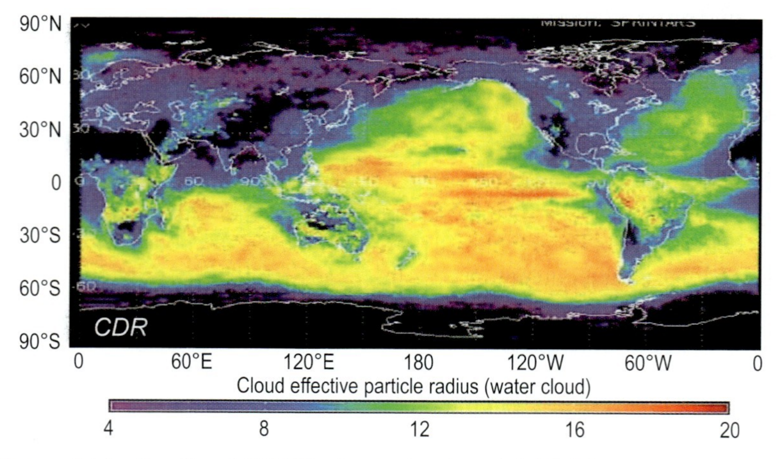

Figure 17.4 Satellite (MODIS)-observed and model (MIROC+SPRINTARS)-simulated global distributions of AOD and CDR for water clouds.

ones. It is interesting to note that a corresponding similarity is also found in the observed and simulated CDR fields; that is, the elongated area of small CDR is much wider in observations than in simulation in the North Pacific and vice versa in the Indian Ocean. This indicates that the northern Pacific aerosol plume in the satellite result seems to be real and responsible for the reduced CDR in this area, even though the aerosol remote sensing is known to overestimate coarse-mode particles attributable to cloud contamination. Clearly, this example shows that a careful analysis is needed to extract the real signature of aerosol–cloud interactions, as both models and satellite systems have their own problems.

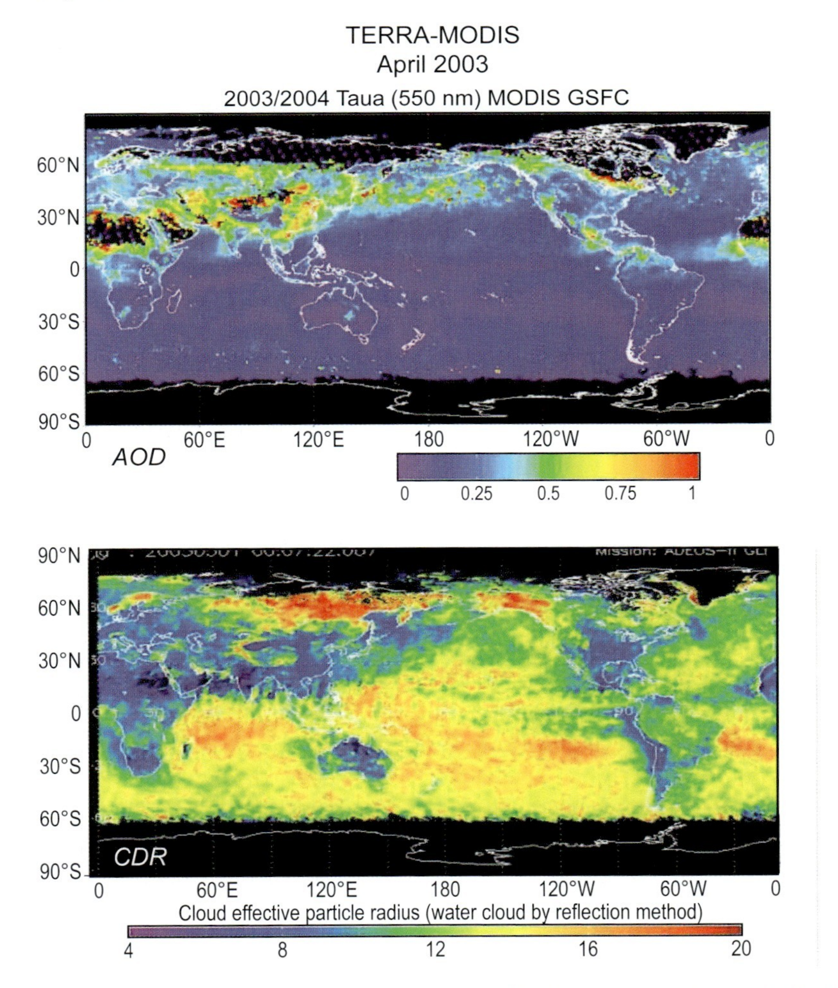

Figure 17.4 (continued) Monthly mean of April 2003 (provided by NASA MODIS team, Takashi Y. Nakajima and Toshihiko Takemura).

Change in the Cloud Microphysical Condition

The strength of aerosol–cloud interactions can be evaluated by the following regression between satellite-observed cloud parameter for water clouds (Y) and columnar particle concentrations (A):

$$Y = a(Y) + b(Y) \log(A), \quad (17.1)$$

where *log* stands for the base 10 logarithm (log_{10}), and $a(Y)$ and $b(Y)$ are regression coefficients to be determined from the analysis. McComiskey and Feingold (2008) discussed that the slopes of various cloud parameters are interrelated for linearly stratified clouds and are useful to evaluate the first indirect forcing of aerosols, which is the cloud forcing when aerosols act as CCN under the condition of a fixed cloud liquid water path (LWP). Many past studies use AOD as a proxy for the aerosol columnar particle concentrations, A.

Some studies use AOD for fine-mode aerosols (AOD-fine) of the NASA's MODIS product as a better index for A. Others use the aerosol index, $AI = \alpha \times AOD$, calculated from satellite-derived AOD and Ångström exponent (α), which is an invariant without large dependence on the size distribution of optically relevant particles (Nakajima et al. 2001; Bréon et al. 2002). This AI correlates fairly well with the column aerosol particle number (N_a) in the optically important particle radius range larger than about 0.05 μm. Some studies directly use N_a, calculated from satellite aerosol observables and an assumed size distribution, the latter being the most relevant parameter for the cloud nucleation process. We should bear in mind, however, that A = AOD may cause artifacts as a result of increasing particle size through adsorbed water vapor at high relative humidities (e.g., the magnitude of slope b might be underestimated when A = AOD.) Other artifacts, as discussed later for each cloud parameter, arise from using a simple regression analysis, as in Equation 17.1, to elucidate the causes and results of the interaction. Here, it is not possible to address all of these differences; however, we believe that it is useful to summarize past results to see the differences and commonalities between them. Future studies should investigate detailed variances of sampling conditions, target clouds, and models to eliminate artifacts, which may exist in our simple comparison.

One consideration, in terms of cloud physics, suggests that a relation between N_a and cloud droplet number (N_c) can be approximated by a form, as in Equation 17.1, with $Y = \log(N_c)$. Table 17.3 and Figures 17.5–17.7 summarize past reported values of the corresponding slope $b(\log N_c) = d\log(N_c)/d\log(A)$ from Equation 17.1, with A = AOD, AI, or N_a (as indicated in the third column of Table 17.3); other frequently used slopes are explained later. In Table 17.3, values for different aerosol concentration indices (AOD, AI, and N_a) are roughly compared without correction, under the assumption that the regression slope does not change when a linear proportionality among them is assumed. Figure 17.5 indicates that $b(\log N_c)$ ranges widely from 0.19–0.7, which is

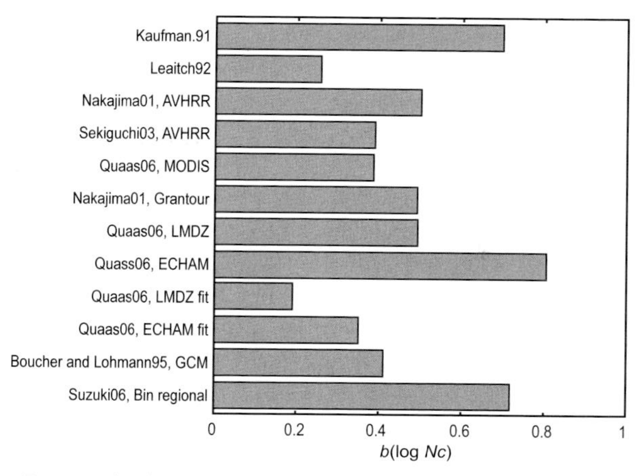

Figure 17.5 Reported values of the regression slope $b(\log N_c)$ (see Table 17.3 for data information).

similar to the range (0.26–1.0) given by McComiskey and Feingold (2008). It has been found, however, that satellite-retrieved values of $b(\log N_c)$ over the global ocean exist in a relatively narrow range, from 0.4–0.5, regardless of the data sources (i.e., AVHRR or MODIS). A large value from aircraft measurements (Kaufman et al. 1995) and a bin-type regional model result (Suzuki et al. 2006) suggest that a strong aerosol–cloud interaction is possible in a favorable condition of cloud nucleation, but insensitive cases (e.g., Leaitch et al. 1992) compensate for these large b values when a global mean is calculated using the tabulated values. Strongly differing values within one model, such as MIROC for various parameterizations (Suzuki et al. 2004) in Table 17.3, demonstrate that there is significant room for any model to be tuned so as to bring model results closer to the observed N_a–N_c relation.

LWP is related to CDR and COT in the following manner:

$$\text{LWP} = 2\,\rho\,\text{CDR} \times \text{COT}/3, \tag{17.2}$$

where ρ is the density of water. This leads to a simple relation between the slopes for CDR and COT (Sekiguchi et al. 2003):

$$b(\log \text{LWP}) = b(\log \text{CDR}) + b(\log \text{COT}). \tag{17.3}$$

Table 17.3 and Figure 17.6 show values of $b(\log \text{CDR})$ and $b(\log \text{LWP})$ reported by past studies, and values of $b(\log \text{COT})$ can be calculated from Equation 17.3. Most satellite values correspond to water clouds, and it has been found that large differences exist among the reported values. This is understandable since the scatterplots from which the slopes have been derived tend to have extremely large dispersions, as reported in the past (e.g., Storelvmo et al. 2006). In our future work, this large dispersion will be studied. The preselection of one type of cloud as a target and a certain analysis method can also effect large differences in results; a serious problem for any satellite sensing lies in the fact

Source[1]	Data	A	$Y = \log$ CDR	log COT	log LWP	$\log N_c$	CF	
(1)	aircraft	*Na*				0.700		
(2)	aircraft	*Na*				0.257		
(3)	POLDER	AI	–0.046					
(4)	MODIS	AOD	0.010	0.150	0.160		0.342*t*	
(5)	MODIS	AOD			0.051		0.244*w*	
(6)	GCM	*Na*				0.410		Globe
(3)	LMDZ	AI	–0.150					
(3)	LMDZ-fit	AI	–0.078					
(3)	LMDZ multi, fit	AI	–0.048					
(7)	AVHRR	*Na*	–0.094	0.257	0.163	0.500		
(8)	AVHRR	*Na*	–0.100	0.156	0.040	0.388	0.086*w*	
(9)	POLDER	AI	–0.078					
(8)	POLDER	*Na*	–0.069					
(10)	POLDER	AI	–0.044	0.203	0.159			
(11)	MODIS	AI	–0.244	0.036	–0.208			
(12)	MODIS	AOD-fine				0.384		
(4)	MODIS	AOD	–0.068	0.142	0.073		0.330*w*	
(5)	MODIS	AOD			0.042		0.299*w*	
(7)	Grantour	*Na*	–0.203			0.490		Global Ocean
(13)	MIROC, Berry	*Na*	–0.187	0.204	0.064			
(13)	MIROC, Khairoutdinov	*Na*	–0.103	0.262	0.227			
(13)	MIROC, Sundqvist	*Na*	–0.178	0.052	–0.104			
(10)	LMDZ	AI	–0.099	0.150	0.051			
(12)	LMDZ	AOD-fine				0.493		
(12)	ECHAM	AOD-fine				0.805		
(12)	LMDZ fit	AOD-fine				0.191		
(12)	ECHAM fit	AOD-fine				0.350		
(11)	GOCART	AI	–0.243	0.017	–0.226			

(10)	POLDER, 30°–60°N	AI			0.113			
(14)	MODIS, AO 30°–60°N	AOD	–0.046	0.069	0.023		0.121w	
(14)	MODIS, AO 5°–30°N	AOD	–0.046	0.081	0.035		0.271w	
(14)	MODIS, AO 20°S–5°N	AOD	–0.124	0.042	–0.081		0.293w	
(14)	MODIS, AO 30°–20°S	AOD	–0.174	0.309	0.135		0.238w	
(4)	MODIS, AO 30°–60°N	AOD					0.445t	Regional Ocean
(4)	MODIS, AO 5°–30°N	AOD					0.433t	
(4)	MODIS, AO 20°S–5°N	AOD					0.471t	
(4)	MODIS, AO 30°–20°S	AOD					0.457t	
(10)	LMDZ, 30–60°N	AI			0.078			
(15)	Regional bin model	*Na*	–0.373	0.330	–0.043	0.720		
(9)	POLDER	AI	–0.035					
(8)	POLDER	*Na*	–0.035					
(10)	POLDER	AI	0.003	0.032	0.035			Global Land
(4)	MODIS	AOD	0.079	0.157	0.236		0.352t	
(5)	MODIS	AOD			0.062		0.210w	
(10)	LMDZ	AI	–0.086	0.105	0.019			
(10)	POLDER, 30°–60°N	AI			0.221			
(10)	LMDZ, 30°–60°N	AI			0.146			Regional Land
(16)	Surface, China S, E & N	AOD					–0.054t	

Table 17.3 Slope b of the regression line between aerosol loading, A, and cloud parameter, $Y = a + b \log(A)$, reported by past studies. Listed regions for (1) and (4) are latitudinal zones of the Atlantic Ocean (AO). Underlined values were not reported originally, but are estimated by the relation of b(log LWP) = b(log CDR)+b(log COT). Values of b(CF) in (5) were estimated from the b(log CF) and mean CF values of (4). The value in (16) is for the total cloud fraction change in the period of 1951–1994. AOD: aerosol optical depth; CDR: cloud droplet radius; COT: cloud optical thickness; LWP: liquid water path; N_c: cloud droplet number, w and t attached to values stand for water cloud fraction and total CF, respectively.

[1] (1) Kaufman et al. (1991), (2) Leaitch et al. (1992), (3) Quaas and Boucher (2005), (4) Myhre et al. (2007), (5) Quaas et al. (2008), (6) Boucher and Lohmann (1995), (7) Nakajima et al. (2001), (8) Sekiguchi et al. (2003), (9) Bréon et al. (2002), (10) Quaas et al. (2004), (11) Matsui et al. (2006), (12) Quaas et al. (2006), (13) Suzuki et al. (2004), (14) Kaufman et al. (2005), (15) Suzuki et al. (2006), (16) Mukai et al. (2008).

that aerosol and cloud information is derived from clear and cloudy pixels, respectively, which are different from each other both spatially and temporarily. Despite these difficulties, however, a decreasing trend of CDR with the aerosol loading has been found over the ocean and is relatively well simulated by models with a slight overestimation.

Over land, a neutral or a slight increasing trend in CDR is observed but is difficult to simulate. A simple Twomey mechanism is not enough to explain this phenomenon, as will be discussed later for the cloud fraction (CF) change. LWP seems to increase over both ocean and land, suggesting that there is a mechanism of the cloud lifetime effect (Albrecht 1989). Further modeling efforts are needed to simulate such an observed tendency. Changes of ECHAM cloud and aerosol schemes lead to –1.9 Wm^{-2}, whereas former versions of ECHAM produce values of –1.0 Wm^{-2} to –2.6 Wm^{-2} for the simulated indirect effect (Table 3 in Lohmann et al. 2007). Suzuki et al. (2004) obtained a large negative slope *b*(log LWP) when the lifetime effect is eliminated from the model by introducing the Sundqvist parameterization ($p = 0$) of the autoconversion rate, $\tau_p \propto Nc^p$, for the cloud to precipitation conversion process. This is because aerosols tend to accumulate in high-pressure areas where a less than average amount of clouds are present. Inclusion of the lifetime effect by Berry's ($p = 1$) or Khairoutdinov's ($p = 1.79$) parameterization makes the system produce more LWP through the cloud lifetime effect. It should be recognized that the cloud lifetime effect makes the LWP insensitive to particle concentrations by

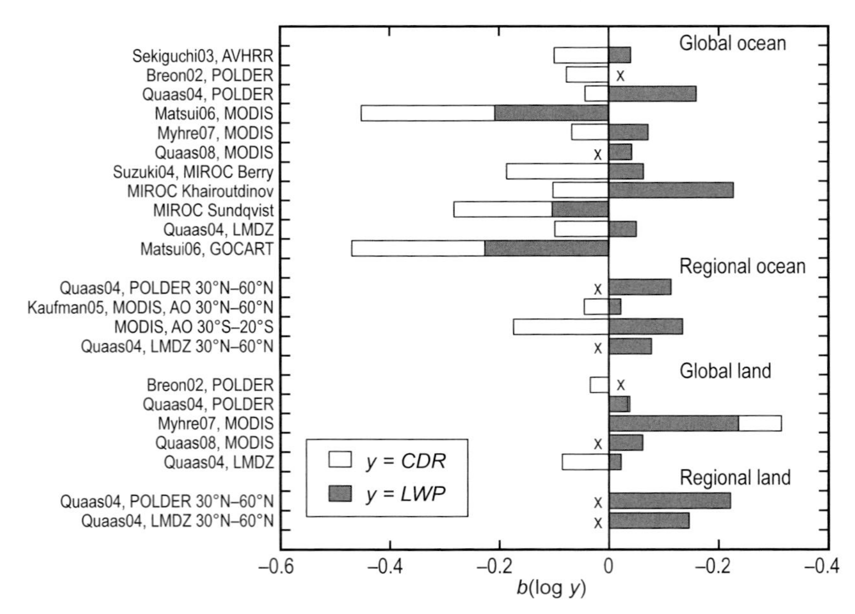

Figure 17.6 Reported values of the regression slope *b*(log CDR) and *b*(log LWP). Data labeled "x" have only one LWP or CDR slope.

canceling the large negative slope in the case of no aerosol–cloud interaction in the autoconversion process by the Sundqvist parameterization. Figure 17.6 suggests a suitable p value exists around 1–1.79, since satellite values over global ocean range from 0.04–0.163 for b(log LWP) other than a large negative value of Matsui et al. (2006), whereas the model result of Suzuki et al. (2004) ranges from 0.064 (for $p = 1$) to 0.227 ($p = 1.79$), respectively. More validation and cloud microphysical modeling is needed, however, to confirm this point. The large negative value by Matsui et al. (2006) was obtained for low clouds with the cloud-top temperature above 273 K. They suggest two possible mechanisms of reduced LWP for such warm clouds under high aerosol loading condition: (a) enhanced evaporation of cloud droplets by light absorption of aerosols and/or by reduced cloud droplet size; (b) a change in the atmospheric stability by drizzle quenching to reduce water supply from the cloud bottom. These mechanisms are diffucult to implement explicitly in GCMs. It is also shown that the difference in target clouds is significant for comparison of the statistics.

Cloud Fraction Change

Cloud fraction can also be altered by aerosol effects. Figure 17.7 shows reported values of the slope of the following regression:

$$\mathrm{CF} = a(\mathrm{CF}) + b(\mathrm{CF}) \log(A). \tag{17.4}$$

In using this, we note that the relation may include an error caused by aerosol-containing pixels when the relative humidity is high. Another important point is that some studies use the water CF whereas others use the total CF for the analysis. In case middle- and upper-layer clouds do not respond to A, the slope b(CF) for water clouds will be similar to that for total clouds, although there are case studies that suggest this assumption is questionable (cf. Figure 17.10).

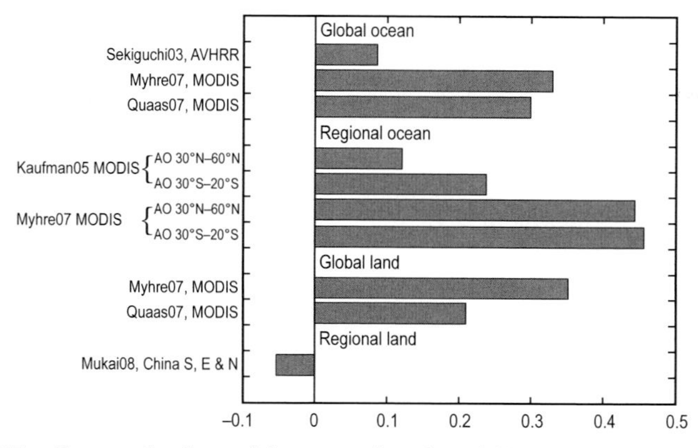

Figure 17.7 Reported values of the regression slope b(CF).

All measurements shown in Figure 17.7 suggest an increasing trend of the CF over the ocean, but there is a large difference between 0.086 by Sekiguchi et al. (2003) and 0.35 by Myhre et al. (2007) and Quaas et al. (2008). Kaufman et al. (2005) reported a significant increase of CF of shallow water clouds with increasing aerosol concentrations that produces an aerosol indirect RF larger than those by first and second indirect effects over the Atlantic Ocean. Myhre et al. (2007) reported a similar analysis for the same regions, but for the total CF, and obtained stronger CF changes than those obtained by Kaufman et al. (2005). Thus, more investigations are needed to identify the causes of the different cloud types. It is interesting to note that clouds over land have globally a positive slope compared to the ocean values. This indicates a significant increase in the large area mean of CF with increasing aerosol concentrations, despite several reports which suggest decreases of the CF attributable to aerosols over some regions (Kaiser 2000; Koren et al. 2004), also shown by the value given by Mukai et al. (2008) for the Chinese region.

Various Indirect Effects of Aerosols

As many researchers have suggested, the cloud field is influenced not only by first and second aerosol indirect effects but also by other processes, such as an atmospheric stability. Matsui et al. (2005, 2006) found that the drizzle particle growth in low-level clouds can be classified by AI and the lower tropospheric stability. In Figure 17.8 we compare their distribution of precipitating/non-precipitating clouds and b(log CDR) of Sekiguchi et al. (2003). The figure indicates that areas of large negative b-values tend to correspond to the location of non-precipitating clouds without much drizzling particles. On the other hand, regions of neutral to positive b(log CDR) values exist in regions with precipitating clouds, which also coincide with strongly active convective regions like ITCZ or SPCZ. This suggests that to simulate b-values better, it is important to model the drizzle-forming or suppression processes in conjunction with the nucleation process and large-scale dynamics that control lower tropospheric stability. High-resolution and bin-type aerosol–cloud particle growth models should be useful for such studies.

Figure 17.9 shows two characteristic COT-CDR scatterplots from AVHRR, which have been observed frequently off the Californian coast and over the North Atlantic (Nakajima and Nakajima 1995). Here, we present corresponding simulation results by a bin-type non-hydrostatic cloud model for low and high CCN conditions (Suzuki et al. 2006). The model successfully simulates the characteristic negative correlation between COT and CDR in the pristine condition with abundant drizzle particles by which the COT decreases when CDR becomes large. For the polluted case, there tends to be a positive correlation between COT and CDR because the COT increases when drizzling is quenched by reduced CDR with increasing CCN. Hence the large negative

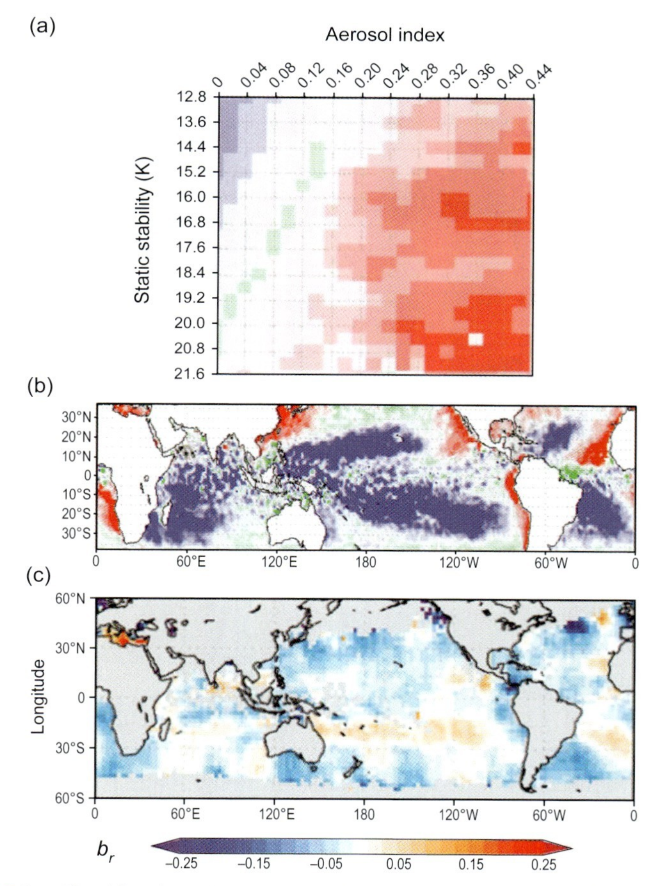

Figure 17.8 Classification of precipitating/non-precipitating clouds from a combined visible and microwave remote sensing (a) and their global distribution for March–June 2000 (after Fig. 3a in Matsui et al. 2004); (b) data from Matsui et al. (2004) is compared with *b*(log CDR) for April, 1990 (after fig. 5 in Matsui et al. 2004); (c) data from Sekiguchi et al. (2003). Red, green, and blue in (a) and (b) indicate non-precipitating, neutral, and precipitating clouds, respectively.

b(log CDR) values in these regions are produced by a drizzle-quenching mechanism that results from CCN input.

In terms of the cloud stratification, it is interesting to study how cloud-top temperature depends on aerosol parameters. Sekiguchi et al. (2003) found the *T14*-temperature, at which CDR exceeds 14 μm (Rosenfeld 2000), decreases with N_a, but they did not find a significant dependence of the cloud-top temperature on N_a. This result looks reasonable because the particle growth is delayed when CCN are added to the cloud system, whereas the cloud-top height is rather controlled by large-scale dynamics. However, Myhre et al. (2007) reported a significant positive correlation between the cloud-top pressure and

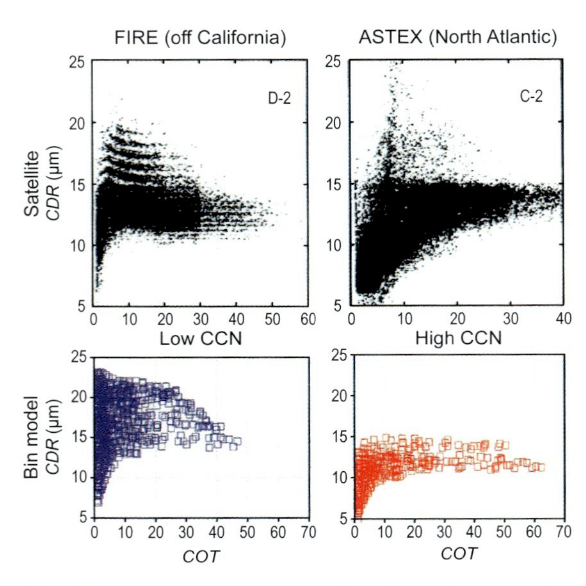

Figure 17.9 COT–CDR correlations from satellite (Nakajima and Nakajima 1995) and a bin-type cloud model (Suzuki et al. 2006) for FIRE and ASTEX experiments.

AOD. Further investigations are needed to resolve the differences in target clouds used in these studies.

Lohmann et al. (2006) point out that high and low AOD cases correspond to very different atmospheric conditions. In addition, a secondary general circulation is induced by a large-scale aerosol direct forcing, which cools the Earth's surface and heats the atmosphere (Menon et al. 2002; Chuang and Ramanathan 2006, 2007; Takemura et al. 2005, 2007). A complication of these mechanisms associated with atmospheric dynamics is that clouds are under the influence of both anthropogenic aerosols and greenhouse gases. Unraveling the mechanisms of cloud change is thus not a simple task, because an increase in anthropogenic aerosols is correlated with an increase in anthropogenic greenhouse gas concentrations. Figure 17.10 demonstrates long-term, surface-observed changes in the total CF from the World Surface Observation data (Japan Meteorological Business Support Center) from 1951–1994 as well as simulated CF changes in three regions of China (Mukai et al. 2008). The simulation results show differences between equilibrium experiments of the MIROC+SPRINTARS model, coupled with a mixed ocean, for preindustrial conditions and for current anthropogenic aerosols (AER), greenhouse gases (GHG), and the combination of both (AER+GHG). Thus the magnitude of the change should not be used for quantitative comparison with observed changes. Figure 17.10 suggests that the large CF increase attributable to the aerosol effect is mostly negated in the model simulation by the CF decrease as a result of the GHG effect in the eastern (E) and northern (N) regions. By comparison, the

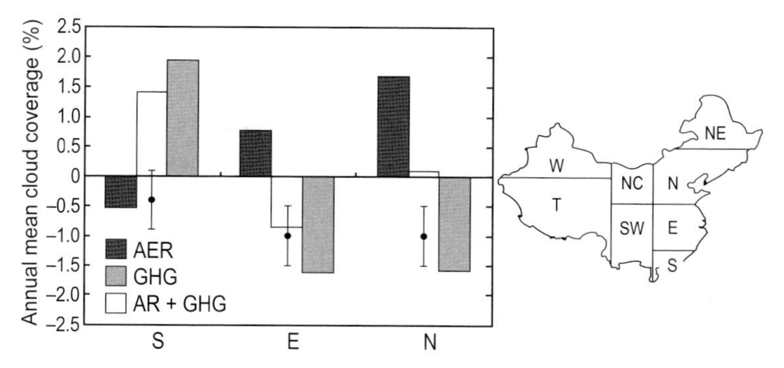

Figure 17.10 Simulated changes in the annual mean cloud coverage (%) attributable to anthropogenic aerosols (AER), greenhouse gases (GHG), and both (AER+GHG) over three Chinese regions: S, E, and N. Closed circles with standard error bar represent the changes in the observed cloud coverage (%) from 1951–1994 (data from Mukai et al. 2008).

southern (S) region exhibits increase/decrease trends that are almost opposite to those of regions E and N. A further complication is that the CF decrease in region S is caused by a decrease in the convective CF, which exceeds the increase of low CF produced by the cloud lifetime effect of anthropogenic aerosols in this region. This large decrease of the convective CF is the result of an enhanced anti-cyclonic circulation caused by the enhanced atmospheric stability and land–ocean temperature gradient, which are, in turn, caused by the large negative aerosol RF at the surface. Convective clouds increase in region N through the counter-circulation of the region S. Thus the change of the total CF, including all types of clouds, results from a very complicated mechanism over this region, as suggested partly (remember that satellite values are mostly for water clouds) by a poor model simulation of *b*(log CDR) and *b*(log LWP) changes over the global land (see Figure 17.6). In addition, the areal mean of the CF change over China becomes very small (< 10 %), much smaller than the large-area averages estimated by satellites. Thus, we need to understand how such a small change in total CF from long-term surface observations and a large change in CF from satellite data can occur, if we are to exact a global understanding of perturbed clouds.

Conclusions and Recommendations

Columnar anthropogenic aerosol amounts and their RF are captured in some detail by observations from space and surface. Model estimates of total AOD are within a range of 0.05 of sun photometer and satellite retrieval estimates (Kinne et al. 2006). Global-scale features of the aerosol–cloud interaction are also estimated from observations and are well simulated by models over the ocean, but perturbed clouds over the land seem to be more difficult to simulate.

Dimming and brightening by aerosol particles have been suggested to alter the surface solar radiation on decadal scales regionally by several W m^{-2}. This perturbation of the regional surface radiation budget seems to have caused noticeable large-scale cloud field changes.

To address the open issues, we recommend the following research efforts:

- Aerosol properties, especially the mixing state of soot particles, should be measured more thoroughly because the anthropogenic AOD and the clear-sky forcing efficiency are smaller in the models than retrieved from observations. The regional aerosol effect and the aerosol-induced perturbation of the vertical structure of the atmosphere depend critically on particulate absorption.
- Active remote measurements should be combined with passive sensing to reduce the difficulty in obtaining simultaneous aerosol and cloud measurements to derive the aerosol–cloud interaction strength. CloudSat, CALIPSO, and future EarthCARE platforms offer a unique opportunity to improve our understanding of these processes. Passive satellite sensors cannot alone discriminate well enough the location in the column aerosol where clouds are actually in the process of being mixed.
- Global climate modeling needs to include more elaborate aerosol and cloud microphysical processes to simulate various aerosol and cloud processes, including large-scale dynamic–hydrologic effects. High-resolution modeling may help us understand continental and convective clouds without the simplifying cumulus parameterizations that have been used in global climate models. At the same time, such simulations require more thorough testing with benchmark datasets to determine the quality of model descriptions of critical processes. The nature of most of the aerosol observations prevents a direct identification of the anthropogenic contributions. In many regions, natural aerosols dominate particulate mass, AOD, and often CCN concentrations as well. Therefore, natural aerosols require as much attention as anthropogenic aerosols when investigating the anthropogenic perturbations of the cloud system.
- Our understanding of long-term trends of atmospheric aerosol concentrations and thus the anthropogenic perturbation is incomplete. Further efforts are required to characterize sufficiently pristine and anthropogenic aerosol levels, and we recommend that this be done by integrating results from different methods. A conceptual or even quantitative model of the temporal and regional evolution of particle composition and CCN concentrations may be helpful to investigate the anthropogenic perturbation of clouds.

Acknowledgments

Figure 17.4 was kindly provided by NASA MODIS team, Takashi Y. Nakajima and Toshihiko Takemura. Model results of anthropogenic aerosol properties have been made accessible thanks to the AeroCom project partners: Olivier Boucher, Shekar Reddy, Philip Stier, Toshihiko Takemura, Gunnar Myhre, Joyce Penner, and Xhiaohong Liu.

References

Albrecht, B. A. 1989. Aerosols, cloud microphysics, and fractional cloudiness. *Science* **245**:1227–1230.

Andreae, M. O. 2007. Aerosols before pollution. *Science* **315(5808)**:50–51.

Bellouin, N., O. Boucher, J. Haywood, and M. S. Reddy. 2005. Global estimate of aerosol direct radiative forcing from satellite measurements. *Nature* **438(7071)**:1138–1141.

Boucher, O., and U. Lohmann. 1995. The sulfate-CCN-cloud albedo effect: A sensitivity study with two general circulation models. *Tellus B* **47**:281–300.

Boucher, O., and M. Pham. 2002. History of sulfate aerosol radiative forcings. *Geophys. Res. Lett.* **29(9)**:22–25.

Bréon, F. -M., D. Tanré, and S. Generoso. 2002. Aerosol effect on cloud droplet size monitored from satellite. *Science* **295**:834–837.

Chameides, W. L., C. Luo, R. Saylor et al. 2002. Correlation between model-calculated anthropogenic aerosols and satellite-derived cloud optical depths: Indication of indirect effect. *J. Geophys. Res.* **107**:4085.

Chung, C. E., and V. Ramanathan. 2006. Weakening of North Indian SST Gradients and the Monsoon Rainfall in India and the Sahel. *J. Climate* **19**:2036.

Chung, C. E., and V. Ramanathan. 2007. Relationship between trends in land precipitation and tropical SST gradient. *Geophys. Res. Lett.* **34**:L16809.

Chung, C. E., V. Ramanathan, D. Kim, and I. A. Podgorny. 2005. Global anthropogenic aerosol direct forcing derived from satellite and ground-based observations. *J. Geophys. Res.* **110**:D24207.

Coakley, J. A., Jr., R. L. Bernstein, and P. A. Durkee. 1987. Effect of ship-track effluents on cloud reflectivity. *Science* **237**:1020–1022.

Forster, P., V. Ramaswamy, Artaxo et al. 2007. Radiative Forcing of Climate Change. In: Climate Change 2007: The Physical Science Basis. Contribution of Working Group I to the Fourth Assessment Report of the Intergovernmental Panel on Climate Change, edited by S. Solomon, D. Qin, M. Manning et al. pp. 129–234. New York. Cambridge Univ. Press.

Goto-Azuma, K., and R. M. Koerner. 2001. Ice core studies of anthropogenic sulfate and nitrate trends in the Arctic. *J. Geophys. Res. Atmos.* **106(D5)**:4959–4969.

Han, Q., W. B. Rossow, and A. A. Lacis. 1994. Near-global survey of effective droplet radii in liquid water clouds using ISCCP data. *J. Climate* **7**:465–497.

Haywood, J., and M. Schulz. 2007. Causes of the reduction in uncertainty in the anthropogenic radiative forcing of climate between IPCC (2001) and IPCC (2007). *Geophys. Res. Lett.* **34(20)**:L20701.

Heidam, N. Z., J. Christensen, P. Wahlin, and H. Skov. 2004. Arctic atmospheric contaminants in NE Greenland: levels, variations, origins, transport, transformations and trends 1990–2001. *Sci. Environ.* **331(1–3)**:5–28.

Kaiser, D. P. 2000. Decreasing cloudiness over China: An updated analysis examining additional variables. *Geophys. Res. Lett.* **27**:2193–2196.

Kaufman, Y. J., O. Boucher, D. Tanré et al. 2005. Aerosol anthropogenic component estimated from satellite data. *Geophys. Res. Lett.* **32(17)**:L17804.

Kaufman, Y. J., R. S. Fraser, and R. L. Mahoney. 1991. Fossil fuel and biomass burning effect on climate: Heating or cooling? *J. Climate* **4**:578–588.

Kaufman, Y. J., I. Koren, L. A. Remer, D. Rosenfeld, and Y. Rudich. 2005. The effect of smoke, dust, and pollution aerosol on shallow cloud development over the Atlantic Ocean. *PNAS* **102**:11,207–11,212.

Kawamoto, K., T. Hayasaka, T. Nakajima, D. Streets, and J.-H. Woo. 2004. Examining the aerosol indirect effect using SO_2 emission inventory over China using SO_2 emission inventory. *Atmos. Res.* **72**:353–363.

Kawamoto, K., T. Hayasaka, I. Uno, and T. Ohara. 2006. A correlative study on the relationship between modeled anthropogenic aerosol concentration and satellite-observed cloud properties over East Asia. *J. Geophys. Res.* **111**:D19201.

Kawamoto, K., and T. Nakajima. 2003. Seasonal variation of cloud particle size as derived from AVHRR remote sensing. *Geophys. Res. Lett.* **30**:1810.

Kawamoto, K., T. Nakajima, and T. Y. Nakajima. 2001. A global determination of cloud microphysics with AVHRR remote sensing. *J. Climate* **14**:2054–2068.

King, M. D., Y. J. Kaufman, D. Tanré, and T. Nakajima. 1999. Remote sensing of tropospheric aerosols from space: Past, present, and future. *Bull. Amer. Meteor. Soc.* **80**:2229–2259.

Kinne, S., M. Schulz, C. Textor et al. 2006. An AeroCom initial assessment optical properties in aerosol component modules of global models. *Atmos. Chem. Phys.* **6**:1815–1834.

Koren, I., Y. J. Kaufman, L. A. Remer, and J. V. Martins. 2004. Measurement of the effect of Amazon smoke on inhibition of cloud formation. *Science* **303**:1342–1345.

Leaitch, W. R., G. A. Isaac, J. W. Strapp, C. M. Banic, and H. A. Wiebe. 1992. The relationship between cloud droplet number concentrations and anthropogenic pollution: Observation and climatic implications. *J. Geophys. Res.* **97**:2463–2474.

Legrand, M., C. Hammer, M. DeAngelis et al. 1997. Sulfur-containing species (methanesulfonate and SO_4) over the last climatic cycle in the Greenland Ice Core Project (central Greenland) ice core. *J. Geophys. Res. Ocean.* **102(C12)**:26,663–26,679.

Liepert, B., and I. Tegen. 2002. Multidecadal solar radiation trends in the United States and Germany and direct tropospheric aerosol forcing. *J. Geophys. Res. Atmos.* **107(D12)**:4153.

Lohmann, U., I. Koren, and Y. J. Kaufman. 2006. Disentangling the role of microphysical and dynamical effects in determining cloud properties over the Atlantic. *Geophys. Res. Lett.* **33**:L09802.

Lohmann, U., P. Stier, C. Hoose et al. 2007. Cloud microphysics and aerosol indirect effects in the global climate model ECHAM5-HAM. *Atmos. Chem. Phys.* **7**:3425–3446.

Matsui, T., H. Masunaga, S. M. Kreidenweis et al. 2006. Satellite-based assessment of marine low cloud variability associated with aerosol, atmospheric stability, and the diurnal cycle. *J. Geophys. Res.* **111**:D17204.

Matsui, T., H. Masunaga, R. A. Pielke, Sr., and W.-K. Tao. 2004. Impact of aerosols and atmospheric thermodynamics on cloud properties within the climate system. *Geophys. Res. Lett.* **31**:L06109.

McComiskey, A., and G. Feingold.2008. Quantifying error in the radiative forcing of the first aerosol indirect effect. *Geophys. Res. Lett.* **35**:L02810.

McConnell, J. R., R. Edwards, G.L. Kok et al. 2007. 20th-century industrial black carbon emissions altered arctic climate forcing. *Science* **317(5843)**:1381–1384.

Menon, S., J. Hansen, L. Nazarenko, and Y. Luo. 2002. Climate effects of black carbon aerosols in China and India. *Science* **297**:2250–2253.

Mishchenko, M. I., I. V. Geogdzhayev, B. Cairns et al. 2007a. Past, present, and future of global aerosol climatologies derived from satellite observations: A perspective. *J. Quant. Spectrosc. Rad. Trans.* **106(1–3)**:325–347.

Mishchenko, M. I., I. V. Geogdzhayev, W. B. Rossow et al. 2007b Long-term satellite record reveals likely recent aerosol trend. *Science* **315(5818)**:1543–1543.

Mukai, M., T. Nakajima, and T. Takemura. 2008. A study of anthropogenic impacts on the radiation budget and the cloud field in East Asia. *J. Geophys. Res.* **113**:D12211.

Myhre, G., F. Stordal, M. Johnsrud et al. 2007. Aerosol–cloud interaction inferred from MODIS satellite data and global aerosol models. *Atmos. Chem. Phys.* **7**:3081–3101.

Nakajima, T., A. Higurashi, K. Kawamoto, and J. E. Penner. 2001. A possible correlation between satellite-derived cloud and aerosol microphysical parameters. *Geophys. Res. Lett.* **28**:1171–1174.

Nakajima, T., M. D. King, J. D. Spinhirne, and L.F. Radke. 1991. Determination of the optical thickness and effective radius of clouds from reflected solar radiation measurements. Part II: Marine stratocumulus observations. *J. Atmos. Sci.* **48**:728–750.

Nakajima, T. Y., and T. Nakajima. 1995. Wide-area determination of cloud microphysical properties from NOAA AVHRR measurements for FIRE and ASTEX regions. *J. Atmos. Sci.* **52**:4043–4059.

Preunkert, S., M. Legrand, and D. Wagenbach. 2001. Sulfate trends in a Col du Dome (French Alps) ice core: A record of anthropogenic sulfate levels in the European midtroposphere over the twentieth century. *J. Geophys. Res. Atmos.* **106(D23)**:31,991–32,004.

Quaas, J., and O. Boucher. 2005. Constraining the first aerosol indirect radiative forcing in the LMDZ GCM using POLDER and MODIS satellite data. *Geophys. Res. Lett.* **32**:L17814.

Quaas, J., O. Boucher, N. Bellouin, and S. Kinne. 2008. Satellite-based estimate of the direct and indirect aerosol climate forcing. *J. Geophys. Res.* **113**:D05204.

Quaas, J., O. Boucher, and F.-M. Bréon. 2004. Aerosol indirect effects in POLDER satellite data and in the LMDZ GCM. *J. Geophys. Res.* **109**:D08205.

Quaas, J., O. Boucher, and U. Lohmann. 2006. Constraining the total aerosol indirect effect in the LMDZ and ECHAM4 GCMs using MODIS satellite data. *Atmos. Chem. Phys.* **6**:947–955.

Quinn, P. K., G. Shaw, E. Andrews et al. 2007. Arctic haze: Current trends and knowledge gaps. *Tellus* **59(1)**:99–114.

Radke, L. F., J. A. Coakley, Jr., and M. D. King. 1989. Direct and remote sensing observations of the effects of ships on clouds. *Science* **246**:1146–1149.

Ramanathan, V., M. V. Ramana, G. Roberts et al. 2007. Warming trends in Asia amplified by brown cloud solar absorption. *Nature* **448(7153)**:575–578.

Richter, A., J. P. Burrows, H. Nuss, C. Granier, and U. Niemeier. 2005. Increase in tropospheric nitrogen dioxide over China observed from space. *Nature* **437(7055)**:129–132.

Rosenfeld, D. 2000. Suppression of rain and snow by urban and industrial air pollution. *Science* **287**:1793–1796.

Schulz, M., C. Textor, S. Kinne et al. 2006. Radiative forcing by aerosols as derived from the AeroCom present-day and preindustrial simulations. *Atmos. Chem. Phys.* **6**:5225–5246.

Sekiguchi, M., T. Nakajima, K. Suzuki et al. 2003. A study of the direct and indirect effects of aerosols using global satellite datasets of aerosol and cloud parameters. *J. Geophys. Res.* **108(D22)**:4699.

Storelvmo, T., J. E. Kristjánsson, G. Myhrel, M. Johnsrud, and F. Stordal. 2006. Combined observational and modeling based study of the aerosol indirect effect. *Atmos. Chem. Phys.* **6**:3583–3601.

Suzuki, K., T. Nakajima, T. Y. Nakajima, and A. Khain. 2006. Correlation pattern between effective radius and optical thickness of water clouds simulated by a spectral bin microphysics cloud model. *SOLA* **2**:116–119.

Suzuki, K., T. Nakajima, A. Numaguti et al. 2004. A study of the aerosol effect on a cloud field with simultaneous use of GCM modeling and satellite observation. *J. Atmos. Sci.* **61**:179–194.

Takemura, T., Y. J. Kaufman, L. A. Remer, and T. Nakajima. 2007. Two competing pathways of aerosol effects on cloud and precipitation formation. *Geophys. Res. Lett.* **34**:L04802.

Takemura, T., T. Nozawa, S. Emori, T. Y. Nakajima, and T. Nakajima. 2005. Simulation of climate response to aerosol direct and indirect effects with aerosol transport-radiation model. *J. Geophys. Res.* **110**:D02202.

Takemura, T., Y. Tsushima, T. Yokohata et al. 2006. Time evolutions of various radiative forcings for the past 150 years estimated by a general circulation model. *Geophys. Res. Lett.* **33**:LI9705.

Wild, M., H. Gilgen, A. Roesch et al. 2005. From dimming to brightening: Decadal changes in solar radiation at Earth's surface. *Science* **308(5723)**:847–850.

Yu, H., Y. J. Kaufman, M. Chin et al. 2006. A review of measurement-based assessment of aerosol direct radiative effect and forcing. *Atmos. Chem. Phys.* **5**:613–666.

Sonia
Kreidenweis

18

The Extent and Nature of Anthropogenic Perturbations of Clouds

Patrick Y. Chuang and Graham Feingold, Rapporteurs

Greg Ayers, Robert J. Charlson, William R. Cotton, Sonia M. Kreidenweis, Zev Levin, Teruyuki Nakajima, Daniel Rosenfeld, Michael Schulz, and Holger Siebert

Introduction

Clouds are a critical component of the climate system. They both influence and are influenced by climate via a myriad of pathways. Anthropogenic perturbations on clouds have been hypothesized and studied for decades, yet despite significant progress, few have been demonstrated and quantified. The inextricable linkages between clouds and other components of the climate system necessitate careful assessment of the nature and extent of anthropogenic perturbations on clouds. This was the purview of our working group. More specifically, we were charged with identifying knowledge gaps that must be bridged in order for progress to be achieved on this important question.

We identified three primary pathways through which anthropogenic influences can perturb clouds; namely, through (a) change in the concentration and composition of aerosols; (b) greenhouse gas warming and consequent changes in the climate system; and (c) land-surface changes, primarily as a result of land use. Because of the expertise in our group, we focused primarily on the more direct means by which aerosols affect clouds, although our discussions did touch on all three types.

It was explicitly recognized that the aerosol–cloud–climate system is tightly coupled; that is, all mechanisms in the system interact. For the sake of simplifying the structure of our discussions, we chose to address separately anthropogenic perturbations in terms of their impact on cloud and aerosol microphysics, radiation, precipitation, atmospheric dynamics, and atmospheric chemistry.

These are two-way interactions. For example, aerosols affect precipitation, but precipitation also affects aerosols (and similarly for the other components). We stress, however, that these pairs of interactions are all tightly coupled and that an anthropogenic perturbation may set off a chain reaction resulting in feedbacks, both positive and negative, to the cloud system and involving all of these categories.

We divided the gaps in understanding into three categories:

1. Conceptual gaps, which represent "big picture" gaps in our understanding of how to proceed.
2. Knowledge/data gaps, which are deficits that could be filled using present-day instruments and data, but for some reason (e.g., lack of resources) have not.
3. Tool gaps or deficits in our ability to make relevant measurements.

We begin with a general discussion related to our current understanding of perturbed clouds and then proceed to each of the specific areas impacted by anthropogenic perturbations. We focus on the three types of gaps as they relate to each of these topics. As with the other reports in this volume, our intent is to highlight our discussions during the Forum. By nature, this report does not comprehensively cover any given topic nor does it encompass all possible topics.

General Discussion on Perturbed Clouds

It was generally felt that although a great deal is known about the physical mechanisms of particle activation, cloud microphysics, and cloud dynamics at the single cloud scale, our level of understanding regarding the nature and the extent of anthropogenic perturbations on *cloud systems* is relatively poor. This may be because perturbations to clouds, especially from aerosol changes, tend to be small relative to the natural variability of a cloud system, or perhaps because the system is self-regulating (e.g., Feingold and Siebert, this volume) and thus perturbations lead to feedbacks (both positive and negative), which makes their manifestation difficult to identify.

Establishing a *causal* relationship, rather than a *correlative* one, between, for example, aerosols and clouds requires us to minimize confounding (i.e., co-varying) factors: we must choose a location and/or regime where meteorological variability is minimal while anthropogenic perturbations are maximal (i.e., an area where the aerosol is ideally uncorrelated with meteorology). The usefulness of time series analysis as an attribution strategy is limited because of these confounding factors (i.e., covariance between aerosols and meteorological factors such as relative humidity; for further discussion, see Stevens and Brenguier, this volume). Other fields of study, such as epidemiology, have experience in addressing such issues, and thus may have tools that are useful to account for such factors.

Tools

Remote Sensing

High-quality (well-calibrated) long-term records (both *in situ* and remotely sensed) are critical to trend detection. A timescale of thirty years is generally regarded as a minimum record length. Satellites provide a critical source of global-scale observations, but are subject to uncertainties that must be accounted for and understood. Satellite overpass times drift over the lifetime of the satellite platform, leading to spurious trends. Moreover, the entire period of satellite observations is not yet even 30 years old. Satellite sensors suffer from degradation over time, lack (in many cases) absolute calibration, and suffer calibration drift. An overlap of sensors to conduct similar measurements is critical to resolve any discrepancies. The lack of an overlap period between ERBE and CERES, along with a significant bias between the two, has led to considerable uncertainty in the radiative budget. Advances in measurement capabilities can be attained through synergistic deployment of instruments (e.g., the A-Train constellation of satellites). Polar-orbiting satellites are constrained to observe any given location once or twice a day, and this greatly limits their usefulness over hourly to daily timescales. Polar-orbiting satellites do not measure clouds at the peak of afternoon convection; for that, we must depend on geostationary satellites with their limited payloads. Geosynchronous satellites have typical resolutions of 1 km and cannot (at the present) constrain many of the critical parameters identified above. They also suffer from edge detection problems. The possibility of a satellite positioned at the Lagrange point was mentioned. Such a satellite would view the entire sunlit hemisphere (with problems near the edge of the field of view) with roughly an 8 km resolution; this is suitable for larger-scale studies, but does not match well with smaller-scale studies.

A lack of co-location of aerosols and clouds viewed from space confounds correlation studies. Whereas subcloud aerosols in updrafts are most pertinent to cloud drop formation, the presence of the cloud generally excludes the ability of most sensors to measure subcloud aerosol properties. This problem can be alleviated using surface remote sensors, but only at limited locations. Aerosols above clouds may also influence cloud properties, but it is difficult to measure aerosols from space accurately above a bright (i.e., cloudy) layer. Finally, because of the complication of surface radiative inhomogeneity, aerosol measurements over ocean locations are for some instruments more reliable than those over land.

Measurement simulators are viewed as a powerful means to compare model output with measurements, including assessment of measurement biases and random error, as well as to optimize the measurement strategy for an observational program, including understanding the added benefit of a given instrument. For example, EarthCARE simulators (Schutgens and Donovan 2004) are being used prior to deployment to optimize the sensor capabilities and exact the most useful dataset. We recommend that future satellite missions include the development of such simulators for distribution to the scientific community. We feel that this is a fruitful direction for future studies to consider.

Integration of Models and Observations

The appropriateness of tools for a given problem depends, among other things, on their spatial and temporal resolution. Figure 18.1 illustrates a conceptual framework for the interaction of different tools (modeling, observations, and laboratory measurements) at various scales. Laboratory measurements provide information on fundamental physical and chemical cloud processes. This information is critical for informing models and field observations. At the smallest length scales, field observations and large eddy simulation (LES) models operate at very similar length and timescales, and resolve aerosol and cloud processes as well as small-scale dynamics. Progress in our understanding at this scale depends on the close interactions between models and observations. One important step is to establish the usefulness of a given model. Generation and testing of new ideas and hypotheses can only be achieved through such cooperation. Further, this strategy is the most effective method to establish the extent to which observed correlations are causal.

Comparison between models and observations should be done to the greatest extent possible using fundamental measured properties (e.g., irradiances rather than albedo; aerosol extinction rather than mass). This will minimize the inconsistencies in assumptions in derived quantities. One example is the comparison between modeled and measured radar reflectivity, which is likely to be much more meaningful if the model actually calculates the sixth moment of the drop size distribution, rather than rely on an empirical relationship between reflectivity and a modeled variable (e.g., liquid water mixing ratio or other empiricisms). Another example would be a comparison between measured and modeled optical depth, which would be greatly facilitated by model calculation of extinction. Confidence in smaller-scale models obtained through this process can feed into larger-scale models in the form of parameterizations. Collaboration between the modeling and observational communities must occur at the full range of scales, if any real progress is to be made.

The quote, "All models are wrong, but some are useful," attributed to George E. P. Box, was raised as an important point for consideration. A suggested modification to this quote is: All models and measurements are wrong, but some are useful. We recognize that all measurements and models are imperfect; still, they can be extremely useful if used and applied appropriately.

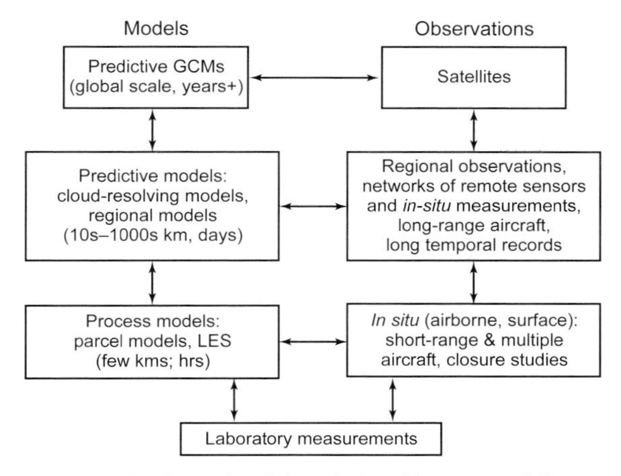

Figure 18.1 Conceptual schematic of the relationship among different subdisciplines for addressing questions regarding anthropogenic perturbations to clouds and precipitation. GCM: general circulation model; LES: large eddy simulations.

The main point is that the quality of data and model output, as well as usefulness for addressing the problem at hand must be given careful consideration to evaluate the strengths and weaknesses.

Predictability

In our discussion of predictability, a number of interesting questions were raised: Is there a limit to the predictably of cloud systems? Does predictability depend on length scale? At the largest scales, for example, precipitation must be constrained by total evaporation and large-scale circulation, which in turn relate to top-of-atmosphere (TOA) and surface energy budgets. These constraints may lead to greater predictability at these scales. At small length and timescales (e.g., single cloud scale), the largest-scale features (such as mean vertical profiles of potential temperature and humidity, subsidence rates, etc.) may be easily predicted because they vary in ways that are fairly well understood (e.g., diurnally with solar heating) on these short timescales. If this exists, such a scenario may grant advantages for prediction at small scales. At intermediate scales, however, it is conceivable that high-frequency variability from smaller scales, as well as large-scale constraints, will play a significant role, thus causing a minimum of predictability. Whether or not this is true remains an open question (cf. Siebesma et al., this volume).

Microphysics

Perturbations on cloud microphysical properties, especially through changes in particle concentration and/or composition, serve as a starting point for discussing many of the subsequent perturbations related to radiation, precipitation, dynamics, and chemistry. The state of cloud condensation nuclei (CCN) measurements and knowledge, as well as anthropogenic perturbations of CCN, has been presented by Kreidenweis et al. (this volume). Another excellent summary is available from McFiggins et al. (2006). Because of this existing body of work and the limited time available, the group did not address anthropogenic perturbations to a number of cloud microphysical processes (e.g., activation, collision–coalescence and evaporation/entrainment) in much detail. This is reflected in the summary below and does not imply an absence of gaps within these important subjects.

Some group members felt that the largest remaining gap in understanding CCN activation is the mass accommodation coefficient of water vapor onto atmospheric aerosol. Recent measurements by Ruehl et al. (2008) suggest that the value is variable, even for the limited number of sampling sites and short periods that were sampled. In contrast, measurements of ice nuclei are limited to a few instruments worldwide and are unable to observe all possible modes of heterogeneous nucleation. This technological gap creates a fundamental difficulty in understanding potential perturbations to glaciated clouds. Bioaerosols are potentially important as ice nuclei, and our understanding of this component remains poor, although some fluorescence-based instruments do exist for detecting a subset of such particles.

Supersaturation (with respect to both water and ice) is not accurately measurable, and thus this important constraint on cloud drop production is not available. This technological gap has been known for a long time, yet it endures in large part because of the difficulty of making such a measurement.

There is a need for long-term, spatially resolved records of CCN spectra. To our knowledge, the only multidecadal record of CCN is from Cape Grim, Tasmania, although a handful of other sites are now instrumented for routine CCN measurement (e.g., Mace Head, Ireland). Thus, we recommend that the creation and maintenance of such a network be considered an important priority for understanding this fundamental perturbation to clouds. The usefulness of CCN proxies, such as the dry-state aerosol accumulation mode number concentration along with some (limited) hygroscopic information, should be considered. A satellite sensor-derived aerosol index (product of aerosol optical depth and Ångström exponent) has also been proposed as a proxy for CCN. This may be qualitatively useful but its quantitative accuracy is highly questionable, since it cannot account for particle humidification (Kapustin et al. 2006). Finally, we suggest further exploration of possibilities to obtain routine CCN and other aerosol measurements from platforms of opportunity, such as ships and commercial aircraft.

Removal Mechanisms

Aerosol-scavenging mechanisms that are associated with clouds and precipitation (also known as wet deposition) dominate aerosol removal. Although generally not as well-studied as other aerosol–cloud interactions, they are important in understanding aerosol budgets at a range of scales (single cloud to global).

The following questions and associated challenges were raised:

1. A perturbation in precipitation by aerosols can lead to a perturbation in aerosol wet deposition. Could this be an important positive feedback? Some boundary layer modeling studies suggest this (e.g., Feingold and Kreidenweis 2002).
2. Model results (e.g., Respondek et al. 1995) show a relationship between precipitation efficiency and scavenging efficiency. Is it possible to make important inferences about precipitation efficiency by measuring scavenging efficiency?
3. Is coalescence scavenging an important mechanism for converting external mixtures of particles to internal mixtures?
4. What are the consequences of aerosol scavenging for the climate system? For example, the scavenging of soot by snow, its deposition to the surface, and associated amplification of snow melt, is an active topic of research.

A number of important tool and data gaps exist. Measurements of precipitation chemistry are routinely made, but are likely inadequate for understanding the lifetime of aerosols and the cycling of organics, dust, and soot. This represents a crucial gap in the understanding of the global aerosol budget. Some existing programs include the European Monitoring Evaluation Programme and the U.S. National Atmospheric Deposition Program. *In-situ* measurement of cloud drop chemical composition is challenging to conduct on appropriately fast timescales, but would be valuable for understanding cloud chemistry. Scavenging efficiency is hard to assess because of volatility of some scavenged species. It was proposed that ice cores represent a potential historical record for deposition, but they do not appear to have been fully exploited.

Radiation

The globally averaged temperature over the last 10,000 years appears to have been stable to within ~1 K, thus constraining, most likely, the globally averaged albedo to within ~1% throughout this period. The implications of this 1% variability are not negligible since it represents about 3.5 W m^{-2}, or 1% of the incoming solar radiation at TOA. Clouds represent roughly two-thirds of the global planetary albedo. This suggests the presence of feedbacks that maintain global cloud albedo, which remain poorly understood. Siebesma et al. (this volume), however, suggest that this is simply a result of the Law of Large Numbers and does not, therefore, require the presence of such feedback mechanisms.

By directly scattering and absorbing radiation, and through modifying cloud drop size, aerosol particles modify the Earth's radiation budget at the top of and within the atmosphere, as well as at the surface. The interaction of aerosols and clouds with radiation modifies heating profiles with feedbacks, which in turn modify cloudiness. Perturbations in land use, or aerosol-induced perturbations in net surface radiation, can alter surface fluxes of sensible and latent heat and moisture, thus perturbing convection and, therefore, cloud formation. Anthropogenic perturbations in cloud albedo have the potential to impose changes in the planetary radiative balance. Local changes have been demonstrated (e.g., ship tracks). At larger scales, however, studies have not been definitive. There was general agreement in the group that perturbations on cloud fraction, precipitation, and the longevity of clouds have not been convincingly demonstrated (cf. Anderson et al., this volume).

Cloud Fraction

The effect of aerosols on cloud fraction is poorly constrained. Not only the magnitude, but also the sign of the effect (i.e., positive or negative) is unclear and may be dependent on the regime. Differentiation between cloudy air and cloud-free air is somewhat arbitrary, and thus cloud fraction is not a well-defined quantity. The presence of cloud haloes as a result of humidified aerosols and small cloud fragments renders the separation between cloudy and cloud-free atmospheres artificial (Koren et al. 2007; Charlson et al. 2007). Thus, aerosol perturbations on cloud fraction are also ill-defined. Furthermore, measurement of cloud fraction depends both on the spatial resolution and detection sensitivity of the instrument. Studies of the size and area distribution of clouds show that small clouds are more numerous, and they contribute more to the cloud fraction than larger clouds, down to the smallest scales resolved by current instruments (Koren et al. 2008). Thus, we suggest that the definition of cloud fraction is dependent on the application. For example, even though small clouds contribute more strongly to cloud fraction, there may exist a size below which smaller, optically thinner clouds can be neglected for the purposes of determining planetary albedo. This might yield a lower bound on the size of clouds relevant to cloud fraction, and thereby an appropriate way to define cloud fraction in this context. This lower bound may be different for other problems, such as longwave radiation. If true, this would imply that the appropriate measurement spatial resolution also depends on the problem under consideration.

Radiative Forcing

For detecting trends in radiation, one possible strategy for overcoming uncertainties in satellite retrievals is to utilize regions where models reproduce observations well; that is, regions where the signal is strongest. Such models can then be used to generate a global distribution of radiation, which suggests the following question: Is albedo best determined by *measurements alone* or through a *hybrid of measurements and models*? Arguments for both sides were presented.

Surface radiation measurements are useful for deriving surface (as opposed to TOA) forcing. In the absence of surface measurements, surface forcing has to be derived from TOA measurements. This conversion requires assumptions about atmospheric absorption. Surface measurements of net irradiance, combined with satellite TOA measurements, obviate the need for such assumptions. Surface supersites that combine remote sensing from a variety of instruments with *in-situ* aerosol sampling and thermodynamic profiles provide extremely useful datasets for comparison with satellites. The high temporal resolution allows for process studies not achievable from space. Examples include sites at Southern Great Plains (U.S.A.), Cabauw (Netherlands), Lindenburg (Germany), Gosan (Korea), and Okinawa (Japan). Lidar networks provide information on the vertical distribution of aerosol, cloud base, and cirrus clouds at high vertical and temporal resolution. New technologies, such as high spectral resolution lidar, enable the measurement of extinction profiles and comparison with aerosol optical depth, but they are limited to thin clouds. Combining lidar with radar, or radar with infrared radiometry, enables retrievals of effective ice crystal radius. This is a good example of measurement synergy. Optimal estimation methods are increasingly being applied to yield best-estimates of physical parameters based on observations from multiple instruments, each with their own strengths and weaknesses (Feingold et al. 2006; Hogan 2007).

The AERONET sun photometer network has developed since the late 1990s into an important research tool aimed at characterizing the global aerosol for climate study purposes (see also Kinne, this volume). Although its primary purpose is to validate remote-sensing products, AERONET serves also to constrain atmospheric aerosol modeling in all aspects (e.g., the international AeroCom initiative). A link to other sun photometer measurement networks (e.g., those assembling under the Global Aerosol Watch Program) is most desirable. It would be useful if these and other measurement systems could be combined with AERONET to address questions related to aerosol–cloud interactions. This includes higher measurement frequencies, which should not pose a problem with current computing capacities. Better documentation of the actual sequence of cloudy and clear-sky scenes would allow better proxies for total cloud cover (see Koren et al. 2007). Additional parameters (e.g., rain amount, rain drop size spectra, cloud base height, and broadband radiative fluxes) should be measured at an increasing number of AERONET stations.

The location of new supersites for aerosol and cloud characterization should be coordinated with long existing AERONET sites. Supersites should ensure that more than one sun photometer is operational at any time to avoid any discontinuity of these basic aerosol observational time series.

Tools

The CERES and ERBE satellite missions have been critical in advancing our understanding of the Earth's radiation budget. Their lack of overlap raises questions, however, about their apparent disagreement and limits their utility. The fact that no replacement for CERES is planned was unanimously viewed as a serious problem, and we recommend strongly that this be remedied as soon as possible.

It was suggested that a satellite at the Lagrange point L1 would be useful for radiation studies, as it would always be pointed at the sunlit side of Earth, yielding time series (including the diurnal cycle) of global albedo, which requires a radiance-flux calculation, potentially resolved down to ~8 km. Such a satellite could also serve as an important constraint on other satellite sensors.

It was noted that the multi-wavelength radiometric data from MODIS, aboard both the Terra and Aqua platforms, are being used for a number of inverse retrievals; among these are the fractional cloud cover, the effective radius of cloud droplets, and the aerosol optical depth. Because of the importance of these quantities to studies such as those conducted by the IPCC, the *specificity* and *uncertainty* of these retrievals are currently being questioned. Although these satellite retrievals are of great importance to global sensing of gross phenomena, such as the annual outbreaks of dust from the Sahara Desert or the smoke from biomass burning in Africa, they may not be fully adequate to quantify accurately the direct and indirect climate forcing by anthropogenic aerosols. Some of the difficulty involved in quantifying the uncertainty of the remote retrievals from MODIS lies in the inherent difficulty of separating clear and cloudy conditions (e.g., over the horizontal extent of a MODIS pixel; Charlson et al. 2007). This open issue remains to be resolved.

Measurements of earthshine (i.e., the sunlight reflected by Earth that illuminates the unlit (by the Sun) side of the Moon) are a relatively inexpensive and simple technique used to derive large-scale albedo of the Earth. Earthshine data from multiple locations could provide a very useful, independent measure of albedo. Such data can also provide temporal information that is not available from polar-orbiting satellites. Comparison of climate model output could be achieved through an "earthshine simulator."

Finally, surface albedo and emissions are fundamental to the production of accurate satellite retrievals of aerosol optical depth. The baseline surface radiation network was also mentioned as a potential source of measurements, but its utility for cloud measurements requires most likely additional calibration and instrumentation effort.

Semi-direct Effect

The semi-direct effect influences clouds over a range of spatial scales. Absorbing aerosol particles generate local heating, and the presence of the semi-direct effect at the appropriate levels can modify the atmospheric stability and suppress vertical motion and cloud formation (Hansen et al. 1997). Alternately, heat added to the planetary boundary layer might raise the temperature and evaporate cloud droplets. We expand our definition to include the fact that the presence of aerosol, particularly absorbing aerosol, is very effective at reducing the downwelling solar radiation at the surface, and consequently, the surface fluxes. The combination of these factors, as well as the (much smaller) microphysical influences on droplet growth, results in significant reduction in cloudiness. This has been observed in biomass burning regions in Brazil (Koren et al. 2004), and satellite imagery strongly suggests effects at the cloud scale as well. Modeling studies suggest that the semi-direct effect might change regional precipitation patterns (Menon et al. 2002). Therefore, an open question is: To what extent do absorbing aerosols modify dynamics, clouds, and precipitation, and might these effects be large enough to influence global circulation?

Precipitation

Clouds are the only entities that deliver precipitation to the Earth's surface. Precipitation is critical to water resources and the global energy cycle. Globally, precipitation must balance global evaporation over sufficiently long timescales. (How long is long enough is unclear, but it may be approximately the residence time of water vapor in the troposphere, which is about one week.) Thus, perturbations in total precipitation amount are constrained by impacts on global evaporation. However, the spatial and temporal *distribution* of precipitation as well as the distribution of precipitation intensity may change significantly. Greenhouse gas warming is predicted to cause increases in evaporation and precipitation on a global scale, as well as changing precipitation patterns on a regional scale. Currently, however, there is not enough evidence to suggest that changes occur on smaller scales. Aerosol perturbations have been linked to precipitation suppression (Warner 1968), but there is no statistically robust proof of aerosol suppression of surface rainfall at scales large enough to be of importance to the hydrological cycle. Moreover, although questions persist about warm rain initiation, greater unresolved issues remain in terms of the ice phase, especially concerning possible anthropogenic perturbations to ice processes (for further discussion, see Levin and Cotton 2008; Ayers and Levin, this volume; Cotton, this volume).

Aerosol perturbations are spatially inhomogeneous on a number of scales. Because their ability to affect precipitation depends on the co-location of precipitation and aerosol, aerosol-related precipitation perturbations are also expected to be spatially inhomogeneous across a variety of scales. The variability in precipitation associated with natural variability is very large relative to the aerosol perturbation, and thus attribution of changes in precipitation to aerosols is extremely challenging. Presently, the sign of the overall aerosol effect is unclear and may not be the same for all cloud regimes and environments. Single cloud studies show strong changes in precipitation in response to aerosol perturbations in particular places, times, or regimes, but response at larger scales remains elusive. Models can help, but important physical knowledge is lacking (particularly in regards to ice microphysics), thus limiting the usefulness of models that might otherwise be adequate to study this problem.

Numerical weather prediction (NWP) models may be a resource that has not been fully exploited for investigating aerosol–precipitation interactions. Over the last forty years, NWP has improved because data accuracy has improved more than microphysical parameterizations. What are the relative benefits to further improvements in defining model initial and boundary conditions versus a better representation of aerosol? Some members in our group stated that there does not appear to be a difference in skill between clean and polluted aerosol conditions, although this may not necessarily imply that aerosol representation is not needed to improve NWP. The question of how much of the residual variability in NWP is the result of aerosols was raised, but no clear answer was proposed. Use of retrospective (30-yr) data to test model improvements in quantitative precipitation forecasting may allow some of these ideas to be tested. One advantage of NWP is that it is performed daily; assimilation methods establish the errors and biases daily, at relatively low cost.

Because precipitation perturbations that result from aerosols are expected to be small relative to natural variability, it is unclear whether we have tools sufficient to the task of detecting such perturbations. Traditional *in-situ* measurement of drops relevant to precipitation (especially precipitation embryos) exhibit large uncertainties. It was suggested that there exist aerosol-dependent biases in radar reflectivities (i.e., the relationship between radar reflectivity Z and rain rate R is aerosol dependent), and therefore they might not be the ideal tool for aerosol–precipitation interactions, although not everyone agreed on this point. Rain gauge data represent an important long-term record with widespread ground stations in populated areas. This dataset is, however, problematic for a variety of reasons (e.g., representativeness, biases). How can this dataset be improved? It is important to note that the rather large number of weather modification studies over the past half-century or more (see Cotton, this volume) has not demonstrated highly significant perturbations in precipitation despite very large local aerosol perturbations.

A number of strategies may be useful in detecting aerosol influences on precipitation. For example, it could be advantageous to capitalize on human-induced variability as a perturbation experiment (e.g., weekly aerosol cycles) wherever possible. Weekly aerosol cycles are observed in some, but not all locations. However, the weekly precipitation cycle is even less clear in the observation record than in the aerosol cycle. Orographic clouds may serve as a useful test bed, with mixed results thus far (for further discussion, see Ayers and Levin, this volume). Single water drainage basin studies are proposed as targets. Consensus was not achieved in terms of the potential for closing the energy and moisture budgets. Some argued that this is the necessary approach, and therefore we should do the best we can, while others stated that a single basin is too large of an area over which energy and moisture budgets can be closed, and thus this strategy cannot succeed at present. This issue remains to be resolved.

Dynamics

Dynamics (i.e., the three-dimensional movement of the atmosphere) are the primary driver of clouds. Anthropogenic perturbations to clouds can affect dynamics through myriad feedbacks on various spatial and temporal scales. The resulting perturbations in dynamics from such feedbacks are, however, poorly understood. Below we present some proposed feedbacks that emanated from our discussions, which we felt were of particular interest and/or importance:

1. Aerosol perturbations can modify precipitation rates. Evaporating precipitation cools and thus stabilizes the subcloud layer.
2. In a field of clouds, perturbations to precipitation can change the strength of downdrafts associated with rain shafts, altering surface convergence and thus subsequent convection. This can occur at a range of scales, from squall lines down to open cell convection in stratocumulus.
3. Suppression of freezing associated with smaller drops formed in polluted clouds results in latent heat release at higher altitudes, leading to cloud invigoration and deepening.
4. Land-use changes can alter the magnitudes of the surface fluxes, as well as the Bowen ratio, with consequences for the strength of convection and cloud base and cloud top heights.
5. Smaller drops associated with polluted clouds may evaporate more readily, promoting increased entrainment mixing and turbulence, with possible ramifications for turbulent collection and precipitation formation.
6. Greenhouse gas warming is predicted to change general circulation patterns, with possibly important consequences for the distribution of clouds as well as the distribution and intensity of precipitation.

Strategies

There was broad discussion about strategies for addressing dynamic feedbacks. It was generally agreed that a team comprising experts in atmospheric dynamics, cloud physics, and aerosols, as well as their models and observations, offers the best hope for success. The ultimate goal should not be merely correlation but causation, for which models appear necessary. We should attempt to look for systems that are naturally simple, but it was unclear which cloud systems best fit this criterion. It was believed that working at the smallest scales plausible for examining dynamic feedbacks held a number of advantages, including the ability to integrate large eddy simulations with observations, as well as in finding the largest signals. However, the minimum spatial and temporal scale at which such feedbacks are observable is unknown. Even with a relatively simple system, it was unclear to us what an appropriate strategy would be, and whether a realistic set of instrumentation would detect any such feedbacks, either because of inherent instrument limitations or because of the lack of an adequate sampling strategy. It was noted that increasing sampling statistics is useful only for reducing random errors, whereas biases (e.g., covariance of aerosol and meteorology; see Stevens and Brenguier, this volume) cannot be reduced in this way. Therefore, we recommend that future studies should identify and study regimes with minimum biases.

Tools

The role of measurements in understanding anthropogenic perturbations to dynamics was viewed primarily as constraints on models (horizontal arrows in Figure 18.1). As the quality and quantity of constraints improve, so too will the degree of our confidence in the model predictions. We had a long discussion on which variables are most important to constrain. The critical ones that we identified fell into both the dynamic and microphysical categories: temperature and relative humidity vertical profiles; large-scale subsidence rates; surface fluxes of moisture, energy, and momentum; winds; cloud liquid water content profiles; cloud liquid water path; cloud base and top height; vertical profiles of rain rate (including below-cloud); cloud fraction; supersaturation with respect to water and ice (particularly in the upper troposphere); and many microphysical properties related to the ice phase. Many of these variables are very difficult to measure accurately, especially over the spatial and temporal scales that might be needed to observe dynamic feedbacks properly. At large scales, the most successful strategy will likely require combining numerous high-resolution satellite products (e.g., gases, aerosol, cloud hydrometeors, radiation), over large spatial and temporal scales, with appropriate models. A number of problems with existing satellite retrievals were identified that confound studies of the coupling between clouds and atmospheric dynamics. In particular, there is a strong bias between surface- and satellite-derived cloud fraction, and satellite- and *in situ*-derived drop sizes in broken cloud fields.

Chemistry

We considered the broader issue of the interactions of atmospheric chemistry and clouds. The average particle cycles through clouds many times before it is removed from the atmosphere. Therefore, the potential for chemical modification of particles is very high (Hoppel et al. 1994). Most anthropogenic pollutants are reduced relative to the oxidizing atmosphere, and thus atmospheric processing leads to increases in polar and therefore water soluble species. This chemical modification has implications for physical, chemical, and optical particle properties and affects subsequent cloud cycles. Cloud particle chemistry can occur in both aqueous and ice phases. Aqueous phase chemistry is likely to be faster because the components have greater mobility. Cloud chemistry also affects gas phase chemistry, partitioning soluble components to the aqueous phase. Aqueous phase chemistry in sea-salt haze may be significant despite its small volume, as a result of spatial and temporal persistence (Sievering et al. 2004). Processing of hydrophobic particles (e.g., dust, soot) in clouds may be an efficient way to increase hygroscopicity. For dust, this could be a mechanism for producing giant CCN. In general, it was thought that quantitative understanding of cloud processing remains a knowledge gap, but one that may be filled with current instrumentation and existing sampling strategies.

Clouds are an important means of transporting boundary layer constituents (trace gases, aerosol) to the free troposphere and lower stratosphere. Perturbed clouds may exhibit changes in their depth of convection as a result of changes in, for example, precipitation or latent heating, which will affect the altitude to which these constituents are transported.

It is well known that the sulfur cycle is strongly perturbed by anthropogenic pollutants. Sulfur is a key component of the atmospheric aerosol and therefore CCN. Oceanic seawater sulfate is highly abundant and thus substrate availability does not limit the production of volatile sulfur compounds (e.g., dimethyl sulfide, carbonyl sulfide). Instead, biology appears to be the determining factor in producing such compounds. Advances have been made in understanding the specific factors that govern such production, but many gaps in knowledge remain. Perturbations to the surface ocean (e.g., temperature, pH, salinity) may affect biological production of such compounds to a degree sufficient to impact the atmospheric sulfur cycle significantly. Production of volatile sulfur compounds is sensitive to sea surface temperature as well as meteorology, setting up potential feedbacks.

New particle formation in the vicinity of clouds is poorly understood and is hard to measure, yet it may represent an important contribution to the global particle number concentration budget. *In-situ* measurements may constitute the only method for constraining this problem because Aitken mode particles are extremely difficult to detect optically. Existing techniques are often insufficiently fast for aircraft measurements, which may make this a measurement gap. However, slower airborne platforms do exist (e.g., dirigibles) and may permit present instrumentation to address this question (see Feingold and Siebert, this volume.)

Organic compounds are an important component of atmospheric aerosol. Models underestimate strongly the amount of secondary organic aerosol relative to measurements. Clouds may contribute to this missing source (e.g., Ervens et al. 2008), which we suggest is a potentially fruitful direction to explore to help resolve this issue. On a global scale, biogenic organic particles appear to be more abundant than anthropogenic particles. Perturbations to precipitation, temperature, sunlight, and thus to the land ecosystem have the potential to alter this source strength (for a recent review, see Möhler et al. 2007).

The deposition of inorganic ions is documented by a number of networks, such as the European Monitoring and Evaluation Programme and the U.S. National Atmospheric Deposition Program. Long-term monitoring on large spatial scales of the organic and black carbon concentrations in rain water is lacking, yet it is important for closing their budgets. The techniques necessary to conduct these measurements exist, so this gap could be filled given appropriate resources.

Finally, if anthropogenic perturbations alter convection and therefore lightning production, this may alter NO_x production. The potential significance of this to photochemistry was unclear to the group and remains to be illuminated.

References

Charlson, R. J., A. S. Ackerman, F. A. M. Bender, T. L. Anderson, and Z. Liu. 2007. On the climate forcing consequences of the albedo continuum between cloudy and clear air. *Tellus B* **59(4)**:715–727.

Ervens, B., A. G. Carlton, B. J. Turpin et al. 2008. Secondary organic aerosol yields from cloud-processing of isoprene oxidation products. *Geophys. Res. Lett.* **35**:L02816

Feingold, G., R. Furrer, P. Pilewskie et al. 2006. Aerosol indirect effect studies at Southern Great Plains during the May 2003 Intensive Operations Period. *J. Geophys. Res.* **111(D5)**:S14.

Feingold, G., and S. M. Kreidenweis. 2002. Cloud processing of aerosol as modeled by a large eddy simulation with coupled microphysics and aqueous chemistry. *J. Geophys. Res.* **107**:D23.

Hansen, J., M. Sato, and R. Ruedy. 1997. Radiative forcing and climate response. *J. Geophys. Res.* **102(D6)**:6831–6864.

Hogan, R. J. 2007. A variational scheme for retrieving rainfall rate and hail reflectivity fraction from polarization radar. *J. Appl. Meteor.* **46(10)**:1544–1564.

Hoppel, W. A., G. M. Frick, J. Fitzgerald, and R. E. Larson. 1994. Marine boundary-layer measurements of new particle formation and the effects nonprecipitating clouds have on aerosol-size distribution. *J. Geophys. Res.* **99(D7)**:14,443–14,459.

Kapustin, V. N., A. D. Clarke, Y. Shinozuka et al. A. 2006. On the determination of a cloud condensation nuclei from satellite: Challenges and possibilities. *J. Geophys. Res.* **111(D4)**:202.

Koren, I., Y. J. Kaufman, L. A. Remer, and J. V. Martins. 2004. Measurement of the effect of Amazon smoke on inhibition of cloud formation. *Science* **303(5662)**:1342–1345.

Koren, I., L. Oreopoulos, G. Feingold, L. A. Remer, and O. Altaratz. 2008. How small is a small cloud? *Atmos. Chem. Phys. Discuss.* **8**:6379–6407.

Koren, I., L. A. Remer, Y. J. Kaufman, Y. Rudich, and J. V. Martins. 2007. On the twilight zone between clouds and aerosols. *Geophys. Res. Lett.* **34(8)**:L08805.

Levin, Z., and W. R. Cotton, eds. 2008. Aerosol pollution impact on precipitaton: A scientific review. The WMO/IUGG International Aerosol Precipitation Science Assessment Group. New York: Springer, in press.

McFiggans, G., P. Artaxo, U. Baltensperger et al. 2006. The effect of physical and chemical aerosol properties on warm cloud droplet activation. *Atmos. Chem. Phys.* **6**:2593–2649.

Menon, S., J. Hansen, L. Nazarenko, and Y. F. Luo. 2002. Climate effects of black carbon aerosols in China and India. *Science* **297(5590)**:2250–2253.

Möhler, O., P. J. DeMott, G. Vali, and Z. Levin. 2007. Microbiology and atmospheric processes: The role of biological particles in cloud physics. *Biogeosciences* **4**:1059–1071.

Respondek, P. S., A. I. Flossmann, R. R. Alheit, and H. R. Pruppacher. 1995. A theoretical study of the wet removal of atmospheric pollutants: The uptake, redistribution, and deposition of (NH4)2SO4 by a convective cloud containing ice. *J. Atmos. Sci.* **52**:2121–2132.

Ruehl, C. R., P. Y. Chuang, and A. Nenes. 2008. How quickly do cloud droplets form on atmospheric particles? *Atmos. Chem. Phys.* **8**:1043–1055.

Sievering, H., J. Cainey, M. Harvey et al. 2004. Aerosol non-sea-salt sulfate in the remote marine boundary layer under clear-sky and normal cloudiness conditions: Ocean-derived biogenic alkalinity enhances sea-salt sulfate production by ozone oxidation. *J. Geophys. Res.* **109**:D19317.

Warner, J. 1968. A reduction in rainfall associated with smoke from sugar-cane fires: An inadvertent weather modification? *J. Appl. Meteor.* **7**:247–251.

19

Global Indirect Radiative Forcing Caused by Aerosols

IPCC (2007) and Beyond

Jim Haywood[1], Leo Donner[2], Andy Jones[1], and Jean-Christophe Golaz[2]

[1]Met Office, Exeter, U.K.
[2]Geophysical Fluid Dynamics Laboratory/NOAA, Princeton University, Princeton, NJ, U.S.A.

Abstract

Anthropogenic aerosols are thought to exert a significant indirect radiative forcing because they act as cloud condensation nuclei in warm cloud-forming processes and ice nuclei in cold cloud-forming processes. Although many of the processes associated with the perturbation of cloud microphysics by anthropogenic aerosols were discussed, IPCC (2007) provided only an estimate of full quantification of the radiative forcing attributable to the first indirect effect (which they referred to as the cloud albedo effect). Here we explain that this approach is necessary if one is to compare the radiative forcing from the indirect effect of aerosols with those from other radiative forcing components such as that from changes in well-mixed greenhouse gases. We also highlight the problems in assessing the effect of anthropogenic aerosols upon clouds under the strict definitions of radiative forcing provided by the IPCC (2007). Although results from global climate models, at their current state of development, suggest that an analysis of indirect aerosol effects in terms of forcing and feedback is possible, a key rationale for the IPCC's definition of radiative forcing, a straightforward scaling between an agent's forcing and the temperature change it induces, is significantly compromised. Feedbacks from other radiative forcings are responses to radiative perturbations, whereas feedbacks from indirect aerosol effects are responses to both radiative and cloud microphysical perturbations. This inherent difference in forcing mechanism breaks down the consistency between forcing and temperature response. It is likely that additional characterization, such as climate efficacy, will be required when comparing indirect aerosol effects with other radiative forcings. We suggest using the radiative flux perturbation associated with a change from preindustrial to present-day composition, calculated in a global climate model using fixed sea surface temperature and sea ice, as a supplement to IPCC forcing.

The Concept of Radiative Forcing

In the latest IPCC report, Forster et al. (2007) define the concept of radiative forcing as follows:

> The radiative forcing of the surface-troposphere system due to the perturbation in or the introduction of an agent is the change in the net irradiance at the tropopause *after* allowing for stratospheric temperatures to readjust to radiative equilibrium, but with surface and tropospheric temperatures and state held fixed at the unperturbed values.

Over the last couple of decades, radiative forcing has proved to be a useful concept because the global mean near surface temperature response, dT, to a particular radiative forcing, dF, may be related to the climate sensitivity, λ, via the relationship:

$$dT = \lambda dF. \tag{19.1}$$

Generally, studies suggest that this relationship appears to be approximately independent of the forcing mechanism (e.g., Meehl et al. 2004; Matthews et al. 2004), which means that the relative importance of many different forcing mechanisms may be quantified and compared.

While λ is approximately independent of the forcing mechanism, it may be quite strongly model-dependent. Additionally, some recent studies have suggested that λ may not be entirely independent of the forcing mechanism, as they may induce different feedbacks, which may lead to a modification of Equation 19.1 via the climate forcing efficacy, ε, (Joshi et al. 2003; Hansen and Nazarenko 2004):

$$dT = \varepsilon_i \lambda_{CO2} dF \tag{19.2}$$

where ε_i is defined as λ_i/λ_{CO2}. The inter-forcing mechanism differences in ε_i appear to be greatest for absorbing aerosols, where absorption of solar radiation induces the so-called "semi-direct effect," but large-scale dynamic feedbacks differ in the models used in assessing ε_i. This results in values that are either larger (e.g., Jacobson 2001) or smaller (e.g., Roberts and Jones 2004; Hansen et al. 2005; Jones et al. 2007) than unity with values that typically range from 0.7–1.3 (Forster et al. 2007).

It is important to realize that to diagnose an indirect radiative forcing that may be compared to other forcing mechanisms, it is the *global* radiative forcing that is related to the *global* temperature change in Equation 19.1 and 19.2. A local radiative forcing does not correspond to a local temperature change because of the myriad of local-, regional-, and global-scale feedback processes that vary over the different regions of the Earth.

Now that the IPCC (2007) definition of radiative forcing and its limitations have been presented, we will discuss the specifics of the indirect aerosol radiative forcing via interactions with clouds.

Radiative Forcing and the Aerosol Indirect Effects

Aerosol particles act as cloud condensation nuclei and can thereby modify the microphysical, macrophysical, and optical properties of clouds. Figure 19.1 summarizes schematically the processes considered in assessing aerosol indirect radiative effects. As depicted, unperturbed clouds reflect a proportion of incident solar radiation back to space; when anthropogenic aerosols are introduced under the assumption of constant cloud liquid water, the cloud is made up of a larger number of smaller droplets, as represented by the albedo effect/1st indirect effect/Twomey effect (Twomey 1977). The change in the cloud droplet radius affects the cloud development in a number of complex ways, some of which are shown in Figure 19.1. Smaller cloud drops may lead to a decrease in the coalescence rate and thus reduce rainfall: decreased rainfall will increase the cloud liquid water content, leading to more developed and therefore deeper clouds (e.g., Pincus and Baker 1994). Reduced rainfall and the increase in cloud liquid water content may also lead to an increase in the lifetime of clouds (Albrecht 1989). All of these effects were considered by Forster et al. (2007) to be encompassed by the labels "cloud lifetime effect/2nd indirect effect/Albrecht effect" in Figure 19.1. Furthermore, two other effects are represented in Figure

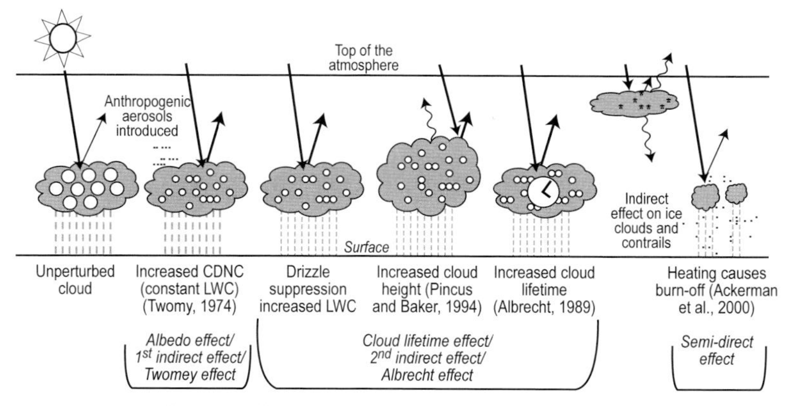

Figure 19.1 A schematic diagram representing the processes considered in assessing aerosol direct and indirect radiative effects (modified from Haywood and Boucher 2001).The small black dots represent anthropogenic aerosols, and the large/small open circles depict large/small cloud droplets, respectively. Ice particles are represented by stars. Solar fluxes are represented by arrows, and the magnitude of the fluxes represented by the width of the arrows. Wavy lines represent infrared fluxes. Precipitation is depicted by dashed vertical gray lines. CDNC: cloud droplet number concentration; LWC: liquid water content.

19.1: the effect of anthropogenic aerosols as ice nuclei and the semi-direct effect, whereby additional solar absorption by anthropogenic aerosols causes heating of the ambient environment and reduces the ambient relative humidity, thereby changing the cloud amount.

Much higher-resolution models, such as cloud-resolving models (CRMs) or large eddy simulation models, are better able to represent the microphysical and dynamical interaction of aerosols and clouds, but they still require subgrid-scale parameterizations of some processes (e.g., entrainment). Their limited domains mean that they are currently only capable of estimating the local indirect radiative forcing for specific case studies. Thus, only general circulation models (GCMs) or chemical transport models (CTMs) are suitable for quantifying the *global* indirect radiative forcing, which can then be compared with other forcing mechanisms such as that attributable to changes in well-mixed greenhouse gases.

The History of Aerosol Indirect Radiative Forcing

In the mid-1990s, when global simulations of aerosol interactions with clouds were in their infancy, a compromise between CTM and GCM modeling efforts was devised to investigate aerosol indirect forcing. CTMs generally did not include realistic radiative transfer codes, which need accurate representation of absorbing gases, treatment of scattering, and accurate representation of surface reflectance, emissivity, and temperature profiles. Similarly, GCMs did not include the detailed chemical processes that were necessary to represent the emission, formation, transportation, microphysical processes, and wet and dry deposition of aerosols and their precursors. Therefore, CTMs were frequently used to model the particulate mass mixing ratios for preindustrial and present-day conditions. These CTM-derived mass mixing ratios were used by the GCMs to simulate the effect of aerosols upon the cloud effective radius by using parameterizations based on either Köhler theory or aircraft observations (e.g., Leaitch et al. 1992; Hegg et al. 1993; Martin et al. 1994). These GCM parameterizations related the particle number or mass concentration to the number of cloud droplets, which itself is related to cloud effective radius. Typically, GCMs diagnosed the first indirect radiative forcing by making two calls to the radiation code: the first call used preindustrial particle concentrations and the second used present-day particle concentrations. "Preindustrial" was used for advancing the model physics at each radiation time-step and the top-of-the-atmosphere (TOA) outgoing shortwave radiation, $SW_{TOA}{}^{\uparrow}{}_{PREIND}$, was diagnosed. "Present-day" was used to modify the cloud effective radius at each radiation time step and determined the TOA outgoing shortwave radiation perturbed by anthropogenic aerosols, $SW_{TOA}{}^{\uparrow}{}_{PRESENT}$. The first indirect effect can then be simply diagnosed as $SW_{TOA}{}^{\uparrow}{}_{PREIND} - SW_{TOA}{}^{\uparrow}{}_{PRESENT}$. Because increased concentrations of anthropogenic aerosol reduce the effective radius of cloud particles and smaller cloud particles reflect more radiation back to space, $SW_{TOA}{}^{\uparrow}{}_{PREIND}$ - $SW_{TOA}{}^{\uparrow}{}_{PRESENT}$ is negative, and the first indirect effect leads to a cooling of the climate system according to Equation 19.1.

There are two contraventions to the definition of radiative forcing of Forster et al. (2007), namely that the change in the net irradiance is frequently derived at the TOA rather than at the tropopause, and the definition does not account for stratospheric adjustment. Neglecting these effects makes very little difference to the derived first indirect radiative forcing when it is derived diagnostically. It is important to realize that the present-day call to the radiative transfer code does not affect the atmospheric heating rates in the model in any way, and therefore the model evolution is entirely unaffected by the perturbation to the particle concentrations. This means that feedback processes are not induced, and cloud, precipitation, sensible and latent heat fluxes, etc. are not affected in diagnosing the first indirect effect in this manner.

GCMs and CTMs have continued their development, and increases in computing power mean that they have to some extent merged: GCMs are now capable of including detailed parameterizations for many types of aerosol which account for the emissions, gas and aqueous phase formation, transportation, and wet and dry deposition of aerosols and their precursors. Therefore, GCMs no longer rely on the off-line particle mass mixing ratios generated by CTMs. As a result, deriving the first aerosol indirect forcing in the manner described above has fallen out of favor. This is because global climate simulations are now driven by emissions, and they explicitly model (albeit in a highly parameterized way) the effect of aerosols upon the microphysical properties of clouds in a far more complete way, and account for the host of second indirect effects, shown schematically in Figure 19.1. Once this approach is taken, diagnosis of the first indirect effect by itself is difficult because separation of the first from the second indirect effects is difficult to achieve in a consistent way. Although the first indirect effect can be approximated by global models without inducing any feedbacks to the climate system (using two calls to the radiation code), when the second indirect effects are invoked, there are necessarily significant fast feedbacks that come into play. For example, each of the second indirect effects represented schematically in Figure 19.1 and represented by GCMs show perturbations to the precipitation. Therefore the state of the atmosphere is not held fixed because surface and tropospheric temperatures respond to changes in, for example, the surface sensible and latent heat flux perturbations caused by the perturbation to the aerosol. Thus, the definition of radiative forcing of Forster et al. (2007) is further compromised. Because we wish, however, to quantify approximately the potential global climatic impact of the second indirect effects, a so-called *quasi-* or *pseudo-indirect forcing* is frequently diagnosed (e.g., Rotstayn and Penner 2001). This "quasi-forcing" is diagnosed in a very different way to the first indirect effect. The TOA net flux (the sum of the short- and longwave fluxes, $SW_{TOA}{}^{\uparrow}$ and $LW_{TOA}{}^{\uparrow}$) in two separate parallel global GCM simulations is diagnosed. The first simulation uses preindustrial aerosol emissions while the second simulation uses present-day aerosol emissions, and the total indirect "quasi-forcing" is diagnosed as the difference between the net top of atmosphere fluxes. Importantly, the simulations are generally carried out in atmosphere-only GCMs with fixed SSTs and sea-ice extents; this means that the temperature response of the model and slow feedbacks associated with changes in global temperatures are inhibited, but the atmospheric state is no longer strictly held fixed (Hansen et al. 2002). Further methods for diagnosing the radiative forcing have been proposed where the land-surface temperatures are also held fixed (e.g., Shine et al. 2003).

At this point, it is worth considering whether the "quasi-forcing" provides an adequate representation of the strict definition of radiative forcing, as defined by Forster et al. (2007). Figure 19.2a (after Ming et al. 2005), shows the radiative forcing from the cloud albedo effect calculated using this definition in the GFDL AM2 (GFDL 2004). To obtain the forcing in Figure 19.2a, a simulation is run with preindustrial aerosols. The fields from this simulation are used to calculate a second set of TOA radiative fluxes, with the only change being the replacement of preindustrial by present-day aerosols in the radiation code. The difference between fluxes with present-day and preindustrial aerosols (the IPCC forcing) is shown in Figure 19.2a. Figure 19.2b shows the "quasi-forcing" from the cloud albedo effect, obtained by integrating AM2 with preindustrial aerosols but using present-day aerosols for purposes of radiative transfer only. The simulation in Figure 19.2b is a preindustrial simulation, except that the cloud albedo corresponds to present-day aerosols. We use this simulation to illustrate the effect of the preindustrial to present-day change in aerosols on cloud albedo while modeling all other cloud processes for preindustrial aerosols. The "quasi-forcing" in Figure 19.2b includes feedbacks caused by radiative, but not microphysical, perturbations associated with anthropogenic aerosols. Many features of the forcing and "quasi-forcing" patterns are similar. Although the amplitudes of the features tend to be larger for the "quasi-forcing," the global means are almost the same. The feedbacks, which are included in the "quasi-forcing," are responses to a radiative perturbation only. This is also true of the feedbacks, which occur in response to forcing by changes in well-mixed greenhouse gases. For cloud-albedo feedback, then, "quasi-forcing" and forcing behave similarly, at least for the GFDL AM2.

The use of the radiative forcing concept becomes problematic when one considers feedbacks associated with microphysical aspects of the aerosol indirect effect. We refer to these feedbacks here as non-albedo effects. Figure 19.2c shows non-albedo "quasi-forcing," obtained by integrating AM2 with present-day aerosols but using preindustrial aerosols for purposes of radiative transfer only. The simulation in Figure 19.2c models all cloud processes, except albedo, using present-day aerosols. We use this simulation to illustrate the effects of the preindustrial to present-day change in aerosols on all cloud processes except albedo, which corresponds to preindustrial aerosols. The most important point is that the magnitude of the non-albedo "quasi-forcing" is a

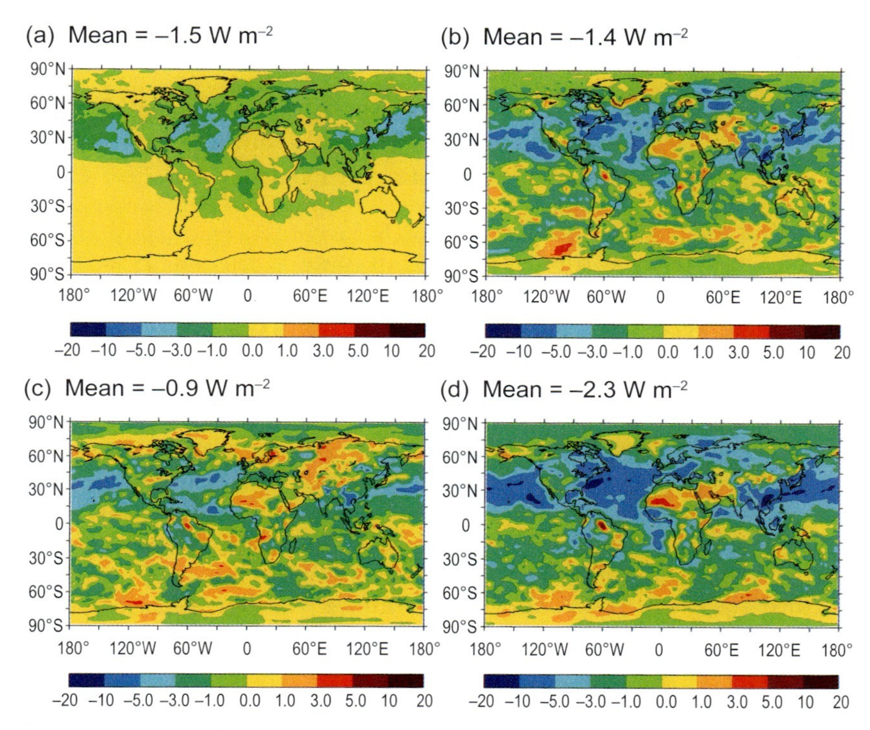

Figure 19.2 (a) Indirect forcing (by cloud albedo effect) using the definition of Forster et al. (2007); (b) quasi-forcing by cloud albedo effect only; (c) quasi-forcing by non-albedo effects only; and (d) quasi-forcing by all indirect effects. The quasi-forcings are derived from ten-year means from parallel GCM simulations where the sea surface temperatures are held fixed. Simulations use the GFDL AM2 model. Figure 19.2a modified from Ming et al. (2005).

large fraction of the magnitude of the albedo forcing, and this is apparent in Figure 19.2d, which shows the total indirect-effect "quasi-forcing." Given the nonlinear nature of the forcing mechanisms, the near additivity of the albedo and non-albedo forcings is noteworthy; at this point we have no way of knowing whether this result is general or particular to this model. The microphysical feedbacks, which produce the "quasi-forcing" in Figure 19.2c are unique among the forcings considered by IPCC and are not associated with instantaneous forcing, as defined by Forster et al. (2007). The relationship between forcing and "quasi-forcing" (and thus between forcing and temperature change) at the global mean will differ for indirect aerosol effects from the relationship for changes in well-mixed greenhouse gases or aerosol direct effects.

A key rationale for use of forcings by IPCC is to scale temperature changes among various agents of climate change. Aerosol indirect effects will not fit into this scaling because unique microphysical feedback mechanisms operate, and the magnitude of the microphysical feedbacks is large relative to the radiative feedbacks.

Rationale of IPCC (2007) for Including Only the First Indirect Effect

IPCC (2007) continues to adopt the approach that only the first aerosol indirect effect can be rigorously defined as a forcing because, in determining any of the second indirect effects, cloud feedbacks necessarily come into play in the climate system. The second indirect effects are therefore considered as feedbacks (responses to the initial anthropogenic perturbation of aerosol via the first indirect effect) to the climate system rather than as radiative forcings. IPCC (2007) continues to rely on global climate models for these estimates, as they are the only tool presently capable of providing global estimates. Arguably, satellite retrievals could provide global estimates of the indirect effects of aerosols (e.g., Nakajima et al. 2001; Brenguier et al. 2000) but an unambiguous determination of a change in cloud reflectance when influenced by anthropogenic aerosols is hampered by large natural variability and/or artifacts in retrievals of cloud liquid water path and/or cloud effective radius caused by absorbing aerosol above cloud (e.g., Haywood et al. 2004). As with all such satellite-based approaches, there is also the question of the degree to which modern-day "clean" conditions are representative of global preindustrial conditions. Figure 19.3 shows the first indirect radiative forcing determined by IPCC (2007), ranging from –0.2 to –1.8 W m^{-2}.

It may be argued that the second indirect effects have to be comprehensively quantified if we are to understand fully the effects of aerosols upon clouds. While there is no doubt that clouds remain one of the most challenging aspects in accurate simulations of climate change, the following argument shows that there is some wisdom in not (yet) including the second indirect effects as radiative forcing mechanisms. This statement stems from the different behavior of the second aerosol indirect effect when assessed by GCM schemes compared with more detailed microphysical large eddy models (LEMs). In GCM parameterizations, the second indirect effect leads to a decrease in precipitation efficiency and therefore increases in cloud water path and fraction. In LEM simulations, however, the response to increases in particle concentration depends on precise details of the situation. In studies of tradewind cumuli, Xue and Feingold (2006) and Xue et al. (2008) found that the competing effects of aerosol-induced precipitation suppression and evaporation enhancement determined whether aerosols increased or decreased cloud fraction, and could be quite regime-dependent. Detailed modeling studies of stratiform clouds by Ackerman et al. (2004) and Wood (2007) showed that the second indirect effect can operate either to enhance or reduce cloud amount, again depending on the precise conditions that prevail (humidity above cloud and cloud-base

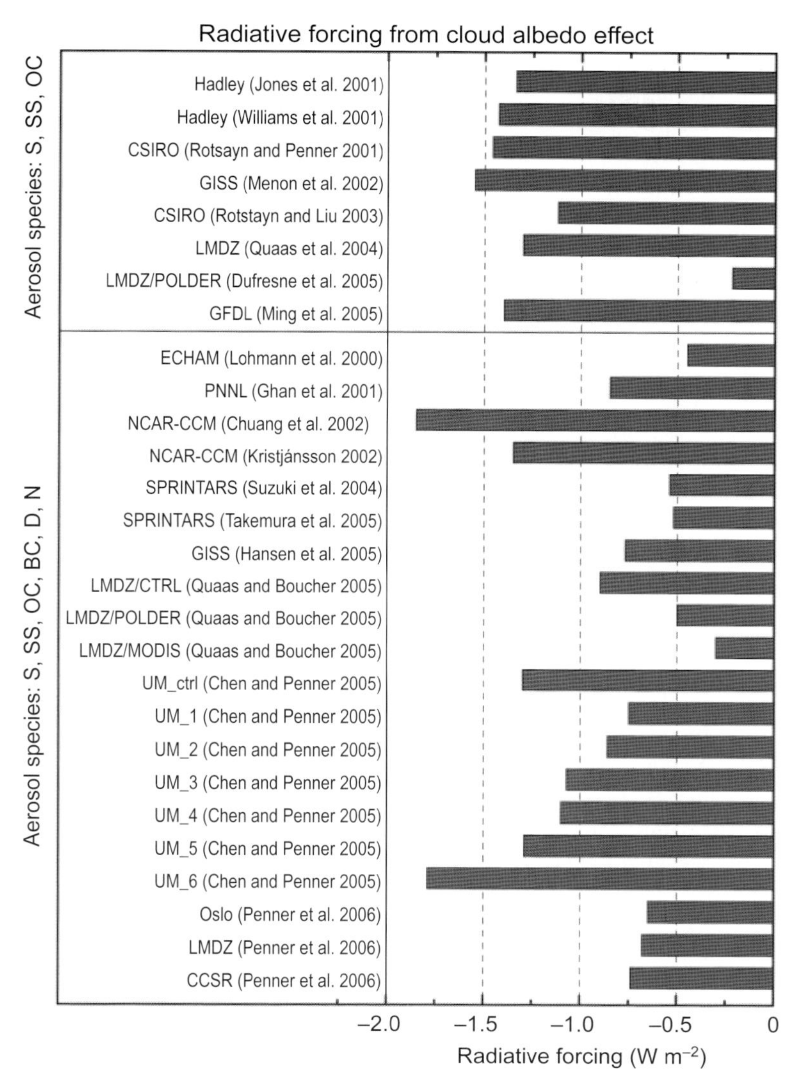

Figure 19.3 Assessment of the first indirect effect. Modified from Forster et al. (2007). BC: black carbon; D: dust; N: nitrate; OC: organic carbon; S: sulfate; SS: sea salt.

height being two of the conditions identified). Studies with CRMs for deep convection indicate that increasing aerosols can either increase or decrease precipitation (van der Heever et al. 2006; Khain and Pokrovsky 2004; Khain et al. 2004; Khain et al. 2005; Lynn et al. 2005), depending on the stability and moisture structure of the flow in which convection develops.

Although global climate models appear to simulate the first-order effect of aerosols on cloud droplet number and precipitation fairly well, the climate impact of this change depends on a host of feedback processes, some of which have only crudely been modeled. This means that in GCMs, which are the only tool available at present for diagnosing the second indirect effect on a global basis, cloud water path or fraction always increases with particle concentration, leading to a cooling of the Earth–atmosphere system. Analysis of the results presented in Forster et al. (2007) shows that in all of the climate models studied, the total indirect effects were more negative than the first indirect effect alone, indicating that the second indirect effect in all the models operated in this manner. In more detailed models, which represent the microphysical, dynamic, and radiative properties of both clouds and aerosols far more explicitly, the second indirect effect can operate in either direction, suggesting that the global impact on the Earth–atmosphere system is unclear.

Limitations of Forcing as a Measure of Indirect Effects

In the previous section, we made a case for viewing forcing as IPCC has traditionally construed it; that is, as an instantaneous flux change produced by a specified change in atmospheric composition. Here, we consider further characteristics of cloud–aerosol interactions which may limit the interpretation of forcing when applied to aerosol indirect effects.

Figure 19.1 presented a traditional view of aerosol direct and indirect effects in GCMs. As noted above, process-level models (e.g., LEM) raise the possibility that, as clouds evolve from an initial state with increased aerosol, evaporation may compete with reduced precipitation and break the sequence depicted in Figure 19.1. A concern for GCM development is that current cloud parameterizations are not able to capture processes in LEMs that could fundamentally alter the response of clouds to aerosols. However, even with current state-of-the-science GCM cloud parameterizations, the response of clouds to increased aerosols can differ fundamentally, depending on atmospheric state. Figure 19.4 shows the change in cloud liquid profiles over two days of integration at high and low particle concentrations, using the single-column model for the GFDL AM2 (GFDL 2004). (A simplified radiation parameterization has been used in these calculations.) For the moist sounding, the clouds and aerosols interact as depicted in Figure 19.1. For the dry sounding, Figure 19.5 shows that the interaction sequence is broken after drizzle is suppressed and liquid water content increased; this, in turn, results in an increase in cloud-top longwave cooling. The associated instability leads to increased entrainment of dry air into the planetary boundary layer and subsequent increased evaporation. Cloud lifetime *decreases* with increasing aerosol particle concentrations, and the sign of the cloud lifetime effect is changed. Figure 19.6 summarizes the sequences of interactions between clouds and radiation occurring in Figure 19.1 and 19.5. The mechanisms shown in Figures 19.5 and 19.6 are by no means the only possibilities for reducing cloud liquid water at high particle concentration.

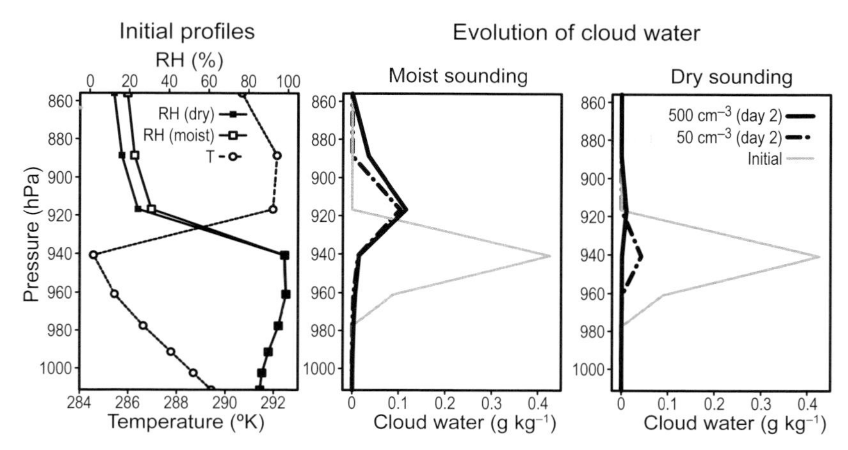

Figure 19.4 Evolution of cloud liquid over two days at high and low aerosol concentrations for two initial relative humidity (RH) profiles (otherwise identical soundings) in the GFDL AM2 single-column model. Cloud liquid is a grid mean, i.e., the product of cloud fraction and in-cloud water mixing ratio.

Among other possibilities is turbulence that results from precipitation suppression (Ackerman et al. 2004; Wood 2005), and all such mechanisms are not included in current GCM parameterizations.

A change in the sign of the cloud lifetime effect does not, by itself, negate the case made earlier about the utility of the forcing concept. The cloud lifetime effect acts as a feedback. Whether the feedback is positive or negative depends on the moisture structure of the atmosphere in which the clouds form in the illustrative case we have just discussed, and probably on additional

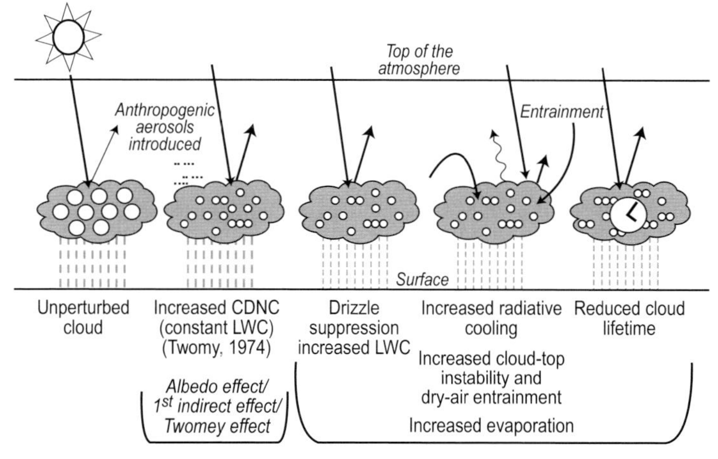

Figure 19.5 A schematic diagram showing how the sequence of indirect effects depicted in Figure 19.1 can change if drier air overlies the planetary boundary layer. Curved lines represent entrainment, while other notation follows Figure 19.1.

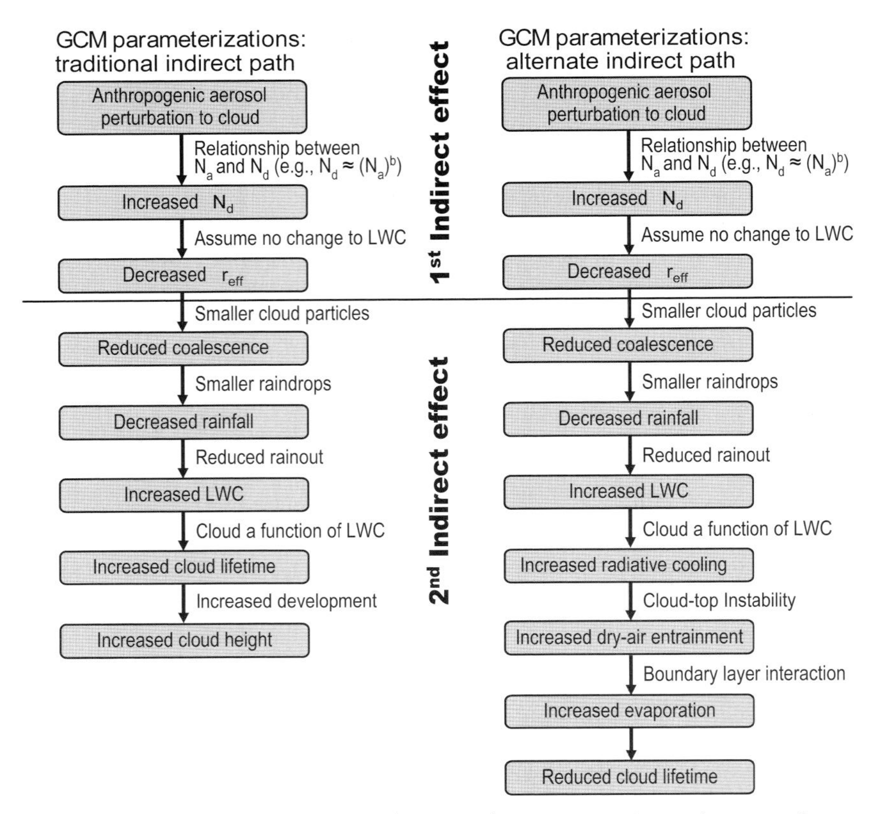

Figure 19.6 Sequence of feedbacks in GCMs in response to increasing aerosol concentration. The GCM paths on the left and right correspond to Figure 19.1. LWC: liquid water content.

factors in other cases. A more interesting situation would arise were a negative cloud lifetime effect to reduce cloud lifetime sufficiently rapidly so that the resulting reduction in reflected shortwave radiation exceeded the increase in reflected shortwave radiation associated with the first indirect effect. In this case, the combination of oppositely signed albedo and lifetime effects would have a net effect of opposite sign to the forcing. This certainly is not consistent with the general view of a feedback amplifying or damping a forcing, although it would still fit within the framework of Equation 19.1 or 19.2, but only if *negative* climate sensitivities or efficacies are allowed. Much of the IPCC rationale for using forcing is that it scales with temperature response in a fairly straightforward manner. The relationship between forcing and temperature response is not straightforward if the magnitude of a cloud lifetime effect exceeds that of an oppositely signed cloud albedo effect.

The use of "quasi-forcing" avoids issues of this type but is likely to exhibit considerable model-dependence. It also clearly (and undesirably) includes some but not all feedbacks operating in the climate system. As a practical matter, the relative merits of forcing and "quasi-forcing" would depend on the prevalence of situations where indirect non-albedo cloud effects oppose the indirect cloud albedo effect and the relative magnitudes of albedo and non-albedo effects. If the prevalence is low or the albedo effects are much larger than the non-albedo effects, the advantages of clear separation of forcing from feedback using the current IPCC approach are very evident. As noted, Forster et al. (2007) report that climate models developed thus far, which include indirect effects, exhibit same-signed global-mean indirect albedo and non-albedo effects. This is true for the GFDL AM2 (Ming et al. 2005), as is evident in Figure 19.2, even though its parameterizations allow oppositely signed non-albedo and albedo effects. The current state of climate modeling of indirect effects, then, does not make the case for departing from the current approach to forcing for indirect aerosol effects. However, a major caveat lies in the poorly developed state of many of the cloud parameterizations on which indirect effects depend. The case shown in Figures 19.1, 19.4, and 19.5 depends strongly on the treatment of cloud radiation and boundary-layer turbulence, both of which are probably represented with limited realism in GCMs. The preliminary conclusions about relative magnitudes and signs of indirect cloud–albedo and non-albedo effects may change with further GCM development.

Conclusion

Indirect aerosol effects challenge the traditional view of forcing in climate models. For liquid clouds alone, several aerosol effects are likely. In state-of-the-science GCMs, these effects can be considered in terms of traditional forcing–feedback analysis, since the global responses to changes in droplet size, which produce the indirect cloud albedo effect, amplify the albedo effect. The concept remains useful if indirect non-albedo effects are oppositely signed to the albedo effects, at least if the magnitude of these effects remains smaller than the cloud albedo effect. (If not, quantities such as climate sensitivity and forcing efficacy need not even remain positive for indirect forcing.)

Even though forcing–feedback analysis remains a viable tool for considering indirect aerosol effects, much of its attractiveness in earlier applications disappears. For example, the climate sensitivity for forcing by indirect effects of aerosols in Equation 19.1 will almost certainly take values considerably different than for forcing by changes in well-mixed greenhouse gases, at least for some models. Although this can be formally treated using the climate forcing efficacy in Equation 19.2, substantial departures of the forcing efficacy from unity are quite possible for indirect forcing. Given the range of results in Figure 19.3, substantial variation of the climate forcing efficacy among models is also likely, at least at the present stage of model development. Note here that feedbacks for well-mixed greenhouse gases are also important. The limitations on the utility of the forcing concept arise because the nature of the feedbacks differs for perturbations of clouds by aerosols and perturbations in well-mixed greenhouse gases.

The case of indirect forcing by aerosols simulated by the GFDL AM2, shown in Figure 19.2, provided an illustrative example. Although it exhibited qualitative similarities in the regional patterns of flux changes for forcing and "quasi-forcing," the regional and global mean magnitudes differed appreciably. These differences are attributable to the presence of strong microphysical feedbacks that result from changes in the sizes of cloud droplets when aerosols increase. Direct aerosol forcing and forcing by changes in well-mixed greenhouse gases are assumed to have feedbacks that respond only to radiative perturbations, whereas both radiative and microphysical feedbacks occur with aerosol indirect effects. The large magnitude of the non-albedo (microphysical) "quasi-forcing" indicates that the temperature response to a forcing whose feedbacks result from a purely radiative perturbation will differ from a forcing whose feedbacks result from both radiative and microphysical perturbations.

As a result, the chief rationale for using forcings in IPCC—to compare cleanly the relative importance of perturbations in atmospheric composition—is likely to be limited for indirect aerosol effects. At least, the climate forcing efficacy will require consideration, along with forcing, to compare indirect aerosol effects with other forcing. This situation introduces complexity but is an inevitable consequence of the vast difference between the forcing mechanism associated with indirect aerosol effects and other forcing mechanisms considered by the IPCC. Although changes in greenhouse gases or direct aerosol effects are assumed to force the atmosphere by changing its radiative properties, indirect aerosol effects force the atmosphere by changing both its radiative and microphysical properties. Feedbacks occur in all cases but can operate through more processes in response to indirect aerosol effects.

"Quasi-forcing" or pseudo-indirect forcing is a strong candidate to supplement IPCC forcing. Its chief drawback is probably the confusion which using the term "forcing" produces. Given that "quasi-forcing" is, in fact, the TOA radiative flux perturbation produced by integrating an atmospheric GCM from preindustrial to present-day composition with fixed SSTs and sea-ice extents, we recommend using the term *radiative flux perturbation* instead. Radiative flux perturbations are likely to scale more closely with global mean surface temperature change than IPCC forcings and, like forcings, remain reasonably easy to calculate, relative to fully coupled GCMs. In addition to incorporating aerosol indirect effects, recent results from Gregory and Webb (2008) suggest that the radiative flux perturbations will include fast cloud feedbacks as well. Invariably, intermodel spread in radiative flux perturbations is likely to be greater than the spread in forcings. Thus, radiative flux perturbation would more accurately capture uncertainty associated with indirect effects than forcing.

Acknowledgment

C. Seman (GFDL) provided formatting assistance for figures.

References

Ackerman, A. S., M. P. Kirkpatrick, D. E. Stevens, and O. B. Toon. 2004. The impact of humidity above stratiform clouds on indirect aerosol climate forcing. *Nature* **432**:1014–1017.

Albrecht, B. 1989. Aerosols, cloud microphysics and fractional cloudiness. *Science* **245**:1227–1230.

Brenguier, J. -L., H. Pawlowska, L. Schueller et al. 2000. Radiative properties of boundary layer clouds: Droplet effective radius versus number concentration. *J. Atmos. Sci.* **57**:803–821.

Chen, Y., and J. E. Penner. 2005. Uncertainty analysis of the first indirect aerosol effect. *Atmos. Chem. Phys.* **5**:2935–2948.

Chuang, C. C., J. E. Penner, J. M. Prospero et al. 2002. Cloud susceptibility and the first aerosol indirect forcing: Sensitivity to black carbon and aerosol concentrations. *J. Geophys. Res.* **107(D21)**:4564.

Dufresne, J.-L., J. Quaas, O. Boucher, S. Denvil, and L. Fairhead. 2005. Contrasts in the effects on climate of anthropogenic sulfate aerosols between the 20th and the 21st century. *Geophys. Res. Lett.* **32**:L21703.

Forster, P., V. Ramaswamy, P. Artaxo et al. 2007. Radiative forcing of climate change. In: Climate Change 2007: The Physical Science Basis. Contribution of Working Group I to the Fourth Assessment Report of the Intergovernmental Panel on Climate Change, edited by S. Solomon, D. Qin, M. Manning et al. pp. 129–234. New York: Cambridge Univ. Press.

GFDL Global Atmosphere Model Development Team. 2004. The GFDL new global atmosphere and land model AM2/LM2: Evaluation with prescribed SST simulations. *J. Climate* **17**:4641–4673.

Ghan, S., N. Laulainen, R. Easter et al. 2001. Evaluation of aerosol direct radiative forcing in MIRAGE. *J. Geophys. Res.* **106(D6)**:5295–5316.

Gregory, J., and M. J. Webb. 2008. Tropospheric adjustment induces a cloud component in CO_2 forcing. *J. Climate* **19(1)**:39.

Hansen, J., and L. Nazarenko. 2004. Soot climate forcing via snow and ice albedos. *PNAS* **101**:423–428.

Hansen, J., M. Sato, L. Nazarenko et al. 2002. Climate forcings in Goddard Institute for Space Studies SI2000 simulations. *J. Geophys. Res.* **107**:D184347.

Hansen, J., M. Sato, R. Ruedy et al. 2005. Efficacy of climate forcings. *J. Geophys. Res.* **110**:D18104.

Haywood, J. M., S. R. Osborne, and S. J. Abel. 2004. The effect of overlying absorbing aerosol layers on remote sensing retrievals of cloud effective radius and cloud optical depth. *Q. J. Roy. Meteor. Soc.* **130**:779–800.

Hegg, D. A., R. J. Ferek, and P. V. Hobbs. 1993. Light scattering and cloud condensation nucleus activity of sulfate aerosol measured over the Northeast Atlantic Ocean. *J. Geophys. Res.* **98**:14,887–14,894.

IPCC. 2007. Climate Change 2007: The Physical Science Basis. Contribution of Working Group I to the Fourth Assessment Report of the Intergovernmental Panel on Climate Change, ed. S. Solomon, D. Qin, M. Manning et al. New York: Cambridge Univ. Press.

Jacobson, M. Z. 2001. Strong radiative heating due to the mixing state of black carbon in atmospheric aerosols. *Nature* **409**:695–697.

Jones, A., D. L. Roberts, and M. J. Woodage. 2001. Indirect sulphate aerosol forcing in a climate model with an interactive sulphur cycle. *J. Geophys. Res.* **106(D17)**:20,293–20,301.

Jones, A., J. M. Haywood, and O. Boucher. 2007. Aerosol forcing, climate response and climate sensitivity in the Hadley Centre climate model. *J. Geophys. Res.* **112**:D20211.

Joshi, M., K. Shine, M. Ponater et al. 2003. A comparison of climate response to different radiative forcings in three general circulation models: Towards an improved metric of climate change. *Climate Dyn.* **20**:843–854.

Khain, A., A. Pokrovsky, M. Pinsky, A. Seifert, and V. Phillips. 2004. Simulation of effects of atmospheric aerosols on deep turbulent convective clouds using a spectral microphysics mixed phase cumulus cloud model. Part I: Model description and possible applications. *J. Atmos. Sci.* **15**:2963–2982.

Khain, A., and A. Pokrovsky. 2004. Simulation of effects of atmospheric aerosols on deep turbulent convective clouds using a spectral microphysics mixed-phase cumulus cloud model. Part II: Sensitivity study. *J. Atmos. Sci.* **24**:2983–3001.

Khain, A., D. Rosenfeld, and A. Pokrovsky. 2005. Aerosol impact on the dynamics and microphysics of deep convective clouds. *Q. J. Roy. Meteor. Soc.* **131**:2639–2663.

Kristjánsson, J. E., A. Staple, J. Kristiansen, and E. Kaas. 2002. A new look at possible connections between solar activity, clouds and climate. *Geophys. Res. Lett.* **29(23)**:2107.

Leaitch, W. R., G. A. Isaac, J. W. Strapp, C. M. Banic, and H. A. Wiebe. 1992. The relationship between cloud droplet number concentration and anthropogenic pollution: Observations and climatic implications. *J. Geophys. Res.* **97**:2463–2474.

Lohmann, U., J. Feichter, J. E. Penner, and W. R. Leaitch. 2000. Indirect effect of sulfate and carbonaceous aerosols: A mechanistic treatment. *J. Geophys. Res.* **105(D10)**:12,193–12,206.

Lynn, B. H., A. P. Khain, J. Dudhia et al. 2005. Spectral (bin) microphysics coupled with a mesoscale model (MM5). Part I: Model description and first results. *Mon. Wea. Rev.* **133**:44–58.

Martin, G. M., D. W. Johnson, and A. Spice. 1994. The measurement and parameterisation of effective radius of droplets in warm stratocumulus clouds. *J. Atmos. Sci.* **51**:1823–1843.

Matthews, H. D., A. J. Weaver, K. J. Meissner, N. P. Gillett and M. Eby. 2004. Natural and anthropogenic climate change: Incorporating historical land cover change, vegetation dynamics and the global carbon cycle. *Climate Dyn.* **22**:461–479.

Meehl, G. A., W. M. Washington, C. M. Ammann et al. 2004. Combinations of natural and anthropogenic forcings in twentieth-century climate. *J. Climate* **17**:3721–3727.

Menon, S., J. Hansen, L. Nazarenko, and Y. Luo. 2002. Climate effects of black carbon aerosols in China and India. *Science* **297**:2250–2253.

Ming, Y., V. Ramaswamy, P. A. Ginoux, L. W. Horowitz, and L. M. Russell. 2005. Geophysical Fluid Dynamics Laboratory general circulation model investigation of the indirect radiative effects of anthropogenic sulphate aerosol. *J. Geophys. Res.* **110**:D22206.

Nakajima, T., A. Higurashi, K. Kawamoto, and J. Penner. 2001. A possible correlation between satellite-derived cloud and aerosol microphysical parameters. *Geophys. Res. Lett.* **28**:1171–1174.

Pincus, R., and M. B. Baker. 1994. Effect of precipitation on the albedo susceptibility of clouds in the marine boundary layer. *Nature* **372**:250–252.

Quaas, J., O. Boucher, and F.-M. Breon. 2004. Aerosol indirect effects in POLDER satellite data and the Laboratoire de Météorologie Dynamique-Zoom (LMDZ) general circulation model. *J. Geophys. Res.* **109**:D08205.

Quaas, J., and O. Boucher. 2005. Constraining the first aerosol indirect radiative forcing in the LMDZ GCM using POLDER and MODIS satellite data. *Geophys. Res. Lett.* **32**:L17814.

Penner, J. E., J. Quaas, T. Storelvmo et al. 2006. Model intercomparison of indirect aerosol effects. *Atmos. Chem. Phys. Discuss.* **6**:1579–1617.

Ramaswamy, V., O. Boucher, J. Haigh et al. 2001. Radiative forcing of climate change. In: Climate Change 2001: The Scientific Basis. Contribution of Working Group I to the Third Assessment Report of the Intergovernmental Panel on Climate Change, ed. J. T. Houghton et al. pp. 349–416. New York: Cambridge Univ. Press.

Roberts, D. L., and A. Jones. 2004. Climate sensitivity to black carbon aerosol from fossil fuel combustion. *J. Geophys. Res.* **109**:D16202.

Rotstayn, L., and J. E. Penner. 2001. Indirect aerosol forcing, quasi forcing, and climate response. *J. Climate* **14**:2960–2975.

Rotstayn, L. D., and Y. Liu. 2003. Sensitivity of the first indirect aerosol effect to an increase of the cloud droplet spectral dispersion with droplet number concentration. *J. Climate* **16**:3476–3481.

Shine, K. P., J. Cook, E. J. Highwood, and M. M. Joshi. 2003. An alternative to radiative forcing for estimating the relative importance of climate change mechanisms. *Geophys. Res. Lett.* **30**:2047.

Suzuki, K., T. Nakajima, A. Numaguti et al. 2004. A study of the aerosol effect on a cloud field with simultaneous use of GCM modeling and satellite observation. *J. Atmos. Sci.* **61**:179–194.

Takemura, T., T. Nozawa, S. Emori, T. Y. Nakajima, and T. Nakajima. 2005. Simulation of climate response to aerosol direct and indirect effects with aerosol transport-radiation model. *J. Geophys. Res.* **110**:D02202.

Twomey, S. A. 1977. The influence of pollution on the shortwave albedo of clouds. *J. Atmos. Sci.* **34**:1149–1152.

van den Heever, S. C., G. G. Carrio, W. R. Cotton, P. J. DeMott, and A. J. Prenni. 2006. Impacts of nucleating aerosol on Florida Storms. Part I: Mesoscale Simulations. *J. Atmos. Sci.* **63**:1752–1775.

Williams, K. D., A. Jones, D. L. Roberts, C. A. Senior, and M. J. Woodage. 2001. The response of the climate system to the indirect effects of anthropogenic sulfate aerosols. *Clim. Dyn.* **17(11)**:845–856.

Wood, R. 2005. Drizzle in stratiform boundary layer clouds. Part I. Vertical and horizontal structure. *J. Atmos. Sci.* **62**:3011–3033.

Wood, R. 2007. Cancellation of aerosol indirect effects in marine stratocumulus through cloud thinning. *J. Atmos. Sci.* **64**:2657–2669.

Xue, H., and G. Feingold. 2006. Large-eddy simulations of trade wind cumuli: Investigation of aerosol indirect effects. *J. Atmos. Sci.* **63**:1605–1622.

Xue, H., G. Feingold, and B. Stevens. 2008. Aerosol effects on clouds, precipitation, and the organization of shallow cumulus convection. *J. Atmos. Sci.* **65**:392–406.

20

Simulating Global Clouds

Past, Present, and Future

William D. Collins[1] and Masaki Satoh[2]

[1]Department of Earth and Planetary Science, University of California, Berkeley, CA, U.S.A.
[2]Center for Climate System Research, University of Tokyo, Kashiwa, Chiba, Japan

Introduction

Clouds have a significant effect on the Earth's heat budget. Changes in clouds affect the temperature change in global warming. This is called cloud feedback and has posed the largest uncertainty in the study of climate sensitivity for almost twenty years (Bony et al. 2006; Soden and Held 2006). Change that occurs in low clouds represents the largest uncertainty (Webb et al. 2006; Bony and Dufresne 2005). Although several hypotheses have been proposed for the feedback stemming from high clouds (Lindzen 1990; Ramanathan and Collins 1991; Lindzen 2001), no consensus has yet been obtained (Lau et al. 1994; Lin et al. 2002). All of the hypotheses relate the changes in high clouds to those in deep convection. Thus it is essential to understand cumulus convection to discuss the high cloud feedback.

We have recognized the uncertainty of cumulus convection for a long time, but have not simulated it in detail. Spatial resolution of current global climate models (GCMs) for climate simulations is on the order of 100 km, by which we cannot resolve cumulus convection. Available computer resources present the main constraint on resolution. GCMs introduce cumulus parameterizations which try to simulate the essential role of cumulus convection on the GCM grid. We begin by describing climate change simulations based on traditional cloud and cumulus parameterizations and the open issues identified in these simulations. Cloud-resolving simulations under idealistic conditions (Wu and Moncrieff 1999; Bretherton 2007; Tompkins and Craig 1999), global simulations with embedded cloud-resolving grids in a GCM grid (Grabowski 2001; Randall et al. 2003; Wyant et al. 2005), and global cloud-resolving simulations that resolve convective systems over the entire globe (Satoh et al. 2008; Miura et al. 2005, 2007) have been started and are the emerging approaches in global process-oriented cloud modeling. We discuss the prospects for new observations for evaluating cloud radiative effects and feedbacks in global models. In addition, we consider the response of cloud systems to an idealized global warming in a global cloud-resolving model (GCRM).

Projected Changes in Clouds and Cloud Effects

In global simulations of climate change with current generation models, cloud amounts are projected to decrease throughout most of the troposphere below 200 hPa and between 50°N and 50°S (IPCC 2007). Cloud cover is projected to increase in the upper extratropical troposphere with robust changes close to 200 hPa. Generally, the amplitude of the spatial patterns of cloud perturbation expand with increasing concentrations of CO_2, but the magnitude of the changes remains less than 4% at 2100 for the canonical SRES A1B emissions scenario.

In part, because the vertical dipole of projected cloud amount changes, the forecast perturbations in column-integrated cloud amount are inconsistent across the multi-model ensemble for most regions on the globe. The models do not concur on either the magnitude or sign of changes in global mean cloud radiative effect, and the size of the cloud radiative effect changes relative to the present-day value is generally less than 10%; this is comparable to various forcings and current uncertainties in the top-of-atmosphere planetary energy budget. Thus, the model projections suggest that observational verification of global cloud radiative effect changes will be a significant challenge. The sign of radiative feedbacks from low clouds and deep convective systems is not consistent across the IPCC multi-model ensemble (Bony and Dufresne 2005). Tropical clouds, mixed-phase clouds and cloud phase, as well as intermodel differences in meridional shifts in storm tracks contribute the divergences in cloud responses (IPCC 2007).

Open Issues in Global Simulation of Perturbed Clouds

Observational and Theoretical Tests

The necessary and sufficient conditions for accurate global projections of clouds and cloud radiative effects are not known. The absence of a complete theory has complicated efforts to construct comprehensive observational tests of cloud parameterizations. It has also complicated efforts to derive these parameterizations through systematic simplification of known cloud physics and process-oriented CRMs. It would be particularly useful to have an analog of the tests for snow, connecting the observed seasonal cycle in snow cover and the simulated snow–albedo feedbacks (Hall and Qu 2006). Some aspects of cloud perturbations from climate change project onto observed intra- and interannual cloud variations (Williams et al. 2006). Whether present-day seasonal variations in clouds should be used as a test for cloud radiative feedbacks in global models has not been definitively determined (Cess et al. 1992; Tsushima et al. 2005).

Scaling Issues and Parametric Uncertainty

Forecasting climate change with skill at local and regional scales has become increasingly necessary to assess societal and environmental impacts. Extreme rainfall events on small temporal and spatial scales constitute some of the most important impacts. Partly in response, the climate modeling community has increased the spatial resolution of their climate forecasts by a factor of five from the first to the fourth IPCC assessment reports (IPCC 2007). However, whether the simulated hydrological cycle, cloud features, and cloud feedbacks improve and converge with increasing resolution remains open. Some models exhibit systematic and apparently nonconvergent variations in the main features of the hydrological cycle (e.g. the Intertropical Convergence Zone, ITCZ), with increasing resolution (Williamson et al. 1995). The same models exhibit large, systematic changes in climate sensitivity, cloud radiative effects and feedbacks, and the probability distributions of rainfall with increasing resolution (Kiehl et al. 2006; Williamson et al. 2007). The parameterizations of clouds and hydrological processes are generally adjusted to produce a realistic simulation at particular model resolution or resolutions. To date, these parameterizations have not usually been designed to yield climate simulations that are invariant to horizontal and vertical truncation scales.

Current parameterizations contain large numbers of free parameters that govern the evolution of micro- and macrophysical cloud properties in relation to the meteorological environment. As is well known, simulated climate radiative feedbacks and climate responses are quite sensitive to variations in these free parameters (Senior and Mitchell 1993; Murphy et al. 2004). It has proven difficult to exclude parameter settings that produce large positive cloud feedbacks and climate sensitivities to increasing greenhouse gas concentrations. Many of the parameters are only loosely constrained by observations or process-oriented modeling and are connected to the physics of cloud formation through scaling arguments.

Global Process-oriented Simulation

Global Cloud-resolving Models and Multiscale Modeling Framework

Traditional climate models cannot explicitly resolve a number of climatically important cloud processes. It would be desirable to design and apply global models that explicitly simulate cloud system organization; relationships among cloud geometry, radiation, and precipitation; and the interactions between clouds and convection. Two new frameworks have been created for process-oriented cloud simulation: a GCRM (Satoh et al. 2008) and the multiscale modeling framework (MMF), originally introduced as super-parameterization (Grabowski 2001; Randall et al. 2003). Features of the GCRM and its simulations for cloud response to idealized climate change are discussed below.

MMF is essentially an enhancement of traditional GCMs to treat interactions between clouds and their mesoscale environment. In MMF-enhanced GCMs, two- or three-dimensional CRMs replace the physical parameterizations at each grid point. The grid-point tendencies in the prognosed fields (e.g., temperature, humidity, and momentum), calculated with the traditional parameterizations, are overwritten by corresponding tendencies computed by the CRMs. MMF is an intermediate step towards a GCRM, and MMFs can be formulated to converge smoothly to a GCRM in the limit of very high spatial resolution. The MMF simplifies the interactions between clouds and their synoptic environment by averaging the boundary conditions applied to the CRMs and the tendencies applied to the host GCM at the scale of individual GCM grid points. Early results from MMF-enhanced GCMs show significant improvements relative to the original GCM in the strength, frequency, and characteristics of the Madden-Julian Oscillation and in the diurnal cycle of convective precipitation (Khairoutdinov et al. 2007). It is not yet known which cloud-scale processes are responsible for the improved fidelity of the climate simulations.

GCRM and MMF represent some of the first attempts to simulate global clouds with process-oriented models. Perhaps the most important question is whether these models can reduce the uncertainty in simulations of climate sensitivity and of cloud radiative feedbacks relative to current GCMs. Since there is only one GCRM and only a few MMF-enhanced GCMs, the characteristics of multi-model ensembles of these systems are still unresolved. The implications of explicit simulation of cloud processes for the coupled climate system are unknown. Although GCRM and MMF eliminate many of the ad hoc parameters governing cloud interactions at the mesoscale, there are still many free parameters for unresolved processes, including turbulence and cloud microphysics. Given the computational expense of the GCRM and MMF, the sensitivity of the simulated climate to these remaining free parameters has not been determined. The convergence of the MMF with increasing spatial resolution of the grid in the host GCM has yet to be determined.

These frameworks could be used as benchmark or reference models for improving GCMs by quantifying the systematic tendency errors caused by conventional cloud parameterizations. The models have not been applied in this manner, and therefore the utility of GCRM and MMF for the improvement of conventional parameterizations has not been established.

Near-term Prospects for Advances in Global Numerical Modeling

The currently available GCRM is a non-hydrostatic icosahedral atmospheric model (NICAM) (Satoh et al. 2008). NICAM runs at horizontal mesh intervals of a few kilometers (3.5km, at most) over the globe using the Earth Simulator[1] high performance computing system. Under these resolutions, deep convective clouds are marginally permitted with representing updraft cores explicitly. Thus far, NICAM has been used for many purposes. A realistic simulation of a Madden-Julian Oscillation event was produced with its internal cloud-system structures (Miura el al. 2007). In addition, NICAM produces realistic climatology close to that observed (Iga et al. 2007). NICAM is now coupled with an aerosol transport radiation model, SPRINTARS (Takemura et al. 2005), to study aerosol direct and indirect effects. Using 7 km mesh global cloud-resolving simulation by NICAM–SPRINTARS, Suzuki et al. (2008) examined vertical profiles of effective cloud particle radius as functions of cloud-top temperature in different regions, an approach similar to that by Rosenfeld (2000). Suzuki and Stephens (submitted) compared timescales of warm rain formation by comparing the GCRM results and CloudSat and MODIS combined data. Suzuki et al. (2008) further analyzed the simulated data to identify aerosol indirect effects on different cloud types and compared with satellite observations. Since the structures of shallow clouds are not resolved by the current GCRM simulations, aerosol indirect effects on shallow clouds depend on subgrid parameterizations, as in GCMs. The advantage of using results of GCRMs is that convective movements are resolved. However, concern remains about the representation of ice phase microphysical processes. Evaluations of indirect effects on deep clouds from satellite observations are hard because contributions of cloud liquid and ice particles are difficult to resolve. This can partly be overcome by using an analysis of GCRM results to guide retrieval algorithms of satellite observations.

In principle, numerical data produced by GCRMs can be used to improve conventional cloud and cumulus parameterizations. For example, shapes of PDFs of cloud fraction using GCRM results can be evaluated and used to improve large-scale cloud schemes in GCMs (Watanabe et al., submitted). By analyzing the spectrum of momentum fluxes, we see resolution dependency and intend to use this to improve cumulus parameterization and momentum transport.

At present, GCRMs do not pose an alternative to current GCMs, but they should be viewed as a complementary tool. GCRMs are not yet coupled in the Earth system model and cannot be used for simulations on a temporal scale of hundreds of years. However, within the coming decade, we should be able to use an Earth system model by replacing current atmospheric GCMs with a

[1] http://www.es.jamstec.go.jp/index.en.html

GCRM and using the results to study climate for time-slice experiments of a few years.

Related to GCMs, the MMFs replace the conventional cloud parameterizations with a CRM in each grid column of a GCM (Grabowski 2001; Khairoutdinov and Randall 2001; Randall et al. 2003). The MMF can explicitly simulate deep convection, cloudiness and cloud overlap, cloud radiation interaction, surface fluxes, and surface hydrology at the resolution of a CRM. In addition, MMFs provide global coverage and two-way interactions between the CRMs and GCMs. MMF could be a natural extension of current cloud-resolving modeling activities, as they simulate reasonable features in the tropics, such as the diurnal cycle of precipitation and intraseasonal oscillations (Randall et al. 2003). They are also used to examine the cloud–aerosol interactive processes by implementing more sophisticated microphysics and coupling with global aerosol transport models. In the future, we expect MMFs to bridge the gap between traditional CRM simulations and current as well as future GCRMs.

Challenges in Observations

Future Prospects for Remote Sensing

One of the principal predictions of climate models is that clouds and cloud effects will evolve in response to anthropogenic climate change (IPCC 2007). However, the projected changes in cloud radiative effects are small compared to the climatological mean cloud radiative effect. Depending on the rate of change, it may take decades for the signals in cloud radiative effect to become sufficiently large for robust detection relative to unforced natural variability and other secular trends in the climate system (e.g. greater forcing by CO_2). Together, the small magnitude and long integration times imply that very stable, and preferably absolutely calibrated, satellite instruments are required to detect cloud radiative effect feedbacks. Although, to date, such instruments have not yet been deployed, stable and accurate observations of cloud radiative effects over several decades would be valuable in the evaluation of global simulations of clouds and climate. Direct measurement of the Earth's radiation field is preferable to calculations of cloud radiative effects based on retrieved cloud properties (Loeb et al. 2007). Spectrally resolved measurements are especially useful to separate and classify the radiative effects from climate forcing and climate response (Goody et al. 1998). None of the proposed satellites offers a stand-alone solution for better quantification of planetary albedo for tests of climate models, since albedo requires adequate sampling over eight independent parameters (Wielicki et al. 1996).

CLARREO (CLimate Absolute Radiance and Refractivity Observatory) offers one possibility, as it is designed to detect the radiative forcing, thermal response, and radiative feedbacks in the Earth's climate system (Space Studies Board 2007). The spectral radiometers in CLARREO are made for absolute calibration against traceable standards to ensure that trends observed in the observations are as free as possible of instrumental artifacts (Keith et al. 2001; Anderson et al. 2004). The payload of CLARREO will include infrared as well as ultraviolet, visible, and near-infrared (UV/VIS/NIR) radiometers. It could be used either as an orbiting calibration facility ("NIST in space") or as an Earth-observing platform in its own right. Although the feasibility for detection of infrared greenhouse gas forcing has been amply illustrated in modeling and satellite studies (e.g., Haskins et al. 1997), it is important to recognize that the utility of the infrared measurements or detection of longwave cloud feedbacks remains unproven (Leroy et al. 2006). The advantages of the UV/VIS/NIR data for detection and estimation of shortwave forcings and feedbacks have not been studied or demonstrated in detail. Open questions include:

1. To what extent is it possible to isolate forcings and feedbacks associated with changes in specific species and processes in the CLARREO measurements?
2. Can the indirect shortwave forcings from aerosol–cloud interactions and the feedbacks from clouds be detected and quantified using CLARREO data?
3. Can changes in and longwave feedbacks from low, middle, and high clouds be detected and quantified using the CLARREO infrared data?

DSCOVR offers another possibiltiy. It has been designed for deployment at Lagrange point L1 from which it would measure the radiation emitted by the sunlight side of the Earth (Valero et al. 1999). DSCOVR would include several single-pixel NIST-advanced radiometers (NISTARs) with a ground-based calibration chain tied directly to primary national standards. These instruments would measure the total solar, near infrared, and infrared radiance field emitted by the Earth in the direction of L1. The inherent stability and traceable calibration of these instruments are ideally suited for the detection of secular trends in the Earth's short- and longwave radiation. Attribution of any observed changes to perturbations in clouds and cloud radiative effects will depend on auxiliary coincident measurements from multichannel imagers onboard DSCOVR.

Upcoming satellite experiments designed to connect clouds, aerosols, and radiative processes include the EarthCARE mission (ESA 2001). EarthCARE is designed to characterize the vertical and horizontal distributions of clouds and aerosols in the Earth's atmosphere. This data will be particularly useful for constraining (a) simulations of radiative energy divergence, (b) the interactions among radiation, aerosols, and clouds, and (c) the spatial distributions and interactions of cloud condensate and precipitation.

Global Cloud-resolving Simulations

Models and Experiments

Here, we discuss the first global warming experiment results using NICAM with a realistic topography. To summarize, a significant increase in cirrus has been seen over the entire subtropics, which has strongly intensified temperature increase (see Figure 20.1). Centralized deep convection supplies more moisture to the tropopause. Such an increase in cirrus has never been found in the same tests by using current GCMs (Ringer et al. 2006). Our results suggest the vital role of cumulus convection for cloud feedback mechanism. Most of the results in this section are based on Tsushima et al. (submitted).

For the experimental setting, we choose the procedure of perpetual July and a perturbed sea surface temperature (SST) of +2K simulation (Cess and Potter 1988). Detailed experimental design and the reproducibility of the control climate follow Iga et al. (2007). For comparison with the current GCM, we conducted the same simulations using an atmospheric GCM of MIROC3.2 (Hasumi and Emori 2004) with a horizontal resolution of 2.8°, 20 levels in the vertical in the atmosphere. The model's climate sensitivity to a doubling of atmospheric CO_2 concentration in the slab ocean experiments is 3.8°C (Ogura et al. 2008). In both models, we incorporated the ISCCP simulator (Schiffer and Rossow 1983; Webb et al. 2001), which diagnoses values of cloud's optical thickness and cloud-top pressure from the models in a manner consistent with the view of satellites from space. Cloud height and optical thickness are important metrics for cloud radiative effects, and it is useful to analyze clouds according to the ISCCP categories when we discuss the change in clouds and the relevance to the radiative field.

We examined the response of cloud fractions and cloud forcing to the idealized global warming experiment, and not cloud feedback directly. Generally, cloud feedback means positive or negative feedback on surface temperature through changes in cloud properties as a response to global warming (Stephens 2005; Bony et al. 2006). To understand cloud feedback mechanism in models,

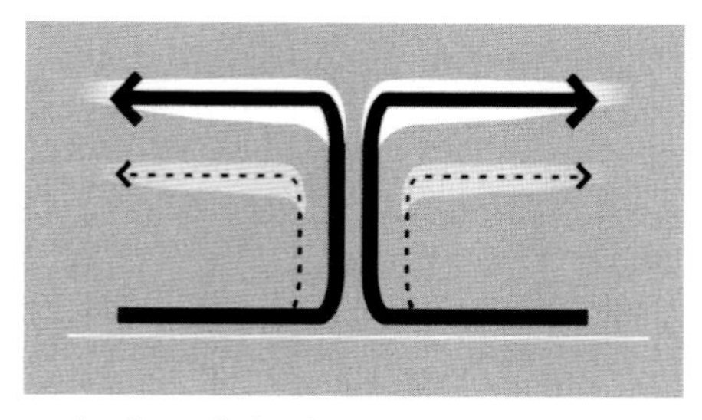

Figure 20.1 Schematic view of cloud responses to the +2K and control experiments using the global cloud-resolving model, showing increase in high clouds detrained from deep convection under warmer condition.

an important intermediate step is to know how cloud changes in idealized warming experiments, such as this study. Changes in cloud fractions of different types of cloud categories simulated by the ISCCP simulator can be used as proxies of various cloud properties. Values of cloud radiative forcing of model results are also used to evaluate cloud feedback parameters, although the difference between cloud feedbacks and cloud radiative forcing poses a limitation (Soden et al. 2004).

Changes in Clouds

Figure 20.2a–c show global distributions of the difference in the frequency of high-, mid-, and low-level clouds determined by the ISCCP simulator between the +2K simulation and control simulation in NICAM. Here we find a significant increase in high-level clouds, especially over the tropics and the subtropics. The same is shown for MIROC (Figure 20.2d–f). The common characteristics of the changes in global mean cloud fractions categorized by optical thickness and cloud-top pressure of the ISCCP simulator clouds in MIROC (Table 20.1 and Figure 20.3) are the same as those in GCM results listed in IPCC AR4 (IPCC 2007; Ringer et al. 2006). Compared to MIROC, the magnitude of the changes in cloud fraction is much larger in NICAM, and the

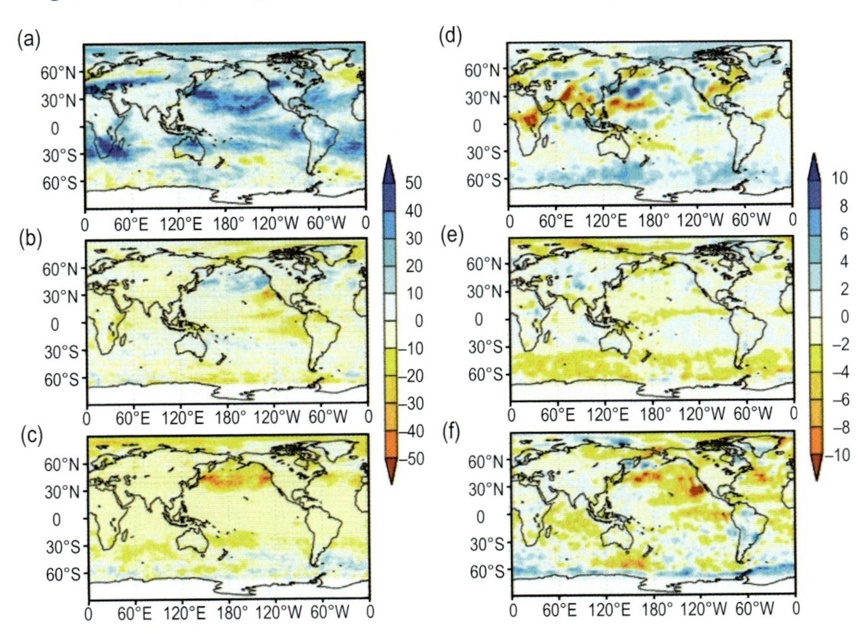

Figure 20.2 Difference of cloud fractions classified by the ISCCP simulator between the control experiment and the +2K experiment. Panels (a), (b) and (c) are high, middle and low clouds for NICAM. Panels (d), (e) and (f) are those for MIROC.

Table 20.1 ISCCP cloud amount changes for the +2K experiment for NICAM and MIROC.

ISCCP cloud categories	Thin		Medium		Thick	
	NICAM	MIROC	NICAM	MIROC	NICAM	MIROC
High	3.17	–0.01	1.51	0.01	0.78	0.11
Middle	–0.20	–0.09	–0.17	–0.24	–0.15	–0.01
Low	–0.15	–0.11	–1.17	–0.46	–0.21	0.10

most significant differences in NICAM are its change in high clouds. Both the magnitude and the spatial patterns of the changes in cloud cover in high clouds are quite different between the two models. Spatial patterns of the changes in mid- and low-level clouds are similar. As summarized by Figure 20.3, the signs of changes in global mean cloud fraction are similar between NICAM and MIROC, except for high thin clouds.

In Figure 20.4a, four colored solid lines show the zonal mean change in the frequency of high clouds in NICAM with different optical thickness categories (thin, medium, thick, and total) classified by the ISCCP simulator. High thin clouds and high thick clouds in the ISCCP are considered to correspond to cirrus clouds and deep convective clouds in their meteorological classification (Rossow and Schiffer 1999), respectively. Hereafter, we will refer to them as cirrus and deep convective clouds. We see that the significant increase in high clouds mostly comes from the change in cirrus. In MIROC (Figure 20.4b), the changes in cirrus are much smaller than those in NICAM. Changes

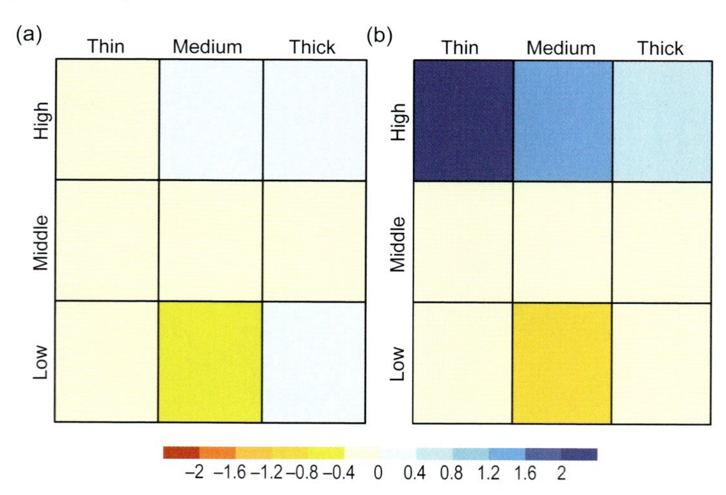

Figure 20.3 Matrix showing difference of global mean cloud fractions classified by the ISCCP simulator between the control experiment and the +2K experiment for NICAM (a) and MIROC (b).

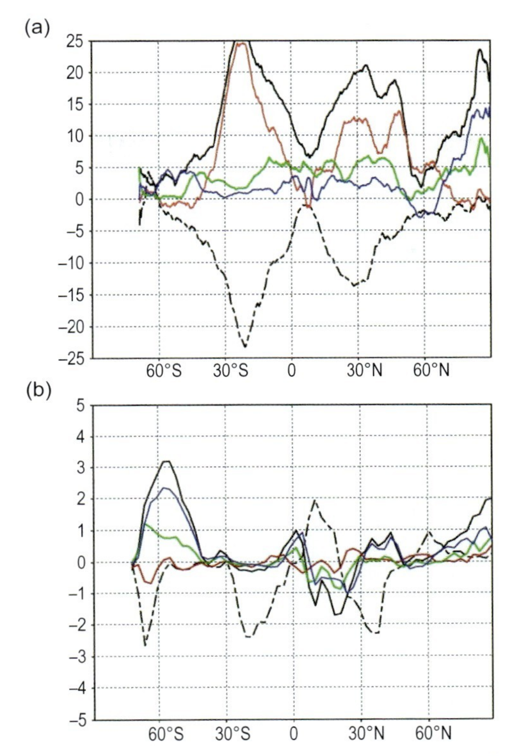

Figure 20.4 Zonally averaged difference of high clouds with different optical thickness (thin: red line, medium: green, thick: blue, total: black, invisible: dashed black) classified by the ISCCP simulator between the control experiment and the +2K experiment for NICAM (a) and MIROC (b).

in other clouds are similar, and most of the change in high clouds comes from the change in deep convective clouds.

A common change in the two models is that the temperature profile (isothermal line) shifts upward because of the temperature increase; the mixed-phase level, where liquid and solid condensate coexist, and the tropopause shifts upward. Together with this upward shift, the distributions of clouds change. A GCM intercomparison of the change in cloud condensate (Tsushima et al. 2006) revealed that the change is predominantly in the liquid condensates' increase in the mixed-phase layer, because the liquid condensate increases according to the phase change with the temperature increase in this level. In NICAM, the magnitude of the increase in ice condensate in the upper troposphere up to the tropopause is also large, and this rise in ice condensate corresponds to the significant increase in cirrus.

It is known that invisible clouds exist throughout the sky; their optical thickness falls below the threshold of the sensor and they are thus not detectable through observations. The threshold thickness is considered to be 0.3 in the ISCCP (Schiffer and Rossow 1983). Although they are invisible in the observational data, the ISCCP simulator in a model can classify these invisible clouds. The black dashed lines in Figure 20.3a, b show the differences in high invisible clouds classified by the ISCCP simulator between the control experiment and the +2K experiment using NICAM and MIROC, respectively. In NICAM, invisible clouds decrease significantly, especially over the subtropics, and the decrease in invisible clouds corresponds well to the increase in cirrus (Figure 20.4a). In MIROC, the variation in invisible clouds does not correspond to changes in high clouds.

One of our interests in the climate sensitivity test using the GCRM is the changes in deep convection and precipitation. Figure 20.5a, b shows these changes over the tropics in NICAM. The spatial pattern of the variation in

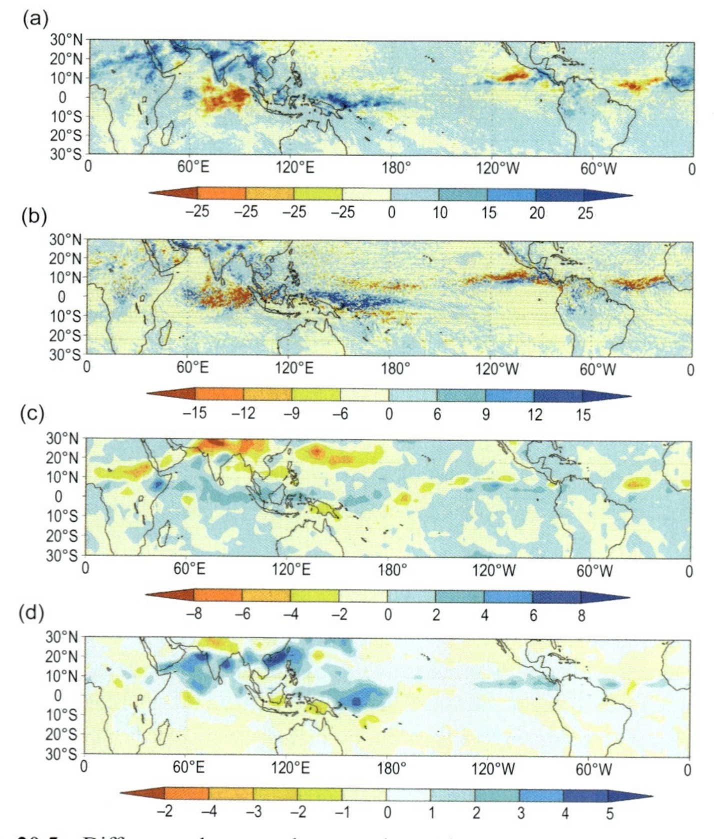

Figure 20.5 Differences between the control experiment and the +2K experiment for NICAM, high thick clouds (a) and precipitation (b), as well as MIROC, high thick clouds (c) and precipitation (d).

precipitation is highly correlated to the change in the deep convective clouds. Deep convective clouds and precipitation decrease over the central and eastern part of the Indian Ocean, but increase over the western part. Over the Pacific and the Atlantic oceans, both deep convective clouds and precipitation increase in the center of the convergence zone, and decrease in the subsidence regions on both sides of the convergence region with increasing precipitation in NICAM. These differences show that the area of the convergence zone becomes centralized; ITCZ and SPCZ become closer.

In MIROC (Figure 20.5c, d), we do not find a clear link between the spatial pattern of the change in precipitation and deep convective clouds. In addition, centralization of the convergence zone is not clear.

Changes in the Radiative Effects of Clouds

Clouds exert two radiative effects on the Earth. They reflect the solar insolation and reduce the warming of the Earth (albedo effect). In addition, they create a greenhouse effect by utilizing the thermal radiation emitted from the Earth, which reduces the radiative cooling of the Earth. Lower clouds have a smaller greenhouse effect, because the emission from the clouds is closer to the emission from the surface temperature. Higher clouds exert a larger greenhouse effect.

Figure 20.6 demonstrates the changes in the radiative effect of clouds on the Earth (albedo effect and greenhouse effect) in NICAM and MIROC. A positive or negative sign indicates that the change warms or cools the Earth,

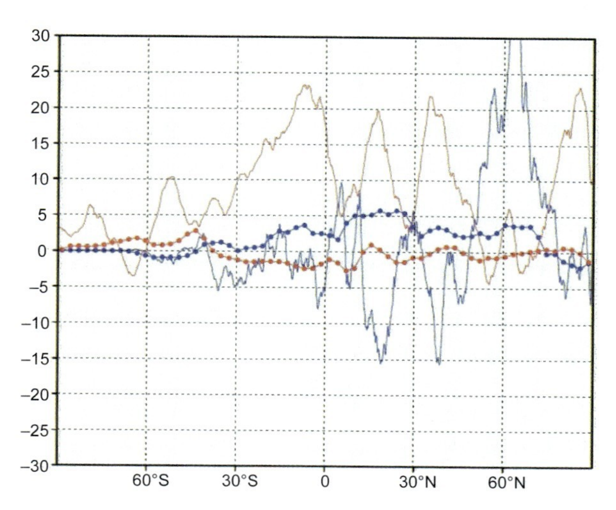

Figure 20.6 Difference in cloud radiative effect (cloud radiative forcing) (W m^{-2}) between the control experiment and the +2K experiment in NICAM (solid line) and in MIROC (solid line with filled circle). Red line indicates greenhouse effect (longwave forcing) and blue line indicates albedo effect (shortwave forcing).

respectively. Although net change is positive in both models, the magnitudes of the changes in both radiative effects in NICAM are much larger than in MIROC. The signs of the changes in the two cloud radiative effects vary between the two models. In MIROC, both the albedo effect and greenhouse effect weaken, with the weakening larger in the albedo effect. This corresponds to the result that the decrease in low cloud is dominant in MIROC. In NICAM, both the albedo effect and greenhouse effect intensify, and the intensification of greenhouse effect is larger than that of albedo effect. This corresponds to the significant increase in high clouds in NICAM.

Net radiative effect of high clouds depends on optical thicknesses. If the optical thickness is thin enough, the greenhouse effect exceeds the albedo effect (Inoue and Ackerman 2002), and an increase in these high thin clouds intensifies the warming.

Relevance between the Changes in Moisture Transport and Cirrus

Centralized deep convection supplies more moisture to the tropopause under the higher SST condition, and high thin clouds, which had invisible optical thickness, become thicker and visible (Figure 20.1). Moisture transport increases in the tropics, despite the decreasing vertical velocity over almost all latitudes. The rise in moisture transport through deep convection increases the detrainment of moisture at the top of the convection until anvils form. This contributes to the increase in the cirrus in the tropopause. Further detailed study is required before a full understanding of the mechanisms involved in the changes in moisture transport is possible.

Uncertainties remain because of our insufficient understanding of the microphysics of clouds. Sensitivity studies of radiative convective experiments over the tropics (not shown here) reveal that the extent of high clouds and the thickening of thin clouds (magnitude of the increase in albedo effect) are sensitive to our model's parameters of microphysics and boundary layer. However, almost all experiments show an increase in the clouds' greenhouse effect.

From these results, the decrease in cirrus and smaller difference in cloud greenhouse effect among GCMs might be attributed to the different treatment of cirrus and cumulus convection, which are represented by large-scale condensation and cumulus parameterization, respectively. In the GCM, whose grid scale is on the order of 100 km, we cannot explicitly resolve the deep convection and detrained anvils associated with them. Our results indicate the necessity of quantifying the amount of detrained moisture from the convection in the subgrid-scale distribution of moisture in a GCM grid in a more consistent manner.

Climate studies using CRMs will become more widespread in the near future. The acquisition of new cloud data with higher dimension and accuracy has already begun (CloudSat, CALIPSO), and subsequent evaluation will be essential to reduce the uncertainty in climate sensitivity simulations.

Summary

Studies of cloud responses using GCRMs have just started. GCRM is a future tool to deepen our understanding of perturbed clouds in the climate system. GCRMs will be supplementary to the *present* methods using conventional GCMs with cloud parameterization. Current GCMs have ambiguities derived from cloud parameterizations, and future GCRMs have ambiguities derived from cloud microphysics schemes. However, by addressing which processes of cloud microphysics affect certain types of cloud properties, we can achieve better results. High cloud properties behave differently in a GCM and GCRM, and this drives us to ask: How is high cloud amount sensitive to cloud processes such as cloud generation processes, precipitation efficiency, or sedimentation of cloud ice? How are model results of high clouds comparable to current satellite observations such as CloudSat and CALIPSO? How can we understand the change in dynamic fields such as narrowing the precipitation regions, increase in transport of water, and relative humidity?

Acknowledgments

M. Satoh thanks the members of NICAM developing team (Y. Tsushima, S. Iga and H. Tomita) for their help in preparing the GCRM part of this chapter and for allowing the authors to use portions of their work. This study was partly supported by the innovative program of climate change for the 21st century in Japan.

References

Anderson, J., J. Dykema, R. Goody, H. Hu, and D. Kirk-Davido. 2004. Absolute, spectrally-resolved, thermal radiance: A benchmark for climate monitoring from space. *J. Quant. Spectrosc. Radiat. Trans.* **85(3-4)**:367–383.

Bony, S., R. Colman, V. M. Kattsov et al. 2006. How well do we understand and evaluate climate change feedback processes? *J. Climate* **19**:3445–3482.

Bony, S., and J.-L. Dufresne. 2005. Marine boundary layer clouds at the heart of tropical cloud feedback uncertainties in climate models. *Geophys. Res. Lett.* **32**:L20806.

Bretherton, C. S. 2007. Challenges in numerical modeling of tropical circulations. In: The Global Circulation of the Atmosphere, ed. T. Schneider and A. H. Sobel, pp. 302–330. Princeton: Princeton Univ. Press.

Cess, R. D., E. F. Harrison, P. Minnis et al. 1992. Interpretation of seasonal cloud–climate interactions using Earth Radiation Budget Experiment data. *J. Geophys. Res.* **97**:7613–7617.

Cess, R. D., and G. L. Potter. 1988. A methodology for understanding and intercomparing atmospheric climate feedback processes in general circulation models. *J. Geophys. Res.* **93**:8305–8314.

ESA. 2001. EarthCARE–Earth clouds, aerosols, and radiation explorer. *ESA Report* SP-1257(1).

Goody, R. M., J. Anderson, and G. North. 1998. Testing climate models: An approach. *Bull. Amer. Meteor. Soc.* **79**:2541–2549.

Grabowski, W. 2001. Coupling cloud processes with the large-scale dynamics using the cloud-resolving convection parameterization (CRCP). *J. Atmos. Sci.* **58**:978–997.

Hall, A., and X. Qu. 2006. Using the current seasonal cycle to constrain snow albedo feedback in future climate change. *Geophys. Res. Lett.* **33**:L03502.

Hasumi, H., and S. Emori, eds. 2004. K-1 coupled model (MIROC) description. K-1 tech. rep. 1, ed. H. Hasumi and S. Emori, 34 pp. Tokyo: Center for Climate System Research, Univ. of Tokyo.

Iga, S., H. Tomita, Y. Tsushima, and M. Satoh. 2007. Climatology of a nonhydrostatic global model with explicit cloud processes. *Geophys. Res. Lett.* **34**:L22814.

Inoue, T., and S. A. Ackerman. 2002. Radiative effects of various cloud types as classified by the split window technique over the eastern subtropical Pacific derived from collocated ERBE and AVHRR data. *J. Meteor. Soc. Japan* **80**:1383–1394.

IPCC. 2007. Climate Change 2007: The Physical Science Basis. Contribution of Working Group I to the Fourth Assessment Report of the Intergovernmental Panel on Climate Change, ed. S. Solomon, D. Qin, M. Manning et al. New York: Cambridge Univ. Press.

Keith, D., and J. Anderson. 2001. Accurate spectrally resolved infrared radiance observation from space: Implications for the detection of decade-to-century-scale climatic change. *J. Climate* **14(5)**:979–990.

Khairoutdinov, M., C. Demott, and D. Randall. 2007. Evaluation of the simulated interannual and subseasonal variability in an AMIP-style simulation using the CSU multi-scale modeling framework. *J. Climate* **21(3)**:413.

Khairoutdinov, M. F., and D. A. Randall. 2001. A cloud-resolving model as a cloud parameterization in the NCAR community climate system model: Preliminary results. *Geophys. Res. Lett.* **28**:3617–3620.

Kiehl, J. T., C. A. Shields, J. J. Hack, and W. D. Collins. 2006. The climate sensitivity of the community climate system model: CCSM3. *J. Climate* **19**:2584–2596.

Lau, K.-M., C.-H. Sui, M.-D. Chou, and W.-K. Tao. 1994. An inquiry into the cirrus-cloud thermostat effect for tropical sea surface temperature. *Geophys. Res. Lett.* **21**:1157–1160.

Lin, B., B. A. Wielicki, L. H. Chambers, Y. Hu, and K. M. Xu. 2002. The iris hypothesis: A negative or positive cloud feedback? *J. Climate* **15**:3–7.

Lindzen, R. S. 1990. Some coolness concerning global warming. *Bull. Amer. Meteor. Soc.* **71**:288–299.

Lindzen, R. S., M.-D. Chou, and A. Y. Hou. 2001. Does the Earth have an adaptive infrared iris? *Bull. Amer. Meteor. Soc.* **82**:417–432.

Loeb, N. G., B. A. Wielicki, F. G. Rose, and D. R. Doelling. 2007. Variability in global top-of-atmosphere shortwave radiation between 2000 and 2005. *Geophys. Res. Lett.* **34**:L03704.

Miura, H., M. Satoh, T. Nasuno, A. T. Noda, and K. Oouchi. 2007. A Madden-Julian Oscillation event simulated by a global cloud-resolving model. *Science* **318**:1763–1765

Miura, H., H. Tomita, T. Nasuno et al. 2005. A climate sensitivity test using a global cloud resolving model under an aqua planet condition. *Geophys. Res. Lett.* **32**:L19717.

Murphy, J. M., D. M. Sexton, D. N. Barnett et al. 2004. Quantification of modeling uncertainties in a large ensemble of climate change simulations. *Nature* **429**:768–772.

Ogura, T., S. Emori, M. J. Webb et al. 2008. Climate sensitivity of the CCSR/NIES/FRSGC AGCM with different cloud modeling assumptions. *J. Meteor. Soc. Japan* **86**:69–79.

Ramanathan, V., and W. Collins. 1991. Thermodynamic regulation of ocean warming by cirrus clouds deduced from observations of the 1987 El Niño. *Nature* **351**:27–32.

Randall, D. A., M. Khairoutdinov, A. Arakawa, and W. Grabowski. 2003. Breaking the cloud-parameterization deadlock. *Bull. Amer. Meteor. Soc.* **84**:1547–1564.

Ringer, M. A., B. J. McAvaney, N. Andronova et al. 2006. Global mean cloud feedbacks in idealized climate change experiments. *Geophys. Res. Lett.* **33**:L07718.

Rosenfeld, D. 2000. Suppression of rain and snow by urban and industrial air pollution. *Science* **287**:1793–1796.

Rossow, W. B., and R. A. Schiffer. 1999. Advances in understanding clouds from ISCCP. *Bull. Amer. Meteor. Soc.* **80**:2261–2288.

Satoh, M., T. Matsuno, H. Tomita et al. 2008. Nonhydrostatic icosahedral atmospheric model (NICAM) for global cloud-resolving simulations. *J. Comp. Phys.* **227**:3486–3514.

Schiffer, R. A., and W. B. Rossow. 1983. The international satellite cloud climatology project (ISCCP): The first project of the World Climate Research Programme. *Bull. Amer. Meteor. Soc.* **64**:779–784.

Senior, C. A., and J. F. B. Mitchell. 1993. Carbon dioxide and climate: The impact of cloud parameterization. *J. Climate* **6**:393–418.

Soden, B. J., A. J. Broccoli, and R. S. Hemler. 2004. On the use of cloud forcing to estimate feedback. *J. Climate* **19**:3661–3665.

Soden, B. J., and I. M. Held. 2006. An assessment of climate feedbacks in coupled ocean–atmosphere models. *J. Climate* **19**:3354–3360.

Space Studies Board. 2007. Earth Science and Applications from Space: National Imperatives for the Next Decade and Beyond. Washington DC: Natl. Academy Press.

Stephens, G. L. 2005. Cloud feedbacks in the climate system: A critical review. *J. Climate* **18**:237–273.

Suzuki, K., T. Nakajima, M. Satoh et al. 2008. Global cloud-system-resolving simulation of aerosol effect on warm clouds. *Geophy. Res. Lett.*, in press.

Takemura, T., T. Nozawa, S. Emori, T. Y. Nakajima, and A. Khain. 2005. Simulation of climate response to aerosol direct and indirect effects with aerosol transport radiation model. *J. Geophys. Res.* **110**:D02202.

Tompkins, A. M., and G. C. Craig. 1999. Sensitivity of tropical convection to sea surface temperature in the absence of large-scale flow. *J. Climate* **12**:462–476.

Tsushima, Y., A. Abe-Ouchi, and S. Manabe. 2005. Radiative damping of annual variation in global mean surface temperature: Comparison between observed and simulated feedback. *Climate Dyn.* **24**:591–597.

Tsushima, Y., S. Emori, T. Ogura et al. 2006. Importance of the mixed-phase cloud distribution in the control climate for assessing the response of clouds to carbon dioxide increase: A multi-model study. *Climate Dyn.* **27**:113–126.

Valero, F. P. J., J. Herman, P. Minnis et al. 1999. Triana: A Deep Space Earth and Solar Observatory. National Academy of Sciences Review.

Webb, M. J., C. A. Senior, S. Bony, and J.-J. Morcrette. 2001. Combining ERBE and ISCCP data to assess clouds in the Hadley Centre, ECMWF and LMD atmospheric climate models. *Climate Dyn.* **17**:905–922.

Webb, M. J., C. A. Senior, D. M. H. Sexton et al. 2006. On the contribution of local feedback mechanisms to the range of climate sensitivity in two GCM ensembles. *Climate Dyn.* **27**:17–38.

Wielicki, B. A., R. Barkstrom, E. F. Harrison et al. 1996. Clouds and the Earth's radiant energy system (CERES): An Earth Observing System experiment. *Bull. Amer. Met. Soc.* **77**:853–868.

Williams, K. D., M. A. Ringer, C. A. Senior et al. 2006. Evaluation of a component of the cloud response to climate change in an intercomparison of climate models. *Climate Dyn.* **26**:145–165.

Williamson, D. L., J. T. Kiehl, and J. J. Hack. 1995. Climate sensitivity of the NCAR Community Climate Model (CCM2) to horizontal resolution. *Climate Dyn.* **11**:377–397.

Wyant, M. C., M. Khairoutdinov, and C. S. Bretherton. 2006. Climate sensitivity and cloud response of a GCM with a superparameterization. *Geophy. Res. Lett.* **33**:L06714.

Wu, X., and M. W. Moncrieff. 1999. Effects of sea surface temperature and large-scale dynamics on the thermodynamic equilibrium state and convection over the tropical western Pacific. *J. Geophys. Res.* **104**:6093–6100.

21

Observational Strategies from the Micro- to Mesoscale

Jean-Louis Brenguier[1] and Robert Wood[2]

[1]Meteo-France, CNRS, Toulouse, France
[2]Department of Atmospheric Sciences, University of Washington, Seattle, WA, U.S.A.

Abstract

In a changing climate, clouds are perturbed through large-scale variations in the general circulation induced by both greenhouse gases and aerosols. Some aerosols perturb cloud microphysical properties as well. The challenge for micro- and mesoscale observational studies of perturbed clouds is thus to establish the links between these two contrasting forcings, in an effort to understand how clouds respond to changes in the general circulation and to quantify how this response might be modulated by changes in their microphysical properties. The two generic classes of micro- to mesoscale observational strategies, the Eulerian column closure and the Lagrangian cloud system evolution approaches, are described using examples of low-level cloud studies, and recommendations are made on how they should be combined with large-scale information to address this issue.

Introduction

There is an apparent paradox in cloud physics: clouds are very diverse, in terms of morphology, depth, cloud base and top heights, horizontal extent, dynamics and microphysics; they exert very different radiative impacts, in terms of the balance between their albedo and their greenhouse effect, as well as very distinct dynamic impacts, in terms of how they contribute to the redistribution of water vapor, sensible, and latent heat in the atmosphere. Cloud spatial and temporal distributions are very heterogeneous; large regions may be devoid of clouds, whereas others can be overcast almost all year round. The lifetimes of cloud systems vary enormously from those that are almost stationary for days to others that are fleeting. Nevertheless—and herein lies the paradox—together they manage to maintain a somewhat constant Earth albedo, close to 0.3, and a fairly constant global balance between their albedo and greenhouse contributions to the climate system.

Such a large-scale equilibrium calls for observational approaches that begin on a global scale and progress down to the microscale to capture the processes responsible for the regulation of the hydrological cycle. However, to date, most observational studies of clouds have focused on the microscale, progressing up to the mesoscale. Such studies have served as our fundamental tools for constructing, from the bottom-up, the current cloud models we use to simulate the climate system. However, model comparison exercises, such as those by the IPCC (2007), clearly show that some key feedback processes are still poorly represented. Moreover, although the models predict a mean cloud radiative forcing in agreement with observations, they also exhibit noticeable biases, with a consistent overprediction of optically thick clouds and an underprediction of optically thin low and middle-top clouds (Zhang et al. 2005). This led the IPCC to conclude that "differences in cloud response are the primary source of inter-model differences in climate sensitivity" (Randall et al. 2007, p. 633).

Better understanding of how clouds react to anthropogenic forcings is therefore a priority for improving the accuracy of climate change projections. Anthropogenic forcings, however, can produce very diverse impacts on the general circulation. The primary greenhouse gases have a long residence time (centuries) and are homogeneously distributed. Aerosols, by contrast, have a short residence time (days to weeks) and, consequently, their spatial distribution is heterogeneous, mainly concentrated in the vicinity of the sources. Greenhouse gases interact primarily with longwave radiation. The net radiative impact of aerosols depends on their chemical composition and the balance between their light scattering and absorbing contributions. Impacts of greenhouse gases on clouds can thus be explored from a global perspective to examine, for instance, how the increase in the column water vapor might be compensated by a damping of the convective mass flux to reduce the fractional increase in precipitation (Held and Soden 2006). Aerosol impacts should also be considered at the regional scale, where they are concentrated. The hypothesized "elevated heat pump" effect (Lau et al. 2006) provides an example of a plausible, but currently unverified, regional-scale aerosol effect which may impact the Indian monsoon at the Himalaya foothills.

This greenhouse perspective suggests that anthropogenic forcings might only perturb clouds by modifying the general circulation (i.e., from the global scale down to the cloud scale). However, aerosols impact also cloud microphysics, since some act as droplet or ice crystal nuclei. Such microphysical changes can propagate up to the cloud scale. The first order response to changing nucleus concentration is an effect upon the cloud albedo by changing the surface area of the droplets, but feedbacks on cloud-scale dynamics must also be considered to understand the response fully. By modifying cloud microphysical and optical properties, aerosol particles thus perturb clouds not only from the global/regional scale downward, via their direct effect, but also upward from the microscale (the aerosol indirect effect). Parameterizations of these processes in global climate models (GCMs) are very crude, partly because they involve nonlinear processes at scales that are not accessible to such models, but also because our knowledge of the various feedbacks of cloud microphysics on cloud dynamics is still limited. For example, high-resolution cloud models suggest that the aerosol impacts result in nontrivial effects, inducing either a decrease or an increase of the cloud liquid water path (LWP), with the sign of the response depending upon poorly understood factors. Hence, the aerosol impacts on clouds remain the most uncertain of the climate forcings in terms of efficacy (Forster et al. 2007, Fig. 2.19).

To improve climate change projections, it is crucial to understand how climate change might impact the spatiotemporal distribution and hence radiative forcing of clouds (from the global scale down). In addition, we must also find out whether clouds respond only to changes in large-scale dynamic forcings, or if microphysical processes might also modulate their response, thereby impacting the hydrological cycle and general circulation (i.e., from the bottom up).

Because of the very large range of scales involved, it has been difficult to connect large-scale observational studies of the hydrological cycle with micro- and mesoscale observations of cloud physics. Using examples from existing and future field studies, we will show how effective the micro- to mesoscale approach was for understanding cloud dynamics and microphysics, and will suggest that it now needs to evolve to progressively larger scales. This mandates a greater degree of multidisciplinarity in the design of future observational studies, in an effort to clarify the interactions between aerosol physics and chemistry, small-scale turbulent dynamics, radiation, and the hydrological cycle at the global scale.

The Ingredients of an Observational Strategy

To begin to determine the magnitude of the aforementioned potential cloud perturbations, the overall objective of the observational approach must be to quantify the respective susceptibilities of cloudiness or, more specifically, cloud radiative properties, δc, to changes in the general circulation and to internal microphysical changes induced by the aerosol. This approach is presented in Stevens and Brenguier (this volume) as:

$$\delta c = \left(\frac{\partial c}{\partial M}\right)_A \left\{\left(\frac{\partial M}{\partial A}\right)_G \delta A + \left(\frac{\partial M}{\partial G}\right)_A \delta G\right\} + \left(\frac{\partial c}{\partial A_\mu}\right)_M \delta A_\mu, \qquad (21.1)$$

where c represents cloudiness, M is the large-scale meteorology, G stands for greenhouse gases, and A for aerosols. A_μ specifically refers to the subset of the aerosols that may impact cloud microphysical properties, namely cloud condensation and ice nuclei (CCN, IN) and absorbing aerosol particles scavenged in hydrometeors.

The term in the braces is the large-scale meteorological forcing, δM, related to the aerosol and greenhouse gas forcings, both of which are the result of integrated radiative forcings on large spatial scales. Thus, from Equation 21.1 we can represent cloud perturbations as:

$$\delta c = \left(\frac{\partial c}{\partial A_\mu}\right)_M \delta A_\mu + \left(\frac{\partial c}{\partial M}\right)_A \delta M, \qquad (21.2)$$
$$\delta c = \lambda_{A_\mu} \delta A_\mu + \lambda_M \delta M,$$

where $\lambda_{A\mu}$ and λ_M are the sensitivities (susceptibilities) of clouds to perturbations in A_μ and M, respectively.

In designing a strategy for an observational program to assess the impacts of aerosols upon clouds, it is thus essential to ask whether variability in observed cloud radiative properties, δc, will be dominated by variability in aerosols or by the varying meteorological forcing. Equation 21.2 shows clearly that to determine the sensitivity of clouds to aerosol perturbations $\lambda_{A\mu}$ through observations, we must first understand the impacts of meteorology upon the cloud system, as expressed by λ_M.

The meteorological sensitivity, λ_M, is poorly known in many cases, and thus it is a major challenge to use observations to determine it. For marine stratocumulus clouds, there has been considerable success in relating cloud properties (e.g., cloud fractional coverage) to the large-scale meteorology using observations (e.g., Klein and Hartmann 1993; Klein et al. 1995; Wood and Bretherton 2006). As Stevens and Brenguier demonstrate (this volume), the meteorological sensitivity is very high in many cases, and this will limit our ability to attribute perturbations in clouds to those in aerosols.

We can further quantify the uncertainties in determining the sensitivity to aerosols by considering the variability across a set of measurements that would be made in a particular observational campaign. If we define σ_x^2 as the observed variance in parameter x across this set of measurements, Equation 21.2 can be used to show that:

$$\sigma_c^2 = \lambda_{A_\mu}^2 \sigma_{A_\mu}^2 + \lambda_M^2 \sigma_M^2 + 2\lambda_{A_\mu} \lambda_M \sigma_{A_\mu} \sigma_M r, \qquad (21.3)$$

where r is the correlation coefficient between the A_μ and M. Equation 21.3 demonstrates that not only do we need to understand the meteorological variability and its impact on the clouds, we also need to understand to what extent the meteorological variability covaries with the aerosol variability. In other words, accounting for the meteorological variability in a dataset (i.e., the second term in Equation 21.3, through knowledge of λ_M) is not sufficient to determine fully the aerosol sensitivity, if one does not also understand how the aerosol properties are tied to the large-scale flow. In addition, we note that such a formulation assumes a one-way cause-and-effect relationship, at least on short timescales, between aerosols and clouds and is a simplistic representation of a tightly coupled system in which clouds and aerosols interact on all scales. For instance, clouds can significantly impact aerosols via scavenging by precipitation, hence contributing to the covariance between aerosol and the meteorology. These mutual interactions raise serious concerns about the numerous and contrasting satellite studies that claim correlations between aerosols and clouds, which are generally interpreted as evidence of aerosols impacting clouds. In fact, quite the opposite may actually be true.

To optimize the chances of distinguishing an aerosol signal from the background meteorological noise, we must first select a place where the aerosol variability is significant, while the variability of the meteorology is minimized. The covariance between the two, however, sets a limit to this strategy: reducing meteorological variability (e.g., by selecting specific weather situations) necessarily reduces the aerosol variability. The second aerosol characterization experiment (ACE-2), which took place in the North Atlantic in June 1997, illustrates this impediment. Within ACE-2, the cloudy column experiment was designed to test the Twomey hypothesis; namely, that the cloud optical thickness, τ, scales with the LWP, W, and the droplet number concentration N as $\tau \propto W^{5/6} N^{1/3}$. The Atlantic Ocean, north of the Canary Islands, offers opportunities to sample air masses flowing around the Azores high that are generally pristine, except when they skim along the European continent (where they become polluted by anthropogenic aerosols). Droplet concentrations observed during the eight case studies ranged from less than 50 cm^{-3} in pristine air masses up to more than 250 cm^{-3} in the most polluted ones (i.e., a factor of five), whereas the LWP ranged from 33 $g\ m^{-2}$ to 77 $g\ m^{-2}$ (i.e., a factor of only two). Such a gap between aerosol and LWP variability allowed a precise validation of the Twomey hypothesis (discussed further below), but the dataset did not provide any evidence of an aerosol impact on the cloud life cycle. Indeed, aerosol and meteorology were closely correlated because polluted air masses had flown over the continent, and had hence experienced greater sensible and lower latent heat fluxes than the pristine oceanic air masses. Overall, the most polluted cases exhibited lower LWP than their pristine counterpart (Brenguier et al. 2003), and it was not feasible to determine whether clouds were thinner as a result of the reduced latent heat fluxes two days ahead over the continent, attributable to direct aerosol effects on the air mass when it moved over the ocean from the continent to the sampling area, or because of an indirect aerosol effect on the cloud layer.

Attempts were made to examine the climatology of precipitation downwind of large cities such as St. Louis, Missouri (Changnon et al. 1971). Careful analysis of the observations with a detailed numerical model indicated, however, that urban land use forced convergence downwind of the city, rather than the presence of greater aerosol concentrations, was the dominant control on the locations and amounts of precipitation in the vicinity of an urban complex (van den Heever and Cotton 2007).

An attractive alternative approach is to select situations where aerosol and meteorological variations are uncorrelated. This has served as the basis of weather modification control experiments for many years. Accidental biomass burning events offer such opportunities, but they are generally not frequent enough to build significant statistics. To improve the statistics, weekend effects have been examined climatologically (Forster and Solomon 2003; Gong et al. 2006). Evidence of a detectable weekly cycle of the diurnal temperature range has been shown, as well as the anthropogenic origin of this cycle. A potential aerosol effect on clouds was investigated by Bell et al. (2008), but may be masked by other atmospheric parameter changes (e.g., heat island effect) that are also correlated to the weekly cycle of anthropogenic activities. The identification of situations where aerosol variability is uncorrelated with meteorology thus remains to be resolved.

For shallow clouds, the issue of sensitivity is particularly acute. Indeed, their liquid water content is typically a few hundredths of the total water content. This explains why deriving their LWP and its temporal evolution from field observations of the thermodynamic fields (i.e. temperature and water vapor) remains beyond our grasp. The magnitude of the energy fluxes that govern the evolution of a cloud-topped boundary layer (e.g., surface fluxes including precipitation, cloud-top entrainment, and the flux divergences of short- and longwave radiation, λ_M) are comparable in magnitude to their potential modulation by the aerosol, for example by suppressing precipitation ($\lambda_{A\mu}$). The susceptibility of marine stratocumulus clouds to aerosols is thus noticeable, as demonstrated by ship tracks.

Quantifying the susceptibilities of marine stratocumulus clouds to the meteorology and aerosol, respectively, is a challenge in the sense that small perturbations of the boundary layer state parameters need to be precisely measured. What facilitates the observation of these clouds, however, is their long synoptic lifetime (although, of course, the cloud element lifetime is only a few minutes), large spatial extension, relative statistical homogeneity at the mesoscale, and their reproducible diurnal cycle.

Beyond the susceptibilities, additional ingredients must also be carefully evaluated when designing a field experiment. This includes the spatial and temporal scales of interest, and the identification of all physical processes which may interfere with the observations. Next we describe two basic categories of observational approach that are particularly useful for the study of the interactions between marine stratocumulus clouds and their environment. The principles involved, however, are applicable to other cloud systems.

Eulerian versus Lagrangian

Closure Experiments

Closure experiments aim to measure the consistency of the atmospheric state parameters with respect to models of the underlying physical processes (Ogren

1995; Quinn et al. 1996). The methodology consists of measuring input parameters to initialize a model and derive output parameters; concomitantly, the control parameters are measured for comparison with model predictions. To illustrate this, we will describe the observational strategy of the ACE-2 Cloudy-Column experiment (Brenguier, Chuang et al. 2000), which was the first field study dedicated entirely to the aerosol indirect effects in extended boundary-layer cloud systems. The hypotheses to be tested can be summarized in the form of three key questions:

1. For specified cloud fields, is the droplet concentration consistent with the predictions of aerosol activation models?
2. Do cloud radiative properties vary with droplet concentration, as anticipated by Twomey (1977)?
3. For a particular value of LWP, is the precipitation rate modulated by the droplet concentration?

Temporal and Spatial Scales

Sampling a single convective cell during its vertical ascent while measuring aerosol properties, vertical velocity, and cloud droplet concentration to examine CCN activation is not feasible using the existing airborne platforms. In addition, radiative transfer raises serious difficulties for single cells because it is, in essence, three dimensional and thus measurements of irradiance performed from above a single convective cell will necessarily be affected by radiation from neighboring cells. Finally, the cycle of precipitation formation in a single convective cell is short (a few tens of minutes), resulting in very heterogeneous drizzle patches below cloud base.

An alternative strategy is to examine the phenomenon at a larger scale at which aerosol properties, turbulence, cloud microphysics, and precipitation are statistically homogeneous (ergodic), and the three-dimensional heterogeneities of the radiation and precipitation fields are smoothed over a large number of cells. Such conditions are often satisfied in boundary-layer marine stratocumulus clouds at a scale of a few tens of kilometers.

Aerosol Activation Closure

This experiment aims to evaluate 0-D kinetic models of CCN activation to predict the cloud droplet concentration (control parameter) from the vertical velocity at cloud base and the physicochemical properties of the aerosol (input parameters) (Guibert et al. 2003; Snider et al. 2003).

Since we are unable to perform a closure experiment on individual CCN activation events, a statistical approach to the problem must be adopted, which necessarily must encompass the spatial variability of the system being studied. Aerosol properties can, far from the aerosol sources, be reasonably assumed to be uniform over the area and the duration of the experiment. Vertical velocity, on the other hand, varies from a few cm s^{-1} up to more than 1 m s^{-1} in the most active cells. Thus, comparison involves the probability distribution function (PDF) of measured droplet concentration and its comparison with the predictions of a CCN activation model initialized with the full spectrum of measured vertical velocities. Figure 21.1 shows the comparison of the deciles of the measured droplet concentration PDF with the predictions of the model initialized successively with the deciles of the measured vertical velocity distribution. This figure demonstrates that the range of concentration variability resulting from vertical velocity fluctuations is broader than the difference between the mean values of a pristine and a polluted case.

Therefore, a closure experiment on CCN activation will not be conclusive if the vertical velocity is not fully constrained by observations. A consistent definition of the cloud droplet concentration used here as a control parameter is also crucial. In fact, the droplet concentration measured in a cloud system is different from the one resulting from CCN activation, even though both are tightly related. After CCN activation is completed, additional processes (e.g., mixing with the environmental dry air and scavenging by precipitation) dilute droplet concentration significantly. Thus, for comparison with a CCN activation model prediction, it is sensible to select only those droplet concentration samples that are not affected by either mixing or precipitation scavenging. In ACE-2, for instance, the droplet concentration after selection was

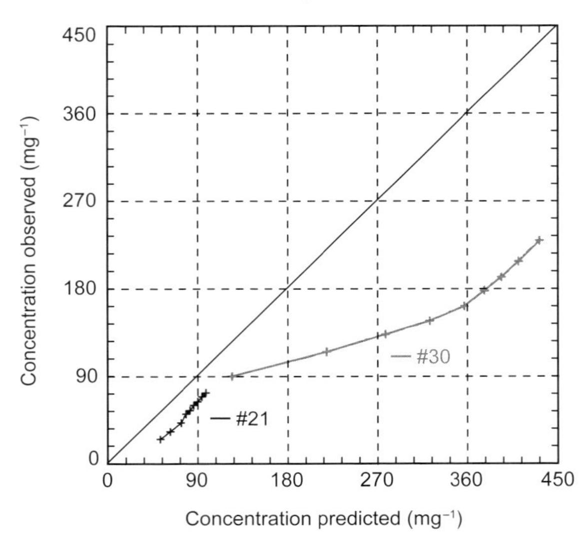

Figure 21.1 Deciles of the observed droplet concentration probability distribution function (PDF) versus predictions of the droplet concentration with a CCN activation model initialized successively with the deciles of the observed vertical velocity PDF for flight 21 (pristine case) and flight 30 (polluted case).

typically 30% higher than the average over all the samples (Pawlowska and Brenguier 2000).

The main limitation of this closure experiment was the incomplete characterization of the size-segregated chemical composition of the particles, which introduces uncertainties in the prediction of their hygroscopic properties. Since the development of airborne mass spectrometers and improved chemical analysis, activation closure has been significantly refined (Conant et al. 2004; Fountoukis et al. 2007). Further work remains to be conducted to characterize the mass accommodation coefficient for small growing particles, and instruments need to be designed to characterize their state of chemical mixture.

Column Closure on Radiative Transfer

This type of experiment aims to corroborate the Twomey hypothesis (i.e., that aerosol-induced microphysical changes are reflected by changes of cloud radiative properties). The input parameters measured *in situ* are the vertical distribution and horizontal variability of the cloud droplet size distribution, and the output parameter is the optical thickness of the cloud layer, derived as the vertical integral of extinction. The control parameter is the optical thickness derived independently from multispectral radiances measured with a second aircraft flying above the cloud layer.

In ACE-2, the statistics of all input parameters were very robust because of the long duration of sampling with two collocated aircraft: one was dedicated to the microphysical fields while the other focused on radiation. Thus, ACE-2 provided the first observational evidence of the Twomey effect in extended cloud systems, e.g., the scaling of the optical thickness with LWP and droplet concentration as anticipated by Twomey (see Fig. 6 in Brenguier, Pawlowska et al. 2000). One serious limitation, however, was that *in-situ* airborne measurements do not provide information on how the microphysical fields sampled at various levels are distributed vertically. The optical thickness used as the output parameter was derived by assuming either random or maximum overlap of the microphysical fields measured *in situ*. The accuracy of the prediction was thus significantly degraded by the overlap uncertainty. Since ACE-2, other closure experiments have been performed using remote-sensing systems that better constrain the vertical organization of the microphysical fields (Feingold et al. 2003).

An alternative approach to radiative closure is to validate the same radiative transfer model but in the inverse mode. Indeed, inverse models are currently used to derive cloud properties from space measurements of multispectral radiances (Nakajima and King 1990). In this approach, radiance measurements are used to derive cloud geometrical thickness, or LWP, and droplet concentration, which are then compared to the ones measured *in situ* (Schüller et al. 2003).

Column Closure on Precipitation

This experiment contributes to the improvement of precipitation formation parameterization in global climate models. Recent field studies suggest that the precipitation rate of stratocumulus clouds, averaged over a large domain containing numerous cloud cells, scales with the mean cloud thickness or LWP and the typical droplet concentration: ACE-2 (Pawlowska and Brenguier 2003), EPIC (Comstock et al. 2004; Wood 2005), and DYCOMS-II (Van Zanten et al. 2005). The nature of such a relationship is a major determinant of the magnitude of the aerosol impact on cloud extent thickness and lifetime. Large eddy simulations (LES) are therefore used to corroborate these observations and quantify the empirical relationship better.

The observations are summarized in Figure 21.2a. For each field campaign, the precipitation rate at cloud base scales well with the cloud thickness and cloud droplet concentration. Each dataset appears, however, to have offsets that mainly reflect measurement biases and differences in the methodology: precipitation rate averaged over the cloud layer (ACE-2) or at cloud base only (EPIC and DYCOMS-2), droplet concentration measured in cloud samples that are not affected by mixing or precipitation scavenging (ACE-2), averaged over the cloud layer (DYCOMS-II), or extrapolated from remote sensing (EPIC), cloud thickness derived from detection of cloud base and top (ACE-2 and DYCOMS-II) or derived from remote sensing of the LWP (EPIC). These discrepancies reveal how sensitive the results can be to the definition of the physical parameters derived from diverse measurement and data processing techniques. In addition, the results demonstrate just how sensitive the precipitation rate is to cloud macrophysical properties, with a doubling of precipitation requiring only a change of ~100 m in cloud thickness. This emphasizes the importance of controlling for meteorological variability when examining microphysical impacts.

Numerical simulations of similar cloud systems were performed with an LES model over a broad range of LWP and CCN concentration values to explore the parameter space of the measurements (Geoffroy et al. 2008). Figure 21.2b–d show the comparison of the model results with the measurements, using the same parameters and scaling laws as in each field campaign, respectively. The similarity between observations from three different areas and the results of numerical simulation suggests that the large-scale relationship between LWP, droplet concentration, and the precipitation rate at cloud base is robust.

Summary and Recommendations

These three examples illustrate different types of closure experiments. Each type has clearly testable hypotheses. When the model is straightforward, such as the 0-D model of CCN activation, the closure experiment follows closely the basic methodology: measured input parameters, numerical simulations, comparison of model predictions with the control parameter. In the second example, the distinction between input and control parameters is less obvious, depending on whether the model is used in the direct or in the inverse mode, such as for satellite 1-D retrieval techniques. The third example, with its 3-D model of stratocumulus clouds, suggests how the technique can be extrapolated to compare relationships between specified physical parameters that have been observed and further simulated with a model over the same parameter space. The three approaches share in common the following methodological rules, which are generally not given sufficient attention in most of the closure studies:

1. Models need to be fully constrained: All parameters, which might impact the prediction of the model to be tested, must be documented according to a level of accuracy consistent with their impact (e.g., as for vertical velocity).
2. Ensure consistency in the definition of the measured and model parameters: The measured and model values of a parameter must be defined over the same spatial and temporal scales. For example, droplet concentration will exhibit significant differences, depending upon whether it is defined as the mean value over a cloud system or the value specifically measured in regions of CCN activation.
3. Redundancy in measurements is highly desirable: Single validation experiments often succeed, whereas redundant controls are more difficult to reconcile, but allow for a higher degree of confidence. For closure to be robust, attempts must be made to combine redundant closures of the same process, such as combining a CCN activation spectrum and a droplet activation closure on the same data set (Snider et al. 2003) or radiation closures on both transmitted and reflected light in the same cloud system (Platnick 2000).

In general, dynamic, thermodynamic, and microphysical properties exhibit important variability and covariability on the kilometer scale (i.e., scales smaller than a typical climate model grid box), and this variability has marked impacts on how aerosol–cloud interaction affects the large-scale properties of clouds. It is important to design a sampling strategy that allows us to characterize further these important subgrid statistical connections between variables (see, e.g., Illingworth and Bony, this volume; Larson et al. 2001, 2002). For instance, airborne measurements in clouds are often optimized by targeting cloud cells along the flight track; however, such an approach introduces bias into the data base (overestimated cloud fraction). Thus, it is crucial to adopt unbiased sampling or provide additional information to reduce potential biases in the data base.

Column closure experiments are useful to validate models of physical processes and their parameterizations for general circulation models, as long as the

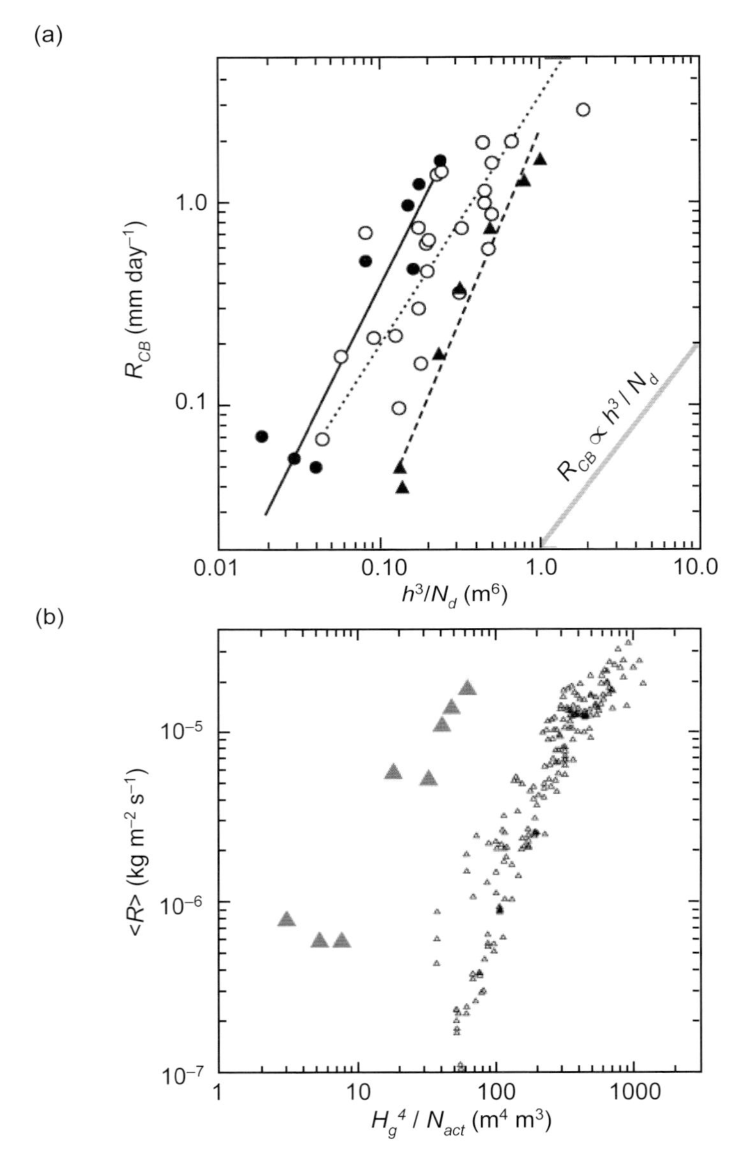

Figure 21.2 (a) Cloud base precipitation rates, R_{CB}, from observational case studies in subtropical marine stratocumulus, plotted against h^3/N_d, where h is the cloud thickness and N_d the droplet concentration. Black circles: Pawlowska and Brenguier (2003), *in-situ* aircraft; white circles: Comstock et al. (2004), radiometric and radar drizzle; triangles: van Zanten et al. (2005), *in-situ* aircraft and radar drizzle. The lines represent linear least-distance regressions to the case studies for each field campaign. Comparison of model predictions (small triangles) with scaling laws derived from: (b) ACE-2 (large triangles): precipitation rate <R> averaged over the cloud layer, cloud thickness H_g

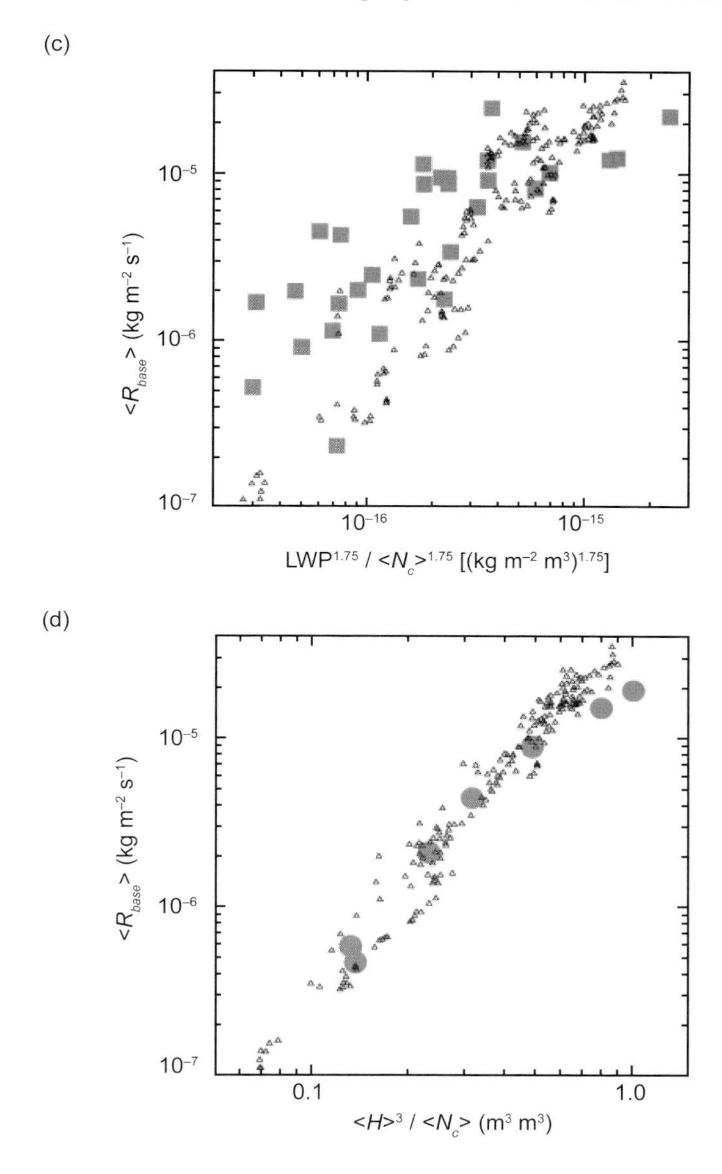

Figure 21.2 (continued) derived from *in-situ* measurements and droplet concentration, N_{act}, derived from samples not affected by mixing or droplet scavenging. (c) EPIC (squares): precipitation rate at cloud base $\langle R_{base} \rangle$, liquid water path (LWP), and mean droplet concentration N_c, derived from surface remote sensing. (d) DYCOMS-II) (circles): precipitation rate at cloud base $\langle R_{base} \rangle$ and cloud thickness $\langle H \rangle$ derived from remote sensing and mean droplet concentration, N_c, derived from *in-situ* measurements.

system is evolving sufficiently slowly. They are not suited, however, for studies of the temporal evolution of the system, which call for a different approach.

Lagrangian Experiments: Linking Small and Large Scales

The Lagrangian Concept

The temporal and spatial scales over which cloud fields evolve and interact with the general circulation in which they are embedded are typically days and thousands of kilometers. Scales involved in the general circulation exceed those accessible to single aircraft flights and limit seriously our ability to understand the two-way interactions between microscale processes and the general circulation.

This need to observe the evolution of meteorological fields and the processes that influence their development over long timescales led to the development of the Lagrangian observational concept, which seeks to make quasi-continuous observations of an air mass (ideally over several days, although as we shall see there are considerable benefits from flights lasting only a few hours) often by employing multiple observation platforms which overlap in time. The Lagrangian observational approach aims to study processes within a frame of reference that moves with the air, an approach necessary to study time-evolving processes in the atmosphere without resorting to making assumptions about poorly constrained advective processes.

Despite the high costs associated with fully realizing the potential of the Lagrangian approach using *in-situ* aircraft observations, a number of such observational studies have been conducted over the last few decades. These studies have yielded a tremendously rich understanding of a wide variety of different processes, as we discuss below.

Lagrangian Boundary Layer and Cloud Evolution

Lagrangian sampling of cloud-related processes on timescales of a day or more has primarily focused upon the marine boundary layer (MBL), because many of the key processes relevant to climate occur there. In addition, the logistics associated with conducting a Lagrangian study in the lowest level are simpler (not least of which is the presence of a solid and uniform boundary at the base of the air mass). Although earlier Lagrangian studies involved balloons and tracers, the first multi-flight Lagrangian experiments devoted to understanding cloud processes were conducted during the Atlantic Stratocumulus Transition Experiment/Marine Aerosol and Gas Experiment (ASTEX/MAGE) in June 1992 (Albrecht et al. 1995; Businger et al. 2006). Two such studies were carried out, each for a period of 36–48 hours (~800 km of spatial advection). During these studies, constant-density balloons floated downwind with the air mass and radioed their GPS-derived locations to a relay of three sampling aircraft. The balloons served as markers for the boundary-layer air mass, which was continually being modified by chemical and energy fluxes at the surface, entrainment of free tropospheric air, wind shear within the boundary layer, horizontal dispersion, chemical reactions, and aerosol transformations. Here we will focus on those aspects most pertinent to clouds.

The ASTEX/MAGE Lagrangian experiments were revolutionary for a number of reasons. By continuously sampling the MBL over a period approaching two days, it was possible to make observations of the transition from a shallow, well-mixed stratocumulus-capped MBL to a much deeper, decoupled MBL containing trade cumuli of less extensive coverage (Bretherton and Pincus 1995). Such air mass transitions are a critical component of the transition from the subtropical to the tropical MBL, and the physical mechanism by which the MBL undergoes this transition was deduced directly as a result of the Lagrangian observations. Previous hypotheses for the transition followed ideas based upon the work on diurnally modulated decoupling of the boundary layer that had been carried out at fixed locations. The ASTEX Lagrangian data were able to demonstrate that although decoupling of the MBL is a key process in the transition, it is actually the increasing latent heat flux as the air mass moves over warmer water, rather than solar radiation, that is the key driver of the subtropical transition (Bretherton and Wyant 1997).

Lagrangian Entrainment Rate Estimates

A second novel aspect of Lagrangian studies is the ability to derive direct estimates of the rate of MBL deepening, and thus the entrainment rate. It is now widely appreciated just how important the entrainment rate is, not only for its physical impacts upon the MBL thermodynamic and cloud climatology (Stevens 2002), but also because entrainment is a critical mediator of the response of cloud field to perturbations in atmospheric aerosols (Ackerman et al. 2004). Because microphysically induced changes (e.g., suppression of drizzle) impact the entrainment rate by changing the MBL turbulent structure, this introduces much longer timescales than the Twomey effect (Wood 2007), which cannot be quantified observationally using measurements from fixed locations.

ASTEX allowed Lagrangian estimates of the entrainment rate over the two days of observations (Bretherton et al. 1995). Recent studies have employed the Lagrangian technique using flight durations of only a few hours. For example, in DYCOMS-II (Stevens et al. 2003), single 6-hr Lagrangian flights using a C-130 aircraft were used to ascertain the entrainment rate by determining the observed rate of boundary layer deepening and subtracting the subsidence rate from reanalysis data (Faloona et al. 2005). This was then compared with the entrainment rate estimated using the standard flux-jump method to provide closure on the time evolution of the entrainment rate that can be used to constrain model ability to represent the entrainment process correctly.

Aerosol–Cloud–Chemistry Interactions using Lagrangians

Although aerosol indirect effects were not a specific focus of the ASTEX Lagrangians, the aerosol and gas phase chemistry measurements taken as part of the MAGE component of the campaign provided the first concrete demonstration of the benefits that the Lagrangian observational concept could bring to our understanding of chemical process rates in the atmosphere (Zhuang and Huebert 1996; Clarke et al. 1996), a need that had been identified since the early 1970s. One might argue that the development of the sampling concept for the early Lagrangian studies led to the realization that aerosol particles and their precursors are themselves a fundamental component of the atmospheric general circulation, which thereby provides further possibilities for interactions between these scales that were not previously appreciated.

Lagrangian experiments that focused specifically upon aerosols and aerosol precursors were conducted as part of the First and Second Aerosol Characterization Experiments (ACE-1, ACE-2) in 1995 and 1997, respectively (Bates et al. 1998; Johnson et al. 2000). They provided a wealth of data on the coupling between aerosols, aerosol precursors, clouds, and the MBL, in both pristine and polluted MBLs. From the ACE-2 Lagrangian studies it has been possible to assemble a conceptual model for continental pollution outbreaks that serves as an invaluable basis for the planning of future sampling of the modified continental plume. Timescale analysis applied to the ACE-2 Lagrangian datasets (Hoell et al. 2000) demonstrated the importance of meteorological factors (e.g., dilution) in controlling the properties of aerosols in the MBL. The example from the third Lagrangian experiment during ACE-2, shown in Figure 21.3, demonstrates the ability of Lagrangian experiments to connect micro- and mesoscale characteristics with those of the large-scale flow.

A Strategy for Future Lagrangian Experiments

Although they have provided invaluable information about air mass transformations and the physical and chemical processes within them, the Lagrangian experiments performed to date have not yet reached their full potential as important strategies for determining how clouds will change in a perturbed climate:

1. Sampling limitations: To date, Lagrangian studies have sampled with one aircraft at any one given time, whereas current state-of-the-art observational techniques require the use of multiple aircraft (e.g., column closure experiments discussed above) to link the aerosol, thermodynamic, and dynamic properties of the atmosphere measured *in situ* with the column cloud radiative properties remotely sensed from above. Adopting this strategy within a Lagrangian experiment would be very powerful, especially if scanning radiometers could be used to

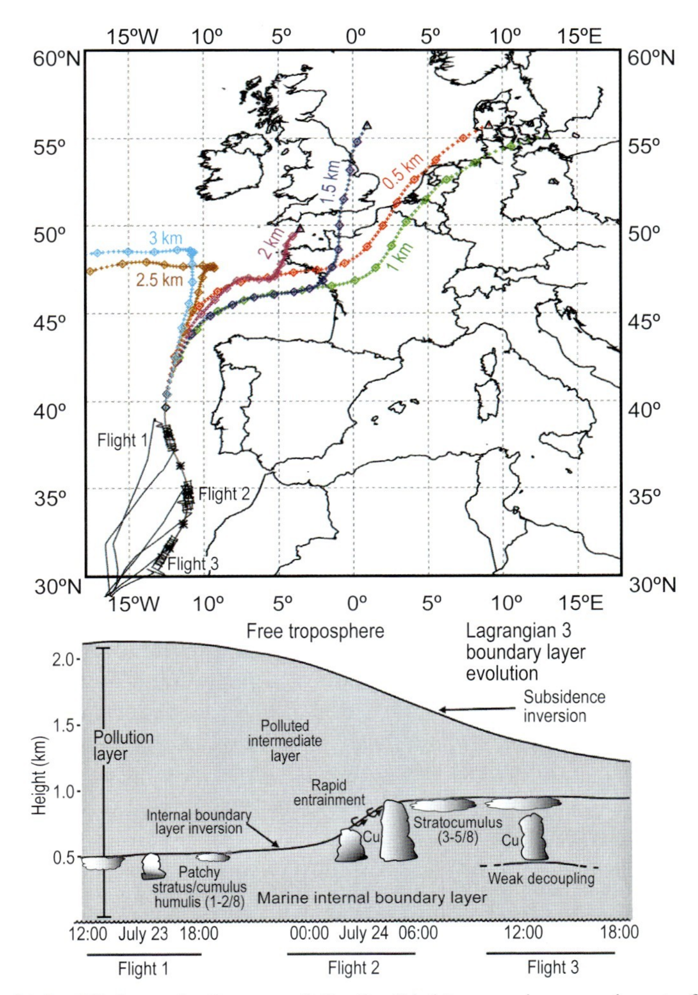

Figure 21.3 Flight tracks (top panel) for the third Lagrangian experiment of ACE-2 (Wood et al. 2000) which sampled a polluted continental outbreak heading southward over the northeast subtropical Atlantic Ocean. Back trajectories, estimated using EC-MWF analyses, are shown ending at different altitudes at the location of the start of the sampling. The three flights allow us to construct a picture of the changing structure of the lowest 2 km of the atmosphere, the essence of which is detailed schematically (bottom panel). Modified from Wood et al. (2000).

broaden the measurements to be representative of an area. Moreover nadir-pointing remote-sensing systems provide crucial information on how microphysical fields that are sampled horizontally by *in-situ* aircraft overlap vertically (Damiani et al. 2006).

2. New sampling techniques: Beyond piloted aircraft, new vectors are becoming available for *in-situ* sampling of boundary-layer clouds at

low speed. Unmanned aerial vehicles have a typical sampling speed of 30 m s^{-1} and they can now fly autonomously a predefined track (Ramanathan et al. 2007). Blimps have been used in the past and new platforms (e.g., the Zeppelin) might be used for Lagrangian studies, although the logistics involved remain a serious obstacle.

3. Interdisciplinary spirit: To characterize the evolution of the cloud layer fully, it is important to have a detailed characterization of all the potential factors controlling it, including the large-scale meteorology, the surface and MBL top boundary conditions (both physical and chemical) as well as the cloud microphysical, thermodynamic, and turbulent information. This is a tall order for a single platform but can be achieved, provided there is effective interdisciplinary communication and adequate financial support. The forthcoming EUCAARI and VOCALS experiments provide good examples of the possible connections that can be made between research communities (oceanography, atmospheric chemistry, cloud physics, and climate dynamics).
4. Synergy between models and observation: The most serious obstacle in the characterization of a boundary layer is our inability to measure the thermodynamic conservative variables (total heat and moisture) and their budgets accurately enough to constrain the resulting LWP. However, LES models, which can simulate the evolution of a boundary layer, suggest observable signatures of the mechanisms that drive its evolution. For instance, Sandu et al. (2008) showed that the impacts of the aerosol on the diurnal cycle of stratocumulus clouds are reflected by observable differences in the vertical velocity variance at cloud top, in the vertical profile of liquid water content, and in the level of turbulent kinetic energy below cloud base.
5. Complementary information: With the current generation of satellite remote sensing, there is a tremendous amount of additional information that can be added from space to multi-day Lagrangian studies. Geostationary satellites, especially those with high-resolution (1 km), multi-channel and high-sampling frequency (15 min for SEVIRI on MSG) spectrometers can provide cloud microphysical, aerosol, and cloud-top temperature information. Polar-orbiting satellites with active remote sensing, such as in the A-Train, provide additional information, at least once a day, to calibrate geostationary satellite observations better. Promising techniques appear to derive accurate temperature, humidity, and boundary-layer depth information from spaceborne GPS limb sounding. A reanalysis focused on clouds, aerosols, and the boundary layer will be an excellent way to incorporate the satellite information into a useable framework. High-resolution numerical modeling could be calibrated using *in-situ* data to “fill in the gaps” in rather the same way that assimilation is used in the numerical weather prediction community.

In addition, aircraft observations do not provide complete information about the large-scale dynamic fields; key variables, such as the large-scale vertical motion field, are completely unknown. These fields are critically important for understanding the interactions between the small and large scales and for allowing a top-down assessment of the entrainment rate that can be compared with bottom-up measurements carried out by integrating over many turbulent eddies. This is currently accomplished by using standard reanalysis products, but these are not generally produced with the needs of the cloud-climate community in mind.

General Recommendations and Conclusions

Understanding how clouds respond to a changing climate requires carefully designed micro- and mesoscale observational experiments. The task of separating the impacts of variability in large-scale meteorology from those caused by changes in aerosol physical and chemical properties is very challenging. This is because cloud and precipitation formation is sensitive to both changes in the large-scale forcings as well as to microphysical aerosol-induced impacts. Moreover, not only the instantaneous relationships between microphysical and macrophysical variables, but also the temporal evolution of the fields must be documented. Thus, it is important to put the observed microphysical changes in perspective with the effects of the large-scale meteorological variability.

Need for Better Connection between Small and Large Scales

Acquiring a better understanding of the interactions between meso- and microscale processes and the large-scale flow requires the provision of accurate boundary conditions for meso- and microscale studies. In recent years, this has been achieved using large-scale meteorological data from reanalysis datasets (e.g., Kistler et al. 2001; Uppala et al. 2005), which have proven extremely useful because they provide gridded data at regular temporal intervals everywhere in the atmosphere.

Current reanalyses place a high value on creating datasets that are homogeneous over a long period of time (40 years), hence preventing the use of the most advanced, but computationally expensive, models as well as limiting assimilation to data that are available over the whole reanalysis period, or at least relatively lengthy fractions of it. Aerosol processes, for instance, are not represented at all in current reanalyses; however, given the recent advances to our observing system, this should be possible. We argue that the "one-size-fits-all" approach is not optimal to address the cloud–climate problem. Indeed, we believe that focused reanalysis systems, which target key processes and assimilate new datasets, would be of major benefit to the observational community. These systems would not necessarily be global, but could instead be run over much smaller spatial and temporal domains at higher resolution. Regional reanalysis (e.g., the North American regional reanalysis) is a move in this direction. Incorporating physical process specialization with geographical specialization would also be a beneficial development. An assimilation system that focuses upon the lower troposphere, including the boundary layer, would greatly benefit our understanding of low cloud systems and their relationship with large-scale meteorological and aerosol forcings.

To optimize boundary conditions, the instrumental setup should be designed to connect the meso- to the microscale. This should rely on a combination of large-scale, though limited and discontinuous, sampling with instrumented aircraft and surface (either ground or ship) stations that provide continuous sampling of the atmosphere flowing above the site.

Need for Consistent Multidisciplinary Approaches

Clouds are perturbed in their dynamics, microphysics, and radiative properties. Field experiments dedicated to this challenging issue must be designed with an instrumental setup that permits the documentation of each source of perturbation and each effect with accuracies consistent with their anticipated contribution or level of response to the forcings. New airborne instruments are now available to document crucial parameters, such as the size-segregated chemical composition of the aerosol. Considerable progress has been made to overcome the sampling limitation of instrumented aircraft by using airborne active remote-sensing systems that provide crucial information about the vertical organization of the cloud systems sampled horizontally by the aircraft (e.g., Damiani et al. 2006). We argue that instead of running numerous separate partial studies that do not consider the critical atmospheric parameters well enough, efforts should be concentrated on large-scale multidisciplinary and multiplatform projects.

Detailed Planning

Most surface remote-sensing systems sample only two-dimensional slices of atmosphere, and instrumented aircraft samples are limited to long lines of data of a few mm^2 cross section each. Their use, over the short duration of an intensive observation period, must therefore be highly optimized; compromises must be made on where to locate surface systems and which flight tracks to allocate to each aircraft. Thus, the preparation of a field experiment involves preliminary studies of the expected phenomena to identify very specific and observable signatures of the physical processes and mechanisms that are supposed to express the impact of the large-scale meteorological and microscale aerosol forcings. The use of large-scale modeling, process modeling, and theory is extremely helpful in designing testable hypotheses which observational datasets can help to refute. A clear and precise definition of the sampling strategy is by far the most important ingredient of a micro- and mesoscale observation program.

Acknowledgments

We thank Bruce Wielicki for discussions from which the ideas for an improved reanalysis focused upon the boundary layer emerged.

References

Ackerman, A. S., M. P. Kirkpatrick, D. E. Stevens, and O. B. Toon. 2004. The impact of humidity above stratiform clouds on indirect aerosol climate forcing. *Nature* **432**:1014–1017.

Albrecht, B. A., C. S. Bretherton, D. Johnson, W. H. Schubert, and A. S. Frisch. 1995. The Atlantic stratocumulus experiment: ASTEX. *Bull. Amer. Meteor. Soc.* **76**:889–904.

Bates, T. S., B. J. Huebert, J. L. Gras, F. B. Griffiths, and P. A. Durkee. 1998. International Global Atmospheric Chemistry (IGAC) project's first Aerosol Characterization Experiment (ACE 1): Overview. *J. Geophys. Res.* **103**:16,297–16,318.

Bell, T. L., D. Rosenfeld, K.-M. Kim et al. 2008. Midweek increase in U.S. summer rain and storm heights suggests air pollution invigorates rainstorms. *J. Geophys. Res.* **113**:D02209.

Brenguier, J. L., P. Y. Chuang, Y. Fouquart et al. 2000. An overview of the ACE-2 CLOUDYCOLUMN closure experiment. *Tellus* **52B**:814–826.

Brenguier, J. L., H. Pawlowska, and L. Schüller. 2003. Cloud microphysical and radiative properties for parameterization and satellite monitoring of the indirect effect of aerosol on climate, PACE Topical Issue. *J. Geophys. Res.* **108(D15)**:8632.

Brenguier, J. L., H. Pawlowska, L. Schueller et al. 2000. Radiative properties of boundary layer clouds: Droplet effective radius versus number concentration. *J. Atmos. Sci.* **57**:803–821.

Bretherton, C. S., P. Austin, and S. T. Siems. 1995. Cloudiness and marine boundary layer dynamics in the ASTEX Lagrangian experiments. Part II: Cloudiness, drizzle, surface fluxes and entrainment. *J. Atmos. Sci.* **52**:2724–2735.

Bretherton, C. S., and R. Pincus. 1995. Cloudiness and marine boundary layer dynamics in the ASTEX Lagrangian experiments. Part I: Synoptic setting and vertical structure. *J. Atmos. Sci.* **52**:2707–2723.

Bretherton, C. S., and M. C. Wyant. 1997. Moisture transport, lower-tropospheric stability, and decoupling of cloud-topped boundary layers. *J. Atmos. Sci.* **54**:148–167.

Businger, S., R. Johnson, and R. Talbot. 2006. Scientific insights from four generations of Lagrangian smart balloons in atmospheric research. *Bull. Amer. Meteor. Soc.* **87**:1539–1554.

Changnon, S. A., F. A. Huff, and R. G. Semonin et al. 1971. An investigation of inadvertent weather modification. *Bull. Amer. Meteor. Soc.* **52**:958–968.

Clarke, A. D., T. Uehara, and J. N. Porter. 1996. Lagrangian evolution of an aerosol column during the Atlantic Stratocumulus Transition Experiment. *J. Geophys. Res.* **101**:4351–4362.

Comstock, K. K., R. Wood, S. E. Yuter, and C. S. Bretherton. 2004. Reflectivity and rain rate in and below drizzling stratocumulus. *Q. J. Roy. Meteor. Soc.* **130**:2891–2919.

Conant, W. C., T. M. VanReken, T. A. Rissman et al. 2004. Aerosol–cloud drop concentration closure in warm cumulus. *J. Geophys. Res.* **109**:D13204.

Damiani, R., G. Vali, and S. Haimov. 2006. The structure of thermals in cumulus from airborne dual-doppler radar observations. *J. Atmos. Sci.* **63**:1432–1450.

Faloona, I., D. H. Lenschow, T. Campos et al. 2005. Observations of entrainment in eastern Pacific marine stratocumulus using three conserved scalars. *J. Atmos. Sci.* **62**:3268–3285.

Feingold, G., L. E. Wynn, D. E. Veron, and M. Previdi. 2003. First measurements of Twomey indirect effect using ground-based remote sensors. *Geophys. Res. Lett.* **30**:1287.

Forster, P., V. Ramaswamy, P. Artaxo, et al. 2007. Changes in atmospheric constituents and in radiative forcing. In: Climate Change 2007: The Physical Science Basis. Contribution of Working Group I to the Fourth Assessment Report of the Intergovernmental Panel on Climate Change, ed. S. Solomon, S. D. Qin, M. Manning et al., pp. 129–234. Cambridge: Cambridge Univ. Press.

Forster, P., and S. Solomon. 2003. Observations of a "weekend effect" in diurnal temperature range. *PNAS* **100**:11,225–11,230.

Fountoukis, C., A. Nenes, N. Meskhidze et al. 2007. Aerosol–cloud drop concentration closure for clouds sampled during ICARTT. *J. Geophys. Res.* **112**:D10S30.

Geoffroy, O., J. L. Brenguier, and I. Sandu. 2008. Relationship between drizzle rate, liquid water path and droplet concentration at the scale of a stratocumulus cloud system. *Atmos. Chem. Phys. Discuss.* **8**:3921–3959.

Gong, D.-Y., D. Guo, and C.-H. Ho. 2006. Weekend effect in diurnal temperature range in China: Opposite signals between winter and summer. *J. Geophys. Res.* **111**:D18113.

Guibert, S., J. R. Snider, and J. L. Brenguier. 2003. Aerosol activation in marine stratocumulus clouds part I: Measurement validation for a closure study PACE topical issue. *J. Geophys. Res.* **108(D15)**:8628.

Held, I. M., and B. J. Soden. 2006. Robust responses of the hydrological cycle to global warming. *J. Climate* **19(21)**:5686–5699.

Hoell, C., C. O'Dowd, S. Osborne, and D. Johnson. 2000. Timescale analysis of marine boundary layer aerosol evolution: Lagrangian case studies under clean and polluted cloudy conditions. *Tellus* **52B**:423–438.

IPCC. 2007. Climate Change 2007: The Physical Science Basis. Contribution of Working Group I to the Fourth Assessment Report of the Intergovernmental Panel on Climate Change, ed. S. Solomon, D. Qin, M. Manning et al. New York: Cambridge Univ. Press.

Johnson, D. W., S. R. Osborne, R. Wood et al. 2000. An overview of the Lagrangian experiments undertaken during the second Aerosol Characterisation Experiment. *Tellus* **52B**:290–320.

Kistler, R., E. Kalnay, W. Collins et al. 2001. The NCEP–NCAR 50-year reanalysis: Monthly means CD-ROM and documentation. *Bull. Amer. Meteor. Soc.* **82**:247–267.

Klein, S. A., and D. L. Hartmann. 1993. The seasonal cycle of low stratiform clouds. *J. Climate* **6**:1588–1606.

Klein, S. A., D. L. Hartmann, and J. R. Norris. 1995. On the relationships among low-cloud structure, sea surface temperature, and atmospheric circulation in the summertime northeast Pacific. *J. Climate* **8**:1140–1155.

Larson, V. E., J.-C. Golaz, and W. R. Cotton. 2002. Small-scale and mesoscale variability in cloudy boundary layers: Joint probability density functions. *J. Atmos. Sci.* **59**:3519–3539.

Larson, V. E., R. Wood, P. R. Field, J.-C. Golaz, and T. H. Vonder Haar. 2001. Small-scale and mesoscale variability of scalars in cloudy boundary layers: One-dimensional probability density functions. *J. Atmos. Sci.* **58**:1978–1994.

Lau, K. M., M. K. Kim, and K. M. Kim. 2006. Asian summer monsoon anomalies induced by aerosol direct forcing: The role of the Tibetan Plateau. *Climate Dyn.* **26**:855–864.

Nakajima, T., and M. D. King. 1990. Determination of the optical thickness and effective radius of clouds from reflected solar radiation measurements, Part I: Theory. *J. Atmos. Sci.* **47**:1878–1893.

Ogren, J. A. 1995. A systematic approach to *in situ* observations of aerosol properties. In: Aerosol Forcing of Climate, ed. R. J. Charlson and J. Heintzenberg, pp. 215–226. Chichester: Wiley.

Pawlowska, H., and J. L. Brenguier. 2000. Microphysical properties of stratocumulus clouds during ACE-2. *Tellus* **52B**:867–886.

Pawlowska, H., and J. L. Brenguier. 2003. An observational study of drizzle formation in stratocumulus clouds during ACE-2 for GCM parameterizations. PACE Topical Issue, *J. Geophys. Res.* **108(D15)**:8630.

Platnick, S. 2000. Vertical photon transport in cloud remote sensing problems. *J. Geophys. Res.* **105**:22,919–22,935.

Quinn, P. K., T .L. Anderson, T. S. Bates et al. 1996. Closure in tropospheric aerosol–climate research: A review and future needs for addressing aerosol direct shortwave radiative forcing. *Contrib. Atmos. Phys.* **69(4)**:547–577.

Ramanathan, V., M. Ramana, G. Roberts et al. 2007. Warming trends in Asia amplified by brown cloud solar absorption. *Nature* **448**:575–578.

Randall, D. A., R. A. Wood, S. Bony et al. 2007. Climate models and their evaluation. In: Climate Change 2007: The Physical Science Basis. Contribution of Working Group I to the Fourth Assessment Report of the Intergovernmental Panel on Climate Change, ed. S. Solomon, S. D. Qin, M. Manning et al., pp. 589–662. Cambridge: Cambridge Univ. Press.

Sandu, I., J. L. Brenguier, O. Geoffroy, O. Thouron, and V. Masson. 2008. Aerosol impacts on the diurnal cycle of marine stratocumulus. *J. Atmos. Sci.*, in press.

Schüller, L., J. L. Brenguier, and H. Pawlowska. 2003. Retrieval of microphysical, geometrical and radiative properties of marine stratocumulus from remote sensing, PACE topical issue. *J. Geophys. Res.* **108(D15)**:8631.

Snider, J. R., S. Guibert, and J. L. Brenguier. 2003. Aerosol activation in marine stratocumulus clouds, part II: Köhler and parcel theory closure studies, PACE topical issue. *J. Geophys. Res.* **108(D15)**:8629.

Stevens, B. 2002. Entrainment in stratocumulus topped mixed layers. *Q. J. Roy. Meteor. Soc.* **128**:2663–2690.

Stevens, B., D. H. Lenschow, G. Vali et al. 2003. Dynamics and chemistry of marine stratocumulus, DYCOMS-II. *Bull. Amer. Meteor. Soc.* **84**:579–593.

Twomey, S. A. 1977. The influence of pollution on the shortwave albedo of clouds. *J. Atmos. Sci.* **34**:1149–1152.

Uppala, S. M., P. W. Kållberg, A. J. Simmons et al. 2005. The ERA-40 reanalysis. *Q. J. Roy. Meteor. Soc.* **131**:2961–3012.

van den Heever, S., and W. R. Cotton. 2007. Urban aerosol impacts on downwind convective storms. *J. Appl. Meteor. Climatol.* **46**:828–850.

Van Zanten, M. C., B. Stevens, G. Vali, and D. Lenschow. 2005. Observations of drizzle in nocturnal marine stratocumulus. *J. Atmos. Sci.* **62**:88–106.

Wood, R. 2005. Drizzle in stratiform boundary layer clouds. Part I: Vertical and horizontal structure. *J. Atmos. Sci.* **62**:3011–3033.

Wood, R. 2007. Cancellation of aerosol indirect effects in marine stratocumulus through cloud thinning. *J. Atmos. Sci.* **64**:2657–2669.

Wood, R., and C. S. Bretherton. 2006. On the relationship between stratiform low cloud cover and lower tropospheric stability. *J. Climate* **19**:6425–6432.

Wood, R., D. W. Johnson, S. R. Osborne et al. 2000. Boundary layer and aerosol evolution during the third Lagrangian experiment of ACE-2. *Tellus* **52B**:401–422.

Zhang, M. H., W. Y. Lin, S. A. Klein et al. 2005. Comparing clouds and their seasonal variations in 10 atmospheric general circulation models with satellite measurements. *J. Geophys. Res.* **110**:D15S02.

Zhuang, L. Z., and B. J. Huebert. 1996. Lagrangian analysis of the total ammonia budget during Atlantic stratocumulus transition experiment marine aerosol and gas exchange. *J. Geophys. Res.* **101**:4341–4350.

22

Observational Strategies at Meso- and Large Scales to Reduce Critical Uncertainties in Future Cloud Changes

Anthony Illingworth[1] and Sandrine Bony[2]

[1]Department of Meteorology, University of Reading, Reading, U.K.
[2]Laboratoire de Météorologie Dynamique, CNRS/UPMC, Paris, France

Abstract

The response of clouds to climate change remains very uncertain. This is attributable to both an incomplete knowledge of cloud physics and to the difficulties that large-scale models have in simulating the different properties of clouds. An observational strategy is proposed to improve the representation of clouds in large-scale models and to reduce uncertainties in the future change of cloud properties. This consists of determining first what key aspects of the simulation of clouds are the most critical, with respect to future climate changes, and then of using specific methodologies and new datasets to improve the simulation of these aspects in large-scale models.

Introduction

After decades of research, the response of clouds to a change in climate, and in particular to a global climate warming induced by anthropogenic activities, remains poorly understood and is still identified as a key source of uncertainty for climate sensitivity estimates. Given the slow progress in this area over the last fifteen years, one may wonder what strategy might help to reduce this uncertainty. There are so many physical processes and cloud properties that need to be better understood, and so many weaknesses in the representation of clouds in climate models.

When proposing an observational strategy, it would be helpful to know whether there is a hierarchy among the different problems; whether there are some priorities among the different processes that need to be better understood, better observed, or better simulated in climate models. Therefore, we think that an observational strategy to reduce uncertainties in cloud–climate feedbacks should be composed of two steps: (a) determine what are the most critical uncertainties, (b) determine how observations might be used to reduce some of these uncertainties.

Critical Uncertainties in the Large-scale Modeling of Clouds and Their Impact on Climate

Key Aspects of the Simulation of Clouds in Large-scale Models

Part of the reason why progress in the representation of clouds in large-scale models has been so slow is that major aspects of the simulated cloud distribution could not be assessed observationally. For instance, the vertical structure of cloud layers, their overlap, the cloud water content, and the cloud water phase are known to play a key role in the radiation budget at the top of the atmosphere (TOA), at the surface, and in the troposphere. Given the lack of reliable and global observations of these quantities, a good agreement between models and observations of TOA radiative fluxes or of the total cloud cover could be obtained with compensating errors. However, differences in the way the radiative balance is achieved can affect the sensitivity of radiative fluxes to a change in climate.

The vertical structure of clouds constitutes one well-known example of a key factor critical for climate studies for which observations have long been lacking. Another example is the cloud phase. Many papers have drawn attention to the vastly different amounts of cloud ice held in various climate models, all of which satisfy the TOA radiation constraint. In large-scale models, the cloud water phase is still commonly diagnosed as a simple function of temperature. As climate warms, the fraction of cloud water may increase at the expense of cloud ice, and owing to differences in the microphysical and radiative properties of liquid and ice clouds, the change in cloud phase contributes to cloud feedbacks. Uncertainties in the diagnostic of the cloud water phase in the current climate thus translate into uncertainties in cloud feedbacks (e.g., Tsushima et al. 2006).

The simplicity of the cloud phase diagnostic in large-scale models has long been justified by the fact that the factors influencing glaciation processes and the presence of supercooled droplets are still poorly understood and loosely constrained by observations. It is thus essential to use new data from satellite or ground-based measurements to provide a better explanation of the factors that influence the cloud phase, and to develop more reliable parameterizations for large-scale models.

In addition to the lack of key measurements, the representation of clouds in large-scale models is complicated by our poor understanding of the large-scale controls of (measured) cloud properties and of the physical processes through which the different cloud properties interact with each other (cf. Bretherton and Hartmann; Grabowski and Petch, both this volume). Such an understanding would require that measurements of cloud properties are done simultaneously for several variables, over a wide range of meteorological situations. Numerical weather prediction (NWP) models may also be very helpful in that regard. If an NWP model is producing the observed cloud amount in the right place at the right time (which has been shown to be the case in Illingworth et al. 2007), then we can be reasonably sure that the meteorological processes which produce the clouds are being well represented. NWP model outputs can then be used to understand how clouds are controlled by meteorological processes.

With the open availability of model simulations to the international climate science community (Meehl et al. 2007), the number of biases of climate models reported in the literature has dramatically increased. However, compared to the hundreds of scientists involved in the analysis of climate simulations, the number of scientists actually working on the development and continuous improvement of physical parameterizations in climate models is fairly small. This may be caused by the fact that institutional structures do not reward this activity enough, especially since it may take a long time and considerable effort to get real improvements, while it is much easier to demonstrate errors in models. In this situation, how should the effort of parameterization improvement be concentrated on the most critical processes?

The same question arises regarding the reduction of uncertainties in the models' projections of the future climate. The Fourth Assessment Report (AR4) (IPCC 2007) reports a large range of global climate sensitivity estimates among climate models. The largest contribution to this spread arises from intermodel differences in cloud feedbacks (Soden and Held 2006). Given the very large number of factors or processes potentially involved in these differences, which ones should we concentrate on to reduce, as efficiently as possible, the uncertainties in future climate change?

Based on different methodologies, recent studies suggest that the response of marine low-level clouds to climate change was the root cause of a large part of intermodel differences in global cloud feedbacks (Bony and Dufresne 2005; Webb et al. 2006; Wyant, Bretherton et al. 2006; Williams and Tselioudis 2007). By using an analysis method based on the stratification of the large-scale tropical circulation into dynamic regimes (Bony et al. 2004) or by analyzing simulations performed with idealized and simplified (aquaplanet) versions of climate models (Medeiros et al. 2008), it was shown that intermodel differences in the tropical cloud response were dominated by the response of clouds in the trade-wind regions. This suggests that an improvement in the representation of shallow convection and trade cumulus clouds is crucial for climate sensitivity. These studies shed light on the "silent majority" of tropical clouds that has locally a less spectacular impact on radiation than deep convective clouds or stratus clouds, but plays a major role for climate sensitivity. This finding will foster further studies focused on the understanding, the simulation, and the evaluation of shallow clouds, and will thus help to reduce this critical uncertainty.

This example shows that by carrying out idealized studies of climate change and by decomposing the global cloud feedback problem into components related to specific physical processes, the problem becomes more tractable and can suggest targeted diagnostics of model-data comparison or of data analysis. Such approaches have the great potential of identifying the processes that are critical for climate change projections. This provides guidance to establish a hierarchy of necessary model developments, helps to fill the gap between climate studies and process studies, and contributes to a better understanding (and thus a better assessment of our confidence) in the models' results. Therefore, such studies should be considered as a key step in the strategy to reduce critical uncertainties associated with the response of clouds to a changing climate.

One caveat associated with this strategy, however, is that processes or cloud regimes that may be missing in all the models or that may be represented equally badly in all the models may not be identified as a key source of uncertainty, as they might actually play an important role in nature. It is thus important to complement this strategy with comparisons of models with observations, and with idealized studies investigating the potential impact of model weaknesses on the simulation of climate.

Studies of this kind are necessary to assess the extent to which some processes contribute more than others to uncertainties in climate change projections. For instance, the inability of large-scale models (NWP or climate models) to simulate accurately the diurnal cycle of convection over tropical land areas is often cited as a concern for the credibility of climate change projections (whatever they are). This bias, which assuredly reveals some weaknesses in the models' representation of physical processes, is likely to be a concern for representing realistically the interactions between local precipitation and vegetation processes for instance. However, for other issues, such as the magnitude of global climate change or the change in some monsoon characteristics, the extent to which this bias actually affects climate model projections has yet to be demonstrated or assessed.

Recent studies show that climate models still exhibit substantial biases in their simulation of the water vapor and temperature distributions in the current climate. However, John and Soden (2007) found no relationship between the biases exhibited by models in the current climate and the magnitude of the water vapor–lapse rate feedback produced in climate change. They interpret this result by the fact that the water vapor feedback depends on the fractional change of humidity, and that this quantity is insensitive to biases in the mean state. Although we cannot exclude the possibility that biases in humidity are associated with biases in cloudiness (such an association may even be expected), this example illustrates the fact that biases in the mean state do not necessarily affect climate change feedbacks.

Similarly, aerosol effects are known to play a key role in the evolution of the 20th century climate through their direct effect on radiation. They have an influence on the formation and radiative properties of clouds at the small or regional scale. It has been proposed that the radiative effects of aerosols at the surface and in the troposphere affect surface temperature and large-scale atmospheric dynamics, and then affect climate phenomena such as the south Asian monsoon (e.g., Ramanathan et al. 2005). Some studies have indicated that the indirect effect of aerosols on liquid water clouds has the potential to affect global-scale climate change. Thus, evaluating and refining the representation of these processes in climate models certainly constitutes an important area of model development. In addition, we know very little about the ability of a very small fraction of the aerosol particles to act as ice nuclei, thus influencing the glaciation of clouds and so affecting the cloud lifetime and precipitation efficiency. Potentially, such processes are very sensitive to small amounts of anthropogenic aerosols and have led to the suggestion that they might affect the development of deep convective clouds.

However, in terms of the understanding and simulation of the tropical or global cloud response to climate change, it is still unclear whether the interaction between aerosols and clouds is of primary importance, compared to the influence of other small-scale (e.g., boundary-layer turbulence and shallow convection) or large-scale (e.g., changes in the large-scale circulation) processes. In the remote trade-wind regions of the Pacific Ocean, for instance, it is unlikely that the physical and radiative properties of trade wind cumuli will be strongly affected by anthropogenic changes in aerosol properties. We can therefore consider, for the moment, that improving the representation of cloud–aerosol interactions in climate models is of lower priority, to reduce the uncertainty in simulated large-scale climate changes, than the improvement of physical processes (e.g., boundary-layer turbulence, atmospheric convection, and radiative transfer).

Observational Strategies that Address These Uncertainties

Cloud Schemes in Large-scale Models

Clouds often form at scales much smaller than the typical size of a grid box in general circulation models (GCMs) and cannot therefore be explicitly predicted in these models. To predict the cloud fraction (together with other cloud properties), many large-scale models diagnose the cloud fraction by using statistical cloud parameterizations. In this approach, subgrid-scale fluctuations of variables (e.g., total water, potential temperature, or vertical velocity) are described by a probability distribution function (PDF) whose statistical moments (mean, variance, and skewness) must be diagnosed based on large-scale prognostic variables plus eventually some subgrid-scale variables predicted by turbulence or convection schemes (e.g., Bony and Emanuel 2001; Tompkins 2002). Here, the cloud fraction and water content are related to the fraction of the PDF above saturation and its first moment, respectively. This PDF can also be used to predict cloud overlap and consequent radiative properties as well as to provide a better description of the development of precipitation. Such an approach is promising to fill the gap between the different cloud scales and to strengthen the physical coupling between the different cloud processes.

A better documentation and understanding of the influence of subgrid-scale processes (e.g., turbulence, convection, gravity waves) on the PDF of large-scale variables for different cloud regimes (e.g., deep convective clouds, shallow clouds, cirrus) is required to guide development or improve statistical cloud parameterizations. For this purpose, modelers often use simulations from cloud-resolving models (CRMs) to get some guidance. This approach might now be developed with the arrival of global CRM simulations and super-parameterizations. This would allow, for instance, the examination and better understanding of how the PDF of different variables relates to small-scale physical processes and interacts with the large-scale environment. One might then investigate why large-scale models fail to simulate middle-level clouds while some CRMs do a better job (e.g., Liu et al. 2001). An important prerequisite for the success of this approach, however, is that CRMs (or large eddy simulations) simulate the subgrid-scale fluctuations accurately. To assess whether it is actually the case, comparisons between observed and simulated fluctuations and cloud distributions produced by high-resolution models on the 100 m to 2 km scale are required. This emphasizes the need to observe the humidity structure of the atmosphere in three dimensions with a high enough resolution.

Evaluating Cloud Properties Simulated by Large-scale Models with Ground-based Data

Ground-based observations, such as those derived from Cloudnet (Illingworth et al. 2007), or ARM-instrumented sites (Mather et al. 1998) have proved useful in evaluating models. Although they lack the global coverage of satellites, they have the advantage of greater spatial and temporal resolution with a more powerful array of remote-sensing instruments. Provided that analysis is restricted to times when winds are high enough to ensure that sufficient amounts of clouds advect past the sensor and a reasonable cross section of the model grid box is sampled, valid comparisons with the model can be made every hour. The Cloudnet study of seven operational models over one year showed that the representation of clouds in a given grid box over the observing site was surprisingly good, but particular biases could be identified. The vertical profile of mean cloud fraction revealed that all models underestimated the occurrence of mid-level cloud. Mean ice water content profiles in the models showed good agreement with observations, and more recent versions of the models captured the observed mean liquid water content well. It is interesting to note that the performance of a mesoscale (12 km resolution) model was not notably better than the same model when run at a global scale with 60 km resolution, although a fairer test would be to carry out the comparisons at the same scale by aggregating the 12 km model data up to the 60 km resolution. It is important to note that the models carry the correct mean values, but this is not the whole story. The PDF of cloud fraction showed that models have fewer completely filled grid boxes than observed and more partially filled grid boxes. Model PDFs of liquid water content were more peaked than observations. The PDFs of ice water content revealed that the one model that had the worst mean value below 7 km had actually the best PDF below 0.1 g m^{-3}; however, because any higher ice water content was considered to be falling snow rather than cloud, the result was a mean value of ice water content that was far too low. Most models have low-level water clouds that drizzle all the time, with the drizzle reaching the ground; observations show the same mean drizzle rate, but a completely different PDF, with occasional bursts of heavier drizzle reaching the ground but usually much lighter drizzle, which evaporates 100 m or so below cloud base. Ground-based studies have also shown (Hogan et al. 2000) that the common assumption of maximum random overlap of clouds is appropriate for clouds shallower than 2 km, but should be modified so that as the clouds become deeper the overlap tends towards maximum. These Cloudnet results, which show that the NWP models have considerable skill in producing clouds at the right time in the right place, suggest that the models are capturing the fundamental meteorological processes that produce the clouds and are correctly locating the regions of ascent. These encouraging results suggest that the assimilation of clouds within NWP models may be feasible, and that such studies could lead to improvements in parameterization schemes. In addition, the success of NWP models give us confidence that climate models should also be representing clouds reasonably well, since they are using essentially the same cloud parameterization schemes.

This example shows how analyzing observations, both in terms of mean values and PDFs, helps to elucidate the processes responsible and should greatly help to improve the physical basis of statistical cloud schemes and to reduce the degree of empiricism in them. Further analysis should be undertaken. This could, for example, investigate if the implicit ice particle sizes in the models are correct, examine if the lack of mid-level clouds in the models is important, and establish the scale and relevance of the errors in the representation of drizzle in low-level clouds. Given the critical uncertainties associated with trade-cumulus clouds (see above), the analysis of long-time series of ground-based data collected in regions covered by such clouds would be very beneficial. Aspects to be investigated would be the values of liquid water content and liquid water path, cloud base, cloud top as a function of the depth of the boundary layer, the formation of any precipitation including small drizzle droplets, and their subsequent fate as they fall below cloud to evaporate or on occasion to reach the ground. A sensitive high-resolution (e.g., 6 m/30 s) ground-based Raman lidar should provide detailed observations of the PDF of water vapor within the boundary layer, which can be combined with the liquid water content within cloud, to provide the observed PDF of total water content.

This PDF, which is the fundamental basis for statistical cloud schemes, can then be compared with that predicted by models. The deployment of the mobile ARM facility in the Azores, from April to December 2009, to observe the springtime overcast stratocumulus regime and the summertime broken trade cumulus, should be particularly fruitful. It may well be that the observations made in recent field projects, such as RICO and BOMEX, are also able to furnish some of the data required to see if climate models are simulating such clouds realistically.

Recent observational campaigns with advanced multiple wavelength and depolarization lidars may provide more information on the ability of aerosol particles to influence cloud properties. In particular, it seems that the size of aerosol particles can be inferred from the lidar backscatter and/or extinction spectrum (Müller et al. 2000, 2001) and the shape from the depolarization ratio, and that large non-spherical particles may act as efficient ice nuclei and promote glaciation. There is some evidence that Saharan dust may be a source of ice nuclei and that when the dust is lofted to high altitudes, such ice nuclei can cross the Atlantic (e.g., DeMott et al. 2003; Ansmann et al. 2008). The degree to which such dust particles promote the glaciation of supercooled clouds and how often this occurs is still questionable, but this is an area of active research which should yield quantitative results. Incorporating such phenomena into climate models will be difficult; the degree to which Saharan dust is lofted is dependent upon the performance of the transport model and the gustiness of the surface winds—a local effect which will be difficult to capture reliably in large-scale models.

Evaluating Cloud Properties Simulated by Large-scale Models with Satellite Data

The evaluation of the clouds in GCMs has long been hampered by the lack of global observations of the vertical structure of clouds. The situation is now radically changing with the arrival of new observations from the A-Train constellation of satellites, including the spaceborne radar (CloudSat) and lidar (CALIOP/CALIPSO) instruments. The observational definition and detection of clouds, however, depends strongly on the type of measurements and sensitivity of sensors, as well as the vertical overlap of cloud layers in the atmosphere. This definition also differs from the definition of a cloud layer in large-scale models or in high-resolution models (e.g., CRMs). Therefore, a raw and direct comparison of cloud products derived from observations with model simulations does not guarantee that apples are not compared with oranges.

To make more meaningful comparisons between models and observations, it is better to use a *simulator* to diagnose from the model outputs some quantities that are directly comparable with observations. Such an approach has been widely used to compare model cloud covers with ISCCP data (e.g., Klein and Jakob 1999; Webb et al. 2001; Zhang et al. 2005). New simulators, which compare the observed radar and lidar backscatter profiles with those profiles calculated from the model parameters, are now under development. First studies using a CloudSat simulator (Bodas-Salcedo et al. 2008) and an ICESAT (Wilkinson et al. 2008) or a CALIPSO (Chepfer et al. 2008) lidar simulator show already how promising the approach is to evaluate the cloudiness simulated by climate models. Biases can now be identified much more clearly and in more detail (in particular, the vertical structure can be documented) than with previous comparisons using passive measurements.

Global comparisons of histograms of CloudSat radar reflectivity (Bodas-Salcedo et al. 2008) as a function of height computed over several months from the Met Office model, which has an implicit exponential ice particle size distribution with an intercept parameter that is a function of temperature and an ice particle density which is inversely proportional to size, appear perhaps initially discouraging. The observed histograms are much smoother than those observed, but they do show that the model is underestimating mid-level clouds. However, when the comparisons are subdivided into geographical regions, they are much more revealing. Over the North Atlantic, the model performance for the ice cloud is quite good, indicating that parameterization of the intercept parameter as a function of temperature performs well. Problems are evident for low cloud: the model has two separate drizzle regimes rather than one. Comparisons over the California stratocumulus region and the tropical Pacific also reveal specific errors in the model. As noted with the ground-based measurements, the lack of mid-level clouds seems to be ubiquitous, and the occurrence of drizzle in low-level clouds seems to be overestimated in the model. Clearly, such an approach has powerful implications, although care is needed to distinguish between the relative influence of cloud and precipitation biases in errors of the simulated radar reflectivities. As with the ground-based observations, comparing the statistics of the mean values of the reflectivity histograms with the observations are just the first step. The next step is to classify the data in terms of different weather regimes; this is accomplished by separating the data according to, for example, large-scale vertical motion and surface stability. Thereafter, those processes which are being poorly represented must be identified to establish if, for example, the lack of mid-level clouds in the models is important for some aspects of the simulated climate.

The computation and interpretation of a radar simulator is reasonably straightforward in that the model holds an implicit size distribution of the ice particles, and for water droplets there is a prescribed droplet size over the ocean and over land, and that in general the attenuation of the radar signal is rather small. Lidar measurements are very sensitive to the presence of cloud particles, and the horizontal and vertical resolutions of the measurements are very high (330 m and 30 m, respectively, for CALIPSO). The analysis of lidar measurements thus constitutes a powerful means of diagnosing the vertical distribution of cloud layers and their overlap. However, the attenuation of lidar signals is much larger than that of radar reflectivities, so that the signal from the satellite can be totally extinguished at low altitudes when thick upper-level clouds are present. The attenuation is related to the observed lidar backscatter through the "lidar ratio," or the ratio of backscatter to extinction, but this lidar ratio is very sensitive to the (unknown) ice particle shape and size. In addition, the penetration of the lidar beam through multiple levels of broken cloud is very sensitive to the degree of cloud overlap; this could be considered as an advantage in that the very sensitivity to the cloud overlap could be regarded as an excellent method of diagnosing if the cloud overlap implied in the model is in fact realistic. In the presence of upper clouds, as in the tropics, the simulated and observed backscatter from lower-level clouds depends on how well the thicker higher-level cirrus clouds are represented. However, a large fraction of tropical oceans are associated with large-scale subsidence in the free troposphere, so boundary-layer clouds are not overlapped by upper-level clouds. In these situations, attenuation problems are minimal and the lidar is able to provide unambiguous returns from cloud top. The lidar simulator is thus particularly useful for studying those (ubiquitous) clouds which are important for the Earth's radiation budget but are often below the sensitivity of radar, such as stratus, stratocumulus, and fair weather cumulus clouds, which we identified earlier as being of crucial importance. Lidar can observe cloud top to 30 m, and thus, for an ensemble of clouds, it should be possible to identify the cloud base of these clouds. Cloud water droplets will generally yield a radar reflectivity too low to be detected from space, so any observed radar reflectivity will indicate the presence of small drizzle droplets or precipitation. If this is combined with inferred values of liquid water path in the cloud and effective radius from passive "MODIS"-type instruments, then the properties of the fair-weather cumulus clouds can be compared in detail with their representation in models. Evaluating the ability of climate models to simulate accurately the geometrical thickness and the precipitation efficiency of shallow-level clouds, together with their variation with natural climate fluctuations, is of paramount importance if we are to have confidence in the simulated response of these cloud properties to climate change and then in the model cloud feedbacks. At high latitudes, the persistent low-level polar clouds should also be well detected by lidar measurements.

Note that the use of simulators is also a way to fill the gap between the different cloud scales since the comparison of the cloud covers predicted at the large scale can be compared to observations derived at a much smaller scale. For example the lidar signals are, in principle, available for each lidar pulse, with a horizontal resolution of 330 m or so for the highly reflecting water clouds and a vertical resolution of 30 m; the radar reflectivity has a horizontal resolution of just 1.1 km and 500 m in the vertical. Unfortunately, when we are considering the representation of tropical broken cumulus clouds, high-resolution observations of the PDF of humidity via Raman lidar do not seem to be possible from space, and, at present, only values of water vapor path integrated over the vertical are available with a horizontal resolution of some 20 km.

Turning to ice clouds, the ice particle size can be derived from the ratio of the radar return (which varies as the sixth power of the particle diameter) to the lidar backscatter signal (which, when corrected for attenuation, depends on the square of the particle diameter). The first stage would be to compare the inferred ice particle size and its variation globally with location and temperature, and then compare it with the particle size, which in most models is prescribed in terms of the temperature alone. Results of this analysis should be available very soon; Delanoë and Hogan (2008) have demonstrated how the errors of the derived products from a combination of active and passive sensors can be obtained using a variational technique. Depending upon the results, one can envisage having a prescribed ice particle size in the models which varies not only with temperature but also with other environmental conditions. The next stage could be to have a double moment scheme to represent the ice particles, as is done in CRMs, provided the particle size in such a scheme could be constrained to agree with the size inferred from the active radar and lidar onboard the satellites.

Supercooled layer clouds can be identified relatively easily from space by their very high lidar backscatter and sharp backscatter gradient at cloud top. Using data from the LITE mission on the space shuttle Hogan et al. (2004) found that around 20% of all clouds between –10°C and –15°C contained supercooled layers. Such thin layer clouds have a much larger radiative impact than ice clouds of the same water content because of their smaller particle size, yet they are scarcely represented in climate models. Quantification of the radiative impact of such clouds on a global scale will soon be possible using CALIPSO data.

In addition, CALIPSO lidar data provides us with high-resolution observations of aerosol backscatter, with the "color ratio" of backscatter at the two wavelengths and depolarization ratio giving us aerosol size and shape information, respectively. The origin of these particles may be desert dust lofted by convection. Clearly, the lidar observations have the potential to quantify the global occurrence of both anthropogenic and natural aerosol but cannot by themselves distinguish the two types. It should be possible to establish just how widespread is the modification by man of the sizes and concentrations of the droplets within liquid water clouds when such clouds are embedded within haze. For ice clouds, lidar returns should reveal the frequency with which dust aerosols (natural or anthropogenic) are being lofted and transported large distances, and whether they are significantly modifying the glaciation rates and ice particle sizes of these high-level clouds. An essential first step is to quantify the magnitude of these effects on a global scale.

Thus far we have discussed the evaluation of NWP models using satellite or ground observations to see if they are producing clouds with the correct average properties and the correct PDF of these properties. This requires several months of data. On a global scale, observations over a few years should be sufficient to establish the characteristics associated with the Madden-Julian Oscillation, interannual variability, and possibly El Niño. The ground-based studies have shown that in regions where there are abundant observations, NWP mesoscale models have skill in producing the right cloud at the right time, and this skill can be evaluated on a monthly basis. In data-sparse regions, this skill is much lower so only an evaluation of the correct statistical properties of the clouds can be achieved. Evaluating the fidelity of clouds in climate models run for many years is more difficult; only the global statistics of mean cloud properties, their PDFs, and the temporal fluctuations of these metrics can be determined. However, experiments in which climate models have been run in a forecast mode indicate that some systematic biases (e.g., in the cloud and humidity fields), noticed in the climate mode, appear in a few days in the model. This suggests that the evaluation of clouds in climate models may be done in part based on short-term experiments and high-frequency observations (e.g., data from field experiments if the model is initialized with large-scale forcings from this experiment). One word of warning is in order: currently, active satellites with active radars and lidars are in sun-synchronous orbits and thus information on the diurnal cycle is limited.

Cloud Feedbacks in a Changing Climate

Cloud and radiative observations are only available for a short period (at best for about 25 years, more generally for just a few years), and no climate variation occurring at this timescale may be considered as an analog of long-term climate change. Until long time series (three decades or more) of cloud and radiation data become available, it is hopeless to assess directly the response of clouds to global climate changes using observations and to compare it with model simulations. This is even more true since establishing long-term trends based on satellite or surface-based measurements is made very difficult by problems such as changes in instrument calibration, or satellite drift in altitude, etc. Once reliable and long time series of observations (of clouds but not only) become available, it might become easier to assess cloud feedback processes directly from observations and then to evaluate cloud feedbacks in climate models. In the meantime, available observational records can be useful in assessing the natural climate variability on various time scales, and also in investigating the physics that controls cloud changes and variability. For this purpose, some approaches have been developed over the last few years that take advantage of the available observations to assess climate model simulations in a way that may be relevant for assessing cloud–climate feedbacks.

Cloud feedbacks are related to the response of clouds to changing climate conditions. To have confidence in the feedbacks produced by climate models, it is therefore not sufficient to evaluate mean cloud properties. Assessing the *sensitivity* of clouds to changing environmental conditions is more likely to be relevant for assessing the realism of the simulated feedbacks.

For this purpose, one approach is to use compositing techniques to assess, in models and in observations, how clouds change in association with dynamic or thermodynamic conditions (e.g., with changes in lower tropospheric stability, in the intensity of large-scale rising or sinking motions in the free troposphere, in humidity and temperature). For this purpose, observations and model simulations are not only compared in terms of geographical distributions but also in terms of covariations between several variables. For instance, recognizing that many cloud properties (in particular, the prominent cloud type) are controlled to a large degree by the large-scale atmospheric circulation, several studies have stratified cloud observations as a function of dynamic regimes (cf. Bretherton and Hartmann, this volume) and then investigated how, for specified dynamic conditions, these cloud properties varied with other environmental conditions, such as surface temperature, static stability, or horizontal advections (Bony et al. 1997; Williams et al. 2003, Bony et al. 2004, Norris and Iacobellis 2005). Other studies have decomposed global cloudiness into a small number of prominent cloud regimes and used this decomposition to understand and assess the response of clouds to long-term climate changes (e.g., Williams and Tselioudis 2007).

Such an approach makes it possible to evaluate simulations from idealized simulations having different geographical distributions of the dynamic features (e.g., aquaplanets) by using observations. Decomposing the large-scale feedback mechanisms or cloud changes in terms of a series of composites also makes it possible to bridge more easily climate studies with process studies. Once a cloud process or a sensitivity is identified as a key component of cloud–climate feedbacks, more detailed investigations using uni-dimensional models, cloud resolving models, or more detailed observations such as those collected during campaigns such as RICO or BOMEX may be performed to explore more deeply the underlying physics.

Cloud–Climate Metrics for Assessing the Relative Reliability of Climate Change–Cloud Feedbacks Produced by Climate Models

With the realization and the open availability of a large coordinated set of climate simulations performed by a large number of climate models (Meehl et al. 2007), the question now arises whether some model results are more reliable than others. Giving more importance (or more weight) to models that seem to perform better in simulating the current climate is sometimes presented as a way to reduce uncertainties in climate projections (Murphy et al. 2004). To address this question, the climate modeling community is currently developing efforts to define a basket of "metrics" to assess the relative merits of the different climate models in reproducing observed features (always remembering that the models may have common errors that need to be identified and corrected, because such errors may offset one another so that some of our present crude criteria for assessing models are satisfied whereas in truth the models are flawed). This effort, in some way, extends to climate models a procedure that has been routinely applied to NWP models for thirty years. However, it raises many questions and concerns.

As provocatively asked during the presentation of the IPCC AR4: "Might the 5th Assessment Report of the IPCC be the end of models democracy?" (IPCC 2007). Indeed, thus far, different climate models have all been treated equally, in terms of their ability to simulate climate change projections. However, we feel that there is a growing desire (and pressure) to rank the different models and to give them different weights depending on their relative ability to reproduce the observed climate.

Certainly, the climate community welcomes the possibility of quantifying, for a large ensemble of climate models and for a wide range of diagnostics, the resemblance between simulations and observations. For example, one may imagine developing metrics focused on the ability of climate models to simulate a realistic diurnal cycle or realistic tropical intraseasonal oscillations. The scrutiny by a very large community of analysts of the simulations performed in support of the IPCC AR4 is already contributing to this very constructively. However, concerns might be expressed regarding the meaning and the future use of these metrics.

As explained above, we still do not know whether some model biases matter more than others for climate change prediction. Common sense suggests that the answer to this question depends on the climate question to be addressed. To assess the reliability of climate model projections in regions dominated by monsoon or ENSO phenomena, one might find useful metrics focused on the simulation of these processes. However, there is some danger in the use of nonspecific metrics based on mean climate features (e.g., mean cloudiness or mean radiative fluxes) to assess the relative reliability of different model estimates of global climate sensitivity (besides, it turns out that climate models producing very different cloud responses to climate change may not be distinguishable in their simulation of mean cloud properties in the current climate). To address this question, we need instead to encourage the development and use of some process-based metrics assessing the ability of climate models to simulate cloud relationships, processes, or composites shown to play a critical role in climate change–cloud feedbacks. Again, analyses and idealized studies of the kind described earlier in this chapter provide some guidance about the processes to be considered in such metrics.

Ways to Reduce Critical Uncertainties in the Prediction of Clouds in a Changing Climate

Comparison of climate simulations with observations reveals a large number of systematic biases in current models. Faced with the long-standing biases of climate models and uncertainties in climate change projections, the optimal way to improve models is still open to question.

Resolution

The increase of the (horizontal and vertical) resolution of large-scale models is often cited as a way to improve model simulations. The experience of many modeling centers indicates that increasing the resolution does reduce some biases, such as the occurrence and strength of midlatitude storms, the simulation of extreme precipitation, or of orographic precipitation. However, it is far from solving all the problems. In particular, the simulation of continental precipitation, of the diurnal cycle, or of the Madden-Julian Oscillations does not improve substantially with resolution. This is the case for many other errors of large-scale models, including the difficulties of representing the cloud processes themselves, such as condensation on aerosol particles, glaciation, the size distribution of the cloud particles and their interaction with radiation, the degree of cloud overlap in the vertical, and the conversion of cloud water into precipitation. We know these are inadequately parameterized and lead to modeled clouds with different characteristics from those indicated by our limited database of cloud observations, but it is unclear if increased resolution will ameliorate the situation. Current models have a higher vertical resolution in the boundary layer and thus can resolve some of the vertical structure of stratocumulus and, to a lesser extent, fair weather cumulus. We know that mid-level clouds are underestimated in nearly all models. Why is this? Is it because we cannot resolve the position of cloud top and base? Is it because the radiation scheme is not called often enough? Is it because the diagnosed phase is incorrect? Is it because there is no turbulent mixing scheme outside the boundary layer? It is also becoming clear that supercooled clouds commonly form and are widespread and persistent and have potentially important radiative effects, but are scarcely represented in the models. Is this also a resolution problem?

Complexity

Another avenue of model development is the increase of complexity. Coupled ocean–atmosphere models are now coupled to complex land-surface schemes, aerosol modules, chemistry, carbon cycle, etc. to form so-called Earth System Models. This allows us to investigate new climate feedbacks (such as carbon–climate feedbacks) but does not reduce the uncertainty in climate change projections. On the contrary, intermodel differences in regional precipitation changes and in climate sensitivity are often amplified by carbon-cycle feedbacks (which are very sensitive to precipitation and climate sensitivity changes) or aerosol feedbacks.

Physical Parameterizations

Improving the physical parameterizations used in large-scale models (in particular, the representation of turbulent, convective cloud processes, and the interaction with radiation) seems to be the most efficient way to reduce uncertainties in model projections of the future climate. However, improving parameterizations is difficult, the number of people actively involved in this work is fairly small at the present, and the progress is slow. National and international funding agencies might play a role in encouraging these activities. Nevertheless, with the arrival of new cloud observations and with the increasing interactions and collaborations between meso- and large-scale modelers, one may expect more progress over the next few years than there has been in the past. From the observations, can we demonstrate that we really need dual moment schemes to represent ice and liquid water, and can we show that such schemes are adequately constrained to lead to improvements? Can observations reliably confirm the existence of large cloud-free regions (see Kärcher and Spichtinger, this volume, and references therein), which are very highly supersaturated with respect to ice, and do we need to adjust our parameterization schemes to take this into account? What level of sophistication is needed in the treatment of aerosols?

Using Cloud-resolving Models instead of Cloud Parameterizations in Climate Models

Now that "super-parameterizations" and "global CRMs" have become available (cf. Grabowski and Petch; Collins and Satoh, both this volume), using CRMs instead of cloud parameterizations might constitute an option to reduce the uncertainty in cloud feedbacks associated with cloud parameterizations. Although these new approaches are promising, they are unlikely, however, to solve the cloud–climate problem issue in the near future for at least for two reasons. First, these approaches are computationally very expensive. Thus, it seems unlikely that ensembles of century-scale simulations can be performed with such models to study changes in the global climate or in climate extremes. Second, the resolution of CRMs is insufficient to resolve boundary-layer turbulence or cloud microphysics and, therefore, parameterizations are still required. The results obtained with these models are likely to depend on these parameterizations and at least on some poorly constrained parameters. This dependence should be explored and quantified before the cloud feedbacks produced by these models can be considered less uncertain than those derived from large-scale models.

On the other hand, sensitivity experiments performed with global CRMs or super-parameterizations can be very instructive to explore the physics of cloud feedbacks and climate sensitivity. It would be very valuable, for example, to understand why an aquaplanet global CRM (Miura et al. 2005) and a GCM embedding a two-dimensional CRM within each grid box instead of a cloud parameterization (Wyant, Khairoutdinov et al. 2006) both predict a climate sensitivity weaker than estimated by most global climate models. It would also be valuable (and computationally cheaper) to perform climate simulations by embedding a CRM or a LES over a limited domain of the Earth instead of globally (e.g., a LES in subtropical regions predominantly covered by boundary-layer clouds). A complementary and constructive (rather than competitive) interaction between large- and mesoscale modeling approaches to study the cloud–climate problem is strongly required.

Conclusion

With the arrival of new and powerful observational datasets, particularly the new space-based active radar and lidar sensors in the "A-Train," we are entering a new era for the evaluation of clouds in large-scale models. Observations with active sensors have already demonstrated that NWP models have skill in representing clouds in the right place and the right time, and have also identified some shortcomings. It will soon be possible to assess key aspects of the simulation of clouds, such as the three-dimensional distribution of cloud layers, the cloud water phase, the cloud precipitation efficiency, and the physical and radiative properties of shallow-level clouds. As these aspects have the potential to affect the response of these crucial shallow tropical clouds to climate change, their evaluation in the current climate under a large variety of environmental conditions will allow us to assess better the realism of their change in the future. Moreover, as high-resolution models are increasingly used to assess and develop physical parameterizations, as well as to investigate cloud-in-climate issues, we recommend that new satellite data be used to evaluate the cloud distributions produced by high-resolution models, including operational NWP models.

We emphasize that the better our physical understanding is of the response of clouds to climate change, the more efficient the strategy for evaluating this response will be. Thus, developing a strategy of evaluation of climate change–cloud feedbacks requires efforts in analyzing and unraveling the physical mechanisms underlying these feedbacks. For this purpose, a promising approach consists of conducting idealized studies using a hierarchy of climate models of different complexities. However, to reduce the uncertainties in cloud–climate feedback processes and improve climate models, it is not sufficient to point out deficiencies in a particular process; physical parameterizations must be improved if we want these deficiencies to be remedied. For this, it is important to keep developing collaborations between the large- and mesoscale cloud communities, as well as between the modeling and observational communities.

The radars and lidars now in space should enable us to observe the global vertical distribution of clouds and aerosols and aid in ascertaining the degree to which aerosols are modifying both warm and cold clouds, as well as the geographic extent of any modification. These measurements should allow us to quantify, for the first time, the effect aerosols are having on the present climate, and hence reduce the large uncertainties in the effects of aerosols on the future climate.

Turning to future satellite-observing systems, we have some concern that the long time series of global monitoring of TOA radiation, which is now being carried out with the CERES sensors, may not continue, although Mega-Tropique may fill the gap but only at low latitudes. We look forward to the launch of the ESA/JAXA EarthCARE mission (in 2013), which will embark a cloud radar and lidar on the same platform. The high spectral resolution lidar should provide direct observations of the optical depths of thin cirrus and aerosols and characterize the aerosol and ice cloud particles. The radar will have improved sensitivity and so should detect more of the high thin ice high clouds as well as the lower-level water clouds, while the Doppler capability should help to characterize the vertical cloud motions and thus contribute to the evaluation of convective parameterization schemes, provide information on ice sedimentation velocities within extensive cirrus decks to inform the model ice schemes, and quantify the drizzling rates in low-level water clouds.

Many small-scale cloud processes remain which must be parameterized in the models; a better understanding of them is needed but cannot be provided from space. Examples include the entrainment and detrainment for both layer and convective clouds; the growth of ice particles from the vapor, their aggregation, riming, and subsequent evaporation; the warm rain coalescence process and the mechanisms that lead to the production and persistence of supercooled layer clouds. Progress can best be provided through detailed observation, whether *in situ* or remotely from the ground, of the evolving physical and dynamic variables. One particularly glaring gap remains: We still have no technique to observe the humidity structure of the atmosphere in three dimensions with a high enough resolution to characterize its PDF within the model grid box even in clear air. To achieve this, when clouds are present, is an even greater challenge.

Acknowledgments

Thoughtful comments from Tony Slingo, Jean-Louis Brenguier, Johannes Quaas, Jon Petch, Graham Feingold, Jost Heintzenberg, Patrick Chuang and an anonymous reviewer are gratefully acknowledged. They were very helpful to clarify some of the issues discussed in this paper.

References

Ansmann, A., M. Tesche, D. Althausen et al. 2008. Ice formation in Saharan dust over northern Africa observed during SAMUM. *J. Geophys. Res.* **113**:D04210.

Bodas-Salcedo, A., M. J. Webb, M. E. Brooks et al. 2008. Evaluation of cloud systems in the Met Office global forecast model using CloudSat data. *Geophy. Res. Abst.* **10**:EGU2008-A-10141.

Bony, S., and J.-L. Dufresne. 2005. Marine boundary layer clouds at the heart of tropical cloud feedback uncertainties in climate models. *Geophys. Res. Lett.* **32**:L20806.

Bony, S., J.-L. Dufresne, H. Le Treut, J.-J. Morcrette, and C. Senior. 2004. On dynamic and thermodynamic components of cloud changes. *Climate Dyn.* **22**:71–86.

Bony, S., and K. A. Emanuel. 2001. A parameterization of the cloudiness associated with cumulus convection: Evaluation using TOGA COARE data. *J. Atmos. Sci.* **58**:3158–3183.

Bony, S., K. M. Lau, and Y. C. Sud. 1997. Sea surface temperature and large-scale circulation influences on tropical greenhouse effect and cloud radiative forcing. *J. Climate* **10**:2055–2077.

Chepfer, H., S. Bony, M. Chiriaco et al. 2008. Use of CALIPSO lidar observations to evaluate the cloudiness simulated by a climate model. *Geophys. Res. Lett.* **35**:L15804.

Delanoë, J., and R. J. Hogan. 2008. A variational scheme for retrieving ice cloud properties from combined radar, lidar and infrared radiometer. *J. Geophys. Res.* **113**:D07204.

DeMott, P. J., K. Sassen, M. R. Poellot et al. 2003. African dust aerosols as atmospheric ice nuclei. *J. Geophys. Res.* **30(14)**:1732.

Hogan, R. J., M. D. Behera, E. J. O'Connor, and A. J. Illingworth. 2004. Estimate of the global distribution of stratiform supercooled liquid water clouds using the LITE lidar. *Geophys. Res. Lett.* **31**:L05106.

Hogan, R. J., and A. J. Illingworth. 2000. Cloud overlap statistics from long-term radar observations. *Q. J. Roy. Meteor. Soc.* **126**:2903–2909.

Illingworth, A. J., R. J. Hogan, E. J. O'Connor et al. 2007. Cloudnet, continuous evaluation of cloud profiles in seven operational models using ground-based observations. *Bull. Amer. Meteor. Soc.* **88(6)**:883–898.

IPCC. 2007. Climate Change. 2007: The Physical Science Basis. Contribution of Working Group I to the Fourth Assessment Report of the Intergovernmental Panel on Climate Change, ed. S. Solomon, D. Qin, M. Manning et al. New York: Cambridge Univ. Press.

John, V. O., and B. J. Soden. 2007. Temperature and humidity biases in global climate models and their impact on climate feedbacks. *Geophys. Res. Lett.* **34**:L18704.

Klein, S. A., and C. Jakob. 1999. Validation and sensitivities of frontal clouds simulated by the ECMWF model. *Mon. Wea. Rev.* **127**:2514–2531.

Liu, C., M. W. Moncrieff, and W. W. Grabowski. 2001. Explicit and parameterized realizations of convective cloud systems in TOGA COARE. *Mon. Wea. Rev.* **129**:1689–1703.

Mather, J. H., T. P. Ackerman, W. E. Clemens et al. 1998. An atmospheric radiation and cloud station in the tropical western Pacific. *Bull. Amer. Meteor. Soc.* **79**:627–642.

Medeiros, B., B. Stevens, I. M. Held et al. 2008. Aquaplanets, climate sensitivity, and low clouds. *J. Climate*, in press.

Meehl, G. A., C. Covey, T. Delworth et al. 2007. The WCRP CMIP3 multimodel dataset: A new era in climate change research. *Bull. Amer. Meteor. Soc.* **88**:1383–1394.

Miura, H., H. Tomita, T. Nasuno et al. 2005. A climate sensitivity test using a global cloud resolving model under an aqua planet condition. *Geophys. Res. Lett.* **32**:L19717.

Müller, D., F. Wagner, U. Wandinger et al. 2000. Microphysical particle parameters from extinction and backscatter lidar data by inversion with regularization. *Exp. Appl. Opt.* **39(12)**:1879–1892.

Müller, D., U. Wandinger, D. Althausen, and M. Fiebig. 2001. Comprehensive particle characterization from 3-wavelength Raman-lidar observations: Case study. *Appl. Opt.* **40(27)**:4863–4869.

Murphy, J. M., D. M. Sexton, D. N. Barnett et al. 2004. Quantification of modeling uncertainties in a large ensemble of climate change simulations. *Nature* **429**:768–772.

Norris, J. R., and S. F. Iacobellis. 2005. North Pacific cloud feedbacks inferred from synoptic-scale dynamic and thermodynamic relationships. *J. Climate* **18:**4862–4878.

Ramanathan, V., C. Chung, D. Kim et al. 2005. Atmospheric brown clouds: Impacts on South Asian climate and hydrological cycle. *PNAS* **102**:5326–5333.

Soden, B. J., and I. M. Held. 2006. An assessment of climate feedbacks in coupled ocean–atmosphere models. *J. Climate* **19**:3354–3360.

Tompkins, A. M. 2002. A prognostic parameterization for the subgrid-scale variability of water vapor and clouds in large-scale models and its use to diagnose cloud cover. *J. Atmos. Sci.* **59**:1917–1942.

Tsushima Y., S. Emori, T. Ogura et al. 2006. Importance of the mixed-phase cloud distribution in the control climate for assessing the response of clouds to carbon dioxide increase: a multi-model study. *Climate Dyn.* **27(2–3)**:113–126.

Webb, M., C. Senior, S. Bony, and J.-J. Morcrette. 2001. Combining ERBE and ISCCP data to assess clouds in three climate models. *Climate Dyn.* **17**:905–922.

Webb, M., C. A. Senior, D. M. H. Sexton et al. 2006. On the contribution of local feedback mechanisms to the range of climate sensitivity in two GCM ensembles. *Climate Dyn.* **27 (1)**:17–38.

Wilkinson, J. M., R. J. Hogan, and A. J. Illingworth. 2008. Use of a lidar forward model for global comparisons of cloud fraction between the ICESat lidar and the ECMWF model. *Mon. Wea. Rev.*, in press.

Williams, K. D., M. A. Ringer, and C. A. Senior. 2003. Evaluating the cloud response to climate change and current climate variability. *Climate Dyn.* **20**:705–721.

Williams, K. D., and G. Tselioudis. 2007. GCM intercomparison of global cloud regimes: Present-day evaluation and climate change response. *Climate Dyn.* **29**:231–250.

Wyant, M. C., C. S. Bretherton, J. T. Bacmeister et al. 2006. A comparison of tropical cloud properties and responses in GCMs using mid-tropospheric vertical velocity. *Climate Dyn.* **27**:261–279.

Wyant, M. C., M. Khairoutdinov, and C. S. Bretherton. 2006. Climate sensitivity and cloud response of a GCM with a superparameterization. *Geophys. Res. Lett.* **33**:L06714.

Zhang, M. H., W. Y. Lin, S. A. Klein et al. 2005. Comparing clouds and their seasonal variations in 10 atmospheric general circulation models with satellite measurements. *J. Geophys. Res.* **110**:D15S02.

23

Aerosols and Clouds in Chemical Transport Models and Climate Models

Ulrike Lohmann[1] and Stephen E. Schwartz[2]

[1]ETH Zurich, Institute for Atmospheric and Climate Science, Zurich, Switzerland
[2]Atmospheric Sciences Division, Brookhaven National Laboratory, Upton NY, U.S.A.

Abstract

Clouds exert major influences on both short- and longwave radiation as well as on the hydrological cycle. Accurate representation of clouds in climate models poses a major problem because of the high sensitivity of radiative transfer and water cycle to cloud properties and processes, an incomplete understanding of these processes, and the wide range of scales over which these processes occur. Small changes in the amount, altitude, physical thickness, and/or microphysical properties of clouds that occur as a result of human influence can exert changes in Earth's radiation budget comparable to the radiative forcing by anthropogenic greenhouse gases, thus either partly offsetting or enhancing the warming due to these gases. Because clouds form on aerosol particles, changes in the amount and/or composition of aerosols affect clouds in various ways. The forcing of the radiation balance due to aerosol–cloud interactions (indirect aerosol effect) has large uncertainties because a variety of important processes are not well understood, precluding their accurate representation in models.

Introduction

Clouds are an extremely important element of Earth's climate system. They are highly reflective in the solar spectrum, yet strongly absorbing in the thermal infrared; consequently, they produce a large impact on Earth's radiation budget. This impact, termed cloud radiative forcing (CRF), has been quantified through satellite observations: globally, on average, clouds decrease the absorption of solar radiation by about 50 W m^{-2} (shortwave CRF) and decrease the upwelling thermal infrared radiation by 30 W m^{-2} (longwave CRF), thus exerting a net CRF of about –20 W m^{-2} (Kiehl and Trenberth 1997). Locally and instantaneously, clouds can reduce absorbed shortwave radiation by as much as 700 W m^{-2}. In addition, clouds play a central role in Earth's hydrological cycle, which is coupled to the energy budget through the release of latent heat that results from water condensation or evaporation. This, in turn, influences atmospheric circulation on a variety of scales.

The nature and extent of these cloud processes may be expected to change in the future in response to changes in concentrations and properties of trace gases and aerosols and resulting changes in climate. Thus, it is imperative for clouds, as well as their radiative and hydrological properties, to be represented accurately in climate models. However, for a variety of reasons, accurate representation of clouds and cloud influences on radiation and hydrology in climate models remains particularly challenging. Key among these reasons are:

- the small fraction of the total water in the cloud that is present in condensed (solid or liquid) phase; this necessitates an accurate representation of both the total water content and temperature governing saturation vapor concentration;
- the complexities associated with the presence of several forms of condensed phase water (liquid, supercooled liquid, ice, mixed);
- the spatial and temporal diversity of cloud microphysical structure, as reflected in the number concentration and size distribution of cloud hydrometeors and the crystal habit of ice clouds; and
- the numerous varieties and morphologies of clouds as well as the resultant complexity of their three-dimensional structure on many scales (see Figure 23.1).

Small changes to macrophysical (coverage, structure, altitude) or microphysical properties (droplet size, phase) can exert substantial effects on climate. For example, a 5% increase of the shortwave cloud forcing, which could result from changes in the nature or amount of the atmospheric aerosol, would be enough to compensate for the increase in greenhouse gases between 1750–2000 (Ramaswamy et al. 2001). Recognition of this has stimulated the development of improved physically based representations of cloud processes, in general, and of aerosol influences on clouds, in particular, for inclusion in climate models. However, despite intensified research, the feedbacks on clouds and cloud processes that result from forcings by increasing greenhouse gases and aerosols remain among the greatest uncertainties in climate modeling projections of future and climate change (Randall et al. 2007). Similarly, understanding the radiative forcing by aerosols through their influences on clouds remains the greatest uncertainty in radiative forcing of climate change over the industrial period (IPCC 2007).

The principal tools for examining prospective consequences of future emissions of greenhouse gases and aerosols on Earth's climate are general circulation models (GCMs). The acronym GCM is also used to denote global climate model, and the terms are often used interchangeably. Global climate models are

Figure 23.1 Complexity of three-dimensional structure of clouds; note penetration of cumulonimbus clouds through thin cirrus layer (courtesy of Y.-N. Lee, Brookhaven National Laboratory).

not only the primary tool for simulating global climate change; they are also used to evaluate the regional effects of anthropogenic emissions on modifying precipitation amounts and distribution. By integrating atmospheric, radiative, oceanic, and land-surface processes on a global scale, global climate models can provide an indication of expected changes in the coupled system, including possible consequences of coupled increases in greenhouse gases and aerosols on atmospheric radiation, clouds, precipitation, and the climate system in general. Here we examine the current state of understanding of aerosol and cloud processes that must be represented in GCMs and the state of such representation. In addition, we identify recent advances and further developments that are needed.

When used to examine aerosol influences on clouds and precipitation, GCMs must accurately represent the macrophysical properties of clouds and precipitation, including their geographical and seasonal variation. Although GCMs have been used to examine the influence of widespread anthropogenic sources of cloud condensation nuclei (CCN) on global climate (i.e., aerosol particles that serve as the nuclei on which cloud droplets form), this has presented numerous problems. First is the issue of scales. Typically, GCM grid cells have a horizontal dimension of 150–250 km and a vertical dimension of hundreds to thousands of meters, over which there can be substantial spatial inhomogeneity. For example, clouds cover often only a small fraction of the volume of a grid cell, necessitating rather ad hoc parameterizations, and the average vertical velocities in a grid cell are very small (~0.01 m s^{-1}), whereas actual vertical velocities, which control cloud formation and the activation of aerosol particles to cloud droplets, might be 1 m s^{-1} or greater. The poor representation of convection is likely a major source of error in modeled liquid and solid water in clouds.

It is clear that there are major disparities among GCMs (see Figure 23.2), even in zonal averages of cloud albedo, which is a major determinant of Earth's

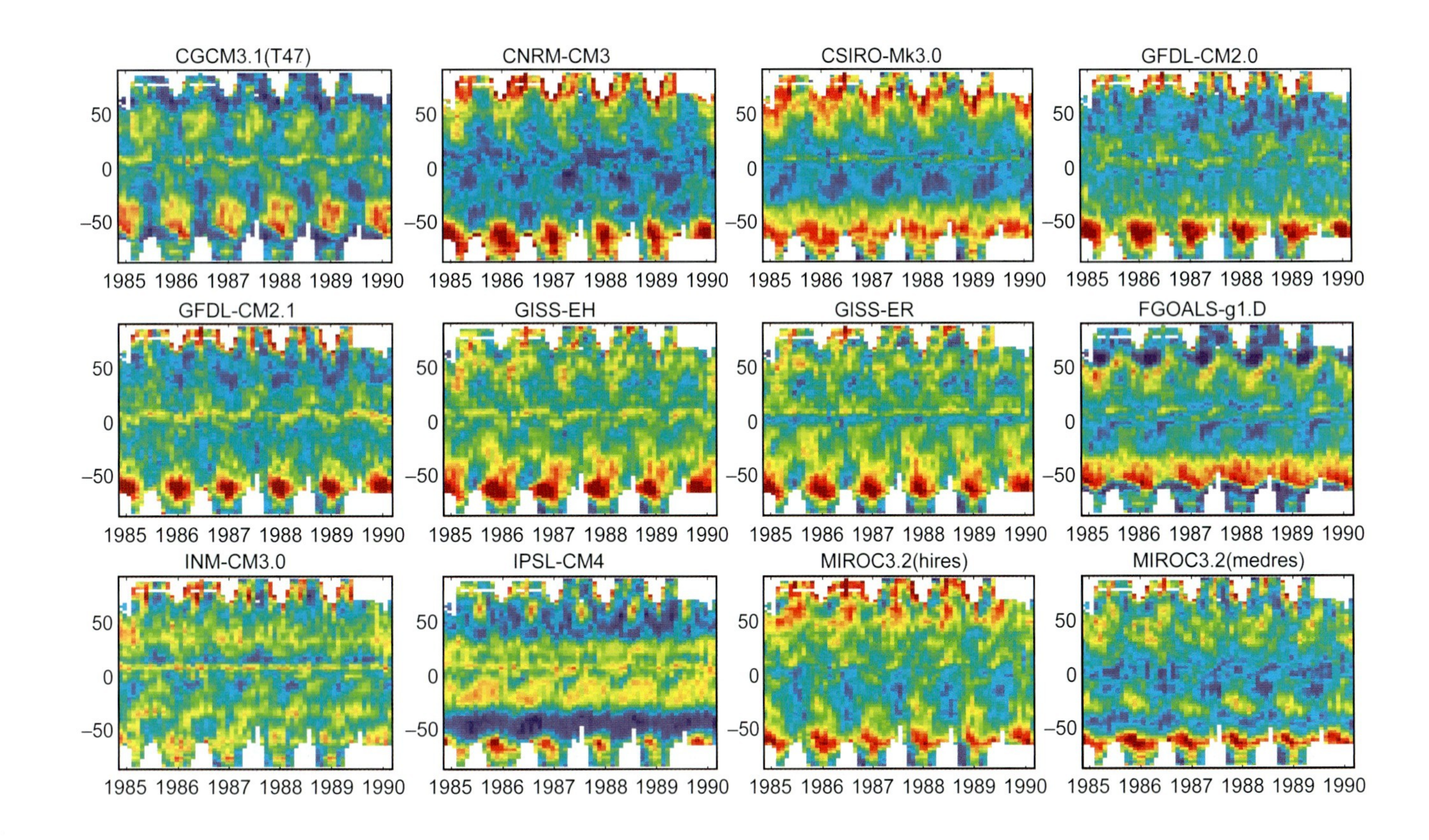
CGCM3.1(T47)
CNRM-CM3
CSIRO-Mk3.0
GFDL-CM2.0
GFDL-CM2.1
GISS-EH
GISS-ER
FGOALS-g1.D
INM-CM3.0
IPSL-CM4
MIROC3.2(hires)
MIROC3.2(medres)
50
0
−50
1985 1986 1987 1988 1989 1990

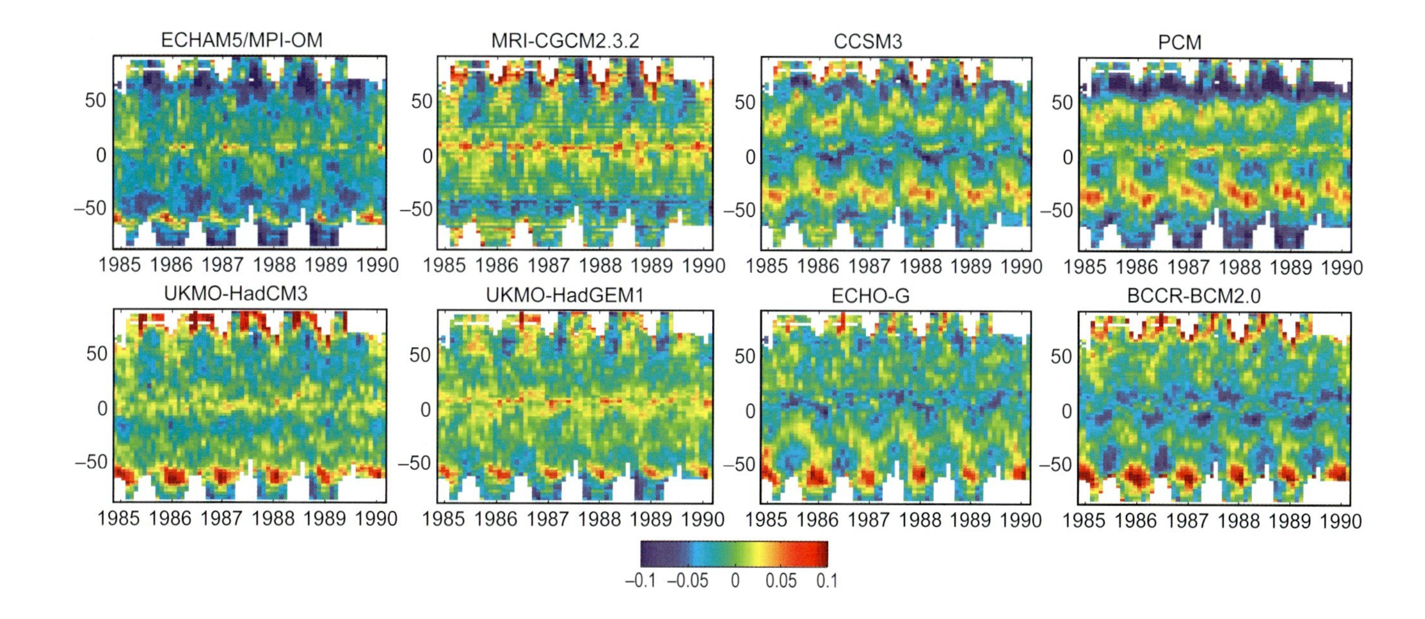

Figure 23.2 Difference between cloud albedo as determined by satellite measurements (ERBE) and twenty global climate models as a function of latitude and time (November 1984 to February 1990). Positive anomalies, where ERBE is higher, are indicated with red, and negative anomalies, where ERBE is lower, with blue colors. Courtesy of Frida Bender; modified from Bender at al. (2006).

radiation budget. In Figure 23.2, each panel corresponds to a different climate model. The model output was obtained from coordinated simulations with twenty different coupled ocean–atmosphere climate models, performed in support of the IPCC Fourth Assessment Report. Clearly these models cannot all be correct. Although space-based measurements can identify models that are doing better or worse, relative to this important cloud variable, such measurements are also difficult, although uncertainties in observations are smaller than intermodel differences.

Cloud microphysical properties are determined by processes such as droplet and crystal nucleation, condensation, evaporation, gravitational settling, and precipitation, all of which operate at the scale of the individual cloud particles or local populations. In contrast, the spatial distribution of clouds is determined by dynamic processes (e.g., turbulence, updrafts, downdrafts, and frontal circulations) and radiative cooling, which operate across meter to global scales. However, these scales are coupled by a variety of processes (e.g., microphysical influences on precipitation development), which in turn affect the release of latent heat below cloud that affects atmospheric stability and vertical motions. Treatment of these processes in climate models and the confidence in this treatment are limited by a lack of understanding and computational resources to represent these on all relevant scales; the latter necessitates development and application of parameterizations, which are inherently scale-dependent. The requirement of accurately representing the many roles of clouds in the climate system and more generally in the biogeochemistry of the planet applies not only to the present atmosphere but also to prior atmospheres (necessary for evaluation of performance of climate models over the instrumental record of the past 150 years or so) and to future atmospheres (necessary for evaluation of the influences of different projected emissions scenarios of greenhouse gases and aerosols). A concern is that each role's common dependence on many of the same cloud properties and processes suggests that errors in simulating one role would produce errors in other roles. Conversely, improving cloud treatment to reduce uncertainty in one role may also reduce uncertainty in other roles. Hence, improving representations and parameterizations of cloud processes in climate models will produce benefits well beyond the simulation of cloud feedbacks and aerosol indirect effects.

Representation of Aerosols in Global-scale Chemical Transport Models and Global Climate Models

Although there are many similarities between treatment of aerosol processes in chemical transport models (CTMs) and global climate models, it is useful to distinguish the two modeling approaches. Global climate models simulate their own meteorology and couple aerosol cycles with clouds, precipitation, and radiation transfer, thereby allowing the projection of future climate under different emissions scenarios. Because climate modeling emphasizes long-term simulation of climate, treatment of aerosol processes in climate models must be greatly simplified. In contrast, with CTMs it is possible to treat aerosol processes and interactions between aerosols (and hydrometeors) and atmospheric chemistry in greater detail. CTMs are often driven by observed meteorology; in such models, the aerosol chemistry and physics do not feed back on the meteorology. CTMs and global climate models need to be driven by observed meteorology to capture detailed aerosol processes and to compare simulated aerosol fields with observations.

Since the pioneering study by Langner and Rodhe (1991), who used a coarse horizontal resolution CTM based on climatological meteorology to represent the global distribution of the mass concentration of sulfate aerosol (without explicit representation of aerosol microphysics), substantial advances have been made in the complexity of treatment of many key processes: aerosol precursor chemistry, aerosol microphysical processes, transport processes, and particle dry and wet deposition. Attempts have recently been undertaken to calculate the aerosol mass concentration as well as the particle number concentration by parameterizing aerosol formation and dynamic processes (e.g., Easter et al. 2004; Stier et al. 2005). An overview of the processes which must be understood and represented in models is given in Figure 23.3.

Most of the earlier studies concerned with the effect of aerosol particles on the climate system took only sulfate particles into account or considered sulfate to be a surrogate for all anthropogenic aerosols. Lately, most major global climate models include also carbonaceous particles, dust, and sea salt (for a synopsis of the state of model development, see Kinne et al. 2006 and the AeroCom model intercomparison project[1]). AeroCom has enabled a comparison of the results of aerosol simulations from more than a dozen modeling groups worldwide. Figure 23.4 provides an example of a comparison for global and annual mean aerosol optical depth and the vertical integral of aerosol extinction coefficient. Although fairly good agreement is demonstrated for most models, it is clear that there are substantial differences in the contributions of the several aerosol species.

A major source of uncertainty in present aerosol modeling is the lack of accurate time-resolved emission inventories. In particular, biogenic sources and emissions from biomass burning are highly uncertain. Both biogenic and biomass burning emissions depend on environmental conditions (e.g., weather) and exhibit high interannual variability, which has not been taken into account by climate studies. Probably the largest uncertainty is associated with organic aerosols, because current measurement techniques cannot identify the many organic species present in primary emissions (Kanakidou et al. 2004). A second issue is that the chemical pathways in the atmosphere are complex and not fully understood. Organic particles result both from primary emission and from

[1] http://nansen.ipsl.jussieu.fr/AEROCOM/

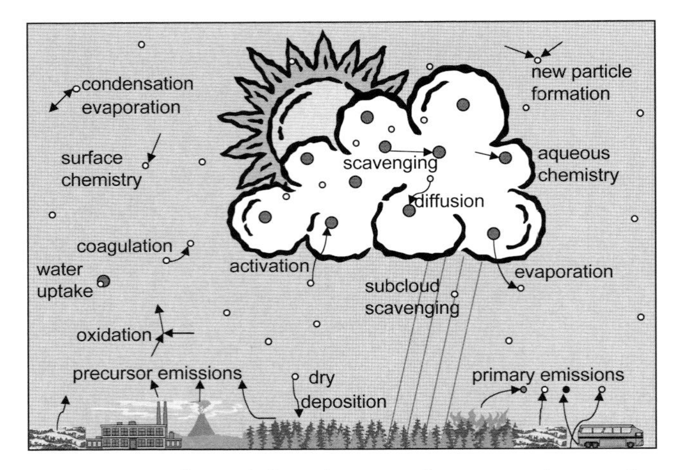

Figure 23.3 Important climate-influencing aerosol processes that must be accurately represented in climate models. Aerosol particles are directly emitted as primary particles and are formed secondarily by oxidation of emitted gaseous precursors. The formation of low-volatility materials results in new particle formation and condensation onto existing particles. Aqueous-phase oxidation of gas-phase precursors within cloud droplets accretes additional mass onto existing particles but does not form new particles. Particles age by surface chemistry, coagulation and condensation. The uptake of water with increasing relative humidity increases particle size, which affects the particle optical properties as well. As clouds form, some fraction of aerosol particles are "activated" to produce cloud droplets. Within clouds, interstitial particles can become attached to cloud droplets by diffusion, and activated particles are combined when cloud droplets collide and coalesce. If cloud droplets evaporate, the particles remain in the atmosphere; if the cloud precipitates, the particles are carried below the cloud to the surface, unless the precipitating particles evaporate completely. Aerosol particles below precipitating clouds can also be removed from the atmosphere through impaction by precipitating drops or through dry deposition to the surface (from Ghan and Schwartz 2007).

gas-to-particle conversion in the atmosphere (secondary production). The total source of these organic particles is therefore a major wildcard in simulations of future scenarios. Advances in measurement techniques for particles are thus of critical importance; one such recent advance is the aerosol mass spectrometer, which permits the development of a systematic measurement database to be developed of general aerosol composition and the identification of primary and secondary organic species (Zhang et al. 2007). Simulating nitrate particles remains problematic because of their semi-volatile nature. In addition to all of the difficulties that exist in developing an understanding of the chemical and microphysical processes, the simulation of aerosol processes in large-scale models is very CPU-time consuming.

There is increasing evidence that individual aerosol particles consist predominantly of a conglomerate of multiple internally mixed chemical substances. In contrast, most global climate models treat aerosols as external mixtures, because internal mixtures have more degrees of freedom, are more complex,

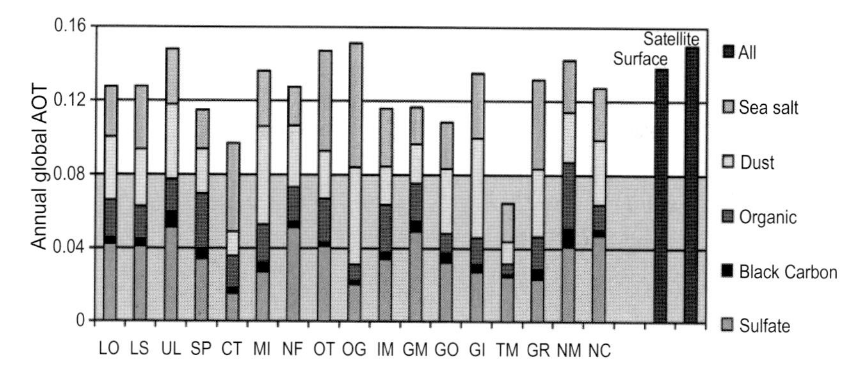

Figure 23.4 Simulated contributions of five aerosol components (seasalt, dust, organic, black carbon, and sulfate) to annual and global mean aerosol optical thickness (AOT), at 550 nm by 17 chemical transport models. For comparison, surface measurements taken by the AERONET network and a composite of satellite measurements are shown. Modified from Kinne et al. (2006).

and impose an additional computational burden. However, the mixing state of aerosol particles (externally vs. internally mixed) influences strongly their optical properties and ability to act as CCN. For example, a slight coating of a particle by only a moderately soluble organic species can drastically increase its ability to act as a CCN. Therefore, treating the degree of mixing properly is essential to represent accurately aerosol processing in global climate models, including aerosol–cloud interactions. Advanced aerosol modules in some global climate models have been expanded to include aerosol mixtures (see Lohmann and Feichter 2005 for references).

Representing the particle size distributions of aerosols and their evolution is also essential. Two kinds of aerosol dynamics models have been developed: modal schemes and bin schemes. Modal schemes represent each aerosol mode with a log-normal distribution of aerosol mass and possibly number. Bin schemes divide the aerosol spectrum into a large number of bins. Typically, modal representations of the aerosol size distribution evolve aerosol number concentration as well as mass concentration.

Representation of Cloud Microphysical Processes in Global Climate Models

Microphysics of Large-scale Clouds

Major improvements have recently been made in the description of cloud microphysics for large-scale models. Although early studies diagnosed cloud amount based on relative humidity, most current global climate models explicitly calculate cloud condensate in large-scale clouds. The degree of sophistication varies from calculating the sum of cloud water and ice to calculating cloud water, cloud ice, snow, and rain as separate species (Lohmann and Feichter 2005). Because the aerosol indirect effect is based on the change in cloud droplet number concentration, some global climate models calculate explicitly cloud droplet number concentrations in addition to the cloud water mass mixing ratio using one of the above described physically based aerosol activation schemes as a source term for cloud droplets. Similarly, the number of ice crystals needs to be calculated in addition to the ice water mass mixing ratio to estimate the effect of aerosols on mixed-phase and ice clouds. Determining the size-dependent sedimentation rate of hydrometeors requires at least a two-moment scheme. Representing size-dependent sedimentation leads to important differences in the cloud vertical structure, cloud lifetime, and cloud optical properties. Two-moment schemes are superior to one-moment schemes provided that the second moment can be treated adequately. Major uncertainties remain, however, in terms of cloud droplet activation and precipitation formation, as discussed below. Theoretically, the best approach would be to use a size-resolved treatment of the cloud microphysics. However, using this approach in a global climate model would be questionable, because treatment of cloud dynamics, including entrainment and advection, is not accurate enough to warrant this level of detail.

Microphysics of Convective Clouds

There is currently a substantial discrepancy between the degree of sophistication in cloud microphysics in large-scale clouds and the very rudimentary treatment of cloud microphysics in convective clouds. This may reflect the fact that stratiform clouds are generally much more susceptible to indirect effects than convective clouds. Recently, however, evidence has emerged to show that biomass burning may affect convective clouds, thus necessitating improvements in the treatment of microphysical processes in convective clouds. In the first global study, Nober et al. (2003) accounted for this effect by decreasing the precipitation efficiency for warm cloud formation in convective clouds, and making it dependent on the cloud droplet number concentration. This approach was taken a step further by Lohmann (2007), who introduced the same microphysical processes (e.g., nucleation, autoconversion, freezing, aggregation) considered in large-scale clouds into convective clouds.

Another option, though considerably more computationally intensive, is to use so-called "super-parameterizations," in which cloud-resolving models (CRMs) are embedded within the normal GCM grid cells, but at only a small fraction of the area of the parent GCM grid cell (e.g., Randall et al. 2003). These models have the capability of calculating cloud-scale vertical velocities and liquid water content (LWC) and thus represent explicitly precipitation processes. They have yet, however, to be applied to aerosol effects on precipitation. If the representation of aerosol and clouds can be improved in such models, or in others through new and innovative techniques for representing subgrid processes, this should increase the accuracy of calculations of the influence of aerosols on the amount and distribution of clouds and precipitation as well as on radiation.

Cloud Droplet Formation

Linking aerosol particles to cloud droplets is a weak point in estimates of the indirect aerosol effects. Accurate treatment of cloud droplet formation requires knowledge of the particle number concentration and size-distributed chemical composition of the aerosol and the vertical velocity on the cloud scale. Parameterizations based on the Köhler theory (Köhler 1923) have been developed to describe cloud droplet formation for a multimodal aerosol. This approach has been extended to include kinetic effects that consider mass accommodation of water at the gas–liquid interface and account for the fact that the largest particles may not have time to grow to their equilibrium size and activate. Competition between natural and anthropogenic aerosol particles, such as between sulfate and sea salt, is also considered (Forster et al. 2007).

Organic carbon is an important constituent of CCN, especially if it is surface active. Facchini et al. (1999) indicate that by lowering the surface tension of surface-active organic particles (e.g., obtained from fog water samples) the cloud droplet number concentration and cloud albedo can be enhanced, leading to a negative forcing as large as ~ -1 W m^{-2}. In contrast, amphiphilic film-forming compounds may retard cloud droplet formation (Feingold and Chuang 2002). Delayed activation enables the growth of larger drops, which have formed earlier, and results in increased dispersion and enhanced drizzle formation. Chemical effects on cloud droplet formation, and thus on the indirect effect, may be as large as the effects of unresolved cloud dynamics (Lohmann and Feichter 2005). Whereas the effect of surface-active organics has recently been included in parameterizations of cloud droplet formation (Abdul-Razzak and Ghan 2004), other effects of organics, such as their film-forming ability have not yet been treated.

Application of parameterizations of cloud drop activation requires estimating cloud-scale vertical velocities in models which do not resolve these cloud scales. Recognizing that this information may not be available, some modelers assume an empirical relationship between modeled sulfate mass concentrations and droplet concentrations (e.g., Boucher and Lohmann 1995), which is equivalent to assuming there is only one single value of cloud updraft velocity for all clouds in the model. Others estimate vertical velocity based on turbulent kinetic energy calculated in boundary layer models (e.g., Lohmann et al. 1999). The latter represents a step in the right direction, but it does not account for the fact that cloudy updrafts are at the tail of the probability density function (PDF) of vertical velocity. Ghan et al. (1997), among others, assumed a normal distribution of vertical velocity with a mean value given by the

GCM grid point mean. They determined the velocity-weighted mean droplet concentration, taking into account the tails of their assumed PDF of vertical velocity. However, observed PDFs of vertical velocity in clouds in the boundary layer are multimodal and are better represented by double-Gaussian PDFs (Larson et al. 2001) with a mean that is a function of the root mean square vertical velocity rather than by a GCM grid point mean (Peng et al. 2005).

Precipitation Formation in Warm Clouds

The influences of precipitation and drizzle processes on cloud lifetime, cloud water content, and cloud radiative properties discussed above cannot be simulated well in current GCM cloud parameterization schemes. For example, the autoconversion rate, which is the rate at which cloud droplets collide and coalesce with each other to form precipitation size drops, is a nonlinear function of the total water condensate. Thus, the mean LWC from a GCM model grid box is essentially meaningless for the representation of precipitation production (e.g., Pincus and Klein 2000). Since the autoconversion bias attributable to horizontal heterogeneity has been found to scale strongly with cloud fractional coverage (Wood et al. 2002), it may be overcome using a parameterization that takes this bias into account. Alternatively, a PDF approach to subgrid modeling may be better in resolving these deficiencies. PDFs of subgrid quantities, such as vertical velocity and liquid water path, are determined from prescribed basis functions in which various moments of the basis functions are calculated in the models (e.g., Pincus and Klein 2000).

Autoconversion of cloud droplets to rain drops is a key process governing the amount and lifetime of clouds in the atmosphere, and must be represented accurately in models from the cloud-resolving to global scale. Even though the mass transfer rate of cloud drops to rain is dominated by accretion in most clouds (Wood 2005), autoconversion is the dominant process in most GCMs because rain is assumed to reach the surface within one model time step. Thus, developing parameterizations for autoconversion suitable for incorporation in large-scale models is an active area of research. Traditional parameterizations are either empirically or intuitively obtained (e.g., Kessler 1969 and Sundqvist 1978) or are derived by curve-fitting detailed microphysical models with simple functions, such as a power law (e.g., Berry 1968; Beheng 1994). These parameterizations lack, however, clear physical bases and have arbitrarily tunable parameters. Furthermore, parameterizations that calculate at least the cloud droplet number concentration in addition to the LWC would be expected to provide much better representation of cloud radiative influences and aerosol effects than existing one-moment schemes.

One promising scheme, which has been derived from theoretical considerations (see Liu et al. 2007 and earlier papers referenced therein), represents the autoconversion rate as the product of a rate function based on the collection efficiency of falling rain drops that describes the conversion rate after the onset of the autoconversion process times a threshold function. The threshold function, unlike that of earlier parameterizations such as the widely used Kessler (1969) parameterization, does not increase abruptly at a critical value of mean droplet mass but instead increases gradually over a range of mean droplet masses that is dependent on the relative dispersion of the cloud droplet size distribution (ratio of standard deviation to mean radius). This dependence captures initiation of the autoconversion by large drops at the high end of the size distribution; its variation for differing values of relative dispersion encompasses prior empirical representations of threshold behavior. This approach yields a strong dependence of autoconversion rate on relative dispersion; for example, for liquid water volume content 0.3 g m^{-3} and cloud droplet number concentration 50 cm^{-3}, as the relative dispersion increases from 0.33 to 1 the characteristic time of autoconversion decreases from 10 hours to 0.1 hour. This parameterization has found application in modeling on regional (Gustafson et al. 2007) and global scales (Rotstayn and Liu 2005), modeling scavenging of soluble gases by precipitation (Garrett et al. 2006), and remote sensing of precipitation (Berg et al. 2006).

Aerosol Indirect Effects

Aerosol particles affect radiative fluxes by scattering solar radiation and absorbing solar and thermal radiation (direct effect). In addition, they interact with clouds and the hydrological cycle by acting as CCN and ice nuclei. For a given cloud LWC, a greater concentration of CCN increases cloud albedo (indirect cloud albedo effect) and is supposed to reduce the precipitation efficiency (indirect cloud lifetime effect), both of which are likely to result in a reduction of the global, annual mean net radiation at the top of the atmosphere (TOA). These effects may be partly offset through the evaporation of cloud droplets attributable to absorbing aerosols (semi-direct effect) and/or by more ice nuclei (glaciation effect). The influences of these processes on radiation at TOA and at the surface and on precipitation are summarized in Table 23.1. The following discussion is based on Denman et al. (2007), which also provides references to the studies noted.

Another aerosol influence on clouds and radiation that may be climatologically important is the enhancement of downwelling longwave radiation from Arctic haze (Blanchet and Girard 1994) and thin Arctic stratus whose longwave optical thickness is augmented by increased droplet concentration (Lubin and Vogelmann 2006).

In addition to raising the number concentration of aerosol particles, there is evidence that increased particle concentrations can broaden the cloud drop size distribution (Liu and Daum 2002). This would have the effect of decreasing aerosol influences on shortwave radiation and inhibiting precipitation development (e.g., Peng and Lohmann 2003).

Table 23.1 Overview of known aerosol indirect effects on net radiative flux (at TOA and at the surface) and on precipitation, and an assessment of level of current scientific understanding. Modified from Denman et al. (2007).

Effect	Cloud albedo effect	Cloud lifetime effect	Semi-direct effect	Glaciation indirect effect	Thermodynamic effect
Cloud types affected	All; greatest for clouds of intermediate optical thickness	All	All	Mixed-phase	Mixed-phase
Process	For same cloud water or ice content more but smaller cloud particles reflect more solar radiation	Smaller cloud particles decrease precipitation efficiency prolonging cloud lifetime	Absorption of solar radiation by absorbing aerosols evaporates cloud particles, increases static stability	Increase in ice nuclei increases precipitation efficiency	Smaller cloud droplets delay freezing and cause supercooled clouds to extend to colder temperatures
Change in net TOA irradiation	–	–	+ / –	+	+ / –
Potential magnitude	medium	medium	small	medium	medium
Scientific understanding	low	very low	very low	very low	very low
Change in surface irradiation	–	–	+ / –	+	+ / –
Potential magnitude	medium	medium	large	medium	medium
Scientific understanding	low	very low	very low	very low	very low
Change in precipitation	N/A	–	–	+	+ / –
Potential magnitude	N/A	small	large	medium	medium
Scientific understanding	N/A	very low	very low	very low	very low

The increase in albedo of liquid water clouds attributable to anthropogenic aerosols has received much attention. Although uncertainties remain regarding the breadth of the cloud drop size distribution, more and probably larger uncertainties are related to aerosol effects on precipitation as well as on mixed- and ice-phase clouds, as discussed below.

Aerosol Effects on Water Clouds and Warm Precipitation

Aerosol particles are hypothesized to lengthen the lifetime of clouds because increased concentrations of smaller droplets lead to decreased drizzle production and reduced precipitation efficiency (Albrecht 1989). It is difficult to devise observational studies that can separate the cloud lifetime from the cloud albedo effect. Thus, observational studies provide estimates of the combined effects. Similarly, climate models cannot easily separate the cloud lifetime indirect effect once the aerosol scheme is fully coupled to a cloud microphysics scheme. Instead, they calculate the combined cloud albedo, lifetime, and semi-direct effect.

GCM studies suggest that, in the absence of giant CCN and aerosol-induced changes in ice microphysics, anthropogenic aerosols suppress precipitation. It should be noted, however, that precipitation would also be suppressed in mixed-phase clouds in which the ice phase plays only a minor role. A decrease in the formation of precipitation leads to increased cloud processing of aerosols. CRM studies have shown that cloud processing can lead to either an increase or a decrease in precipitation in subsequent cloud cycles, depending on the size and concentration of activated CCN (e.g., Feingold and Kreidenweis 2002). When the actual cloud lifetime is analyzed in CRM simulations, an increase in aerosol concentration—from very clean to strongly anthropogenically influenced situations—does not increase cloud lifetime, even though precipitation is suppressed (Jiang et al. 2006). This results from competition between precipitation suppression and enhanced evaporation of the more numerous smaller cloud droplets at high cloud droplet concentration. Giant sea-salt nuclei, on the other hand, may override the precipitation suppression effect of the large number of small CCN.

Aerosol Impacts on Mixed-phase Clouds

GCM studies suggest that if, in addition to mineral dust, hydrophilic black carbon particles are assumed to act as ice nuclei at temperatures between 0° and –35°C, then increases in aerosol number concentration from preindustrial to present times may have resulted in greater glaciation of supercooled stratiform clouds and an increase in the amount of precipitation via the ice phase. This process could decrease the global mean cloud cover, leading to enhanced absorption of solar radiation. Whether the glaciation effect or the warm cloud lifetime effect is larger depends on the chemical nature of the particles (Lohmann and Diehl 2006).

Simulations of precipitation from single-cell mixed-phase convective clouds suggest a reduction for various background aerosol concentrations when particle concentrations are increased. Khain et al. (2005) postulated that smaller cloud droplets, such as those affected by human activities, would change the thermodynamics of convective clouds. More but smaller droplets would reduce the production of rain in convective clouds. When these droplets freeze, the associated latent heat release would then result in more vigorous convection and more precipitation. In a clean cloud, on the other hand, rain would have depleted the cloud so that less latent heat is released when the cloud glaciates, resulting in less vigorous convection and less precipitation. For a thunderstorm in Florida, in the presence of Saharan dust, the simulated precipitation enhancement lasted only two hours, after which precipitation decreased as compared with clean conditions. This highlights the complexity of the system and indicates that the sign of the global change in precipitation attributable to aerosols is not yet known. Note that microphysical processes can only change the temporal and spatial distribution of precipitation, whereas the total amount of precipitation can only change if evaporation from the surface changes.

Subgrid-scale Variability and Radiative Transfer

Model inaccuracy can result from treating clouds as simple parallel homogeneous clouds. This would be the case if clouds are uniformly distributed over grid cells, as is conventional. Such treatment can overestimate the Twomey effect by up to 50% (Lohmann and Feichter 2005). One way to obviate this problem is to treat grid cells as being nonuniform. A PDF-based approach can be used here to account for subgrid-scale variability in cloud cover and cloud condensate in radiative transfer through inhomogeneous cloud fields (e.g., Pincus and Klein 2000).

Aerosol Impacts on Cirrus Clouds

The influence of aerosols from aircraft emissions on cirrus cloud extent and properties has received considerable attention because these particles are emitted at an altitude where clouds can exert a strong radiative influence. This subject is examined by Kärcher and Spichtinger (this volume) and Denman et al. (2007) and is thus not reviewed here.

Global Climate Model Estimates of the Total Anthropogenic Aerosol Effect

The total anthropogenic aerosol effect, as defined here, consists of the direct effect, semi-direct effect, indirect cloud albedo effect, and cloud lifetime effect for warm clouds. The total anthropogenic aerosol effect is obtained by calculating the difference between a multiyear simulation with present-day aerosol emissions and a simulation representative for preindustrial conditions, in which anthropogenic emissions are turned off. It should be noted that the representation of the cloud lifetime effect in global climate models is essentially one of changing the autoconversion of cloud water to rainwater.

The radiative forcing that results from the indirect cloud albedo effect attributable to anthropogenic aerosols is estimated from global models as –0.7 W m^{-2}, with a 90% confidence range of –0.3 to –1.8 W m^{-2} (Forster et al. 2007). Feedbacks that result from the cloud lifetime effect, semi-direct effect, or aerosol–ice cloud effects can either enhance or reduce the cloud albedo effect. Climate models estimate the total aerosol effect (direct plus indirect effects) on the TOA net radiation since preindustrial times to be –1.2 W m^{-2}, with a range of –0.2 to –2.3 W m^{-2} (Figure 23.5 and Denman et al. 2007). The range of the total aerosol effect from different models cannot easily be compared to the range of the indirect cloud albedo effect alone because different model simulations entered these various compilations.

All models agree that the total aerosol effect is larger over the northern hemisphere than over the southern hemisphere (Figure 23.5), consistent with emissions of anthropogenic aerosols and precursor gases being much greater in the northern hemisphere. This effect has not been seen, however, in satellite data (Han et al. 1998; Schwartz 1988), suggesting that either dynamic influences on the liquid water path mask such an effect or that the models do not represent aerosol–cloud interactions realistically. The values of the northern hemisphere total aerosol effect vary between –0.5 and –3.6 W m^{-2}; in the southern hemisphere they range between slightly positive to –1.1 W m^{-2}; and the average southern/northern hemisphere ratio is 0.3. Estimates of the ocean/land partitioning of the total aerosol effect vary from 0.03 to 1.8, with an average value of 0.7. Although the combination of ECHAM4 model results with POLDER satellite estimates suggests that the total aerosol effect should be larger over oceans, combined estimates of the LMD and ECHAM4 models with MODIS satellite data reach the opposite conclusion. The average total aerosol effect over the ocean of –1 W m^{-2} agrees with estimates between –1 to –1.6 W m^{-2} from AVHRR/POLDER (Denman et al. 2007).

Estimates of the total aerosol effect from global climate models are generally larger than those estimated from inverse approaches, which constrain the indirect aerosol effect to be between –0.1 and –1.7 W m^{-2} (Anderson et al. 2003; Hegerl et al. 2007). The estimated total anthropogenic aerosol effect is now lower than was stipulated in IPCC's Third Assessment Report and

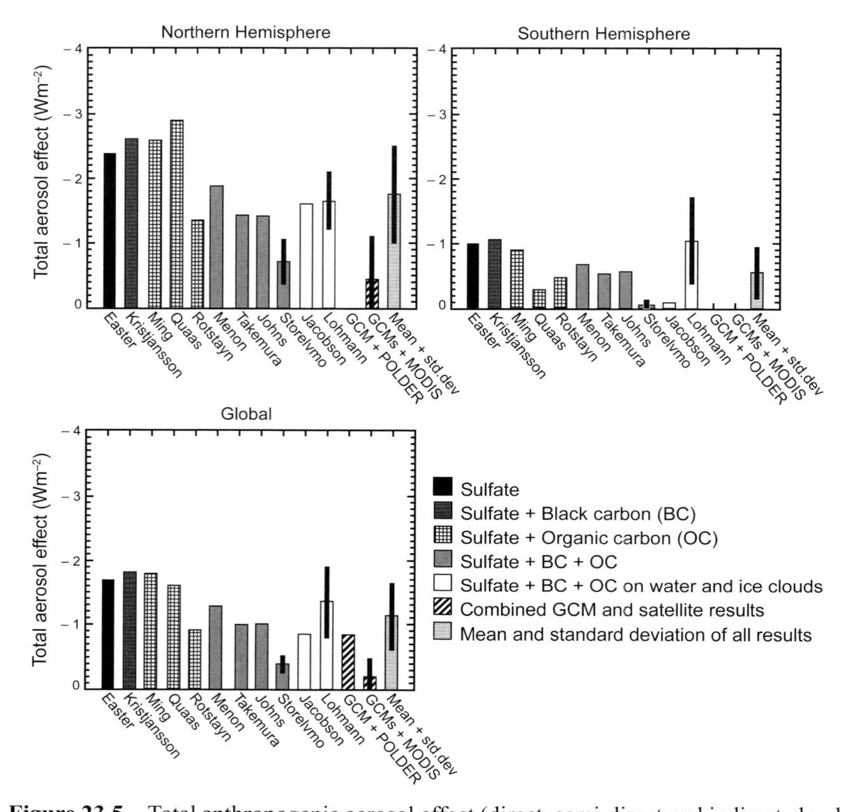

Figure 23.5 Total anthropogenic aerosol effect (direct, semi-direct and indirect cloud albedo and lifetime effects) in 12 global climate models and two determinations from satellite observations in global mean, over the northern and southern hemispheres, over oceans and over land, and the ratio over oceans/land. Anthropogenic aerosol effect is defined as the change in net radiation at TOA from preindustrial times to the present day resulting from anthropogenic emissions of aerosols and aerosol precursors. Patterns denote different anthropogenic species whose forcings were examined and the cloud types affected; all are for water clouds except as indicated.

this is attributable to improvements in cloud parameterizations. Still, large uncertainties remain.

The influence of aerosols on evapotranspiration and precipitation is also quite uncertain, with model results for the change in global average precipitation ranging from almost no change to a decrease of 0.13 mm day^{-1} (5 cm yr^{-1}), with much greater changes locally. Decreases in precipitation are larger when the atmospheric GCMs are coupled to mixed-layer ocean models, where the sea surface temperature and, hence, the evaporation from the ocean is also allowed to vary, than in models in which the sea surface temperature is held constant (Denman et al. 2007). The decrease in evapotranspiration results primarily from decreases in solar radiation at the surface, as a result of increased

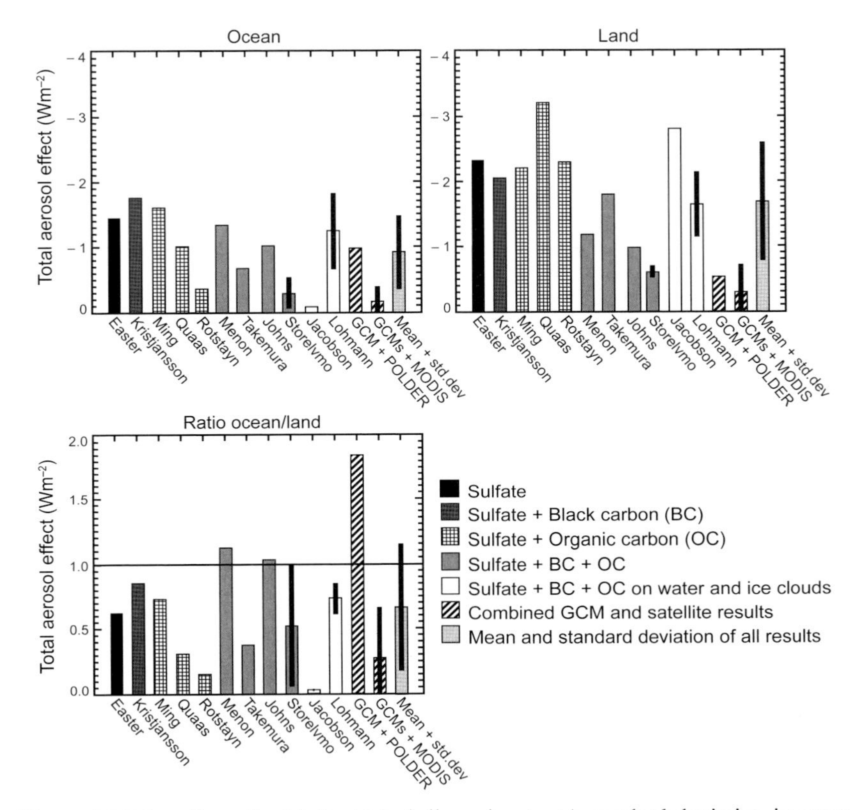

Figure 23.5 (continued) Vertical black lines denote ±1 standard deviation in cases of multiple simulations and/or results. Modified after Denman et al. (2007).Vertical black lines denote ±1 standard deviation in cases of multiple simulations and/or results. Modified after Denman et al. (2007).

aerosol optical depth and optically thicker clouds. The decrease in solar radiation at the surface is then partly balanced by a decreased latent heat flux, which results in a reduced global mean precipitation rate (e.g., Liepert et al. 2004).

Cloud Feedbacks in Climate Models and Their Influence on Climate Sensitivity

The energy balance model of Earth's climate system is useful to assess the influences of particular processes on global mean surface temperature (T) and to compare these influences across different climate models. Within this energy balance framework, the time-dependent change in heat content of the climate system, ΔQ, is related to radiative forcing, ΔF, and the change in T, ΔT, as:

$$\frac{d}{dt}(\Delta Q) = \Delta F - \frac{1}{\lambda}\Delta T, \tag{23.1}$$

where λ is the equilibrium climate sensitivity, as is readily seen by considering a system in a new equilibrium in response to a forcing ΔF, for which $d(\Delta Q)/dt = 0$. Hence,

$$\lambda = \frac{\Delta T_{eq}}{\Delta F}, \tag{23.2}$$

where ΔT_{eq} is the temperature difference between two equilibrium states. At present, climate sensitivity is not well constrained in climate models or in empirical analyses. According to the IPCC's Fourth Assessment Report (IPCC 2007), the likely range of global equilibrium temperature increase for the doubling of CO_2, ΔT_{2x}, lies between 2.0 and 4.5 K; values below 1.5 K are considered very unlikely. This sensitivity range is in agreement with values exhibited by current climate models (Figure 23.6). Since a doubling of CO_2 causes direct radiative forcing of about 3.7 W m^{-2} (Forster et al. 2007), the range of 2.0–4.5 K for a doubling of CO_2 corresponds to climate sensitivity between 0.54 and 1.22 K/(W m^{-2}). Roe and Baker (2007) point out that the upper range of the climate sensitivity is relatively insensitive to decreases in uncertainties associated with the underlying climate processes. In this context, we note that some recent experiments with CRMs embedded into climate models (Miura et al. 2005; Wyant et al. 2006) suggest a lower climate sensitivity, with values of 0.44 and 0.41 K/(W m^{-2}), respectively (ΔT_{2x} 1.6 and 1.5 K). These findings point to the strong influence of the treatment of clouds on modeled climate sensitivity.

Cloud feedback is the response of CRF to a change in global temperature. Key questions involve the nature and extent of CRF changes in a greenhouse-warmed world: As the climate warms, will longwave CRF increase (positive feedback) or decrease (negative feedback)? Similarly, will shortwave CRF increase (negative feedback) or decrease (positive feedback)? Current climate models produce a wide variety of cloud feedbacks ranging from weakly negative to strongly positive, depending on the relative magnitude of different cloud feedback mechanisms (Bony et al. 2006; Webb et al. 2006). As seen in Figure 23.6, much of the model-to-model difference in climate sensitivity results from differences in cloud feedback.

CRF depends largely on the spatial and temporal distribution of clouds and their radiative properties, which are determined by their microphysical characteristics such as size distribution of droplets and ice crystals. These cloud properties are controlled by cloud-forming processes (e.g., cooling through rising air or radiative cooling) and by cloud-dissipating processes (e.g., precipitation, sinking motion, and mixing with dry air). Thus, cloud feedbacks involve changes in the spatial distribution of clouds and their microphysical properties that result from alterations in processes that form and dissipate

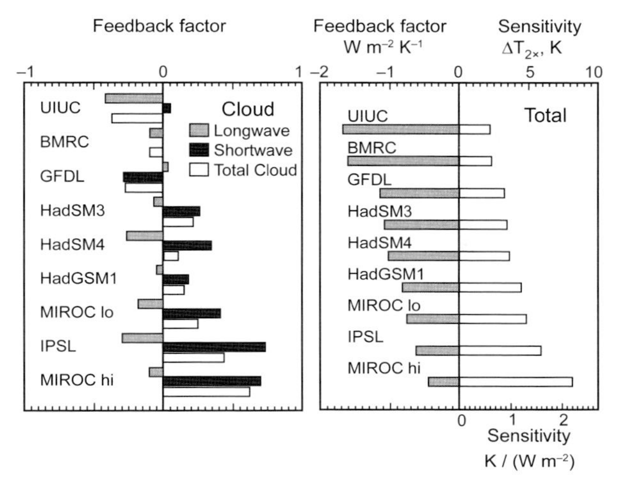

Figure 23.6 Influence of cloud feedback on total feedback and sensitivity of current GCMs. Left panel shows total cloud feedback and long- and shortwave components in nine current climate models. Right panel shows total feedback factor and sensitivity expressed as the inverse of the total feedback factor, in units of K/(W m^{-2}), and as the equilibrium increase in global mean surface temperature that would result from a doubling of CO_2, $\Delta T_{2\times}$, evaluated as –3.7 W m^{-2} upon the total feedback factor. Modified from Webb et al. (2006).

clouds. Uncertainty arises because it is not clear how cloud microphysical properties and cloud horizontal/vertical distributions respond to alterations in controlling variables as the climate changes. In contrast to CRF, cloud feedback cannot be measured directly; it can only be determined from GCM simulations. Consequently, confidence in estimates of cloud feedback can only be assessed by using observations to evaluate simulations of cloud microphysical and radiative properties, cloud distribution, and radiative forcing in a variety of conditions that span the range expected under climate change. Therefore, atmospheric process research on cloud feedbacks focuses on how these cloud features depend on processes that form and dissipate clouds under a variety of conditions.

Closely related to cloud feedback is water-vapor feedback. Water vapor is the most important greenhouse gas in Earth's atmosphere. Consistent with basic physics, all climate models show an increase in the amount of atmospheric water vapor with rising global mean surface temperature. However, the amount and spatial distribution of the resultant radiative forcing differ considerably. This water-vapor feedback is strongly connected to the cloud feedback because water vapor is the source of condensed phase water in clouds, and clouds remove water from the atmosphere when they precipitate. Clouds also influence evapotranspiration, which is the source of water vapor to the atmosphere.

Conclusions and Outlook

Clouds play a crucial role in determining Earth's energy balance. This must be accurately represented in climate models, if the models are to be used with confidence to project future climate change. Much of the variation in climate sensitivity observed in current climate models can be attributed to differences in the treatment of clouds, as evidenced by model-to-model variation in cloud feedback. Microphysical processes are now recognized to exert strong influences on cloud dynamics and that the influences of clouds on short- and longwave radiation must be understood and accurately represented in climate models. In addition, the strong coupling between aerosols and clouds has large influences on climate and climate change. An increased atmospheric aerosol burden alters the microphysical properties of clouds and influences short- and longwave radiation and the locus and intensity of precipitation.

One of the better understood influences of aerosols on clouds is a reduction of the amount of solar radiation absorbed by Earth–atmosphere system, as quantified by net shortwave radiation at the TOA and a similar decrease in shortwave radiation reaching the surface. The negative radiative forcing of anthropogenic aerosols competes with greenhouse gas warming as a forcing of climate change and in altering evaporation and precipitation. Although much has been learned about these effects, they are not understood be well enough to be fully represented in climate models. None of the transient climate model simulations conducted thus far accounts for all of the known aerosol–cloud interactions; thus the net effects of aerosols on clouds and climate deduced from global climate models cannot be considered conclusive. Therefore, the cloud feedback and sensitivity of Earth's climate system remain highly uncertain. One reason is that aerosol–cloud interactions take place on microscale and thus are at best crudely represented in GCMs.

In terms of aerosols and clouds, the principal areas for future development in global climate models are twofold. The treatment of clouds themselves requires improvement in all aspects, if models are to represent accurately cloud feedbacks in a greenhouse-warmed world. A good representation of cloud dynamics, including entrainment, is especially important for the representation of convective clouds and boundary layer clouds. Cloud microphysical processes are important for the conversion of cloud particles into precipitation-size particles. The treatment of aerosol–cloud interactions needs to be improved as well, if aerosol radiative forcing is to be accurately quantified.

Acknowledgment

S. Schwartz's work at Brookhaven National Laboratory was supported by the U.S. Dept. of Energy's Atmospheric Science Program and Atmospheric Radiation Measurement Program (Office of Science, OBER) under Contract No. DE–AC02–98CH10886.

References

Abdul-Razzak, H., and S. J. Ghan. 2004. Parameterization of the influence of organic surfactants on aerosol activation, *J. Geophys. Res.* **109**:D03205.

Albrecht, B. 1989. Aerosols, cloud microphysics, and fractional cloudiness. *Science* **245**:1227–1230.

Anderson, T. L., R. J. Charlson, S. E. Schwartz et al. 2003. Climate forcing by aerosols: A hazy picture. *Science* **300**:1103–1104.

Beheng, K. D. 1994. A parameterization of warm cloud microphysical conversion processes. *Atmos. Res.* **33**:193–206.

Bender, F. A. M., H. Rodhe, R. J. Charlson, A. M. L. Ekman, and N. Loeb. 2006. 22 views of the global albedo: Comparison between 20 GCMs and two satellites. *Tellus* **58**:320–330.

Berg, W., T. L'Ecuyer, and C. Kummerow. 2006. Rainfall climate regimes: The relationship of regional TRMM rainfall biases to the environment. *J. Appl. Meteor. Climatol.* **45(3)**:434.

Berry, E. X. 1968. Modification of the warm-rain process. Proc. 1st. Natl. Conf. Weather Modification, pp. 81–88. Albany: *Amer. Meteor. Soc.*

Blanchet, J.-P., and E. Girard. 1994. Arctic greenhouse effect. *Nature* **371(6496)**:383.

Bony, S., R. Colman, V. M. Kattsov et al. 2006. How well do we understand and evaluate climate change feedback processes? *J. Climate* **19**:3445–3482.

Boucher, O., and U. Lohmann. 1995. The sulfate-CCN-cloud albedo effect: A sensitivity study using two general circulation models. *Tellus* **47B**:281–300.

Denman, K. L., G. Brasseur, A. Chidthaisong et al. 2007. Couplings between changes in the climate system and biogeochemistry. In: Climate Change 2007: The Physical Science Basis. Contribution of Working Group I to the Fourth Assessment Report of the IPCC, ed. Solomon, S., D. Qin, M. Manning et al., pp. 499–588. New York. Cambridge Univ. Press.

Easter, R. C., S. J. Ghan, Y. Zhang et al. 2004. MIRAGE: Model description and evaluation of aerosols and trace gases. *J. Geophys. Res.* **109**:D20210.

Facchini, M. C., M. Mircea, S. Fuzzi, and R. J. Charlson. 1999. Cloud albedo enhancement by surface-active organic solutes in growing droplets. *Nature* **401**:257–259.

Feingold, G., and P. Y. Chuang. 2002. Analysis of the influence of film-forming compounds on droplet growth: Implications for cloud microphysical processes and climate. *J. Atmos. Sci.* **59**:2006–2018.

Feingold, G., and S. M. Kreidenweis. 2002. Cloud processing of aerosol as modeled by a large eddy simulation with coupled microphysics and aqueous chemistry. *J. Geophys. Res.* **107(D23)**:4687.

Forster, P., V. Ramaswamy, P. Artaxo et al. 2007. Radiative forcing of climate change. In: Climate Change 2007: The Physical Science Basis. Contribution of Working Group I to the Fourth Assessment Report of the IPCC, ed. by S. Solomon, D. Qin, M. Manning et al., pp. 129–234. New York: Cambridge Univ. Press.

Garrett, T. J., L. Avey, P. I. Palmer et al. 2006. Quantifying wet scavenging processes in aircraft observations of nitric acid and cloud condensation nuclei. *J. Geophys. Res.* **111**:D23S51.

Ghan, S. J., L. R. Leung, R. C. Easter, and H. Abdul-Razzak. 1997. Prediction of cloud droplet number in a general circulation model. *J. Geophys. Res.* **102**:21,777–21,794.

Ghan, S. J., and S. E. Schwartz. 2007. Aerosol properties and processes: A path from field and laboratory measurements to global climate models. *Bull. Amer. Meteor. Soc.* **88**:1059–1083.

Gustafson, W. I., Jr., E. G. Chapman, S. J. Ghan, R. C. Easter, and J. D. Fast. 2007. Impact on modeled cloud characteristics due to simplified treatment of uniform cloud condensation nuclei during NEAQS 2004. *Geophys. Res. Lett.* **34**:L19809.

Han, Q., W. B. Rossow, J. Chou, and R. M. Welch. 1998. Global survey of the relationships of cloud albedo and liquid water path with droplet size using ISCCP. *J. Climate* **11**:1516–1528.

Hegerl, G. C., F. W. Zwiers, P. Braconnot et al. 2007. Understanding and attributing climate change. In: Climate Change 2007: The Physical Science Basis. Contribution of Working Group I to the Fourth Assessment Report of the IPCC, ed. S. Solomon, D. Qin, M. Manning et al., pp. 665–744. New York: Cambridge Univ. Press.

IPCC. 2007. Climate Change 2007: The Physical Science Basis. Contribution of Working Group I to the Fourth Assessment Report of the Intergovernmental Panel on Climate Change, ed. S. Solomon, D. Qin, M. Manning et al. New York: Cambridge Univ. Press.

Jiang, H., H. Xue, A. Teller, G. Feingold, and Z. Levin. 2006. Aerosol effects on the lifetime of shallow cumulus. *Geophys. Res. Lett.* **33(14)**:L14806.

Kanakidou, M., J. Seinfeld, S. Pandis et al. 2004. Organic aerosol and global climate modelling: A review. *Atmos. Chem. Phys.* **4**:5855–6024.

Kessler, E. 1969. On the Distribution and Continuity of Water Substance in Atmospheric Circulation. Meteor. Monogr., vol 10. Boston: Amer. Meteor. Soc.

Khain, A. P., D. Rosenfeld, and A. Pokrovsky. 2005. Aerosol impact on the dynamics and microphysics of convective clouds. *Q. J. Roy. Meteor. Soc.* **131(611)**:2639–2663.

Kiehl, J. T., and K. E. Trenberth. 1997. Earth's annual global mean energy budget. *Bull. Amer. Meteor. Soc.* **78**:197–208.

Kinne, S. et al. 2006. An AeroCom initial assessment: Optical properties in aerosol component modules of global models. *Atmos. Chem. Phys.* **6**:1815–1834.

Köhler, H. 1923. Zur Kondensation des Wasserdampfes in der Atmosphäre, erste Mitteilung. *Geophys.* **2**:1–15.

Langner, J., and H. Rodhe. 1991. A global three-dimensional model of the global sulfur cycle. *J. Atmos. Chem.* **13**:225–263.

Larson, V. E., R. Wood, P. R. Field et al. 2001. Small-scale and mesoscale variability of scalars in cloudy boundary layers: One-dimensional probability density functions. *J. Atmos. Sci.* **58**:1978–1996.

Liepert, B. G., J. Feichter, U. Lohmann, and E. Roeckner. 2004. Can aerosols spin down the water cycle in a warmer and moister world? *Geophys. Res. Lett.* **31**:L06207.

Liu, Y., and P. H. Daum. 2002. Indirect warming effect from dispersion forcing. *Nature* **419**:580–581.

Liu, Y., P. H. Daum, R. McGraw, M. Miller, and S. Niu. 2007. Theoretical formulation for autoconversion rate of cloud droplet concentration. *Geophys. Res. Lett.* **34**:L116821.

Lohmann, U. 2007. Global anthropogenic aerosol effects on convective clouds in ECHAM5-HAM. *Atmos. Chem. Phys.* **7**:14,639–14,674.

Lohmann, U., and K. Diehl. 2006. Sensitivity studies of the importance of dust ice nuclei for the indirect aerosol effect on stratiform mixed-phase clouds. *J. Atmos. Sci.* **63**:968–982.

Lohmann, U., and J. Feichter. 2005. Global indirect aerosol effects: A review. *Atmos. Chem. Phys.* **5**:715–737.

Lohmann, U., J. Feichter, C. C. Chuang, and J. E. Penner. 1999. Predicting the number of cloud droplets in the ECHAM GCM. *J. Geophys. Res.* **104**:9169–9198.

Lubin, D., and A. M. Vogelmann. 2006. A climatologically significant aerosol longwave indirect effect in the Arctic. *Nature* **439**:453–456.

Miura, H., H. Tomita, T. Nasuno et al. 2005. A climate sensitivity test using a global cloud resolving model under an aqua planet condition. *Geophys. Res. Lett.* **32**:L19717.

Nober, F. J., H. F. Graf, and D. Rosenfeld. 2003. Sensitivity of the global circulation to the suppression of precipitation by anthropogenic aerosols. *Global Planetary Change* **37**:57–80.

Peng, Y., and U. Lohmann. 2003. Sensitivity study of the spectral dispersion of the cloud droplet size distribution on the indirect aerosol effect. *Geophys. Res. Lett.* **30(10)**:1507.

Peng, Y., U. Lohmann, and W. R. Leaitch. 2005. Importance of vertical velocity variations in the cloud droplet nucleating process of marine stratus clouds. *J. Geophys. Res.* **110**:D21213.

Pincus, R., and S. A. Klein. 2000. Unresolved spatial variability and microphysical process rates in large scale models. *J. Geophys. Res.* **105**:27,059–27,065.

Ramaswamy, V., O. Boucher, J. Haigh et al. 2001. Radiative forcing of climate change. In: Climate Change 2001: The Scientific Basis. Contribution of Working Group I to the Third Assessment Report of the IPCC, ed. J. T. Houghton, Y. Ding, D.J. Griggs et al., pp. 349–416. New York: Cambridge Univ. Press.

Randall, D., M. Khairoutdinov, A. Arakawa, and W. Grabowski. 2003. Breaking the cloud parameterization deadlock. *Bull. Amer. Meteor. Soc.* **84**:1547–1564.

Randall, D. A., R. A. Wood, S. Bony et al. 2007. Climate models and their evaluation. In: Climate Change 2007: The Physical Science Basis. Contribution of Working Group I to the Fourth Assessment Report of the IPCC, ed. S. Solomon, D. Qin, M. Manning et al., pp. 591–662. New York: Cambridge Univ. Press.

Roe, G. H., and M. B. Baker. 2007. Why is climate sensitivity so unpredictable? *Science* **318**:629–632.

Rotstayn, L. D., and Y. Liu. 2005. A smaller global estimate of the second indirect aerosol effect. *Geophys. Res. Lett.* **32**:L05708.

Schwartz, S. E. 1988. Are global cloud albedo and climate controlled by marine phytoplankton? *Nature* **336**:441–445.

Stier, P., J. Feichter, S. Kinne et al. 2005: The aerosol–climate model ECHAM5–HAM. *Atmos. Chem. Phys.* **5**:1125–1156.

Sundqvist, H. 1978. A parameterization scheme for non-convective condensation including prediction of cloud water content. *Q. J. Roy. Meteor. Soc.* **104**:677–690.

Webb, M. J., C. A. Senior, D. M. H. Sexton et al. 2006. On the contribution of local feedback mechanisms to the range of climate sensitivity in two GCM ensembles. *Climate Dyn.* **27**:17–38.

Wood, R. 2005. Drizzle in stratiform boundary layer clouds. Part II: Microphysical aspects. *J. Atmos. Sci.* **62**:3034–3050.

Wood, R., P. R. Field, and W. R. Cotton. 2002. Autoconversion rate bias in stratiform boundary layer cloud parameterizations. *Atmos. Res.* **65**:109–128.

Wyant, M., M. Khairoutdinov, and C. Bretherton. 2006. Climate sensitivity and cloud response of a GCM with a superparameterization. *Geophys. Res. Lett.* **33**:L06714.

Zhang, Q., J. L. Jimenez, M.R. Canagaratna et al. 2007. Ubiquity and dominance of oxygenated species in organic aerosols in anthropogenically-influenced northern hemisphere midlatitudes. *Geophys. Res. Lett.* **34**:L13801.

24

Current Understanding and Quantification of Clouds in the Changing Climate System and Strategies for Reducing Critical Uncertainties

Johannes Quaas, Rapporteur

Sandrine Bony, William D. Collins, Leo Donner, Anthony Illingworth, Andy Jones, Ulrike Lohmann, Masaki Satoh, Stephen E. Schwartz, Wei-Kuo Tao, and Robert Wood

Abstract

To date, no observation-based proxy for climate change has been successful in quantifying the feedbacks between clouds and climate. The most promising, yet demanding, avenue to gain confidence in cloud–climate feedback estimates is to utilize observations and large-eddy simulations (LES) or cloud-resolving modeling (CRM) to improve cloud process parameterizations in large-scale models. Sustained and improved satellite observations are essential to evaluate large-scale models. A reanalysis of numerical prediction models with assimilation of cloud, aerosol, and precipitation observations would provide a valuable dataset for examining cloud interactions. The link between climate modeling and numerical weather prediction (NWP) may be exploited by evaluating how accurate cloud characteristics are represented by the parameterization schemes in NWP models.

A systematic simplification of large-scale models is an important avenue to isolate key processes linked to cloud–climate feedbacks and would guide the formulation of testable hypotheses for field studies.

Analyses of observation-derived correlations between cloud and aerosol properties in combination with modeling studies may allow aerosol–cloud interactions to be detected and quantified. Reliable representations of cloud dynamic and physical processes in large-scale models are a prerequisite to assess aerosol indirect effects on a large scale with confidence.

To include aerosol indirect effects in a consistent manner, we recommend that a "radiative flux perturbation" approach be considered as a complement to radiative forcing.

Are There Observational Proxies in the Present-day Climate for Future Cloud Perturbations?

Climate sensitivity, defined as the equilibrium change in global mean temperature in response to a doubling in the atmospheric CO_2 concentration, is still a very uncertain quantity (e.g., Randall et al. 2007). The primary reason for the spread in climate sensitivities, as simulated by different global climate models, is the difference in the representation of cloud processes and cloud–climate feedbacks (Soden and Held 2006; Dufresne and Bony 2008). Cloud processes determine the amount and distribution of precipitation, which is a key parameter for land–atmosphere interactions, for carbon–climate feedbacks, and for climate impact studies (e.g., water resources). Moreover, the representation of cloud microphysical properties is critical for the simulation of interactions between clouds and aerosols. Improving how cloud processes are represented in global climate models is thus of paramount importance, not only for estimates of climate sensitivity but also for projections of future climate change and their use in impact studies.

Limitations of Observational Proxies

Since cloud and radiation observations are available on a global scale for only a climatically short time period (at best, ca. 25 years), our ability to observe cloud–climate feedbacks directly is limited. Thus the question arises as to whether there might be some observational proxies in the present-day climate for future cloud perturbations. The dynamic and thermodynamic forcings of clouds that occur on shorter timescales do not correspond directly to the dynamic and thermodynamic changes that are expected to occur with global warming. Therefore, neither the seasonal cycle, which is useful to constrain the snow/ice albedo feedback (Hall and Qu 2006), nor El Niño/La Niña events can be considered as good analogues of long-term climate changes for the analysis of cloud–climate feedbacks. This was confirmed by the FANGIO study (Cess et al. 1990), which found no direct relationship between cloud seasonal variations and the cloud response to climate change. Based on a study comparing the response of clouds simulated by a climate model in the case of a volcanic eruption (Mt. Pinatubo) or of a doubling CO_2 experiment (Yokohata et al. 2005), it appears that volcanic eruptions may also not be considered as useful proxies for future cloud changes, despite the success of studies constraining the water vapor feedback (Soden et al. 2002). Thus, no direct observational proxy for future cloud changes has yet been identified.

A better understanding of the physical processes that control cloud–climate feedbacks may, however, help uncover new pathways to exploit existing observational datasets for the understanding of cloud–climate feedbacks.

Options for the Exploitation of Observational Data

Compositing techniques can be used to compare models and observations in a way that can be relevant for evaluating cloud–climate feedbacks (see Illingworth and Bony, this volume). These methodologies permit an assessment of how clouds change with dynamic and thermodynamic conditions in the present-day climate (e.g., Bony et al. 2004). In these studies, it is essential to eliminate, as much as possible, the dynamic contribution to observed relationships between cloud properties and temperature. Other studies have decomposed the global cloudiness into a small number of prominent cloud regimes and used this decomposition to understand and assess the response of clouds to long-term climate changes (e.g., Williams and Tselioudis 2007). Of course, a perturbed climate will be characterized by both perturbed thermodynamics and dynamics, both of which are a challenge to model.

Rapid changes in clouds as a response to instantaneous CO_2 doubling have been identified by Gregory and Webb (2008) in a slab model ensemble. Changes in radiative cooling in the atmosphere and fast surface responses, which are associated with the CO_2 change, operate on timescales of months; that is, before variations in longer-term sea surface temperature (SST) are established. Such changes constitute a substantial and, in some cases, dominant fraction of longer-term changes. They may be detectable in the satellite record of changes of cloud forcing. However, issues related to confounding changes in the atmosphere (e.g., attributable to aerosols) may complicate detection of a signal.

The fluctuation dissipation (FD) theorem states that the transient behavior (and sensitivity) of dynamic systems to perturbations can be determined from their natural variability. This theorem has been adapted from statistical physics, where it applies to a wide array of classical and quantum mechanical systems. It provides a potential mechanism for estimating or constraining climate sensitivity from short-term unforced variations in the climate system (Leith 1975; Schwartz 2007). To date, however, there is no rigorous determination of whether the FD theorem is applicable to the Earth's climate and, if so, which fields and fluctuations constrain climate sensitivity. The prospect that the theorem might apply to spectral features of the Earth's radiation field is under active investigation.

Process-based Evaluation of Large-scale Models to Gain Confidence in Cloud Feedback Determination

Cloud–climate feedbacks can be simulated with general circulation models (GCMs). To gain confidence in the results from these simulations, the reliability of the physics of climate models (e.g., the representation of turbulence, convection, aerosols, and clouds) must be improved (Illingworth and Bony, this volume). For this purpose, a well-recognized methodology, which forms the basis of the GEWEX cloud system studies, can be employed: the physics of climate models within a single-column framework are compared with observations from field experiments and/or LES or CRM simulations driven by observed forcings (Browning et al. 1993; Randall et al. 2003). The resulting parameterizations are then evaluated in a full 3-D GCM with global datasets to assess whether an improvement of cloud representation has been achieved. Such an approach can be very powerful in pointing out deficiencies in model parameterizations and in improving model parameterizations.

Satellite observations provide particularly well-suited datasets to evaluate GCM processes, because of their global coverage and the availability of large statistics (Illingworth and Bony, this volume). Recent spaceborne active remote-sensing data providing information about the vertical distribution of cloud-related quantities (CALIPSO, CloudSat), as well as precipitation radar (TRMM) and advanced passive instrument retrievals, are especially valuable to enhance the understanding of processes and enable the evaluation of climate models. This, in turn, has enhanced our understanding of processes and enabled the evaluation of climate models. The development and application of satellite simulators in climate models that predict sensor responses assures the comparability of simulations with actual observations (Webb et al. 2001; Bodas-Salcedo et al. 2008; Chepfer et al. 2008). A persistent shortcoming, however, is the lack of high-resolution water vapor retrievals and its probability density function (PDF), which forms the basis of statistical cloud parameterization schemes and has a large impact on cloud processes. Potentially, differential absorption and Raman lidar technology should help improve this situation in the future.

The general consistency of climate models with the global atmospheric models used for NWP suggests that it is possible to evaluate climate models in the NWP framework. Comparison of the representation of clouds in NWP models (or GCMs run in NWP mode) with observations (e.g., satellite data or networks of instrumental sites such as Cloudnet; Illingworth et al. 2007) can identify strengths and shortcomings in NWP parameterization schemes. Parameter values that are applied in parameterization schemes used in an NWP model run with data assimilation may be rejected if they lead to systematic and unrealistic tendencies in short-term forecasts (Rodwell and Palmer 2007). This technique might be adapted to identify features or errors in short-term cloud simulations that map directly onto features in long-term cloud radiative feedbacks.

Systems to predict the climate evolution on a decadal timescale are currently under development. Once in place, they may permit hindcast simulations to be performed, from which information about cloud–climate feedbacks could be inferred.

Despite the fundamental importance of physical parameterizations to improve climate model simulations and reduce uncertainties in climate projections, very few people are actively involved in the development, evaluation, and improvement of physical parameterizations. Progress in developing reliable model-based projections of future climate change could be greatly enhanced by these activities, for which considerable time and energy are required.

Essential Observations

The Oklahoma Atmospheric Radiation Measurement (ARM) site, which has been operating for 15 years, provides an example of detailed ground-based cloud process observations over long time series. Although these measurements contribute greatly to the improvement of process understanding and the development of process representations in models, studies that provide reliable information about cloud–climate feedbacks derived directly from these data are still lacking. One reason might be that the available data relate only to a specific, individual location. An ensemble of dedicated ground-based measurement sites positioned over a large spatial scale and over a long time period could resolve this issue. Such sites might be less well equipped than ARM or Cloudnet sites. However, as a minimum requirement, they would need to observe some main cloud properties (e.g., cloud-base and cloud-top heights, integrated liquid water path (LWP), radiative cooling and fluxes, precipitation) together with thermodynamic properties derived from radiosondes. Maintaining such an ensemble for a few years (to encompass, e.g., El Niño and La Niña events) could provide the opportunity to explore relationships between cloud properties, large-scale atmospheric circulation, and thermodynamic stratification of the atmosphere (e.g., low-level tropospheric static stability, low-level humidity gradients).

For cloud–climate feedback studies, a sustained long-term monitoring of the Earth's radiation budget (ERB) is essential. Loeb et al. (2007) suggest that with the current measurement stability of broadband radiometers, such as CERES, a record of about ten years of consistent observations would allow for the detection of some cloud–climate feedbacks. Since a substantial part of cloud feedbacks act on longer timescales, sustained ERB observation is necessary. The commitments to operate broadband radiometers onboard the EarthCARE and NPOESS NPP platforms are important, as are the contributions of instruments monitoring ERB at high temporal resolution. However, for the latter, spatial coverage is limited (e.g., GERB on MSG and ScaRaB on Mega-Tropique). In addition, data are not able to be intercompared as a result of possible instrument failures, suggesting the need for an absolutely calibrated

ERB instrument. Retrievals with such an instrument at any two times could then be compared to estimate the temporal change in the ERB.

A dataset of reanalyses, including the assimilation of satellite observations of cloudy atmospheres, may prove particularly useful for the investigation of cloud–climate feedbacks.

Summary

Cloud–climate feedbacks represent a major uncertainty in projections of climate change. Proxies for climate change (e.g., seasonal cycle, interannual variability, volcanic eruptions) have thus far been found to be unfit to quantify cloud feedbacks. Improved understanding of the cloud–climate feedback processes may lead to a better exploitation of existing data. Compositing techniques to separate feedback mechanisms, as well as the combination of model results with observational data, have already enhanced our understanding of cloud feedbacks. New analysis methods of cloud responses to forcing in models, and of natural variability, may yield further insights. A very promising, though demanding pathway is to improve cloud process parameterizations in large-scale models through the use of observations and LES or CRM, as well as by exploiting NWP experiences. Sustained observations from existing and additional satellites and ground-based sites are needed to support these approaches.

What Are the Limitations of Bottom-up Diagnoses that Require Top-down Approaches?

Limitations of Current Small- to Mesoscale Observations

Traditional field programs have advanced our process-level understanding of the climate system and the representation of cloud processes in detailed models. Still, it has proven far more challenging to apply the field observations to constrain the parameterizations and large-scale properties of clouds in models. Challenges arise from the disparities in spatial scales between point measurements and climate models (the so-called "process-parameterization gap") as well as from disparities in temporal scales between short-term field programs and long-term climate change.

To engage climate model development with field measurements, we suggest that sensitivity studies with large-scale models be used more frequently to guide the design of field programs. For example, intermodel differences across ensembles of climate models could be used to identify processes, cloud systems, or aerosol–cloud interaction processes that represent major sources of uncertainty in simulations of climate perturbations. These differences could then be targeted for detailed field programs. Consideration should be given to methods that bridge the spatial and temporal gaps, so that field measurements could be used to test climate sensitivity.

Simplification of Large-scale Models to Guide Process Understanding

Over the last decade, the climate modeling community has increased the complexity of models to address critical scientific questions by adding more processes or components (e.g., aerosols, chemistry, carbon cycle, dynamic vegetation) and increasing the model resolution. It appears, however, that this did not yield a better understanding of cloud–climate feedbacks, because the diagnosed uncertainty in climate sensitivity has not decreased with time. Large-scale models appear most useful not only as a basis for the quantification of climate sensitivity but also as a framework for advancing ideas and understanding on how the climate works and responds to external perturbations. Such models can be used to develop concepts, which in turn must then be explored, or isolated, through simplifications.

A better understanding of the physical mechanisms underlying the cloud–climate feedbacks produced by climate models could be useful in designing a strategy to evaluate these feedbacks using observations. By simplifying (rather than complicating) models and conducting idealized experiments, we may be able to pinpoint the main critical processes, to prioritize them, and to test resulting ideas or theories. Interpretation frameworks could then be proposed to assist our understanding of the physics of and intermodel differences in cloud–climate feedbacks.

One approach to simplify GCMs may be to reduce the complexity of the large-scale boundary conditions (e.g., aquaplanet versions of the models, even for CRMs or super-parameterizations), to reduce the dimensionality of the system (e.g., 2-D or 1-D model versions derived from the 3-D model), or even to remove some processes (i.e., replace complex microphysical schemes with simpler ones). Models of intermediate complexity, such as the quasi-equilibrium tropical circulation model (Neelin and Zeng 2000), might also be used. Simple conceptual models (e.g., 2-box models) may be viewed as the ultimate step to this simplification process.

The extent to which simplified models are useful in reproducing and interpreting complex model results should be tested and quantified by analyzing, for a given model, how the cloud–climate feedbacks compare across the hierarchy of model complexities. An ideal situation would be for each GCM to be associated with a suite of simpler or more idealized model versions to support the analysis and the understanding of its results.

Ensembles of Process Model Simulations to Assess Large-scale Cloud Feedbacks

Systematic comparisons between process models (LES and CRM) and large-scale models may be used to analyze and improve cloud parameterizations in an effort to assess cloud–climate feedbacks on a large scale (discussed above). Idealized sensitivity experiments (e.g., +2 K SST perturbations or perturbations applied to cloud properties) might be conducted using LES or CRMs (i.e., in the absence of any interaction with the large-scale circulation) and compared to GCM simulations allowing for this interaction.

Comparing the response of clouds at these different scales would enable us to understand the response of clouds to a perturbation at a specific scale as well as the various controlling factors involved. Global CRMs or superparameterization models constitute integrating tools of small and large spatial scales and may play a particularly important role in this regard (Collins and Satoh, this volume).

Summary

For observational studies to be successful, key mechanisms of cloud–climate feedbacks need to be identified. This can be accomplished by simplifying large-scale models. Systematic comparisons between process and large-scale models offer the best opportunity to improve the representation of cloud processes in models and to analyze cloud feedback processes.

Is the Uncertainty Range of Radiative Flux Perturbations by Aerosols Related to Systematic Errors in Climate State? What Aspects of Clouds Must Be Represented to Treat Cloud–Aerosol Interactions with Fidelity?

Uncertainties in Estimates of Radiative Flux Perturbations

Aerosol forcing, as inferred from representing the geographical distribution of anthropogenic aerosols and their radiative influences in climate models (i.e., the so-called forward estimates of forcing), has been consistently larger than "inverse" estimates, whereby aerosol forcing is inferred from the total forcing required to obtain the observed temperature change over the industrial period for the generally accepted range of climate sensitivity, observed ocean heat uptake, and given anthropogenic greenhouse gas forcing (Anderson et al. 2003). Given the large uncertainties in processes connected to both aerosol forcings and climate sensitivity, it is not obvious that this discrepancy can simply be attributed to an overestimate of the aerosol forcings by forward estimates.

In a multimodel comparison, Denman et al. (2007) analyzed the forcing of anthropogenic aerosols with respect to the aerosol compounds and aerosol–cloud interaction processes considered (i.e., whether or not aerosol interactions with mixed-phase and ice clouds are taken into account). If sulfate or sulfate and black carbon are solely used, they found that global mean forcing is larger than if organics and aerosol interactions with mixed-phase and ice clouds are included as well. However, the variations within a given aerosol compound category are at least as large as the differences between different categories.

Calculations of the total aerosol forcing depend critically on estimates of both present-day and preindustrial aerosol (precursor) emissions. In particular, the limited knowledge of preindustrial aerosol concentrations introduces considerable uncertainty about the radiative forcing by aerosols. This is manifest in our lack of knowledge about the physical processes that determine the unperturbed cloud droplet number concentration (CDNC) for use in global climate model parameterizations. Observations show that CDNC, even in very clean regions, falls rarely below some lower bound (approximately 10 cm^{-3}).

Cloud Aspects Critical for Aerosol Indirect Effect Quantifications

Cloud Albedo

The sensitivity of cloud albedo to perturbations in cloud condensation nuclei, and thus the strength of the Twomey effect (Twomey 1974), depends on the value of cloud albedo. To simulate aerosol indirect effects with fidelity, we must thus be able to simulate realistically the distribution of cloud albedo (Feingold and Siebert, this volume). Satellite data of this quantity exist, allowing for an observation-based evaluation of the distribution.

Boundary Layer Moisture Budget

Process modeling and theoretical studies suggest that for stratocumulus, the cloud thickness response to perturbations in CDNC may, in many cases, act to oppose the changes in albedo caused by the Twomey effect alone. The dynamic and microphysical processes responsible for this depend critically on surface relative humidity. In general, the total aerosol indirect effect may be strongly sensitive to the cloud macrophysical and boundary layer properties (see Brenguier and Wood; Feingold and Siebert; and Stevens and Brenguier, all this volume).

Cloud Subgrid-scale Variability

The aerosol and microphysical processes that determine cloud radiative properties occur on scales much smaller than those resolved by GCMs, and they are nonlinear. Thus, the application of GCM-resolved moisture and motion fields will produce unrealistic estimates of aerosol indirect effects. A general approach to address this problem is to construct PDFs for subgrid velocity and moisture distributions (Lohmann and Schwartz, this volume). Aerosol and microphysical processes are then evaluated using the PDF vertical velocity and moisture.

Microphysical Processes in Ice and Mixed-phase Clouds

Aerosols may have an important influence on the optical properties of ice clouds as well as on microphysical processes in mixed-phase clouds. To simulate such effects realistically, the various ice crystal nucleation pathways and mixed-phase microphysical processes must be accurately represented in the model (Lohmann and Schwartz, this volume).

Summary

"Inverse" estimates (i.e., aerosol forcings inferred from the observed global warming) are hampered by the uncertainty about climate sensitivity. Biases in modeled cloud fields, limited process representation, missing aerosol–cloud interaction processes, and uncertainties in the unperturbed aerosol distribution are the main sources of uncertainties for simulated aerosol forcings ("forward" estimates). Key cloud aspects necessary for a realistic representation of aerosol–cloud interactions in large-scale models are the distribution of cloud albedo, boundary layer moisture budget, subgrid-scale variability of moisture and vertical updraft velocity, as well as ice crystal nucleation and mixed-phase cloud microphysical processes. In general, the quantification of perturbations of cloud properties by anthropogenic aerosols in large-scale models is highly dependent on the formulation of the cloud process representations.

Is It Possible to Design an Observational Program to Detect and Quantify Aerosol Indirect Effects?

Limits to Existing Observation-based Studies

Despite many observational campaigns, no large body of observational evidence exists to provide a positive correlation between cloud albedo and aerosol perturbations on a large scale. Although it has been established that an increase in aerosol concentration results usually in an increase of CDNC, this does not necessarily lead to an increase in cloud albedo, since the LWP of the cloud also changes. Although the LWP may be responding to aerosol-induced changes, it may be determined to a first order by differences in large-scale meteorological forcings.

Ship Track Analyses

Studies of plumes that interact with clouds, as typified by ship tracks, provide a natural laboratory to sample a partial derivative with respect to aerosol concentration or cloud droplet concentration, unlike, for example, a regional-scale plume of polluted air, in which the meteorology co-varies with the aerosol properties. Heat and moisture are also injected into the atmosphere within ship exhaust, but these perturbations can be regarded as negligible after the plume spreads over a few kilometers in width. However, ship tracks occur only in one type of cloud (marine stratocumulus), and they appear to be more likely in shallow boundary layers (Coakley et al. 2000). Despite these limitations, analyses of ship tracks might be exploited to yield further insights into indirect effects of aerosols, since they reveal information about the important processes (e.g., entrainment) that control the cloud response to aerosol perturbations (see Cotton, this volume).

Statistical Relationships

Correlations between cloud properties (cloud droplet radius, CDNC) and column aerosol concentration (aerosol index or aerosol optical depth, or a proxy such as hemispheric or land–sea contrasts) from satellite data may be used to infer some clues about aerosol–cloud interactions (Nakajima and Schulz, this volume). Since it is, however, impossible with a passive satellite instrument to quantify the relevant cloud and aerosol parameters simultaneously at a particular location, these analyses rely on the assumption that aerosol properties in clear scenes are similar to those in near-by cloudy situations. Long-term measurements from ground-based sites are necessary to establish similar correlations (Feingold and Siebert, this volume).

Approaches for Better Use of Observations

Analysis of Time History in Satellite Retrievals

Observations and analyses from a combination of various satellite instruments (e.g., from the A-Train constellation) provide a comprehensive dataset for studying global cloud and aerosol properties, their radiative effects, and their interrelationship. It is critical to determine whether the correlations imply causation (e.g., whether correlations between cloud albedo and aerosol optical depth from adjacent clear-sky regions constitute evidence for aerosol indirect effects). Correct construction and interpretation of the correlations require methods for identifying measurements with similar physical and chemical tendencies over the lifetime of the clouds prior to measurement. These tendencies are not available from the instantaneous observations collected by polar-orbiting satellites. However, the tendencies could be constructed in a chemical and meteorological assimilation of satellite observations. Analyses of satellite data might be enhanced to include the tendencies of physical and chemical environmental factors that have been shown to govern cloud–aerosol interactions.

Model–Data Comparisons to Infer Causalities

To explore the extent to which observed correlations between cloud and aerosol properties represent causality, similar correlations can be computed in models. Consistency between a process model and observations and sensitivity studies with the model, with and without relevant processes, provide confidence that causality can be inferred from observed correlations (Feingold and Siebert, this volume).

Careful Choice of Analyzed Situations

The use of natural meteorological variability is a means of testing the ability of a model to represent correctly the key physics of the aerosol indirect effects. Careful choice needs to be made regarding possible locations where this might be beneficial. The system should have some simple and effective proxies for meteorological control, the meteorological variability must not greatly swamp the essential aerosol signal, the albedo of the clouds in the system need to be potentially susceptible to perturbations in aerosols, and the clouds in the system should have sufficient horizontal homogeneity such that their properties (microphysical and macrophysical) are detectable from space. To some degree, marine stratocumulus satisfies these criteria. Spaceborne retrievals of CDNC, LWP, and cloud cover, and possibly other variables (e.g., drizzle from Cloudsat), can be used for model evaluation. CDNC could serve as a proxy for the aerosol variability. In principle, models could be used to calculate the radiances actually measured by satellites.

Ensembles of LES/CRM Model Simulations

The understanding of aerosol–cloud interactions at a large scale might be advanced by simulating broad ranges of meteorological situations with process models. LESs are highly idealized, limited domain (order of 10 km) simulations. When embedding LES within a regional model, mesoscale forcing is provided to the finer LES grid. Two-way nesting allows for communication between grids. In a different approach, meteorological parameters, such as SST and large-scale subsidence, may be varied along with aerosol perturbations to explore the relative importance of aerosol and meteorology.

A multiscale modeling framework, which replaces the conventional cloud parameterizations with a CRM in each grid-column of a general circulation model, or global CRMs are ideal tools to apply CRMs to a large variety of meteorological situations for all regions of the globe. These could be used to investigate the effects of indirect aerosols on a large scale (Collins and Satoh; Grabowski and Petch, both this volume).

Improving Large-scale Model Parameterizations

It has been a long-standing goal to use field programs, process models, and satellite observations effectively to improve GCM parameterizations, which are of critical importance to represent the effects of indirect aerosols accurately. Still, this remains a daunting challenge. A basic strategy to achieve this goal is first to develop confidence in process models (e.g., LES, CRMs). Field campaigns can play a central role in doing this. Next, process models must be used in the parameterization process. Finally, satellite observations, including statistical analysis of covariances, provide a means to evaluate the GCM with its parameterization.

Ice and Mixed-phase Clouds

Much of the previous discussion centered on the link from aerosols to the CDNC and albedo of warm clouds. Whereas warm clouds (extensive stratus or stratocumulus sheets especially over the oceans) are important for the radiation balance, mixed-phase and ice clouds are more important in terms of the possible effect of aerosols on precipitation. However, there are much larger uncertainties associated with the processes acting in these clouds.

Summary

Various studies of observational data have established statistical relationships between cloud microphysical properties and aerosol concentrations consistent with an assumed aerosol indirect effect. However, covariance of aerosol concentrations and meteorological cloud-controlling factors leads to a variety of responses beyond the enhancement of CDNC at increased aerosol concentrations. Time history in 3-D data is needed to overcome some of the issues. To interpret correlations found in observational data, process model studies can establish causality, whereas process models applied to larger scales can be used to comprehend aerosol effects beyond instantaneous influences. A better understanding of models at the process scale may help us improve parameterizations in large-scale models.

Is the Concept of Forcing for Aerosols Useful? If Not, What Is a Viable Definition of Forcing?

Shortcomings of IPCC's Radiative Forcing Concept

The forcing summary graph put forth by the IPCC (2007) includes only the direct forcing and Twomey aerosol indirect effect. It does not reflect the full magnitude or uncertainty of cloud-mediated aerosol forcing. However, some effects (e.g., the aerosol "cloud lifetime effect,") for which confidence in understanding is limited do not necessarily have the smallest magnitude, and therefore they cannot be omitted. The inclusion of such uncertain effects emphasizes the imperative need to narrow the gap in our understanding.

Forcing, as used by IPCC, is well-defined for the cloud albedo effect when simulated in a simplified manner, but it cannot capture the microphysical responses to aerosol-related changes in cloud particle number. For well-mixed greenhouse gases, a consistent scaling between forcing and climate change is a reasonable expectation, but this is less so for agents such as aerosols, which are not well-mixed. Aerosol–cloud interactions break this consistency further (Haywood et al., this volume).

The Radiative Flux Perturbation as an Alternative

Standard radiative forcing is computed as the change in radiation fluxes at the tropopause attibutable to an external perturbation, with tropospheric temperature and humidity profiles held fixed, but allowing for fast (order of months) temperature adjustment in the stratosphere. In contrast, radiative flux perturbation allows for rapidly responding tropospheric meteorological fields to adjust as well. The radiative flux perturbation concept does amount to "fixed-SST forcing" (Shine et al. 2003; Hansen et al. 2005; Forster et al. 2007) or "quasi-forcing" (Rotstayn and Penner 2001).

The radiative flux perturbation is likely to yield a more consistent scaling with climate change than forcing. It is also likely to convey more intermodel uncertainty than forcing, thus including some information about the level of scientific uncertainty.

Recent work suggests that one of the consequences of adopting a flux perturbation approach for the assessment of greenhouse gas forcing is that the flux perturbation may include a significant cloud response component (Gregory and Webb 2008; Andrews and Forster 2008). This work further suggests that the cloud–climate feedback, when strictly defined as being the response of clouds to increases in surface temperature, may actually be quite small. Thus, uncertainty in the proportionality between radiative flux perturbation and global mean change in surface temperature would probably be less than for the proportionality between forcing and global mean temperature change. Of course, this inclusion of response to forcing in the radiative flux perturbation does not reduce overall uncertainty in, for example, expected warming at the end of the 21st century.

The proposal to adopt the radiative flux perturbation concept for aerosol radiative impacts implies the convolution of direct and indirect effects, rapid cloud responses, and other rapid climate responses in a single measure. This convolution is similar to the mixing of forcing and response in the actual climate system. Observational estimates of flux perturbations should be based on total derivatives of regional and global planetary albedo with respect to measurable extrinsic aerosol properties, such as optical depth. Since the flux perturbation is based upon fixed SSTs, it might be useful to composite oceanic observations with similar spatial patterns of SST but varying measures of aerosol loading. Since the flux perturbation includes the radiative adjustment of the troposphere to anthropogenic aerosol perturbations, it may be necessary to combine observations over the appropriate radiative dynamic timescales.

Aerosol indirect effects other than the Twomey effect are difficult to define. The distinction between first, second, and semi-direct effects is obsolete, and many more cloud–aerosol interaction "effects" may be defined. Thus, the replacement of various ill-defined aerosol indirect effects by just the combined radiative flux perturbation quantification appears to be an advantage of this concept.

Open Issues

Whether radiative flux perturbations exhibit the same additivity as practiced for traditional radiative forcings remains unknown. The relationships between flux perturbations and traditional forcings for greenhouse gases and aerosols need to be quantified across the ensemble of climate models used for IPCC.

To calculate forcing (as precisely defined), GCM simulations using a double-call to the radiation scheme need to run typically for only one year in order to sample the full seasonal cycle. In contrast, simulations to calculate the flux perturbation must last considerably longer, typically from 5–10 years. This is because the meteorology usually differs between control and perturbed simulations in calculating the flux perturbation, and thus one has to run the simulations for a number of years to obtain a good signal-to-noise ratio.

One difficulty with the flux perturbation is that usually only SSTs and sea-ice extents are prescribed. This means that although feedbacks via surface temperature over ocean areas are eliminated, those via changes in land-surface temperature are not; neither are other land-surface responses, such as changes in soil moisture or snow cover. Ideally, land-surface temperatures would also be prescribed, but the complex formulation of land-surface parameterization schemes in current GCMs makes this difficult and may not be possible for all models. Thus, we suggest the flux perturbation in the form of "fixed-SST forcing."

Summary

We recommend that the community quantify the effects of aerosols on climate using the concept of radiative flux perturbations, known in the literature as quasi-forcing or fixed-SST forcing. Although extremely useful for well-mixed greenhouse gases, the concept of forcing has limited utility and conceptual rigor when applied to cloud–aerosol interactions. Since the comparison of forcings from various radiative species has been very useful for scientific and policy-oriented applications, we recommend, in addition, that the community compute the effects of all anthropogenic agents using the radiative flux perturbation approach. The traditional IPCC bar chart for forcings could be complemented by a corresponding bar chart for radiative flux perturbations. The application of the flux perturbation concept includes the advantage of a more intuitive observational assessment possibility, a tighter definition of feedbacks as a pure response to surface temperature warming, and a more plausible definition of aerosol effects.

References

Anderson, T. L., R. J. Charlson, S. E. Schwartz et al. 2003. Climate forcing by aerosols: A hazy picture. *Science* **300**:1103–1104.

Andrews, T., and P. M. Forster. 2008. CO_2 forcing induces semi-direct effects with consequences for climate feedback interpretations. *Geophys. Res. Lett.* **35**:L04802.

Bodas-Salcedo, A., M. J. Webb, M. E. Brooks et al. 2008. Evaluation of cloud systems in the Met Office global forecast model using CloudSat data. *Geophy. Res. Abst.* **10**:EGU2008-A-10141.

Bony, S., J.-L. Dufresne, H. Le Treut, J.-J. Morcrette, and C. Senior. 2004. On dynamic and thermodynamic components of cloud changes, *Climate Dyn.* **22**:71–86.

Browning, K. A., et al. 1993. The GEWEX Cloud System Study (GCSS). *Bull. Amer. Meteor. Soc.* **74**:387–399.

Cess, R. D., G. L. Potter, J. P. Blanchet et al. 1990. Intercomparison and interpretation of climate feedback processes in 19 atmospheric general circulation models. *J. Geophys. Res.* **95(D10)**:16,601–16,615.

Chepfer, H., S. Bony, M. Chiriaco et al. 2008. Use of CALIPSO lidar observations to evaluate the cloudiness simulated by a climate model. *Geophys. Res. Lett.* **35**:L15804.

Coakley, J. A., P. A. Durkee, K. Nielsen et al. 2000. The appearance and disappearance of ship tracks on large spatial scales. *J. Atmos. Sci.* **57**:2765–2778.

Denman, K. L., G. Brasseur, A. Chidthaisong et al. 2007. Couplings between changes in the climate system and biogeochemistry. In: Climate Change 2007: The Physical Science Basis. Contribution of Working Group I to the Fourth Assessment Report of the Intergovernmental Panel on Climate Change, ed. S. Solomon, D. Qin, M. Manning et al., pp. 499–588. New York: Cambridge Univ. Press.

Dufresne, J.-L., and S. Bony. 2008. An assessment of the primary sources of spread of global warming estimates from coupled ocean–atmosphere models. *J. Climate,* in press.

Forster, P., V. Ramaswamy, Artaxo et al. 2007. Radiative Forcing of Climate Change. In: Climate Change 2007: The Physical Science Basis. Contribution of Working Group I to the Fourth Assessment Report of the Intergovernmental Panel on Climate Change, edited by S. Solomon, D. Qin, M. Manning et al., pp. 129–234. New York: Cambridge Univ. Press.

Gregory, J. M., and M. J. Webb. 2008. Tropospheric adjustment induces a cloud component in CO_2 forcing. *J. Climate.* **21(1):**58–71.

Hall, A., and X. Qu. 2006. Using the current seasonal cycle to constrain snow albedo feedback in future climate change. *Geophys. Res. Lett.* **33**:L03502.

Hansen, J., M. Sato, R. Ruedy et al. 2005. Efficacy of climate forcings. *J. Geophys. Res.* **110**:D18104.

Illingworth, A. J., R. J. Hogan, E. J. O'Connor et al. 2007. Cloudnet: Continuous evaluation of cloud profiles in seven operational models using ground-based observations. *Bull. Am. Meteor. Soc.* **88**:883–898.

IPCC. 2007. Climate Change 2007: The Physical Science Basis. Contribution of Working Group I to the Fourth Assessment Report of the Intergovernmental Panel on Climate Change, ed. S. Solomon, D. Qin, M. Manning et al. New York: Cambridge Univ. Press.

Leith, C. E. 1975. Climate response and fluctuation dissipation. *J. Atmos. Sci.* **32**:2022–2026.

Loeb, N. G., B. A. Wielicki, W. Su et al. 2007. Multi-instrument comparison of top of the atmosphere reflected solar radiation, *J. Climate* **20**:575–591.

Neelin, J. D., and N. Zeng. 2000. A quasi-equilibrium tropical circulation model—formulation. *J. Atmos. Sci.* **57**:1741–1766.

Rodwell, M. J., and T. N. Palmer. 2007. Using numerical weather prediction to assess climate models. *Q. J. Roy. Meteor. Soc.* **133**:129–146.

Rotstayn, L. D., and J. E. Penner. 2001. Indirect aerosol forcing, quasi forcing, and climate response. *J. Climate* **14**:2960–2975.

Schwartz, S. E. 2007. Heat capacity, time constant, and sensitivity of Earth's climate system. *J. Geophys. Res.* **112**:D24S05.

Shine, K. P., J. Cook, E. J. Highwood, and M. M. Joshi. 2003. An alternative to radiative forcing for estimating the relative importance of climate change mechanisms. *Geophys. Res. Lett.* **30**:2047.

Soden, B. J., and I. M. Held. 2006. An assessment of climate feedbacks in coupled ocean–atmosphere models. *J. Climate* **19**:3354–3360.

Soden, B. J., R. T. Wetherald, G. L. Stenchikov, and A. Robock. 2002. Global cooling after the eruption of Mount Pinatubo: A test of climate feedback by water vapor. *Science* **296**:727.

Twomey, S. A. 1974. Pollution and the planetary albedo. *Atmos. Environ.* **8**:1251–1256.

Webb, M., C. Senior, S. Bony, and J.-J. Morcrette. 2001. Combining ERBE and ISCCP data to assess clouds in the Hadley Centre, ECMWF and LMD atmospheric climate models. *Climate Dyn.* **17**:905–922.

Williams, K. D., and G. Tselioudis. 2007. GCM intercomparison of global cloud regimes: Present-day evaluation and climate change response. *Climate Dyn.* **29**:231–250.

Yokohata, T., S. Emori, T. Nozawa et al. 2005. Climate response to volcanic forcing: Validation of climate sensitivity of a coupled atmosphere–ocean general circulation model. *Geophys. Res. Lett.* **32**:L21710.

Abbreviations

ACE	Aerosol Characterization Experiment: 1, southern hemisphere; 2, North Atlantic Regional; Asia, Asian Pacific Regional
ACES	Artificial Cloud Experimental System
ACTOS	Airborne Cloud Turbulence Observation System
AeroCom	Aerosol Comparisons between observations and models
AERONET	AErosol RObotic NETwork
AGCM	Atmospheric General Circulation Model
AI	Aerosol Index
AIDA	Aerosol Interactions and Dynamics in the Atmosphere
AIM	Aerosol Inorganics Model
Air NEPM	National Environmental Protection Measure for ambient Air quality
AIRS	Atmospheric InfraRed Sounder
AMMA	African Monsoon Multidisciplinary Analysis
AMSR-E	Advanced Microwave Scanning Radiometer – EOS
AMSU-A	Advanced Microwave Sounding Unit-A
AnP	Ångstrom Parameter
AOD	Aerosol Optical Depth
AOT	Aerosol Optical Thickness
ARM	Atmospheric Radiation Measurement
ASTEX	Atlantic Stratocumulus Transition EXperiment
ATEX	ATmosphere EXplosive
AVHRR	Advanced Very High Resolution Radiometer
BAMEX	Bow Echo And MCV EXperiment
BASC	Board of the National Research Council
BOMEX	Barbados Oceanographic and Meteorological EXperiment
BRE	Bridger Range Experiment
BSOA	Biogenic Secondary Organic Aerosols
BSRN	Baseline Surface Radiation Network
CALIOP	Cloud–Aerosol Lidar with Orthogonal Polarization
CALIPSO	Cloud–Aerosol Lidar and Infrared Pathfinder Satellite Observation
CAM	Community Atmosphere Model
CAPE	Convectively Available Potential Energy
CAPS	Cloud Aerosol and Precipitation Spectrometer
CAS	Cloud and Aerosol Spectrometer
CCD	Charge Coupled Device
CCN	Cloud Condensation Nuclei
CCNC	Cloud Condensation Nucleus Counter

CCOPE	Cooperative COnvective Precipitation Experiment
CDNC	Cloud Droplet Number Concentration
CDR	Cloud Droplet Radius
CERES	Clouds and the Earth's Radiant Energy System
CF	Cloud Fraction
CFDC	Continuous Flow Diffusion Chamber
CFDE	Centre de Formation et de Documentation sur l'Environnement industriel
CIN	Center Ice Nuclei
CLARREO	CLimate Absolute Radiance and REfractivity Observatory
CNES	Centre National d'Etudes Spatiales
CNRS	Centre National de la Recherche Scientifique
CMOS	Canadian Meteorological and Oceanographic Society
COT	Cloud Optical Thickness
CPR	Cloud-Profiling Radar
CPU	Central Processing Unit
CRF	Cloud Radiative Forcing
CRM	Cloud-Resolving Model
CSRM	Cloud-System Resolving Model
CTM	Chemical Transport Model
CVI	Counterflow Virtual Impactor
DEE	Double End Element
DJF	December, January, February
DLR–IPA	Deutsches Zentrum für Luft und Raumfahrt–Institut für Physik der Atmosphäre
DMA	Differential Mobility Analyzer
DNS	Direct Numerical Simulations
DRE	Direct Radiative Effect
DSCOVR	Deep Space Climate Observatory
DUSEL	Deep Underground Science and Engineering Laboratory
DYCOMS-II	Dynamics and Chemistry of Marine Stratocumulus, Phase II
EARLINET	European Aerosol Research LIdar NETwork
EarthCARE	Earth Clouds, Aerosols, and Radiation Explorer
ECHAM	ECMWF model, Hamburg Version
ECMWF	European Centre for Medium-range Weather Forecasts
EDB	ElectroDynamic Balance
EECRA	Extended Edited Cloud Report Archive
EIS	Estimated Inversion Stability
EMEP	European Monitoring and Evaluation Programme
ENSO	El Nĩno Southern Oscillation
EOS	Earth Observing System
EPA	Environmental Protection Agency
EPIC	East Pacific Investigation of Climate
ERB	Earth's Radiation Budget

ERBE	Earth Radiation Budget Experiment
ERBS	Earth Radiation Budget Satellite
ESA	European Space Agency
EUCAARI	EUropean integrated project on aerosol cloud climate air quality interactions
EULAG	EULerian LAGrangian
EXA	EXtended Area
FACE	Florida Area Cumulus Experiment
FANGIO	Feedback ANalysis for GCM Intercomparison and Observation
FD	Fluctuation Dissipation
FINCH	Fast Ice Nucleus CHamber
FIRE	First International Radiation Experiment
FRH	Freezing Relative Humidity
FSSP	Forward Scattering Spectrometer Probe
GADS	Global Aerosol Data Set
GARP	Global Atmospheric Research Program
GATE	GARP Atlantic Tropical Experiment
GAW	Global Atmospheric Watch
GCRM	Global Cloud-Resolving Model
GCSS	GEWEX Cloud System Study
GEBA	Global Energy Budget Archive
GERB	Geostationary ERB
GEWEX	Global Energy and Water Cycle Experiment
GFDL AM2	Geophysical Fluid Dynamics Laboratory Atmospheric Model version 2
GFSC	Goddard Space Flight Center
GHG	GreenHouse Gas
GISS	Goddard Institute for Space Studies
GLI	GLobal Imager
GOCART	Global Ozone Chemistry Aerosol Radiation Transport
GOES	Geostationary Operational Environmental Satellite
GPS	Global Positioning System
HALO	High Altitude and Long-range Aircraft
HIAPER	High-performance Instrumented Airborne Platform for Environmental Research
HIPLEX	HIgh PLains EXperiment
HIRS	High-resolution Infrared Radiometer Sounder
HSRL	High Spectral Resolution Lidar
H-TDMA	Humidity Tandem Differential Mobility Analyzer
IAPSAG	International Aerosol Precipitation Science Assessment Group
ICESAT	Ice, Cloud, and land Elevation SATellite
IGAC	International Global Atmospheric Chemistry

IMPROVE	Interagency Monitoring of PROtected Visual Environments
IN	Ice Nuclei
IOCI	Indian Ocean Climate Initiative
IPCC	Intergovernmental Panel on Climate Change
ISCCP	International Satellite Cloud Climatology Project
ISCCP FD	International Satellite Cloud Climatology Project Flux Dataset
ITCZ	Intertropical Convergence Zone
IUGG	International Union of Geodesy and Geophysics
IWC	Ice Water Content
JAXA	Japan Aerospace Exploration Agency
JJA	June, July, August
KWAJEX	KWAJalein EXperiment
LACIS	Leipzig Aerosol Cloud Interaction Simulator
LCL	Lifting Condensation Level
LES	Large Eddy Simulations
LIS	Lightning Imaging Sensor
LITE	Lidar In-Space Technology Experiment
LMD	Laboratoire de Météorologie Dynamique du CNRS
LMDZ–INCA	Laboratoire de Météorologie Dynamique Zoom–INteraction of Chemistry and Aerosol
LSCE	Laboratoire des Science du Climat et de l'Environnement
LTS	Lower Tropospheric Stability
LW	LongWave
LWC	Liquid Water Content
LWP	Liquid Water Path
MAGE	Marine Aerosol and Gas Experiment
MAM	March, April, May
MAST	Monterey Area Ship Tracks experiment
MBL	Marine Boundary Layer
MCV	Mesoscale Convective Vortex
METROMEX	METROpolitan Meteorological EXperiment
MIROC	Model for Interdisciplinary Research On Climate
MISR	Multiangle Imaging Spectroradiometer
MJO	Madden-Julian Oscillation
MLS	Microwave Limb Sounder
MMF	Multiscale Modeling Framework
MODIS	MODerate resolution Imaging Spectroradiometer
MPI-HAM	Max-Planck-Institute, Hamburg
MPL-Net	MicroPulse Lidar Network
MRI	Meteorological Research Institute
MSA	MethaneSulfonic Acid
MSG	Meteosat Second Generation
MTF	Modulation Transfer Function

MVD	Median Volume Diameter
NAS	National Academy of Sciences
NASA	National Aeronautics and Space Administration
NCAR	National Center for Atmospheric Research
NCEP	National Centers for Environmental Protection
NCRF	Net Cloud Radiative Forcing
NEPM	National Environmental Protection Measure
NHRE	National Hail Research Experiment
NICAM	Non-hydrostatic ICosahedral Atmospheric Model
NIES	National Institute for Environmental Studies
NIR	Near-Infrared Radiation
NIST	National Institute of Standards and Technology
NISTARs	NIST Advanced Radiometers
NOAA	National Oceanic and Atmospheric Administration
NPOESS NPP	National Polar-orbiting Operational Environmental Satellite System Preparatory Project
NRC	National Research Council
NWP	Numerical Weather Prediction
NSW	New South Wales
OAP	Optical Array Probe
OLR	Outgoing Longwave Radiation
OPAC	Optical Properties of Aerosols and Clouds
ORBIMAGE	ORBital IMAGing Corporation
PATMOS	Pathfinder Atmosphere
PBL	Planetary Boundary Layer
PDF	Probability Density Function
PFR	Precision Filter Rad
PFs	Precipitation Features
PID	Proportional–Integral–Derivative
PIV	Particle Imaging Velocimetry
PMS	Particle Measuring System
POA	Primary Organic Aerosol
POLDER	POLarization and Directionality of the Earth's Reflectances
PR	Precipitation Radar
PRE-STORM	Preliminary Regional Experiment for STormscale Operational and Research Meteorology
RF	Radiative Forcing
RH	Relative Humidity
RICO	Rain In Cumulus over Ocean
RPF	Rainfield Per Flash
SACZ	South Atlantic Convergence Zone
SAFARI	Southern African Regional Science Initiative
ScaRaB	Scanner for Radiation Budget
SCAR-B	Smoke, Clouds and Radiation–Brazil

SCRF	Surface Cloud Radiative Forcing
SeaWiFS	Sea-viewing Wide Field-of-view Sensor
SEVIRI	Spinning Enhanced Visible and InfraRed Imager
SLH	Spectral Latent Heating
SOCEX	Southern Ocean Cloud Experiment
SON	September, October, November
SPARC	Stratospheric Processes And their Role in Climate
SPEC HVPS	Straton Park Engineering Corporation High Volume Precipitation Spectrometer
SPCZ	South Pacific Convergence Zone
SPRINTARS	Spectral Radiation-Transport Model for Aerosol Species
SRES	Special Report on Emissions Scenarios
SST	Sea Surface Temperature
STP	Standard Temperature and Pressure
SW	ShortWave
TAPM	The Air Pollution Model
TAR	Third Assessment Report (IPCC)
TARFOX	Tropospheric Aerosol Radiative Forcing Observation Experiment
TIROS	Television Infrared Observation Satellites
TOA	Top-Of-Atmosphere
TOGA-COARE	Tropical Ocean Global Atmosphere Coupled Ocean Atmosphere Response Experiment
TOMS	Total Ozone Mapping Spectrometer
TRMM	Tropical Rainfall Measuring Mission
TWC	Total Water Content
UAV	Unmanned Airborne Vehicle
UCLA	University of California, Los Angeles
UIO-CTM	Chemistry Transport Model, Univ. of Oslo
UMI	IMPACT at University of Michigan
UV	UltraViolet
VOCALS	Vamos Ocean-Cloud-Atmosphere-Land Study
WFOV	Wide-Field-Of-View
WHO	World Health Organization
WMO	World Meteorological Organization
ZINC	Zürich Ice Nucleation Chamber
ZSR	Zdanovski–Stokes–Robinson

Name Index

Abbatt, J. P. D. 240, 250, 306
Abdul-Razzak, H. 541
Ackerman, A. S. 3, 127, 134, 182–185, 239, 326, 355, 453, 458, 461, 501
Ackerman, B. 375
Ackerman, S. A. 482
Adams, P. J. 259
Aitken, J. 320
Akiyama, T. 117, 119, 122
Albrecht, B. 3, 108, 112, 134, 181–184, 190, 278, 321, 420, 453, 500, 545
Allan, J. D. 66
Allan, R. P. 32
Alpert, P. 25, 377, 378, 395
Anderson, J. G. 144, 475
Anderson, T. L. 2, 5, 7, 41, 68, 127, 287, 288, 440, 547, 564
Andreae, M. O. 3, 59, 61–64, 127, 128, 296, 302, 312, 371, 409
Andrews, T. 570
Ansmann, A. 94, 249, 518
Anthes, R. A. 340
Aoki, K. 42
Arblaster, J. 390
Archeluta, C. M. 249
Arkin, P. A. 223
Arrhenius, S. A. 11
Asa-Awuku, A. 301, 309
Atlas, D. 348
Austin, P. H. 303
Ayers, G. 8, 52, 110, 134, 321, 369, 371, 381, 384, 385, 386, 394, 443, 445
Ayyalasomayajula, S. 163

Back, L. E. 221, 222
Bailey, M. 154, 248
Baker, M. B. 185, 453, 550
Barsugli, J. J. 222
Bates, B. C. 392
Bates, T. S. 127, 502
Bauer, P. 96
Baumann, R. 247
Baumgardner, D. 75–78, 85, 90, 248
Beheng, K. D. 542
Bellon, G. 181
Bellouin, N. 412
Bell, T. L. 3, 492
Bender, F. A. M. 535
Bengtsson, L. 32, 208, 254
Benz, S. 152, 160
Ben-Zvi, A. 342
Berg, W. 543
Berger, W. H. 350
Berry, E. X. 542
Bertram, A. K. 246
Betts, A. K. 181, 182, 202
Bigg, E. K. 254, 346, 379, 381, 382, 384
Bilde, M. 294, 298, 299
Birmili, W. 4
Biswas, K. R. 346
Biter, C. J. 77
Bjerknes, J. 175
Black, J. F. 340
Blake, D. 176
Blanchet, J.-P. 543
Bodas-Salcedo, A. 519, 560
Boe, B. A. 344
Boers, R. 189, 323, 382, 383
Bögel, W. 247
Bony, S. 10, 21, 200, 227, 276, 469, 470, 476, 497, 511, 513, 516, 523, 550, 557–560
Borys, R. D. 370, 375, 376
Bösenberg, J. 43
Boucher, O. 404, 409, 419, 453, 459, 541
Bougeault, P. 183
Bowen, E. G. 346
Bradley, R. S. 350
Braham, R. R., Jr. 201, 342, 373, 378
Brenguier, J.-L. 6, 10, 79, 85, 173, 179, 183, 189, 269, 276, 278, 285, 301, 325, 434, 446, 458, 487–496, 565
Bréon, F.-M. 131, 416, 419
Brest, C. L. 30
Bretherton, C. S. 6, 7, 20, 112, 131, 137, 176, 182, 185, 217, 221–225, 228, 229, 238, 269, 301, 469, 490, 501, 513, 523
Broekhuizen, K. 308
Brooks, S. D. 304
Browning, K. A. 560

Brown, S. J. 116
Bryson, R. A. 351
Budyko, M. 352
Bundke, U. 158
Bunz, H. 153, 160
Burkhardt, U. 259
Burnet, F. 179
Businger, S. 500
Byers, H. R. 201

Cadeira, K. 353
Cai, W. 390
Cantrell, W. 240
Carlin, B. 251
Carras, J. N. 381
Carrió, G. G. 243
Cess, R. D. 471, 476, 558
Chahine, M. 81
Chameides, W. L. 413
Changnon, S. A., Jr. 373, 491
Charlson, R. J. 1, 23, 134, 142, 176, 324, 433, 440, 442
Charney, J. 188
Chen, J.-M. 331
Chen, T. 244, 249, 252
Chen, Y. 459
Chepfer, H. 519, 560
Cho, H. 358
Chosson, F. 179
Chuang, C. C. 459
Chuang, P. Y. 9, 291, 301, 309, 433, 493, 541
Chung, C. E. 412
Chylek, P. 242
Clarke, A. D. 330, 502
Clayton, H. H. 176
Clegg, S. L. 60, 298
Coakley, J. A., Jr. 3, 131, 371, 413, 567
Cober, S. G. 77, 78, 87–89, 98
Colberg, C. A. 250
Collins, W. D. 10, 199, 209, 212, 226, 230, 273, 469, 526, 557, 564, 569
Comstock, K. K. 189, 496, 498
Conant, W. C. 323, 495
Connolly, P. J. 243
Cooper, W. A. 342, 344
Cotton, W. R. 8, 134, 321, 339–348, 350–354, 370, 373–376, 393, 433, 443, 444, 491, 567
Coulier, P. J. 320
Cox, S. K. 257
Craig, G. C. 199, 205, 207, 208, 228, 258, 469
Crutzen, P. J. 352, 354, 358
Cruz, L. A. 185
Cui, Z. 203, 243
Cziczo, D. J. 240, 242, 243, 255

d'Almeida, G. A. 40
Damiani, R. 503, 506
Daum, P. H. 543
Davidovits, P. 152
Davis, E. 159
Davis, M. 189
Dean, S. M. 239, 247, 259
Deardorff, J. W. 177
Delanoë, J. 521
DeMott, P. J. 152, 161, 241, 242, 249, 302, 303, 358, 518
Denman, K. L. 190, 543, 544–549, 565
Dennis, A. S. 342, 344, 346
Dentener, F. 51
Dessens, J. 348
Dessler, A. E. 239
Detwiler, A. 358
Deuzé, J. L. 41
Devasthale, A. 3
Diehl, K. 242, 546
Di Michele, S. 96
Diner, D. J. 141
Dobbie, S. 239, 244
Donner, L. 451, 557
Douville, H. 3
Dubovik, O. 42
Dufresne, J.-L. 21, 200, 227, 459, 470, 513, 558
Duft, D. 152, 159
Durkee, P. A. 286, 354, 371
Dusek, U. 66, 67
Dye, J. E. 90
Dyson, F. J. 352

Early, J. T. 353
Easter, R. C. 537
Eklund, D. L. 348
Emanuel, K. A. 516

Emori, S. 116, 476
Ervens, B. 322, 329, 448
Evan, A. T. 30
Eyring, V. 253

Facchini, M. C. 295, 541
Faloona, I. 501
Farley, R. D. 346
Federer, B. 348
Feichter, J. 539, 540, 541, 546
Feingold, G. 8, 112, 131, 184, 185, 191, 299, 301, 319, 326–329, 333, 334, 416, 417, 433, 434, 439, 441, 448, 458, 495, 541, 545, 565–568
Field, P. R. 85, 98, 133, 248, 252, 255
Fierer, N. 313
Flanagan, P. 303
Fletcher, N. H. 370, 385, 394
Flueck, J. A. 345
Forster, P. 142, 412, 413, 452–459, 460, 463, 489, 492, 541, 547, 550, 570
Fountoukis, C. 495
Frederiksen, C. S. 390
Frederiksen, J. S. 390
Fridlind, A. M. 243
Friis-Christensen, E. 350
Fu, Q. 135, 251
Fugal, J. P. 164
Fusina, F. 244, 254
Fuzzi, S. 62

Gagin, A. 344
Gao, R. S. 246
Gardiner, B. A. 248
Garner, S. 210
Garrett, T. J. 129, 334, 392, 543
Garstang, M. 3, 341, 348
Gary, B. L. 239, 247
Gasparini, R. 311
Gassó, S. 134
Gavish, M. 302
Gayet, J.-F. 77, 98, 130, 133, 242, 255
Gentry, R. C. 348
Geoffroy, O. 189, 496
Gerber, H. 75, 109, 189
Gettelman, A. 236, 238, 246, 253
Ghan, S. J. 459, 538, 541
Gierens, K. 238, 251, 252, 257
Gilgen, H. 24
Gill, P. S. 309
Girard, E. 543
Givati, A. 376–378, 395
Golaz, J.-C. 451
Goldreich, Y. 378
Gong, D.-Y. 142, 492
Goodman, B. M. 351
Goody, R. M. 144, 474
Goto-Azuma, K. 403
Govindasamy, B. 353
Grabowski, W. W. 6, 9, 197, 205–207, 210, 230, 258, 269, 270, 273, 288, 301, 469, 472, 474, 513, 526, 569
Graham, N. E. 226
Grant, L. O. 343
Graßl, H. 3, 134
Gregory, J. M. 220, 464, 559, 570
Griffith, D. A. 376
Guibert, S. 493
Gultepe, I. 98, 99, 133, 306
Gunn, R. 321
Gupta, S. K. 25
Gustafson, W. I., 543
Gu, Y. 251

Haag, W. 238, 240, 242, 246, 249–254
Hahn, C. J. 23
Hall, A. 470, 558
Hallett, J. 75, 154, 248
Haman, K. E. 75
Han, Q. 3, 333, 413, 547
Hand, J. 43
Hansen, J. 184, 329, 443, 452, 456, 459, 570
Harries, J. E. 26
Hartmann, D. L. 6, 7, 20, 29, 112, 115, 127, 131, 137, 176, 199, 217, 219, 221, 225, 228, 301, 490, 513, 523
Haskins, R. D. 144
Hasumi, H. 476
Havens, B. S. 357
Haynes, J. M. 113
Haywood, J. 9, 413, 451, 458, 570
Hegerl, G. C. 547
Hegg, D. A. 309, 330, 454
Heidam, N. Z. 404
Heimbach, J. A. 343, 344

Heintzenberg, J. 1, 63, 130, 269, 303
Held, I. M. 208, 219, 254, 469, 488, 513, 558
Hendricks, J. 250
Hennessy, K. J. 386
Herman, M. 81
Herzegh, P. H. 236
Hess, M. 40
Heymsfield, A. J. 77, 88, 240, 248, 252, 255
Higgins, R. W. 118
Hindman, E. E., II 372
Hobbs, P. V. 157, 236, 329, 342, 343, 346, 372, 376, 392
Hoell, C. 502
Hoffman, R. N. 3
Hogan, R. J. 258, 441, 517, 521
Holben, B. N. 42
Holton, J. R. 236
Hoppel, W. A. 447
Hori, M. 294, 299
Hosseini, A. S. 153
Houghton, H. G. 320, 322, 323
Houze, R. A., Jr. 113, 136, 202
Howard, L. 173, 174
Howell, W. E. 320
Hoyle, C. R. 240, 241, 246
Hudson, J. G. 66, 67
Huebert 502

Iacobellis, S. F. 523
Iga, S. 473, 476
Illingworth, A. J. 10, 276, 497, 511, 513, 516, 557, 559, 560
Imbrie, J. 350
Imbrie, K. P. 350
Immler, F. 252
Imre, D. G. 159
Inoue, T. 482
Isaac, G. A. 73, 75, 78, 83, 85, 86, 88, 97–99, 127, 342, 344

Jacobowitz, H. 31
Jacobson, M. Z. 259, 452
Jakob, C. 21, 519
Jayaweera, K. 303
Jeffery, C. A. 303
Jensen, E. J. 239–241, 244, 246, 257
Jensen, J. B. 75
Jiang, H. 327, 328, 333, 355, 545
Jin, M. 375
Jirak, I. L. 376
John, V. O. 514
Johnson, D. W. 502
Jonas, P. R. 244
Jones, A. 451, 452, 459, 557
Jones, C. 118
Joos, H. 247, 259
Joshi, M. 452

Kahn, R. 41, 92
Kaiser, D. P. 422
Kanakidou, M. 62, 537
Kandler, K. 66, 67
Kapustin, V. N. 438
Kärcher, B. 7, 153, 235–243, 246–259, 269, 281, 282, 285, 291, 301, 303, 358, 526, 546
Karl, T. R. 115
Kaufman, Y. J. 41, 83, 134, 405, 406, 413, 417, 419, 422
Kawamoto, K. 413
Keith, D. W. 144, 475
Kerr, R. A. 342
Kessler, A. 394
Kessler, E. 272, 542, 543
Kew, S. F. 258
Khain, A. 109, 203, 243, 459, 546
Khairoutdinov, M. F. 210, 230, 232, 472, 474, 527
Kharin, V. V. 116
Kiehl, J. T. 135, 228, 471, 531
King, M. D. 42, 109, 111, 414, 495
King, W. D. 83, 84, 343
Kingsmill, D. E. 88
Kinne, S. 4, 37, 41, 46, 47, 63, 67, 68, 79, 127, 128, 141, 405, 412, 425, 441, 537, 539
Kinnison, D. E. 253
Kistler, R. 505
Klein, H. 43
Klein, S. A. 29, 112, 131, 176, 179, 219, 225, 490, 519, 542, 546
Klemp, J. B. 203
Klett, J. D. 59, 61, 189, 238, 370
Kley, D. 246

Knapp, K. R. 24
Knippertz, P. 118, 122
Knollenberg, R. G. 75, 77, 88
Knopf, D. A. 249, 250
Knutson, T. R. 136
Koehler, K. 302, 306
Koenig, L. R. 346
Koerner, R. M. 403
Köhler, H. 151, 320, 541
Köhler, M. 244
Koop, T. 240, 247–250, 303, 308, 313
Köpke, P. 40
Korczyk, P. 163
Koren, I. 23, 324, 329, 422, 440–443
Korolev, A. V. 75–78, 85, 91, 97, 133
Kreidenweis, S. M. 8, 62–68, 78, 128, 151, 280, 291–295, 299, 310, 322, 433, 438, 439, 545
Kristjánsson, J. E. 459
Krüger, O. 3, 134
Krummel, P. B. 382
Kuang, Z. 210
Kubar, T. L. 221
Kuhn, P. M. 357
Kummerow, C. D. 41, 245

Lamarck, J.-B. 173
Lance, S. 77
Landsberg, H. H. 378
Langner, J. 39, 537
Larson, K. 115, 219
Larson, V. E. 497, 542
Latham, J. 176, 355
Lau, K.-M. 469, 488
Lauer, A. 259
Law, K. 249
Lawson, R. P. 75, 85, 248, 342, 344
Leaitch, W. R. 417, 419, 454
Legrand, M. 403
Lehmann, K. 269
Leisner, T. 152, 159
Leith, C. E. 559
Lensky, I. M. 110, 111
Leovy, C. B. 23
Leroy, S. 144
Levin, Z. 8, 52, 110, 134, 321, 341–345, 354, 369–372, 393, 433, 443, 445
Lewellen, D. C. 181
Lewellen, W. S. 181, 183
Liepert, B. G. 351, 404, 549
Liljegren, J. C. 95
Lilly, D. K. 178
Lin, B. 469
Lin, J. C. 372
Lindzen, R. S. 199, 469
Liou, K. N. 237, 251, 358
Liu, C. 516
Liu, X. 253, 259
Liu, Y. 459, 542, 543
Lock, A. P. 179
Loeb, N. G. 28, 139, 474, 561
Lohmann, U. 11, 236, 240–242, 247, 253–259, 370, 409, 419, 420, 424, 459, 531, 539–543, 546, 557, 566
Lott, F. 247
Lu, M. -L. 355
Lubin, D. 543
Lunn, G. W. 343
Luo, Z. Z. 24, 245
Lynch, D. K. 235–241, 244, 253
Lynn, B. H. 459
Lyons, T. J. 390

MacPherson, J. I. 75
MacVean, M. K. 179
Magee, N. 153
Malinowski, S. P. 163
Malm, W. 43
Manabe, S. 3, 136, 188
Manes, A. 378
Mann, M. E. 349
Manton, M. J. 381, 384, 389
Marcolli, C. 298, 302, 304, 305
Marland, G. 352
Marsh, N. D. 3
Marsham, J. 239
Martin, G. M. 454
Martin, J. E. 118, 122
Martins, J. V. 134
Martonchik, J. V. 41
Mason, B. J. 108, 342, 343, 370
Masunaga, H. 5, 107, 111, 127, 143, 223
Mather, G. K. 346, 370, 372
Mather, J. H. 516
Matsui, T. 112, 419, 421–423
Matthews, H. D. 452

Mauger, G. 131, 132
Mazin, I. P. 78
McComiskey, A. 416, 417
McConnell, J. R. 403, 404
McFarlane, J. F. 247
McFarquhar, G. M. 248
McFiggans, G. 59, 61–64, 68, 293, 301, 308, 310, 312
McGuirk, J. P. 118
McIntyre, S. 349
McKitrick, R. 349
McMurry, P. H. 62, 157
Medeiros, B. P. 229, 513
Medina, J. 308
Meehl, G. A. 452, 513, 523
Meerkötter, R. 3
Mehta, A. 32
Menon, S. 22, 424, 443, 459
Mertes, S. 4, 255, 302
Mesinger, F. 348
Mesinger, N. 348
Michelsen, M. L. 199
Mielke, P. W., Jr. 342–344
Mikhailov, E. 62, 65
Miller, M. 247
Miller, R. L. 176
Miloshevich, L. M. 252
Ming, Y. 456, 457, 459, 463
Minnis, P. 358
Mishchenko, M. I. 404, 405
Mitchell, J. F. B. 471
Mitchell, R. M. 323
Miura, H. 209, 469, 473, 527, 550
Mochida, M. 308
Moeng, C.-H. 179
Möhler, O. 149, 153, 160, 240, 246, 302, 306, 448
Moncrieff, M. W. 258, 469
Mooney, M. L. 343
Morrison, H. 206
Mukai, M. 419, 422, 424, 425
Müller, D. 518
Murphy, D. M. 240, 246, 247
Murphy, J. M. 471, 523
Murray, B. J. 246
Murray, F. W. 346
Murty, A. S. R. 346
Myhre, G. 419, 422, 423
Nakajima, T. 3, 8, 109–111, 133, 134, 287, 401, 413, 414, 422, 433, 458, 495, 567
Nakajima, T. Y. 109, 110, 415, 422, 424
Nakamura, K. 108, 113
Nazarenko, L. 452
Neelin, J. D. 223, 563
Neggers, R. A. 182
Nenes, A. 157, 301
Neumann, J. 344
Newell, R. E. 238
Nghiem, S. V. 135
Nicholls, J. D. 177
Nicholls, N. 388, 389, 392
Nicholls, S. 108, 181
Nickerson, E. C. 345
Nigam, S. 176
Nilsson, B. 49
Ninomiya, K. 117, 119, 122
Nishizawa, T. 43
Nober, F. J. 540
Norment, H. G. 85
Norris, J. R. 2, 17, 21, 23–25, 30, 127, 131–134, 523
Notholt, J. 241

O'Dowd, C. D. 323
Ogren, J. A. 492
Ogura, T. 476
Ohmura, A. 24, 25
Ohring, G. 28
Onasch, T. B. 304
Orville, H. 331, 342

Pallé, E. 26
Palmer, T. N. 560
Paltridge, G. W. 179
Pandis, S. N. 59, 61
Park, S. 23, 181
Parrish, J. L. 88
Parsons, M. T. 304
Pawlowska, H. 189, 495, 496
Pearson, R. 177
Peng, Y. 542, 543
Penner, J. E. 135, 455, 459, 570
Perry, K. D. 330
Petch, J. C. 6, 9, 197, 206, 230, 269, 270, 273, 288, 301, 526, 569

Peter, Th. 246, 256
Peters, M. E. 223–225
Petters, M. D. 62–67, 128, 291–295, 299, 310
Pfister, L. 241
Pham, M. 404
Philander, S. G. H. 176
Phillips, B. B. 321
Phillips, V. 206
Pielke, R. A., Sr. 340–342, 350, 351, 353
Pincus, R. 179, 185, 453, 501, 542, 546
Pinsky, M. B. 109
Pitman, A. 3, 390
Platnick, S. 131, 286, 323, 497
Plougonven, R. 239
Polacheck, T. 395
Porch, W. M. 354
Pöschl, U. 4, 37, 58, 62, 63, 127
Potter, G. L. 476
Prenni, A. J. 306
Preunkert, S. 403, 404
Pruppacher, H. R. 59, 61, 162, 238, 370

Qu, X. 470, 558
Quaas, J. 11, 419, 422, 459, 557
Quinn, P. K. 404, 493

Radke, L. F. 132, 157, 183, 354, 413
Raes, F. 62, 63
Raga, G. B. 75
Ramanathan, V. 18, 20, 199, 226, 412, 424, 469, 504, 515
Ramaswamy, V. 532
Randall, D. 21, 22, 177, 183, 205, 230, 469, 472, 474, 488, 532, 540
Rangno, A. L. 342, 343
Rapp, A. D. 199
Rauber, R. M. 276
Ray, D. K. 3
Redelsperger, J.-L. 204
Reisin, T. 346
Respondek, P. S. 439
Reynolds, D. W. 344
Richter, A. 405
Ridgway, W. 181, 182
Riedi, J. 92
Ringer, M. A. 476, 477
Rissler, J. 64, 293
Roberts, D. L. 452
Roberts, G. 77, 157, 309
Robinson, A. B. 349
Robock, A. 351, 359
Rodhe, H. 39, 537
Rodwell, M. J. 560
Roe, G. H. 550
Rogers, D. C. 75, 77, 84, 152, 158, 161
Rokicki, M. L. 346
Rose, D. 53, 60, 62, 64–68, 75, 309
Rosenfeld, D. 3, 59, 61–67, 110, 111, 128, 183, 296, 302, 312, 342, 345, 348, 371, 376–378, 384–388, 395, 423, 433, 473
Rossow, W. B. 23, 28, 49, 218, 236, 245, 476, 478, 480
Rotstayn, L. D. 74, 98, 100, 208, 370, 392, 455, 459, 543, 570
Rudolf, B. 348
Ruehl, C. R. 63, 309, 323, 438
Ryan, B. F. 343

Sandu, I. 186, 191, 504
Santer, B. D. 208, 254
Sassen, K. 242, 249, 257, 358
Satoh, M. 10, 209, 212, 230, 273, 469, 472, 473, 526, 557, 564
Satoh, W. D. 569
Saunders, C. P. R. 153
Saunders, P. M. 185
Savic-Jovcic, V. 185
Sax, R. I. 342, 343
Schaefer, V. J. 370
Schemenauer, R. S. 89, 342
Schiffer, R. A. 23, 218, 236, 245, 476, 478, 480
Schmidt, K. S. 73, 127, 251
Schrems, O. 252
Schüller, L. 495
Schulz, M. 8, 46, 50, 109, 127, 133, 134, 287, 401, 404, 405, 408, 409, 413, 433, 567
Schumacher, C. 113
Schumann, U. 239, 358
Schütz, L. 66, 67
Schwartz, S. E. 11, 13, 531, 547, 557, 559, 566
Scinocca, J. F. 247

Scorer, R. 3, 173
Scott, B. C. 346
Scott, W. D. 329
Seager, R. 222
Seidel, D. J. 208
Seifert, P. 242
Seinfeld, J. H. 59, 61, 259, 355
Sekiguchi, M. 3, 417, 419, 422, 423
Senior, C. A. 471
Shaw, R. 149
Sheets, R. C. 348
Shige, S. 122
Shine, K. P. 456, 570
Shipley, S. T. 94
Short, D. A. 108, 113
Shutts, G. J. 211
Siebert, H. 8, 75, 112, 131, 291, 299, 301, 319, 433, 434, 448, 565–568
Siebesma, A. P. 7, 182, 269, 437, 440
Sievering, H. 447
Silverman, B. A. 341, 342, 345, 346
Simpson, J. S. 342, 345, 348
Slingo, A. 2, 17, 20, 74, 98, 100, 127
Slingo, J. 176, 182, 183, 198, 232
Smith, I. N. 386
Smith, M. H. 176
Smith, P. L. 342, 348
Smith, R. B. 239
Smith, R. N. B. 183
Smolarkiewicz, P. K. 210, 238
Snider, J. R. 309, 493, 497
Soden, B. J. 188, 219, 254, 469, 477, 488, 513, 514, 558, 559
Solomon, S. 142, 237, 492
Sorjamaa, R. 295
Spichtinger, P. 7, 153, 235, 238, 239, 246, 251, 252, 256, 257, 269, 281, 282, 285, 291, 301, 303, 526, 546
Spracklen, D. V. 55
Squires, P. 73, 97, 183, 184, 371
Starr, D.'O.C. 240, 244, 257
Stephens, G. L. 41, 79, 93, 113, 236, 243, 245, 473, 476
Stetzer, O. 158, 255
Stevens, B. 6, 173, 178–182, 185, 269, 276, 278, 285, 301, 434, 446, 489, 490, 501, 565
Stier, P. 259, 537
Stolzenburg, M. R. 157
Storelvmo, T. 417
Stowe, L. 39
Strapp, J. W. 79, 89–91
Stratmann, F. 6, 52, 149, 161, 269, 280–284
Ström, J. 129, 238, 241, 246, 247, 254
Stroud, C. A. 308
Stubenrauch, C. J. 238, 245
Sukarnjanasat, W. 346
Sun, B. 23, 350
Sundqvist, H. 183, 185, 542
Super, A. B. 343, 344
Suppiah, R. 386, 387
Surowiecki, J. 395
Susskind, J. 32
Suzuki, K. 109, 113, 417, 419–422, 424, 459, 473
Svenningsson, B. 294, 298, 299
Svensmark, H. 3, 350

Tabazadeh, A. 241
Takayabu, Y. 5, 107, 121, 127, 143, 223
Takemura, T. 414, 415, 424, 459, 473
Tanre, D. 41
Tao, W.-K. 557
Taraniuk, I. 301, 309
Targino, A. C. 303
Tarmy, B. H. 340
Tegen, I. 404
Textor, C. 41, 46, 62
Thomson, D. W. 135
Thornton, B. F. 236
Thorpe, A. J. 203
Tiedtke, M. 239
Timball, B. 390
Tompkins, A. M. 183, 199, 205, 207, 208, 228, 238, 252, 258, 469, 516
Topping, D. O. 295
Torres, O. 39
Travis, D. J. 3
Treffeisen, R. 238
Trenberth, K. E. 118, 122, 135, 531
Tselioudis, G. 21, 513, 523, 559
Tsigaridis, K. 306–308
Tsonis 342
Tsushima,Y. 471, 476, 479, 512
Tukey, J. W. 342, 343

Turton, J. D. 177, 181
Turvey, D. E. 379, 380, 382, 384
Twohy, C. H. 84, 303
Twomey, S. A. 3, 112, 131, 183, 184, 188, 320–323, 371, 372, 379, 380, 383–385, 453, 461, 493, 565

Uppala, S. M. 505

Valero, F. P. J. 144, 475
Vali, G. 302, 305
van den Heever, S. 374, 375, 491
Van Zanten, M. C. 189, 332, 496, 498
Vecchi, G. A. 222
Verver, G. 2569
Vestin, A. 308, 311
Vogelmann, A. M. 543
Vohl, O. 162
Voigt, C. 236
Vonnegut, B. 302

Waliser, D. E. 226
Wang, C. 203
Wang, P. H. 236, 239
Wang, S. 325, 326
Wang, Z. 257
Warner, J. 131, 183, 321, 371, 372, 383–386, 443
Warren, S. G. 23
Watts, R. G. 358
Weaver, C. J. 177
Weaver, C. P. 21
Webb, M. J. 220, 464, 469, 476, 513, 519, 550, 551, 559, 560, 570
Weber, R. J. 330
Webster, P. J. 243
Weisman, M. L. 203
Welton, E. J. 43
Wendisch, M. 133, 251
Westbrook, C. D. 252
Wetherald, R. T. 3, 188
Wex, H. 64, 149, 151
Whiteway, J. 240
Wielicki, B. A. 26, 27, 474
Wild, M. 25, 30, 134, 404
Wilkinson, J. M. 519
Williams, D. J. 381
Williams, K. D. 459, 471, 513, 523, 559
Williamson, D. L. 471
Wilson, C. T. R. 320
Winker, D. M. 41, 79, 95
Wojciechowski, T. A. 183
Wong, A. P. S. 390
Wong, T. 27, 29
Wood, R. 10, 131, 176, 179, 191, 229, 278, 285, 332, 458, 461, 487, 490, 496, 501, 503, 542, 557, 565
Wood, S. E. 246, 250
Woodley, W. L. 342, 345, 347
Worthington, R. M. 239
Wu, X. 469
Wyant, M. C. 181, 231, 469, 501, 513, 527, 550
Wylie, D. 32

Xie, P. 223
Xu, K.-M. 183, 205
Xue, H. 185, 191, 326, 327, 333, 334, 458

Yamagata, S. 165
Yanai, M. 122
Yang, G.-Y. 198, 232
Yin, J. H. 222, 254
Yin, Y. 346, 347
Yoh, S. 183
Yokohata, T. 558
Young, K. C. 342, 346
Yu, H. 407, 408, 412, 413

Zeng, N. 223, 563
Zhang, F. 239
Zhang, M. H. 223, 231, 236, 251, 253, 488, 519
Zhang, Q. 538
Zhang, Y. 25, 28, 179
Zhao, C. 334
Zhong, W. 244
Zhu, P. 222
Zhuang, L. Z. 502
Ziese, M. 161
Zipser, E. J. 120, 122
Zobrist, B. 250, 302, 306
Zuidema, P. 324, 333

Subject Index

accommodation coefficient 166, 167, 170, 280
  mass 153, 301, 309, 310, 322, 438
accumulation mode 37–41, 48–50, 54, 307
ACE-2 325, 332, 491, 493, 494, 495, 496, 502, 503
ACE-Asia 38
activation 150, 166, 168 , 280, 296–301, 307, 308, 312 331, 438
ACTOS 284
Adelaide plume 384–386
adiabatic cooling 160, 217, 221, 229
AeroCom 41, 47, 50, 54, 405–410, 537
AERONET 42–50, 67, 102, 139, 441, 442, 539
aerosol 4, 22, 38, 39, 53, 67, 138, 142, 183, 329–331, 351, 433, 488, 489
  anthropogenic 17, 22, 135, 425, 451
  as CCN 52–72, 61–72, 175, 293
  forcing 407–411
  global modeling 46, 536–539
  ground-based monitoring 41, 56
  hygroscopic properties of 127, 128
  indirect effects 250, 251, 422–425, 451–465, 543–546, 566–569
  large-scale changes 402–413
  secondary organic 66, 294, 329
  vertical variations in 96–106
aerosol optical depth (AOD) 37, 38, 42–46, 50–72, 83, 132, 134, 139, 371, 404–410, 413, 549
  global distribution of 414–415
  maps 40, 45–47, 51–53
AIDA 159, 160, 167
AIM 60, 298
aircraft measurements 79, 84, 93–95, 130
AIRS 81, 86, 92, 97, 137
Aitken mode particles 330, 447
albedo 11, 18, 22, 92, 131, 134, 320–325, 533–535, 565
  effect 451, 453, 456, 461, 463, 481, 482, 487, 543, 544, 547
  engineering changes to 340
  of low clouds 175, 186
  snow feedbacks 470
  susceptibility 323, 331–336
Albrecht effect 453
Amazon 114 , 372
AMMA 38, 140, 204, 206
Ångström parameter 37, 42, 44, 82, 83, 139
  maps 46–78
anvil cirrus 237–240, 243, 244, 248, 251, 254, 281
Aqua satellite 26
ARM 102, 139, 276, 518, 561
ASTEX 424, 500, 501
asymmetry factor 37, 38
  maps 51, 53
ATEX 288
A-Train 41, 95, 101, 102, 112, 232, 257, 435, 518, 527, 567
AVHRR 39, 44, 92, 413

Baiu Front 119, 120
BAMEX 202
BASC 385, 393
Baseline Surface Radiation Network (BSRN) 25, 404
Bergeron–Findeisen process 153, 251, 280, 281
biomass burning 25, 40, 41, 54, 66, 142, 294, 385, 405, 443, 492
BOMEX 288, 518
Bony-binning 231

CALIOP 95, 518
CALIPSO 10, 41, 79, 93, 129, 137, 138, 257, 288, 426, 483, 518–521, 560
Canadian Convair 580 84, 93–95
Canadian Freezing Drizzle Experiment III 86–88, 97
CAPS 130
CCOPE 202, 203
ceilometer 82, 93–106
CERES 28, 138, 139, 435, 442, 528, 561

cirrus 2, 7, 235, 237, 252, 253, 482
aerosol effects on 240–243, 281–284
anvil 237–240, 243, 244, 248, 251, 254, 281
contrails 237–240, 244, 251, 254, 279, 357, 358
ice initiation and growth 249, 250
life cycle 245–247
measurement techniques 255–257
modeling techniques 257–259
seeding 357, 358
CLARREO 474, 475
Clausius–Clapeyron rate 115, 116
climate engineering 339, 340, 352–360
climate sensitivity 13, 33, 135, 200, 227, 452, 471, 511–514, 549–551, 564
defined 21, 558
estimated 188
inter-model differences in 488
Climax experiments 343
closure experiments 308–312, 492–499
cloud chambers 163, 164, 168
cloud–climate metrics 523, 524
cloud condensation nuclei (CCN) 4, 52–75, 97, 201, 298, 299, 370, 451
anthropogenic perturbations 127, 291, 408–413, 438
closure studies 308–312, 492–499
counters 62, 64, 156–158, 166
giant 323, 334, 356, 370, 372–374, 393, 447, 545
global maps 55–57
cloud droplets. *See* droplet
cloudiness 23, 182–185, 489
impact of meteorology 133, 175, 187
large-scale climatology of 218
net radiative effect of 220
role of aerosol 132, 133, 175
cloud lifetime 143, 144, 285, 321, 324, 327–329, 335, 420, 460, 546
effect 453, 461, 462, 543–547, 570
Cloudnet 139, 276, 516
cloud optical thickness 18, 21, 30, 108, 110, 114, 221, 323, 413, 422, 482, 491
Cloud Particle Imager 85
cloud-profiling radar (CPR) 93, 112, 113
cloud radiative forcing (CRF) 18, 21, 22, 218, 223, 226, 531
TOA 7, 407–413
cloud-resolving models (CRMs) 137, 183, 202, 210, 230, 277, 454, 516, 540, 557
global 10, 273, 274, 470–473, 476, 480, 483, 516, 526, 527, 564
weakness of 206, 207
CloudSat 10, 41, 79, 93–95, 112, 129, 137, 138, 257, 288, 331, 426, 473, 483, 518, 519 , 560
cloud seeding 8, 134, 340, 348, 353, 354, 358
altostratus 356, 357
glaciogenic 341–345
hygroscopic 346, 347
political placebos 360
cloud-top temperature 24, 110, 114, 221
coalescence 108, 109, 152–156, 270, 325, 327, 331, 334, 371, 386, 438
coarse mode 37, 41, 48, 49, 54
collision 108, 111, 129, 152, 155, 156, 270, 325, 331, 334, 438
communicating results 360, 393–395
condensation 111, 536
continuous flow diffusion chamber 158, 167
contrails 237–240, 244, 251, 254, 279, 357, 358
convective-radiative quasi-equilibrium 198, 203–205
critical supersaturation 292–300, 309
cumulonimbus 2, 201, 202, 270
cumulus 2, 181–183, 279
convection 469, 476, 482, 513
glaciogenic seeding of 342

decoupling 180, 181, 186
deep convection 6, 19, 20, 122, 197–213, 219, 226–229, 243, 280, 459, 469, 476, 480, 482
in atmospheric GCMs 208–210
detrainment 155, 330
diabatic heating 221
Diamond Dust 235
differential mobility analyzer 155, 157
Doppler radar 82, 93, 143

droplet
collision efficiency 189
formation 52, 58–60, 151, 155, 198, 201, 292–297, 305, 314, 335, 541
freezing 152, 154
number concentration 301, 565
radius 108–114
global distribution of 414–415
regional changes in 22
size distribution 108–111, 495
–turbulence interactions 150, 155, 160, 169, 170
dry particle composition 293–296
DSCOVR 139, 144, 475
DYCOMS-II 177, 189, 276, 288, 332, 496, 501

EARLINET 42, 43, 102
EarthCARE 426, 436, 475, 528, 561
Earthshine method 139
Earth simulator 209
Earth's radiation budget 5, 17–34, 243, 440, 531
controlling factors 108
difficulties in measuring 20
ERBE 139, 435, 442, 535
ERBS 27–29, 32
geostationary 26, 561
long-term monitoring of 561
satellite measurements of 26–36, 139
ECHAM 247, 252, 547
ECMWF 138
El Chichón 352
electrodynamic balance 156, 159, 166
El Niño 19, 26, 118, 522, 558, 561
emissivity 18–20, 19, 20, 82, 174
entrainment 155, 156, 169, 179, 270, 284, 325, 438, 445, 501
EPIC 288, 496
estimated inversion strength 176, 229
EUCAARI 504
Eulerian column closure 487, 492–499
evaporation 327, 336, 438, 458, 536
Explorer 7 26
Extended Edited Cloud Report Archive 23, 24
extreme rainfall 107, 115–120, 128, 471
Baiu Front 119, 120
defined 116

FACE experiments 345
FINCH (fast ice nucleus chamber) 158, 167
fires 286, 289, 404
biomass 25, 40, 41, 54, 66, 142, 294, 385, 405, 443, 492
sugar cane 383–384, 385–386
fixed-SST forcing 570, 571, 572
fluctuation dissipation theorem 559

GATE 137, 202, 206, 288
GAW 43, 441
Gayet's cloud nephelometer 130
general circulation models (GCMs) 143, 454–456, 532, 560
atmospheric 208–212
feedbacks in 462
limits to 140, 217, 288
GEWEX cloud system studies 142, 560
GFDL 456, 457, 460, 464
giant CCN 323, 334, 356, 370, 372–374, 393, 447, 545
glaciation effect 543–545
Global Aerosol Data Set 40
global brightening 25, 426
global climate models 21–36, 136, 278, 279, 531–553
cloud feedbacks in 549–551
connecting scales 271–276
microphysical processes 539–543
representation of aerosols in 41, 536
global cloud-resolving models 10, 273, 274, 470–473, 476, 480, 483, 516, 526, 527, 564
global dimming 25, 351, 426
global warming 13, 22, 340, 351, 354
detection of 26
response of clouds to 17, 22, 135–140
Glomap simulations 55
gravitational settling 536
greenhouse gases 10, 206, 254, 270, 348–351, 425, 488, 489, 532, 537,

greenhouse gases (*continued*) 570–572
forcing 17, 220, 273, 390, 392, 551
greenhouse warming 13, 339, 351, 370, 433, 445, 481, 482, 487, 551
countering 8, 356, 357

Hadley circulation 114, 135, 208, 211
Hallett–Mossop effect 143, 153, 168, 281
HALO 256, 257
HIAPER 256, 257
HIPLEX-1 342
H-TDMA 32, 62, 64, 156, 157, 166, 295
hygroscopic growth 4, 37, 150, 157, 166, 168, 170, 280
hygroscopicity 58–68, 61, 64–67, 68, 127, 128, 294–298, 303, 306, 310, 311, 447

IAPSAG 59
ice nucleation 5, 143, 150, 152, 167, 235, 249, 250, 323
heterogeneous 154, 167, 241, 255, 301–306, 312
homogeneous 255, 301–305, 308, 313
ice nuclei (IN) 128–130, 152, 201, 451
counters 156, 158
measuring 74–78, 97
ice particle size distributions 255
ICESAT simulator 519
industrial emissions 39, 127, 142, 354
regional changes in 404
infrared iris hypothesis 199
in-situ measurements 73–79, 108, 141
aircraft 79, 84, 93–95, 130
cirrus 237, 238, 241, 244, 254, 281
limits to 83, 129, 191, 495
probes 85–91
units 98–101
unmanned airborne vehicle 74, 80, 357, 504
invisible clouds 479, 480, 482
IOCI 390, 392
IPCC 10, 21, 59, 116, 188, 203, 208, 212, 320, 385, 451, 453, 458–460, 470, 471, 474, 477, 488, 532, 570
Fourth Assessment Report 413, 513, 524, 536, 550
ISCCP (International Satellite Cloud Climatology Project) 23, 24, 28–32, 218, 231, 241
simulator 476–478, 480

King LWC probe 90
Köhler theory 59, 60, 64
KWAJEX 137

laboratory studies 6, 149, 309
cloud simulations 154–165
for cirrus 255–258
LACIS (Leipzig aerosol cloud interaction simulator) 159, 161, 166, 167
Lagrangian experiments 487, 500–505
large eddy simulation 141–143, 202, 230, 251, 256–258, 277, 436, 437, 557
EULAG model 257
of stratocumulus 325, 331–332
UCLA model 179, 180
latent heating 221
lidar 41, 81, 82, 93–106, 143, 519, 520, 528
liquid water path 18, 131, 312, 320, 323–326, 331, 420, 421, 458, 489
LITE 257, 521
Little Ice Age 349, 351
longwave cloud radiative forcing 18, 19, 24, 32, 220–222, 227
anomalies 23, 27, 29
low clouds 173–192, 519
defined 174, 175
lower tropospheric stability 131, 132, 176, 223, 225, 228

Madden-Julian Oscillation 118, 209, 270, 473, 522, 525
Magnus-effect sailing ships 355, 356
marine stratocumulus 21, 174, 183, 218
mass accommodation coefficient 153, 301, 309, 310, 322, 438

MAST 286
Medieval Climate Optimum 349, 351
Mega-Tropique 528, 561
mesoscale convective systems 107, 113, 119, 122, 128, 202, 204
Meteorological Research Institute (MRI) 159, 161
meteorology 130–133, 138, 285, 489, 490, 491
  of low clouds 176–182
Meteosat 26, 285
METROMEX 373, 374, 375, 393
Michigan Technological University cloud chamber 164
microwave radiometry 82, 95, 111
mid-visible aerosol properties 38–40, 44, 47
MIROC 414, 477–482
MISR 41, 92
mixing 325, 326, 335
MODIS 41, 44, 92, 113, 129, 137, 138, 220, 372, 405, 414, 442, 473, 547
MPL-Net 42, 43, 102
Mt. Isa 384
Mt. Pinatubo 24, 26, 241, 352
multiscale modeling framework 210, 211, 258, 471–474, 569

NASA Icing Research Tunnel 90
National Hail Research Experiment 202
Nevzorov probes 130
NICAM 473, 476–482
NIES 42, 43
Nimbus 7 26, 27
NIST 475
NISTARs 144
numerical modeling 200–202, 211
numerical weather prediction 514
  reanalyses 32, 138
  model 275, 276, 444, 513, 522, 527, 557, 560

observational record 23, 143, 144
  current needs 137–139, 199, 200
  EECRA (Extended Edited Cloud Report Archive) 23, 24
  GADS (Global Aerosol Data Set) 24
  GEBA (Global Energy Balance Archive) 24
  ice cores 403, 404
  NASA Earth Observing System 26
observational strategies 487–507, 511–529
Oklahoma–Kansas PRE-STORM 202
optical array probe 85, 91
orographic ratio 377–379

particle imaging velocimetry 164
particle measuring system (PMS) 75, 85, 87–90 96
particle size 292, 295, 303, 307–311
PATMOS 31, 32
Peters-Bretherton model 223–226
phase state 292, 294, 298, 299, 302, 304, 306, 310
polarimetry 82
POLDER 41, 92, 129, 413, 547
pollution 320, 321, 334, 379–384, 491, 502
  air 369–395
  control measures 403
  effects on rainfall 373–375
precipitation 5, 8, 107–115, 143, 175, 334, 369, 370, 422, 423, 443–445, 480, 491, 496, 536
  aerosol effects on 134, 321, 331, 370
  anthropogenic effects on 372–375
  Australia 386–393
  cloud-base 189, 190
  extreme rainfall 107, 115–120, 128, 471
  intermodel differences 525
  orographic clouds 375–378
  shallow rainfall 108
    butterfly pattern 5, 107, 112–115, 123
  suppression 325, 336, 385, 458, 460
  susceptibility 331, 336
  TRMM radar 113, 114, 560

probability distribution function 99, 220, 231, 252, 253, 272, 494, 515–518, 541, 566
quasi-forcing 455, 462–464

radar 81, 92, 93, 191, 528
cloud-profiling 93, 112, 113
Doppler 82, 93, 143
precipitation radar 113, 114, 560
radiative forcing 251, 252
anthropogenic 22
cloud (CRF) 18, 21, 22, 218, 223, 226, 531
definition of 451, 452
shortcomings of IPCC 570
indirect 451–465
longwave 18, 19, 24, 32, 220–222, 227
shortwave 18, 33, 220, 221, 227
TOA 407–413
rainyield per flash (RPF) values 121, 122
Raoult effect 59, 60, 62
remote sensing 81, 102, 139, 435, 506
of aerosols 44, 45
future prospects 474, 475
instrumentation 79–106
limits to 129
measurements 73, 91–96
units of 98–101
satellite 108, 320, 325
RICO 276, 288, 518
riming 153, 154

SAFARI 38
Santa Ana Winds 286
ScaRaB 26, 27, 561
sea surface temperature (SST) 198, 199, 208, 226, 559
fixed-SST forcing 570–572
secondary organic aerosols 66, 294, 329
semi-direct effect 454, 544, 547
shallow clouds. *See* low clouds
shallow convection 122, 123, 124
shallow rainfall. *See* precipitation
ship tracks 1, 131, 132, 286, 289, 321, 354, 355, 567
satellite images of 183, 184
shortwave cloud radiative forcing 18, 19, 33, 220, 221, 227
anomalies 23, 29
single-scattering albedo 37, 38, 42–46, 49, 50, 82
maps 46–72, 48, 51, 53
retrievals 83
SKYNET 42, 43, 47
SOCEX 382, 383
solar radiation 18, 24, 25, 236, 244, 454
altering 353
semi-direct effect 452
TOA 139, 220–222, 284–285, 439, 512, 528, 543, 544, 547
SPARC Water Vapor Assessment Report 246
spectral latent heating 122–124
splintering 153, 168
SPRINTARS 414, 473
stratocumulus 3, 21, 108, 177–180, 185–187, 270, 286
hygroscopic seeding of 355
large eddy simulations of 325, 331–332
marine 21, 174, 183, 218
precipitation rate 496–499
sun photometers 39, 42–47, 53, 81–83
super-parameterization 210, 212, 230, 232, 273, 274, 472, 526, 540, 564
surface tension 295–298, 303, 310, 322
synoptic-scale systems 107, 122, 124

TARFOX 38
Terra satellite 26, 92
thermostat hypothesis 199
Thunderstorm Project 201
TIROS 1 26
TOA radiation 139, 220–222, 284–285, 407–413, 512 , 528, 543, 544, 547
TOGA-COARE 137, 204, 206
TOMS 39, 44
TRMM 26, 93, 113, 122, 257, 372, 560
intense thunderstorms 120, 121
precipitation radar 113, 114, 560
tropical plumes 107, 118, 119

turbulence 154, 156, 163, 169, 229, 270–274, 326, 445, 536
in cirrus 239, 240
Twomey effect 242, 251, 371, 453, 461, 495, 546, 565, 570

Univ. of Mainz wind tunnel 159, 162
Univ. of Manchester cloud chamber 168
unmanned airborne vehicle (UAVs) 74, 80, 357, 504

van't Hoff factor 60, 65, 322
vertical motion 217–223, 237, 256, 272, 288, 319, 329
vertical structure 512, 518
visible and infrared imagery 82, 91
VOCALS 140, 276, 504
volcanic activity 24, 26, 241, 351, 352, 360, 384, 558, 562

Walker circulation 114, 136, 218, 222
warm clouds 331–334, 371, 372, 421
formation 1, 58, 68, 127, 128, 308–312
particle precursors 292–295
rain formation in 451, 473, 542–545
Warsaw University cloud chamber 163
weather
engineering 340–348, 358, 359
hazards 201, 202
modification 134, 285, 370, 492
severe 6, 120, 121, 211
wet deposition 314, 404, 439, 455
whole-earth monitoring 139, 144
wind tunnels 79, 83, 90, 159, 162, 169

Zdanovski-Stokes-Robinson assumption 65
Zürich ice nucleation chamber (ZINC) 158, 167